W9-CRB-244

V. Hessel, S. Hardt, H. Löwe

**Chemical Micro Process
Engineering**

Related Titles

V. Hessel, S. Hardt, H. Löwe

Chemical Micro Process Engineering

Processing, Applications and Plants

2004
ISBN 3-527-30998-5

W. Ehrfeld, V. Hessel, H. Löwe

Microreactors

New Technology for Modern Chemistry

2000
ISBN 3-527-29590-0

W. Menz, J. Mohr, O. Paul

Microsystem Technology

2001
ISBN 3-527-29634-4

J. G. Sanchez Marcano, Th. T. Tsotsis

Catalytic Membranes and Membrane Reactors

2002
ISBN 3-527-30277-8

T. Gh. Dobre, J. G. Sanchez Marcano

Chemical Engineering

Modelling, Simulation and Similitude

2004
ISBN 3-527-30607-2

K. Sundmacher, A. Kienle (Eds.)

Reactive Distillation

Status and Future Directions

2003
ISBN 3-527-30579-3

S. P. Nunes, K.-V. Peinemann (Eds.)

Membrane Technology

in the Chemical Industry

2001
ISBN 3-527-28485-0

Volker Hessel, Steffen Hardt, Holger Löwe

Chemical Micro Process Engineering

Fundamentals, Modelling and Reactions

WILEY-VCH

WILEY-VCH Verlag GmbH & Co. KGaA

Chemistry Library

Dr. Volker Hessel
Dr. Steffen Hardt
Dr. Holger Löwe
IMM – Institut für Mikrotechnik Mainz GmbH
Carl-Zeiss-Straße 18–20
55129 Mainz
Germany

Cover Illustration

Upper left: Production- and pilot-scale gas/gas counter-flow heat exchanger comprising microstructured channel arrays. The device (including flanges about 36 kg heavy and 54 cm long), made of stainless steel, is designed for gas throughput in the range of m^3/min at 100 mbar pressure drop for a power of about 10 kW. The internals consist of a stack of microstructured plates having multi-channel arrays of a channel width of 2 mm, depth of 250 µm, and length of 240 mm. Totaling, 6685 micro channels are operated in parallel in this device. The flange-type connection allows installation in large-scale industrial plants (IMM Mainz-Hechtsheim, Germany).

Center: CFD simulation of streamlines of a liquid flow in a caterpillar micro mixer. This device utilizes the split-recombine principle leading to distributive mixing. It is seen that by multiple repetition of this principle the entanglement of the streams increases (IMM Mainz-Hechtsheim, Germany).

Lower right: Cross-flow catalyst screening device with multiple short mini-fixed beds. The fixed-bed catalyst section is fed by bifurcation-channel flow architectures that serve for flow equipartition. This device is a typical example for the class of smart chip reactors, widely employed for analytical-chemistry, kinetic studies and process/catalyst screening purposes on a lab-scale level, and is fabricated using MEMS technology based on silicon micromachining (Courtesy K. S. Jensen, MIT Cambridge, USA).

This book was carefully produced. Nevertheless, authors and publisher do not warrant the information contained therein to be free of errors. Readers are advised to keep in mind that statements, data, illustrations, procedural details or other items may inadvertently be inaccurate.

Library of Congress Card No.: Applied for.
A catalogue record for this book is available from the British Library.

Bibliographic information published by Die Deutsche Bibliothek
Die Deutsche Bibliothek lists this publication in the Deutsche Nationalbibliografie; detailed bibliographic data is available in the internet at http://dnb.ddb.de.

© 2004 Wiley-VCH Verlag GmbH & Co. KGaA, Weinheim

Printed in the Federal Republic of Germany.
Printed on acid-free paper.

Composition Manuela Treindl, Laaber
Printing betz-druck GmbH, Darmstadt
Bookbinding Buchbinderei J. Schäffer GmbH & Co. KG, Grünstadt

ISBN 3-527-30741-9

Preface

Carrying out chemical reactions in volumes as small as possible is *a priori* not a completely new idea. In the beginnings of chemical experimentation, dating back to the age of alchemy, chemical substances like sulphuric acid or ammonia were much more valuable than gold, and very small reaction vessels were used to economize on the precious materials. When analytical chemistry was established as a second, independent discipline, the desire to make do with ever less material was very strong in order to avoid consuming large portions of the product for analysis. Establishing increasingly sensitive analytical techniques has therefore been one of the most significant driving forces in analytics research.

The beginning of the industrial age saw a substantial increase in demand for basic materials and chemicals, and the chemical industry was established to satisfy these demands for high production volumes. The tall and impressive silhouettes of modern chemical plants dominate industrial estates, visible from afar as symbols for the vast capabilities and capacities of today's chemical industry. Without this industry and its equipment of enormous proportions, our economic wealth would be quite inconceivable.

Bearing all this in mind, what is the purpose of Chemical Micro Process Technology?

Conventionally, the development of chemical manufacturing processes takes place subsequently *via* a sequence of different intermediate stages. Approaching the final process design, the reaction volume is successively increased from laboratory scale to reaction vessel dimensions suitable for production outputs of several kilotons *per annum*. This procedure, known as "scale-up", is expensive and time-consuming. During the scale-up, new and previously unencountered problems often crop up and have to be solved. It may even occur that the complete development process has to be re-initiated in order to cirumvent severe obstacles. Furthermore, the developed industrial process is laid out for a specific, predefined throughput, a fact which constrains the later flexibility of production significantly.

The solution of these problems is based on a simple idea: the developed laboratory-scale process is used for manufacturing of a chemical product by parallelization of many small units. Although promising great advantages over scale-up, this procedure, denoted "numbering-up", is not trivial by far. It cannot be carried out in a simple way due to the tremendous technological effort necessary: a chemical plant with hundreds or even thousands of small-scaled vessels, stirrers, heaters, pumps, etc. would be impractical. A new way of engineering and new technologies had to be developed to combine the advantages of lab-scale processing with the necessities associated with production-scale throughput. First steps into this direction have been taken, and despite some remaining throughput restrictions, first successes have become visible. Also, economical and ecological reasons create increasing demand for further steps in process intensification and sustainable development.

The present book is devoted to both the experimentally tested micro reactors and micro reaction systems described in current scientific literature as well as the corresponding processes. It will become apparent that many micro reactors at first sight "simply" consist of a multitude of parallel channels. However, a closer look reveals that the details of fluid dynamics or heat and mass transfer often determine their performance. For this reason, besides the description of the equipment and processes referred to above, this book contains a separate chapter on modeling and simulation of transport phenomena in micro reactors.

Using specific examples of gas-phase, gas/liquid and liquid-phase reactions, the advantages of microstructured reactors are highlighted in comparison to conventional equipment. At the same time, known problems are pointed out and some processes are listed for which micro reactors so far failed to show superior performance. Furthermore, the book is conceived as a compendium. Processes, microstructured reactors and chemical reactions are described in an integrated manner, providing in each case the relevant original citations. Equipped with the data given in this book, readers will be able to identify the most suitable reactor to successfully perform a given chemical reaction on the micro scale.

By now, Chemical Micro Process Technology has been established as an independent discipline, bringing forth over 1500 publications in the last few years, and an end is not foreseeable. The surge of scientific cognitions encouraged the authors to write this book, which should provide a deeper insight into this new and fascinating subject.

We are very grateful to those who helped this project become reality. In particular, we would like to mention K. Bouras, T. Hang, C. Mohrmann, and L. Widarto, who prepared electronic versions of many of the figures appearing in this book. We also wish to thank C. Mohrmann and L. Widarto for handling the copyright transfer formalities and T. Hang for taking pictures of some of IMM's micro devices. A special thanks goes to B. Knabe and R. Schenk for helping us with literature retrieval. Last but not least, we are indebted to K. S. Drese and F. Schönfeld for the thorough checking of parts of our manuscript.

Mainz, November 2003   *The authors*

Contents

List of Symbols and Abbreviations

$\langle \ldots \rangle$	Ensemble average
A	Cross sectional area
A	Coefficient matrix
a	Chemical species
A'	Coefficient matrix
$a_{\mathrm{p}}, a_{\mathrm{i}}$	Numerical coefficients
ASE	Advanced silicon etching
a_{spec}	Internal surface area
ATR	Attenuated total reflection
$a_l^{u_i}$	Numerical coefficient
b/a	Channel aspect ratio
$b_{\mathrm{i,P}}$	Source term in Navier-Stokes equation
$(b_i)_k$	Transformation vector
Bo	Bodenstein number
Boc	tert.-Butyloxycarbonyl
c	Species concentration
Ca	Capillary number
c_{a}	Concentration of species a
CAD	Computer aided design
c_{b}	Concentration of species b
CCD	Charge-coupled device
c_{f}	Fluid specific heat
CFD	Computational fluid dynamics
CGC	Constrained-geometry catalyst
c_{i}	Concentration at node i
c_{i}	Concentration of species i
$C_{\mathrm{i\pm1/2}}$	Flux limiter
c_{p}	Specific heat
CSTR	Continuous-stirred tank reactor
CVD	Chemical vapor deposition

$\bar{c}$	Concentration averaged over the cross section of a tube
$\tilde{c}_i^e(\omega)$	Laplace transform of the effective concentration field
$\tilde{c}$	Smoothed volume-fraction function
d	Bubble diameter
d	Typical length scale
D	Diffusion constant
D	Channel diameter
D	Distance
Da	Damköhler number
DBU	1,8-Diazabicyclo[5.4.0]-undec-7-ene
DCC	1,3-Dicyclohexylcarbodiimide
DCM	Dichloromethane
D_e	Dispersion coefficient
D_h	Hydraulic diameter
D_i	Species diffusion constant inside a pore
Dmab	4-[N-(1-(4,4-Dimethyl-2,6-dioxocyclohexylidene)-3-methylbutyl)-3 amino]benzyl
DMAP	4-Dimethylamino pyridine
DMF	N,N-Dimethyl formamide
DNDA	N,N'-Dialkyl-N,N'-dinitro-urea
DRIE	Deep reactive ion etching
DSMC	Direct simulation Monte Carlo method
D_e^{cur}	Dispersion coefficient in curved ducts
e	Channel depth
E	Activation energy
e	Thermal energy density
E	Magnitude of electric field strength
EDCI	3-Ethyl-1-(3-dimethylaminopropyl)-carbodiimide hydrochloride
EDDA	Ethylenediamine diacetate
EDL	Electric double layer
E_i	Electric field strength
EMA	Effective-medium approximation
EOF	Electroosmotic flow
f	Indicates the fluid phase
F	Number of molecules per unit area and time hitting a surface
F	Cost function
f_0	Maxwell distribution
FCT	Flux-corrected transport
FDM	Finite-difference method
FEM	Finite-element method
FEP	Fluorinated ethylene propylene
FFMR	Falling film micro reactor

Fmoc	9-Fluorenylmethoxycarbonyl
f	Friction factor
FTIR	Fourier transform infrared
FVM	Finite-volume method
GC	Gas chromatography
GHSV	Gas hourly space velocity
g_i	Gravity vector
GPC	Gel permeation chromatography
h	Channel height
h	Perturbation function
HPLC	High performance liquid chromatography
i, j, k, l, m, n	Summation indices
ID	Inner diameter
IR	Infrared
J_i	Thermodynamic flux
K	Dean number
k	Thermal conductivity
k	Reaction rate constant
k	Heat transfer coefficient
K	Permeability
k_0	Pre-exponential factor of Arrhenius equation
k_B	Boltzmann constant
k_L	Specific interface in gas/liquid systems
$k_l\alpha$	Mass-transfer coefficient
k_n	Time-dependent dispersion coefficient
Kn	Knudsen number
K_w	Reaction rate constant
L	Characteristic length scale of the flow domain
L	Length of a tube
l	Length of a plug
L	Length of a channel
L	Length of a channel segment
Lab-Chip	Lab-on-a-chip
LBC	Laboratory column
LC	Liquid chromatography
LC-MS	Liquid chromatography coupled wit mass spectrometry
L_{hy}	Hydrodynamic entrance length
LIGA	German acronym for lithography, electroforming, moulding (Lithograpie, Galvanik, Abformung)
LPCVD	Low pressure chemical vapour deposition

L_s	Slip length
L_{s0}	Reference slip length
L_{th}	Thermal entrance length
m	Molecular mass
MBC	Micro bubble column
MCR	Multi-component reaction
MD	Molecular dynamics
MS	Mass spectrometry
MSE	Micro-strip electrodes
$\dot{m}$	Mass flow rate
n	Coordinate normal to a wall
N	Number of molecules
n_a	Molar amount of a
n_i	Unit vector normal to an interface
n_i	Outward normal vector
n_i	Number of moles of species i
NIR	Near infrared
Nml	Standard milliliter
NMR	Nuclear magnetic resonance
NPT	Normal pressure and temperature
Nu	Nusselt number
OAOR	Oxidation and outgassing reduction
P	Grid node
p	Pressure
p	Partial pressure
P	Poincaré map
P	Channel perimeter
PDE	Partial differential equations
PDMS	Polydimethylsiloxane
Pe	Peclet number
Pe^*	Modified Peclet number containing the Taylor-Aris dispersion
PLIC	Piecewise-linear interface construction
PMMA	Poly methylmethacrylate
Pr	Prandtl number
PTFE	Poly tetrafluorethylene
PVD	Physical vapour deposition
Q_f	Orthogonal subspace
$\dot{q}$	Heat source
$\dot{q}_v$	Heat source due to viscous dissipation

R	Tube radius
r	Source term due to chemical reactions
r	Distance between molecules
R	Gas constant
R	Mean radius of curvature of a channel
r	Radial coordinate of a tubular geometry
Re	Reynolds number
R_{ij}	Matrix defining how a specific reaction contributes to a change in concentration of the chemical species involved
r_j	Rate of the jth reaction
R_s	Radius of curvature along an interface
RTD	Residence time distribution
$\bar{r}$	Mean radius of a pore

s	Indicates the solid phase
s_{abs}	Adsorption probability at an active site of the surface
Sc	Schmidt number
s_{des}	Site-specific desorption probability
SDS	Sodium dodecyl sulphate
SEM	Scanning electron microscopy
s_i	Unit vector
S_{ij}	Surface of a computational cell
S_{kat}	Surface area of a catalyst
SLIC	Single-line interface construction
slpm	Standard liter per minute
SOI	Silicon-on-insulator
SPOS	Solid-phase organic chemistry
STP	Standard temperature and pressure
S_Φ	Source term
$\dot{S}$	Entropy generation per unit time
$(S_\Phi)_P$	Value of a source term at node P

T	Temperature
T_c	Critical temperature
TEM	Transmission electron microscopy
THF	Tetrahydrofurane
t_i	Unit vector
TOF	Turnover frequency
TOF-MS	Time-of-flight mass spectrometry

u	Magnitude of velocity
u	Line velocity
u	Typical velocity scale
U	Mean flow velocity
u_i	Flow velocity vector

u_{max}	Maximum velocity
u_p	Velocity at the wall
UV	Ultraviolet
u_i^{int}	Velocity of an interface
$\bar{u}$	Average velocity
u_i^m	Velocity field at time step m
V	Volume flow
Vi	Viscous number
V_{ij}	Volume of computational cell (i,j)
VIS	Visible
V_{kl}	Interaction potential between molecules k and l
VOF	Volume-of-Fluid
w	Channel width
W, P, E	Computational nodes
W_c	Micro channel width
w_c	Channel width
WGS	Water-gas-shift reaction
W_{ij}	Transport coefficient
x_i	Spatial coordinate vector
X_i	Thermodynamic force
$x_i^{(k)}$	Spatial coordinate i of particle k
$(\dot{y}_s)_j$	Expansion coefficient for chemical reaction kinetics
z	Coordinate along the axis of a pore
Z_{eff}	2×2 tensor related to the slip flow on a grooved surface
Δp	Pressure drop
Δx	Grid spacing
Φ	Field quantity
Γ	Diffusivity
Λ_e	Effective thermal conductivity tensor
Λ_{ij}	Kinetic coefficients
Ψ	Electric potential
α	Heat transfer coefficient between a fluid and a solid
α	Aspect ratio of a channel
β	Dimensionless parameter representing a pseudo-homoge-neous reaction
δ_{ij}	Kronecker symbol
ε	Porosity
ε	Dielectric constant

ε	Energy scale
γ	Ratio of specific heats
γ	Liquid/vapor surface tension
γ_{SL}	Solid/liquid surface tension
γ_{SV}	Solid/vapor surface tension
$\dot{\gamma}_c$	Critical shear rate
κ	Local curvature of an interface
λ	Thermal conductivity
λ	Mean free path of a gas molecules
λ_a	Eigenvalue
λ_e	Effective thermal conductivity
λ_f	Fluid thermal conductivity
λ_{nc}	Correction factor accounting for the non-circularity of a channel
λ_s	Solid thermal conductivity
$\lambda_{s,eff}$	Effective thermal conductivity
μ	Dynamic viscosity
μEDM	Micro electro discharge machining
μTAS	Micro-total-analysis system
ν	Critical exponent
ν_i	Stoichiometric coefficient of species i
θ	Contact angle
θ	Angle to the main flow direction
θ	Surface coverage
θ_a	Advancing contact angle
θ_d	Dynamic contact angle
θ_r	Receding contact angle
ρ	Density
ρ_f	Fluid density
ρ_n	Residual
σ	Correlation length for density fluctuations in a fluid
σ	Interfacial tension
σ	Liquid conductivity
σ	Range of a potential
σ_T	Thermal accommodation coefficient
σ_v	Tangential momentum accommodation coefficient
τ	Intrinsic time scale
τ_{ij}	Stress tensor
ξ_i	Computational space curvilinear coordinates
ζ	Zeta potential

1
A Multi-faceted, Hierarchic Analysis
of Chemical Micro Process Technology

To give a thorough, rational review of the field of chemical micro-process technology itself, one ideally would like to follow a deductive analysis route, pursuing a bottom-up approach. First, one may provide a definition of micro reactors, then search for the impacts on the engineering of chemical processes, and try to propose routes for exploitation, i.e. applications. Alternatively, for a less comprehensive, but more in-depth description, one could use a top-down approach starting with a selected application and try to design an ideal micro reactor for this.

Such ideal scenarios could not be followed in this book and are not practised in any other publications worldwide so far, since the developments and the theory in the field are not complete. Since mere deduction, therefore, could not be followed, the way of narration chosen gives a multi-faceted, hierarchic analysis of the field of chemical micro processing.

At four hierarchic levels, groups of entities are given in such a way that both the groups and the entities have casual relationships to each other, i.e. adapt a certain linear order (Figure 1.1). This main row may have side arms, comprising entities of minor importance. Each entity as well as the mutual relationships between the entities are not described in a complete, analytical fashion (for the reasons given above), but are outlined in a more heuristic or phenomenological way by discussing multiple facets thereof. Each entity is discussed in a corresponding section within this chapter; the facets are given in a sub-section. Thereby, the analysis given here has a kind of encyclopedia format.

At a first level, *fundaments* related to the use of micro-process devices are discussed. The nature of these fundaments is considered in Section 1.2.

At a second level, a *definition* or *conception* of a micro reactor, the so-called *micro-reactor differentiation*, is provided by analysis of its constituting properties. Conceptual approaches that arose from this definition are highlighted in the following such as *numbering-up, the multi-scale concept* and *process intensification*. This is accompanied by supplying auxiliary information and definitions on *interrelated topics* such as green chemistry, micro total analysis systems, and others. At the end of description of the first level, information on the micro-reactor history and *forums/ organizations* acting in the field is given.

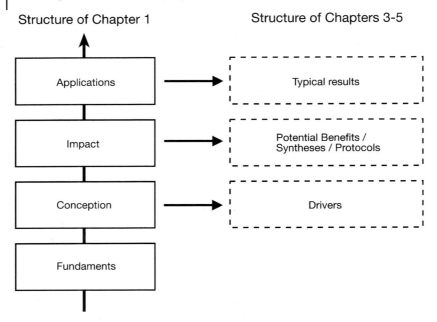

Figure 1.1 Entity hierarchy: four-level architecture of this first chapter resulting in the topics implication – conception – impact – applications. This logic is kept in three of the four following chapters, being sub-divided into drivers – potential benefits/syntheses/protocols – typical results.

A micro reactor according to the logic given here uses the fundaments to permit *impacts* on several entities, all related to chemical micro processing. Further, the new way of processing has impacts on mankind and environment. These various kinds of impacts provide altogether a third level, i.e. a third group of entities. Some of the impacts are in a consecutive relationship, some are in a parallel position. The nature of these fundaments is disclosed in Section 1.2.

Besides naming the third level 'impact', one could also term it *cognition* about what the conception actually can do.

At a fourth and final level, the impacts determine which *applications* are derived from the conception (or from the micro reactor itself).

The four-levelled approach is a guideline for the configuration of three of the four following chapters, describing experimental achievements made with micro reactors. The conception transfers to the *drivers* that trigger investigations with micro reactors. The impacts are evident when cognising on the potential benefits and developing a route to achieve this, i.e. choosing proper *experimental protocols*. The applications are specified and corroborated by *typical results*. This strategy was used for the three experimentally based chapters on gas-phase, liquid-phase and liquid/liquid-phase and gas/liquid-phase reactions. Of course, this could not be applied to modeling and simulation; hence Chapter 2 has a different structure.

1.1
Micro-reactor Differentiation and Process Intensification

1.1.1
Structure or Being Structured? Miniature Casings and Micro Flow

Micro reactors use *chemical micro processing*, which is continuous flow through regular domains with characteristic dimensions much below those applied in conventional apparatus, typically in the sub-millimeter range. There is a mutual relationship between structure and a process medium which is structured. Geometrically defined flow demands precisely structured small units for *casing* it. It is this synergetic interplay between geometric confinement and novel ways of processing that leads to the beneficial properties of micro reactors to be discussed below.

Today's micro-fabrication techniques have contributed substantially to providing miniature casings for chemical micro processing with much wider flexibility and accessibility (see, e.g., [1–4]). Besides simple *confinement*, today's precise manufacturing creates micro-flow elements of virtually any shape. Shaping in this way provides miniature passive elements, such as interdigital structures, which assume a specific *function*, e.g. for multi-lamination mixing, when passed by fluid flow. Accordingly, the high standard and variety of today's microfabrication techniques allow the provision of tailor-made micro-reaction units, possibly with a flexibility higher than for macroscopic reactors. Combined with the aid of modern simulation and the speed of modern screening approaches to determine optimum experimental parameters, there is definitely potential for a steep learning curve. If this potential is brought into application, micro reactors may affect the operation of chemical processes in a way not previously known in chemistry and, with huge value to the development of conventional apparatus, to an extent also not known in chemical engineering.

1.1.2
Symmetry and Unit Cells

A favorable feature of many micro flows is regularity or symmetry, i.e. manifold repetition of internal flow structures as a constituting element. First, this relates to flows in channels or in platelets (with channel arrays) parallel to each other. Second, this flow symmetry may even concern the flow within one micro channel, e.g. regarding the radial extension, the axial evolution, or both. To illustrate this abstract consideration by examples, focused and normal multi-lamination patterns and slug-micro flow are flows of high symmetry (Figure 1.2). The latter flow pattern is composed of a row of alternating segments, consisting of immiscible liquids or gas/liquid segments.

When symmetry is given, there is no need to consider the flow in a micro device or even a micro channel as a whole, but favorably one can refer to only a small part of it and regard this as a true functional region. A first-approach description for slug flow, according to the above consideration, may therefore refer to the repeti-

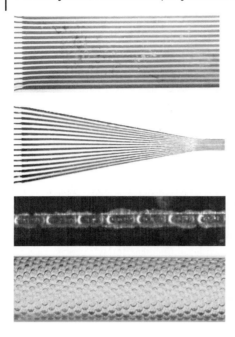

Figure 1.2 Normal and focused multi-lamination flow patterns, slug flow composed of gas/liquid segments ('Taylor flow'), and ordered foam flow ('hexagon flow') (from top to bottom).

tion unit only, i.e. to a single slug. This confined processing region may be termed unit-flow cell or unit cell.

Owing to the repetitive structure of the flow domain, a rational design of micro reactors is alleviated (see also the discussion on first-principle modeling approaches in Section 1.4.4.3). Instead of having to consider the complete system, often the study of a representative unit cell is sufficient to gain valuable information on the behavior of the full set-up. In practical terms, this leads to a more thorough involvement of simulation/modeling in future micro-reactor investigations which, in turn, will demand more microfluidic experiments in advance of the engineering and design phase of microstructured devices.

1.1.3
Process Design Dominates Equipment Manufacture and Choice

Knowing about ideal unit cells has the potential to turn the situation around in process development (Figure 1.3). First, the process is 'designed' for a very confined volume, the unit cell. Here, modern simulation techniques undoubtedly have a key role (see Chapter 2). The choice of the best unit-sized 'process design' is followed by choosing proper micro reactor elements and architectures, then repeating them and, in this way, constituting the total micro-reactor device. As the net result of this novel approach in process development, the reaction (or the process) becomes pre-eminent over the reactor, simply because the best reactor allows the reaction to be in the limelight.

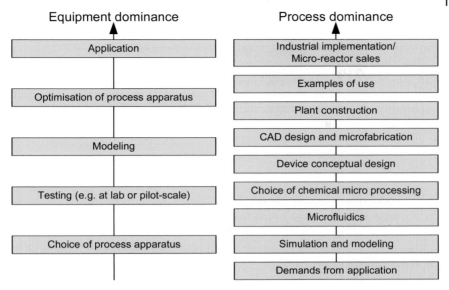

Figure 1.3 Today's bottom-up (equipment dominance) and sketch of future top-down (process dominance) approaches in process development.

1.1.4
Micro-reactor and Chemical-micro-processing Differentiation

Many authors regard chemical micro engineering as a separate, novel discipline (see, e.g., [5]), the term micro-reactor engineering being used already in an early citation [6]. What is essentially different about micro reactors, i.e. what differentiates micro-reactor and chemical-micro-processing?

Older definitions ascribe to micro reactors their manufacture as a distinguishing feature; a micro reactor accordingly is termed a device that is, at least partly, made by microfabrication or modern precision-engineering techniques (see, e.g., [3]). As a result, micro reactors have sub-millimeter internal dimensions. This definition was very exact and useful at that time, however, in the light of modern investigations, it seems to be somewhat too narrow. What if a device is composed of 'micro channels', albeit not being made by microfabrication techniques? Then undoubtedly one could rely on virtually the same micro-flow processing as with microfabricated devices. Should such a device not be termed a micro reactor? In this sense, the older definitions were technology-driven; therefore, in the following it is aimed at finding a more application-driven definition.

Micro-reactor differentiation
A micro reactor is a casing for performing reactions that was designed or selected to induce and exploit deliberately micro-flow phenomena, i.e. flow guidance and flow processing with characteristic dimensions much below that of conventional apparatus, typically below the sub-millimeter range.

In the same way, any other micro-flow device performing other processing functions can be defined. By custom, all such micro-flow devices are subsumed under the term 'micro reactor', although strictly this is not correct.

How, then, do we define chemical-micro processing (or micro-reaction technology, micro-chemical process engineering, or whatever substitute naming is used)? Is that synonymous with simply using a micro reactor for an application, i.e. replacing conventional equipment by novel micro reactors? It is not; in lieu of the conceptions made in Section 1.1.3, this is an essential, but not a commensurate condition.

> *Chemical-micro-processing differentiation*
> Chemical micro processing is based on 'process design' in a unit cell before manufacturing new or selecting existing micro-processing equipment, composed of a multitude of such cells. This results in having tailored processing equipment at the micro-flow scale.

A few micro reactors, especially those having large throughput, may fulfil the criteria of *process intensification* (PI) (see Section 1.1.6). Chemical micro processing can then be a branch of process intensification. More often, process intensification will rely on novel, specifically designed reaction equipment of larger scale such as spinning-disk reactors [7], i.e. tools which neither are micro devices nor have micro-flow equipment. Accordingly, micro-flow devices, PI reactors, and conventional apparatus constitute processing equipment of varying scale, the so-called *multi-scale concept* (see Section 1.1.7).

Owing to their normal size of equipment and of processing volumes, PI devices usually cannot refer to a unit-flow description as given for micro reactors, since they are dealing with non-micro flows. Instead, an analysis of their performance is usually based on a more gross or global description of the flows. In turn, they have the advantage of bridging the gap from conventional chemical apparatus to micro-flow devices. Thus, while still exhibiting improved performance, PI devices may also meet requirements of productivity, costs and reliability in a way which micro reactors cannot satisfy.

1.1.5
Numbering-up

1.1.5.1 Progressive Increase in Capacity by Addition of Modules

There are opinions that micro-flow processing, the repetition of unit cells (see Sections 1.1.2 and 1.1.3), not only holds at the laboratory scale, but may also be applied to – smaller or larger – production as well [5]. This is achieved by numbering-up of the elements, i.e. by multiple repetitions of small, precisely structured regions. In this way, unit-cell processing should be maintained at enlarged to process-unit size, although capacity is increased significantly – very different from conventional processing. By numbering-up, potentially the pilot-scale processing may be by-passed [5] and a phased (gradual) increase in production volumes may be achieved by progressively bringing on-stream micro reactor modules.

1.1.5.2 **Internal vs. External Numbering-up: Scaling-out of Elements or Devices**

Numbering-up can be performed in two ways (Figure 1.4). *External numbering-up* is referred to as the connection of many devices in a parallel fashion [8] (a similar, but less elaborate, definition was already provided in [9, 10]; see also [11] for a realized industrial example). A device in the sense as it is used here is defined as a functional element, e.g. a micro-mixing flow configuration such as an interdigital

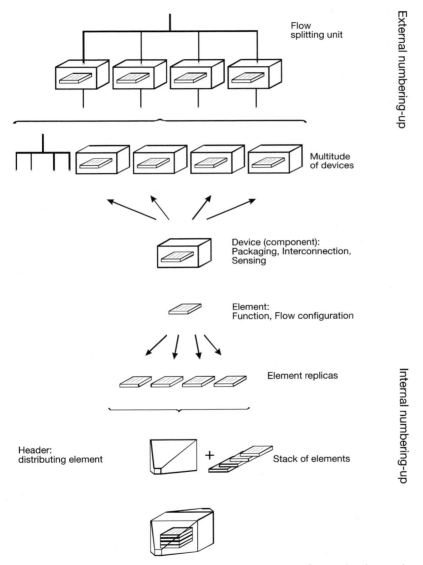

Figure 1.4 Schematic representation of the generic conceptions of external and internal numbering-up [8].

Figure 1.5 Photograph of the liquid-flow splitting unit for liquid/liquid processing with three tanks and six separation-layer micro mixers [8].

feed array with mixing chamber attached, which is encased and has outer connections for fluid supply and withdrawal. Combining of devices is thus achieved via their outer connections which most often follow commercial standards. Accordingly, conventional tubing with standard process-control equipment may be used here, but can suffer from fluid equipartition problems as mentioned in [11]. For this reason, a first specialty fluid-splitting tool was recently proposed and developed. The construction of this six-fold liquid splitting unit is given in Figure 1.5.

External numbering-up is numbering-up in the truest sense. Virtually, the complete fluid path is repeated. This resembles also the real meaning of scaling-out.

Internal numbering-up means the parallel connection of the functional elements only, rather than of the complete devices. These elements are grouped in a new way, usually as a stack, and are encompassed in a new housing. This housing typically contains one flow manifold and one collection zone, most often having a simple design like a header of diffusor. Although often overlooked, internal numbering-up actually is state of the art. To name only a few realized examples, a mixer array with 10 parallel interdigital units [12], a gas/liquid contactor array with 10 packed beds [13], and micro heat exchangers with hundreds of parallel platelets [14] (see [15] for a recent development) have been realized (Figure 1.6). The internal numbering-up of the latter type of device, for instance, permits a throughput of up to 7 t h^{-1} water flow [14]. Hence one device, of a size of a shoebox up to a computer, may be sufficient for a complete production.

External numbering-up benefits from true repetition of the fluidic path, and hence preserves all the transport properties and hydrodynamics, determined in advance for a single-device operation. This is commonly believed to result in a

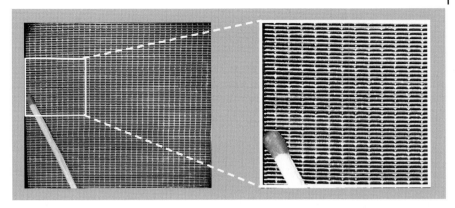

Figure 1.6 Photograph and detailed view of 6685 parallel micro channels which are numbered-up internally to give a micro-flow heat exchanger of large capacity [15].

faster time-to-market of the process development. The disadvantage is the need for a sophisticated monitoring and control system, in particular to achieve fluid distribution. Another drawback stems from the unfavorable ratio of housing (i.e. reactor) material to active internal volume. At the laboratory scale (single-device operation), this does not matter, whereas at a production scale it is almost not practicable. It is easy to imagine that even only a few tens of parallel connected devices with all their tubing attached do not present really elegant, compact engineering solution. Furthermore, the costs for micro-reactor fabrication are raised considerably owing to the increasing material demand and, more especially, to the need to structure the housings multiple times.

In contrast, internal numbering-up provides, as the existing examples demonstrate, compact reactor architecture at reasonably high throughput. Meanwhile, advanced fabrication exists, reducing the costs of such production devices more and more to an acceptable level. Fluid distribution is manageable, in particular using modern simulation methods. The disadvantage is that possibly a slight redesign of the fluidic path, sometimes even of the functional element, has to be made. This, however, can be avoided by the taking into account during design of the initial single device how the internally numbered-up production device may be arranged later.

Obviously, internal numbering-up has more arguments in favor than external numbering-up. Hence why propose a liquid-splitting tool for external numbering-up? There are still good reasons to do so. First, not all micro-reactor processes are amenable to internal numbering-up, e.g. for fluid dynamic reasons. Concerning the latter, the application reported for the first external numbering-up device refers to the field of precipitating organic synthesis and in some way stands also for crystallization. The special feature of the corresponding micro-flow processing is tri-layered droplet formation causing delayed mixing [16, 17]. The droplet can be made using a single separation-layer micro mixer; the same processing cannot, however,

Table 1.1 Some thoughts on features of chemical processing which favor the choice of either external or internal numbering-up. Provided that more in-depth knowledge is gained, such a (modified) list may provide selection criteria for choosing the right splitting concept when looking for scaling-out.

External numbering-up	Internal numbering-up
Low degree of parallelism	High degree of parallelism
Favoring multi-phase processing, including powder making	Favoring single-phase processing
Bubble/droplet/film formation or movement; surface flows	Stream guidance; volume (channel) flows
Specialty, complex processes	Unit operations; standard reactions
Precious products; functional chemicals	Bulk chemicals; fine chemicals
Slow reactions/processes, i.e. demand for laminar-flow post-processing, e.g. by delay-loop	Fast reactions/processes, i.e. no demand for laminar-flow post-processing, process completed within device
Hazardous processes	Safe processes

be simply made by an internal parallel arrangement. As a rule of thumb, such solid/liquid processes and, in a wider sense, multi-phase processes in general may restrict (or at least favor) parallelization to external numbering-up. These and more thoughts in favor of or against a numbering-up concept are given in Table 1.1.

In addition, when carrying out a process in the explosive regime, the whole fluidic path possibly has to be set below a certain characteristic dimension to guarantee inherent safety (see, e.g., [18]). Hence the processing of hazardous reactions may demand an optimized, well-understood specialty device, rather than restarting developments at the level of internal numbering-up. The use of any conventional tubing or any larger-scale own-constructed splitting unit (such as the one here) may be prohibitive, as it does not fulfil the stringent safety requirements.

Finally, external numbering-up may be chosen for simple, practical reasons, if the degree of parallelism does involve only low numbers, e.g. not exceeding 10 devices. Developing an internally numbered-up device will demand development costs and time and, therefore, may be inefficient at a low degree of parallelism. Meanwhile, since micro reactors are available commercially and tentatively will become cheaper, the rapid availability of a small series of devices is no longer impossible.

1.1.5.3 Issues to be Solved; Problems to be Encountered

The numbering-up concept demands achieving absolutely uniform flow equipartition by placing special headers in front of the parallel micro channels [5]. Although this has been solved for tubes in conventional multi-tubular reactors, the expenditure for equipartition in micro channel stacks is assumed to be higher, since the small channels may have more relative differences in structural precision. In addition, their sensitivity to particulate clogging may be higher, requiring a filtration system. Fouling inside the channels may be even more detrimental, since there may be major difficulties with removal. Heat management, already an issue at the device scale, will demand more elaborate solutions for a scale-out device, taking into account the heat management of each subunit. Another issue also refers to the multi-unit architecture, namely sensing and controlling all of the subunits in a reliable and robust manner. Here, a first approach based on valves, flow meters, power supply, and control circuitry was proposed by industry and academia [19, 20].

The existence and the success of scaled-up industrial plants using micro or mini reactors reported in the open literature [11, 21], however, gives hope for the possibility overcoming all the problems mentioned above, at least for selected applications.

1.1.5.4 Limits of Mini- and Micro Plants for Scale-up

There are not only limitations concerning the achievement of numbering-up, but also concerning the right use of a numbered-up micro-reactor plant.

In this context, the simulation of complete plant behavior, including recycling and energy management, at the mini-plant or micro-plant stage, is critically reviewed by Zlokarnik [22]. Mini plants describe homogeneous-phase processes adequately; scale-dependent processes can be processed satisfactorily. However, Zlokarnik admonishes that mini-plants in the general case are not suited to develop scale-up rules for scale-dependent processes. This holds even more for micro plants, since they will lose their performance upon an increase in device size, i.e. scale-up (even when keeping the relative internal dimensions). (Parallel) numbering-up is hence the only right means to scale-out micro devices.

1.1.5.5 First Large-capacity Numbered-up Micro-flow Devices Reported

Having reviewed the limitations predicted, one should have a look at the achievements that already have been made with numbering-up.

Numbering-up within one device, i.e. internal numbering-up, is no longer a dream, but is reality. For a long time, the Forschungszentrum Karlsruhe has reported on parallel micro heat exchangers that have a productivity of up to $7 \, t \, h^{-1}$ liquid flow per passage, leading to a heat-transmission power of 200 kW, at an outer volume of only $27 \, cm^3$ [23].

The Institut für Mikrotechnik Mainz GmbH (IMM) recently introduced a production- and pilot-scale gas/gas counter-flow heat exchanger comprising microstructured channel arrays [15] (see Figure 1.6). This stainless-steel device is equipped with industrial flanges, and is about 36 kg in weight and 54 cm long. It is operated at gas throughput in the range of $m^3 \, min^{-1}$ at 100 mbar pressure drop, having a power capacity of about 10 kW at a thermal efficiency above 95%. The internals consist of a stack of microstructured plates having multi-channel arrays of a channel width of 2 mm, depth of 250 µm, and length of 240 mm. In total, 6685 micro channels are operated in parallel in this device. The flanges allow installation in large-scale industrial plants.

1.1.5.6 **First Complete Test Station for Multiple-micro-reactor testing**

DuPont and MIT built a so-called *turnkey multiple-micro-reactor test station* providing reactors with fluidic distribution and control components in a scaleable fashion (Figure 1.7) [2, 19]. A detailed design description for a whole plant concept starting from one generic idea is provided. A key feature of this concept is the use of standard components originating from the semiconductor industry for micro chemical processing, either in the original version or partially modified. This holds for, e.g., the so-called *known good die sockets* of Texas Instruments and for the special type of arrangement chosen that was oriented at the concept of *printed circuit boards*. Available microtechnical standard components such as MEMS-based mass flow controllers of Redwood Microsystems and further commercial equipment such as *programmable logic controllers* of Siemens are also integrated. With respect to these monitoring and controlling functions, the time-scale of temperature change is given exemplarily. Temperature is reported to be changeable within about 3 ms, allowing real-time algorithms. Besides the use of standard components, the micro-reactor test station is characterized by showing up technical solutions for a fluid manifold, the essential component for numbering-up.

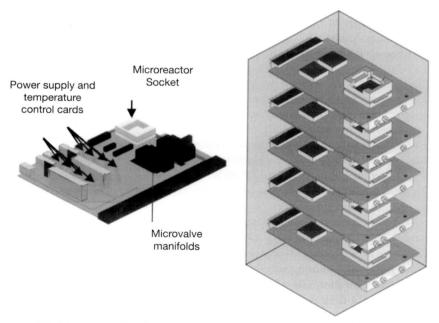

Figure 1.7 Schematic outline of the constructional principle of the multiple-micro-reactor test station [2].

1.1.6
Process Intensification

1.1.6.1 Definitions

Process intensification is a concept aimed at notable improvements in chemical processing. Ramshaw, one of the pioneers in the field, defines process intensification as a design approach for making dramatic reductions in the physical size of a plant while achieving a given production objective [24] (see also [25] and [5]). A volume reduction of 100-fold or more was regarded as dramatic by Ramshaw [24], whereas newer publications, with the achievements of the last decade, give two as a realistic number [25]. The size reduction in equipment, which is typically a plant or a system, can come from shrinking the parts of the equipment, i.e. device miniaturization, or from omitting complete parts, i.e. unit operations, from the equipment.

Stankiewicz and Moulijn extend the definition of process intensification from size reduction of apparatus to dramatic improvements in key features of chemical processing [25]. Substantial decreases in the following process-engineering properties are listed:

- ratio of equipment size to production capacity
- energy consumption
- waste production

Therefore, in this definition process intensification encompasses both novel apparatus and techniques which are designed to bring dramatic improvements in production and processing (Figure 1.8) [25]. As a result, safe, cheaper, compact, sustainable (environmentally friendly), and energy-efficient technologies are obtained.

Process intensification means following engineering methods and using appropriate apparatus; improvements in materials, e.g. catalysts, or chemical paths, are not regarded as falling into this category [25]. The list of engineering methods includes the integration of operations such as reaction and separation [26] (multifunctional reactors), the use of alternative energy sources such as ultrasound, and new process-control methods such as unsteady operation. There is a tendency that process intensification will no longer be based on pure unit operations, but rather will involve hybrid forms. The impact on equipment is seen to be even more prominent; it is said that some types of apparatus will disappear from plants because of process intensification [25].

Phillips of the BHR Group, UK, provides a compact definition of process intensification, saying it is '... a design philosophy whereby the fluid dynamics in a process are matched to its chemical, biological and/or physical requirements, ...' [27]. In this way, significant benefits are gained, those listed above.

1.1.6.2 Matching Fluidics to Physico-chemical Requirements of a Reaction

Following the latter definition, the BHR Group <www.bhrgroup.co.uk> gives straightforward goals reflecting how they envisage a chemical process becoming

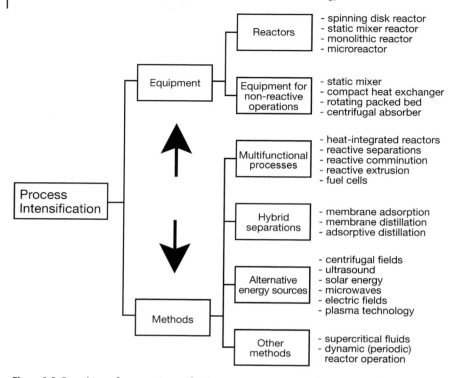

Figure 1.8 Branching of process intensification into engineering methods and corresponding equipment [26].

ideal. They propose to match variaous dynamic features with the physico-chemical requirements of the reaction:

- match reaction rate with mixing rate
- match mechanism with flow pattern
- match reaction time with residence time
- match exothermicity with heat transfer

1.1.6.3 Relationship of and Difference between of PI and Micro-reaction Technology

Micro-reaction technology can be one of the tools that process intensification may use [5]. Hence chemical micro processing and process intensification have a share, where the former supplies devices for the latter concept or purpose. However, both chemical micro processing and process intensification also cover unique aspects that the other field does not comprise.

Strictly, chemical micro processing, in addition to being a device field based on micro channels, is a means to use micro flows, which is oriented not at one, but rather at a multitude of purposes. Process intensification, also strictly, is a concept (but specifying no concrete means) and apparatus for a specific purpose (see above).

Hence chemical micro processing may also be applied for purposes other than PI, e.g. screening, which is not related to production and, accordingly, does not fall into the category of process intensification (see the definition given above).

1.1.6.4 Process Intensification Achieved by Use of Micro Reactors

As examples of micro-channel process intensification and the respective equipment, in particular gas/liquid micro reactors and their application to toluene and various other fluorinations and also to carbon dioxide absorption can be mentioned [5]. Generally, reactions may be amenable to process intensification, when performed via high-temperature, high-pressure, and high-concentration routes and also when using aggressive reactants [5].

1.1.7
The Multi-scale Concept

Hessel and Löwe report on hybrid, i.e. multi-scale, approaches which are currently most often favored for micro-reactor plant construction, simply for practical time and cost reasons [9, 10]. In addition, such an approach allows one to fit micro reactors in existing industrial, producing and academia, measuring environments. The micro reactor is only used where it is really needed and costs for changing the processing are kept to a minimum in such a way (Figure 1.9).

The hybrid concept is generically different from the monolithic concept, such as practiced for Lab-on-a-Chip designs, aiming at integrating all processing and controlling functions into one miniature device. It stands to reason that the monolithic concept has advantages for analytics and screening, whereas the hybrid (multi-scale) concept is demanded for production and synthesis of preparative quantities. A review on concrete multi-scale architecture, including the first commercial ones (Figure 1.10), is given in [9, 10].

The hybrid approach seems to be the more pragmatic procedure [9, 10]. The impressive results gathered so far, especially on the industrial side, substantiate that this – and so far only this – concept is not lacking innovative character besides pragmatism [11, 28]. Set-ups with many micro-reactor components, pointing to a monolithic concept, will find use when each of these components on its own or the interplay between all of them gives advantages. This holds, for instance, for applications in the field of automobiles and consumer care. Here, space and weight savings are important technical drivers, thereby demanding miniaturization of all components. In turn, chemical applications typically target other issues. In any case, one should be aware that the fluidic peripherals of many of today's micro-reactor plants still take a considerably large portion of the total plant space.

The time of the hybrid approach has come already [9, 10]. It allows one without delay to analyze the advantages of micro reactors, especially facing today's industrial time demands. The monolith approach needs more time for development, but can be built on the progress achieved so far. The respective developments will certainly gain an additional impetus when more and better tailor-made applications especially for this concept are identified.

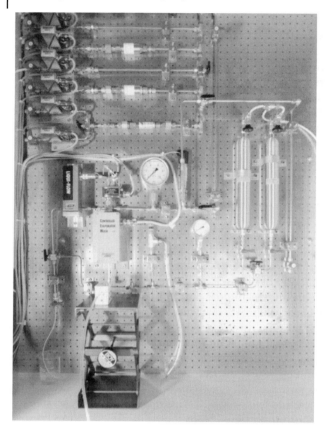

Figure 1.9 Hybrid, multi-scale micro-reactor plant for catalyst testing for propane steam reforming [15].

Figure 1.10 Hybrid, multi-scale micro-reactor plant, a commercial product of mgt mikroglas technik AG, Mainz, Germany [9].

1.1.8
A Word of Caution on the Probability of a Deductive Analysis

Needless to say, all conclusions drawn in Sections 1.1.1–1.1.7 are ideal-case considerations of an abstract nature aimed at showing the maximum potential of chemical micro processing, and the ideas behind. In reality, a performance less than ideal (but often better than conventional) may be found, at least initially, e.g. for reasons of imperfect exhibition of flow patterns or due to limits of micro flow compared with existing technology. This reality description is given in Chapters 3–5.

1.1.9
Other Concepts Related to or Relevant for Chemical-Micro Processing

1.1.9.1 μTAS: Micro Total Analysis Systems
Micro Total Analysis Systems (μTAS) are chip-based micro-channel systems that serve for complete analytics. The word 'Total' refers to the monolithic system character of the devices, integrating a multitude of miniature functional elements with minimal dead volumes. The main fields of application are related to biology, pharmacology, and analytical chemistry. Detailed applications of μTAS systems are given in Section 1.9.8. Recently, μTAS developments have strongly influenced the performance of organic syntheses by micro flow (see, e.g., [29]). By this, an overlap with the micro-reactor world was made, which probably will increase more and more.

There is no doubt that the description of μTAS investigations could fill several volumes on their own. In the following chapters, only those applications with relevance to synthetic chemistry are presented. For reviewing all other μTAS applications, actually the majority, the reader is referred to reviews in the field (pioneering literature [30–40]; up-to-date literature [41–44]; essays [45]; see also [46] for a description of a consortium) and to the proceedings of the μTAS conferences.

1.1.9.2 Green Chemistry
Haswell and Watts use the term *Green Chemistry* in way similar to the definitions for methods for process intensification (see Section 1.1.6) [47]. This refers to optimum use of material, energy, and consequent waste management, with an appreciation of the environmental resources and the need for reducing pollution. While traditionally waste was reduced at the end of the pipeline, a change is needed to reduce waste at the source. In this context, micro-reaction devices fulfil the criteria of sustainability in the way described below, e.g. because they have high selectivity, have efficient heat management, and use high reactant concentrations, reducing solvent volumes.

1.1.9.3 Sustainable Development and Technology Assessment
Jischa gives an overview of sustainable development and technology assessment, particularly from the German point of view and reviewing the German initiatives

over the last four decades [48]. Technology assessment is the older term, essentially of similar meaning to sustainable development, mainly used nowadays. Concerning sustainable development, no unambiguously accepted, detailed definition can be given, but it is related to planned resource use and focused resource development. As relevant problem areas, world-population growth, provision of energy and raw materials, and environmental destruction are seen. Micro reactors certainly are of relevance to the latter two areas. Consequently, sustainable development is concerned with environmental, social, and economic aspects.

One coalition of 165 international companies united by a shared commitment to sustainable development is the *World Business Council for Sustainable Development* (WBCSD; <www.wbscd.org>). The three pillars economic growth, ecological balance, and social progress constitute the idea of sustainable development. This movement is joined by companies from more than 30 countries and from more than 20 major industrial sectors. The aims of WBCSD are business leadership, policy development, best practice, and global outreach. The mission of WBCSD is to be a catalyst for a change towards sustainable development. Key projects refer to accountability and reporting, advocacy and communication, capacity building, energy and climate, and sustainable livelihoods. As cross-cutting themes are considered corporate social responsibility, eco-efficiency, ecosystems, innovation/technology, risk, and sustainability and markets.

1.1.9.4 Microfluidic Tectonics (μFT)

Moorthy and Beebe subsume under this term '… the fabrication and assembly of microfluidic components into a universal platform, in which one starts with a "blank state" (shallow cavity) and proceeds to shape microchannels and components within the cavity by liquid-phase photopolymerization …' [49]. Tectonics is the science and art of assembling or shaping of materials during construction. μFT is seen as new strategy to generate organic and biomimetic designs for microfluidic systems. Especially the large freedom in design and fabrication on an *ad hoc* basis are seen as benefits. Hydrogels are cross-linked polymers that can adsorb and move water by swelling and have responsive functions and were considered as construction material of particular interest.

1.1.9.5 Compact Flow-through Turbulent Reactors, also Termed Microreactor (MR) Technology

For a longer time than the onset of chemical micro processing, the term micro reactor was also used for small-scale testing continuous-flow equipment without micro channels and micro flow (see, e.g., [50]). Instead, turbulent flow was usually utilized. By this, numerous benefits over existing batch technology could be figured out, which are similar to the 'real' micro reactors. This stems from process-engineering implications which are basically the same for the two classes of micro reactors: high mass and heat transfer, shortening of residence time, and so on. Particularly interesting is that economic data are available for the use of the first type of equipment, perhaps providing valuable hints for the future industrial use of micro reactors.

Table 1.2 Key data for production of polymers in the compact reactor/mixer/heat exchanger, termed micro reactor, by cationic polymerization yielding propylene, piperylene, butylenes, etc. [50].

Process parameter	Micro reactor	Batch reactor in CIS	Batch reactor of Amoco USA
Reactor volume [m^3]	0.02–0.06	1.5–4.0	22.7
Output on raw stock [m^3 h^{-1}]	5–10	2	9
Residence time [h]	10^{-3}	1	2.5
Metal consumption [t]	0.05	7.5	40
Specific expenses [relative units]:			
– raw stock	0.84	1	1
– catalyst	0.6–0.8	1	1
– electric energy	0.75	1	–
– water	0.6–0.85	1	–

As an example, a compact high-throughput turbulent flow reactor/mixer/heat exchanger was used for various applications, including polymer and rubber manufacture [50]. Further applications refer to emulsification and intensified heat exchange.

Typical properties (Table 1.2) of these micro reactors preserved the basic characteristics of the existing processes, reduced the reactor volume with even slightly increased output and distinctively reduced residence time, thereby having a very large specific output [50]. In addition, the quality of the final product was improved, a high process flexibility was achieved, less raw material and catalyst were needed, specific expenses were reduced, and gains in economy of the process were achieved. Fewer staff members were needed for process operation.

Other types of non-micro-channel, non-micro-flow 'micro reactors' were used for catalyst development and testing [51, 52]. A computer-based micro-reactor system was described for investigating heterogeneously catalyzed gas-phase reactions [52]. The 'micro reactor' is a Pyrex glass tube of 8 mm inner diameter and can be operated up to 500 °C and 1 bar. The reactor inner volume is 5–10 ml, the loop cycle is 0.9 ml, and the pump volume adds a further 9 ml. The reactor was used for isomerization of neopentane and n-pentane and the hydrogenolysis of isobutane, n-butane, propane, ethane, and methane at Pt with a catalyst.

1.1.9.6 **Supramolecular Aggregates, Also Termed Micro Reactors**
Supramolecular organic structures such as micelles, vesicles, liposomes, or monolayers were often also termed micro reactors when being used for reaction processing (Figure 1.10) [53]. Hollow structures created by inorganic crystallites such as zeolites were named and used in a similar way [54]. All these represent reaction vessels of nanometer dimensions, spatially and parametrically segregated from the homogeneous solution. For the latter reason, these 'nano reactors', which is probably the more appropriate term, can change reactions in a way not accessible to real micro reactors. They create unique, localized chemical environments

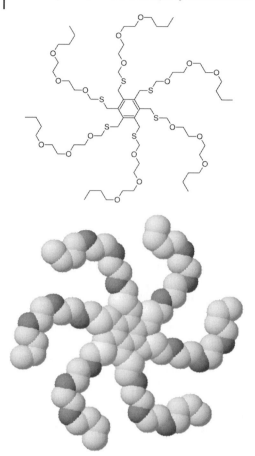

Figure 1.11 Hexapus, a multi-polar amphiphile providing a supramolecular structure [55]. Similar types of multi-polar amphiphiles were used as nano-reactors for changing the course of reactions [56].

with, e.g., modified solubility, solvent polarity, and reactant concentration. In addition, they may exhibit steric interactions not given in bulk solution such as hindered rotation in the case of molecules encapsulated in the zeolites' cavities.

In [53], segregated catalyst and polymer particles act as 'micro reactors' where the polymerization process takes place. Each particle is an individual reactor with its own energy and material balance. During polymerization, the catalyst particles undergo a change in volume by a factor of 10^3–10^4, thereby generating the corresponding polymer particles. The particle size distributions of catalyst and polymer are the same.

1.1.10
Some Historical Information on Micro-reactor Evolution

Hessel and Löwe give some information on the 'pre-history' of micro reactors by mentioning that this approach is not fundamentally novel as the foundation of the American Microchemical Society in 1934 is a relevant factor (Figure 1.12) [57, 58]. Wörz (BASF, Ludwigshafen, Germany) overstates this argument when making the – undoubtedly correct – remark that 'Microreactors are not really new, because every capillary tube is a microreactor' [28]. It is, however, also a matter of fact that a proper transformation of this system of thought into practice was pushed to the sidelines and at times is still being pushed.

The developments at the Kernforschungszentrum Karlsruhe (KfK), now the Forschungzentrum Karlsruhe, in the late 1980s and early 1990s, belong to the earliest, if not the first, reported activities on micro reactors [59–62]. In particular, Schubert et al. report on probably the first activities within industry documented in the open literature in 1993 [63]. KfK's activities were isolated until about 1994 (albeit being intense in themselves), i.e. not many other reports of other groups followed this example.

Accordingly, it is valid to claim the joint onset of micro-reaction technology in 1995 and to try to name those who were the pathfinders in this development (see, e.g., [3] for a description of such an opinion). Regarding the birth of micro-reaction

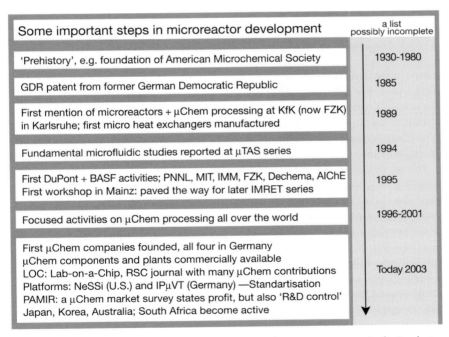

Some important steps in microreactor development	a list possibly incomplete
'Prehistory', e.g. foundation of American Microchemical Society	1930-1980
GDR patent from former German Democratic Republic	1985
First mention of microreactors + µChem processing at KfK (now FZK) in Karlsruhe; first micro heat exchangers manufactured	1989
Fundamental microfluidic studies reported at µTAS series	1994
First DuPont + BASF activities; PNNL, MIT, IMM, FZK, Dechema, AIChE First workshop in Mainz: paved the way for later IMRET series	1995
Focused activities on µChem processing all over the world	1996-2001
First µChem companies founded, all four in Germany µChem components and plants commercially available LOC: Lab-on-a-Chip, RSC journal with many µChem contributions Platforms: NeSSi (U.S.) and IPµVT (Germany) —Standartisation PAMIR: a µChem market survey states profit, but also 'R&D control' Japan, Korea, Australia; South Africa become active	Today 2003

Figure 1.12 Timetable for the micro-reactor evolution, marking important steps in the two last decades.

technology as a discipline, many researchers refer to the year 1995, when the first Workshop on Microsystems Technology for Chemical and Biological Microreactors was held in Germany [5]. A section on the pre-history of micro reactors, covering MEMS (*Micro ElectroMechanical Systems*) fabrication and the start of µTAS developments, is provided in, e.g., [5]. In [57, 58], it is stressed in the same sense that 'Microfabrication techniques have made headway with the total development of chemical microprocessing ...'.

The status of micro reactor and µTAS developments from the beginnings to 1998 is given in [64]. Selected micro devices and applications are presented, and also key players, the foundations and the main drivers for undergoing such development.

A similar overview of German activities in 1998 is provided in [65]. The German working party on micro-reaction technology is presented, a federal funding program is introduced, and three top-priority research projects are presented. They show the activities at the edge of acceptance by industry.

Wegeng et al. give an overview of US and Japan funding programs in 1996 and plans for future funding [1]. They also name interest groups, funding programs and conferences in the field of MEMS at that time. They then identify major players in Europe in 1996 and, finally, comment on German activities which were then just starting the series of *International Conferences on Micro-REaction Technology* (IMRET).

1.1.11
Micro-reactor Consortia/Forums

1.1.11.1 The Laboratory on a Chip Consortium (UK)

<www.labonachip.org.uk>
This is said to be UK's largest collaborative research project on miniaturized chemistry, mainly concerning the use of chip technology. Members are universities, vendors and user companies (see also [45]). The consortium initiates a large, multidisciplinary collaboration, which combines academic excellence, user steering, and manufacturer input.

The project is managed through the Laboratory of the Government Chemist in Teddington, UK, and is part of the British government's Foresight Link program [45]. The cost of the Lab-on-a-Chip project was £ 3.2 million. Two key tasks are the exploration of reactions and processes on a micro-scale and the commercialization of the results.

1.1.11.2 MicroChemTec and IPµVT (D)

<www.microchemtec.de>
The strategic project *Modulare Mikroverfahrenstechnik* (modular micro-chemical engineering) is funded in the framework of the German program Mikrosystemtechnik 2000+ by the German ministry for education and science, BMBF. The project was started in October 2001 and lasts 3 years.

It was initiated by work results of an industrial platform on micro chemical engineering, the so-called *IndustriePlattform MikroVerfahrensTechnik* (IPµVT). This forum combines industrial users and commercial providers of the technology, meeting in Frankfurt/Main at the Dechema organization. This forum has several reporting sub-groups which each dealing with special topics such as future device development (e.g. standardization) and chemical and biological applications.

The main aims of MicroChemTec are the development of a unit construction kit for micro reactors, definition of standardized interfaces, investigations of modules on the market for their suitability for affiliation in the unit construction kit, documentation for this purpose, and demonstration of functioning of the concept with the example of selected unit operations or processes.

1.1.11.3 NeSSI (USA)

<www.cpac.washington.edu/NeSSI/NeSSI.htm>
NeSSI is a non-affiliated international *ad hoc* group composed of over 250 people (and growing) including end-users such as Dow and ExxonMobil, and manufacturers such as A+, ABB Analytical, Siemens, Fisher-Rosemount, Swagelok, Tescom, Parker-Hannifin, Circo and many others who are looking to modularize and miniaturize process analyzer sample system components. NeSSI operates under the sponsorship and umbrella of CPAC (Center for Process Analytical Chemistry) at the University of Washington in Seattle. The membership focus of NeSSI has been on manufacturers who are willing to supply parts and components for this initiative and also end-users who are willing to do 'in-house' testing. There are now several NeSSI sampling manifold designs which conform (approximately) to the SEMI 1.5" manifold. An ISA standard is concurrently being drafted, called SP76; this, however, is not an integral part of NeSSI.

NeSSI's driver is to simplify and standardize sample system design. There is also a huge opportunity to adapt the emerging class of 'lab-on-a-chip' sensors to a miniature/modular 'smart' manifold which could fundamentally change the way in which industry does process analysis.

1.1.11.4 Micro Chemical Process Technology, MCPT (J)

<www.mcpt.jp>
The Research Association of Micro Chemical Process Technology (abbreviation: MCPT) was established in July 2002 with the aim to conduct experimental research on micro-chemical process technology through the collaborative work of its members. The association has been implementing R&D together with its members from 29 different firms of various industries including the chemical industry, the precision machinery industry, and the electrical equipments and electronics industry. In addition, the participation of national, public, and private universities as well as independent administrative institutions has been greatly appreciated.

With the support of the Ministry of Economy, Trade and Industry (METI) and the New Energy and Industrial Technology Development Organization (NEDO),

The Research Association of Micro Chemical Process Technology will establish micro-chemical plant technology and microchip technology, which is the core of micro-chemical process technology. By integrating these two technologies in an effort to establish micro-chemical process technology as the fundamental technology that is common to all technologies of the industrial world, the association will continue to dedicate itself to the R&D of the systematization of micro-chemical process technology.

Concrete aims of MCPT are to implement experimental research on high-efficiency micro-chemical process technology, to apply micro-spaces to chemical reactions and analysis technologies by utilizing semiconductor production technologies that have developed rapidly in recent years. Furthermore, the work conducted is said to contribute to the establishment of high-efficiency chemical plants, which is expected to reduce resource and energy consumption in the chemical industry. In a similar way, contributions to the development of high-speed analysis technologies using minute quantities are foreseen, which is expected to create a large-scale market of life-science products such as diagnostic chips, etc. for household and other uses. As a sum of all these targets, it is hoped to establish a fundamental technology beneficial to the industrial world, through the systematization of the aforementioned knowledge and technologies.

1.1.11.5 CPAC Micro-reactor Initiative (USA)

<www.cpac.washington.edu/micro-reactor/cpac_micro.htm>
The Center for Process Analytical Chemistry (CPAC) in Seattle (WA, USA) released a white paper specifying its capabilities on micro reactors and trying to initiate collaborations to fill vacancies in this portfolio. For these reasons, the CPAC Micro-reactor initiative will focus on the acquisition and operation of commercially available micro-reactor systems as test beds for the testing and development of analytical sensing technologies, process integration, and process control issues. CPAC also envisions the platform as a valuable educational tool in the chemistry and chemical engineering disciplines, particularly as part of the engineering unit operations laboratory section of chemical engineering. The three main aspects for implementing the CPAC Micro-reactor initiative are:

- acquire commercially available micro-reactor platforms
- select appropriate chemistries for test applications running on the micro-reactor platforms
- operate the micro-reactor platforms as test beds for research and testing.

1.2
Consequences of Chemical Micro Processing

1.2.1
Limits of Outlining Top-down Impacts for Micro Reactors

The field of chemical micro process engineering is moving fast, having impacts on chemical process industry, academia, markets, and society (see Section 1.3), and is starting to bring the technical progress made into application (see Section 1.9). Hence it is not surprising that not all impacts of this field have so far been fundamentally described, ideally in the format of a bottom-up scenario (see Section 1.4.3), and that some applications still remain underdeveloped. In the first 10 years, developments focused on establishing fabrication technology, to a considerable extent by microtechnology development, and on chemical feasibility studies. Detailed, e.g. mechanistic, scientific and technical analyses, as well as practical scenarios, e.g. cost analysis of a process, are left to future studies.

Nevertheless, based on our own considerations and on expert opinions given in the literature, it is aimed in the following to provide as far as possible a comprehensive overview.

1.2.2
Categories of 'Micro-reactor Fundaments and Impacts'

Many authors have described the consequences of chemical micro processing (see Section 1.4.3) in reviews. When following the multi-faceted hierarchic structure proposed at the beginning of this chapter, one is able to specify the entities grouped at four levels by studying the expert information supplied in the reviews (Figure 1.13). As a result, at the first level physical and chemical basics (see Sections 1.3 and 1.4), named briefly fundaments in the following, have to be described. At the second level is micro-reactor development, which has impacts on chemical engineering, process engineering, process results, society/ecology and the economy (see Sections 1.5–1.8). Knowing the process results, one is able to evaluate the potential for the use of micro reactors for applications in various fields (see Section 1.9).

Some reviews provide information on impacts of various types, whereas others focus on one aspect in detail. Most descriptions cover physical fundaments and the impact on chemical engineering; the impact on process results, although indispensable for the field of micro reaction engineering, has so far not been described in a systematic and comprehensive manner (probably for this more progress in experimental results is needed). Process engineering visions originally developed for transportable bench-scale plants composed of small conventional devices, but also relevant and adjustable to micro reactors, were given at very early stage of development. A vision of the socio-economic aspects of the field is given in a few specialist articles.

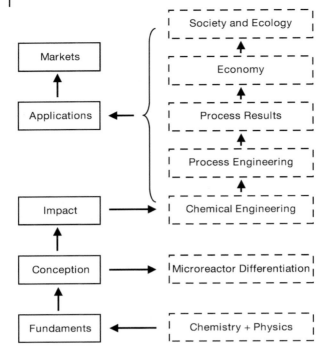

Figure 1.13 Hierarchic structure of entities that are related to using micro reactors – fundaments, impacts, and applications. An analysis and specification of the entities and their relation gives a multi-faceted knowledge on chemical micro processing, the content of this chapter.

1.2.3
Comprehensive Reviews and Essays

At an early stage of development, Wegeng et al. gave a near-comprehensive view covering all of the above-mentioned fundaments and impacts (with the exception of the chemical one) [1]. Naturally, information on applications was limited, hence the review had to be visionary in many parts, but it made correct predictions, as we now know. Much later and based on the much wider knowledge available then, Hessel et al. provided a series of reviews of comprehensive and extended character, full of details on applications and up-to-date references [9, 10, 57, 58, 66]. The two CIT and CET series by these authors contain particularly a section on plant concepts with regard to micro reactors. Gavriilidis et al. gave similar information in a compact manner in one review, included a section on modeling, and reported in detail on chemical applications [5].

The first book on the field aims at combining all this information and to group it according to the way of processing [4].

1.2.4
Reviews and Essays on Physical Fundaments and the Impact on Chemical Engineering and Process Engineering

Ehrfeld et al. gave a particularly systematic description of the physical fundaments, the resulting fundamental advantages, generic device concepts, and some early-stage, relevant applications [3]. Ehrfeld et al. also reviewed microfabrication techniques and in a phenomenological way described many kinds of devices [67–70]. Wegeng et. al also reported at an early stage physical fundaments, microfabrication methods needed for micro-reactor manufacture, and the first micro-flow devices [1]. Later, the first ideas on industrial implementation and commercialization were also sketched by Ehrfeld et al. [71]. Further, at an early stage of development, Jäckel [72, 73] and Lerou et al. [74] each gave a description of the physical fundaments and the impact on process engineering from an industrial perspective, touching also upon economic aspects. Jensen reported on chemical engineering aspects, but also summarized the microfabrication issues, considering especially chip micro reactors [2, 75, 76]. One reference addresses the impact on process engineering in a compact form [75].

1.2.5
Reviews and Essays on the Impact on Process Results, Society/Ecology and the Economy

Haswell et al. prepared very detailed reviews on the use of chemical micro processing in chip-based micro reactors for organic synthesis [29, 77] and for chemistry in general [78], covering also green chemistry aspects [47]. Other reviews also concern the chemical field [79–81]. An especially detailed review on the chemical fundaments was recently provided by Jähnisch et al. [82]. Schwalbe et al. gave a first insight into the relationship between potential-energy profiles of the reaction course and setting proper micro-reactor operation [81]. Löwe et al. gave information on the chemical misuse of micro reactors [83]. Rinard et al. focused on socio-ecologic-economic analysis with strong consideration of the social aspects of future chemical processing [84–86]. Hessel et al. discussed hybrid/multi-scale plant concepts and the state of market implementation [9, 10, 57, 58, 66, 86, 87].

1.2.6
Reviews and Essays on Application Topics and Microfabrication

Concerning one application only and the design concepts involved, reviews were given on micro mixing [88], micro heat exchange [14], energy [64, 89, 90], and extra-terrestrial applications [91, 92]. Some reviews give in-depth information on selected micro-fabrication techniques: electroforming in the framework of the LIGA process for metal micro reactors [93], µEDM [94–96], specialty-tool milling [61], and ceramic moulding [97–100]. In [3], a summary and many references on micro-fabrication technologies applied are given.

1.2.7
Reviews and Essays on Institutional Work

Other reviews explicitly concern summaries of the work of the author's group, supplemented by a short introduction in the field: see [75, 76, 101] for MIT work, [87, 93, 102, 103] for IMM work, [104] for TU Eindhoven work, [105] for Sandia National Laboratories work, [106] for PNNL work, and [23] for the work of the Forschungszentrum Karlsruhe.

Reviewing all these experts' analyses, in the following it is aimed to give a summary of these considerations, supplemented by our own or otherwise newly presented thoughts. Special attention is given to achieving a thorough separation and classification of the fundaments and impacts according to their origin.

1.3
Physical and Chemical Fundaments

1.3.1
Size Reduction of Process Equipment

The most obvious, trivial consequence of micro-device miniaturization is a

- decrease in volumes, e.g. for reaction flow-through chambers or for interconnects.

Many other, less obvious physical consequences of miniaturization are a result of the scaling behavior of the governing physical laws, which are usually assumed to be the common macroscopic descriptions of flow, heat and mass transfer [3, 107]. There are, however, a few cases where the usual continuum descriptions cease to be valid, which are discussed in Chapter 2. When the size of reaction channels or other generic micro-reactor components decreases, the surface-to-volume ratio increases and the mean distance of the specific fluid volume to the reactor walls or to the domain of a second fluid is reduced. As a consequence, the exchange of heat and matter either with the channel walls or with a second fluid is enhanced.

Generally, whenever fluids are processed in a confined space, two different types of phenomena are observed: surface and volume effects. An example of a surface effect is a heterogeneously catalyzed reaction occurring at the walls of the vessel, whereas the motion of a fluid due to gravitational forces would be described as a volume effect. In brief, it can be stated that the surface effects gain in importance compared with the volume effects when the size of a reactor decreases. In particular, the reduction of length scale leads to a

- decrease in diffusion distances, e.g. for mixing and heat transfer
- increase in respective gradients, e.g. concentration or temperature
- increase in specific surfaces, e.g. when using films or catalysts
- increase in specific interfaces, e.g. in multi-phase flow.

A more specific discussion of the importance of various hydrodynamic phenomena when the length scale is reduced is given in the following section.

1.3.2
Scaling Effects Due to Size Reduction: Hydrodynamics

In order to assess the importance of various physical phenomena relevant to standard and micro-process technology, it is helpful to compare two scenarios of typical processes. In standard process technology, a generic reactor is the stirred vessel which is used for liquid/liquid and gas/liquid reactions. When the goal is to process a gas/liquid mixture, the impeller blades may contain spargers for introduction of gas bubbles. A schematic representation of such a device is shown on the left in Figure 1.14 and a generic micro reactor design on the right. In such a device, either liquid/liquid or gas/liquid reactions can be conducted, depending on the types of fluids fed to the two inlets (gas-phase reactions are not considered in this section). The figure indicates the transport of gas bubbles through the reaction channel. In contrast to the standard process, which is of batch type, the micro reactor allows the fluids to be proceeded in a continuous-flow manner.

In general, a multitude of different phenomena of flow, heat and mass transfer occur during a liquid/liquid or gas/liquid reaction. Rather than discussing all relevant effects, which would be a tremendous task, the focus of this section is solely on flow phenomena in either single-phase aqueous systems or air/water systems. The purpose is to highlight some of the relevant scaling laws to be taken into account when reducing the size of process equipment.

To be specific, a liquid with the properties of water at 20 °C and 1 atm, i.e. a density of $\rho_l = 998.21$ kg m^{-3}, a dynamic viscosity of $\mu_l = 1.002 \cdot 10^{-3}$ Pa s and a surface tension of $\sigma = 72.7 \cdot 10^{-3}$ N m^{-1}, is considered. For the gas, the properties of air under the same conditions are assumed ($\rho_g = 1.188$ kg m^{-3}, $\mu_g = 18.24 \cdot 10^{-6}$ Pa s). The velocity scale u was taken to be 1 m s^{-1} for the stirred vessel and 0.01 m s^{-1} for the micro reactor, and the fluids are assumed to be subject to a gravi-

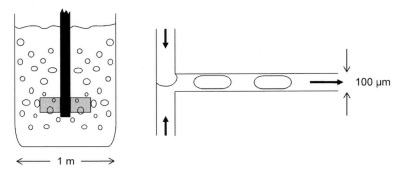

100 µm

1 m

Figure 1.14 Schematic representation of a stirred vessel (left) and a T-shaped micro reactor (right). Both devices can be used for liquid/liquid and for gas/liquid reactions. The length scales indicate typical physical dimensions.

tational acceleration $g = 9.81$ m s^{-2}. The central quantity the present analysis is based on is the relevant length scale L in both types of reactors. For the stirred vessel, a diameter of 1 m is assumed, while the channel diameter of the T-shaped micro reactor is fixed at 100 μm. When gas/liquid flows are considered, a second length scales comes into play, which is the bubble diameter. Characteristic bubble diameters in heavily agitated tanks are about 500 μm [108], whereas in a micro reactor the dimension of the bubbles is typically the same as the channel diameter. Alternatively, the T-shaped micro reactor might also be used to process droplets or slugs instead of bubbles.

In Table 1.3, some of the most important dimensionless groups characterizing the flow in both the standard macroscopic vessel and the micro reactor are listed. If not stated otherwise, the geometric dimensions (1 m and 100 μm) were used as the length scale L. Likewise, the properties of water were normally used to compute the expressions in the second column. The evaluation of the Bond number in the second line shows that in both the macro and micro devices, gravitational forces can usually be neglected when compared with surface-tension forces. However, the Froude number indicates that in these two cases gravitational forces dominate over inertial forces. The product of both numbers gives the Weber number, listed

Table 1.3 Comparison of miscellaneous dimensionless groups characterizing different hydrodynamic regimes in macroscopic vessels and micro reactors.

Name	*Formula*	*Phyical interpretation*	*Typical macro value*	*Typical micro value*	*Comments*
Bond number	$Bo = \dfrac{(\rho_l - \rho_g)\, L^2\, g}{\sigma}$	$\dfrac{\text{gravitational force}}{\text{surface-tension force}}$	$3 \cdot 10^{-2}$	$1 \cdot 10^{-3}$	Relevant for bubble (droplet) flows. Length scale: bubble diameter
Capillary number	$Ca = \dfrac{\mu\, u}{\sigma}$	$\dfrac{\text{viscous force}}{\text{surface-tension force}}$	$1 \cdot 10^{-2}$	$1 \cdot 10^{-4}$	Relevant for bubble (droplet) flows
Froude number	$Fr = \dfrac{\rho_g\, u^2}{(\rho_l - \rho_g)\, g\, L}$	$\dfrac{\text{inertial force}}{\text{gravitational force}}$	$2 \cdot 10^{-1}$	$1 \cdot 10^{-4}$	Relevant for bubble (droplet) flows. Length scale: bubble diameter
Reynolds number	$Re = \dfrac{\rho\, L\, u}{\mu}$	$\dfrac{\text{inertial force}}{\text{viscous force}}$	$1 \cdot 10^{6}$	1	
Weber number	$We = \dfrac{\rho_g\, L\, u^2}{\sigma}$	$\dfrac{\text{inertial force}}{\text{surface-tension force}}$	$8 \cdot 10^{-3}$	$2 \cdot 10^{-7}$	Relevant for bubble (droplet) flows. Length scale: bubble diameter

in the last line. Hence, for gas/liquid flows the ranking of surface-tension, gravitational and inertial forces is the same in both types of devices, with surface-tension forces being the most important. Viscous forces, represented by the capillary number, are of about the same importance as gravitational forces in such reactors. Naturally, the perhaps surprising similarities between the macro- and micro reactors disappear when not the gas/liquid interface of the bubbles, but the interface between the bulk liquid and the surrounding gas in the stirred vessel is considered. In that case the relevant length scale is 1 m rather than 500 μm, with the result that gravitational and inertial forces dominate over surface-tension forces.

The most pronounced difference between the stirred vessel and the micro reactor is to be seen in the Reynolds number, which is the ratio of inertial and viscous forces. The Reynolds number is a dimensionless group relevant for both single-phase and two-phase flows. In a macroscopic vessel, viscous forces are usually completely negligible compared with inertial forces. This is no longer true in micro reactors, where both are of the same order of magnitude. As a result, flow in macroscopic devices is turbulent in most cases, whereas usually laminar flows are found in micro devices. This fact bears far-reaching consequences for process modeling. Except for a few special cases, turbulent flows require special heuristic models which incorporate the effects of stochastic velocity fluctuations [109]. The range of validity of these models is often limited and it is often not clear which model is best suited for a specific problem. As a result, the predictive power of, e.g., methods of computational fluid dynamics, is often not sufficient for a rational design of complex macroscopic reactors. In contrast, first-principle modeling techniques can be applied for flow prediction in micro reactors. In this context, the term 'first principle' refers to the fact that the fundamental equations for flow, heat and mass transfer are solved directly, without any reference to heuristic models. Hence, the predictive power of micro reactor models is often higher than their counterparts for macroscopic devices are able to achieve. In such a way, a rational design of micro reactors and even a virtual prototyping process seem within reach.

1.3.3
Chemical Fundaments

The majority of syntheses described in the literature belong to the class of homogeneous organic reactions. These are typical bulk reactions and the underlying chemistry is typically performed at the molecular level. Hence it is not likely that micro channels of a typical size of a few micrometers or more will have an impact on the reaction course, unlike nanometer-sized reaction vessels such as micelles or zeolites, which e.g. may hinder rotation, impact on electron density distribution or change the state of solvent environment. The majority of industrially practised processes belong to heterogeneous reactions, the chemistry of which is also not likely to be changed.

However, it is known that, even when using construction materials only (no functional polymer resin or catalyst), bulk reactions can change to surface reactions with the surface acting as a real 'reactant'. Here, the functional groups of the surface act as reactants. Such findings have only recently been identified (see Section 1.6.10).

1.4
Impact on Chemical Engineering

1.4.1
Basic Requirements on Chemical Engineering from an Industrial Perspective

Arguing from an industrial point of view, Wörz et al. list three basic tasks which an industrial reactor has to fulfil [110–112].

- provision of the residence time needed for reaction
- efficient heat removal or supply
- provision of a sufficiently large interface (for multi-phase reactions)

Wörz et al. describe in detail why micro reactors can give advantages concerning all three tasks [110–112]. They illustrate their analysis by two application examples, concerning a liquid/liquid reaction and a catalyzed gas-phase reaction, both representing industrial chemical production processes.

Note that the tasks defined by Wörz et al., when matched perfectly to the reaction requirements, exactly correspond to the criteria introduced by the BHR Group for defining process intensification (see Section 1.1.6.2). CPC give a similar selection of basic (micro-) reactor tasks [113].

1.4.2
Top-down and Bottom-up Descriptions

The conclusion of Wörz et al. given above is the concentrated result of a number of heuristic bottom-up descriptions, having access to micro-reaction technology, knowing about the corresponding impact on chemical engineering, and selecting by this an application of interest. In total, all these descriptions provide valuable expert opinions and thus are cited in Section 1.4.4. To approach the topic from an industrial and sales view, the study of these conclusions is indispensable. However, the current status of chemical-engineering know-how allows an even more in-depth analysis. Based on a systematic, non-heuristic analysis, a top-down description of the impact of chemical engineering will also be provided. It would be very desirable to have such top-down descriptions also for the process-engineering, chemical, socio-ecologic and economic impacts. However, it stands to reason that the complexity of these topics currently or even hardly ever will not allow such an approach. There is definitely a chance for the first two types of impacts, but more results are needed and an in-depth methodology has to be developed going beyond the scope of this book.

1.4.3
A Top-down Description of Chemical Engineering Impacts

Most of this section is based on the work of Jean-Marc Commenge of CNRS, Nancy, France [114], to whom we are indebted.

1.4.3.1 A Case Study on Gas-phase Reactions

In an economics context, the question has to be raised for which applications micro process technology offers a superior performance to conventional processes. An attempt to answer this question should include various aspects such as reactor efficiency, reliability, maintenance costs, and expenditures for investment and plant operation. However, a complete analysis assessing the benefits and drawbacks of micro process technology in a holistic context has rarely been given. Most reviews are based on a bottom-up approach, describing the achievements of specific operation units and deriving implications for specific applications in chemical engineering. Rather than that, a deductive or top-down approach starts with the requirements related to a specific process and seeks to derive a design allowing one to fulfil these requirements, potentially subject to various constraints and aiming at the minimization of certain cost functions. As an example, the requirement could be to produce a specific quantity of a chemical species per unit time in a continuous-flow process, where a limit on the pressure drop is given and the mole fraction of an undesired by-product should be as small as possible.

One of the most far reaching analyzes along these lines of thought was given by Commenge [114] in the context of gas-phase reactions in continuous-flow processes. Specifically, he analyzed four different aspects of micro reaction devices, namely the *expenditure in mechanical energy*, the *residence-time distribution*, *safety in operation*, and the *potential for size reduction* when the efficiency is kept fixed.

1.4.3.2 Energy Gain from Microstructuring

A standard way of conducting gas-phase reactions in a continuous-flow process is through the use of fixed-bed reactors. A fixed-bed reactor contains a dense packing of catalyst pellets through which the process gas is guided. The chemical species contained in the gas phase diffuse to the pellets and inside their pores where a reaction occurs. Alternatively, a heterogeneously catalyzed gas-phase reaction may be conducted in a micro-channel reactor. In such a set-up, the process gas is guided through a multitude of parallel micro channels which are coated with a catalytically active porous layer. The heat management in corresponding reactors may be done via heating or cooling gas channels, which may be arranged in alternating layers with the micro channels. Alternatively, heating cartridges in thermal contact with the reactor monolith may be employed.

The two different concepts are depicted schematically in Figure 1.15. The fixed bed is assumed to have a cross-section S_F and a height H_F and is filled with non-deformable spherical particles with diameter d_F, where the density of the packing (particles per m^3) is denoted by n_F. The micro-channel reactor has a cross-section S_M and a height H_M and comprises channels of diameter d_M with a specific density (number of channels per m^2) n_M.

In order to compare the micro-channel and the fixed-bed reactor, the design and operation parameters should be adjusted in such a way that certain key quantities are the same for both reactors. One of those key quantities is the porosity ε, defined as the void fraction in the reactor volume, i.e. the fraction of space which is not occupied by catalyst pellets or channel walls. The second quantity is the specific

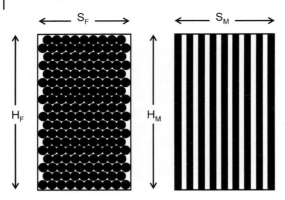

Figure 1.15 Schematic representation of a fixed-bed reactor (left) and a micro-channel reactor (right) [114].

surface area a (surface area/reactor volume), which, in the case of the fixed bed, is defined via the surface area of the catalyst pellets. Finally, the gas flows should be adjusted in such a way that the mean residence time τ is the same for both reactors. Specifically, for equal gas flows, this means that the cross-sections and the heights of the reactors need to be the same. From the equivalence of porosities and specific surface areas, a relation between the pellet size and the micro-channel diameter is obtained:

$$d_{\mathrm{F}} = \frac{3}{2} \frac{1 - \varepsilon}{\varepsilon} d_{\mathrm{M}} . \tag{1}$$

For this purpose, cylindrical channels have been assumed. In randomly packed fixed beds the porosity is about 0.4, from which the relationship $d_{\mathrm{F}} = 2.25\, d_{\mathrm{M}}$ is obtained. Since the focus is on heterogeneously catalyzed gas-phase reactions, it is important to not only ensure comparable conditions from a hydrodynamic point of view, but also as far as chemical reaction kinetics is concerned. Therefore, it is assumed that both reactors contain the same amount of catalyst.

In order to verify that the fixed bed and the micro-channel reactor are equivalent concerning chemical conversion, an irreversible first-order reaction $A \rightarrow B$ with kinetic constant k_{s} was considered. For simplicity, the reaction was assumed to occur at the channel surface or at the surface of the catalyst pellets, respectively. Diffusive mass transfer to the surface of the catalyst pellets was described by a correlation given by Villermaux [115].

$$\frac{k_{\mathrm{d}}\, d_{\mathrm{F}}}{D} = 2 + 1.8\, \mathrm{Re}^{1/2}\, \mathrm{Sc}^{1/3} \tag{2}$$

where k_{d} denotes the mass transfer coefficient and D the molecular diffusivity of the chemical species. The Reynolds number Re and the Schmidt number Sc are given by

$$\text{Re} = \frac{u\,\rho\,d_{\text{F}}}{\mu}\,; \quad \text{Sc} = \frac{\mu}{D\,\rho}\,, \tag{3}$$

where u, ρ and μ are the velocity, density, and dynamic viscosity of the gas, respectively. In the context of heterogeneously catalyzed reactions, an important quantity characterizing the interplay between reaction and diffusion dynamics is the Damköhler number, defined as

$$\text{Da} = \frac{k_{\text{s}}\,d_{\text{F}}}{D}\,. \tag{4}$$

For fast reactions Da becomes large. Based on that assumption and standard correlations for mass transfer inside the micro channels, both the model for the micro-channel reactor and the model for the fixed bed can be reformulated in terms of pseudo-homogeneous reaction kinetics. Finally, the concentration profile along the axial direction can be obtained as the solution of an ordinary differential equation.

For a specific comparison of the two different reactor types, channels of 300 μm diameter were considered. The equivalent pellet size for that case is 675 μm. As a characteristic quantity, the conversion at the reactor exits was computed for different flow velocities and a range of Damköhler numbers spanning three orders of magnitude. The results for the two different reactor types obtained in such a way were practically indistinguishable. This suggests that the different reactors considered in this study are equivalent as far as chemical conversion is concerned.

In continuous flow systems, the expenditure in mechanical energy necessary to run a process is directly proportional to the pressure drop over the system. Hence the pressure drop is an important figure determining the operating costs of a device. After having verified the chemical equivalence of the two reactor types introduced above, the question arises of whether using a micro-channel reactor instead of a fixed-bed reactor allows a decrease in the pressure drop. In order to estimate the pressure drop in the fixed-bed reactor, the Carman–Kozeney hydraulic diameter model (see, e.g., [116]) was used:

$$\frac{\Delta P}{L} = \frac{36\,h_{\text{k}}\,\mu}{d_{\text{F}}^2}\frac{(1-\varepsilon)^2}{\varepsilon^2}\,u \tag{5}$$

where ΔP denotes pressure drop, L the length of the fixed bed, h_{k} the Kozeney coefficient, and u the average gas velocity in the pore system. For the micro-channel reactor, the pressure drop is obtained from the usual Poiseuille flow profile. Accordingly, a comparison of the two reactor types yields

$$\frac{(\Delta P/L)_{\text{F}}}{(\Delta P/L)_{\text{M}}} = \frac{h_{\text{k}}}{2}\,. \tag{6}$$

For a fixed bed of spherical particles, the Kozeney coefficient lies between 4.5 and 5. Hence the pressure drop in the fixed bed is up to 2.5 times larger than that in the micro-channel reactor, for otherwise comparable conditions. This suggests that micro-structured reactors bear a great potential for energy gain when compared with fixed-bed technology.

1.4.3.3 Residence-time Distributions

When a number of competing reactions are involved in a process, and/or when the desired product is obtained at an intermediate stage of a reaction, it is important to keep the residence-time distribution in a reactor as narrow as possible. Usually, a broadening of the residence-time distribution results in a decrease in selectivity for the desired product. Hence, in addition to the pressure drop, the width of the residence-time distribution is an important figure characterizing the performance of a reactor. In order to estimate the axial dispersion in the fixed-bed reactor, the model of Doraiswamy and Sharma was used [117]. This model proposes a relationship between the dispersive Peclet number:

$$Pe'_A = \frac{u \, d_F}{D_{A,F}} \tag{7}$$

the modified Reynolds number:

$$Re' = \frac{\varepsilon \, u \, \rho \, d_F}{\mu} \tag{8}$$

and the Schmidt number. In the definition of the dispersive Peclet number, $D_{A,F}$ denotes an effective diffusion coefficient which contains the effects of axial dispersion in the fixed bed. The relationship is given as

$$\frac{1}{Pe'_A} = \frac{0.5}{1 + \dfrac{3.8}{Re' \, Sc}} + \frac{0.3}{Re' \, Sc} \, . \tag{9}$$

For axial dispersion in the micro-channel reactor, the usual relationships from Taylor–Aris theory were employed. In order to assess the performance of both reactor types, the widths of two initially delta-like concentration tracers are compared after they have passed through the flow domain. The results of this comparison are displayed in Figure 1.16.

The figure shows the ratio of the widths of initially delta-like concentration tracers at the reactor exits as a function of the micro-channel Peclet number for different values of the porosity. Taking a value of $\varepsilon = 0.4$ as standard, it becomes apparent that the dispersion in the micro-channel reactor is smaller than that in the fixed-bed reactor in a Peclet number range from 3 to 100. Minimum dispersion is achieved at a Peclet number of about 14, where the tracer width in the micro-channel reactor is reduced by about 40% compared with its fixed-bed counterpart. Hence the conclusion may be drawn that micro-channel reactors bear the potential of a narrower residence time than fixed-bed reactors, where again it should be stressed that reactors with equivalent chemical conversion were chosen for the comparison.

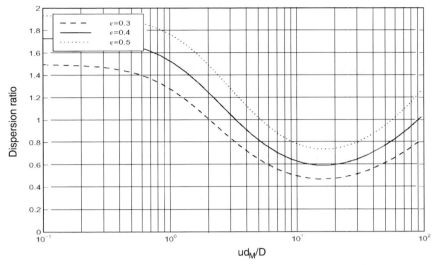

Figure 1.16 Ratio of the width of concentration tracers at the exit of the micro-channel reactor to the corresponding quantity in the fixed-bed reactor as a function of the micro-channel Peclet number $u\,d_M/D$, as obtained in [114].

1.4.3.4 Heat Transfer: Safety in Operation

Micro-structured reactors offer a gain in operational safety for at least two different reasons. In the first place, chain reactions may be quenched when molecules of the gas phase collide with the channel walls. In many cases the channel walls may act as a catalyst, with the result that reactions at the surface are different from the reactions occurring in the bulk, or at least the speed of various elementary reactions is modified. In micro devices, the specific surface area is larger than in macroscopic vessels, associated with an increasing importance of surface reactions and, potentially, enhanced safety due to quenching of chain reactions. Secondly, thermal explosions may also be prevented. The driving mechanism behind thermal explosions is a reaction rate rapidly increasing with temperature. However, due to the very efficient heat transfer between the fluid phase and the reactor housing in micro devices, thermal energy may be subtracted so rapidly that a significant temperature rise is avoided and a thermal explosion can be suppressed. The following discussion focuses exclusively on the second effect, since, owing to the variety of possible chain reaction mechanisms, a sufficiently general treatment of such safety hazards does not seem to be feasible.

In the following, the impact of the micro-channel diameter on the temperature rise due an exothermic gas-phase reaction is investigated. For simplicity, a homogeneous reaction A → B of order n with kinetic constant k is considered. Inside the micro channel, the time evolution of the radially averaged species concentration c and temperature T is governed by the equations

$$\frac{dc}{dt} + k\,c^n = 0 \tag{10}$$

$$\frac{dT}{dt} + \frac{4\,h}{\rho\,d_M\,c_p}\,(T - T_W) = -\frac{k\,c^n\,\Delta H}{\rho\,c_p} \tag{11}$$

where h is the heat transfer coefficient between the fluid and the channel walls, c_p the specific heat of the gas, T_W the temperature of the wall (assumed to be constant), and ΔH the reaction enthalpy. For the gas passing through the micro channel, the conversion between time and space coordinates is done using the average flow velocity. The temperature dependence of the kinetic constant is given by Arrhenius' law:

$$k = k_0\,\exp\left(-\frac{E_a}{R\,T}\right) \tag{12}$$

where k_0 is a pre-exponential factor, E_a the activation energy, and R the gas constant. The temperature rise in the channel due to the chemical reaction is governed by two different time-scales, a reaction time-scale t_r and a heat-exchange time-scale t_{he}:

$$t_r = \frac{1}{k_0\,\exp\left(-\dfrac{E_a}{R\,T_W}\right)}\;;\quad t_{he} = \frac{\rho\,d_M\,c_p}{4\,h}, \tag{13}$$

where a first-order reaction was assumed.

When t_{he} is large compared with t_r, the reaction gas can be heated considerably before the heat is subtracted through the reactor housing. In the opposite case, hardly any temperature rise is observed, since heat is removed rapidly through the channel walls.

In order to derive specific numbers for the temperature rise, a first-order reaction was considered and Eqs. (10) and (11) were solved numerically for a constant-density fluid. In Figure 1.17 the results are presented in dimensionless form as a function of t_r/t_{he}. The y-axis represents the temperature rise normalized by the adiabatic temperature rise, which is the increase in temperature that would have been observed without any heat transfer to the channel walls. The curves are differentiated by the activation temperature, defined as $T_a = E_a/R$. As expected, the temperature rise approaches the adiabatic one for very small reaction time-scales. In the opposite case, the temperature rise approaches zero. For a non-zero activation temperature, the actual reaction time-scale is shorter than the one defined in Eq. (13), due to the temperature dependence of the exponential factor in Eq. (12). For this reason, a larger temperature rise is found when the activation temperature increases.

In order to show how specific guidelines for the reactor layout can be derived, the maximum allowable micro-channel radius giving a temperature rise of less than 10 K was computed for different values of the adiabatic temperature rise and different reaction times. For this purpose, properties of nitrogen at 300 °C and 1 atm and a Nusselt number of 3.66 were assumed. The Nusselt number is a dimensionless heat transfer coefficient, defined as

$$Nu = \frac{h\,d_M}{\lambda} \tag{14}$$

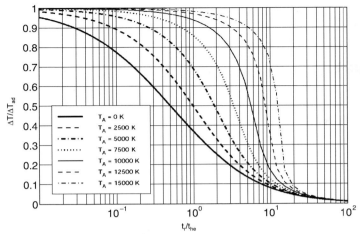

Figure 1.17 Normalized temperature rise for a first-order exothermic reaction as a function of the ratio of reaction and heat-exchange time-scale, obtained from [114]. Different activation temperatures are considered.

where λ is the thermal conductivity. Furthermore, the activation energy was set to zero. The results of that calculation are depicted in Figure 1.18. Each curve corresponds to a fixed channel radius, ranging from 100 µm to 1 mm, and a calculated temperature rise of 10 K. The plane is spanned by the adiabatic temperature rise

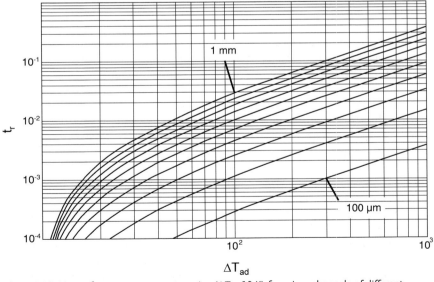

Figure 1.18 Lines of constant temperature rise ($\Delta T = 10$ K) for micro channels of different radius in a plane spanned by the adiabatic temperature rise of the reaction and the reaction time-scale, obtained from [114]. The properties of nitrogen at 300 °C and 1 atm and a Nusselt number of 3.66 were used for this calculation.

(*x*-axis) and the reaction time-scale (*y*-axis). When one of the two criteria is relaxed, i.e. either the reaction time-scale is increased or the adiabatic temperature rise is reduced, successively larger channel diameters are sufficient to limit the temperature rise to 10 K.

In order to exemplify the potential of micro-channel reactors for thermal control, consider the oxidation of citraconic anhydride, which, for a specific catalyst material, has a pseudo-homogeneous reaction rate of $1.62 \, s^{-1}$ at a temperature of 300 °C, corresponding to a reaction time-scale of 0.61 s. In a micro channel of 300 μm diameter filled with a mixture composed of N_2/O_2/anhydride (79.9 : 20 : 0.1), the characteristic time-scale for heat exchange is $1.4 \cdot 10^{-4}$ s. In spite of an adiabatic temperature rise of 60 K related to such a reaction, the temperature increases by less than 0.5 K in the micro channel. Examples such as this show that micro reactors allow one to define temperature conditions very precisely due to fast removal and, in the case of endothermic reactions, addition of heat. On the one hand, this results in an increase in process safety, as discussed above. On the other hand, it allows a better definition of reaction conditions than with macroscopic equipment, thus allowing for a higher selectivity in chemical processes.

1.4.3.5 Potential for Size Reduction

On shrinking the size of micro-channel reactors by reducing the channel dimensions, a number of characteristic quantities such as pressure drop and the degree of chemical conversion are affected. In order to permit a meaningful comparison of the reactor geometry with a scaled geometry, it is important to keep one or a few key quantities fixed and study the influence of size reduction on the remaining quantities. One strategy is to compare reactors with equal efficiency for heat exchange or chemical conversion. In order to study the problem in a generalized framework, it is helpful to define a characteristic time-scale τ_{op} for specific operations such as heat exchange or reaction. A process can be characterized by its number of transfer units

$$NTU = \frac{\tau}{\tau_{op}} \tag{15}$$

where $\tau = L/u$ is the mean residence time in the reactor, defined via the channel length L and the average velocity u. When the number of transfer units is kept fixed, the efficiency of a reactor for specific operations, for example heat transfer or chemical conversion, is also fixed. Comparing reactors of different scales with fixed chemical efficiency ensures that the product concentrations at the channel exits are the same.

In Table 1.4, the characteristic time-scales for selected operations are listed. The rate constants for surface and volume reactions are denoted by k_s and k_v, respectively. Furthermore, the Sherwood number Sh, a dimensionless mass-transfer coefficient and the analogue of the Nusselt number, appears in one of the expressions for the reaction time-scale. The last column highlights the dependence of τ_{op} on the channel diameter d_M. Apparently, the scale dependence of different operations varies from $(d_M)^0$ to $(d_M)^2$. Owing to these different dependences, some op-

Table 1.4 Characteristic time-scale and length scale-dependence for selected operations [114].

Type of operation	Characteristic time-scale τ_{op}	Dependence on channel diameter
Heat exchange	$\dfrac{\rho\, c_p (d_M)^2}{4\, \lambda\, Nu}$	$\sim (d_M)^2$
Heterogeneously catalyzed reaction		
• Diffusion limited	$\dfrac{(d_M)^2}{4\, D\, Sh}$	$\sim (d_M)^2$
• Reaction-rate limited	$\dfrac{d_M}{4\, k_s}$	$\sim d_M$
Homogeneous reaction	$\dfrac{1}{k_v}$	constant

erations are better suited for miniaturization than others, as will be discussed in the following.

In cases where the operation time-scale is independent of the channel diameter, as for a homogeneous reaction, it is necessary to keep the residence time fixed when downscaling a reactor in order keep the efficiency constant. When the flow-rate Q_{tot} of the process gas is given, this means that a reduction in the channel diameter has to be accompanied by an increase in the channel length L or the number of channels N, according to

$$N\, L\, (d_M)^2 = \text{constant} \tag{16}$$

More favorable for miniaturization are processes with an operation time-scale proportional to d_M or $(d_M)^2$. For a linear dependence on the channel diameter, the product $N\, L\, d_M$ is conserved under the conditions described above. This means that with shrinking d_M and for fixed efficiency, the reactor volume decreases proportionally with the channel diameter. For a quadratic dependence of the operation time-scale with channel diameter, the product $N\, L$ is conserved and the reactor volume decreases as the channel diameter squared.

In some cases, it may not be desirable to reduce the volume of a reactor, and rather a decrease of pressure drop or channel length may be the goal. In Table 1.5, the dependence of several characteristic quantities on channel diameter is given, where the efficiency and at least one specific quantity is kept fixed in each line.

The comparison given above shows that a reduction in channel dimensions offers some substantial benefits in cases where surface reactions are involved or efficient heat and mass transfer are needed. One important conclusion to be drawn is that a decrease in the channel diameter at fixed efficiency does not necessarily mean an increase in pressure drop. Rather, the pressure drop can be kept constant

Table 1.5 Dependence of the number of micro channels N, their length L, the cross-sectional area of the reactor S and the pressure drop ΔP on the micro-channel diameter, when the efficiency (i.e. a fixed number of transfer units) and at least one specific characteristic quantity are kept fixed in each line. Three cases with operation time-scales varying as $(d_M)^0$, d_M, and $(d_M)^2$ are considered [114].

τ_{op}	τ	V	N	L	S	ΔP
constant	constant	constant	constant	$\sim(d_M)^{-2}$	$\sim(d_M)^2$	$\sim(d_M)^{-6}$
			$\sim(d_M)^{-2}$	constant	constant	$\sim(d_M)^{-2}$
			$\sim(d_M)^{-3}$	$\sim d_M$	$\sim(d_M)^{-1}$	constant
$\sim d_M$	$\sim d_M$	$\sim d_M$	constant	$\sim(d_M)^{-1}$	$\sim(d_M)^2$	$\sim(d_M)^{-5}$
			$\sim(d_M)^{-1}$	constant	$\sim d_M$	$\sim(d_M)^{-3}$
			$\sim(d_M)^{-2}$	$\sim d_M$	constant	$\sim(d_M)^{-2}$
			$\sim(d_M)^{-5/2}$	$\sim(d_M)^{3/2}$	$\sim(d_M)^{-1/2}$	constant
$\sim(d_M)^2$	$\sim(d_M)^2$	$\sim(d_M)^2$	constant	constant	$\sim(d_M)^2$	$\sim(d_M)^{-4}$
			$\sim(d_M)^{-2}$	$\sim(d_M)^2$	constant	constant

when the number of channels is increased, while still reducing the total reactor volume [for operation time-scales proportional to d_M and $(d_M)^2$]. At the same time, the channel length decreases. When accepting a constant or only slightly decreasing reactor volume, the pressure drop can even be reduced. However, it should be kept in mind that for a large number of micro channels the flow distribution manifold itself might occupy a considerable volume.

In practice, the process regime will often be less transparent than suggested by Table 1.4. As an example, a process may neither be diffusion nor reaction-rate limited, rather some intermediate regime may prevail. In addition, solid heat transfer, entrance flow or axial dispersion effects, which were neglected in the present study, may be superposed. In the analysis presented here only the leading-order effects were taken into account. As a result, the dependence of the characteristic quantities listed in Table 1.5 on the channel diameter will be more complex. For a detailed study of such more complex scenarios, computational fluid dynamics, to be discussed in Section 2.3, offers powerful tools and methods. However, the present analysis serves the purpose to differentiate the potential inherent in decreasing the characteristic dimensions of process equipment and to identify some cornerstones to be considered when attempting process intensification via size reduction.

1.4.3.6 Proposing a Methodology for Micro-reactor Dimensioning and Layout

In the following, a methodology is derived allowing dimensioning of a micro-channel reactor according to specifications and requirements of a specific process which should be implemented on a micro-reactor platform. The methodology rests on the assumptions made in the previous paragraphs. Specifically, gas-phase reactions are considered. It should be pointed out that in general the micro reactor is only a sub-unit in a more complex set up. The question of global process design is still more involved, with a complex interdependence between the performance of the sub-units and that of the complete system. In an immediate manner, the present analysis can only give hints on the dimensioning of a single multi-channel reactor and not on the design of a complex plant. However, as the component and the system level are closely intertwined, it is hoped that this analysis also contains some valuable information for the process designer.

Step 1: Dimensioning of the channel diameter

The micro-channel diameter d_M plays a crucial role for heat and mass transfer. When chemical reactions are involved, the channel diameter should be chosen small enough to render the time-scales for heat and mass transfer smaller than the reaction time-scale. A listing of the most relevant time-scales is provided in Table 1.6, where the crosses indicate for which step of the design process the time-scales play a role. In addition to heat- and mass-transfer limitations, there might be constraints on the maximum allowable temperature change in the channel. Hence the adiabatic temperature rise of the reaction also has to be taken into account. A dimensioning of the channel diameter along these lines might result in either very wide or very narrow channels. Wide channels may not give a satisfactory flow equipartition from the distribution manifold to the multi-channel reactor. In that case, the channel diameter may be decreased beyond the limits derived from heat- and mass-transfer considerations. On the other hand, very narrow channels may induce a prohibitive pressure drop. In that case, the number of micro channels could be increased.

Step 2: Dimensioning of the channel length

After the channel diameter has been fixed, the length of the channels has to be determined. Usually in a process, the number of transfer units as defined in Eq. (15) is fixed, which means that the residence time in the micro channel has to be considered as given. Via the equation $\tau = L/u$ this induces a relationship between the channel length and the flow velocity The average flow velocity can be determined be demanding that the broadening of a concentration tracer while being transported through the channel is as small as possible. Based on the expression for the axial dispersion coefficient obtained from Taylor–Aris theory, it is easy to show that this occurs at a Peclet number of $8\sqrt{3}$. With a correspondingly adjusted flow velocity and a given value for the residence time, the channel length can be determined. Again, it should be noted that channel dimensions derived from that rationale may lead to an elevated pressure drop. In such cases, a compromise between narrow residence-time distributions and a low pressure drop should be envisaged.

Step 3: Dimensioning of the channel walls

During the time the process gas spends in a micro channel, heat is transferred from the gas to the channel walls or vice versa. If in an exothermic reaction the time-scale for heat conduction in the channel walls is larger than the residence time of the fluid, considerable temperature gradients will build up along the walls and an isothermicity assumption will not be valid. Hence, in the case of predominantly longitudinal heat transfer, it is advisable to adjust the heat conduction time-scale in a proper way by providing channel walls of sufficient thickness. In a dynamic regime where temperature ramping becomes important (e.g. for start-up), it is advisable to take into account the time-scale for reactor heat-up which reflects the thermal inertia due to the heat capacity of the construction material.

Step 4: Determining the number of micro channels

After the channel diameter and length and the flow velocity have been fixed, the number of micro channels determines the total throughput and product yield. In applications focused on chemical production, the number of channels is then simply given by throughput requirements. In applications focused on research and development, such as kinetic measurements, a small number of channels might be preferable, since flow equipartition and data analysis can be more difficult when the number of channels is large.

Table 1.6 Characteristic quantities to be considered for micro-reactor dimensioning and layout. Steps 1, 2, and 3 correspond to the dimensioning of the channel diameter, channel length and channel walls, respectively. Symbols appearing in these expressions not previously defined are the effective axial diffusion coefficient D_A, the density ρ_w, thermal conductivity λ_w, specific heat $c_{p,w}$ and total cross-sectional area S_w of the wall material, the total process gas mass flow $\dot{m}$, and the reactant concentration c_0 [114].

Quantity	Expression	Step 1	Step 2	Step 3
Residence time	$\dfrac{L}{u}$		×	
Heat-exchange time-scale	$\dfrac{\rho\, c_p\, (d_M)^2}{4\,\lambda\, \mathrm{Nu}}$	×		
Radial diffusion time-scale	$\dfrac{(d_M)^2}{4\, D}$	×	×	
Axial diffusion time-scale	$\dfrac{L^2}{D_A}$		×	
Time-scale of homogeneous reaction (1st order)	$\dfrac{1}{k_v}$	×	×	
Time-scale of heterogeneous reaction (1st order)	$\dfrac{d_M}{4\, k_s}$	×	×	
Time-scale for reactor heat-up	$\dfrac{\rho_w\, c_{p,w}\, S_w\, L}{\dot{m}\, c_p}$			×
Time-scale for longitudinal heat conduction	$\dfrac{S\, L^2\, \rho_w\, c_{p,w}}{S_w\, \lambda_w}$			×
Adiabatic temperature rise	$\dfrac{\Delta H\, c_0}{\rho\, c_p}$	×		

Step 5: Layout of the flow distribution system

When the number of micro channels is fixed, a flow distribution system for feeding the process gas to the different channels has to be devised. A major requirement for such a system is a narrow residence-time distribution. There are in principle two different ways in which a flow distribution system can effect a broadening of the residence time distribution. First, if flow equipartition over the multi-channel domain cannot be achieved, the mean flow velocities and thus the residence times vary over the different channels. Secondly, the flow distribution system itself may induce a considerable broadening of the residence-time distribution, for example when recirculation zones are present. Among the design concepts for a distribution manifold are a comparatively wide flow coupler interfacing with the inlet and the multi-channel domain, and a tree-like network with successive bifurcations. In the first case, flow equipartition is achieved via the large pressure barrier of the micro channels. In the second case, the geometry of the flow distribution network itself ensures an equipartition over the micro channels. When the number of channels is very large, it may be preferable to split a reactor into various sub-units. In that case, flow equipartition is a two-level problem: first, the flow is distributed between the different sub-units, followed by a distribution manifolds feeding the multi-channel flow domain.

1.4.4
A Bottom-up Description of Chemical Engineering Impacts

1.4.4.1 Mixing

The increase in mass transfer by micro-channel processing is frequently cited in the literature [5].

Confined flows typically exhibit laminar-flow regimes, i.e. rely on a diffusion mixing mechanism, and consequently are only slowly mixed when the diffusion distance is set too large. For this reason, in view of the potential of microfabrication, many authors pointed to the enhancement of mass transfer that can be achieved on further decreasing the diffusional length scales. By simple correlations based on Fick's law, it is evident that short liquid mixing times in the order of milliseconds should result on decreasing the diffusion distance to a few micrometers.

In many cases, speed-up of mixing by miniaturization is achieved by lamellar flow configurations. When the width of the fluid lamellae is reduced, the diffusional length scale decreases and mixing is speeded up. In order to increase the mixing speed even further, hydrodynamic focusing can be employed to create fluid lamellae of a width even smaller than defined by the inlet geometry of a mixer [118]. When a high throughput is required, several lamellar streams are usually combined, giving periodic multi-lamellar arrangements (see Figure 1.19) [119, 120]. As an alternative to multi-lamellar mixers, a number of other concepts have been developed. In split–recombine mixers, two fluid streams are successively split and recombined in such a way that, at least in an ideal situation, multi-lamellar arrangements are created [121]. In another class of mixing devices, channels with corrugated walls, recirculating flows are induced which increase the mixing effi-

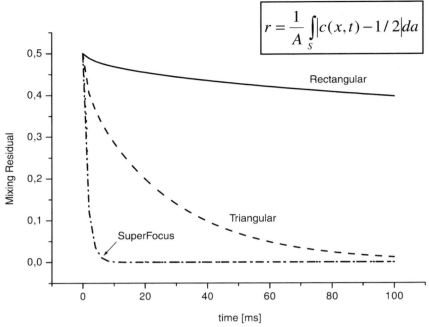

$$r = \frac{1}{A} \int_{S} |c(x,t) - 1/2| da$$

Figure 1.19 Semi-analytical calculation of the speed of mixing in an advanced interdigital micro mixer, named SuperFocus, and comparison with previously developed interdigital micro mixers [120].

ciency beyond the value found for two co-flowing, parallel lamellae [122]. Even in the laminar regime, chaotic flow patterns can be induced by superposing different vortex structures in channel flow. It has been shown that such chaotic micro mixers are very efficient at low Reynolds numbers [123]. Some micro-mixing principles have been proven to be much faster than the mixing observed in large-scale vessels or even in many specialized high-speed mixers (see, e.g., [3, 107, 124]).

Apart from this enhancement of mixing speed, a very uniform spatial distribution of mixing is typically achieved with micro mixers. Due to the laminar flow patterns, the flow field is uniquely determined in every part of the domain. As shown in Figure 1.20, a uniform mixing quality can be attained by multiply repeated flow geometries created by repetition of a unit flow cell. With suitable provisions taken, a very narrow mixing-time distribution is found in micro mixers. This stands in contrast to the turbulent flow characteristic of macroscopic equipment, accompanied by a chaotic velocity field with stochastic fluctuations. The turbulent vortices, the so-called eddies, occur on various length scales and with different intensity [125]. The eddies effect a stretching and folding of liquid volumes, finally creating very thin lamellae which mix by diffusion. Due to the stochastic nature of that process and variations of turbulence intensity, turbulent mixing is not uniform and displays a comparatively broad distribution of mixing times.

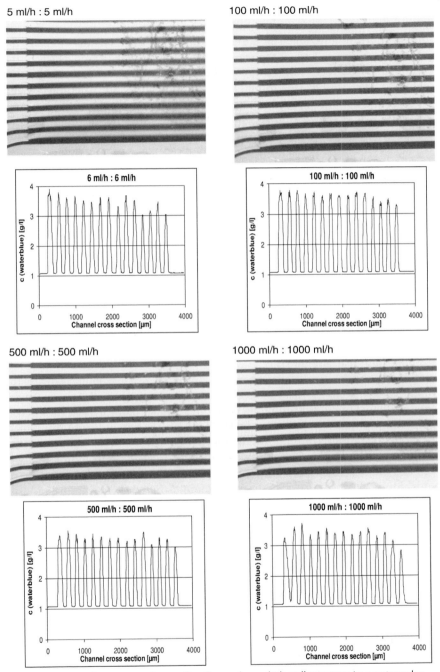

Figure 1.20 Periodic concentration profile of a regular multi-lamellar pattern in a rectangular interdigital micro mixer, determined by photometric-type analysis [119].

The spatio-temporal variations of the concentration field in turbulent mixing processes are associated with very different conditions for chemical reactions in different parts of a reactor. This scenario usually has a detrimental effect on the selectivity of reactions when the reaction time-scale is small compared with the mixing time-scale. Under the same conditions (slow mixing), the process times are increased considerably. Due to mass transfer inhibitions, the true kinetics of a reaction does not show up; instead, the mixing determines the time-scale of a process. This effect is known as mixing masking of reactions [126].

Mixing-masked reactions should develop differently when processed by conventional and micro mixers, for reasons of both global mixing speed and associated spatio-temporal concentration evolution. For instance, different solvent environments may be generated for a short period, impacting on the reactant strength, e.g. on the nucleophilicity or base strength of nucleophiles or on the electron distribution of aromatics. This will affect side reactions or even the nature of the product itself, e.g. by determining the position of reaction at a multi-functional target molecule or the atom moiety that actually attacks a bivalent reactant (for instance, by attack of the C or N moiety for CN^- anions, giving either nitriles or isonitriles) [126].

Mixing masking, i.e. influences of micro mixing on the reaction course, has been determined for a number of reactions, including azo coupling as a prominent process [126]. Since the selectivity of organometallic reactions is known to be particularly sensitive to the feed ratio of the reactants, a deliberate variation of global feed ratios was performed using micro mixers in an industrial study. It was found that the contribution of side reactions increased steeply on moving away from the ideal reactant ratio [11].

Speed-up of mixing is known not only for mixing of miscible liquids, but also for multi-phase systems the mass-transfer efficiency can be improved. As an example, for a gas/liquid micro reactor, a mini packed-bed, values of the mass-transfer coefficient $K_L a$ were determined to be 5–15 s^{-1} [2]. This is two orders of magnitude larger than for typical conventional reactors having $K_L a$ of 0.01–0.08 s^{-1}. Using the same reactor filled with 50 μm catalyst particles for gas/liquid/solid reactions, a 100-fold increase in the surface-to-volume ratio compared with the dimensions of laboratory trickle-bed catalyst particles (4–8 mm) is found.

1.4.4.2 Heat Transfer

The increase in heat transfer by micro-channel processing is frequently cited in the literature [5]. A number of authors cite overall heat-transfer coefficients, which serve to underline the high efficiency of heat transfer in micro-channel devices. The highest one measured so far was reported by Brandner et al. using micro-fin heat exchangers, amounting to 56 000 W m^{-2} K^{-1} using water as test fluid [23]. Heating rates may be another figure of merit to make evident the high thermal performance of micro devices. A heating rate of 680 000 K s^{-1} was reported for gas streams heated to 850 °C [23].

One of the most often cited advantages of micro reactors is their prevention of hot spots by strongly enhanced heat dissipation. Hence higher dosing of reactants, i.e. higher reactant concentration and/or higher catalyst loading, may be possible

Table 1.7 Numerical example illustrating the increased heat transfer on miniaturization of industrial reactors [110].

Scale of reactor	Typical volume/typical internal dimension	Normalized specific surface
Production	$30\ m^3$	1
Laboratory	$1\ l = 10^{-3}\ m^3$	30
Micro	$30\ \mu m$	3000

when dealing with exothermic reactions [124], thereby enhancing productivity. Also for endothermic reactions, improved heating may be advantageous, since thermal ramping can be achieved faster, reducing the pre-heating time to a minimum and avoiding overshoots. In contrast, the slower heating for conventional equipment can lead to undesired side products [124]. For a methane chlorination reaction, the origin of this is seen in a more undefined radical formation during the slower thermal ramping, having in addition a thermal overshoot [127].

For oxidative propane dehydrogenation in a fixed bed, hot spots ranging from 3 to 100 K were detected [128, 129]. Even when using diluted gases, the hot spots were as large as 20 K. Using the micro-channel reactor, isothermal processing was achievable for nearly all conditions applied, giving a maximum temperature increase of 2 K [128, 129].

Wörz et al. give a numerical example to illustrate the much better heat transfer in micro reactors [110–112]. Their treatment referred to the increase in surface area per unit volume, i.e. the specific surface area, which was accompanied by miniaturization. The specific surface area drops by a factor of 30 on changing from a 1 l laboratory reactor to a $30\ m^3$ stirred vessel (Table 1.7). In contrast, this quantity increases by a factor of 3000 if a $30\ \mu m$ micro channel is used instead. The change in specific surface area is 100 times higher compared with the first example, which refers to a typical change of scale from laboratory to production.

Wörz et al. stress the possibility of carrying out very fast reactions with large reaction heat in micro reactors [110]. They often use the terms 'isothermal operation' or 'isothermicity' to describe adequately the carrying out of a reaction with a heat that is taken out of the processing volume immediately upon release. In practice, they often refer to a temperature increase of 1–2 °C as a limit for fulfilling the criterion of isothermal operation.

1.4.4.3 Microfluidics

The key attribute of flows in micro devices is their laminar character, which stands in contrast to the mostly turbulent flows in macroscopic process equipment. Owing to this feature, micro flows are *a priori* much more accessible to a model description than macro flows and can be described by first-principle approaches without any further assumptions. In contrast, for the simulation of turbulent flows usually a number of semi-heuristic models are applied, and in many situations it is not clear which description is most adequate for the problem under investigation. As a result, it stands to reason to assume that a rational design of micro reactors based on simulation approaches is in some sense more realistic than for their macroscopic counterparts.

Apart from obvious features such as laminarity, there are speculations that flows in micro channels exhibit a behavior deviating from predictions of macroscopic continuum theory. In the case of gas flows, these deviations, manifesting themselves as, e.g., velocity slip at solid surfaces, are comparatively well understood (for an overview, see [130]). However, for liquid flows on a length scale above 1 μm, there is no clear theoretical foundation for deviations from continuum behavior. Nevertheless, various unexpected phenomena such as friction factors deviating from the continuum prediction [131–133] have been reported. A more detailed discussion of this still unsettled matter is given in Section 2.2. At any rate, one has to be careful here since it may be that measurements in small systems lack precision, essentially because of the incompatibility of analysis in a confined space and with large measuring equipment.

1.4.5
Fouling

Wegeng et al. mentioned fouling as a major problem, decreasing micro-fluidic system reliability [1]. Active control over fouling was demanded in their review, but no strategies to overcome it were given.

Hessel and Löwe state that fouling is a term which is only said in a whisper among 'micro chemical engineers' [9, 10]. Nonetheless, there is no denying of the fact that it is a frequent experience during handling of microfluidic components. The impact of fouling in micro reactors is over- and underestimated concurrently, this statement not being contradiction as it includes the opinions of different parties. Fouling still attracts too little attention in the technical–scientific exploration of micro systems. It is observed mainly when processing is not sufficiently adapted to the needs of fluid operation in micro channels. By choice of proper parameters, laboratory operation (i.e. from several tens of minutes to hours) of micro reactors becomes possible in many cases and hence usually does not pose a problem. For selected, albeit so far only a few, processes, considerably longer operation times have been achieved up to processing times that are demanded for chemical production.

In some cases, a simple increase in characteristic dimensions is sufficient to suppress micro-channel plugging [9, 10]. In the first euphoria about the new technical capabilities, miniaturization of processing equipment was possibly flogged to death. The motto therefore is: *as small as (beneficially) needed, but not as small as possible.*

As a kind of specialty solutions for the real 'hard' cases where fouling is intense and unavoidable, IMM first proposed ideas to develop special micro mixers for fouling-intense reactions and conducted feasibility tests, among them very fast organic reactions with spontaneous precipitation such as the amidation of acetyl chloride in THF [134]. The Forschungszentrum Karlsruhe developed special anti-fouling coatings in cooperation with partners [135].

1.5
Impact on Process Engineering

1.5.1
Laboratory-scale Processing

1.5.1.1 Provision of a Multitude of Innovative Reactor Designs

Jensen stresses the great flexibility in reactor design that can be achieved by means of microfabrication, in particular when parallel, MEMS-based mask processes are followed, having a multitude of miniature designs on one mask [75]. This should '... invigorate the innovative nature of reactor design ...' and allow one to overcome '... the tedium of stirred tanks, and tubular and trickle bed reactors'. Micro-chemical engineering as such an approach will probably not only provide many more reactors of different, distinct designs, but will also facilitate their testing and to modeling. If this can be put in an optimization loop, it stands to reason that much better performing reactor designs than we know now will become available. How this may be achieved was also shown by Jensen et al., who have provided an idea of an advanced test station where many micro reactors can be tested in parallel [19].

1.5.1.2 Quality of Information – More Accurate and In-depth

Another implication of micro reactors for chemical-process engineering involves the provision of accurate, in-depth and reliable experimentation data. This is exemplarily discussed for high-throughput screening.

Micro reactors permit high-throughput screening of process chemistries under controlled conditions, unlike most conventional macroscopic systems [2]. In addition, extraction of kinetic parameters from sensor data is possible, as heat and mass transfer can be fully characterized due to the laminar-flow conditions applied. More uniform thermal conditions can also be utilized. Further, reactor designs can be developed in this way that have specific research and development functions.

1.5.1.3 Quantity of Information – Speed of Experimentation

Besides quality of information, quantity of data still is a decisive factor in laboratory-scale development.

Following in the footprints of screening for drug discovery, the smallness of micro reactors recommends their use for high-throughput testing [2]. Today's laboratory testing suffers from high costs of reactants and safety concerns. In addition, the introduction of new ideas is hindered by the risk and high capital costs of scale-up, once a lead candidate has been found by laboratory testing. Micro reactors have the potential to overcome pilot-scale testing and redesign. This should be of particular advantage if only small quantities are produced, such as for the pharmaceutical industry. This allows scheduled, gradual investment in new chemical production facilities.

1.5.1.4 **Shrinkage of Total System**

One implication of micro reactors on chemical-process engineering concerns the shrinkage of the total system. This is exemplarily discussed for catalyst testing.

For catalyst testing, conventional small tubular reactors are commonly employed today [2]. However, although the reactors are small, this is not the case for their environment. Large panels of complex fluidic handling manifolds, containment vessels, and extended analytical equipment encompass the tube reactors. Detection is often the bottleneck, since it is still performed in a serial fashion. To overcome this situation, there is the vision, ultimately, to develop PC-card-sized chip systems with integrated microfluidic, sensor, control, and reaction components [2]. The advantages are less space, reduced waste, and fewer utilities.

1.5.1.5 **Integratability of Sensing and Other Functions**

Jensen highlights the potential of integrating multiple functions in one small micro reactor. In this context, he suggests that micro-reactor engineering must go beyond exploiting high transport rates due to their small dimensions, and rather integrate sensing and separation functions [75]. He points out, 'It was the integrated circuit that created the microelectronics revolution, not the transistor itself'. With increased integration, packaging rather than only microstructuring becomes the issue.

Concerning function integration, for example, micro-flow membrane reactors can exhibit similar process intensification, as shown already for their large-scale counterparts [75]. Separation columns for proteomics, immobilizing enzymes, utilize the large surface-to-volume ratios. Surface tension differences can guide and transport liquids selectively.

To realize micro reactors with sensing and surface-control structures, the underlying fabrication methods have to go beyond MEMS and classical micro machining [75]. Plastic and glass manufacture and also novel innovative methods such as soft lithography have to be developed further and applied.

Hessel and Löwe agree on the need of future micro-sensor development, but in a kind of state-of-the-art analysis also discuss today's solutions for monitoring of micro-channel flow [9, 10]. The integration of appropriate conventional sensing, e.g. for temperature monitoring, is a non-ideal but practical, simple solution. The simple transition from a non-transparent to a glass material allows visual inspection of the processing for the operator. A multitude of information is supplied by this smart change without a newly integrated sensing function.

As an alternative to both classical sensing and integration of micro sensors, they also provide examples of contact-free in-line measurements of temperature or concentration, e.g. to characterize transport processes in micro reactors, using advanced, conventional apparatus (Figure 1.21) [9, 10]. Having this in mind, Hessel and Löwe argue in favor of caution when dealing with sensing/monitoring solutions. They come to the conclusion that new developments are definitely desired, albeit we do not need them for every problem. There is no reason why pressure determination, for example, could not still be done in a classical way. The advantages of micro-sensor-based pressure measurements are, at least, not obvious.

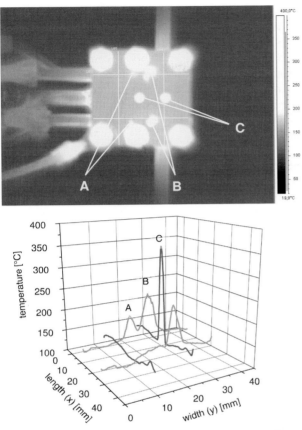

Figure 1.21 Thermographic image taken during heating-up of a gas-phase reactor (top). A multitude of information is supplied on the spatio-temporal temperature evolution at the outside of the reactor housing and via conduits even within the micro reactor (bottom) [15].

The integration of sensing and other functions in a micro-flow system requires either monolithic, on-chip or hybrid, multi-scale approaches. Concerning the latter, Hessel and Löwe discuss the lack of compatibility of today's fluidic interfaces and report on a German project team developing a standard for micro-reactor interconnection [9, 10].

1.5.2
Industrial Process Development and Optimization

1.5.2.1 Information on Industrial Large-scale Chemical Manufacture: Time to Market
There is common agreement that with the aid of micro reactors, process development can be speeded up considerably [5]. First industrial examples on the implementation of micro-reaction technology for pilot-scale and production seem to con-

firm this assumption [11, 21, 110, 136]. Besides providing the mere facts on the specific process developments made, the descriptions of Wörz et al. give vivid examples of the value of micro reactors for process development from an industrial perspective [110].

Felcht reports that the testing of industrial-scale processes can be performed with low expenditure by using micro reactors, since this should result in a faster time to market of the development [137]. He also sees uses for micro reactors at the laboratory scale as a means of high-throughput screening and model examinations such as fast determination of reaction kinetics.

1.5.2.2 Pharmaceutical and Organic Synthesis Process Development

The organizational format in current industrial chemistry impacts on the choice of process development and the respective tools. In a contrasting scenario, a new tool with much improved performance could impact the way of approaching chemical synthesis (Figure 1.22). In this context, Schwalbe et al. describe how micro reactors can change current trends in organizing industry's chemistry development [81]. Whereas the established process is made in several steps by several groups, comprising discovery, functional chemistry, chemical development, and process development until market entry, the micro-reactor process ideally should have only

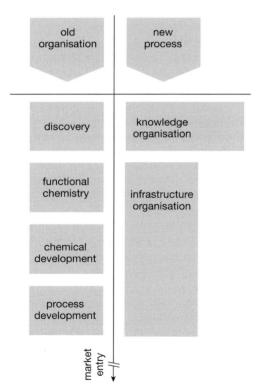

Figure 1.22 Organizational format in current industrial chemistry and proposal for a change in the future which may be induced by use of micro reactors [81].

two steps, termed knowledge and infrastructure organization (see [81] for an explanation). As a result, faster total development and faster market entry are claimed to be achievable.

1.5.2.3 Approval by Public Authorities

According to an older German study, the time-to-market of new chemical products, besides optimizing the internal research and development as outlined above, could profit from faster administrative approval by the public authorities [138]. For example, the phase of getting the concession has a share of about 20% of the overall time for process development, amounting on average to 7.5 years in Germany. Process development in other countries is faster, e.g. 5.0 and 5.7 years in Japan and the USA, respectively.

Although it is unclear at present how the public authorities would react if micro reactors were largely employed for production purposes, it stands to reason that in principle faster approval could be achieved in this way. Approval is often connected with assigning the equipment under consideration to certain reactor categories which typically are defined, among other features, by the reactor internal volume. Due to their small internal volumes and their large fraction of construction material, micro reactors, as a general rule, do not fall into the categories usually considered for tedious and thorough approval. In addition, considering their safety features and new ways of environmental protection (e.g. complete encapsulation [139]), faster approval can certainly be realized. However, one should not be too starry-eyed. Approval processes in the chemical industry have a long tradition and have previously never been faced with an abrupt, quasi-revolutionary change in processing apparatus. Therefore, they will not be changed overnight, but more likely in an evolutionary way.

1.5.3
Pilot-stage Processing and Centralized Production

1.5.3.1 Production as a Challenge for Micro Reactors

There is essentially no other question which more vigorously splits the micro reactor and even the chemical community into groups of complete stances. There are even commentators who consider the proof of production feasibility of micro reactors as the central issue to be solved, like a *conditio sine qua non*. As an extreme view, they claim to see no other *raison d'être* for micro reactors or at least none of similar importance.

Among the chemical community, three groups can be identified. One group of experts strongly doubt that micro reactors can be used at all for production. While this is most of all the point of view of those engaged in chemical producing manufacture (maybe for reasons of not waking up competitors), micro reactor researchers themselves are torn between conservative skepticism and optimistic views, the latter saying it is not practiced now, but will come anyway.

The chemical manufacturers defend their negative view by referring to the scaling law, which predicts that processing equipment becomes more cost-efficient the larger it is, referring to the 'seven tenths law' [139]. However, an opposite position can be taken as well, saying that the economy of scale is now applied to plant manufacture (through production of large numbers of small units) rather than to plant operation itself [139]. In addition, the chemical manufacturers raise objections if fouling problems can be solved and if generally reliability can be set to a level acceptable for automated industrial processing.

Wörz et al. give an 'in principle, yes' answer, but point at the serious problems to be encountered [110–112]. They say that production in micro reactors will be the exception rather than the rule. In view of the background of the authors coming from a large-base chemical provider, they assume that production has to occur in hundreds or even millions of separate parallel streams. Accordingly, they refer to the problems of uniformly splitting one main stream into a multitude of streams and also then regard fouling as a major problem. Cleaning procedures after shutdown of the system, common for laboratory equipment, are said to be 'inconceivable' for industrial production.

1.5.3.2 Micro Reactors as Information Tools for Large-scale Production

One of the most interesting theorems of Wörz et al. is that they see a serious potential for micro reactors to permit small-scale production of some different sort [110–112]. Micro channels serve as an ultra-precise measuring tool, whereas production is done in channels about 10 to 100 times larger, i.e. millimeter-sized channels. The limit of tube diameter of industrial production reactors is reported to be 2 cm; hence any new reactor of smaller characteristic dimensions bears some potential for improvement. Wörz et al. conclude with the remark that the above strategy 'could be the most important result' of their studies [110–112].

Jensen is of the same opinion, pointing out that replication of micro reactors used in the laboratory avoids costly redesign and pilot-plant experiments, shortening the development time from laboratory to commercial production [75]. He sees particular uses of that for the fine-chemical and pharmaceutical industries with production of a few metric tons per year. A peculiar feature of such a numbering-up approach is that it allows for '... scheduled, gradual investment in new chemical production facilities without committing to a large-scale manufacturing facility from the outset' [75]. Accordingly, mass fabrication of individual micro-reactor components and subsequent integration could challenge the traditional centralized economy of scale practiced in the chemical industry. This manufacturing-based approach, albeit not implemented so far for micro-reaction technology, is today's practice in the electronics and automotive industries.

1.5.3.3 Micro reactors for Specialty-chemicals Production

Felcht reports on Degussa's activities in cooperation with partners from academia and industry to develop innovative industrial-scale micro-structured reactors for making large-tonnage products by liquid- and gas-phase reactions [137]. The aim is to make the potential of micro reactors more widely available for a larger variety of processes, naturally with focus on Degussa's fine and specialty chemical productions. Also, the aim is to circumvent traditional problems of scale-up.

Figure 1.23 Large-scale reactor with microstructured internals tested at the industrial site of Degussa and developed jointly by a project team of several partners [137].

One of the reactors that Degussa and the plant manufacturer Krupp-Uhde are currently investigating in the framework of a government-funded project has remarkably large outer dimensions and is being tested for pilot-scale feasibility (see Figure 1.23).

1.5.3.4 Intensification of Transport – Reduction of Equipment Size

Many publications refer to the use of micro reactors for process intensification, with all the implications related to this definition – safety, cost reduction, high productivity rate, environmental friendliness, energy efficiency and so on [5, 25, 104]. A particular feature of interest is the reduction of the equipment size (see also Section 1.4.3.5 for a systematic top-down description of this topic).

Calculations predict that improved heat transfer for reacting systems in micro-channel heat-exchanger reactors could lead to considerable size reduction of the equipment, by enhancing the degree of product formation per micro channel (see Figure 1.24) [140, 141]. This was exemplarily shown for a fast, high-temperature

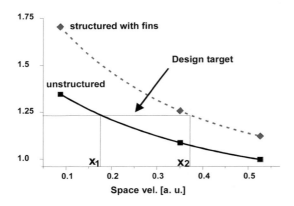

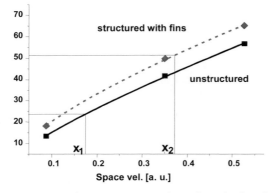

Figure 1.24 Outlet concentration and specific molar flux of product (P1) in arbitrary units as a function of the space velocity of a feed component.

gas-phase industrial reaction, the nature of which was not disclosed, with a hypothetical catalyst. For thick (e.g. 100 μm) catalyst layers, mass-transfer limits in the porous structure are observed, reducing catalyst effectiveness. Accordingly, for the same performance, the catalyst mass can be reduced when using micro reactors as compared with standard fixed-bed technology. This comes together with a reduction in the total device size; for the application investigated this was an order-of-magnitude reduction. Concerning heat transfer, the use of micro fins gave improvements, i.e. an increase in Nusselt number by a factor of five. Similar findings of heat-transfer enhancement with fin-type microstructures have been reported [14, 23, 142, 143].

Substantial heat-transfer intensification was also described for a special micro heat exchanger reactor [104]. By appropriate distribution of the gas-coolant stream, the axial temperature gradient can be decreased considerably, even under conditions corresponding to very large adiabatic temperature rises, e.g. of about 1400 °C.

A detailed characterization of micro mixing and reaction performance (combined mixing and heat transfer) for various small-scale compact heat exchanger chemical reactors has been reported [27]. The superior performance, i.e. the process intensification, of these devices is evidenced and the devices themselves are benchmarked to each other.

1.5.4
Distributed, On-Site Production

1.5.4.1 An Existing Distributed Small-scale Plant for Phosgene Synthesis
When discussing distributed small-scale production of hazardous materials, one should be aware that such plants already exist, albeit not using micro-channel technology. The Ciba-Geigy phosgene plant produces phosgene on demand [144]. It is said to be a very safe apparatus, as phosgene is neither liquefied nor stored. The plant contains only about 10 kg of gaseous phosgene and is encapsulated by a double mantle. On-line monitoring of phosgene is applied. In case of an incident, a two-fold washing equipment is provided, able to quench the whole phosgene content. The rated output can be varied from 10 to 100%. Product selectivity exceeds 97.5%. A 5–10 min start-up time is needed. As waste, only sodium chloride and sodium carbonate are produced. Waste water in an amount of only 3 m^3 per month is generated for an annual production of 5000 t of phosgene.

1.5.4.2 Distributed Manufacturing – A Conceptual Study of Future Scenarios
Concerning process-engineering impacts of micro reactors, many authors argue for distributed production owing to size and weight advantages of these devices and to their modularity and hence flexibility for multi-purpose syntheses [1, 5, 104]. Felcht outlines the potential of external numbering-up of micro-reactor devices for adjusting the capacity to the quantity needed at decentralized sites, albeit saying that this refers to a limited number of devices [137].

As a possible chemical process for distributed manufacture, the production of toxic feedstock gases is claimed [1, 139]. Benson and Ponton were among the first

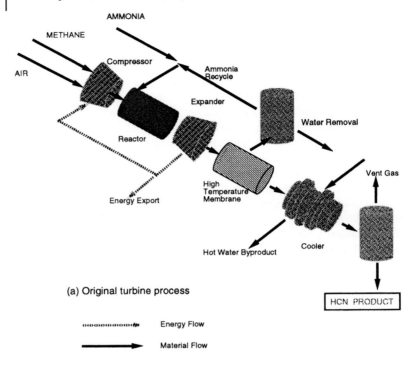

(a) Original turbine process

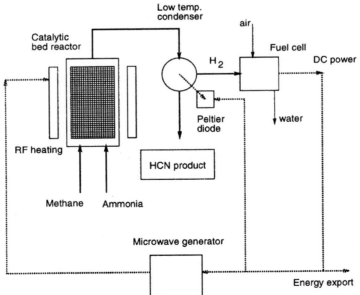

Figure 1.25 Explosion drawing and schematic of a future plant for distributed chemical manufacturing [139].

to figure out some details of a concept of distributed manufacturing [139] (see [145] for the comments of trade press), which will be referred to in detail in Volume 2 of this book series (more information is also provided in this volume; see Section 1.7.3). Basically, they refer to small, transportable plants which are fed with reactants 'over the fence', hence using only non-hazardous, generally available materials by normal piping or standard transport (Figure 1.25).

The authors provide selection criteria, by which the suitability of a process for a distributed production can be assessed [139]. These are explicitly given for the categories of feedstock, processes, customer products, and waste products. This is completed by a list of suitable device types for distributed production such as plate heat exchangers, pressure and temperature swing units, electrostatic dispersers, and membrane units. The various operations often rely on the use of electricity and therefore are said to be particularly suited for operation at the mini scale.

1.5.4.3 Central Role of Control Systems and Process Models

Control systems will play a key role in future distributed plants [139, 145]. As a rule of thumb, plants will be smaller and simpler, but the control systems will be much more advanced, of a standard not known today. Plant personnel for operation and managing will ultimately no longer be required, except for start-up, shutdown, and services. This is a shift from a regulatory to a 'servo' role, supported by a sophisticated sequence control. Control is needed for safety issues, operability, and product quality control. Sensors have a central role to provide the information needed for control and modeling and simulation is needed for process models.

Although 'active' safety is provided by the control systems mentioned above, 'passive' safety is an additional important feature of a distributed plant. Due to the low inventory, even a total release of the reaction volume or an explosion would create no significant impact on the environment [139]. To prevent such scenarios, a total containment of the plant is envisaged; it needs to be 'sealed for life'. Hydrogen cyanide synthesis and chlorine 'point-of-sale' manufacture are two examples for safety-sensitive distributed syntheses.

1.5.4.4 Off-shore Gas Liquefaction

Off-shore gas liquefaction of natural gas on oil platforms is often mentioned as an example of on-site production [5]. Natural gas today is flared as its transportation is not economic. Conversion on-site by conventional reactors suffers from the large weight of the equipment. Gavriilidis et al. state that 5 kg of support are needed to support 1 kg of equipment on an oil platform [5]. Light-weight micro reactors have the potential to use the precious area of the platforms better (Figure 1.26).

1.5.4.5 Energy Generation and Environmental Restoration

Wegeng and Drost give five examples for distributed processing applications, which relate to the fields of energy generation and environmental restoration [106]. The first example concerns distributed heat pumps, designed for military purposes. A lightweight, portable unit serves for cooling in airtight, protective clothing under bio-hazardous conditions. Compared with industrial absorption-cycle heat pumps,

Figure 1.26 Large-capacity heat exchanger, e.g. for use on oil platforms, constructed from diffusion-bonded plate stacks comprising a vast number of millimeter-sized channels. This apparatus was manufactured by Heatric (Poole, UK).

the volume is said to be reduced by a factor of about 60 when relying on micro-channel-based heat-transfer performance.

The second example describes distributed, mobile and portable power-generation systems for proton-exchange membrane (PEM) fuel cells [106]. A main application is fuel processing units for fuel cell-powered automobiles; it is hoped that such processing units may be achieved with a volume of less than 8 l.

The third example is compact cleanup units for waste treatment, mainly in consideration of the numerous radioactive sites, stemming from cold-war military developments [106]. The Hanford, Washington, USA, site with a multitude of seriously contaminated tank wastes is among them. Due to the unknown character of most polluting species, the installation of a central waste-treatment facility is said to be not the best and most inexpensive solution. Rather, small modular units, able to be individually adapted to various separation tasks, which are inserted into the tanks and perform cleanup on site, are seen as the proper solution.

The fourth example, the use of chemical processing on Mars for producing a propellant, is presented in Section 1.9.7 [106]. The fifth and last example describes the use of distributed systems for global carbon dioxide management, aiming at reducing the greenhouse effect [106]. The main issue here is the installation of gas-absorption equipment for CO_2 capture at central, fossil-fuel power plants.

1.5.4.6 Desk-top Pharmacies, Home Factories and More

Other suggestions on distributed processing reach beyond the capabilities of today's micro reactors, and hence are of more visionary character [5]. These include desktop pharmacies, home-recycle factories, mobile factories, house-water treatment plants, processing cereal crops at the combine, purification of blood in the body, recycling of plastics in the collection vehicle and more [5].

1.5.4.7   Production of Chemical Weapons?

One (at least theoretical) variant of distributed processing by micro reactors, although by no means useful or desirable, is to use them for making chemical weapons at various sites. This possibility for the use of micro reactors has been discussed [83], particularly with respect to their employment as tools for terrorist attacks and to facilitate the clandestine manufacture of chemical agents. At the moment, highly skilled personnel are still needed for micro-reactor fabrication. However, the emerging worldwide capabilities of microfabrication techniques will render future processing in micro reactors no longer unique or proprietary. One day they will be available in regular workshops. Thus, it cannot be ruled out completely that people with allegiance to a terrorist organization could misuse micro-reactor technology by construction at their site or simply by purchasing these devices.

Among the hazardous chemical weapons scheduled class 1–3, methyl isocyanate becoming more and more important as a precursor [83]. This is just one among a number of substances which could be made via micro-reactor synthesis. Especially in the case of so-called binary weapons, where two relatively harmless substances are mixed to give a weapon, on-site mixing is demanded; this can be accomplished with high performance by micro reactors. Pocket-sized miniature plants can neither be monitored nor detected.

1.5.4.8   Standardization

Standardization is seen as advantageous for distributive production by micro reactors, as it is supposed to decrease development times and to provide higher flexibility by modularity [5]. Hessel and Löwe discuss the current lack of compatibility of suitable fluidic interfaces for hybrid, multi-scale micro-reactor construction (see Section 1.1.7) and report on a German project team developing a standard for micro-reactor interconnection (Figure 1.27) [9, 10].

1.5.5
The Shape of Future Plants/Plant Construction

1.5.5.1   The Outer Shape of Future Chemical Manufacture Plants

Many authors have addressed this important topic. The visions extend from shoe box-sized construction-kit plants, back bone-based micro-device assemblies, to much larger pilot and production plants of a shape not too different from now. Most authors follow the last path. In the end, as a consequence of process intensification, the equipment of the plants will become smaller in addition to the plant itself. A nice drawing of how such plant with process-intensification equipment may look is given in [25] as a future plant vision (Figure 1.28). Since one type of equipment will be replaced after another, it may take some time until we see a noticeable change in the external shape of a plant. As a governing principle of plant construction, this implies inserting micro-channel reactors in an existing plant environment. This is termed a hybrid [9, 10] or multi-scale approach (see Section 1.1.7). Naturally, micro devices with larger throughputs, hence larger internal components and so larger external dimensions, will be the first to be chosen.

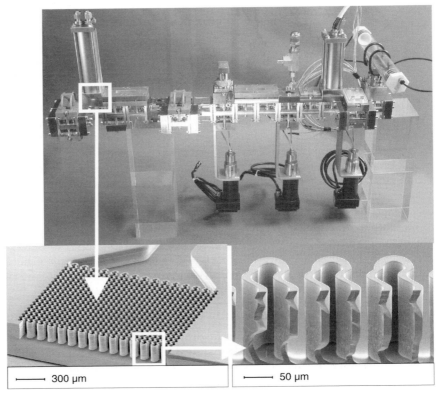

├─────────── 300 μm

├─────────── 50 μm

Figure 1.27 Standardized fluidic backbone for connecting micro-reactor components from different suppliers, yielding a small, compact 'micro-plant' with multiple functions. This backbone not only has a fluidic path, but also contains an electric bus system.

Figure 1.28 Illustration of a future plant using process-intensification equipment, aiming at giving the corresponding shape or perception. Compared with today's plants, a reduction in size is predicted [25].

1.5.5.2 Today's Shape of Micro-reactor Bench-scale Plants: Monolith vs. Hybrid/Multi-scale? Specialty vs. Multi-purpose?

The realization of complete bench-scale micro reactor set-ups is certainly still in its infancy. Nevertheless, the first investigations and proposals point at different generic concepts. First, this stems from the choice of the constructing elements for such set-ups. Either microfluidic components can be exclusively employed (the so-called monolithic concept) or mixed with conventional components (the so-called hybrid or multi-scale concept). Secondly, differences concerning the task of a micro-reactor plant exist. The design can be tailor-made for a specific reaction or process (specialty plant) or be designated for various processing tasks (multi-purpose plant).

Hessel and Löwe provide examples of hybrid, i.e. multi-scale, approaches, including the first commercial systems (see Section 1.1.7) [9, 10]. Such approaches are currently most often favored for micro-reactor construction, simply for practical time and cost reasons. In addition, such an approach allows one to fit micro reactors in existing industrial and academic environments for production and measurement. The micro reactor is only used where it is really needed, and in this way costs of changing the processing are kept to a minimum.

A micro reactor concept proposed by MIT and DuPont on the basis of electronic circuits is the most prominent among the examples listed for the hybrid approach [19, 101]. The so-called *turnkey multiple micro-reactor test station* relies on the use of standard components originating from the semiconductor industry for microchemical processing, the construction being oriented at the concept of *printed circuit boards*.

Micro mixer-tube reactor set-ups also belong to the hybrid category. Institutes such as the Forschungszentrum Karlsruhe [146] and IMM [147] have presented both monolithic and hybrid concepts, giving many variants of the latter option. These set-ups are often highly specialized, often only developed for one specific purpose. Commercial suppliers such as mgt mikroglas technik (see [148]) or CPC (see Figure 1.29 and [81]) developed hybrid concepts which usually target a broader range of applications (<www.mikroglas.de>; <www.cpc-net.com>, [113]).

Figure 1.29 CPC micro-reaction system CYTOS [9].

The hybrid approach seems to be the more pragmatic procedure and its time has come already. It allows one to analyze the advantages of micro reactors without delay, especially facing today's industrial time demands. The impressive results gathered so far, especially on the industrial side, substantiate that this – and so far only this – concept is not lacking in innovative character besides pragmatism [11, 28]. Set-ups with many micro-reactor components, pointing to a monolithic concept, will find use when each of these components on their own or the interplay between all of them gives advantages. This needs more time for development, but can be built on the progress achieved so far.

1.5.5.3 Methodology of Micro/Mini-plant Conception

Rinard dedicated his research to a detailed analysis of methodological aspects of a micro-reactor plant concept which he also termed mini-plant production [85] (see also [4, 9, 10] for a commented, short description). Important criteria in this concept are JIT (*just-in-time*) production, zero holdup, inherent safety, modularity and the KISS (*keep it simple, stupid*) principle. Based on this conceptual definition, Rinard describes different phases in plant development. Essential for his entire work is the pragmatic way of finding process solutions, truly of hybrid character [149] (miniaturization only where really needed). Recent investigations are concerned with the scalability of hybrid micro-reactor plants and the limits thereof [149]. Explicitly he recommends jointly using micro- and meso-scale components.

1.5.5.4 Highly Integrated Systems

Jensen emphasizes the potential of system integration to combine many functions on one chip [101] (see Section 1.5.1.5). Chemical detection is often the rate-limiting step in many chemistry investigations for gathering product information. Macroscopic test systems may be replaced by PC card-sized micro-chemical systems consisting of integrated microfluidic, sensor, control, and reaction components. The advantages are obvious: namely less space requirements, less utilities, and less waste production. This is seen as the step towards high-throughput screening of process chemistries under controlled conditions.

Jensen gives several examples for his present highly integrated chip systems [101], including a gas-phase reactor, a liquid-phase reactor, a catalyst-testing reactor, and a packed-bed multi-phase reactor. In addition, he provides the vision of a multiple micro-reactor test station (see Section 1.5.5.2).

1.6
Impact on Process Results

1.6.1
Selection Criteria for Chemical Reactions for Micro Reactors

Lerou et al. mention the following criteria that render reactions suitable for investigations in micro reactors [74]:

- fast
- homogeneous
- catalytic
- photochemical
- high temperature
- hazardous

1.6.2
Conversion, Selectivity, Yield

1.6.2.1 Conversion

Owing to increased mass transfer and the use of aggressive reaction conditions (e.g. increase in temperature), high conversions can be achieved in micro reactors.

Such improvements in conversion were reported for the oxidation of ethanol by hydrogen peroxide to acetic acid. This is a well-studied reaction, carried out in a continuous stirred-tank reactor (CSTR). Near-complete conversion (> 99%) at near-complete selectivity (> 99%) was found in a micro-reaction system [150]. Processing in a CSTR resulted in 30–95% conversion at > 99% selectivity.

For the Michael addition of 2,4-pentanedione enolate to ethyl propiolate, improvements in conversion were determined. This example serves also to demonstrate that proper process conditions are mandatory to have success with micro-reactor processing. A conversion of only 56% was achieved when using electroosmotically driven flow with a two-fold injection, the first for forming the enolate and the second for its addition to the triple bond (batch synthesis: 89%) [151]. Using instead a stopped-flow technique to enhance mixing, a conversion of 95% was determined.

1.6.2.2 Selectivity

Wörz et al. stress a gain in reaction selectivity as one main chemical benefits of micro-reactor operation [110] (see also [5]). They define criteria that allow one to select particularly suitable reactions for this – fast, exothermic (endothermic), complex and especially multi-phase. They even state that by reaching regimes so far not accessible, maximum selectivity can be obtained [110]. Although not explicitly said, 'maximum' refers to the intrinsic possibilities provided by the elemental reactions of a process under conditions defined as ideal; this means exhibiting isothermicity and high mass transport.

Consecutive reactions can reduce selectivity. Among them, partial oxidations or partial hydrogenations undergo several consecutive reaction steps during the course of the reaction. Usually, an intermediate, rather than the final product is the value target product. In this context, it was proven that micro reactors can exhibit high selectivity to such intermediates. For the hydrogenation of *cis,trans,trans*-1,5,9-cyclododecatriene to cyclododecene, the performance of the micro reactor was benchmarked against three types of fixed beds, two of them being composed of the same catalyst as employed for the micro reactor. Monitoring cyclododecene yield versus selectivity, it was found that the yield decreases from 62–64% within the range of conversions investigated when using the conventional granules. For the

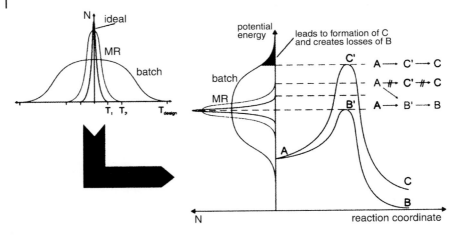

Figure 1.30 Temperature/energy distributions correlated to a generic potential energy curve [81].

foil fragments, a nearly constant selectivity of 73% was found. Wire pieces and the micro-channel reactor both give notably better results than the foils; the micro-channel reactor is slightly better than the wires.

Similar findings were made for liquid-phase reactions such as the naphthalene nitration in micro reactors. The nitro group can be introduced only once per naphthalene molecule or multiple times; up to four nitro groups can ultimately be attached to the aromatic core. For micro-reactor processing, mainly mono- and dinitro products were obtained [152]. For batch processing of naphthalene, a wider range of products are found containing many isomers of the above-mentioned species, but also tri- or tetranitrated products. In the micro reactor, even at 50 °C and using a large excess of nitrating agent, high selectivity was maintained, as revealed by the high degree of mononitronaphthalenes in the product mixture [152].

Side reaction can reduce selectivity as well. In this context, Schwalbe et al. gave the first insight into the relationship between potential-energy profiles of the reaction course and setting proper micro-reactor operation, particularly to enhance selectivity [81]. Using generic, known profiles under kinetic and thermodynamic control, a first mechanistic analysis is presented on how selectivity will be affected on changing the temperature. Normal distributions of temperatures within a micro reactor and a batch are given, the first being much smaller. These temperatures can be related to energy values, thereby revealing the respective energy distributions for the batch and the micro reactor. Two such curves were superimposed on a potential energy curve under kinetic control with a main and a side reaction (Figure 1.30). In this way, it becomes evident that a high-energy energy fraction given in the batch can induce the side reaction (having a higher value of activation energy). In contrast, the micro reactor is virtually only supplying the energy needed for passing the transition state, related to the activation-energy curve. Hence the micro reactor does not induce side reactions to any great extent.

The main achievement of Schwalbe et al. is to have initiated such considerations in micro-reaction technology, known in conventional chemistry for decades, and to have pointed out the theoretical value of such reaction mechanism-based organic micro-reactor processing [81].

Selectivity may also come from reducing the contribution of a side reaction, e.g. the reaction of a labile moiety on a molecule which itself undergoes a reaction. Here, control over the temperature, i.e. the avoidance of hot spots, is the key to increasing selectivity. In this respect, the oxidative dehydrogenation of an undisclosed methanol derivative to the corresponding aldehyde was investigated in the framework of the development of a large-scale chemical production process. A selectivity of 96% at 55% conversion was found for the micro reactor (390 °C), which exceeds the performance of laboratory pan-like (40%; 50%; 550 °C) and short shell-and-tube (85%; 50%; 450 °C) reactors [73, 110, 112, 153, 154].

1.6.2.3 Yield

As it was shown before that conversions and selectivities can be increased, usually not at the expense of each other, it stand to reason that micro reactors provide high yields. For example, the Suzuki coupling of 4-bromobenzonitrile and phenylboronic acid gives a yield of 62% for micro-flow processing which is about six times higher than with batch processing (10%) at comparable process conditions [155].

As a second example, several Hantzsch syntheses using diverse ring-substituted 2-bromoacetophenones and 1-substituted-2-thioureas are given. For these reactions, comparative and better yields were achieved when using a micro-mixing tee chip reactor as compared with conventional laboratory batch technology. The increase in yield amounted to about 10–20% [156, 157].

Finally, yield improvements were also reported for industrial process developments. For the Merck Grignard process, a yield of 95% was obtained by a micro mixer-based process, while the industrial batch process (6 m^3 stirred vessel) had only a 72% yield (5 h, at −20 °C) [11]. The laboratory-scale batch process (0.5 l flask; 0.5 h, at −40 °C) gave an 88% yield.

1.6.3
Reaction Time – Reaction Rate

1.6.3.1 Reaction Time

Many reports confirm notable reductions in reaction times when carrying out reactions under micro flow conditions. Concerning β-dipeptide synthesis, for example, a comparison between batch and micro-reactor processing was made for the reaction of Dmab-β-alanine and Fmoc-L-β-homophenylalanine [158]. While the micro reactor gave a 100% yield in 20 min, only about 5% was reached with the batch method. Even after 400 h, only 70% conversion was achieved.

Most often, such enormous improvements are discussed in a classical way following conventional organic chemistry descriptions, e.g. providing the experimental protocol and briefly giving the results. This is usually not followed by a chemical-engineering explanation. Thus it remains unclear to what extent the batch protocols relied on actually providing much more residence time than kinetically needed. Everyone knows the expression 'stirring overnight' to complete reactions. In the future, we need real kinetic data to compare batch and micro-reactor performance. However, 'stirring overnight' is common practice and organic reactions are typically conducted over hours, if not for days. There is a lot of evidence that micro-reactor processing is definitely faster in most cases (see [29] for an overview). One just does not know which share of this has to be attributed to improvements of mass and heat transfer and how much of the reactions were simply processed too long, in a kind of chemist's tradition.

1.6.3.2 Reaction Rate

Often the micro-reactor data were able to confirm literature values for true reaction rates. For ethylene oxide synthesis, a reaction rate of $4.8 \cdot 10^{-5}$ mol s^{-1} m^{-2} was observed [159]. This figure, when corrected to compensate different experimental conditions, compares with a literature value of $1.7 \cdot 10^{-6}$ mol s^{-1} m^{-2}; then, a value of $1.9 \cdot 10^{-6}$ mol s^{-1} m^{-2} is obtained for the micro-reactor processing.

In some cases, higher effective reaction rate constants were reported for micro-reactor processing, usually without giving a chemical-engineering explanation for this fact. Often, this is associated with somehow suboptimal processing at the macro scale. Accordingly, not all improvements reported really refer to a generic advantage of the micro reactor; sometimes, the latter just facilitates processing improvements which in principle are achievable also with conventional equipment (but, in fact, were not achieved so far). An example of this type is the Kumada–Corriu reaction between 4-bromoanisole and phenylmagnesium bromide. For this reaction, observed rate constants were determined [160]. For high-flow-rate processing ($33.3 \, \mu$l min^{-1}), an observed rate constant of 0.033 1 s^{-1} was obtained. This amounts to a rate enhancement of 3300-fold compared with the value for batch processing. This was explained by being able to feed the reaction solution inside the pores of a Merrifield resin, thereby dramatically increasing the available reaction surface in comparison with traditional processing, relying on the outer surface only.

1.6.4
Space–Time Yield

Given large conversions and sufficiently short contact times, space–time yields can be very high in micro reactors, in particular since the 'space' itself, i.e. the micro-flow-through chamber, inevitably is small. It stands to reason that improvements in space–time yields may be obtained for any fast organic reaction which conventionally is carried out in batch very slowly, e.g. for reasons of control over heat release, and now is performed at high throughput in a micro mixer/heat exchanger system.

When the space–time yield is referred to the total reactor volume (and not only to the micro-channel volume), the large share of 'inactive' construction material has to be taken into account. Consequently, the space–time yields per micro-channel volume have to differ by orders of magnitude, e.g. more than a factor of 1000, from those of conventional reactors, to have a gross result. So far, only a few data have been provided in the open literature, although many such processes, also from the industrial side, have been developed subsequently.

By an industrial investigation of a gas-phase reaction, the chlorination of alkanes, thermal management (faster temperature ramping, avoidance of overshoots) was improved and, hence, control over radical formation was exerted. As a result, a significant increase in space–time yield to about 430 g h^{-1} l^{-1} was achieved using a hybrid micro-reactor plant compared with the conventional performance of 240 g h^{-1} l^{-1} [127, 161].

Even for the well-known ethylene oxide formation, improvements in space–time yield were reported. A value of 0.78 t/(h m^3) using an oxidative modified silver was obtained, which exceeds considerably the industrial performance of 0.13–0.26 t h^{-1} m^{-3} [159].

However, these investigations also point out that we need a proper definition of space–time yields for micro reactors. This refers to defining what essentially the reaction volume of a micro reactor is. Here, different definitions lead to varying values of the respective space–time yields. Following another definition of this parameter for ethylene oxide formation, a value of only 0.13 t h^{-1} m^{-3} is obtained – still within the industrial window [159, 162, 163].

Data are also available for space–time yields of reactions where completely new reaction paths or regimes were followed. In some cases, this was associated with operation in the explosion regime. An example of this type is the direct fluorination of aromatics using elemental fluorine. For toluene fluorination, space–time yields higher by an order of magnitude were found for the falling film micro reactor and the micro bubble column compared with the laboratory bubble column [164]. The space–time yields for the micro reactors ranged from about 20 000 to 110 000 mol of monofluorinated product h^{-1} m^{-3}. The ratio with respect to this quantity between the falling film micro reactor and the micro bubble column was about 2. The performance of the laboratory bubble column was in the order of 40–60 mol of mono-fluorinated product h^{-1} m^{-3}.

1.6.5
Isomerism

1.6.5.1 Cis–Trans Isomerism of Double Bonds
The products of reactions generating double bonds can exhibit positional isomerism, as rotation of the moieties, given before the reaction, is now prevented. This is usually referred to as cis/trans or, as a complementary description, Z/E isomerism. There are first hints that cis/trans or Z/E ratios of the products with micro-reactor processing differ from the corresponding data for conventional processing.

An example of such an impact is the Wittig reaction. For the formation of double bonds from 2-nitrobenzyltriphenylphosphonium bromide and methyl 4-formyl-benzoate, it was determined that the ratio of cis and trans products (Z/E ratio) can be changed by simply adjusting the voltages in an electroosmotic-flow driven chip

[78]. The Z/E ratio is also strongly influenced by moving the reactants separately or as a pre-mixed solution. For a 1 : 1 ratio of the reactants, the Z/E ratio changed from 2.35–3.00 (pre-mixed) to 0.82–1.09 (not pre-mixed, separate movement) [165]. In a subsequent extended study it was found that relatively small voltage changes of the order of only 100 V were needed for large changes in the Z/E ratio. For instance, changing the voltage from 694 to 494 V for one channel decreased the Z/E ratio from 2.30 to 0.57.

For a Michael addition, however, the same isomeric ratio (99% trans : 1% cis) was observed for micro-reactor and batch operations [151].

The results clearly show that more results are needed to confirm the validity of the impact of micro reactors on the regioisomerism of substituted aromatics. Also, an explanation is needed of whether the effect can be confirmed.

1.6.5.2 Regioisomerism in Condensed Aromatics
The substitution of condensed aromatic rings is possible at various sites. This leads to regioisomerism, already when the first substituent is introduced. There are first hints that the distributions between regioisomers of condensed aromatics differ when conducted in a micro reactor as compared with conventional processing. The reason for this is not understood; even suggestions on this are lacking in the literature.

For the nitration of naphthalene a ratio of 1- to 2-mononitronaphthalene of about 20 : 1 is found in industrial processes. This ratio is dramatically increased to more than 30 by using micro reactors [166].

1.6.5.3 Regioisomerism in Aromatics with One Substituent
When aromatics, even single-ring compounds such as benzene, have one substituent already, the introduction of the next also gives rise to regioisomerism. In addition, the first one guides the introduction of the second by steric and electronic effects.

In this context, the second-stage nitration of 1-mononitronaphthalene was investigated. The isomeric ratio of the two regioisomers, 1,5-dinitro- to 1,8-dinitro-naphthalene, was constant at 1 : 3.5 for macroscopic batch reactors, whereas it changes to 1 : 2.8 in micro reactors [166].

For toluene fluorination, the impact of micro-reactor processing on the ratio of ortho-, meta- and para-isomers for monofluorinated toluene could be deduced and explained by a change in the type of reaction mechanism. The ortho-, meta- and para-isomer ratio was 5 : 1 : 3 for fluorination in a falling film micro reactor and a micro bubble column at a temperature of –16 °C [164, 167]. This ratio is in accordance with an electrophilic substitution pathway. In contrast, radical mechanisms are strongly favored for conventional laboratory-scale processing, resulting in much more meta-substitution accompanied by uncontrolled multi-fluorination, addition and polymerization reactions.

1.6.5.4 Keto–Enol Isomerism
Hardly any work has been done on reactions that give both enol and keto forms of a product. One short note is given below, indicating possible changes in micro-

reactor operation by optimization of process parameters. Much more information is needed here and the first results have to be analyzed with care.

A Grignard reaction between cyclohex-2-enone and diisopropyl magnesium chloride was carried out in a micro reactor yielding a keto and enol product each [168]. In this context, 14 different reaction conditions were investigated in 14 hours. By software-supported process optimization (factorial design), the initial yield of 49% could be improved to 78% with a simultaneous increase in keto/enol isomer ratio (A : B) of 65 : 35 to 95 : 5.

1.6.6
Optical Purity

1.6.6.1 Enantiomeric Excess (*ee*)
The very first investigations on this topic pointed out that a similar degree of optical purity is achievable for some reactions in microreactor as compared to conventional processing. Hence there is no reason not to investigate a chiral reaction in a micro reactor; the feasibility has been proven.

For example, the hydrogenation of methyl (*Z*)-α-acetamidocinnamate gives a chiral product when conducted in the presence of a chiral diphosphine catalyst. The enantiomeric excess data for micro-reactor and batch operation are in line when performed under similar conditions [169]. A very high reproducibility of determining data on enantiomeric excess was reported [170]. In addition, the *ee* distribution was quite narrow; 90% of all *ee* data were within 40–48% [170].

1.6.6.2 Racemization
The very first investigations on this topic indicated that racemization can be monitored in micro reactors and that the degree of racemization seems not to be higher than in conventional organic synthesis. For dipeptide formation from the pentafluorophenyl ester of (*R*)-2-phenylbutyric acid and (*S*)-α-methylbenzylamine, racemization of 4.2% was found [158]. At higher concentration (0.5 M instead of 0.1 M), a higher degree of racemization was found (7.8%).

1.6.7
Reaction Mechanism

From all that we know, reactions in micro reactors still have to be considered as bulk reactions, i.e. they follow all the whole known rules which we know for conventional synthesis. In particular, we expect the same reaction mechanisms to occur. However, there may be exceptions to this rule.

1.6.7.1 Preferring One Mechanism Among a Multitude
Micro reactors can have a distinct influence on which reaction path is undergone, if there is close competition between several reaction mechanisms, which may be steered by, e.g., temperature control. This is nothing else than the selectivity impact already mentioned above (see Section 1.6.2). As one example for this impact,

the electrophilic path could be favored for direct fluorination using elemental fluorine at the expense of the undesired radical path by providing better temperature control and decreasing residence times [164, 167].

1.6.7.2 Tuning Bulk Reactions to Surface Control

Another exception to the known mechanisms of conventional chemistry may arise when dominance of surface reactions is achieved in micro reactors. This holds for all catalytic reactions on solid contacts. Beyond that, it was shown that some formerly homogeneous bulk reactions may become heterogeneous when carried out in a micro reactor owing to the very large surface-to-volume ratio [155, 171, 172].

1.6.8
Experimental Protocols

Experimental protocols are amenable to change by using micro-flow conditions.

1.6.8.1 Residence Time

One commonly found feature is a reduction in process times, simply because flow conditions allow a faster sequence of all necessary processing steps such as mixing and completion of reaction and promote reaction by optimized transport properties.

1.6.8.2 Reaction Temperature

Another feature often reported is an increase in reaction temperature from cryogenic conditions below or to ambient temperature, without losing selectivity. Sometimes even selectivity is increased in this way. Most often, such improved performance was found for fast organometallic reactions, probably the most prominent example being the Grignard reaction of Merck which was transferred to industrial production in the final stage.

The industrial Merck process had to use equipment with a surface-to-volume ratio of $4 \ m^2 \ m^{-3}$; the corresponding figure for the laboratory-scale process was $80 \ m^2 \ m^{-3}$ and that for the micro reactor was $10 \ 000 \ m^2 \ m^{-3}$. Accordingly, the residence time had to be increased from 0.5 to 5 h to allow less heat generation per unit time for the large-scale process. As a consequence, the contribution of side and follow-up reactions is larger. In addition, micro-channel operation at $-10 \ °C$ causes lower energy expenditure and costs than the former batch processing at $-20 \ °C$.

This is, at a first sight, against chemists' intuition, since the extent of side reactions is usually the larger, the higher the temperature is set, as they have higher activation energies. However, by going to room-temperature operation, simultaneously the residence time is considerably reduced. There is definitely a need for more in-depth analyses to understand better the chemical-engineering background of the favorable room-temperature processing.

1.6.8.3 Type of Reactants and Auxiliary Agents

Among the most striking changes reported is that auxiliary agents such as bases may no longer be needed, since functional groups on the surface of the micro channels may take this role. As the specific surface area is much increased in micro channels, effects that are not visible on a macro scale may become important upon miniaturization. Hence, the addition of aggressive reactants may be superfluous, i.e. milder reaction conditions may be applied elegantly, since the species they create are made at the channel surface in high local concentration, even though the total amount is very low compared with the bulk reaction material.

In this context, the esterification of 4-(1-pyrenyl)butyric acid with an alcohol to the corresponding ester was investigated [171]. Without the presence of sulfuric acid no reaction to the ester was found in the micro reactor. On activating the surface by a sulfuric acid/hydrogen peroxide mixture, however, a yield of 9% was achieved after 40 min at 50 °C. On making the surface hydrophobic by exposure to octadecyltrichlorosilane, no product formation was observed. Using silica gel in a laboratory-scale batch experiment resulted in conversion, but substantially lower than in the case of the micro reactor. The yield was no higher than 15% (40 min; $0.1\ \mu l\ min^{-1}$), while the best micro reactor result was 83% (40 min; $0.1\ \mu l\ min^{-1}$).

In a similar way, elimination of the need to add a base was reported for Suzuki couplings. Conventionally, such base addition is needed for Suzuki couplings to activate the boronic acid group. Surprisingly, there is no need for addition of a base when performing Suzuki coupling in a glass chip micro reactor [155, 172]. This was explained as being due to the local generation of a base at a heterogeneous site of the micro-channel wall. Under the action of the voltage for electroosmotic micro-flow processing, water can be converted to hydroxide ions. The high specific surface area in the micro reactor probably accelerates this process. Although the corresponding hydroxide concentrations may be low in bulk, they potentially can be large at the catalyst surface where these species enrich. As a consequence, the Suzuki coupling can be performed without a base in micro reactors. Testing with the same process parameters does not lead to any conversion in a batch reactor. Here, the addition of a base is essential.

1.6.9
Safety Profits

1.6.9.1 Share of Safety-relevant Industrial Processes

Approaching the subject from a practical point of view, how many plants actually are performing safety-relevant chemical processes? In an older study, the share of safety-relevant plants in the German state of Rhineland-Palatinate are listed [173]. It is assumed here that the trend given there still is valid.

From 1980 to 1993, the number of safety-relevant plants increased from 65 to 772 [173]. They constituted slightly more than a quarter of all plants (3472) at that time. Concerning big chemical plants, selected as falling into the category of a special law [173], 230 out of 346 plants were safety-relevant; 163 of them, in addition, had special obligations.

1.6.9.2 Safe Micro-reactor Operations in the Explosive Regime or for Otherwise Hazardous Processes

Many examples of safe processing in micro reactors have been reported. Among them, the formation of the poisonous hydrogen cyanide is often mentioned [5]. Rinard and Saha refer to the non-oxidative Degussa variant [174] of this synthesis [85], while a micro reactor has performed the oxidative formation of hydrogen cyanide [175] via the Andrussov process [176, 177].

Another reaction with hazardous potential is the synthesis of methyl isocyanate from methylformamide, which was investigated using a micro reactor in industry [74].

Many examples of the use of safe micro-reactor operation in the explosive regime have been given, including ethylene oxide [159, 162, 163, 178] and maleic anhydride [179] formations, the oxyhydrogen reaction [180, 181], the synthesis of explosive endoperoxides [182] and the Hock phenol process [183]. Concerning the prominent oxyhydrogen reaction, inherent safety was ascribed to hydrogen/oxygen mixtures in the explosive regime when guided through channels of sub-millimeter dimensions under ambient-pressure conditions, based on an analysis of the thermal and kinetic explosion limits [18]. This was confirmed by experiments in a quartz micro reactor [18] and other measurements in steel micro reactors [180, 181]. The reason for safe operation is seen in improved thermal control (better removal of reaction heat; avoidance of hot spots) and in a wall-induced quenching of radical chains (flame-arrestor effect).

An impressive example of the impact of miniaturization on the explosion limit has been given for the oxyhydrogen reaction [18]. For a conventional reactor of 1 m diameter, explosive behavior sets in at 420 °C at ambient pressure (10^5 Pa). An explosion occurs at about 750 °C, when the reactor diameter is decreased to about 1 mm. A further reduction to 100 μm shifts the explosive regime further to higher pressures and temperatures.

1.6.10
New Process Regimes

Micro reactors can open up new process regimes [5, 18, 70, 147, 180, 184–186]. Actually, the term 'new' is used in this respect in the micro-reactor literature with ambiguous meanings. There are at least three types of 'new' processes:

- essentially novel processes that were not known before realization in a micro reactor [187–192];
- processes that were principally known before, but behaved much worse or otherwise differently when carried out in conventional equipment [164, 167, 193–197];
- processes that are known, but bear threat to equipment and life when being carried out in conventional equipment or are otherwise considered as prohibitive [18, 147, 180, 184–186].

1.6.10.1 **Essentially Novel Processes**

Keeping in mind the controversial discussion on 'new physics' in micro reactors [198], we certainly have to be at least as careful when introducing or claiming essentially novel chemical processes. A thorough scientific consideration is required for an exact definition and differentiation here that is beyond the scope of this book. So far, no deep-rooted scientific work has been published analyzing the origin of the novelty of chemistry under micro-channel processing conditions.

1.6.10.2 **Known Processes that Become Entirely Better or Otherwise Different**

It is certainly more philosophical question as to whether a known process that becomes entirely better or otherwise different is to be called new or old with a new complexion. It is – chemically speaking – even more difficult when 'better' or 'otherwise different' refer to undertaking a new elemental pathway, i.e. to following a different reaction mechanism.

Some authors even speak of 'inaccessible regimes' when processes are operated with maximum selectivity [110]. As pointed out above, there is no clear definition of what a new process regime is and what just a change of process parameters is. Hence the definitions given here may be taken as a guideline until there are better ones.

An example of operation in new process regimes is ammonia oxidation which was carried out in membrane-carrying micro reactors [188]. Dependent on triggering heat removal via change of membrane properties such as heat conduction and thickness, the ignition–extinction behavior is completely different from the normal process. Hysteresis-type behavior, typical of conventional operation, vanishes for certain specifications of the membrane reactor. Not only does the processing itself change, it is accompanied by a difference in product spectra. Operation without hysteresis ('no-loop') gives mainly dinitrogen oxide (N_2O) as product, whereas in operation with hysteresis ('loop'), the 'conventional' products nitrogen and nitrogen oxide (NO_2) are obtained instead.

1.6.10.3 **Processes Known, but not Used for Safety Reasons**

Some processes have been known and investigated for a long time, even over decades, but are not used for industrial production or even for common laboratory-scale practice. Although promising, safety reasons may restrict widespread permission to use them or render them inconvenient. The processes are investigated but at extreme dilution or in complex special apparatus.

For example, direct fluorinations with elemental fluorine are kept under control in this way, at very low conversion and by entrapping the molecules in a molecular-sieve reactor. As with some other aromatic substitutions they can proceed by either radical or electrophilic paths, if not even more mechanisms. The products are different then; this may involve position isomerism, arising from different substitution patterns, when the aromatic core already has a primary substituent; further, there may be changed selectivity for undefined addition and polymeric side products (Figure 1.31). It is justified to term this and other similar reactions 'new', as the reaction follows new elemental paths and creates new products or at least new

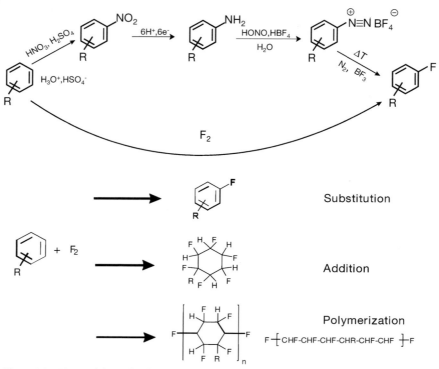

Figure 1.31 Electrophilic and radical paths in direct-fluorination chemistry leading to substitutions, additions and polymerizations with the example of toluene as substrate. The aromatic substitutions give rise to defined, characteristic substitution patterns.

product mixtures; however, the reactants used are the same and so are (some of the) products.

Other well-known examples of processes that fall into the category discussed in this chapter are many oxidation reactions that have extended explosive regimes. Among them, the reaction between oxygen and hydrogen is renowned to be particularly dangerous, as is evident from the extremely large explosion range of oxygen contents from 4 to 96%. The reported explosions for this process were more than vigorous and hindered not only any use of this process, but even most scientific investigations on this topic. Recently, some micro-reactor processing was reported to exert control over this reaction. For instance, even up to 1200 °C safe operation was achieved at relatively large volume flows [18, 147, 180, 184–186]. Somehow, 'the beast was tamed' [199] and acquired a completely new appearance.

According to the information given above, several reasons exist for why promising processes actually do not find application. Literature descriptions refer especially to the following reasons:

• safety precautions [124]
• insufficient turnover rates [124].

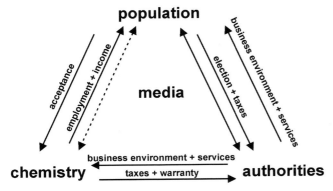

Figure 1.32 'Control circuit' with three cornerstones, creating the socio-economic perception of chemistry (redrawn from [200]).

1.7
Impact on Society and Ecology

The awareness of chemistry in general and its socio-ecologic perception is characterised by some kind of 'control circuit' having three parts, namely population, public authorities, and chemistry itself (Figure 1.32) [200] (see also the same three-cornered relationship between technology push, market pull, and social demands given by Felcht [137]). As a type of information exchange and creating platform, media serve between these parts. There are mutual relationships between each of the cornerstones with the other of the control circuit. As a result, a state is reached that is stable for a time, but gradually amenable to changes, i.e. the socio-ecologic perception of chemistry as a technology for mankind. This view has an influence on the technology itself, besides economics and margins. Hence it is worthwhile to look for changes that especially a technology as innovative as micro reactors may induce here.

1.7.1
The 'Control Circuit' for Chemical Micro Processing

Micro-reaction technology has just left its infancy and started commercialization. Consequently, we do not know about the social response to micro reactors in general, i.e. the 'micro-reactor control circuit' in the sense outlined above has not yet developed, but we do know something of the response of chemical industry and customers to what the micro-reactor offers.

Hessel and Löwe try to take the view of the user [9, 10]. They conclude that micro reactors have long since become a constituent part of today's chemical R&D activities. For them it is not surprising that much news of successes is not disseminated in a branch that profits largely from patents and keeping knowledge secret. Hence all essays on this topic, such as this volume, can only provide a fraction of the information which is free to be communicated.

Figure 1.33 Micro-reactor images as teaching material in German school books.

Hessel and Löwe further comment that cautions optimism meanwhile has replaced the euphoria with which many developments in the 'hype 1990s' were associated [9, 10] (a description of the industrial standpoint was given, e.g., by Oroskar et al. [201]). The suppliers of micro reactors and their components have to comply with their promise to purchase in line with market requirements. Seriousness on the part of the customer is also required concerning the application targets when approaching the supplier. Large, often too large, expectations were aroused during the start phase or were waiting to be awoken. At present, it is very clear that micro reactors are not a panacea for processes that suffer from conversion or selectivity problems. On the contrary, really bad results may be obtained if the process does not fit their capability.

1.7.2
Social Acceptance via Education and Awareness

The fact that the 'control circuit' for micro chemical processing has not been established, as outlined in Section 1.7.1, can be seen as chance to have a deep, positive impact on society in a way not known before. Education and awareness of the technology can be used as one main driver for developing such circuit.

Micro-reaction technology is a classic example of a novel discipline generated at the interface of existing disciplines; hence this new discipline demands special teaching at universities, presented by specialized lecturers [57, 58]. It is the multi-disciplinary merging of chemistry, physics, chemical engineering, and mechanical engineering, to name just a few. In this way, chemical micro processing may become a novel, individual branch of research and education organization within universities.

In fact, this movement has started. Some universities now provide training courses and give lectures on this aspect [202, 203]. Sooner or later, this will also be taught at schools. In the state of Baden-Württemberg in Germany, school books now show photographs of micro reactors (Figure 1.33) [204]. As a result, society may be much more aware of the 'revolutions' of miniature chemical equipment that it was in the past concerning the use of remote, large-scale chemical apparatus. This may be accompanied by a much increased social acceptance of chemistry in general as it can exhibit a modern, dynamic change in a similar way to microelectronics. Once we are using micro-reactor equipment for our household and personal needs, we will no longer feel reluctant about chemistry in general. This may even stimulate further interest in chemical developments in a way such as is given now only for selected scientific topics such as cancer or AIDS drug development and gene therapy.

1.7.3
Ecologic Acceptance via Environmental Acceptability

However, it is not only knowledge transfer on the technology itself which impacts on society; microchemical processing will have repercussions on our wealth, health, and environment. From the late 1960s (first prognoses on limits of production increase, e.g. given by the Club of Rome) to the early 1980s (peace and ecology movement), the awareness of the environmental acceptability of new and existing production processes rose in society. This becomes more and more important when establishing new ways of production.

In this context, Benson and Ponton declare that while the chemical industry has made considerable achievements in reactor performance, safety and control, comparable to those in the microelectronics business, this success is by no means evident to the public, in deep contrast to the latter [139]. It is said that this is mainly and in a way simply due to the visual recognition of chemical production plants. From a distance and for somebody outside the field, the chemical plants of the late 1940s and the early 1990s look virtually similar, whereas one is able immediately to see the big differences in, e.g., television sets and automobiles. Hence it is not evident that notable improvements were made over the decades.

Following this, it stands to reason that micro reactors can aid in changing the perception of chemical plants. Usually their outer shape is small and compact, at least compared with most existing equipment. At best, the outer shape of the whole plant is visibly diminished. Even if this does not arise owing to employing multi-scale architecture of almost unchanged outer shape (see Section 1.1.7), one could accentuate the new tool in various ways, including the press and TV. If desired, micro reactors could stand for a paradigm change; more likely and realistically, they could visualize the enormous improvements in industry in recent years, particularly referring to process intensification.

Nonetheless, no change will happen without external stimuli. Benson and Ponton list the following:

- environmental pressures on the process industries
- electronic point-of-sale (EPOS) demands for just-in-time (JIT) production
- increased emphasis on product quality and consistency
- expansion of plants into the Asian and African regions leading to a demand for smaller distributed plants, also with a high degree of reliable automation
- economics of scale will be 'disproved' if significant technology improvements and computer control are achieved. This especially holds for directly converting raw material into valuable products.

Benson and Ponton propose, based on this analysis, the concept of distributed manufacturing [139, 145], which will be referred to in detail in Volume 2 of this book series. Basically, they refer to small, transportable plants which are fed with reactants 'over the fence', hence using only non-hazardous, generally available materials by normal piping or standard transport. If an aggressive chemical is needed, it has to be made from environmentally friendly base materials as an intermediate on-site. Needless to say, effluents have to be completely harmless, plant operation has to be intrinsically safe, and the plant should be clean and quiet.

The scenario of Benson and Ponton seems to be for the remote future. However, this is no quality rating at all, since we all need such long-term views for new innovations. Moreover, parts of the predictions certainly have already been addressed in the work published so far. It should be a guideline or an ideal for promoting chemical micro-process engineering.

1.7.4
Environmental Restoration

While the discussion give above was more generically oriented, this will be underlined here by a practical example.

Wegeng mentions the use of micro reactors for the cleanup of environmental contamination [1]. In particular, he refers to downwell groundwater cleanup by micro-chemical separations and conversions such as destruction of organics.

In a further paper, Wegeng and Drost give several examples of distributed processing applications, which refer to the fields of energy generation and environmental restoration [106]. Concerning the latter, compact cleanup units for waste treatment are mentioned, mainly in consideration of the numerous radioactive sites, stemming from cold-war military developments [106] (see Section 1.5.4.5 for more information). As another example, the use of distributed systems for global carbon dioxide management aimed at reducing the greenhouse effect [106] (see Section 1.5.4.5) is mentioned.

1.7.5
The Micro-reactor Echo in Trade Press and Journal Cover Stories

Micro reactors have attracted the attention of the press and journals (Figure 1.34). This 'echo' reflects the perception of how the technology is seen by journalists and experts in the field, thereby influencing the opinion of the interested society. It is worth – not only from a marketing perspective – having a closer look at this and attempting to draw some conclusions concerning one's own future developments.

Before showing some examples of press releases and their content, we shall briefly shortly sum up all the information given. Most frequently the relationship of microreaction technology to the development of microelectronics is cited, suggesting a similar success story. Expectations are created that some day micro reactors will be mass fabricated at low cost in a similar way. In addition, it is believed that compactness can be achieved as for the integration of functions in the microelectronics world. In this context, often the vision of a shoebox-sized plant or a plant on a desk is given.

Micro reactors have also frequently been compared with nature and nature's reactors such as cells, organelles, organisms, etc. In turn, the cell is often taken as a model for micro reactors. It is stated that nature decided to not change the size of the cells throughout evolution, e.g. when comparing the small animal the mouse with the very large elephant; both have cells of the same size range, the elephant just has much more than the mouse. This 'numbering-up concept of nature' is seen to resemble the approach of parallelizing micro reactors. From this, it is analyzed if a similar numbering-up concept can be provided for technical reactors. In this context, it is often questioned if there will be a production supplying huge quantities of micro reactors. On this and other topics typically a number of expert

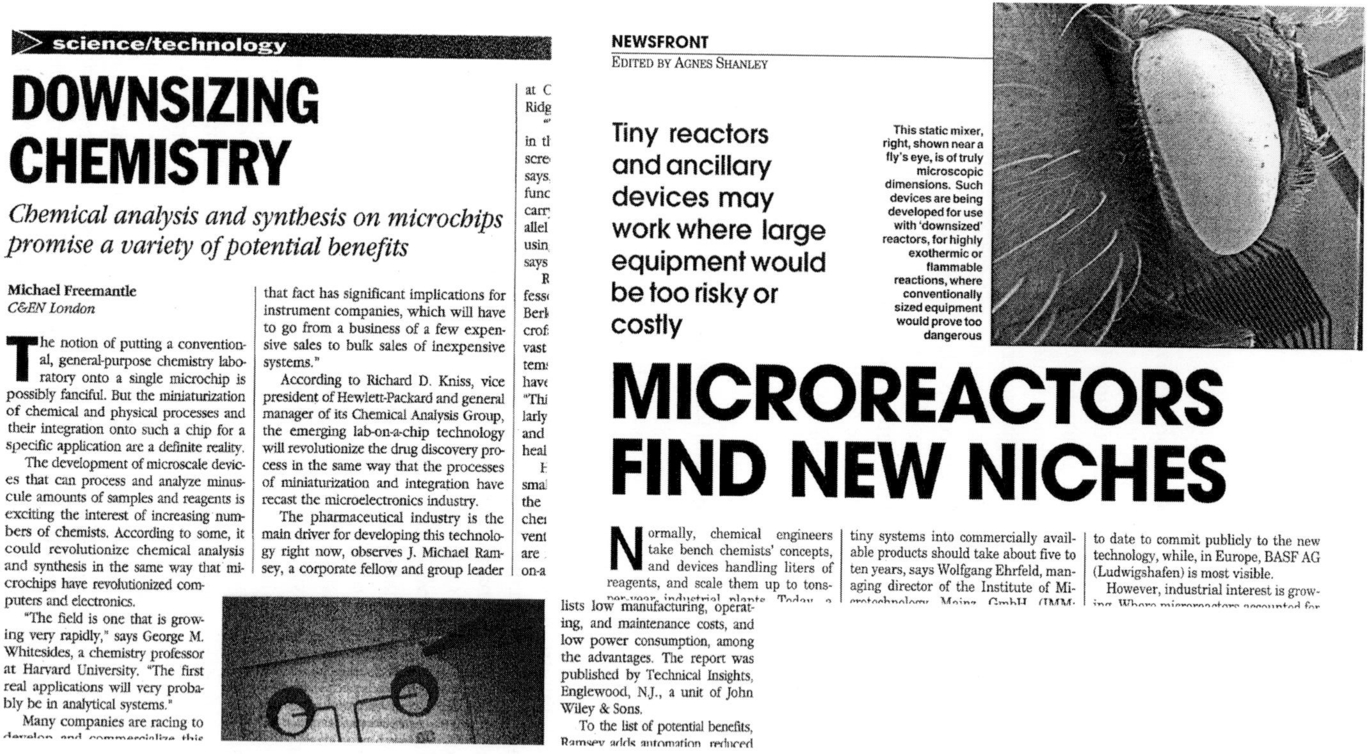

▲ science/technology

DOWNSIZING CHEMISTRY

Chemical analysis and synthesis on microchips promise a variety of potential benefits

Michael Freemantle
C&EN London

The notion of putting a conventional, general-purpose chemistry laboratory onto a single microchip is possibly fanciful. But the miniaturization of chemical and physical processes and their integration onto such a chip for a specific application are a definite reality.

The development of microscale devices that can process and analyze minuscule amounts of samples and reagents is exciting the interest of increasing numbers of chemists. According to some, it could revolutionize chemical analysis and synthesis in the same way that microchips have revolutionized computers and electronics.

"The field is one that is growing very rapidly," says George M. Whitesides, a chemistry professor at Harvard University. "The first real applications will very probably be in analytical systems."

Many companies are racing to develop and commercialize this

that fact has significant implications for instrument companies, which will have to go from a business of a few expensive sales to bulk sales of inexpensive systems."

According to Richard D. Kniss, vice president of Hewlett-Packard and general manager of its Chemical Analysis Group, the emerging lab-on-a-chip technology will revolutionize the drug discovery process in the same way that the processes of miniaturization and integration have recast the microelectronics industry.

The pharmaceutical industry is the main driver for developing this technology right now, observes J. Michael Ramsey, a corporate fellow and group leader

at C Ridg "
in t
scre
says.
func
carr
allel
usin
says
R
fess
Berl
crof
vast
tem
have
"Thi
larly
and
heal
l
sma
the
chea
vent
are
on a

NEWSFRONT
EDITED BY AGNES SHANLEY

Tiny reactors and ancillary devices may work where large equipment would be too risky or costly

MICROREACTORS FIND NEW NICHES

Normally, chemical engineers take bench chemists' concepts, and devices handling liters of reagents, and scale them up to tons-per-year, industrial plants. Today, a report lists low manufacturing, operating, and maintenance costs, and low power consumption, among the advantages. The report was published by Technical Insights, Englewood, N.J., a unit of John Wiley & Sons.

To the list of potential benefits, Ramsey adds automation, reduced

tiny systems into commercially available products should take about five to ten years, says Wolfgang Ehrfeld, managing director of the Institute of Microtechnology, Mainz, GmbH (IMM,

to date to commit publicly to the new technology, while, in Europe, BASF AG (Ludwigshafen) is most visible.

However, industrial interest is growing. When microreactors proved for

This static mixer, right, shown near a fly's eye, is of truly microscopic dimensions. Such devices are being developed for use with 'downsized' reactors, for highly exothermic or flammable reactions, where conventionally sized equipment would prove too dangerous

Figure 1.34 Title pages of two trade-press articles on micro reactors [45, 226].

opinions are quoted. The question concerning production is ultimately related to the marketability of the technology. Start-up enterprises selling micro reactors and market studies/prognoses are hence also in the focus of the press articles.

Finally, the topic of standardization ('plug-and-play') has gained interest. Will there be modular flexible plants in the future, for multi-purpose and on-site production? Concerning the latter, the chemical manufacture of hazardous and explosive substances has attracted readership over many years.

Besides all these comparisons, press releases cite and explain the technological benefits of micro reactors, which were described in Section 1.3, often in a vivid description. This includes the following items:

- safety gains
- process intensification
- the on-site production of, e.g., dangerous materials
- use of new process regimes
- general advantages of micro reactors such as enhancing transport
- highly parallel screening.

In the following, selected trade-press and cover-story releases (mostly from German sources, but not exclusively) are presented. These releases are given by the headline, the source (name of journal), the time of release, and a list of key contents, completed by a citation. The citations are listed in the sequence of their appearance.

Bulk chemicals on the drop, The Economist, June 2003
Better chemistry through confinement; economy of scale; limits and expenditure of scale-up; micro reactors origin from microfabrication; expert opinions; general advantages of micro flow; hybrid construction; industry's efforts; chip micro reactors; numbering-up; vision of computer-like chemical workstation [205].

'Numbering up' small reactors, Chemical Engineering News, June 2003
Expert opinions; general advantages of micro flow; market situation; modular construction; time-to-market; industrial implementation and experience; off-the-shelf catalogue sales; numbering-up; problem hurdles; filament reactors; state of knowledge acquisition [206].

Chemieküche für Zwerge, VDI Nachrichten, April 2003
Safety; risk of incidents; miniature plants; novel process routes; industrial and academic examples of use; HTE; standardization; LEGO-type plants [207].

Tausend Kanäle für eine Reaktion, Chemische Rundschau, February 2003
Industrial and institutional expert opinions; general advantages of micro flow; safety; work of institutes; particle precipitation; pilot-scale operation; challenges; process control; plugging; miniature sensing and controlling; emulsification; market situation [204].

Vor dem Sprung in die Produktion, Chemie Produktion, December 2002
Prognoses on speed of implementation; PAMIR market study; industry's demands; numbering-up risks; expert opinions; Clariant pigment micro-reactor production; smallness not an end in itself; general advantages of micro flow; industrial process development and optimization; share of reactions suited for micro reactors; hybrid approach; standardized interfaces; start of industrial mass production of micro reactors?; unit construction kit [208].

Die Fabrik auf dem Chip, Spektrum der Wissenschaft, October 2002
Miniaturization and modularization of parts of future chemical apparatus; general advantages of micro flow; expert opinions; specialty and fine chemical applications; leading position of German technology; flexible manufacture; large-capacity micro reactors; reformers for small-capacity applications; compatible and automated micro-reaction systems; process-control systems; temperature and pressure sensors [209].

Chemiefabrik im Schuhkarton, Handelsblatt, June 2002
Shoebox-sized lab-on-a-chip laboratories; personal drug manufacture; general advantages of micro flow; Merck's production; nitrations; HTS; parallel catalyst testing; turnkey bench-scale test station; standardization; cube-like modules [210].

Kleine Reaktoren mit großer Zukunft, Chemische Rundschau, April 2002
PAMIR study; large commercial potential; large industrial interest; market volume; standardization; strategic cooperations; time horizon; potential for pharmaceuticals and fine chemistry; Clariant pilot with caterpillar mixer [211].

Les premiers pas des microréacteurs, Industries et Techniques, October 2001
Numbering-up; use for analytics and screening; faster and improved production; smart processes; process intensification; 'intelligent' reactor and new process regimes; dominance of surface over gravity forces; precise control of process conditions; control over selectivity; direct fluorination; distribution problems during numbering-up; CFD modeling; interdisciplinary field; onset of industrialization of micro reactors; market considerations; selected devices for combustion, power-generation (reforming); outer-space applications [212].

IMRET 3: 3rd International Conference on Microreaction Technology, CIT, 2000
This article gives a summary on the topics and selected presentations of the 3rd International Conference on Microreaction Technology and draws conclusions. Among the topics of the conference were design and production of micro-flow devices and microfluidics. Further topics of major concern are micro reactors for production processes, for energy generation and storage, and for biotechnology. In addition, a conference section was devoted to commercialization of the technology [213] (see also [214]).

Mikroreaktoren für die chemische Synthese, Nachrichten aus der Chemie, May 2000
Chip technology initiates quest for small structures; better temperature control on
the small scale; fast mixing by diffusion; several kg productivity per day; no novel,
but better chemistry; perfect control over process parameters; corresponding increase in selectivity; basic micro-reactor functions; selected examples of use; micro
reactors as routine tools in the laboratory; first start-up companies [113].

Gezähmte Chemie im Mikroreaktor, VDI Nachrichten, June 2000
Micro-reactor enterprises; shape and material variety of micro reactors; selectivity
gains and new project regimes; direct fluorination; faster process development;
BASF investigations; safety increase; speed-up of catalyst development; production for fine chemistry and pharmacy; numbering-up; first industrial examples for
micro-reactor production [215].

Wozu Mikroreaktoren?, Chemie in Unserer Zeit, April 2000
Innovation – advocates and opponents; origin from microtechnology; list of
microfabrication techniques; selectivity and efficiency as main driver for industrial
implementation; special properties and general advantages of micro reactors; proc-
ess-development issues; BASF investigations on liquid/liquid and gas-phase reactions; micro reactors as ideal measuring tools; production in micro reactors as
exception, the rule will be transfer to mm-sized channels [111].

Herrscher über die Temperatur, Chemie Technik, 1999
Interview with Wörz/BASF in a special on heat exchangers giving expert opinion
on: compact heat exchangers, feasibility and problems of large-scale implementation of micro reactors; measuring tool for process optimization; exotic status?; scale-
up; unit-construction kit; industrial implementation in 5 years [216].

Chemical reduction, Chemist, 1999
Situation like microelectronics decades ago; impetus by analytical chemistry; lab-
on-a-chip – biological applications; microfabrication and micro devices; scale out;
input–output board; fast and hazardous reactions; plug-and-play modules; interconnects; non-linear synthesis; growth of scientific community; industry's response;
selected key players and their activities [217].

Microchemical systems:
status, challenges, and opportunities, AIChE Journal – Perspective, October 1999
Advances in microelectronics; reduction in size and integration of multiple
functions; MEMS development as base; not only reaction, but also separation
and analytics based on µTAS advances; general advantages of micro reactors;
more aggressive reaction conditions and novel process regimes; monitoring for
replacement of failed reactor unit; numbering-up; scheduled, gradual investment; biological screening, DNA amplification; manipulating processing, in
particular steering thermal control; mixing by novel means; many innovative reactor designs in parallel; more focusing on separation and surface forces; fabrica-

tion beyond classical MEMS; multiple packaged reactors with integrated sensing [75].

Chemie im Kleinen, VDI Nachrichten, May 1999
Smallness of micro-flow components; safety gains; tool for kinetics evaluation; process development for large-scale processes; polymerization; combinatorial catalyst screening; hydrogen via reforming [218].

Downsizing chemistry, Chemical Engineering News, February 1999
Miniaturization of processes is reality, that of complete laboratories not; revolution as in microelectronics possible; joint development of Caliper and Hewlett-Packard; revolutionizing drug discovery; combinatorial screening; huge gains in throughput by vast numbers of analysis systems on one chip; DNA diagnostics; other advantages of µTAS; low-inventory process synthesis and other further uses; increase in performance at small scale; lab-on-a-chip definitions; materials discussion; microfabrication techniques; mixing demonstrating small-scale process intensification; electro-osmotic flow; surface-directed flow on patterned surfaces; miniaturized total analysis systems; selected µTAS devices and applications; electrophoresis-on-a-chip; selected examples for electrophoretic separations [45].

The world in figures: industries, The Economist, January 1999
Original citation: 'Miniature chemical reactors will pave the way for the future. These reactors will cut today's monster chemical plants down to the size of a car, with huge financial and environmental gains' [219].

Process Miniaturization Second International Conference, CATTECH, December 1998
Steep progress in microelectronics in the past; key players; topics of IMRET 2; general advantages of micro flow; energy, safety, process development, combinatorial catalyst testing, lab-on-a-chip biological applications; anodically oxidized catalyst supports as alternatives to non-porous supports [220].

Process miniaturization: 2nd Second International Conference on Microreaction Technology, Chemie-Ingenieur-Technik, October 1998
Number of participants of IMRET 2; steering organizations and initiators of IMRET 2; excellent resonance on IMRET 2 in the framework of AIChE spring meeting; topics of IMRET 2; summary of selected presentations; investigations beyond feasibility and laboratory stage [221].

Sicherer, effizienter, flexibler, Verfahrenstechnik, 1998
Today's use of microtechnical products; microfabrication techniques; general advantages of micro flow; parallelization for screening; steep transport gradients; plant safety; numbering-up; industrial response; outlook on market [222].

Das Chemielabor im Mikrochip, Blick durch die Wirtschaft, December 1997
Chemtel glass chip of Orchid Biocomputer, Princeton; 144 cells for parallel processing; matchbox-sized system with many devices; micro pumps with no movable
parts; 10 nl internal volume; carrying out of different reactions in parallel fashion;
complete chemistry laboratory en miniature; 10 000 cells as future-development
task [223].

1st International Conference on Microreaction Technology, CIT, August 1997
This article gives a summary on the topics and selected presentations of the 1st
International Conference on Microreaction Technology and draws conclusions [224].

Daumengroßes Labor aus Aluminium-Folie, Blick durch die Wirtschaft, June 1997
Heterogeneous gas-phase micro reactor; micro-fabrication of this device; anodic
oxidation of aluminum to porous catalyst support; vision of complete small laboratory; numbering-up; development of new silicon device [225].

Microreactors find new niches, Achema Daily/Chemical Engineering, June 1997
Conclusions on IMRET1; micro-reactor exhibitors at ACHEMA 1997; expert opinions; industry's commitment; general advantages of micro flow; views on commercialization; extended list of leading institutes' and companies' activities; topological approach; numbering-up [226].

Chemielabor auf dem Mikrochip, Blick durch die Wirtschaft, May 1997
Lab-on-a-chip; protein separation; DuPont's investigations; general advantages of
μTAS; DARPA foundation of military biological sensor development; MEMS components [223].

Teufelszeug im Griff, Wirtschaftswoche, April 1997
DuPont's phosgene synthesis in a micro reactor; BASF's vitamin precursor synthesis; developments in the bio field; prognosis on market volume in 2000 [227].

Microreactors find new niches, Chemical Engineering, March 1997
Leading players; expert opinions; conclusions on IMRET1; general advantages of
micro flow; views on commercialization; extended list of leading institutes' and
companies' activities; topological approach; numbering-up [228].

Die Natur der Chemie, FUTURE (Hoechst Magazin), August 1996
Vision of large-scale production in shoebox-sized plants; nature and plant cells as
model for micro reactors; sustainable development; central role of catalysis; general advantages of micro flow; use of clean raw materials; minimization of waste;
the next step in the sequence acetylene-to-ethylene chemistry: ethane chemistry;
renewable resources; combinatorial chemistry; intelligent and creative solutions
[229].

Chemie in neuen Dimensionen, Chemie Produktion, August 1996
Plant cells as model for micro reactors; hidden process regimes opened by micro reactors; general advantages of micro flow; intelligent chemo chip as vision; DuPont's and BASF's pioneering efforts; µTAS; safe processing; distributed manufacture; need for knowledge base [230].

Mikrotechnik – Anwendungsmöglichkeiten in der chemischen Industrie,
CHEManager, November 1995
Interview with Jäckel/BASF: general advantages of micro flow; safety expenditure; multi-phase processing; analytics; process development; heat release of exothermic reactions – hot spot; limits of industrial tube-bundle, single-tube, and short-bundle reactors; micro reactors can overcome these limits; high heat-transfer coefficients; analytics of small volumes; combinatorial testing; mobile sensing systems for inspection of tubes; formulations for agricultural chemistry; production feasible? [72].

Ätzende Wolken, Wirtschaftswoche, June 1995
Nature as model for micro-reactor development; general advantages of micro flow; onset of industrial interest; micro heat exchanger; vision of methanol-fuel reforming; costs still too prohibitive [231].

Small but environmentally friendly, The Chemical Engineer, March 1993
Huge increases in technology in the past; distributed manufacturing in small-scale plants; miniaturization of processes; domestic methanol plant; point-of-sale chlorine; simpler and cheaper plants; economy of plant manufacture; process control and automation; start-up and shut-down; sensor demand [145].

1.7.6
The Micro-reactor Echo in Newspaper Press and Magazines

Chemiker sind der Zelle auf der Spur, Handelsblatt, August 2000
Plant cells as model for micro-reactor development; availability of micro-flow devices; German leadership; first production applications; BASF's motivation; spotting for DNA arrays; materials for micro reactors; Merck production plant; smallness for efficiency, but not an end in itself [232].

Ultra-Hochdurchfluss-Tests in der Arzneimittelforschung unverzichtbar,
Frankfurter Allgemeine Zeitung, June 2000
Miniaturization and parallelization key approaches for drug development; apparatus for combinatorial chemistry; UHTS; 1536 titer-plate format; modular construction of apparatus; applications of UHTS; fine-chemical synthesis by micro reactors; numbering-up; nature as model; general advantages of micro flow; vision of 'plants-on-a-desk' [233].

Chemische Technik findet im Fingerhut statt, Handelsblatt, November 1999
Nature as model; pharmaceutical industry as pathfinder; foxglove-sized micro-flow components; general advantages of micro reactors; direct fluorination; transport

intensification; BASF process development; Merck and Aventis production; production in micro reactors for niche applications, if problems are solved; micro sensors for monitoring of industry's processes; household applications; chemical 'micro nose'; integrated vision needs unit construction kits [234].

Winzige Reaktoren mit Höchstleistung, Handelsblatt, November 1998
General advantages of micro reactors; pioneering efforts of Forschungszentrum Karlsruhe (KfK); development of micro heat exchangers/reactors; micro only where needed – hybrid approach; good temperature control enhances yield and safety and reduces waste; details on cube-like device; details on power capacity (200 kW) and throughput (7000 l h^{-1}); 300 000 t yr^{-1} in shoebox-sized reactor; materials and fabrication; flame arresting effect; new process regimes; boom in micro-reaction technology; KfK as pathfinder [235].

Mikroreaktoren sind so klein wie ein Fingerhut, Handelsblatt, May 1998
Steep progress in microelectronics, sensor and analytical techniques in the past; transport intensification for catalysis; first catalytic micro reactors available; partial oxidation to acrolein; partial hydrogenation to cyclododecene; anodically oxidized catalyst supports as alternatives to non-porous supports; study group on micro reactors at Dechema; safety, selectivity, high pressure; exclusion of using particle solutions; limited experience with lifetime of micro reactors [236].

Chemiefabrik in der Größe eines Chips, Handelsblatt, May 1996
Vision of shoe box-sized micro reactors; plant cells as model for micro-reactor development; cost, performance, and safety advantages; LIGA process; numbering-up; safety processing of hazardous substances [237].

Sichere Chemie in Mikroreaktoren, Frankfurter Allgemeine Zeitung, December 1995
Plant cells as model for micro-reactor development; micro-fabrication techniques; DuPont's investigations; DECHEMA's initiation of micro-reactor platform; BASF's investigations; general advantages of micro flow [238].

1.8
Impact on Economy

1.8.1
Market Development/Commercial Implementation

1.8.1.1 A Historical Description of the Interplay between Technology Push and Market Pull

Hessel et al. describe the state of the market implementation of micro reactors, as opposed to the scientific development (Figure 1.35) [239].

The development of chemical micro processing was, among other influences, strongly promoted by continuing manufacturing and offering of micro devices, as

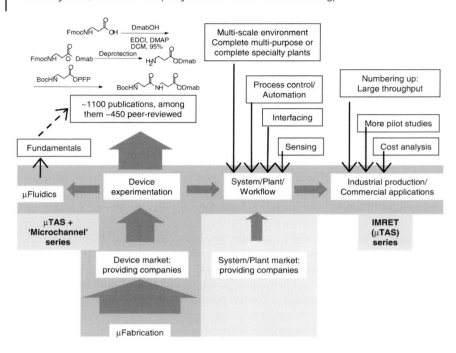

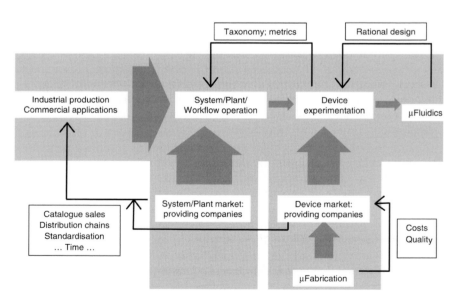

Figure 1.35 Current state of scientific and market implementation of micro reactors and optimistic future scenario. Micro reactors are today a device market and are technology driven (top). The future scenario is under market pull and will ask for more plant/total-system solutions (bottom).

a consequence of emerging microfabrication capabilities [239]. The rising, now even commercial provision of micro devices and their plants allowed a much larger scientific community to test predictions which were made in the early phase of the development, claiming several characteristic means of process intensification by chemical micro processing. Consequently, more and more journal articles report on scientific results in this field and often benchmark it to conventional equipment. By this means, a number of improvements over existing technology were indeed identified. Thus, this first phase of development with chemical processing micro devices was pushed, and so guided, by technology and the growing possibilities for its use.

Meanwhile, the chemical industry, in particular specialty and functional chemical producers, regard the technology to be mature enough to perform extensive industrial laboratory-scale testing and the first few reports on a transfer to pilot or production scale have appeared [239]. Recognizing the industrial demand, a market emerges. Institutes and companies, mostly spin-offs of the institutes, sell a range of chemical micro-processing devices off-the-shelf, for use in organic synthesis and heterogeneous catalysis, and also for various types of mixing and heat exchange. This offer is complemented by the supply of turn-key and application-specific plants, experimentation services, simulation/modeling services, and consulting.

In view of the technology push stated above, one can recognize, particularly in the last 2 years, a promising pull by the market in terms of increasing inquiries [239]. However, these requests are still partly motivated by the scientific success of the technology, gaining attraction, and only to a smaller extent by real business drivers, i.e. business units of the companies that profit from the use of micro devices. Technical obstacles due to the radical change in the way of chemical processing. It will also take some time to establish the new skills needed, to provide the investment, and to change habits that have been exercised for decades. Thus, penetration of micro devices into chemical production lines is a slow process, needing several more years.

1.8.1.2 PAMIR – A Market Study Giving First Insight

Prognoses on the amount and the volume of micro-structured devices are published regularly within the framework of a NEXUS study (see, e.g., [240]). It is, however, striking that in none of these studies do micro-structured reactors appear. Such information is given in a separate market study, named PAMIR ('Potential and Applications of MIcroReaction technology') and dedicated solely to micro-reactor technology, which was carried out by Yole Developpement (Lyon, France) and IMM (Mainz, Germany) [211, 241]. The study claims that the present market volume related to the sales of micro-structured reactors and services for their application are at approximately US $ 35 million per year (Figure 1.36). The study confirms that small- and medium-sized enterprises (SMEs) can make a turnover entirely or in parts of their organization already with their technology today. This also corresponds to the fact that a small number of medium-sized enterprises define micro-structured reactors as their core business, while bigger companies and

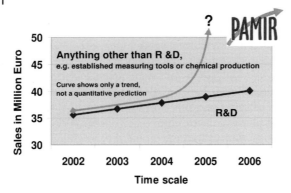

Figure 1.36 Sales-volume forecast for the field of micro-reaction technology for the next few years. A projection is given on when industrial production will set in [241].

also chemical plant manufacturers do not have them in their portfolio yet, but instead observe the market segment intensely. It is interesting to point out that the current suppliers of micro-structured reactors are nearly exclusively situated in Germany.

The study is based on interviews with about 100 selected companies and institutes/universities. Representatives from companies made up 70% of all interviewees. The study initially predicted a small increase in worldwide turnover according to general economic development. It predicts at the same time, however, that an interaction of an increasing acceptance and a significant improvement of the technical suitability of micro-structured reactors (e.g. for chemical production) could lead to an amplified steep increase of the market turnover also in the short term.

1.8.1.3 Market Evaluation

Some concrete conclusions on market evaluation for micro reactors are given by the PAMIR study, described above (see Section 1.8.1.2). In this context, it is worth reviewing general essays on the market evaluation of microsystems technology, since in a broader view they follow the same analysis and come to the same conclusions.

It is a general phenomenon of developments in microtechnology in the 1990s that they were rather technology-driven, hence a proper market evaluation was much behind the first technical breakthroughs [242]. As a result, unsuited market segments were often addressed and, more important, unsaleable products were released; in turn, real markets and their products were sometimes overlooked. This was accompanied by estimating the market potential too large and, in a sense, too euphoric.

Staudt and Krause commented in 1998, based on the results of a study by the Institut für Angewandte Innovationsforschung, Bochum (Institute for Applied Innovation Research), that many micro-systems technical companies still have problems in the analysis and evaluation of application changes and markets for microsystems technology [242]. The marketing problems also refer to bringing their own developments to an end and to establishing a manufacturing line for real products, once a prototype has been made. The causes lie partly in deficits in the competence of the acting personnel, but are also related to the complexity of the interdisciplinary topic microsystems technology. In this context, Steg mentions the term 'system innovation', describing a competence (of a country, society, industry, etc.) to bring new innovative technologies rapidly into many existing market sectors [243]. In the same reference, the different market-implementation strategies and the competitiveness of microsystems technology developments in Europe, Japan and the USA are described.

The technological revolution which microsystem technology brings is fascinating for those who develop; however, it is more frightening for those who are potential users, since it creates fears that the systems are too far from the well-known, marketable products of our daily experience and more generally it does not reach our common perception [242]. Consequently, a lot of future markets have not been identified as possible customers have not expressed interest and possible suppliers have not provided corresponding offers. Government funding tends more and more to select those projects which work on market implementation and thus search for potential needs of the technology.

At the turn of the century, this situation partly changed. By economic selection, only those developments that were market-oriented survived. In addition, most commentators in the field such as the trade press realized that they had to change their attitude on reporting a proper outlook, i.e. what can be realistically expected in the coming years; hence, euphoria changed to (sober) realism [242].

1.8.1.4 Start-up Companies and User–Supplier Platforms

The development of start-up companies in the field of micro-systems technology, including microfluidics, has been reviewed by Wicht et al. [244]. Different types of business models are presented as well as the corresponding growth models. The development of start-ups in Germany in the last 10 years is presented and the product offer is discussed. This is compared with the situation in other European countries. Finally, information on problems and opportunities for the start-ups is provided.

Within the NEXUS activity, aimed at strengthening the interaction of European industry and institutes in microsystems technology, user–supplier clubs (USC) are formed as one means for joint developments (for the USC for CAD tools see [240]; for the MikroWebFab see [245]). This should serve to promote the industrial uptake, by bridging the gap also to new potential users.

Concerning such user–supplier relationships, see also [102] for IMM's view on the role of strategic alliances in the field of micro-reaction technology.

1.8.2
Device Fabrication and Quality Control

1.8.2.1 Cost Estimation from Mass-manufacture Scenarios for Chip-based Microfabrication

Schlaak gives a description of the choice of fabrication technologies suited for a basic function – electronic, mechanical, and optical – and provides an equation for calculating costs for multi-level wafer stacks and hybrid integrated devices [239]. The conclusions drawn hold for classical chip manufacture and can certainly be applied to micro reactors fabricated via this route. They give a first estimate of how costs develop when, e.g., the device area is enlarged or the construction material is changed (which usually involves changing the fabrication route). It is a mass-manufacture scenario and will not hold when only a few micro reactors are made. Naturally, it also cannot take into account design development costs and testing for proper functioning such as leak tightness.

Wegeng et al. refer to the low-cost, mass production of microstructures from metals, ceramics, and plastics as a crucial element for widespread application [1]. Micro technologies, they say, are generally conducive for mass production; however, this has so far only been proven for the field of microelectronics.

1.8.2.2 Quality Control

Off-the-shelf catalogue sales of micro reactors have just started [15]. With an increasing number of commercial products, quality control will become more important. Brandner et al. describe quality control for micro heat exchangers/reactors at the Forschungszentrum Karlsruhe [23]. All manufacturing steps are accompanied by quality control and documentation. Leak rates (down to 10^{-10} mbar l s^{-1} for He) and overpressure resistance (up to 1000 bar at ambient temperature) are measured. Under standardized conditions, the mean hydraulic diameter is determined. Dynamic tests supplement this quality control.

1.8.3
Cost Savings for the Chemical Industry

So far, no scientific extrapolation has been published on the cost savings for the chemical industry when using micro reactors. Industrial experience is also not known; at least, it has not been communicated. Thus, one is bound to rely on expert opinions given in the press and trade press. Mostly these come from suppliers of the technology, aiming to convince industry of the benefits of their systems, by prognosis of a return on investment. Considering the pharmaceutical R&D efforts of the order of US $ 50 billion worldwide, CPC/Mainz sees a potential for an increase in profit of more than US $ 15 billion if micro reactors are implemented consequently [246].

1.9
Application Fields and Markets for Micro Reactors

1.9.1
Transportation/Energy

Many authors mention the use of micro reactors for fuel processing as one of the most promising fields [1, 104]. Wegeng et al. point at using this micro-fuel processing for transportation [1]. The placement of reformers under the hood of an automobile for converting liquid hydrocarbons to hydrogen is explicitly mentioned.

Accordingly, serious commercially oriented attempts are currently being made to develop special gas-phase micro and mini reactors for reformer technology [91, 247–259]. This is a complex task since the reaction step itself, hydrogen formation, covers several individual processes. Additionally, heat exchangers are required to optimize the energy balance and the use of liquid reactants demands micro evaporators [254, 260, 261]. Moreover, further systems are required to reduce the CO content to a level that is no longer poisonous for a fuel cell. Overall, three to six micro-reactor components are typically needed to construct a complete, ready-to-use micro-reformer system.

Since space and weight savings are essential technical goals in reformer development, it is not sufficient to manufacture only one or a few components in miniature. Long and extensive research was carried out especially by research groups at Forschungszentrum Karlsruhe [248], Battelle (Richland, USA) [262], and PNNL (Richland, USA) [247, 250–256, 260, 263]. The PNNL work is governed by the strategy of employing micro reactors for mobile tasks concerned with energy supply and temperature control [255, 264], in particular focusing on outer space and military applications [90, 91]. As an example of outer space research, fuel generation in micro reactors was tested, e.g. by using Sabatier and reverse water-gas shift reactions for future Mars missions [265, 266].

Mini reformers undoubtedly are of interest for the automobile industry and have attracted considerable attention despite the considerable technical problems to be solved. Enlarging the scope of possible applications, mini reformers with a power supply notably higher or lower than needed for car use are currently being evaluated. This includes battery replacement, e.g. as an energy source for laptop computers (see also [262]) or energy supply for households without mains connection (stand-alone supply). Among the components developed are a micro evaporator (volume 0.3 l) for a 50 kW fuel cell [260], a micro heater (weight 200 g) for 30 W power and 85% efficiency [267], and portable energy sources with 10–100 W power (base unit of 21 cm length) [255, 268]. A compact steam reforming reactor (total size 4 l) for automobile applications achieves a conversion of 90% concerning isooctane and serves for the supply of a 50 kW fuel cell [247] (see also [269] for an industrial description of a micro-flow reformer development and testing for the automotive industry).

Fuel processing is not discussed in a separate, detailed chapter within this book; this will be done in Volume 2.

1.9.1.1 How Far is the Development? A Critical Review

Already in 1996, Wegeng et al. admonished that for the development of miniaturized reformers, e.g. especially facing the provision of hydrogen as fuel for transportation, several main issues have to be solved which probably will need a considerable time; among them are the reduction of the CO level, the high energy intensity of conventional reforming processes, the need for thermal integration, and the slow dynamic response of the reformers [1]. Within the last 7 years, after this first comprehensive analysis in 1996, there have been considerable scientific achievements concerning reformer and catalyst development. However, these investigations still appear to cover mainly one or a few aspects only, rather than providing a complete system approach. No flow sheets have been published on the total micro-channel reforming process, making evident the advantageous integration or reaction and heat transfer features [1]. However, with regard to a total-system approach for mobile applications, a conceptual design for a fuel cell-based portable power system of size 8–10 cm^3 was proposed [255] (see also [5]).

A growing number of research groups are active in the field. The activity of reforming catalysts has been improved and a number of test reactors for fuel partial oxidation, reforming, water-gas shift, and selective oxidation reactions were described; however, hardly any commercial micro-channel reformers have been reported. Obviously, the developments are still inhibited by a multitude of technical problems, before coming to commercialization. Concerning reformer developments with small-scale, but not micro-channel-based reformers, the first companies have been formed in the meantime (see, e.g., <www.hyradix.com>) and reformers of large capacity for non-stationary household applications are on the market.

1.9.2
Petrochemistry

Wegeng et al. mention the use of efficient, compact reactors for the conversion of methane to syngas, making natural gas an alternative to petroleum [1]. They see more potential in using the high-temperature partial oxidation of methane as a fast process than the slow steam reforming. In the latter case, short residence times may be utilized due to the speed of the reaction, permitting high productivity for compact reactor sizes. An estimative calculation of the capacity of a 1 m^3 reactor for the partial oxidation of methane predicts a capacity of 1 000 000 ft^3 of hydrogen per day.

1.9.2.1 How Far is the Development? A Critical Review

Petrochemistry has always been a topic in the discussion of possible micro-reactor applications. However, reported micro-reactor developments have not yet entered the field. This may be due to the large gap between the size of current micro reactors and that required for petrochemistry. Already the first demonstrators probably need to be of considerable size. It would not be surprising if an industry that is used to handling very large-scale equipment was the latest to enter such a new emerging field as micro-chemical processing.

1.9.3
Catalyst Discovery and Optimization via High-throughput Screening

Micro reactors can be used as part of the equipment for high-throughput screening [5, 104]. Schouten et al. even say, 'Microreactors are the natural platform for parallel screening and high-throughput testing ... of new catalysts ...' [104]. Micro reactors are also regarded as perfect tools for investigating intrinsic kinetics [104, 270]. Generally, a specific workflow for micro reactors needs to be developed on the basis of combinatorial methodologies to exploit its potential fully (see [271] for an example focusing on minimizing signal dispersion for a serial HTE apparatus).

Several reports underline the ability of micro reactors to perform catalyst testing for process development [187, 272–274] and for high-throughput screening [275–279]. Kinetic data have been extracted and the capability of both fast serial [272, 273], and parallel [275–279] screening has been demonstrated. The range of reactions investigated is large, covering, e.g., hydrogenations of double bonds [272, 273], ammonia oxidation [187, 280–282], and phosgene generation [274].

In-depth knowledge on the catalysts themselves, in addition to the reactions, could be gained, e.g. concerning turnover frequencies [280–282] or the internal composition, morphology, and microstructure of the catalyst [283–285]. Catalyst coating processes are of a much higher standard than some years ago [283–285]. Specialized micro reactors are available for kinetic data acquisition, such as a cross-flow, mini-short-bed reactor [286, 287]. Criteria on judging mass- and heat-transfer limits have been proposed recently, allowing an in-depth analysis of the suitability of these tools and reaction conditions to give reliable kinetic data [181, 286, 287].

It should be noted that ultimately the methods developed for catalyst HTE testing can also be applied to the discovery of new process parameters and new materials.

1.9.3.1 How Far is the Development? A Critical Review

The above given conclusions make it very clear. The investigations concerning catalyst testing and screening are of a high standard today, at least from a scientific point of view. These developments are, besides the organic investigations (see Section 1.9.5), actually one of the major drivers pushing developments in micro-reaction technology.

Many high-ranked catalysis experts have opened their research to micro-reactor studies and have become active. Engineering and catalysis journals have their eye on the field. Thus, cross-border expertise is approaching chemical micro processing. Virtually all chemical-engineering conferences have a micro-reactor session. One can say that micro reactors are now accepted and in a positive sense being 'absorbed' by the catalysis community.

BASF activities have shown that one can make money by investigating gas-phase reactions for process development. If Degussa and Krupp-Uhde have built a very large micro reactor with outer dimensions of more than 1 m, obviously there must be an economic reason for that.

Not to raise too much euphoria, it should be remarked that still much more work has to be done to obtain real, well-founded achievements compared with conventional attempts. The known micro-reactor investigations are a nice piece of work, but still have some kind of feasibility character. However, there is so far no outstanding achievement that can really be regarded as a breakthrough. As an outstanding result, something similar to that achieved by developing the maleic anhydride process (Contractor process), as a selective oxidation in the gas phase, can be envisaged. Here, reactor engineering, catalyst development, heat recycling, phase separation, and many more go hand in hand.

1.9.4
Bulk Chemicals and Commodities

There is not much to be said about the use of micro reactors for bulk chemicals and commodities. Wörz et al. are so far the only ones who have disclosed their work on the potential of micro-structured reactors for the optimization of chemical processes performed on a large scale of industrial relevance [110, 112, 154, 288–290]. This included a fast exothermic liquid/liquid two-phase reaction, which was used for the industrial production of a vitamin intermediate product, and a selective oxidation reaction for an intermediate, a substituted formaldehyde derivative.

1.9.4.1   How Far is the Development? A Critical Review
Wörz has often stressed the importance of using micro reactors as information tools for large-scale production, giving precise information otherwise not achievable [28, 110, 112]. This information is transferred and applied at the production stage, without changing the equipment from conventional to micro, but only affecting the setting of process parameters. In written form, there has been not a very large response to these ideas, i.e. not many follow-up articles by other industrial authors from the major chemistry players have appeared. This stands in obvious contradiction to the large potential provided and the clear evidence given so far by Wörz et al. and others. Maybe it is done in a hidden way due to the secrecy needs of industry (see, e.g., a patent on the use of a micro reactor for one step of the large-scale Hock process [183]). Probably, the large cost pressure in the field of bulk chemicals and commodities does not allow a search for novel, innovative means in laboratory-scale development. Having in mind, however, that Wörz stated that BASF "... would have saved many years of production at low yields" [28]. Thus, considering that obviously his investigations led to enormous cost savings of BASF processes, one might dare to question if this attitude is really fully correct.

1.9.5
Fine Chemicals and Functional Chemicals

1.9.5.1   Fine Chemicals – Drivers and Trends
Specialty chemicals divide into fine and functional chemicals (Figure 1.37). Fine-chemicals manufacture today is undergoing a substantial change. Thus, before

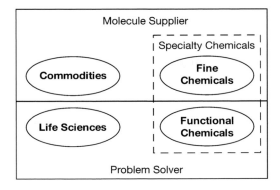

Figure 1.37 Diverse fields belonging to the class of specialty chemicals (redrawn from [137]).

having a look at the state-of-the-art of micro reactor use for fine chemicals, it is worth considering the changes that are occuring in the field. They will create – and already have created – a demand for using micro-reactor technology.

The classes of fine chemicals and functional chemicals are subsumed under the term specialty chemicals. This field is undergoing major changes at present and may need to employ micro-channel technology. ICI, for example, sees the following as current key trends and challenges for specialty chemicals [291]:

1. a move from molecule synthesis towards efficient delivery for the industry's products
2. requirements for sustainable products and services
3. trend to greatly increased regulation of chemicals
4. globalization of products, but at the same time growth in mass customization of products
5. multi-functionality of products as increasing requirement.

Points 1 and 5 refer to the increased importance of functional chemicals [291]. Owing to the wide parameter space determining functionality (not only molecular diversity), this demands much higher flexibility and speed in the preparation of new samples during the research phase. The behavior of complex molecular mixtures needs to be understood. In particular, product application, formulation, and blending skills need to be developed and acquired. In a more remote vision, this demands on-site distributed manufacture of functional chemicals such as paints and similar products.

Points 2–4 are related to green chemistry (see Section 1.1.6), sustainable development (see Section 1.1.6), and process intensification (see Section 1.1.9) which may need micro reactors as a preferred tool [291]. All these efforts are due to rising pressure from customers, regulators, and non-government organizations (NGOs). Once accepted as an unavoidable barrier, the only change for the chemical companies is to face this challenge and see it as an opportunity for profitable growth. One feature of sustainable development may be the upgrading of waste streams to first-quality products and high-efficiency processing of all raw materials.

Globalization creates a need to source complex products with a high degree of consistency across the world [291]. This, in turn, demands a well-defined and transferable process technology. A massive, local customization requires product availability in great variety close to the customer, demanding process intensification, modular operations, transportable plant, and fast response and product changeover. Multifunctionality demands a wider space in formulations and a chemistry set to deliver 'dial-in functionality', which needs assembly from a consistent set of basic materials.

1.9.5.2 Fine Chemicals – State of the Art of Micro-reactor Use

Fine chemicals are most commonly made using a large variety of organic (and inorganic) reactions. Hence the use of micro reactors for fine chemicals requires in-depth information about their performance with regard to several classes of (organic) reactions In this context, one should be aware of the huge steps forward that the performance of organic reactions in micro reactors has made in the past 5 years or so, which are summarized below.

Many of the known chemical syntheses such as Wittig [29, 165], Knoevenagel [29], aldol [292], Ugi [29, 293], Michael addition [29], Hantzsch [29, 156, 157], Diels–Alder [294], Azo coupling [136, 182], Suzuki coupling [29, 155] or enamine [29, 295] reactions (Table 1.8), to name but a few, have been carried out successfully in mi-

Table 1.8 Organic reactions conducted in a micro-chip reactor and details of the corresponding experimentation (redrawn from [29]).

Reaction	Chip material	Solvent	Conversion (%)	Comments
Suzuki	Glass	Aq. THF	67	EOF
Kumada coupling	Polypropylene	THF	60	Syringe pump
Aldol	Glass	THF	100	EOF
Nitration	Glass	Benzene	65	EOF
Wittig	Glass	MeOH	39–59	EOF
Enamine	Glass	MeOH	42	EOF
Ugi four-component coupling	Glass	MeOH		EOF
Peptide synthesis	Glass	DMF	100	EOF
Synthesis of pyridazinones	Glass	EtOH/AcOH	30	EOF
Synthesis of amides from amines and acid chlorides	Glass	DCM	77	EOF
Diazo coupling	Glass	MeOH, MeCN	37, 22	EOF
Aminothiazole synthesis	Glass	NMP	58–100	EOF
Knoevenagel	Glass	MeOH/H_2O	59–68	EOF
Hantzsch thiazole synthesis	Glass	NMP	58–100	EOF
Michael addition	Glass	EtOH	95–100	EOF
S_N2 alkyl halide	Glass	DMF/H_2O	25	EOF
Dehydration	Glass/PDMS	EtOH	85–95	EOF or syringe pump
Photochemical	Silicon/quartz	Me_2CHOH	60	Syringe pump
Phase transfer	Glass	EtOAc	100	Syringe pump
Fluorination	Ni or Cu	Nitrogen gas	90–99	Syringe pump
Fluorination	Silicon/Pyrex	MeOH	80	Syringe pump

cro-structured reactors and in most cases with improved results compared with macro-structured reactors (see also [29]). Fluorinations [164, 167, 193, 197, 296, 297], chlorinations [127], nitrations [152, 298–302], hydrogenations [303–306], and oxidations [18, 147, 150, 159, 162, 163, 178–180, 186, 307, 308] have been described in numerous examples. The range of chemical conversions includes many known chemical reaction mechanisms such as additions [151, 159, 162, 163, 307], eliminations [129, 304, 309–313], nucleophilic substitutions on aliphatics [314], electrophilic substitutions on aromatics [152, 164, 167, 298, 299], cycloadditions, radical polymerizations [21, 315] and more (see also a selection in [4, 316]).

The potential described above is not only of a scientific nature, as all these publications impressively prove. Indeed, it has been applied to bench-scale, pilot-scale and production-scale processing. Examples refer to the fields of Grignard reactions [11], polymerizations [21], boron chemistry [317], and azo pigment generation [136].

Fine-chemical companies have definitely shown interest in micro-reaction technology (see also the commitment in [137]) and have formed their own task forces for this purpose. The increasing number of patents is further proof of the beginning commercial use of micro reactors (see, e.g., [318–321]).

1.9.5.3 Functional Chemicals

Following a theoretical analysis of distributed small-plant manufacture, Benson and Ponton define assessment criteria for processes suitable for such processing [139]. Since micro reactors are one of the favorite and natural tools for distributed manufacture, this selection list also defines micro-reactor applications. In this context, the authors, probably in one of the first regular citations, emphasize that formulation processes, especially those with multiple ingredients, are particularly suited for distribution. The making of paint 'on-site' is referred to as an already existing way to do so. It stands to reason to augment the scope from formulations to functional chemicals.

Meanwhile, many investigations have paved the ground for the use of micro reactors with functional chemicals. Among these is work on emulsification [322–326], foaming [327, 328], creaming [329], and particle formation [17, 136, 330–332].

Noteable are recent studies on the generation of polymer particles as carriers for controlled drug release [333] and of cationic solid lipid micro-particles as synthetic carriers for the targeted delivery of macromolecules to phagocytic antigen-presenting cells [334]. The industrial interest, although rarely disclosed, is evident from the patents filed in the field (see, e.g., [335, 336]).

1.9.5.4 How far is the Development? A Critical Review

The PAMIR study predicts fine chemicals as one of the first fields where commercialization of micro-reactor technology will start [241, 337]. The small quantities, sometimes only in the kilogram range, and the large margins make the substitution of existing processes by micro reactors attractive. Currently, we can see the first signs of this, e.g. the work of Merck, Siemens-Axiva, Clariant and others.

The corresponding scientific investigations have advanced considerably in recent years. Together with the field of catalyst testing and screening (see Section

1.9.3), the conducting of organic reactions which could be applied to fine chemicals has made a huge step forward.

However, in contrast to the field of catalysis, not many high-ranked organic chemistry experts have so far opened their research to micro-reactor studies and have become active (for some exceptions see, e.g., [29, 47, 338–341]). Organic synthesis journals and conferences have yet not recognized to a great extent micro reactors, an exception being [82, 342]. This is, however, not true for researchers oriented towards analytical chemistry. In conjunction with µTAS developments, more and more work is being done in that area.

The organic chemists in academia still stick to their flask glassware. Here, certainly, some time is needed and education has to be provided. Micro-chemical engineering, as the name indicates, still remains a domain of the engineering society. Nonetheless, the fine-chemical companies have accepted micro reactors; the push will come from the industry side.

1.9.6
Cosmetics and Foods

There is one report on the use of micro reactors for cosmetics [323] and none for foods. Of all the topics mentioned, these have the smallest number of reports.

1.9.6.1 How Far is the Development? A Critical Review
In view of the prognoses of the PAMIR market study, predicting at least a certain interest for these fields, this is astonishing, in particular when compared with the great interest stimulated in the field of reaction engineering. The surprise becomes even larger when reviewing also the large potential forecast for functional chemicals (see Section 1.9.5), which is related to cosmetics and foods with regard to the type of processing. From the authors' own experience, it can be said that industrial investigations are being undertaken, but not being published.

1.9.7
Extra-terrestrial Processing

Gavriilidis et al. review the use of lightweight, compact chemical systems for application in the space sector [5], referring to original work done in this field [91]. The latter work was done for the NASA In Situ Resource Utilization (ISRU) program, designated for future missions to Mars. The investigations were aimed at the In Situ Propellant Production (ISPP) by converting the carbon dioxide of the Martian atmosphere to propellants and oxygen for the return trip using stored hydrogen, thereby decreasing the launch mass. Highly energy efficient micro-channel systems use extensive recuperation and energy cascading to minimize thermal and electrical wastage.

ISPP units are not the only micro device units of interest for space applications; micro fuel cells, compact cleanup units for water treatment, portable heating and cooling units and devices for chemical processing and mining are considered [91].

1.9.7.1 How Far is the Development? A Critical Review

Significant challenges exist when applying micro reactors to outer space; the latter will require adaptations, e.g. concerning fluid flow. Energy management has to be improved to yield the predicted major benefits for spacecraft size and payloads [91]. The investigations are long-term oriented, i.e. can profit from more funding in the coming years, as the first manned Mars mission will probably not come sooner than in 10 years. Nonetheless, at that time results will be demanded, as for such large, strategic developments no delays will be accepted.

1.9.8
Chemical Analysis, Analyte Separation, Assays and Further Diverse Applications in the Bio Field

μTAS components and systems exhibit sensor and analytical separation functions. DNA analysis, performance of polymerase chain reactions, clinical assays for pH, enzymes, proteins, oxygen etc., trace pollution monitoring and other sorts of biological analyzes are at the focus of recent developments [5]. Another reference lists environmental monitoring (including speciation), clinical monitoring, and quality control in production processes as applications of μTAS equipment in chemical analysis [30].

Since further reviewing μTAS could easily fill a separate book and is mostly concerned with biochemical applications, it was excluded from this book. Therefore, the description of these applications is beyond the scope of this chapter. The reader is referred to original reviews [31, 32].

Auroux et al. give an up-to-date description of the application of μTAS components and systems, including cell culture and cell handling, immunoassays, DNA separation and analysis, polymerase chain reactions, and sequencing [43] (see also [44] for a description of the μTAS components and systems).

Major drivers for using micro-channel chip systems for chemical analysis are low reagent consumption, low energy consumption, high reliability, good robustness also in the hands of personnel not trained in analytical chemistry, and low maintenance requirements [30].

Concrete applications of micro reactors for chemical analysis, albeit so far not a core application, have been described [5]. Among other uses in chemical analysis, micro devices for gas chromatography, infrared spectroscopy, and photoacoustic detection are mentioned.

1.9.8.1 How Far is the Development? A Critical Review

μTAS applications are very developed, at least from a scientific and technical point of view. For some of the applications given above, even commercial chips are available. A detailed description and a far-reaching analysis of these are not the topic of this book, as outlined earlier.

So far, micro-reactor developments apart from μTAS sector have not been used to a great extent for analytical purposes. Merging this technology base with that of μTAS devices, e.g. for hybrid assemblies, could have some potential.

References

1 WEGENG, R. W., CALL, C. J., DROST, M. K., *Chemical system miniaturization,* in Proceedings of the AIChE Spring National Meeting, pp. 1–13 (25–29 February **1996**), New Orleans, USA.

2 JENSEN, K. F., *Microreaction engineering – is small better?* Chem. Eng. Sci. **56** (**2001**) 293–303.

3 EHRFELD, W., HESSEL, V., HAVERKAMP, V., *Microreactors, Ullmann's Encyclopedia of Industrial Chemistry,* Wiley-VCH, Weinheim (**1999**).

4 EHRFELD, W., HESSEL, V., LÖWE, H., *Micro-reactors,* Wiley-VCH, Weinheim (**2000**).

5 GAVRIILIDIS, A., ANGELI, P., CAO, E., YEONG, K. K., WAN, Y. S. S., *Technology and application of microengineered reactors,* Trans. IChemE. **80/A**, 1 (**2002**) 3–30.

6 ONDREY, G., *Microreactor engineering: birth of a new discipline?* Chem. Eng. **5** (**1995**) 52.

7 OXLEY, P., BRECHTELSBAUER, C., RICARD, F., LEWIS, N., RAMSHAW, C., *Evaluation of spinning disk reactor (SDR) technology for the manufacture of pharmaceuticals,* Ind. Chem. Res. **39** (**2000**) 2175–2182.

8 SCHENK, R., HESSEL, V., HOFMANN, C., KISS, J., LÖWE, H., SCHÖNFELD, F., *Numbering-up of micro devices: a first liquid-flow splitting unit,* Chem. Eng. Technol. **26**, 12 (**2003**) 1271–1280.

9 HESSEL, V., LÖWE, H., *Micro chemical engineering: components – plant concepts – user acceptance: Part III,* Chem. Eng. Technol. **26**, 5 (**2003**) 531–544.

10 HESSEL, V., LÖWE, H., *Mikroverfahrenstechnik: Komponenten – Anlagenkonzeption – Anwenderakzeptanz – Teil 3,* Chem. Ing. Tech. **74**, 4 (**2002**) 381–400.

11 KRUMMRADT, H., KOPP, U., STOLDT, J., *Experiences with the use of microreactors in organic synthesis,* in EHRFELD, W. (Ed.), *Microreaction Technology: 3rd International Conference on Microreaction Technology, Proc. of IMRET 3,* pp. 181–186, Springer-Verlag, Berlin (**2000**).

12 EHRFELD, W., GOLBIG, K., HESSEL, V., LÖWE, H., RICHTER, T., *Characterization of mixing in micromixers by a test reaction: single mixing units and mixer arrays,* Ind. Eng. Chem. Res. **38**, 3 (**1999**) 1075–1082.

13 LOSEY, M. W., SCHMIDT, M. A., JENSEN, K. F., *Microfabricated multiphase packed-bed reactors: characterization of mass transfer and reactions,* Ind. Chem. Res. **40** (**2001**) 2555–2562.

14 SCHUBERT, K., BRANDNER, J., FICHTNER, M., LINDER, G., SCHYGULLA, U., WENKA, A., *Microstructure devices for applications in thermal and chemical process engineering,* Microscale Therm. Eng. **5** (**2001**) 17–39.

15 IMM, Institut für Mikrotechnik Mainz GmbH, unpublished results.

16 SCHÖNFELD, F., RENSINK, D., *Simulation of droplet generating by mixing nozzles,* Chem. Eng. Technol. **26**, 5 (**2003**).

17 SCHENK, R., HESSEL, V., WERNER, B., ZIOGAS, A., HOFMANN, C., DONNET, M., JONGEN, N., *Micromixers as a tool for powder production,* Chem. Eng. Trans. **1** (**2002**) 909–914.

18 VESER, G., *Experimental and theoretical investigation of H_2 oxidation in a high-temperature catalytic microreactor,* Chem. Eng. Sci. **56** (**2001**) 1265–1273.

19 QUIRAM, D. J., JENSEN, K. F., SCHMIDT, M. A., RYLEY, J. F., MILLS, P. L., WETZEL, M. D., ASHMEAD, J. W., BRYSON, R. D., DELANEY, T. M., KRAUS, D. J., MCCRACKEN, J. S., *Development of a turnkey multiple microreactor test station,* in Proceedings of the 4th International Conference on Microreaction Technology, IMRET 4, pp. 55–61 (5–9 March **2000**), AIChE Topical Conf. Proc., Atlanta, USA.

20 QUIRAM, D. J., RYLEY, F., ASHMEAD, J. W., BRYSON, R. D., KRAUS, D. J., MILLS, P. L., MITCHELL, R. E., WETZEL, M. D., SCHMIDT, M. A., JENSEN, K. F., *Device level integration to form a parallel microfluidic reactor system,* in RAMSEY, J. M., VAN DEN BERG, A. (Eds.), *Micro Total Analysis Systems,* pp. 661–663, Kluwer Academic Publishers, Dordrecht (**2001**).

21 BAYER, T., PYSALL, D., WACHSEN, O., *Micro mixing effects in continuous radical polymerization,* in EHRFELD, W. (Ed.),

Microreaction Technology: 3rd International Conference on Microreaction Technology, Proc. of IMRET 3, pp. 165–170, Springer-Verlag, Berlin (**2000**).

22 ZLOKARNIK, M., *Scale-up and Miniplants*, Chem. Ing. Tech. **75**, 4 (**2003**) 370–375.

23 BRANDNER, J., BOHN, L., SCHYGULLA, U., WENKA, A., SCHUBERT, K., *Microstructure devices for thermal and chemical process engineering*, in Proceedings of the Japan Chemical Innovation Institute (JCII) (23 May **2001**), Tokyo, Japan.

24 RAMSHAW, C., *The incentive for process intensification*, in Proceedings of the 1st Int. Conf. Process Intensification for Chem Ind., p. 1 (**1995**), BHR Group, London.

25 STANKIEWICZ, A. I., MOULIJN, J. A., *Process intensification: transforming chemical engineering*, Chem. Eng. Prog. **1** (**2000**) 22–34.

26 STANKIEWICZ, A. I., *Reactive separations for process intensification: an industrial perspective*, Chem. Eng. Proc. **42** (**2003**) 137–144.

27 PHILLIPS, C. H., *Development of a novel compact chemical reactor-heat exchanger*, in GREEN, A. (Ed.), *Proc. of 3rd Int. Conf. on Process Intensification for the Chemical Industry, BHR Group Conference Series*, Vol. 38, pp. 71–87, Professional Engineering Publishing (**1999**).

28 WÖRZ, O., *Microreactors as tools in chemical research*, in MATLOSZ, M., EHRFELD, W., BASELT, J. P. (Eds.), *Microreaction Technology – IMRET 5: Proc. of the 5th International Conference on Microreaction Technology*, pp. 377–386, Springer-Verlag, Berlin (**2001**).

29 FLETCHER, P. D. I., HASWELL, S. J., POMBO-VILLAR, E., WARRINGTON, B. H., WATTS, P., WONG, S. Y. F., ZHANG, X., *Micro reactors: principle and applications in organic synthesis*, Tetrahedron **58**, 24 (**2002**) 4735–4757.

30 VAN DER LINDEN, W. E., *Chances of µTAS in Analytical Chemistry*, in VAN DEN BERG, A., BERGFELD, P. (Eds.), *Micro Total Analysis System*, pp. 29–35, Kluwer Academic Publishers, Dordrecht (**1994**).

31 BERGVELD, P., *The challenge of developing µTAS*, in VAN DEN BERG, A., BERGFELD, P. (Eds.), *Micro Total Analysis System*, pp. 1–4, Kluwer Academic Publishers, Dordrecht (**1994**).

32 MANZ, A., VERPOORTE, E., REYMOND, D. E., EFFENHAUSER, C. S., BURGGRAF, N., WIDMER, H. M., *µ-TAS: miniaturized total chemical analysis systems*, in Proceedings of the Micro Total Analysis System Workshop, pp. 5–27 (Nov. **1994**), Enschede, The Netherlands.

33 MANZ, A., VERPOORTE, E., RAYMOND, D. E., EFFENHAUSER, C. S., BURGGRAF, N., WIDMER, H. M., *µTAS for biochemical analysis*, in VAN DEN BERG, A., BERGFELD, P. (Eds.), *Micro Total Analysis System*, pp. 5–27, Kluwer Academic Publishers, Dordrecht (**1994**).

34 KARUBE, I., *µTAS for biochemical analysis*, in VAN DEN BERG, A., BERGFELD, P. (Eds.), *Micro Total Analysis System*, pp. 37–46, Kluwer Academic Publishers, Dordrecht (**1994**).

35 FLUITMAN, J. H., VAN DEN BERG, A., LAMMERICK, T. S., *Micromechanical components for µTAS*, in VAN DEN BERG, A., BERGFELD, P. (Eds.), *Micro Total Analysis Systems*, pp. 73–83, Kluwer Academic Publishers, Dordrecht (**1995**).

36 WIDMER, H. M., *A survey of the trends in analytical chemistry over the last twenty years, emphasizing the development of TAS and µTAS*, in VAN DEN BERG, A., BERGFELD, P. (Eds.), *Proceedings of the 2nd International Symposium on Miniaturized Total Analysis Systems, Analytical Methods and Instrumentation, Special Issue µTAS '96*, pp. 1–27, Kluwer Academic Publishers, Dordrecht (**1996**).

37 VAN DEN BERG, A., BERGVELD, P., *Development of µTAS concepts at the MESA Research Institute*, in WIDMER, E., VERPOORTE, E., BANARD, S. (Eds.), *Proceedings of the 2nd International Symposium on Miniaturized Total Analysis Systems, Analytical Methods and Instrumentation, Special Issue µTAS '96*, pp. 9–15, Basel (**1996**).

38 RAMSEY, J. M., *Miniature chemical measurement systems*, in WIDMER, E., VERPOORTE, E., BANARD, S. (Eds.), *Proceedings of the 2nd International Symposium on Miniaturized Total Analysis Systems, Analytical Methods and Instrumentation, Special Issue µTAS '96*, pp. 24–27, Basel (**1996**).

39 LI, P. C. H., HARRISON, D. J., *Transport, manipulation, and reaction of biological cells on-chip using electrokinetic effects*, Anal. Chem. **69**, 8 (**1997**) 1564–1568.

40 VAN DEN BERG, A., VAN AKKER, E., OOSTENBROEK, E., TJERKSTRA, W., BARSONY, I., *Technologies and micro-structures for (bio)chemical microsystems*, in EHRFELD, W. (Ed.), *Microreaction Technology – Proc. of the 1st International Conference on Microreaction Technology, IMRET 1*, pp. 91–103, Springer-Verlag, Berlin (**1997**).

41 ZORBAS, H., *Miniatur-Durchfluß-PCR: Ein Durchbruch?* Angew. Chem. **111**, 8 (**1999**) 1121–1124.

42 KNAPP, M. R., KOPF-SILL, A., DUBROW, R., CHOW, A., CHIEN, R.-L., CHOW, C., PARCE, J. W., *Commercialized and emerging Lab-on-a-Chip applications*, in RAMSEY, J. M., VAN DEN BERG, A. (Eds.), *Micro Total Analysis Systems*, pp. 7–9, Kluwer Academic Publishers, Dordrecht (**2001**).

43 AUROUX, P. A., IOSSIFIDIS, D., REYES, D. R., MANZ, A., *Micro total analysis systems: 2. Analytical standard operations and applications*, Anal. Chem. **74**, 12 (**2002**) 2637–2652.

44 REYES, D. R., IOSSIFIDIS, D., AUROUX, P. A., MANZ, A., *Micro total analysis systems: 1. Introduction, theory, and technology*, Anal. Chem. **74**, 12 (**2002**) 2623–2636.

45 FREEMANTLE, M., *Downsizing Chemistry*, Chem. Eng. News **77**, 2 (**1999**) 27–36.

46 http://www.labonachip.org.uk, The laboratory on a chip consortium, 22 February **2001**.

47 HASWELL, S. T., WATTS, P., *Green chemistry: synthesis in micro reactors*, Green Chem. **5** (**2003**) 240–249.

48 JISCHA, M. F., *Sustainable development and technology assesment*, Chem. Eng. Technol. **21**, 8 (**1998**) 629–636.

49 MOORTHY, J., BEEBE, D. J., *Organic and biometric designs for microfluidic systems*, Anal. Chem. **75**, 7 (**2003**) 293–301.

50 BERLIN, A. A., ZAIKOV, G. E., MINSKER, K. S., *Microreactor (MR) technology*, Polym. News **22** (**1997**) 291–292.

51 KOCHLOEFL, K., *Anwendung von Mikroreaktoren bei der Entwicklung und Prüfung von technischen Katalysatoren*, Chemie-Technik **4** (**1975**) 443–447.

52 ZIMMERMANN, C., HAYEK, K., *Computerunterstütztes Mikroreaktorsystem zur Untersuchung heterogen katalysierter Gasreaktionen*, Chem. Ing. Tech. **63**, 1 (**1991**) 68–71.

53 BÖHM, L. L., FRANKE, R., THUM, G., *The microreactors as a model for the description of the ethylene polymerization with heterogeneous catalysts*, in KAMINSKY, W., SINN, H. (Eds.), *Transition metals and organometallics as catalysts for olefin polymerization*, pp. 391–403, Springer-Verlag, Berlin (**1988**).

54 PILLAI, C. N., *Zeolites as microreaction vessels*, Indian J. Technol. **30**, 2 (**1992**) 59–63.

55 NICKON, A., SILVERSMITH, E. F., *Organic Chemistry: the Name Game*, Pergamon Press, New York (**1987**).

56 SUCKLING, C. J., *Host–guest binding by a simple detergent derivative: tentacle molecules*, J. Chem. Soc., Chem. Commun. (**1982**) 661–662.

57 HESSEL, V., LÖWE, H., *Micro chemical engineering: components – plant concepts – user acceptance: Part I*, Chem. Eng. Technol. **26**, 1 (**2003**) 13–24.

58 HESSEL, V., LÖWE, H., *Mikroverfahrenstechnik: Komponenten – Anlagenkonzeption – Anwenderakzeptanz – Teil 1*, Chem. Ing. Tech. **74**, 2 (**2002**) 17–30.

59 SCHUBERT, K., BIER, W., LINDER, G., SEIDEL, D., *Herstellung und Test von kompakten Mikrowärmeüberträgern*, Chem. Ing. Tech. **61**, 2 (**1989**) 172–173.

60 BIER, W., KELLER, W., LINDER, G., SEIDEL, D., SCHUBERT, K., *Manufacturing and testing of compact micro heat exchangers with high volumetric heat transfer coefficients*, ASME, DSC-Microstructures, Sensors, and Actuators **19** (**1990**) 189–197.

61 SCHUBERT, K., BIER, W., LINDER, G., SEIDEL, D., *Profiled microdiamonds for producing microstructures*, Ind. Diamond Rev. **50**, 5 (**1990**) 235–239.

62 BIER, W., KELLER, W., LINDER, G., SEIDEL, D., SCHUBERT, K., MARTIN, H., *Gas-to-gas heat transfer in micro heat exchangers*, Chem. Eng. Process. **32**, 1 (**1993**) 33–43.

63 BIER, W., GUBER, A., LINDER, G., SCHALLER, T., SCHUBERT, K., *Mechanische Mikrofertigung – Verfahren und Anwendungen*, KfK Ber. **5238** (**1993**) 132–137.

64 BRENCHLEY, D. L., WEGENG, R. S.,
*Status of microchemical systems develop-
ment in the United States of America*, in
EHRFELD, W., RINARD, I. H., WEGENG,
R. S. (Eds.), *Process Miniaturization: 2nd
International Conference on Microreaction
Technology, IMRET 2, Topical Conf.
Preprints*, pp. 18–23, AIChE, New
Orleans, USA (**1998**).

65 BASELT, J. P., FÖRSTER, A., HERMANN, J.,
*Microreaction technology: focusing the
German activities in this novel and
promising field of chemical process engi-
neering*, in EHRFELD, W., RINARD, I. H.,
WEGENG, R. S. (Eds.), *Process Miniaturi-
zation: 2nd International Conference on
Microreaction Technology, IMRET 2,
Topical Conf. Preprints*, pp. 13–17,
AIChE, New Orleans, USA (**1998**).

66 HESSEL, V., LÖWE, H., *Micro chemical
engineering: components – plant concepts –
user acceptance: Part II*, Chem. Eng.
Technol. **26**, 4 (**2003**) 391–408.

67 EHRFELD, W., HESSEL, V., MÖBIUS, H.,
RICHTER, T., RUSSOW, K., *Potentials and
realization of micro reactors*, in EHRFELD,
W. (Ed.), *Microsystem Technology for
Chemical and Biological Microreactors,
DECHEMA Monographs*, Vol. 132,
pp. 1–28, Verlag Chemie, Weinheim
(**1996**).

68 EHRFELD, W., LÖWE, H., HESSEL, V.,
RICHTER, T., *Anwendungspotentiale für
chemische und biologische Mikroreaktoren*,
Chem. Ing. Tech. **69**, 7 (**1997**) 931–934.

69 EHRFELD, W., GOLBIG, K., HESSEL, V.,
KONRAD, R., LÖWE, H., RICHTER, T.,
*Fabrication of components and systems for
chemical and biological microreactors*, in
EHRFELD, W. (Ed.), *Microreaction
Technology – Proc. of the 1st International
Conference on Microreaction Technology,
IMRET 1*, pp. 72–90, Springer-Verlag,
Berlin (**1997**).

70 EHRFELD, W., HESSEL, V., KIESEWALTER,
S., LÖWE, H., RICHTER, T., SCHIEWE, J.,
*Implementation of microreaction techno-
logy in process engineering*, in EHRFELD,
W. (Ed.), *Microreaction Technology: 3rd
International Conference on Microreaction
Technology, Proc. of IMRET 3*, pp. 14–34,
Springer-Verlag, Berlin (**2000**).

71 EHRFELD, W., HESSEL, V., LÖWE, H.,
*Extending the knowledge base in micro-
fabrication towards themical engineering
and fluid dynamic simulation*, in
Proceedings of the 4th International
Conference on Microreaction
Technology, IMRET 4, pp. 3–20
(5–9 March **2000**), AIChE Topical Conf.
Proc., Atlanta, USA.

72 JÄCKEL, K.-P., *Mikrotechnik – Anwendungsmöglichkeiten in der chemischen
Industrie*, CHEManager **11** (**1995**) 8.

73 JÄCKEL, K. P., *Microtechnology: Applica-
tion opportunities in the chemical industry*,
in EHRFELD, W. (Ed.), *Microsystem
Technology for Chemical and Biological
Microreactors, DECHEMA Monographs*,
Vol. 132, pp. 29–50, Verlag Chemie,
Weinheim (**1996**).

74 LEROU, J. J., HAROLD, M. P., RYLEY, J.,
ASHMEAD, J., O'BRIEN, T. C., JOHNSON,
M., PERROTTO, J., BLAISDELL, C. T.,
RENSI, T. A., NYQUIST, J., *Microfabricated
mini-chemical systems: technical feasibility*,
in EHRFELD, W. (Ed.), *Microsystem
Technology for Chemical and Biological
Microreactors, DECHEMA Monographs*,
Vol. 132, pp. 51–69, Verlag Chemie,
Weinheim (**1996**).

75 JENSEN, K. F., *Microchemical systems:
Status, challenges, and oportunities*,
AIChE J. **45**, 10 (**1999**) 2051–2054.

76 JENSEN, K. F., AJMERA, S. K.,
FIREBAUGH, S. L., FLOYD, T. M., FRANZ,
A. J., LOSEY, M. W., QUIRAM, D.,
SCHMIDT, M. A., *Microfabricated
chemical systems for product screening and
synthesis*, in HOYLE, W. (Ed.), *Automated
Synthetic Methods for Specialty Chemicals*,
pp. 14–24, Royal Society of Chemistry,
Cambridge (**2000**).

77 HASWELL, S. J., MIDDLETON, R. J.,
O'SULLIVAN, B., SKLETON, V., WATTS, P.,
STYRING, P., *The application of micro
reactors to synthetic chemistry*, Chem.
Commun. (**2001**) 391–398.

78 HASWELL, S. J., *Miniaturization – What's
in it for chemistry*, in VAN DEN BERG, A.,
RAMSAY, J. M. (Eds.), *Micro Total Analysis
System*, pp. 637–639, Kluwer Academic
Publishers, Dordrecht (**2001**).

79 DEWITT, S., *Microreactor for chemical
synthesis*, Curr. Opin. Chem. Biol. **3**
(**1999**) 350–356.

80 DE MELLO, A., WOOTTON, R., *But what is
it good for? Applications of microreactor*

technology for the fine chemical industry, Lab Chip **2** (2002) 7N–13N.

81 SCHWALBE, T., AUTZE, V., WILLE, G., *Chemical synthesis in microreactors,* Chimia **56**, 11 (2002) 636–646.

82 JÄHNISCH, K., HESSEL, V., LÖWE, H., BAERNS, M., *Chemie in Mikrostruktur-reaktoren,* Angew. Chem. **44**, 3 or 4 (2004) in press.

83 LÖWE, H., HESSEL, V., MÜLLER, A., *Microreactors – prospects already achieved and possible misuse,* Pure Appl. Chem. **74**, 12 (2003).

84 RINARD, I. H., *Miniplant design methodology,* in EHRFELD, W., RINARD, I. H., WEGENG, R. S. (Eds.), *Process Miniaturization: 2nd International Conference on Microreaction Technology, IMRET 2, Topical Conf. Preprints,* pp. 299–312, AIChE, New Orleans (**1998**).

85 SAHA, N., RINARD, I. H., *Miniplant design methodology: a case study – manufacture of hydrogen cyanide,* in Proceedings of the 4th International Conference on Microreaction Technology, IMRET 4, pp. 327–333 (5–9 March **2000**), AIChE Topical Conf. Proc., Atlanta, USA.

86 KÜPPER, M., HESSEL, V., LÖWE, H., STARK, W., KINKEL, J., MICHEL, M., SCHMIDT-TRAUB, H., *Micro reactor for electroorganic synthesis in the simulated moving bed-reaction and separation environment,* Electrochim. Acta **48** (2003) 2889–2896.

87 HESSEL, V., HARDT, S., LÖWE, H., *Chemical processing with microdevices: Device/plant concepts, selected applications and state of scientific/commercial implementation,* Chem. Eng. Comm., Special edition – 6th Italian Conference on Chemical and Process Engineering, ICheaP-6 **3** (2003) pp. 479–484.

88 LÖWE, H., EHRFELD, W., HESSEL, V., RICHTER, T., SCHIEWE, J., *Micromixing technology,* in Proceedings of the 4th International Conference on Microreaction Technology, IMRET 4, pp. 31–47 (5–9 March **2000**), AIChE Topical Conf. Proc., Atlanta, USA.

89 BRENCHLEY, D. L., WEGENG, R. S., DROST, M. K., *Development of micro-chemical and thermal systems,* in Proceedings of the 4th International Conference

on Microreaction Technology, IMRET 4, pp. 322–326 (5–9 March **2000**), AIChE Topical Conf. Proc., Atlanta, USA.

90 WEGENG, R. S., DROST, M. K., BRENCHLEY, D. L., *Process intensification through miniaturization of chemical and thermal systems in the 21. century,* in EHRFELD, W. (Ed.), *Microreaction Technology: 3rd International Conference on Microreaction Technology, Proc. of IMRET 3,* pp. 2–13, Springer-Verlag, Berlin (**2000**).

91 WEGENG, R. S., *Application of microreactors in space,* in Proceedings of Microreaction Technology – IMRET5: Proceedings of the 5th International Conference on Micro-reaction Technology (27–30 May **2001**), Strasbourg, France.

92 WEGENG, R. S., TEGROTENHUIS, W. E., RASSAT, S. D., BROOKS, K. P., STENKAMP, V. S., SANDERS, G. B., *Progress in microreactor technology for NASA in-situ propellant production plant on mars,* in Proceedings of Microreaction Technology – IMRET5: Proceedings of the 5th International Conference on Microreaction Technology (27–30 May **2001**), Strasbourg, France.

93 LÖWE, H., EHRFELD, W., SCHIEWE, J., *Micro-electroforming of miniaturized devices for chemical applications,* in SCHULTZE, W., OSAKA, T., DATTA, M. (Eds.), *Electrochemical Microsystem Technologies,* pp. 245–268, Taylor & Francis, London (**2002**).

94 WOLF, A., EHRFELD, W., LEHR, H., MICHEL, F., RICHTER, T., GRUBER, H., WÖRZ, O., *Mikroreaktorfertigung mittels Funkenerosion,* F & M, Feinwerktechnik, Mikrotechnik, Meßtechnik **105**, 6 (1997) 436–439.

95 RICHTER, T., EHRFELD, W., WOLF, A., GRUBER, H. P., WÖRZ, O., *Fabrication of microreactor components by electro discharge machining,* in EHRFELD, W. (Ed.), *Microreaction Technology – Proc. of the 1st International Conference on Microreaction Technology, IMRET 1,* pp. 158–168, Springer-Verlag, Berlin (**1997**).

96 WOLF, A., LEHR, H., NIENHAUS, M., MICHEL, F., GRUBER, H. P., EHRFELD, W., *Combining LIGA in EDM for the generation of complex microstructures in*

hard materials, in Proceedings of Progress in Precision Engineering and Nanotechnology, 9-IPES/UME4 Conf. '97, pp. 657–660 (26–30 May **1997**), Braunschweig, Germany.

97 KNITTER, R., GÜNTHER, E., ODEMER, C., MACIEJEWSKI, U., *Ceramic microstructures and potential applications,* Microsystem Technol. **2** (**1996**) 135–138.

98 KNITTER, R., BAUER, W., FECHLER, C., WINTER, V., RITZHAUPT-KLEISSL, H.-J., HAUSSELT, J., *Ceramics in microreaction technology: materials and processing,* in EHRFELD, W., RINARD, I. H., WEGENG, R. S. (Eds.), *Process Miniaturization: 2nd International Conference on Microreaction Technology, IMRET 2, Topical Conf. Preprints,* pp. 164–168, AIChE, New Orleans (**1998**).

99 GÖHRING, D., KNITTER, R., *Rapid manufacturing keramischer Mikroreaktoren,* Keram. Z. **53**, 6 (**2001**) 480–484.

100 BAUER, W., KNITTER, R., *Formgebung keramischer Mikrokomponenten,* Galvanotechnik **90**, 11 (**1999**) 3122–3130.

101 JENSEN, K. F., *Microchemical systems for synthesis of chemicals and information,* in Proceedings of the Japan Chemical Innovation Institute (JCII) (23 May **2001**), Tokyo, Japan.

102 HESSEL, V., LÖWE, H., STANGE, T., *Micro chemical processing at IMM – from pioneering work to customer-specific services,* Lab Chip **2** (**2002**) 14N–21N.

103 LÖWE, H., HESSEL, V., *Chemical micro process engineering – how a concept became mature for chemical industry,* in Proceedings of the South African Chemical Engineering Congress (3–5 September **2003**), Sun City, South Africa.

104 SCHOUTEN, J. C., REBROV, E., DE CROON, M. H. J. M., *Challenging prospects for microstructured reaction architectures in high-throughput catalyst screening, small scale fuel processing, and sustainable fine chemical synthesis,* in Proceedings of the Micro Chemical Plant – International Workshop, pp. L5 (25–32) (4 February **2003**), Kyoto, Japan.

105 LINDNER, D., *The µChem Lab^{TM} project: micro total analysis system R&D at Sandia National Laboratories,* Lab Chip **1** (**2001**) 15N–19N.

106 WEGENG, R. S., DROST, M. K., *Opportunities for distributed processing using micro chemical systems,* in EHRFELD, W., RINARD, I. H., WEGENG, R. S. (Eds.), *Process Miniaturization: 2nd International Conference on Microreaction Technology, IMRET 2, Topical Conf. Preprints,* pp. 3–9, AIChE, New Orleans (**1998**).

107 LÖWE, H., EHRFELD, W., *State of the art in microreaction technology: concepts, manufacturing and applications,* Electrochim. Acta **44** (**1999**) 3679–3689.

108 PERRY, R. H., GREEN, D. W., *Perry's Chemical Engineers' Handbook,* 7th ed., McGraw-Hill, New York (**1997**).

109 POPE, S. B., *Turbulent Flows,* Cambridge University Press, Cambridge (**2000**).

110 WÖRZ, O., JÄCKEL, K.-P., RICHTER, T., WOLF, A., *Microreactors – a new efficient tool for reactor development,* Chem. Eng. Technol. **24**, 2 (**2001**) 138–143.

111 WÖRZ, O., *Wozu Mikroreaktoren?* Chem. Unserer Zeit **34**, 1 (**2000**) 24–29.

112 WÖRZ, O., JÄCKEL, K.-P., RICHTER, T., WOLF, A., *Mikroreaktoren – Ein neues wirksames Werkzeug für die Reaktorentwicklung,* Chem. Ing. Tech. **72**, 5 (**2000**) 460–463.

113 AUTZE, V., GOLBIG, K., KLEEMANN, A., OBERBECK, S., *Mikroreaktoren für die chemische Synthese,* Nachr. Chem. **48** (**2000**) 683–685.

114 COMMENGE, J.-M., *Réacteurs microstructurés: hydrodynamique, thermique, transfert de matière et applications aux procédés,* PhD thesis, Institut National Polytechnique de Lorraine, Nancy (**2001**).

115 BENNETT, C. O., MYERS, J. E., *Momentum, heat and mass transfer,* McGraw Hill, New York (**1962**).

116 SCHEIDEGGER, A. E., *The physics of flow-through porous media,* 3rd ed., University of Toronto Press, Toronto (**1974**).

117 DORAISWAMY, L. K., SHARMA, M. M., *Heterogeneous reactions – analysis, examples and reactor design,* Wiley – Interscience Publications, New York (**1984**).

118 KNIGHT, J. B., VISHWANATH, A., BRODY, J. P., AUSTIN, R. H., *Hydrodynamic focussing on a silicon chip: mixing nanoliters in microseconds,* Phys. Rev. Lett. **80**, 17 (**1998**) 3863.

119 HESSEL, V., HARDT, S., LÖWE, H.,
SCHÖNFELD, F., Laminar mixing in
different interdigital micromixers – Part I:
Experimental characterization, AIChE
J. 49, 3 (2003) 566–577.

120 HARDT, S., SCHÖNFELD, F., Laminar
mixing in different interdigital micromixers – Part 2: Numerical simulations,
AIChE J. 49, 3 (2003) 578–584.

121 SCHWESINGER, N., FRANK, T., WURMUS,
H., A modular microfluic system with an
integrated micromixer, J. Micromech.
Microeng. 6 (1996) 99–102.

122 MENGEAUD, V., JOSSERAND, J., GIRAULT,
H. H., Mixing processes in a zigzag
microchannel: finite element simulations
and optical study, Anal. Chem. 74 (2002)
4279–4286.

123 STROOCK, A. D., DERTINGER, S. K. W.,
AJDARI, A., MEZIC, I., STONE, H. A.,
WHITESIDES, G. M., Chaotic mixer for
microchannels, Science 295, 1 (2002)
647–651.

124 LÖWE, H., HESSEL, V., Miniaturization as
a concept in chemical technology, in
Proceedings of the 3rd European
Conference of Chemical Engineering,
ECCE, published on CD
(26–28 June 2001), DECHEMA,
Nürnberg, Germany.

125 GAWEDZKI, K., Turbulence under a
magnifying glass, NATO ASI Ser. B Phys.
364 (1997) 123.

126 ROESSLER, A., RYS, P., Selektivität
mischungsmaskierter Reaktionen:
Wenn die Rührgeschwindigkeit die
Produktverteilung bestimmt, Chem.
Unserer Zeit 35, 5 (2001) 314–322.

127 BAYER, T., HEINICHEN, H., LEIPPRAND,
I., Using micro heat exchangers as
diagnostic tool for the process optimization
of a gas phase reaction, in Proceedings of
the VDE World Microtechnologies
Congress, MICRO.tec 2000,
pp. 493–497 (25–27 September 2000),
VDE Verlag, Berlin.

128 STEINFELDT, N., BUYEVSKAYA, O. V.,
WOLF, D., BAERNS, M., Comparative
studies of the oxidative dehydrogenation of
propane in micro-channels reactor module
and fixed-bed reactor, in SPIVEY, J. J.,
IGLESIA, E., FLEISCH, T. H. (Eds.), Stud.
Surf. Sci. Catal., pp. 185–190, Elsevier
Science, Amsterdam (2001).

129 STEINFELDT, N., DROPKA, N., WOLF, D.,
Oxidative dehydrogenation of propane in a
micro-channel reactor-kinetic measurements, modeling and reactor simulation,
in Proceedings of Microreaction
Technology – IMRET 5: Proceedings of
the 5th International Conference on
Microreaction Technology (27–30 May
2001), Strasbourg, France.

130 GAD-EL-HAK, M., The fluid mechanics of
microdevices, J. Fluids Eng. 121 (1999)
5–33.

131 PFAHLER, J., HARLEY, J., BAU, H., ZEMEL,
J., Liquid transport in micron and
submicron channels, Sens. Actuators A
21–23 (1990) 431–434.

132 MALA, G. M., LI, D., Flow characteristics
of water in microtubes, Int. J. Heat Fluid
Flow 20 (1999) 142–148.

133 QU, W., MALA, G. M., LI, D., Pressure-
driven water flows in trapezoidal silicon
microchannels, Int. J. Heat Mass Transfer
43 (2000) 353–364.

134 LÖWE, H., HESSEL, V., RUSSOW, K.,
Progress in chemical microreaction
technology, in Proceedings of the 6th
World Congress of Chemical
Engineers, paper No. 1373, published
on CD (23–27 September 2001),
Melbourne, Australia.

135 FICHTNER, M., BENZINGER, W., HASS-
SANTO, K., WUNSCH, R., SCHUBERT, K.,
Functional coatings for microstructure
reactors and heat exchangers, in EHRFELD,
W. (Ed.), Microreaction Technology: 3rd
International Conference on Microreaction
Technology, Proc. of IMRET 3,
pp. 90–101, Springer-Verlag, Berlin
(2000).

136 WILLE, C., AUTZE, V., KIM, H., NICKEL,
U., OBERBECK, S., SCHWALBE, T.,
UNVERDORBEN, L., Progress in transferring microreactors from lab into
production – an example in the field of
pigments technology, in Proceedings of
the 6th International Conference on
Microreaction Technology, IMRET 6,
pp. 7–17 (11–14 March 2002), AIChE
Pub. No. 164, New Orleans.

137 FELCHT, U.-H., The future shape of process
industries, Chem. Eng. Technol. 25, 4
(2002) 345–355.

138 Faktor Zeit auf dem Prüfstand, Europa
Chemie, 21 July (1994) 1.

139 BENSON, R. S., PONTON, J. W., *Process miniaturization – a route to total environmental acceptability?* Trans. Ind. Chem. Eng. **71**, A2 (**1993**) 160–168.

140 HARDT, S., EHRFELD, W., HESSEL, V., VANDEN BUSSCHE, K. M., *Strategies for size reduction of microreactors by heat transfer enhancement effects,* in Proceedings of the 4th International Conference on Microreaction Technology, IMRET 4, pp. 432–440 (5–9 March **2000**), AIChE Topical Conf. Proc., Atlanta, USA.

141 HARDT, S., EHRFELD, W., HESSEL, V., VANDEN BUSSCHE, K. M., *Strategies for size reduction of microreactors by heat transfer enhancement effects,* Chem. Eng. Commun. **190**, 4 (**2003**) 540–559.

142 SCHUBERT, K., BRANDNER, J., *New designs for microstructured devices for thermal and chemical process engineering,* in Proceedings of the Micro Chemical Plant – International Workshop, pp. L7 (42–53) (4 February **2003**), Kyoto, Japan.

143 BRANDNER, J., FICHTNER, M., SCHYGULLA, U., SCHUBERT, K., *Improving the efficiency of micro heat exchangers and reactors,* in Proceedings of the 4th International Conference on Microreaction Technology, IMRET 4, pp. 244–249 (5–9 March **2000**), AIChE Topical Conf. Proc., Atlanta, USA.

144 STOJANOVIC, I., *Sicherer Umgang mit Phosgen,* Chemie-Technik **21**, 2 (**1992**) 64–68.

145 *Small but environmentally formed,* Chem. Engineer (11 March **1993**) s6–s7.

146 LOHF, A., LÖWE, H., HESSEL, V., EHRFELD, W., *A standardized modular microreactor system,* in Proceedings of the 4th International Conference on Microreaction Technology, IMRET 4, pp. 441–451 (5–9 March **2000**), AIChE Topical Conf. Proc., Atlanta, USA.

147 HAAS-SANTO, K., GÖRKE, O., SCHUBERT, K., FIEDLER, J., FUNKE, H., *A microstructure reactor system for the controlled oxidation of hydrogen for possible application in space,* in MATLOSZ, M., EHRFELD, W., BASELT, J. P. (Eds.), *Microreaction Technology – IMRET 5: Proc. of the 5th International Conference on Microreaction Technology,* pp. 313–320, Springer-Verlag, Berlin (**2001**).

148 FREITAG, A., DIETRICH, T. R., *Glass as a material for microreaction technology,* in Proceedings of the 4th International Conference on Microreaction Technology, IMRET 4, pp. 48–54 (5–9 March **2000**), AIChE Topical Conf. Proc., Atlanta, USA.

149 SHINNAR, R., RINARD, I., *Appropriate scales for μ-RT and hybrid/μ-RT systems,* in Proceedings of the Microreaction Technology – IMRET5: Proceedings of 5th International Conference on Microreaction Technology (27–30 May **2001**), Strasbourg, France.

150 KRAUT, M., NAGEL, A., SCHUBERT, K., *Oxidation of ethanol by hydrogen peroxide,* in Proceedings of the 6th International Conference on Microreaction Technology, IMRET 6, pp. 352–356 (11–14 March **2002**), AIChE Pub. No. 164, New Orleans.

151 WILES, C., WATTS, P., HASWELL, S. J., POMBO-VILLAR, E., *1,4-Addition of enolates to α,β-unsaturated ketones within a micro reactor,* Lab Chip **2** (**2002**) 62–64.

152 ANTES, J., TUERCKE, T., MARIOTH, E., SCHMID, K., KRAUSE, H., LOEBBECKE, S., *Use of microreactors for nitration processes,* in Proceedings of the 4th International Conference on Microreaction Technology, IMRET 4, pp. 194–200 (5–9 March **2000**), AIChE Topical Conf. Proc., Atlanta, USA.

153 WÖRZ, O., JÄCKEL, K. P., *Winzlinge mit großer Zukunft – Mikroreaktoren für die Chemie,* Chem. Techn. **26**, 131 (**1997**) 130–134.

154 WÖRZ, O., JÄCKEL, K. P., RICHTER, T., WOLF, A., *Microreactors, a new efficient tool for optimum reactor design,* in EHRFELD, W., RINARD, I. H., WEGENG, R. S. (Eds.), *Process Miniaturization: 2nd International Conference on Microreaction Technology, IMRET 2, Topical Conf. Preprints,* pp. 183–185, AIChE, New Orleans (**1998**).

155 SKELTON, V., GREENWAY, G. M., HASWELL, S. J., STYRING, P., MORGAN, D. O., *Micro-reactor synthesis: synthesis of cyanobiphenyls using a modified Suzuki coupling of an aryl halide and aryl boronic acid,* in EHRFELD, W. (Ed.), *Microreaction Technology: 3rd International Conference on Microreaction Technology, Proc. of*

IMRET 3, pp. 235–242, Springer-Verlag, Berlin (2000).

156 GARCIA-EGIDO, E., WONG, S. Y. F., *A Hantzsch synthesis of 2-aminothiazoles performed in a microreactor system*, in RAMSEY, J. M., VAN DEN BERG, A. (Eds.), *Micro Total Analysis Systems*, pp. 517–518, Kluwer Academic Publishers, Dordrecht (2001).

157 GARCIA-EGIDO, E., WONG, S. Y. F., WARRINGTON, B. H., *A Hantzsch synthesis of 2-aminothiazoles performed in a heated microreactor system*, Lab Chip 2 (2002) 31–33.

158 WATTS, P., WILES, C., HASWELL, S. J., POMBO-VILLAR, E., *Solution phase synthesis of β-peptides using micro reactors*, Tetrahedron 58, 27 (2002) 5427–5439.

159 KESTENBAUM, H., LANGE DE OLIVERA, A., SCHMIDT, W., SCHÜTH, F., EHRFELD, W., GEBAUER, K., LÖWE, H., RICHTER, T., *Silver-catalyzed oxidation of ethylene to ethylene oxide in a microreaction system*, Ind. Eng. Chem. Res. 41, 4 (2000) 710–719.

160 HASWELL, S. J., O'SULLIVAN, B., STYRING, P., *Kumada–Corriu reactions in a pressure-driven microflow reactor*, Lab Chip 1 (2001) 164–166.

161 HEINICHEN, H., *Kleiner Maßstab – große Wirkung: Mikrowärmeaustauscher zur Verfahrensoptimierung*, Chemie-Technik 30, 3 (2001) 89–91.

162 KESTENBAUM, H., LANGE DE OLIVERA, A., SCHMIDT, W., SCHÜTH, H., EHRFELD, W., GEBAUER, K., LÖWE, H., RICHTER, T., *Synthesis of ethylene oxide in a catalytic microreactor system*, Stud. Surf. Sci. Catal. 130 (2000) 2741–2746.

163 KESTENBAUM, H., LANGE DE OLIVEIRA, A., SCHMIDT, W., SCHÜTH, F., GEBAUER, K., LÖWE, H., RICHTER, T., *Synthesis of ethylene oxide in a catalytic microreacton system*, in EHRFELD, W. (Ed.), *Micro- reaction Technology: 3rd International Conference on Microreaction Technology, Proc. of IMRET 3*, pp. 207–212, Springer-Verlag, Berlin (2000).

164 JÄHNISCH, K., BAERNS, M., HESSEL, V., EHRFELD, W., HAVERKAMP, W., LÖWE, H., WILLE, C., GUBER, A., *Direct fluorina- tion of toluene using elemental fluorine in gas/liquid microreactors*, J. Fluorine Chem. 105, 1 (2000) 117–128.

165 SKELTON, V., GREENWAY, G. M., HASWELL, S. J., STYRING, P., MORGAN, D. O., WARRINGTON, B. H., WONG, S., *Micro reaction technology: synthetic chemical optimization methodology of Wittig synthesis enabling a semi- automated micro reactor for combinatorial screening*, in Proceedings of the 4th International Conference on Microreaction Technology, IMRET 4, pp. 78–88 (5–9 March 2000), AIChE Topical Conf. Proc., Atlanta, USA.

166 LÖBBECKE, S., ANTES, J., TÜRCKE, T., MARIOTH, E., SCHMID, K., KRAUSE, H., *The potential of microreactors for the synthesis of energetic materials*, in Proceedings of the 31st Int. Annu. Conf. ICT, Energetic Materials – Analysis, Diagnostics and Testing (27–30 June 2000), Karlsruhe, Germany.

167 HESSEL, V., EHRFELD, W., GOLBIG, K., HAVERKAMP, V., LÖWE, H., STORZ, M., WILLE, C., GUBER, A., JÄHNISCH, K., BAERNS, M., *Gas/liquid microreactors for direct fluorination of aromatic compounds using elemental fluorine*, in EHRFELD, W. (Ed.), *Microreaction Technology: 3rd International Conference on Microreaction Technology, Proc. of IMRET 3*, pp. 526–540, Springer-Verlag, Berlin (2000).

168 TAGHAVI-MOGHADAM, S., KLEEMANN, A., OVERBECK, S., *Implications of microreactors on chemical synthesis*, in Proceedings of the VDE World Microtechnologies Congress, MICRO.tec 2000, pp. 489–491 (25–27 September 2000), VDE Verlag, Berlin, EXPO Hannover.

169 DE BELLEFON, C., TANCHOUX, N., CARAVIEILHES, S., GRENOUILLET, P., HESSEL, V., *Microreactors for dynamic high throughput screening of fluid–liquid molecular catalysis*, Angew. Chem. 112, 19 (2000) 3584–3587.

170 DE BELLEFON, C., PESTRE, N., LAMOUILLE, T., GRENOUILLET, P., *High-throughput kinetic investigations of asymmetric hydrogenations with micro- devices*, Adv. Synth. Catal. 345, 1+2 (2003) 190–193.

171 BRIVIO, M., OOSTERBROEK, R. E., VERBOOM, W., GOEDBLOED, M. H., VAN DEN BERG, A., REINHOUDT, D. N.,

Surface effects in the esterification of 9-pyrenebutyric acid within a glass micro reactor, Chem. Commun. (**2003**) 1924–1925.

172 GREENWAY, G. M., HASWELL, S. J., MORGAN, D. O., SKELTON, V., STYRING, P., *The use of a novel microreactor for high troughput continuous flow organic synthesis*, Sens. Actuators B: Chem. **63**, 3 (**2000**) 153–158.

173 RINGELSBACHER, G., *Gefährdungpotential verringern*, Chem. Ind. Rheinland-Pfalz (**1995**) 15–18.

174 ENDTER, F., *Die technische Synthese von Cyanwasserstoff aus Methan und Ammoniak*, Chem. Ing. Tech. **30**, 5 (**1958**) 305–310.

175 HESSEL, V., EHRFELD, W., GOLBIG, K., HOFMANN, C., JUNGWIRTH, S., LÖWE, H., RICHTER, T., STORZ, M., WOLF, A., WÖRZ, O., BREYSSE, J., *High temperature HCN generation in an integrated Micro-reaction system*, in EHRFELD, W. (Ed.), *Microreaction Technology: 3rd International Conference on Microreaction Technology, Proc. of IMRET 3*, pp. 152–164, Springer-Verlag, Berlin (**2000**).

176 ANDRUSSOW, L., *Über die katalytische Oxidation von Ammoniak-Methan-Gemischen zu Blausäure*, Angew. Chem. **48**, 37 (**1935**) 593–604.

177 ANDRUSSOW, L., *Blausäuresynthese und die schnell verlaufenden katalytischen Prozesse in strömenden Gasen*, Chem. Ing. Tech. **27**, 8/9 (**1935**) 469–472.

178 KURSAWE, A., HÖNICKE, D., *Comparison of Ag/Al- and Ag/α-Al$_2$O$_3$ catalytic surfaces for the partial oxidation of ethene in microchannel reactors*, in MATLOSZ, M., EHRFELD, W., BASELT, J. P. (Eds.), *Microreaction Technology – IMRET 5: Proc. of the 5th International Conference on Microreaction Technology*, pp. 240–251, Springer-Verlag, Berlin (**2001**).

179 KAH, S., HÖNICKE, D., *Selective oxidation of 1-butene to maleic anhydride – comparison of the performance between microchannel reactors and fixed bed reactor*, in MATLOSZ, M., EHRFELD, W., BASELT, J. P. (Eds.), *Microreaction Technology – IMRET 5: Proc. of the 5th International Conference on Microreaction Technology*, pp. 397–407, Springer-Verlag, Berlin (**2001**).

180 HAGENDORF, U., JANICKE, M., SCHÜTH, F., SCHUBERT, K., FICHTNER, M., *A Pt/Al$_2$O$_3$ coated microstructured reactor/heat exchanger for the controlled H$_2$/O$_2$-reaction in the explosion regime*, in EHRFELD, W., RINARD, I. H., WEGENG, R. S. (Eds.), *Process Miniaturization: 2nd International Conference on Microreaction Technology, IMRET 2, Topical Conf. Preprints*, pp. 81–87, AIChE, New Orleans (**1998**).

181 GÖRKE, O., PFEIFER, P., SCHUBERT, K., *Determination of kinetic data in the isothermal microstructure reactor based on the example of catalyzed oxidation of hydrogen*, in Proceedings of the 6th International Conference on Microreaction Technology, IMRET 6, pp. 262–274 (11–14 March 2002), AIChE Pub. No. 164, New Orleans.

182 WOOTTON, R. C. R., FORTT, R., DE MELLO, A. J., *A microfabricated nano-reactor for safe, continuous generation and use of singlet oxygen*, Org. Proc. Res. Dev. **60** (**2002**) 187–189.

183 WEBER, M., TANGER, U., KLEINLOH, W., *Method and device for production of phenol and acetone by means of acid-catalyzed, homogeneous decoposition of cumol-hydroperoxid*, WO 01/30732, Phenol-chemie GmbH, Priority: 22.10.99.

184 CHATTOPADHYAY, S., VESER, G., *Detailed simulations of catalytic ond non-catalytic ignition during H$_2$-oxidation in a micro-channel reactor: isothermal case*, in Proceedings of the ChemConn-2001, pp. 1–6 (December 2001), Chennai, India.

185 VESER, G., FRIEDRICH, G., FREYGANG, M., ZENGERLE, R., *A simple and flexible micro reactor for investigations of heterogeneous catalytic gas reactions*, in FROMENT, G. F., WAUGH, K. C. (Eds.), *Reaction Kinetics and the Development of Catalytic Processes*, pp. 237–245, Elsevier Science, Amsterdam (**1999**).

186 VESER, G., FRIEDRICH, G., FREYGANG, M., ZENGERLE, R., *A modular microreactor design for high-temperature catalytic oxidation reactions*, in EHRFELD, W. (Ed.), *Microreaction Technology: 3rd International Conference on Microreaction Technology, Proc. of IMRET 3*, pp. 674–686, Springer-Verlag, Berlin (**2000**).

187 SRINIVASAN, R., HSING, I.-M.,
BERGER, P. E., JENSEN, K. F.,
FIREBAUGH, S. L., SCHMIDT, M. A.,
HAROLD, M. P., LEROU, J. J., RYLEY, J. F.,
*Micromachined reactors for catalytic
partial oxidation reactions*, AIChE J. **43**,
11 (**1997**) 3059–3069.

188 JENSEN, K. F., HSING, I.-M.,
SRINIVASAN, R., SCHMIDT, M. A.,
HAROLD, M. P., LEROU, J. J., RYLEY, J. F.,
*Reaction engineering for microreactor
systems*, in EHRFELD, W. (Ed.),
*Microreaction Technology – Proc. of the 1st
International Conference on Microreaction
Technology, IMRET 1*, pp. 2–9, Springer-
Verlag, Berlin (**1997**).

189 JENSEN, K. F., FIREBAUGH, S. L., FRANZ,
A. J., QUIRAM, D., SRINIVASAN, R.,
SCHMIDT, M. A., *Integrated gas phase
microreactors*, in HARRISON, J., VAN DEN
BERG, A. (Eds.), *Micro Total Analysis
Systems*, pp. 463–468, Kluwer Academic
Publishers, Dordrecht (**1998**).

190 QUIRAM, D. J., HSING, I.-M., FRANZ,
A. J., SRINIVASAN, R., JENSEN, K. F.,
SCHMIDT, M. A., *Characterization of
microchemical systems using simulations*,
in EHRFELD, W., RINARD, I. H.,
WEGENG, R. S. (Eds.), *Process
Miniaturization: 2nd International
Conference on Microreaction Technology,
IMRET 2, Topical Conf. Preprints*,
pp. 205–211, AIChE, New Orleans
(**1998**).

191 FRANZ, A. J., QUIRAM, D. J.,
SRINIVASAN, R., HSING, I.-M.,
FIREBAUGH, S. L., JENSEN, K. F.,
SCHMIDT, M. A., *New operating regimes
and applications feasible with
microreactors*, in EHRFELD, W., RINARD,
I. H., WEGENG, R. S. (Eds.), *Process
Miniaturization: 2nd International
Conference on Microreaction Technology,
IMRET 2, Topical Conf. Preprints*,
pp. 33–38, AIChE, New Orleans (**1998**).

192 FRANZ, A. J., AJMERA, S. K., FIREBAUGH,
S. L., JENSEN, K. F., SCHMIDT, M. A.,
*Expansion of microreactor capabilities
trough improved thermal management and
catalyst deposition*, in EHRFELD, W. (Ed.),
*Microreaction Technology: 3rd
International Conference on Microreaction
Technology, Proc. of IMRET 3*, pp. 197–
206, Springer-Verlag, Berlin (**2000**).

193 DE MAS, N., JACKMAN, R. J., SCHMIDT,
M. A., JENSEN, K. F., *Microchemical
systems for direct fluorination of aromatics*,
in MATLOSZ, M., EHRFELD, W., BASELT,
J. P. (Eds.), *Microreaction Technology –
IMRET 5: Proc. of the 5th International
Conference on Microreaction Technology*,
pp. 60–67, Springer-Verlag, Berlin
(**2001**).

194 JÄHNISCH, K., BAERNS, M., HESSEL, V.,
HAVERKAMP, V., LÖWE, H., WILLE, C.,
*Selective reactions in microreactors –
fluorination of toluene using elemental
fluorine in a falling film microreactor*, in
Proceedings of the 37th ESF/EUCHEM
Conference on Stereochemistry (13–19
April 2002), Bürgenstock, Switzerland.

195 JÄHNISCH, K., EHRICH, H., LINKE, D.,
BAERNS, M., HESSEL, V., MORGEN-
SCHWEIS, K., *Selective gas/liquid-reactions
in microreactors*, in Proceedings of the
Inten. Conference on Process
Intensification for the Chemical
Industry (13–15 October 2002),
Maastricht, The Netherlands.

196 DE MAS, N., GÜNTHER, A., SCHMIDT,
M. A., JENSEN, K. F., *Microfabricated
multiphase reactors for the selective direct
fluorination of aromatics*, Ind. Eng.
Chem. Res. **42**, 4 (**2003**) 698–710.

197 DE MAS, N., *Heat effects in a microreactor
for direct fluorination of aromatics*, in
Proceedings of the 6th International
Conference on Microreaction
Technology, IMRET 6, pp. 184–185
(11–14 March 2002), AIChE Pub.
No. 164, New Orleans.

198 HERWIG, H., *Flow and heat transfer in
micro systems: is everything different or just
smaller?* Z. Angew. Math. Mech. **82**
(**2002**) 579–586.

199 http://www.fluoros.co.uk/f2chemicals,
Fluorine: The beast tamed?, 26 March
1999.

200 THÜRING, P., *Chemie, Medien, Bevöl-
kerung und Behörden, ein gesellschafts-
politischer Regelkreis und seine Funktions-
weise beim Störfall*, Chimia **49**, 11 (**1995**)
425–429.

201 OROSKAR, A. R., VANDENBUSSCHE, K.,
ABDO, S. F., *Intensification in microstructured unit operations: performance
comparison between mega and microscale*,
in MATLOSZ, M., EHRFELD, W., BASELT, J.

P. (Eds.), *Microreaction Technology – IMRET 5: Proc. of the 5th International Conference on Microreaction Technology*, pp. 153–163, Springer-Verlag, Berlin (2001).

202 HALBRITTER, A., KLEMM, W., LÖWE, H., ONDRUSCHKA, B., SCHOLZ, P., SCHUBERT, K., *Experimental determination of heat transfer coefficients in micro heat exchangers*, in Proceedings of the 6th International Conference on Microreaction Technology, IMRET 6, pp. 241–246 (11–14 March 2002), AIChE Pub. No. 164, New Orleans.

203 ONDRUSCHKA, B., SCHOLZ, P., GORGES, R., KLEMM, W., SCHUBERT, K., HALBRITTER, A., LÖWE, H., *Mikrowärme-übertrager im Chemisch-technischen Praktikum*, Chem. Ing. Tech. **74**, 11 (2002) 1577–1582.

204 DONNER, S., *Tausend Kanäle für eine Reaktion*, Chem. Rundschau, 11 February (2003).

205 *Bulk chemicals by the drop*, The Economist, 19 June 2003.

206 FREEMANTLE, M., *'Numbering up' small reactors*, Chem. Eng. News **81**, 24 (2003) 36–37.

207 BODDERAS, E., *Chemieküche für Zwerge*, VDI Nachr., 11 April (2003) 26.

208 GEIPEL-KERN, A., *Vor dem Sprung in die Produktion – Trendbeitrag Mikroreaktions-technik*, Chem. Produktion (2002) 28–30.

209 ASCHENBRENNER, N., *Die Fabrik auf dem Chip*, Spektrum der Wissenschaft October (2002) 80–82.

210 LANGE, E., *Chemiefabrik im Schuhkarton*, Handelsblatt **105**, 5 June (2002) 6.

211 KIESEWALTER, S., *Kleine Reaktoren mit grosser Zukunft*, Chem. Rundschau, 5 April (2002).

212 LYON, P., *Les premiers pas des micro-réacteurs*, Industries et Techniques **830** (2001) 80–83.

213 RICHTER, T., LANGER, O.-U., *Tagungsbericht: 3rd International Conference on Microreaction Technology*, Chem. Ing. Tech. **72** (2000) 142–144.

214 EUL, U., RICHTER, T., *Small is useful – IMRET 3, 3rd International Conference on Microreaction Technology in Frank-furt/M.* GIT **43**, 6 (2000) 588–589.

215 *Gezähmte Chemie im Mikroreaktor*, VDI Nachr., 2 June (2000) 13.

216 *Herrscher über die Temperatur*, Chemie Technik **28**, 2 (1999) 64.

217 *Chemical reduction*, The European Chemist **1** (1999) 17–18.

218 SCHAMARI, U., *Chemie im Kleinen*, VDI Nachr. 19 (1999) 27.

219 *The world in 1999*, The Economist (1999) 80.

220 RIBERIO, F., *Process miniaturization – second international conference*, CATTECH **2**, 2 (1996) 124–126.

221 RICHTER, T., *Process Miniaturization: 2nd International Conference on Microreaction Technology*, Chem. Ing. Tech. **70**, Tagungsberichte (1998) 1355–1357.

222 EHRFELD, W., HESSEL, V., STANGE, T., *Sicherer, effizienter, flexibler- Bedeutung der Mikrotechnik für die Verfahrens-technik*, Verfahrenstechnik **32**, 12 (1998) 14–16.

223 *Das Chemielabor im Mikrochip*, Blick durch die Wirtschaft, 1 December (1997).

224 LANGER, O.-U., RICHTER, T., *1st Inter-national Conference on Microreaction Technologie, Tagungsbericht*, Chem. Ing.Tech. **69** (1997) 1026–1027.

225 *Daumengroßes Labor aus Aluminiumfolie*, Blick durch die Wirtschaft, 3 June (1997).

226 CHOPEY, N. P., ONDREY, G., PARKINSON, G., *Microreactors find new niches*, Achema Daily, 9 June (1997) 1–4.

227 JOPP, K., *Teufelszeug im Griff*, Wirtschaftswoche, 24 April (1997) 102.

228 SHANLEY, A., *Microreactors find new niches*, Chem. Eng. **3** (1997) 30–33.

229 VENNEN, H., *Die Natur der Chemie*, FUTURE (Hoechst Magazin) (1996).

230 *Chemie in neuen Dimensionen*, Chemie Produktion 8 (1996) 22–24.

231 PETERS, R.-H., *Wie Brötchen*, Wirt-schaftswoche, 1 June (1995) 94–105.

232 JOPP, K., *Chemiker sind der Zelle auf der Spur*, Handelsblatt, 8 November (2000).

233 *Ultra-Hochdurchfluss-Tests in der Arzneimittelforschung unverzichtbar*, Frankfurter Allgemeine Zeitung, 26 June (2000).

234 VON DER WEIDEN, S., *Chemische Technik findet im Fingerhut statt*, Handelsblatt, 3 November (1999) B 21.

235 SCHUBERT, K. M., *Winzige Reaktoren mit Höchstleistung*, Handelsblatt, 25 November (1998) 50.

236 *Mikroreaktoren sind so klein wie ein Fingerhut,* Handelsblatt, 6 May (**1998**) 25.

237 DETTMANN, J. B., *Chemiefabrik in der Größe eines Chips,* Handelsblatt, 15 May (**1996**).

238 VENNEN, H., *Sichere Chemie in Mikroreaktoren,* Frankfurter Allgemeine Zeitung, 20 December (**1995**).

239 SCHLAAK, H. F., *Fabrication technologies and economic aspects for components in microtechnology,* in Proceedings of the VDE World Microtechnologies Congress, MICRO.tec 2000, pp. 649–653 (25–27 September **2000**), VDE Verlag, Berlin, EXPO Hannover.

240 SALOMON, P., *User needs in design tools for microsystems and microreactors – a NEXUS survey,* in Proceedings of the VDE World Microtechnologies Congress, MICRO.tec 2000, pp. 639–643 (25–27 September **2000**), VDE Verlag, Berlin, EXPO Hannover.

241 KIESEWALTER, S., RUSSOW, K. M., STANGE, T., BALSALOBRE, C., BOULON, P., PROVENCE, M., *PAMIR – a market survey on Potential and Applications of MIcroReaction technology,* in Proceedings of the 6th International Conference on Microreaction Technology, IMRET 6, pp. 135–138 (11–14 March 2002), AIChE Pub. No. 164, New Orleans.

242 STAUDT, E., KRAUSE, M., *Nach einer euphorischen Phase dominiert jetzt Realismus,* Handelsblatt, 24 November (**1998**) 2.

243 STEG, H., *Wachsender weltweiter Wettbewerb um Winzlinge,* Handelsblatt, 24 November (**1998**), 2.

244 WICHT, H., ELOY, J.-C., ROBINET, C., LE FLOCH, C., *From research to industry: start-up companies in microsystems,* in Proceedings of the VDE World Microtechnologies Congress, MICRO.tec 2000, pp. 639–643 (25–27 September **2000**), VDE Verlag, Berlin, EXPO Hannover.

245 ZECHBAUER, U., *Entweder bekommen alle einen Auftrag oder eben keiner,* Spektrum der Wissenschaft (**2002**) 84–85.

246 NIEDER, O., *Branche mit Mikroreaktor ködern,* Mainzer-Rhein-Zeitung, 7 February (**2001**).

247 FITZGERALD, S. P., WEGENG, R. S., TONKOVICH, A. L. Y., WANG, Y., FREEMAN, H. D., MARCO, J. L., ROBERTS, G. L., VANDERWIEL, D. P., *A compact steam reforming reactor for use in an automotive fuel processor,* in Proceedings of the 4th International Conference on Microreaction Technology, IMRET 4, pp. 358–363 (5–9 March **2000**), AIChE Topical Conf. Proc., Atlanta, USA.

248 PFEIFER, P., FICHTNER, M., SCHUBERT, K., LIAUW, M. A., EMIG, G., *Microstructured catalysts for methanol-steam reforming,* in EHRFELD, W. (Ed.), *Microreaction Technology: 3rd International Conference on Microreaction Technology, Proc. of IMRET 3,* pp. 372–382, Springer-Verlag, Berlin (**2000**).

249 WHYATT, G. A., TEGROTENHUIS, W. E., WEGENG, R. S., PEDERSON, L. R., *Demonstration of energy efficient steam reforming in microchannels for automotive fuel processing,* in Proceedings of the Microreaction Technology – IMRET5: Proceedings of the 5th International Conference on Microreaction Technology, pp. 303–312 (27–30 May 2001), Strasbourg, France.

250 TONKOVICH, A. L. Y., CALL, C. J., JIMENEZ, D. M., WEGENG, R. S., DROST, M. K., *Microchannel heat exchangers for chemical reactors,* in Proceedings of the AIChE Symposium Heat Transfer, pp. 119–125 (September **1996**), AIChE No. 310, Houston, TX.

251 TONKOVICH, A. L. Y., ZILKA, J. L., POWELL, M. R., CALL, C. J., *The catalytic partial oxidation of methane in a microchannel chemical reactor,* in EHRFELD, W., RINARD, I. H., WEGENG, R. S. (Eds.), *Process Miniaturization: 2nd International Conference on Microreaction Technology, IMRET 2, Topical Conf. Preprints,* pp. 45–53, AIChE, New Orleans (**1998**).

252 TONKOVICH, A. L. Y., JIMENEZ, D. M., ZILKA, J. L., LAMONT, M. J., WANG, J., WEGENG, R. S., *Microchannel chemical reactors for fuel processing,* in EHRFELD, W., RINARD, I. H., WEGENG, R. S. (Eds.), *Process Miniaturization: 2nd International Conference on Microreaction Technology, IMRET 2, Topical Conf. Preprints,* pp. 186–195, AIChE, New Orleans (**1998**).

253 TONKOVICH, A. L., ZILKA, J. L., LaMONT, M. J., WANG, Y., WEGENG, R., *Microchannel chemical reactor for fuel processing applications – I. Water gas shift reactor,* Chem. Eng. Sci. **54 (1999)** 2947–2951.

254 TONKOVICH, A. L., FITZGERALD, S. P., ZILKA, J. L., LaMONT, M. J., WANG, Y., VANDERWIEL, D. P., WEGENG, R., *Microchannel chemical reactor for fuel processing applications. – II. Compact fuel vaporization,* in EHRFELD, W. (Ed.), *Microreaction Technology: 3rd International Conference on Microreaction Technology, Proc. of IMRET 3,* pp. 364–371, Springer-Verlag, Berlin (**2000**).

255 DAYMO, E. A., VANDERWIEL, D. P., FITZGERALD, S. P., WANG, Y., ROZMIAREK, R. T., LaMONT, M. J., TONKOVICH, A. L. Y., *Microchannel fuel processing for man portable power,* in Proceedings of the 4th International Conference on Microreaction Technology, IMRET 4, pp. 364–369 (5–9 March **2000**), AIChE Topical Conf. Proc., Atlanta, USA.

256 ALLEN, W. L., IRVING, P. M., THOMPSON, W. J., *Microreactor system for hydrogen generation and oxidative coupling of methane,* in Proceedings of the 4th International Conference on Microreaction Technology, IMRET 4, pp. 351–357 (5–9 March **2000**), AIChE Topical Conf. Proc., Atlanta, USA.

257 IRVING, P. M., LLOYD ALLEN, W., HEALEY, T., THOMSON, W. J., *Catalytic micro-reactor systems for hydrogen generation,* in MATLOSZ, M., EHRFELD, W., BASELT, J. P. (Eds.), *Microreaction Technology – IMRET 5: Proc. of the 5th International Conference on Microreaction Technology,* pp. 286–294, Springer-Verlag, Berlin (**2001**).

258 KARNIK, S. V., HATALIS, M. K., KOTHARE, M. V., *Palladium based micro-membrane for water gas shift reaction and hydrogen gas separation,* in MATLOSZ, M., EHRFELD, W., BASELT, J. P. (Eds.), *Microreaction Technology – IMRET 5: Proc. of the 5th International Conference on Microreaction Technology,* pp. 295–302, Springer-Verlag, Berlin (**2001**).

259 DELSMAN, E., REBROV, J., DE CROON, M. H. J. M., SCHOUTEN, J., KRAMER, G. J., COMINOS, V., RICHTER, T., VEENSTRA, T. T., VAN DEN BERG, A., COBDEN, P., DE BRUIJN, F. A., FERRET, C., D'ORTONA, U., FALK, L., *MiRTH-e: Micro-reactor technology for hydrogen and electricity,* in MATLOSZ, M., EHRFELD, W., BASELT, J. P. (Eds.), *Microreaction Technology – IMRET 5: Proc. of the 5th International Conference on Microreaction Technology,* pp. 368–375, Springer-Verlag, Berlin (**2001**).

260 ZILKA-MARCO, J., TONKOVICH, A. L. Y., LaMONT, M. J., FITZGERALD, S. P., VANDERWIEL, D. P., WEGENG, R. S., *Compact microchannel fuel vaporizer for automotive applications,* in Proceedings of the 4th International Conference on Microreaction Technology, IMRET 4, pp. 301–307 (5–9 March **2000**), AIChE Topical Conf. Proc., Atlanta, USA.

261 TURNER, C., SHAW, J., MILLER, B., BAINS, V., *Vapour stripping using a micro-contactor,* in Proceedings of the 4th International Conference on Microreaction Technology, IMRET 4, pp. 106–113 (5–9 March **2000**), AIChE Topical Conf. Proc., Atlanta, USA.

262 JONES, E., HOLLADAY, J., PERRY, S., ORTH, R., ROZMIAREL, B., HU, J., PHELPS, M., GUZMAN, C., *Sub-watt power using an integrated fuel processor and fuel cell,* in MATLOSZ, M., EHRFELD, W., BASELT, J. P. (Eds.), *Microreaction Technology – IMRET 5: Proc. of the 5th International Conference on Microreaction Technology,* pp. 277–285, Springer-Verlag, Berlin (**2001**).

263 MATSON, D. W., MARTIN, P. M., TONKOVICH, A. L. Y., ROBERTS, G. L., *Fabrication of a stainless steel microchannel microcombustor using a lamination process,* in Proceedings of the SPIE Conference on Micromachined Devices and Components IV, pp. 386–392 (September **1998**), SPIE, Santa Clara, CA.

264 DROST, K., FRIEDRICH, M., *A micro-technology-based chemical heat pump for portable and distributed space conditioning applications,* in EHRFELD, W., RINARD, I. H., WEGENG, R. S. (Eds.), *Process Miniaturization: 2nd International Conference on Microreaction Technology, IMRET 2, Topical Conf. Preprints,* pp. 318–322, AIChE, New Orleans (**1998**).

265 VanderWiel, D. P., Zilka-Marco, J. L., Wang, Y., Tonkovich, A. Y., Wegeng, R. S., *Carbon dioxide conversions in microreactors*, in Proceedings of the 4th International Conference on Microreaction Technology, IMRET 4, pp. 187–193 (5–9 March 2000), AIChE Topical Conf. Proc., Atlanta, USA.

266 TeGrotenhuis, W. E., Wegeng, R. S., VanderWiel, D. P., Whyatt, G. A., Viswanathan, V. V., Schielke, K. P., Sanders, G. B., Peters, T. A., *Microreactor system design for NASA in situ propellant production plant on mars*, in Proceedings of the 4th International Conference on Microreaction Technology, IMRET 4, pp. 343–3350 (5–9 March 2000), AIChE Topical Conf. Proc., Atlanta, USA.

267 Drost, M. K., Wegeng, R. S., Martin, P. M., Brooks, K. P., Martin, J. L., Call, C., *Microheater*, in Proceedings of the 4th International Conference on Microreaction Technology, IMRET 4, pp. 308–313 (5–9 March 2000), AIChE Topical Conf. Proc., Atlanta, USA.

268 Palo, D., Rozmiarek, R., Steven, P., Holladay, J., Guzman, C., Wang, Y., Hu, J., Dagle, R., Baker, E., *Fuel processor development for a soldier-portable fuel cell system*, in Matlosz, M., Ehrfeld, W., Baselt, J. P. (Eds.), *Microreaction Technology – IMRET 5: Proc. of the 5th International Conference on Microreaction Technology*, pp. 359–367, Springer-Verlag, Berlin (2001).

269 Hermann, I., Lindner, M., Winkelmann, H., Düsterwald, H. G., *Microreaction technology in fuel processing for fuel cell vehicles*, in Proceedings of the VDE World Microtechnologies Congress, MICRO.tec 2000, pp. 447–453 (25–27 September 2000), VDE Verlag, Berlin, EXPO Hannover.

270 de Bellefon, C., *Application of microdevices for the fast investigation of catalysis*, in Proceedings of the Micro Chemical Plant – International Workshop, pp. L3 (9–17) (4 February 2003), Kyoto, Japan.

271 Pennemann, H., Hessel, V., Kost, H.-J., Löwe, H., de Bellefon, C., *Investigations on pulse broadening for transient catalyst screening in gas/liquid systems*, AIChE J. (2003) 34.

272 Besser, R. S., Ouyang, X., Surangalikar, H., *Hydrocarbon hydrogenation and dehydrogenation reactions in microfabricated catalytic reactors*, Chem. Eng. Sci. **58** (2003) 19–26.

273 Surangalikar, H., Ouyang, X., Besser, R. S., *Experimental study of hydrocarbon hydrogenation and dehydrogenation reactions in silicon microfabricated reactors of two different geometries*, Chem. Eng. J. **90**, 4140 (2002) 1–8.

274 Ajmera, S. K., Losey, M. W., Jensen, K. F., Schmidt, M. A., *Microfabricated packed-bed reactor for phosgene synthesis*, AIChE J. **47**, 7 (**2001**) 1639–1647.

275 Zech, T., Hönicke, D., *Efficient and reliable screening of catalysts for microchannel reactors by combinatorial methods*, in Proceedings of the 4th International Conference on Microreaction Technology, IMRET 4, pp. 379–389 (5–9 March 2000), AIChE Topical Conf. Proc., Atlanta, USA.

276 Zech, T., Hönicke, D., Klein, J., Schunk, S., Demuth, D., *A novel system architecture for high-throughput primary screening of heterogeneous catalysts*, in Proceedings of the Microreaction Technology – IMRET5: Proceedings of the 5th International Conference on Microreaction Technology (27–30 May 2001), Strasbourg, France.

277 Zech, T., Schunk, S., Klein, J., Demuth, D., *The integrated materials chip for high-throughput experimentation in catalysis research*, in Proceedings of the 6th International Conference on Microreaction Technology, IMRET 6, pp. 32–36 (11–14 March 2002), AIChE Pub. No. 164, New Orleans.

278 Müller, A., Drese, K., Gnaser, H., Hampe, M., Hessel, V., Löwe, H., Schmitt, S., Zapf, R., *A combinatorial approach to the design of a screening reactor for parallel gas phase catalyst screening*, Chim. Oggi **21**, 9 (**2003**) 60–68.

279 Müller, A., Drese, K., Gnaser, H., Hampe, M., Hessel, V., Löwe, H., Schmitt, S., Zapf, R., *Fast preparation and testing methods using a microstructured modular reactor for parallel gas phase catalyst screening*, Catal. Today **81** (2002) 377–391.

280 REBROV, E. V., DE CROON, M. H. J. M., SCHOUTEN, J. C., *Design of a micro-structured reactor with integrated heat-exchanger for optimum performance of highly exothermic reaction,* Catal. Today **69** (2001) 183–192.

281 REBROV, E. V., DE CROON, M. H. J. M., SCHOUTEN, J. C., *Development of the kinetic model of platinum catalyzed ammonia oxidation in a microreactor,* Chem. Eng. J. **90** (2002) 61–76.

282 REBROV, E. V., DUINKERKE, S. A., DE CROON, M. H. J. M., SCHOUTEN, J. C., *Optimization of heat transfer characteristics, flow distribution, and reaction processing for a microstructured reactor/heat-exchanger for optimal performance in platinum catalyzed ammonia oxidation,* Chem. Eng. **93** (2003) 201–216.

283 ZAPF, R., BECKER-WILLINGER, C., BERRESHEIM, K., HOLZ, H., GNASER, H., HESSEL, V., KOLB, G., LÖB, P., PANNWITT, A.-K., ZIOGAS, A., *Detailed characterization of various porous alumina based catalyst coatings within microchannels and their testing for methanol steam reforming,* Trans IChemE **81**, A (2003) 721–729.

284 WUNSCH, R., FICHTNER, M., GÖRKE, O., HAAS-SANTO, K., SCHUBERT, K., *Process of applying Al₂O₃ coatings in micro-channels of completely manufactured micro-structured reactors,* Chem. Eng. Technol. **25**, 7 (2002) 700–703.

285 HAAS-SANTO, K., FICHTNER, M., SCHUBERT, K., *Preparation of microstructure compatible porous supports by sol–gel synthesis for catalyst coatings,* Appl. Catal. A **220** (2001) 79–92.

286 AJMERA, S. K., DELATTRA, C., SCHMIDT, M. A., JENSEN, K. F., *Microfabricated differential reactor for heterogeneous gas phase catalyst testing,* J. Catal. **209** (2002) 401–412.

287 AJMERA, S. K., DELATTRE, C., SCHMIDT, M. A., JENSEN, K. F., *Microfabricated cross-flow chemical reactor for catalyst testing,* Sens. Actuators **82**, 2–3 (2002) 297–306.

288 WÖRZ, O., JÄCKEL, K. P., RICHTER, T., WOLF, A., *Microreactors, new efficient tools for optimum reactor design, Micro-technologies and Miniaturization, Tools, Techniques and Novel Applications for the BioPharmaceutical Industry,* IBC Global Conferences, London (**1998**).

289 RICHTER, T., WOLF, A., JÄCKEL, J.-P., WÖRZ, O., *Mikroreaktoren, ein neues wirksames Werkzeug für die Reaktor-entwicklung,* Chem. Ing. Tech. **71** (**1999**) 973–974.

290 MAINZ, A., EIJKEL, J. C. T., *Miniaturization and chip technology,* Pure Appl. Chem. **73** (2001) 1555–1561.

291 *The challenge for speciality chemicals,* Crystal Faraday Workshop, 20–21 March, 2003, Warrington, UK, oral presentation, ICI.

292 WILES, C., WATTS, P., HASWELL, S. J., POMBO-VILLAR, E., *The aldol reaction of silyl enol ethers within a micro reactor,* Lab Chip **1** (2001) 100–101.

293 SKELTON, V., HASWELL, S. J., STYRING, P., WARRINGTON, B., WONG, S., *A micro-reactor device for the Ugi four component condensation (4CC) reaction,* in RAMSEY, J. M., VAN DEN BERG, A. (Eds.), *Micro Total Analysis Systems,* pp. 589–590, Kluwer Academic Publishers, Dordrecht (**2001**).

294 FERNANDEZ-SUAREZ, M., WONG, S. Y. F., WARRINGTON, B. H., *Synthesis of a three-member array of cycloadducts in a glass microchip under pressure driven flow,* Lab Chip **2** (2002) 170–174.

295 SANDS, M., HASWELL, S. J., KELLY, S. M., SKELTON, V., MORGAN, D., STYRING, P., WARRINGTON, B., *The investigation of an equilibrium dependent reaction for the formation of enamines in a microchemical system,* Lab Chip **1** (2001) 64–65.

296 CHAMBERS, R. D., SPINK, R. C. H., *Microreactors for elemental fluorine,* Chem. Commun. 10 (**1999**) 883–884.

297 CHAMBERS, R. D., HOLLING, D., SPINK, R. C. H., SANDFORD, G., *Elemental fluorine Part 13. Gas–liquid thin film reactors for selective direct fluorination,* Lab Chip **1** (2001) 132–137.

298 ANTES, J., TÜRCKE, T., MARIOTH, E., LECHNER, F., SCHOLZ, M., SCHNÜRER, F., KRAUSE, H. H., LÖBBECKE, S., *Investigation, analysis and optimization of exothermic nitrations in microreactor processes,* in MATLOSZ, M., EHRFELD, W., BASELT, J. P. (Eds.), *Microreaction Technology – IMRET 5: Proc. of the 5th International Conference on Microreaction Technology,* pp. 446–454, Springer-Verlag, Berlin (**2001**).

299 Burns, J. R., Ramshaw, C., *A microreactor for the nitration of benzene and toluene*, in Proceedings of the 4th International Conference on Microreaction Technology, IMRET 4, pp. 133–140 (5–9 March 2000), AIChE Topical Conf. Proc., Atlanta, USA.

300 Dummann, G., Quitmann, U., Gröschel, L., Agar, D. W., Wörz, O., Morgenschweis, K., *The capillary-microreactor: a new reactor concept for the intensification of heat and mass transfer in liquid–liquid reactions*, Catalysis Today, Special edition – 4th International Symposium on Catalysis in Multiphase Reactors, CAMURE IV **78–79**, 3 (2002) pp. 433–439.

301 Burns, J. R., Ramshaw, C., Harston, P., *Development of a microreactor for chemical production*, in Ehrfeld, W., Rinard, I. H., Wegeng, R. S. (Eds.), *Process Miniaturization: 2nd International Conference on Microreaction Technology, IMRET 2, Topical Conf. Preprints*, pp. 39–44, AIChE, New Orleans (1998).

302 Burns, J. R., Ramshaw, C., *Development of a microreactor for chemical production*, Trans. Inst. Chem. Eng. **77**, 5/A (1998) 206–211.

303 Födisch, R., Reschetilowski, W., Hönicke, D., *Heterogeneously catalyzed liquid-phase hydrogenation of nitro-aromatics using microchannel reactors*, in Proceedings of the DGMK-Conference on the Future Role of Aromatics in Refining and Petrochemistry, pp. 231–238 (1999), Erlangen, Germany.

304 Surangalikar, H., Besser, R. S., *Study of catalysis of cyclohexene hydrogenation and dehydrogenation in a microreactor*, in Proceedings of the 6th International Conference on Microreaction Technology, IMRET 6, pp. 248–253 (11–14 March 2002), AIChE Pub. No. 164, New Orleans.

305 Wiessmeier, G., Hönicke, D., *Heterogeneously catalyzed gas-phase hydrogenation of cis,trans,trans-1,5,9-cyclododecatriene on palladium catalysts having regular pore systems*, Ind. Eng. Chem. Res. **35** (1996) 4412–4416.

306 Wiessmeier, G., Hönicke, D., *Microfabricated components for heterogeneously catalyzed reactions*, J. Micromech. Microeng. **6** (1996) 285–289.

307 Kursawe, A., Hönicke, D., *Epoxidation of ethene with pure oxygen as a model reaction for evaluating the performance of microchannel reactors*, in Proceedings of the 4th International Conference on Microreaction Technology, IMRET 4, pp. 153–166 (5–9 March 2000), AIChE Topical Conf. Proc., Atlanta, USA.

308 Kursawe, A., Dietzsch, E., Kah, S., Hönicke, D., Fichtner, M., Schubert, K., Wiessmeier, G., *Selective reactions in microchannel reactors*, in Ehrfeld, W. (Ed.), *Microreaction Technology: 3rd International Conference on Microreaction Technology, Proc. of IMRET 3*, pp. 213–223, Springer-Verlag, Berlin (2000).

309 Rouge, A., Spoetzl, B., Gebauer, K., Schenk, R., Renken, A., *Microchannel reactors for fast periodic operation: the catalytic dehydration of isopropanol*, Chem. Eng. Sci. **56** (2001) 1419–1427.

310 Ciu, T., Fang, J., Maxwell, J., Gardner, J., Besser, R., Elmore, B., *Micromachining of microreactor for dehydrogenation of cyclohexane to benzene*, in Proceedings of the 4th International Conference on Microreaction Technology, IMRET 4, p. 488 (5–9 March 2000), AIChE Topical Conf. Proc., Atlanta, USA.

311 Cao, E., Yeong, K. K., Gavriilidis, A., Cui, Z., Jenkins, D. W. K., *Microchemical reactor for oxidative dehydrogenation of methanol*, in Proceedings of the 6th International Conference on Microreaction Technology, IMRET 6, pp. 76–84 (11–14 March 2002), AIChE Pub. No. 164, New Orleans.

312 Maurer, R., Claivaz, C., Fichtner, M., Schubert, K., Renken, A., *A microstructured reactor system for the methanol dehydrogenation to water-free formaldehyde*, in Proceedings of the 4th International Conference on Microreaction Technology, IMRET 4, pp. 100–105 (5–9 March 2000), AIChE Topical Conf. Proc., Atlanta, USA.

313 Zheng, A., Jones, F., Fang, J., Cui, T., *Dehydrogenation of cyclohexane to benzene in a membrane reactor*, in Proceedings of the 4th International Conference on Microreaction Technology, IMRET 4,

pp. 284–292 (5–9 March **2000**), AIChE Topical Conf. Proc., Atlanta, USA.

314 Herweck, T., Hardt, S., Hessel, V., Löwe, H., Hofmann, C., Weise, F., Dietrich, T., Freitag, A., *Visualization of flow patterns and chemical synthesis in transparent micromixers*, in Matlosz, M., Ehrfeld, W., Baselt, J. P. (Eds.), *Microreaction Technology – IMRET 5: Proc. of the 5th International Conference on Microreaction Technology*, pp. 215–229, Springer-Verlag, Berlin (**2001**).

315 Pysall, D., Wachsen, O., Bayer, T., Wulf, S., *Verfahren und Vorrichtung zur kontinuierlichen Hestellung von Polymerisaten*, DE 19816886, Aventis Research & Technologies GmbH, Priority: 17.04.98.

316 Hessel, V., Löwe, H., *Mikroverfahrenstechnik: Komponenten – Anlagenkonzeption – Anwenderakzeptanz – Teil 2*, Chem. Ing. Tech. **74**, 3 (**2002**) 185–207.

317 Hessel, V., Löwe, H., Hofmann, C., Schönfeld, F., Wehle, D., Werner, B., *Process development of a fast reaction of industrial importance using a caterpillar micromixer/tubular reactor set-up*, in Proceedings of the 6th International Conference on Microreaction Technology, IMRET 6, pp. 39–54 (11–14 March 2002), AIChE Pub. No. 164, New Orleans.

318 Wehle, D., Dejmek, M., Rosenthal, J., Ernst, H., Kampmann, D., Trautschold, S., Pechatschek, R., *Verfahren zur Herstellung von Monochloressigsäure in Mikroreaktoren*, DE 10036603 A1, Priority: **27.07.2000**.

319 Dietz, E., Weber, J., Schnaitmann, D., Wille, C., Unverdorben, L., Brychcy, B., *Verfahren zur Feinverteilung von organischen Pigmenten durch Fällung*, EP 1195413 A1, Priority: 14.09.2001.

320 Dietz, E., Weber, J., Schnaitmann, D., Wille, C., Unverdorben, L., Brychcy, B., *Verfahren zur Herstellung von flüssigen Pigmentpräparationen*, EP 1195414 A1, Priority: 14.09.2001.

321 Dietz, E., Weber, J., Schnaitmann, D., Wille, C., Unverdorben, L., Brychcy, B., *Verfahren zur Feinverteilung von Pigmenten*, EP 1195415 A1, Priority: 14.09.2001.

322 Haverkamp, V., Ehrfeld, W., Gebauer, K., Hessel, V., Löwe, H., Richter, T., Wille, C., *The potential of micromixers for contacting of disperse liquid phases*, Fresenius' J. Anal. Chem. **364** (**1999**) 617–624.

323 Mahe, C., Tranchant, J. F., Burgold, J., Schwesinger, N., *A microstructured device for the production of emulsions on demand*, in Proceedings of the 6th International Conference on Microreaction Technology, IMRET 6, pp. 159–167 (11–14 March 2002), AIChE Pub. No. 164, New Orleans.

324 Bayer, T., Heinichen, H., Natelberg, T., *Emulsification of silicone oil in water. Comparison between a micromixer and a conventional stirred tank*, in Proceedings of the 4th International Conference on Microreaction Technology, IMRET 4, pp. 167–173 (5–9 March **2000**), AIChE Topical Conf. Proc., Atlanta, USA.

325 Sugiura, S., Nakajima, M., Seki, M., *Monodispersed droplet formation caused by interfacial tension from microfabricated channel array*, in Matlosz, M., Ehrfeld, W., Baselt, J. P. (Eds.), *Microreaction Technology – IMRET 5: Proc. of the 5th International Conference on Microreaction Technology*, pp. 252–261, Springer-Verlag, Berlin (**2001**).

326 Kobayashi, I., Nakajima, M., Kikuchim, Y., Chun, K., Fujita, H., *Micromachined straight-through silicon microchannel array for monodispersed microspheres*, in Matlosz, M., Ehrfeld, W., Baselt, J. P. (Eds.), *Microreaction Technology – IMRET 5: Proc. of the 5th International Conference on Microreaction Technology*, pp. 41–48, Springer-Verlag, Berlin (**2001**).

327 Mathes, H., Plath, P. J., *Generation of monodisperse foams using a microstructured static mixer*, in Proceedings of the Tunisian–German Conference of Smart Systems and Devices, submitted for publication (27–30 March 2001), Hammamet, Tunisia.

328 Ganan-Calvo, A., Gordillo, J. M., *Perfectly monodisperse microbubbling by capillary flow focusing*, Phys. Rev. Lett. **87**, 27 (**2001**) 4501–4504.

329 Schiewe, J., Ehrfeld, W., Haverkamp, V., Hessel, V., Löwe, H., Wille, C.,

ALTVATER, M., RIETZ, R., NEUBERT, R., *Micromixer based formation of emulsions and creams for pharmaceutical applications,* in Proceedings of the 4th International Conference on Microreaction Technology, IMRET 4, pp. 467–477 (5–9 March **2000**), AIChE Topical Conf. Proc., Atlanta, USA.

330 SCHENK, R., HESSEL, V., JONGEN, N., BUSCAGLIA, V., GUILLEMET-FRITSCH, S., JONES, A. G., *Nanopowders produced using microreactors, Encyclopedia of Nanoscience and Nanotechnology,* in press **(2004)**.

331 JONGEN, N., DONNET, M., BOWEN, P., LEMAITRE, J., HOFMANN, H., SCHENK, R., HOFMANN, C., AOUN-HABBACHE, M., GUILLEMET-FRITSCH, S., SARRIAS, J., ROUSSET, A., VIVIANI, M., BUSCAGLIA, M. T., BUSCAGLIA, V., NANNI, P., TESTINO, A., HERGUIJUELA, J. R., *Development of a continuous segmented tubular flow reactor and the "scale-out" concept – in search of perfect powders,* Chem. Eng. Technol. **26**, 3 **(2003)** 303–305.

332 PENTH, B., *New non-clogging microreactor,* in Proceedings of Microreaction Technology – IMRET5: Proceedings of the 5th International Conference on Microreaction Technology (27–30 May 2001), Strasbourg, France.

333 FREITAS, S., WALZ, A., MERKLE, H. P., GANDER, B., *Solvent extraction employing a static micromixer: a simple, robust and versatile technology for the microencapsulation of proteins,* J. Microencapsulation **20**, 1 **(2003)** 67–85.

334 ERNI, C., SUARD, C., FREITAS, S., DREHER, D., MERKLE, H. P., WALTER, E., *Evaluation of cationic solid lipid microparticles as synthetic carriers for the targeted delivery of macromolecules to phagocytic antigen-presenting cells,* Biomaterials **23** **(2002)** 4667–4676.

335 HILDEBRAND, G., TACK, J., HARNISCH, S., *Method for producing morphologically uniform micro and nanoparticles using micromixers,* WO 00/72955, Schering AG, Priority: **26.05.1999**.

336 EISENBEISS, F., KINKEL, J., *Verfahren zur Herstellung von Perlpolymerisaten,* DE 19920794, Merck Patent GmbH, Darmstadt, Priority: **06.05.1999**.

337 STANGE, T., KIESEWALTER, S., RUSSOW, K., HESSEL, V., *PAMIR: Potential and applications of microreaction technology – a market survey,* IMM Institut für Mikrotechnik Mainz GmbH and Yole Developpement Lyon **(2002)**.

338 SKELTON, V., GREENWAY, G. M., HASWELL, S. J., STYRING, P., MORGAN, D. O., WARRINGTON, B. H., WONG, S. Y. F., *The preparation of a series of nitrostilbene ester compounds using micro-reactor technology,* Analyst **126 (2001)** 7–10.

339 GARCIA-EGIDO, E., SPIKMANS, V., WONG, S. Y. F., WARRINGTON, B. H., *Synthesis and analysis of combinatorial libraries performed in an automated micro-reactor system,* Lab Chip **3 (2003)** 67–72.

340 SUGA, S., OKAJIMA, M., FUJIWARA, K., YOSHIDA, J.-I., *Cation Flow method: a new approach to conventional and combinatorial organic syntheses using electrochemical microflow systems,* J. Am. Chem. Soc. **123**, 32 **(2001)** 7941–7942.

341 FUKUYAMA, T., SHINMEN, M., NISHITANI, S., SATO, M., RYU, I., *A copper-free Sonogashira coupling reaction in ionic liquids and its application to a microflow system for efficient catalyst recycling,* Org. Lett. **4**, 10 **(2002)** 1691–1694.

342 JÄHNISCH, K., HESSEL, V., LÖWE, H., BAERNS, M., *Chemistry in microstructured reactors,* Angew. Chem. Int. Ed. **(2004)** in press.

2
Modeling and Simulation of Micro Reactors

2.1
Introduction

In the field of conventional, macroscopic process technology, modeling and simulation approaches are now used on a routine basis to design and optimize processes and equipment. Many of the models employed have been developed for and carefully adjusted to specific processes and reactors and allow one to predict flow as well as heat and mass transfer, sometimes with a high degree of accuracy. In comparison, modeling and simulation approaches for micro reactors are more immature, but bear great potential for even more reliable computer-based process engineering, as will be discussed below. In general, the purposes of computer simulations are manifold, such as feasibility studies, optimization of process equipment, failure modeling or modeling of process data. For each of these tasks within the field of chemical engineering, simulation methods have been applied successfully in the last few years.

Micro reactors are developed for a variety of different purposes, specifically for applications which require high heat and mass transfer coefficients and well-defined flow patterns. The spectrum of applications includes gas and liquid flow as well as gas/liquid or liquid/liquid multi-phase flow. The variety and complexity of flow phenomena clearly pose major challenges to the modeling approaches, especially when additional effects such as mass transfer and chemical kinetics have to be taken into account. However, there is one aspect which makes the modeling of micro reactors in some sense much simpler than that of macroscopic equipment: the laminarity of the flow. Typically, in macroscopic reactors the conditions are such that a turbulent flow pattern develops, thus making the use of turbulence models [1] necessary. With turbulence models the stochastic velocity fluctuations below the scale of grid resolution are accounted for in an effective manner, without the need to model explicitly the time evolution of these fine details of the flow field. Heat and mass transfer processes strongly depend on the turbulent velocity fluctuations, and for this reason the accuracy of the turbulence model is of paramount importance for a reliable prediction of reactor performance. However, there is no model available which is capable of describing turbulent flow phenomena in a universal manner and is computationally inexpensive at the same time. For this reason, simulation approaches for micro reactors, while usually not necessitating turbulence models, offer some potential to make predictions with a degree of accuracy unparalleled by models of macroscopic reactors.

When comparing processes in micro reactors with those in conventional systems, a few general differences can be identified:

- Flow in microstructures is usually laminar (as mentioned above), in contrast to the turbulent flow patterns on the macro scale.
- The diffusion paths for heat and mass transfer are very small, making micro reactors ideal candidates for heat or mass transfer-limited reactions.
- The surface-to-volume ratio of microstructures is very high. Hence surface effects are likely to dominate over volumetric effects.
- The share of solid wall material is typically much higher than in macroscopic equipment. Hence solid heat transfer plays an important role and has to be accounted for when designing micro reactors.

Although the absence of turbulence simplifies many modeling tasks, the predominance of surface effects introduces additional complications, especially in the case of multi-phase flow. Some of the fundamental mechanisms of, e.g., dynamic wetting and spreading phenomena are not yet well understood, thus adding some degree of uncertainty to the modeling of these processes. As more and more practical applications of micro fluidic systems emerge, research in the field of fluidic surface and interfacial phenomena receives additional impetus. It is thus hoped that in the coming years refined models for microfluidic multi-phase systems will be formulated and will add an additional degree of predictability to flow phenomena in micro reactors.

The purpose of this book is to give an overview of reactions, micro reactor designs and simulation methods generic to chemical micro process technology. Although, on the one hand, such a thematic area is delimited from conventional process technology, on the other hand there exists a borderline to the fields of Lab-on-a-Chip (Lab-Chip) and micro total analysis (µTAS) systems. Both Lab-Chip and chemical micro process technology utilize microfluidics, in the one case in order to implement bioanalytical assays and in the other for chemical production. However, there are a couple of differences as far as the principles of microfluidics generic for each application area are concerned. As an example, in chemical micro process technology gas flows are of major importance, whereas Lab-Chip technology is mainly concerned with liquid micro flows. Another example relates to electrokinetic flows, which are rarely used in micro process technology, but have found widespread applications in Lab-Chips. The discussion of micro flow modeling and simulation techniques in the following sections mainly refers to those thematic areas which are of relevance for chemical micro process technology. This means that little attention is paid to some specific topics such as electrokinetic flows, which are almost exclusively related to Lab-Chip and µTAS systems.

2.2
Flow Phenomena on the Micro Scale

When facing the task of formulating a model of a specific microfluidic system, the question arises of whether the conventional macroscopic equations describing fluid flow, heat and mass transfer are still valid on the micro scale. Systems for chemical processing rarely contain structures with dimensions less than 10 μm; the relevant length scale is often in the range of 100 μm. The purpose of this section is to identify the boundaries beyond which the macroscopic description ceases to be valid, in terms of length scales and fluid properties. It will turn out that in many cases the usual macroscopic descriptions can be applied and that genuine micro scale effects mainly appear in gas-phase systems, whereas for many practical applications liquids in micro channels can be described by the usual continuum models. The case of multi-phase systems with free-surface flow requires some special discussion. It should be pointed out that fluid mechanics on the micro scale is a research field of its own. The following sections can only give an overview of the most important effects and provide a general framework for the formulation of models. More detailed discussions can be found elsewhere [2, 3].

2.2.1
Gas Flows

A standard approach to modeling transport phenomena in the field of chemical engineering is based on convection–diffusion equations. Equations of that type describe the transport of a certain field quantity, for example momentum or enthalpy, as the sum of a convective and a diffusive term. A well-known example is the Navier–Stokes equation, which in the case of compressible media is given as

$$\rho\left(\frac{\partial u_i}{\partial t} + u_j \frac{\partial u_i}{\partial x_j}\right) = -\frac{\partial p}{\partial x_i} + \rho\, g_i + \frac{\partial}{\partial x_k}\left[\mu\left(\frac{\partial u_i}{\partial x_k} + \frac{\partial u_k}{\partial x_i} - \frac{2}{3}\delta_{ik}\frac{\partial u_j}{\partial x_j}\right)\right], \qquad (1)$$

where u_i denotes the ith component of the fluid velocity, ρ and μ density and dynamic viscosity, p pressure, g_i the ith component of the gravity vector and δ_{ij} the Kronecker symbol. Derivatives are taken with respect to space and time coordinates and the Einstein convention of summation over repeated indices is assumed throughout this book. Summation indices always appear as lower-case letters i, j, k, l, m and n, if not explicitly stated otherwise. The Navier–Stokes equation can be viewed as a transport equation for the momentum of the fluid. There are two different mechanisms for momentum transport. The first mechanism, convection, is modeled by the last term on the left-hand side of Eq. (1). The convective term describes transport of local fluid momentum along the streamlines of the flow, co-moving with the particles of the fluid. In the co-moving frame of reference, the local momentum density changes either due to a pressure difference (first term on the right-hand side), gravitational forces (second term) or viscous dissipation (last term). The dissipation term describes diffusive transport of momentum due to thermal motion of the particles.

The Navier–Stokes equation defines a set of three relations for four unknown quantities, u_1, u_2, u_3 and p. Another equation is needed to close the set, which is the equation of mass conservation:

$$\frac{\partial \rho}{\partial t} + \frac{\partial}{\partial x_i} (\rho\, u_i) = 0 \,. \tag{2}$$

Similar convection–diffusion equations to the Navier–Stokes equation can be formulated for enthalpy or species concentration. In all of these formulations there is always a superposition of diffusive and convective transport of a field quantity, supplemented by source terms describing creation or destruction of the transported quantity. There are two fundamental assumptions on which the Navier–Stokes and other convection–diffusion equations are based. The first and most fundamental is the continuum hypothesis: it is assumed that the fluid can be described by a scalar or vector field, such as density or velocity. In fact, the field quantities have to be regarded as local averages over a large number of particles contained in a volume element embracing the point of interest. The second hypothesis relates to the local statistical distribution of the particles in phase space: the standard convection–diffusion equations rely on the assumption of local thermal equilibrium. For gas flow, this means that a Maxwell–Boltzmann distribution is assumed for the velocity of the particles in the frame-of-reference co-moving with the fluid. Especially the second assumption may break down when gas flow at high temperature or low pressure in micro channels is considered, as will be discussed below.

The principle quantity determining the flow regime of gases and deviations from the standard continuum description is the Knudsen number, defined as

$$\mathrm{Kn} = \frac{\lambda}{L} \,. \tag{3}$$

Kn is the ratio of two length scales, the mean free path of the gas molecules λ and a characteristic length scale of the flow domain L, for example the channel diameter. For molecules interacting as hard spheres of diameter d, the mean free path is given as (see, e.g., [4])

$$\lambda = \frac{k_B\, T}{\sqrt{2}\, \pi\, p\, d^2} \,, \tag{4}$$

where k_B is the Boltzmann constant, T, p are temperature and pressure and d is the hard-sphere diameter. When Kn > 1, a gas molecule is more likely to collide with the channel wall than with another molecule. As the transport of momentum or enthalpy is to a large extent governed by the collisions between molecules, major changes in the flow behavior are expected when the Knudsen number exceeds 1. This may happen when gas flow through narrow channels is considered, but also when the temperature is high and/or the pressure is low.

Based on the Knudsen number, four different flow regimes can be distinguished [5]:

- continuum flow with no-slip boundary conditions (Kn $\leq 10^{-2}$);
- continuum flow with slip boundary conditions ($10^{-2} < $ Kn $\leq 10^{-1}$);
- transition flow ($10^{-1} < $ Kn ≤ 10);
- free molecular flow (Kn > 10).

In the first two cases the Navier–Stokes equation can be applied, in the second case with modified boundary conditions. The computationally most difficult case is the transition flow regime, which, however, might be encountered in micro-reactor systems. Clearly, the defined ranges of Knudsen numbers are not rigid; rather they vary from case to case. However, the numbers given above are guidelines applicable to many situations encountered in practice.

2.2.1.1 Slip Flow Regime

For applications in the field of micro reaction engineering, the conclusion may be drawn that the Navier–Stokes equation and other continuum models are valid in many cases, as Knudsen numbers greater than 10^{-1} are rarely obtained. However, it might be necessary to use slip boundary conditions. The first theoretical investigations on slip flow of gases were carried out in the 19th century by Maxwell and von Smoluchowski. The basic concept relies on a so-called slip length L_s, which relates the local shear strain to the relative flow velocity at the wall:

$$u_{gas} - u_{wall} = L_s \frac{\partial u_{gas}}{\partial y} \bigg|_{Wall} . \tag{5}$$

For this purpose a coordinate frame is introduced with the x-coordinate in the flow direction and the y-coordinate perpendicular to the wall (see Figure 2.1). For isothermal conditions the above relation was rigorously derived by Maxwell from the kinetic theory of monoatomic gases. The slip length enters as an empirical parameter containing information on how the gas molecules interact with the wall. There are two limiting cases for this interaction. In the first case, called 'specular reflection', the tangential velocity component of a molecule is conserved and the normal component is inverted. Thus, the momentum transfer of the molecule to the wall in flow direction vanishes. Such an interaction can be imagined to occur when the walls are 'smooth'. In the case of a more intense interaction, or synonymously 'rough' walls, the molecule can transfer longitudinal momentum to the walls. The limiting case where the molecule has lost the 'memory' on its velocity before the collision with the wall and is reflected randomly to all angles is called 'diffuse reflection'. The two limiting cases are illustrated in Figure 2.1.

When the fractions of molecules reflected specularly and diffusively are known, the slip length can be determined, as shown by Maxwell. Maxwell introduced a tangential momentum accommodation coefficient defined as

$$\sigma_v = \frac{\tau_i - \tau_r}{\tau_i - \tau_w} , \tag{6}$$

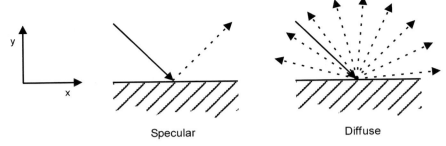

Figure 2.1 Velocity vectors of a gas molecule for specular (left) and diffuse (right) reflection.

where τ_i denotes the average tangential momentum of the incoming molecules, τ_r that of the reflected molecules and τ_w the tangential momentum of a molecule co-moving with the wall. In the case of specular reflection, σ_v assumes a value of zero. Based on the accommodation coefficient, the slip length is given by

$$L_s = \frac{2 - \sigma_v}{\sigma_v} \lambda \, . \tag{7}$$

Apparently, the slip length diverges when $\sigma_v \to 0$. In practice, the shear strain in Eq. (5) will approach zero in such a case, thus leaving the velocity jump finite. Several experimental results on σ_v have been reported [6, 7], most of them indicating values between 0.8 and 1.0, compatible with 'rough' walls.

The physical reason for the velocity slip is the fact that close to the wall the gas is not in thermal equilibrium. For the same reason, a temperature jump is induced, and a more detailed investigation based on the kinetic theory of gases shows that heat transfer and momentum transfer are coupled. Expressions for velocity slip and temperature jump valid in the case of non-isothermal conditions are given by

$$u_{gas} - u_{wall} = \frac{2 - \sigma_v}{\sigma_v} \lambda \frac{\partial u_{gas}}{\partial y} + \frac{3}{4} \frac{\mu}{\rho T_{gas}} \frac{\partial T_{gas}}{\partial x} \, , \tag{8}$$

$$T_{gas} - T_{wall} = \frac{2 - \sigma_T}{\sigma_T} \frac{2 \gamma}{\gamma + 1} \frac{\lambda}{Pr} \frac{\partial T_{gas}}{\partial y} \, . \tag{9}$$

All derivatives are to be evaluated at the wall. In these expressions, μ and ρ are the viscosity and density of the gas, γ is the ratio of specific heats of the gas and the wall material and the Prandtl number of the gas is defined as

$$Pr = \frac{c_p \mu}{k} \, , \tag{10}$$

with the specific heat c_p and the thermal conductivity k. The thermal accommodation coefficient σ_T is defined similarly to the tangential momentum accommodation coefficient. An interesting effect which can be derived from the second term on the right-hand side of Eq. (8) is *thermal creep*. In a microfluidic channel, an axial heat flux through the wall material induces a flow in the opposite direction.

Relations as those defined by Eqs. (8) and (9) can be viewed as a special case of a low-Knudsen number expansion which in case of isothermal conditions is given as [8]

$$u_{\text{gas}} - u_{\text{wall}} = \frac{2 - \sigma_v}{\sigma_v} \left(\text{Kn} \frac{\partial u_{\text{gas}}}{\partial n} + \frac{\text{Kn}^2}{2} \frac{\partial^2 u_{\text{gas}}}{\partial n^2} + \ldots \right), \tag{11}$$

where n denotes a unit vector normal to the wall. As will be discussed below, the Navier–Stokes equation can be obtained from an expansion of the Boltzmann equation up to first order in Kn. Although higher-order expansions of the flow boundary conditions as indicated in Eq. (11) are more accurate in the slip-flow and the transition-flow regime, using such boundary conditions is of little use in combination with the Navier–Stokes equation which is of lower order accuracy.

2.2.1.2 Transition Flow and Free Molecular Flow

In the transition flow and free molecular regime, the use of the Navier–Stokes equation with slip-flow boundary conditions is no longer justified. A general framework for computing gas flows at arbitrary rarefaction is provided by the Boltzmann equation. One might argue that the transition flow regime is rarely reached in micro reactors. However, the first microfluidic systems with sub-micron structures have already been reported, although for liquid-flow applications [9–12]. As nanotechnology becomes more mature, more applications in the field of nanofluidics might open up. Apart from of hypothetical future developments, a well-known component of chemical reaction systems requires an understanding of transport processes in the transition of the free molecular regime: porous media. Often, catalyst materials used for heterogeneously catalyzed gas-phase reactions contain pores with diameters in the nanometer range. Such nanoporous materials are beginning to be used in the field of micro process technology [13, 14]. A brief survey of models for transition and free molecular flows is thus justified.

A detailed derivation and discussion of the Boltzmann equation can be found in many standard statistical physics textbooks, for example [15]. The Boltzmann equation is based on a distribution function $f(x_i, u_i, t)$ defined such that $f(x_i, u_i, t) \, d^3r \, d^3u$ represents the average number of molecules in a phase space volume element around (x_i, u_i), i.e. the distribution function is related to the probability of finding a molecule at spatial coordinates x_i with velocity components u_i. For molecules of mass m in an external force field with components $F_i(x_j)$ the equation takes the form

$$\frac{\partial f}{\partial t} + u_i \frac{\partial f}{\partial x_i} + \frac{F_i}{m} \frac{\partial f}{\partial u_i} = J(f, f^*). \tag{12}$$

The right-hand side of this equation is the so-called collision integral, an integral over terms quadratic in the distribution function (in the case of binary collisions) containing the molecular scattering cross-section. The collision integral describes the rearrangement of the distribution function due to molecular collisions. The terms on the left-hand side describe the collision-less evolution of the molecular distribution due to external forces, for example gravitation. For two reasons the Boltzmann equation is much more difficult to solve than the Navier–Stokes equation. On the one hand, it is an integro-differential equation containing a coupling of points in phase-space separated by a finite distance. On the other hand, the unknown function to be solved for is a function not only of the spatial coordinates, but also of the velocity coordinates.

In order to reduce the complexity of the problem, several approximation schemes have been developed. In the BGK model, the collision integral is replaced by a simple local term ensuring that the well-known Maxwell distribution is reached at thermal equilibrium [16]. The linearization method assumes that the phase space distribution is given by a small perturbation h on top of a (local) Maxwell distribution f_0 (see, e.g., [17, 18]):

$$f(x_i, u_i, t) = f_0[1 + h(x_i, u_i, t)] \,. \tag{13}$$

In order to establish a link with the Navier–Stokes equation, the distribution function can be expanded in a series of powers of the Knudsen number as

$$f = f^{(0)} + \mathrm{Kn}\, f^{(1)} + \mathrm{Kn}^2\, f^{(2)} + \dots . \tag{14}$$

When this so-called Chapman–Enskog expansion [19] is inserted into the Boltzmann equation, a series of equations of different approximation order in Kn is obtained. To lowest order, the Euler equation, i.e. the Navier–Stokes equation in the limit of vanishing viscosity, is recovered. The next approximation level yields the Navier–Stokes equation, in the second-order approximation a generalization of the Navier–Stokes equation is obtained, the so-called Burnett equations. The Burnett equations could be employed to describe gas flows in the transition regime; they are, however, rarely used in practice, owing to their complexity and uncertainties related to the implementation of boundary conditions. The standard approach for computing transition flows is the direct simulation Monte Carlo (DSMC) method, which will be outlined below. In addition, approaches aiming at solving the Boltzmann equation directly, partially making use of approximations as in the BGK model, are used to describe transition flows [20].

In order to obtain a qualitative view of how the transition regime differs from the continuum flow or the slip flow regime, it is instructive to consider a system close to thermodynamic equilibrium. In such a system, small deviations from the equilibrium state, described by thermodynamic forces X_i, cause thermodynamic fluxes J_i which are linear functions of the X_i (see, e.g., [15]):

$$J_i = \sum_j \Lambda_{ij}\, X_j \tag{15}$$

As an example, J_i can be a heat flux and X_i a temperature gradient. The thermodynamic fluxes determine the irreversible time evolution of a system to thermodynamic equilibrium, e.g. a temperature difference can be equalized by a heat flux. In general, the kinetic coefficients Λ_{ij} are non-zero for $i \neq j$, implying several so-called cross-diffusion effects. As an example, a concentration gradient can induce a heat flux or a temperature gradient a mass flow. The coupling between the velocity field and the temperature field of Eq. (8) is a special case of mass flow being induced by a temperature gradient. When rarefaction effects are absent, the non-diagonal terms of the kinetic-coefficients matrix can be neglected in many cases. This is no longer true in the transition flow regime, where cross-diffusion effects become equally important as the diagonal contributions to Eq. (15) [21]. In the transition flow regime, gradients of field quantities such as temperature and concentration induce fluxes of other quantities to which they are linked by the kinetic coefficients to a similar extent as observed for the self-induced fluxes. Thus, the coupling between different field quantities becomes much stronger than in the continuum flow regime.

Nowadays, the DSMC method has become the standard approach for simulation of gas flows in the transition regime [4]. Rather than modeling gas flow by a set of differential or integro-differential equations, this method is based on tracking the trajectories and interactions of gas molecules directly. DSMC is a time marching approach based on a time discretization with steps smaller than the collision time of the molecules. The computational domain is divided into cells with a size smaller than the molecular mean free path. Particles are initialized inside the computational domain. In one time step, a particle moves along its velocity vector, potentially penetrating into a neighboring cell. Inside each cell, the particles can undergo collisions with other particles, changing their velocity vectors. In a sense, DSMC attempts to model the physical scenario of molecular motion interrupted by collisions with other molecules directly. However, even in small systems of a few microns extension, the number of gas molecules is tremendously high. A modeling approach following the trajectories of all of these molecules would typically require computational resources exceeding the capabilities of most of today's computers. For this reason, each particle in a DSMC simulation represents a whole ensemble of molecules. Specifically, a DSMC algorithm comprises a repeated sequence of the following steps:

- *Particle motion*
  Particles are moved along their current velocity vectors without undergoing interactions for a time Δt which is chosen smaller than the mean collision time. If a particle hits the domain boundary, its velocity vector is modified according to the corresponding boundary condition (for example specular or diffuse reflection if a particle hits a wall);

- *Indexing and tracking of particles*
  Attached to each particle is information on the computational cell in which it is located. During a time step a particle might enter a neighboring cell. It is then necessary to update the cell index of the particle.

- *Collision of particles*
  Based on the molecular collision cross-section, a particle might undergo a collision with another particle in the same cell. In a probabilistic process collision partners are determined and velocity vectors are updated according to the collision cross-section. Typically, simple parametrizations of the cross-section such as the hard-sphere model for monoatomic gases are used.

- *Sampling of macroscopic quantities*
  Macroscopic quantities of interest such as pressure, density or average velocity are obtained by sampling over the particle distribution within each cell. Usually sampling from only one simulation run is not sufficient to obtain quantities with an acceptable noise level; instead results are obtained by averaging over many parallel runs.

DSMC simulations have been employed to study flow in micro channels [22–24]. The number of simulated particles usually lies in the range of several hundred thousand to several million. Despite this apparently high number, there still exist considerable problems with statistical noise. For the specific application of micro channel flows, it is very difficult to extract reliable velocity profiles at low Mach numbers. The reason for these problems lies in the different velocity scales of the flow and the single molecules. For low-speed flows the molecular velocity exceeds the flow velocity by orders of magnitude, thus making the extraction of a small non-zero average from a largely fluctuating ensemble very difficult. By the statistical nature of the DSMC approach, the errors are inversely proportional to the square root of the number of simulated molecules. Hence an increase of the number of molecules effects only a slight reduction in the statistical errors. The simulation of low-speed flows by DSMC is a field of ongoing research. A special DSMC scheme has been developed where the particles carry information not only on their microscopic velocity but also on the velocity of a macroscopic ensemble represented by the particle [23, 25]. Alternatively, non-linear filters have been used for post-processing DSMC results and extracting macroscopic quantities from a noisy ensemble [24]. Such novel approaches help to reduce the noise level considerably and allow one to obtain more reliable results for low-speed flows.

In Figure 2.2 DSMC results of Karniadakis and Beskok [2] and results obtained with the linearized Boltzmann equation are compared for channel flow in the transition regime. The velocity profiles at two different Knudsen numbers are shown. Apparently, the two results match very well. The fact that the velocity does not reach a zero value at the channel walls ($Y = 0$ and $Y = 1$) indicates the velocity slip due to rarefaction which increases at higher Knudsen numbers.

When the channel diameter becomes very small, it might no longer be justified to assume smooth channel walls since, depending on the microfabrication technology used, the surface roughness cannot be neglected on the length scale of the channel diameter. One of the first studies of rarified gas flows in channels with rough surfaces based on the DSMC method was performed by Sun and Faghri [26]. They set up a model geometry of a channel with step structures on its surface, as

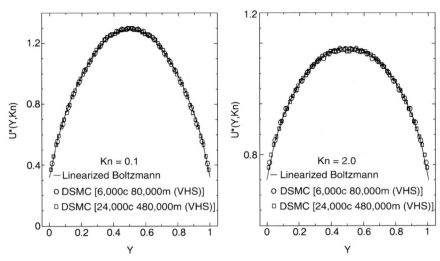

Figure 2.2 Non-dimensionalized velocity distribution across a channel for Kn = 0.1 (left) and Kn = 2.0 (right), taken from [2]. The results were obtained by DSCM using two different collision cross-sections and by solution of the linearized Boltzmann equation.

depicted in Figure 2.3. The model geometry is described by three parameters D_h, s and ε, where the ratio ε/D_h controls the magnitude of the surface roughness. The DSMC simulations were performed for Knudsen numbers between 0.02 and 0.08 and the results were expressed as a friction factor depending on the Knudsen number and the geometry parameters. For very small values of ε/D_h, the friction factor of a smooth channel obtained from continuum theory with slip flow boundary conditions is reproduced. A deviation from the smooth channel results occurs at $\varepsilon/D_h = 5\%$, where the friction factor starts to increase owing to the effects of surface roughness. When varying the Knudsen number it is found that the surface roughness has a larger effect on low Kn flows compared with flows at high Kn.

Owing to the computational cost of the DSMC method, it is advisable to use continuum approaches such as the Navier–Stokes equation wherever possible. In many conceivable situations, there are regions inside of the flow domain where the use of slip boundary conditions in combination with the Navier–Stokes equation is justified, whereas other regions require the use of the DSMC method. A simple example is given by a narrow channel with a high pressure drop. At the entrance of the channel, the pressure and the density are high and the molecular mean free path is comparatively small. As the pressure decreases towards the exit of the channel, the mean-free path increases and the Knudsen number might reach values in the transition regime. An optimized method for such problems would be based on a subdivision of the flow domain into a continuum flow and a transition flow region and would consist of a Navier–Stokes or Euler solver coupled with a DSMC solver. These methods have been developed in the last few years [27, 28] and enhance the applicability of DSMC approaches.

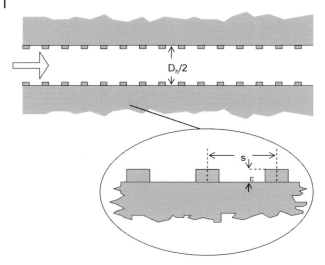

Figure 2.3 Model geometry used for the study of surface-roughness effects on rarefied gas flows in micro channels.

2.2.2
Liquid Flows

The Navier–Stokes equation [Eq. (1)] provides a framework for the description of both liquid and gas flows. Unlike gases, liquids are incompressible to a good approximation. For incompressible flow, i.e. a constant density ρ, the Navier–Stokes equation and the corresponding mass conservation equation simplify to

$$\frac{\partial u_i}{\partial t} + u_j \frac{\partial u_i}{\partial x_j} = -\frac{1}{\rho} \frac{\partial p}{\partial x_i} + g_i + \frac{1}{\rho} \frac{\partial}{\partial x_j} \left(\mu \frac{\partial}{\partial x_j} u_i \right), \qquad (16)$$

$$\frac{\partial u_i}{\partial x_i} = 0 . \qquad (17)$$

In some microfluidic applications liquid is transported with a comparatively low velocity. In such cases, a liquid volume co-moving with the flow experiences inertial forces which are small compared with the viscous forces acting on it. The terms appearing on the left-hand side of Eq. (16) can then be neglected and the *creeping flow* approximation is valid

$$\frac{\partial}{\partial x_j} \left(\mu \frac{\partial}{\partial x_j} u_i \right) - \frac{\partial p}{\partial x_i} + \rho \, g_i = 0 . \qquad (18)$$

An indicator of the validity of the creeping flow approximation is the dimensionless Reynolds number:

$$\mathrm{Re} = \frac{u\, d\, \rho}{\mu},\qquad(19)$$

where d is a length scale characteristic for the flow under consideration. The Reynolds number reflects the ratio of inertial and viscous forces and is possibly the most important dimensionless group used to characterize flow fields. It should be small for the creeping flow approximation to be applicable [29]. Eq. (18) is much simpler to solve than the full Navier–Stokes equation, as it is linear in the velocity. As a special scheme suited for such types of problems, the boundary element method allows one to reformulate a fluid dynamics problem in the creeping flow regime for a specific flow domain as a problem defined on its boundary [30] and permits a fast and efficient solution of Eq. (18) in many cases. Hence, whenever attempting to solve a micro flow problem, it is useful to check the validity of the creeping flow approximation.

Similarly to the case of gas flows discussed above, the question of the range of applicability of continuum models for liquid flow arises. While the kinetic theory of gases provides clear indications of the limits of continuum models and of the onset of rarefaction effects on the micro scale, there is no general framework explaining possible deviations of liquid flow phenomena from their macroscopic behavior. The concept of mean free path ceases to be useful for liquids, as the molecules interact with their neighbors in a permanent manner (in a sense, the molecular mean free path is zero for liquids). A few groups have conducted experiments on liquid flows in microstructures and measured pressure drop and heat transfer coefficients, with largely contradictory results. Pfahler et al. [31] measured the pressure drop in channels with depths ranging from 0.5 to 50 µm and found indications for a reduced viscosity. Peng et al. [32] studied liquid flow through rectangular micro channels of hydraulic diameters between 133 and 367 µm. They found that the transition from laminar to turbulent flow occurred at smaller Reynolds numbers than in macroscopic set-ups and increased or reduced friction factors, depending on the aspect ratio of the channel. The same group studied the heat transfer characteristics in micro channels and again found indications of a transition to turbulent flow at smaller Reynolds numbers than in conventional situations [33]. Other groups measured friction factors in good agreement with predictions of the classical theory [34, 35] or higher friction factors than in conventional set-ups [36, 37]. A compilation of the published results does not seem to leave room for any simple conclusions of general importance. Furthermore, the physics of Newtonian liquids does not suggest any universal mechanisms by which the flow behavior in channels of several hundred microns width could differ considerably from macroscopic behavior at the same Reynolds number. However, in some specific situations liquid micro flows exhibit a behavior which does not occur on the macro scale, but which can be reproduced experimentally and explained by theoretical models. A brief overview of such effects is given in the following.

2.2.2.1 Boundary Slip of Liquids

During the last decade, it was confirmed that not only gases, but also liquids can exhibit boundary slip. For liquids the slip length is defined as the distance behind the interface at which the flow velocity extrapolates to zero [38]. Non-Newtonian liquids such as polymer melts often show very pronounced slip-flow effects, as was discovered by Migler et al. [39]. They used a Couette-flow apparatus and an evanescent wave with a penetration depth of the order of 100 nm to photobleach a small liquid volume in the close vicinity of a solid surface. From the fluorescence-recovery curve of their evanescent wave-induced fluorescence experiment, they could determine the slip length, which was larger than 100 μm in some cases. Measurements of slip flow on hydrophobic surfaces were carried out by Baudry et al. [40]. They used an oscillating microscopic sphere in the vicinity of a planar surface and measured the response force due to the oscillations. The slip length determined in such a way was 38 nm. Recently, it was discovered that not only hydrophobic, but also hydrophilic surfaces exhibit boundary slip. By moving a silica sphere with a radius of 10 μm relative to a flat surface by means of a piezo-driven cantilever beam, the slip flow characteristics of a Newtonian liquid were measured via the drag force exerted on the sphere [41]. By comparison with exact solutions of the Stokes equation, a slip length of up to 20 nm was determined, and the slip length was found to increase with shear rate. Other researchers found values of comparable magnitude [42]. In terms of the values characteristic for some microfluidic systems, the shear rates in these experiments were moderate, reaching 8000 s^{-1} at the maximum. Shear rates in that range are characteristic for flow velocities around 1 m s^{-1} in 100 μm wide channels. Such a comparison indicates that boundary slip of liquids might be of some importance for microfluidic systems, especially when high-speed flows in narrow channels are considered.

For a long time, a unified framework for modeling liquid slip-flow effects such as given by Eq. (5) in case of gases has been missing. In 1997, Thompson and Trojan published a result which might be a major step towards a unified description of wall boundary conditions for liquid flows [43]. Their method was based on molecular dynamics (MD) simulations, an approach which is briefly described below. The MD method is similar to the DSMC approach for gases, i.e. the trajectories of single particles are computed. Unlike in DSMC, the particles never travel freely, but permanently move in a force field generated by the other particles. Thompson and Trojan used a well-established model for the intermolecular and molecule–wall interactions, which relies on a shifted Lennard–Jones potential (the fundamentals of molecular interactions are explained in [44], among others). They showed that the slip length in a nanoscale Couette flow geometry varied as a function of the parameters of their interaction potential. The most efficient momentum transfer, or conversely, the smallest slip length was found when the intermolecular distance of the solid wall material equalled that of the liquid. At a critical shear rate the slip length was found to diverge. The results of all of their simulations can be expressed in the following form:

$$L_s = L_{s0} \left(1 - \frac{\dot{\gamma}}{\dot{\gamma}_c} \right)^{-\frac{1}{2}},$$

(20)

with the shear rate given by $\dot{\gamma} = \partial u / \partial y$, as in Eq. (5). L_s is the slip length, and a reference slip length L_{s0} and a critical shear rate $\dot{\gamma}_c$ enter as parameters. In contrast to Eq. (5), the slip length depends non-linearly on the shear rate. When the shear rate reaches a critical value, the slip length diverges, thus indicating that for high-speed flows in small channels deviations from the macroscopic no-slip boundary conditions might be found even on the micrometer scale. It is hoped that in the coming years more information on characteristic values of the free parameters of Eq. (20) will be collected, thus allowing reliable and accurate predictions of high-speed flows in microstructures. Typical numbers reported so far indicate that the slip length of liquid flows in micro scale geometries is small, but not insignificant.

2.2.2.2 Electric Double Layers

Another interfacial phenomenon which is already present in macroscopic flows, but plays a significant role only on the micro scale, is the formation of electric double layers (EDL) [45]. The first theoretical description of EDL formation in liquids was given by Helmholtz in the 19th century. Many surfaces have the ability to bind ions from the liquid phase. A glass surfaces acquires a negative potential when immersed in water. As a consequence, the liquid phase assumes a net positive charge. The positive ions are attracted to the immobilized negative surface charges by electrostatic forces and stay in the vicinity of the surface. However, the electrostatic attraction is counterbalanced by thermal motion of the ions and a polarization layer, the so-called Debye layer, of finite thickness develops. A sketch of the charge distribution in the vicinity of the solid surface is presented in Figure 2.4. The immobilized negative charges are partly balanced by a layer of positive

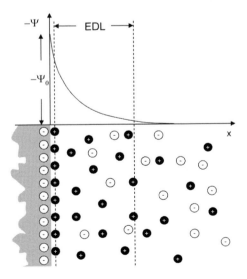

Figure 2.4 Sketch of an electric double layer next to a negatively charged solid surface. Through balance of thermal motion and electrostatic forces a rapidly decaying electric potential Ψ develops inside the liquid phase.

ions sticking to the surface, the so-called Stern layer, followed by an electric double layer of ions dissolved in the liquid. The resulting potential Ψ decays rapidly as a function of distance from the surface.

The EDL charge distribution can be modeled by a Poisson–Boltzmann equation (see, e.g., [46]). In many practical cases values for the layer thickness between 1 and 100 nm are obtained [47].

In comparison, the width of most microfluidic channels is much larger. Hence, whenever EDL effects have to be taken into account in models of microfluidic systems, it is a common strategy to incorporate them as a flow boundary condition rather than resolving the electric double layer explicitly. As a net charge close to the solid surface is created, an external electric field exerts a force on the liquid and can be used to drive a flow. The corresponding transport mechanism is termed *electroosmotic flow* and is used as a fluidic actuation principle in a number of microfluidic systems developed recently [48–50]. The velocity distribution of electroosmotic flow in a capillary has an approximate plug-flow character [51]. This makes electroosmotic flow an interesting principle for microfluidics, as hydrodynamic dispersion are minimized and residence-time broadening and back-mixing is avoided. As mentioned above, in models of electroosmotic flow phenomena the physics of the Debye layer is often incorporated as a flow boundary condition [45]:

$$u_p = -\frac{\zeta \, \varepsilon \, E}{\mu}, \tag{21}$$

where u_p is the velocity at the wall, ε is the dielectric constant of the liquid and μ its viscosity. The zeta potential ζ is equal to the potential drop in the electric double layer (i.e. the total potential drop minus the potential drop in the Stern layer) and the flow is driven by an electric field E along the wall. Such an effective implementation of body forces exerted by the electric field avoids resolving the thin electric double layer explicitly and thus helps to limit the computational complexity of a model.

The formation of an electric double layer close to a solid/liquid interface might offer an explanation for micro channel friction factors or heat transfer coefficients deviating from their macroscopic values. Most surfaces carry electric charges, and a net charge is found within the EDL in the mobile liquid phase. When liquid is forced through a micro channel, the molecules drag these charges along with them, thus inducing an electric current. In this way, an electric potential, the so-called streaming potential, builds up between the channel inlet and outlet [52]. On the other hand, the streaming potential induces an electric current against the flow direction, and finally an equilibrium configuration is found where the two currents compensate each other. The drag force due to the current induced by the streaming potential causes an increased friction factor as compared to a situation without EDL. This effect was studied by Yang et al. [53] for rectangular micro channels. They solved the Poisson–Boltzmann and the enthalpy equations with a finite-difference scheme and the Navier–Stokes equation with a Green's function technique. The channel width and height considered was in the range between 20 and 40 μm. Depending on the solute concentration, they found the friction factor to be significantly increased owing to the presence of an EDL. At the same time, the heat transfer coefficient is reduced by a significant amount. However, the deviations from the corresponding values without EDL rapidly diminish when the channel size increases.

2.2.2.3 Nano Flows

The effects considered above, boundary slip and the formation of electric double layers, are surface effects occurring at the interface of the liquid with a solid. Owing to the large surface-to-volume ratio, they are of potential importance in micro systems but usually negligible in macroscopic systems. Despite these surface effects, the continuum description provided by the Navier–Stokes and other continuum equations remains valid on the micro scale. Only when liquids are confined in nanoscale flow geometries do they show a behavior which can no longer be described with continuum models. Most of the information on nanoflows has been obtained with the help of MD simulations. In the close vicinity of a solid/liquid interface, the liquid molecules usually show a certain ordering which results in density oscillations. Experimental studies of molecular ordering within a few nanometers of a solid surface have been done utilizing the X-ray reflectivity of the molecules [54]. The results of such an experiment are displayed in Figure 2.5. The figure shows the density of oxygen atoms contained in water and exhibits a layer adsorbed to the surface (a) and an ordering of molecules close to the surface (b) which results in density oscillations. Such density oscillations could be reproduced in MD simulation studies [55, 56] and belong to a class of phenomena which are clearly beyond the scope of continuum theories.

Within the MD approach the trajectory of each molecule in the force field generated by the other molecules is computed [57]. For this purpose, Newton's equations of motion are solved:

$$m \frac{d^2 x_i^{(k)}}{dt^2} = \sum_{l \neq k} \frac{\partial V_{kl}}{\partial x_i^{(k)}}, \tag{22}$$

where $x_i^{(k)}$ is spatial coordinate i of particle k, V_{kl} is the interaction potential between molecules k and l and m is the molecular mass. In many simulations the Lennard–Jones interaction potential:

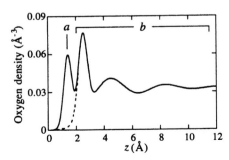

Figure 2.5 Density oscillations of water close to the solid surface, as reported in [54]. The figure shows molecules adsorbed to the surface (a) and in the liquid phase (b).

$$V_{kl}(r) = 4\,\varepsilon\left[\left(\frac{r}{\sigma}\right)^{-12} - \left(\frac{r}{\sigma}\right)^{-6}\right] \tag{23}$$

or slight modifications thereof are used. In this expression, r is the distance between molecules k and l, ε is an energy scale and σ determines the range of the potential. The first term in brackets is a short-range repulsive and the second term a longer range attractive contribution. Potentials of Lennard–Jones type are suited to describe the interaction between neutral, spherical molecules. At the beginning of a simulation, a number of molecules are initialized at specific positions with random velocities sampled from a Maxwell distribution. To compute the time evolution of the system, Eq. (22) is integrated, usually with an explicit, higher order time step algorithm. The integration time steps have to be chosen significantly smaller than the intrinsic time-scale of the problem which is given by

$$\tau = \sqrt{\frac{\sigma^2\,m}{\varepsilon}}\ . \tag{24}$$

The intrinsic time-scale also sets a lower limit to the time the system needs to equilibrate, which is the minimum time the simulation has to run in order to sample average quantities from the molecular ensemble. With no further actions taken, the computational cost of an MD simulation of an ensemble of N molecules is of order $O(N^2)$, as the interactions of each molecule with all other molecules have to be taken into account. However, the Lennard–Jones potential decays as r^{-6}, which means that the contributions of distant molecules can be neglected. This idea has been implemented in algorithms making use of lists of neighboring molecules [58], thereby reducing the computational cost of MD simulations. MD has been used to study laminar channel flow, fluid interfaces and wetting phenomena, among others [57]. Nevertheless, MD simulations remain very challenging. State-of-the-art simulations on high-performance computing platforms are based on numbers of particles in the range of a few billion [59], still not enough for tasks such as following the motion of the $3 \cdot 10^{10}$ water molecules contained in a volume of 1 μm^3.

2.2.3
Multiphase Flows

Several 'unexpected' phenomena have been reported for multiphase systems in milli and micro scale geometries, especially for evaporation of liquids (for an overview, see [60]). Evaporation often occurs via nucleate boiling, and when the bubble size is of the same order of magnitude as the dimension of a fluidic structure, the mechanism of heat removal is modified as compared with, e.g., the classical pool boiling. Viewed from that angle it is not surprising that the conventional correlations for evaporation heat transfer coefficients are not valid in narrow channels or tubes [60]. More difficult to understand are the results of Peng and Wang [61–63], who observed that bubble formation in micro evaporator channels is suppressed, even though in the manifold behind the channel outlet bubbles were observed, indicating that the process was in the nucleate boiling regime.

In general terms, the phenomena described above belong to the class of phase transitions and critical phenomena in confined spaces. From the field of statistical physics, some far-reaching results applying to such problems are known. One fruitful concept used in statistical physics is the correlation length (see, e.g., [64]). The correlation length describes how a local field quantity evaluated at one point in space is correlated with the same quantity at another point. As an example, the correlation length σ for density fluctuations in a fluid is defined via

$$\langle \rho(x_i)\, \rho(x_i + s_i) \rangle \propto \exp(-s/\sigma), \tag{25}$$

where ρ is the density, s the magnitude of s_i and $\langle ... \rangle$ denotes the thermodynamic average. Naturally, the above definition only refers to systems where correlations decay exponentially as a function of distance. When a thermodynamic system approaches the critical temperature T_c for a phase transition, the correlation length diverges with a critical exponent ν [65]:

$$\sigma \propto | T - T_c |^{-\nu} . \tag{26}$$

Hence, close to the critical point thermodynamic quantities at comparatively distant spatial locations become correlated. Especially in the case of liquid micro flows close to a phase transition, these considerations suggest that the correlation length and not the molecular diameter is the length scale determining the onset of deviations from macroscopic behavior.

2.2.3.1 Phase Transitions in Confined Spaces

The theoretical foundation for describing critical phenomena in confined systems is the finite-size scaling approach [64], by which the dependence of physical quantities on system size is investigated. On the basis of the Ising Hamiltonian and finite-size scaling theory, Fisher and Nakanishi computed the critical temperature of a fluid confined between parallel plates of distance D [66]. The critical temperature refers to, e.g., a liquid/vapor phase transition. Alternatively, the demixing phase transition of an initially miscible liquid/liquid mixture could be considered. Fisher and Nakashini found that compared with free space, the critical temperature is shifted by an amount

$$\Delta T_c \propto D^{-\frac{1}{\nu}} . \tag{27}$$

With the critical exponent being positive, it follows that large shifts of the critical temperature are expected when the fluid is confined in a narrow space. Evans et al. computed the shift of the critical temperature for a liquid/vapor phase transition in a parallel-plates geometry [67]. They considered a maximum width of the slit of 20 times the range of the interaction potential between the fluid and the solid wall. For this case, a shift in critical temperature of 5% compared with the free-space phase transition was found. From theoretical considerations of critical phenomena in confined geometries, not only a shift of the critical point, but also other effects as a shift in pressure with respect to the bulk fluid, are found [68]. Such theoretical results are supported by experiments on phase transitions in micro scale geometries. One of the first experimental results was obtained by Jacobs et al. [69], who studied the demixing phase transition of a binary liquid mixture in a narrow slit. They found that for a slit width of 3 µm the critical temperature is shifted by about 10 mK.

The results reported above indicate that small but measurable effects are induced when a fluid close to criticality is confined in a micro scale geometry. The reason for these effects are the long-range correlations building up in the vicinity of the critical point, i.e. the fluid starts to 'feel' the presence of the wall boundaries even at some distance away from the walls. Although, owing to their smallness, most of the effects discovered so far seem to be of limited relevance for practical applications in microfluidics, the principal mechanism of long-range correlations leaves room for significant deviations from free-space behavior on the micro scale.

2.2.3.2 Wetting and Spreading Phenomena

When the length scale of a fluidic system is reduced, surface phenomena become more and more important. On the macroscale, the physics of interfaces, moving contact lines, wetting and spreading is often masked by volumetric effects. When surface effects start to dominate, details come to light which can often be ignored in macroscopic systems but which have a measurable, sometimes decisive impact on the dynamics on the micro scale. Considerable attention has been drawn to the physics of contact lines, i.e. the boundary between immiscible fluid phases, for example liquid and vapor, and a solid surface. The static contact angle θ of the fluid interface with the solid surface is determined by Young's equation (see, e.g., [47]):

$$\cos\theta = \frac{\gamma_{SV} - \gamma_{SL}}{\gamma}, \tag{28}$$

where γ is the liquid/vapor surface tension and γ_{SV} and γ_{SL} the solid/vapor and solid/liquid free energy per unit area, respectively. Experimentally, in many situations it is observed that the contact angle is not unique, but it rather lies in some interval

$$\theta_r < \theta < \theta_a \tag{29}$$

(for an overview, see [70]). The advancing contact angle θ_a is found at the advancing front of a droplet slowly moving over a solid surface and the receding contact angle θ_r is measured at the receding front, as shown in Figure 2.6. For a droplet at rest any contact angle between θ_a and θ_r might be found, depending on the history of the dynamic evolution of the droplet, a phenomenon termed *contact angle hysteresis*.

It has been shown that contact angle hysteresis might arise as a result of inhomogeneities of the surface wetted by the liquid phase or surface roughness [70]. When surface roughness plays a considerable role, the observed contact angle may depend on the exact position of the contact line with respect to the microscopic or

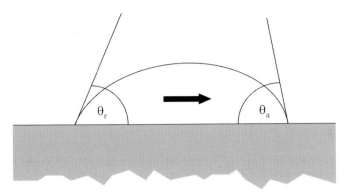

Figure 2.6 Advancing and receding contact angle for a droplet slowly
moving from left to right on a solid surface.

nanoscopic structure of the surface. In a simple model with parallel grooves, different contact angles are observed depending on whether the contact line lies close to
a valley or close to a crest of the surface structure [70]. Inhomogeneities might be
due to chemical contamination distributed over the surface in a random fashion.
Assuming a dilute distribution of such contaminations, equations for the advancing and receding contact angle have been derived [71].

While the above refers mainly to the static limit, new effects come into play when
a moving contact line, i.e. spreading, is considered. It has been observed experimentally that the contact angle of a moving contact line θ_d, the dynamic contact
angle, deviates from the corresponding static value θ_s. As an example, for a completely wettable surface (i.e. $\theta_s = 0$), a relationship of the form

$$u = \text{constant} \; \frac{\sigma \, \theta_d^3}{\mu} \tag{30}$$

was found (see [72] and references therein), where u is the line velocity and μ and σ
are the viscosity and surface tension of the liquid. In theoretical descriptions of
spreading phenomena, the motion of a wedge-like liquid profile over a solid surface is considered within the Stokes-flow regime. De Gennes [73] derived an equation which is valid for small but finite static contact angles θ_s:

$$u = \text{constant} \; \frac{\sigma}{\mu} \theta_d \, (\theta_s^2 - \theta_d^2). \tag{31}$$

Perhaps the most complete description was given by Cox [74], who derived an
expression for the dynamic contact angle which is also valid in the case of large
angles. The Cox prediction was tested experimentally and a good agreement between theory and experiment was found.

Many of these results, most of them more than 10 years old, have attracted some
renewed interest recently. In some µTAS and Lab-Chips small liquid volumes are
transported as plugs in micro channels. Owing to the smallness of the volumes,

surface effects play a dominant role and a profound knowledge of wetting and spreading phenomena is essential for designing systems which are able to process a large number of samples in a parallel way.

2.3
Methods of Computational Fluid Dynamics

In the previous section, an overview of the physical models suitable to describe micro flow phenomena was given. Considering that the relevant length scale in micro reactors is often of the order of 100 µm, the macroscopic continuum models for momentum, heat and mass transfer can be employed in many cases. The first corrections to the macroscopic description typically appear as modifications of the flow boundary condition, such as the introduction of a finite slip length for gas flows. Hence, in many cases, the problem of simulating a micro reactor is, at least formally, equivalent to the problem of solving the corresponding continuum transport equations with suitable boundary conditions. Numerical solution strategies for the partial differential equations (PDE) describing momentum, heat and mass transfer, termed 'Computational fluid dynamics' (CFD), are the main subject of this section. Such an approach, based on a fundamental set of physical models and thorough mathematical methods allowing high-accuracy solutions to be obtained, gives the most detailed insights into flow phenomena in micro reactors and the most reliable prediction of micro-reactor performance.

CFD methods have found widespread applications in conventional chemical process technology. In addition, another class of models has frequently been used, known as lumped-element or macro models. With such types of models no detailed description of flow phenomena is attempted, but rather an integral description of certain functional units used in a chemical process. Corresponding functional units could be pipes, mixers, separators and certain types of reactors. With macro models the transformation and exchange of matter and energy in and between the functional units are described. The functional units are characterized by a number of ports by which they interact with their environment, very similar to the elements of an electronic circuit. As an example, a pipe has an inlet and an outlet port, and the volume flow inside depends on the difference in pressure between inlet and outlet. The details of the flow field inside the pipe are not taken into account explicitly, only in an integral manner via the relationship between pressure drop and volume flow. From this example it is apparent that the accuracy of macro models strongly depends on the quality of the characteristic diagrams and characteristic curves determining the relationship between the response of the system and the excitation at the ports. Typically, several functional units are combined in a conservative network, i.e. a model which guarantees that fundamental quantities such as the volumetric flow rate can distribute over the network, but are conserved as a function of time. By their descriptive nature, relying on integral relationships, macro models are well suited to complex problems allowing a large number of degrees of freedom to be subsumed into a comparatively simple characteristic diagram. An overview of macro models used in the field of chemical process technology is given by Hlavacek et al. [75].

While the use of lumped elements is not necessarily related to a loss in accuracy, it bears two major disadvantages: reduced flexibility and an at best indirect link with the underlying physical models. The reduced flexibility becomes obvious when a number of different designs or geometries of a micro reactor have to be considered. In general, the formulation of accurate macro models is a very expensive process, making a number of validation experiments necessary. When design modifications are taken into account, it is *a priori* unclear how the corresponding macro model has to be modified. In the worst case, the whole expensive process of model formulation and validation has to be repeated. Related to that is the indirect link to physical models. When physical parameters such as density or viscosity are modified, the consequences for the macro model description are often difficult to overlook. Sometimes it is even unclear if a macro model remains valid or if new phenomena occur which would require the use of a refined model. For these reasons, macro models have not yet found extensive use in the field of micro process technology, a fact underpinned also by the application examples presented in this chapter. Instead, standard methods for the simulation of micro reactors are based on solution strategies for the partial differential equations of the underlying physical models.

The three most widespread methods for solving the transport equations for momentum, matter and heat are the finite-difference [76], the finite-element [77, 78] and the finite-volume [79] approaches. In computational fluid dynamics, typically equations of convection–diffusion type:

$$\frac{\partial(\rho\,\Phi)}{\partial t} + \frac{\partial(\rho\,u_i\,\Phi)}{\partial x_i} = \frac{\partial}{\partial x_i}\left(\Gamma\frac{\partial\Phi}{\partial x_i}\right) + S_\Phi \tag{32}$$

are to be solved. In this expression ρ and u_i are the density and the velocity of the fluid and Γ is diffusivity. The equation describes the transport of a field quantity Φ by convection (second term on the left-hand side) and diffusion (first term on the right-hand side), supplemented by a source term S_Φ. As an example, the incompressible Navier–Stokes equation could be considered. In that case Φ would be a component of the linear momentum per unit volume. The three methods indicated above are based on different strategies to obtain approximate solutions Φ. All of the methods rely on a discretization of the computational domain, i.e. a numerical grid. The different types of grids will be discussed later in Section 2.3.3. In order to classify the two most important types beforehand, a structured (left) and an unstructured (right) grid are displayed in Figure 2.7. A 2-D structured grid consists of quadrilateral elements and is, apart from a few special cases, topologically equivalent to a rectangular grid. A 2-D unstructured grid has a more general connectivity of computational cells and can be constructed from triangular or quadrilateral elements or a mixture of these.

Within the finite-difference method (FDM), the derivative terms appearing in Eq. (32) are approximated by finite-difference expressions at each grid node. As an

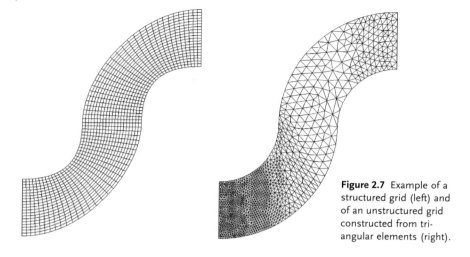

Figure 2.7 Example of a structured grid (left) and of an unstructured grid constructed from triangular elements (right).

example, on a one-dimensional grid a derivative of a function at point i could approximately be expressed via the values of this function at points $i - 1$ and $i + 1$ via a Taylor expansion. By these means a partial differential equation is transformed into a system of algebraic equations for the field values at the grid nodes. The advantage of the FDM is that it is very easy to implement on a rectangular or on a structured grid. However, on more complex grid geometries a FDM formulation can be challenging. A major disadvantage is the fact that the FDM is in general not a *conservative* discretization scheme. Often, the governing transport equations conserve local flow quantities such as mass or momentum. If a numerical approximation scheme does not fulfil the same conservation laws, serious deviations from the exact solution may be induced, as is often found within the FDM.

The finite-volume method (FVM) is closely linked with the FDM in so far as a partial differential equation is transformed into a system of algebraic equations for the field values at the grid nodes. However, within the FVM the system of algebraic equations is obtained by integrating the differential equation over so-called control volumes embracing the grid nodes and expressing the corresponding volume and surface integrals by the field values at the grid nodes. In this way the conservation laws are enforced and a conservative discretization scheme is obtained. The FVM has been applied also to complex grid geometries.

The finite-element method (FEM) is based on shape functions which are defined in each grid cell. The unknown function Φ is locally expanded in a basis of shape functions, which are usually polynomials. The expansion coefficients are determined by a Ritz–Galerkin variational principle [80], which means that the solution corresponds to the minimization of a functional form depending on the degrees of freedom of the system. Hence the FEM has certain optimality properties, but is not necessarily a conservative method. The FEM is ideally suited for complex grid geometries, and the approximation order can easily be increased, for example by extending the set of shape functions.

In computational fluid dynamics only the last two methods have been extensively implemented into commercial flow solvers. Especially for CFD problems the FVM has proven robust and stable, and as a conservative discretization scheme it has some built-in mechanism of error avoidance. For this reason, many of the leading commercially available CFD tools, such as CFX4/5, Fluent and Star-CD, are based on the FVM. The outline on CFD given in this book will be based on this method; however, certain parts of the discussion also apply to the other two methods.

2.3.1
Fundamentals of the Finite–volume Method

The FVM is based on a computational grid with nodes being embraced by control volumes. The values of Φ at the grid nodes are the unknowns to be solved for. An example of a 2-D computational grid with rows labeled by an index j and columns by i is displayed in Figure 2.8. It shows a grid node P embraced by a control volume (shaded in gray) and surrounded by four other nodes $N(orth)$, $S(outh)$, $W(est)$, $E(ast)$. The essential step in deriving a FV approximation is the integration of Eq. (32) over control volumes. For a stationary problem (i.e. a vanishing time derivative) the resulting equation is

$$\int_{S_{ij}} \rho\, u_i\, \Phi\, n_i\, dS = \int_{S_{ij}} \Gamma \frac{\partial \Phi}{\partial x_i} n_i\, dS + \int_{V_{ij}} S_\Phi\, dV . \qquad (33)$$

where V_{ij} denotes the volume of cell (i,j), S_{ij} its surface and n_i the components of an outward normal vector. The surface integrals are obtained from volume integrals via the Gauss theorem. The volume integral appearing on the right-hand side of the equation can easily be approximated:

$$\int_{V_{ij}} S_\Phi\, dV \approx (S_\Phi)_P\, V_{ij} , \qquad (34)$$

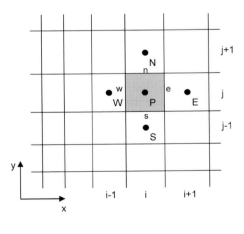

Figure 2.8 2-D FVM grid with a control volume embracing a node P. The grid node is surrounded by four neighboring nodes N, S, W and E.

where $(S_\Phi)_P$ denotes the value of the source term at node P. As only the nodal values of the unknown function should enter the FVM equations, the surface integrals have to be expressed in approximate form via Φ_P, Φ_N, Φ_S, Φ_W and Φ_E. The surface integrals can be split into four contributions from the n, s, w, and e faces. In approximate form, the contribution from the east face to the surface integral of the convective term can be written as

$$
\int_{(S_{ij})_e} \rho\, u_x\, \Phi\, dS \approx (\rho\, u_x\, \Phi)_e\, (S_{ij})_e \, .
\tag{35}
$$

A priori, neither the value of Φ nor the values of density and velocity are known at the faces of the control volume. They have to be determined via interpolation from their values at neighboring nodes. A simple approximation would be

$$
\Phi_e \approx \frac{1}{2}(\Phi_P + \Phi_E)
\tag{36}
$$

in the case of an equidistant grid. Similar equations can be used to approximate the derivative $\partial\Phi/\partial x_i$ appearing in the diffusive term. Expressions as such Eq. (36) induce a nearest-neighbor coupling of grid nodes: the value Φ_P depends on the field values at the neighboring nodes N, S, W and E. In three spatial dimensions, two more nodes, $H(igh)$ and $L(ow)$, are coupled to P. In total, such a nearest-neighbor coupling leads to a system of algebraic equations of the form

$$
a_P\, \Phi_P = \sum_i a_i\, \Phi_i + b_P \, ,
\tag{37}
$$

where the summation is over neighboring nodes. The coefficients a_i depend on the approximation scheme used to determine the field values on the control volume faces, while b_P depends on the form of the source terms appearing in Eq. (33). In matrix form, the system of equations can be written as

$$
A\, \Phi = b \, ,
\tag{38}
$$

where Φ represents the vector of unknowns, A the matrix of coefficients appearing in Eq. (37) and b the vector of source terms.

So far the function Φ has not yet been specified; it could represent a concentration, temperature or velocity field. In the first two cases the velocity field entering Eq. (33) would be assumed as given, it could be determined from a previous FVM simulation run. Eq. (37) then represents a set of coupled linear algebraic equations for the values of Φ at the grid nodes which can be solved by the usual iterative methods for sparse systems of linear equations. When Φ stands for a component of the velocity field itself, the situation is more complex. In that case Eq. (37) is a discretized version of the Navier–Stokes equation with coefficients a_P, a_i depending on the unknown velocity components themselves. The problem then reduces to solving a set of non-linear, coupled algebraic equations which is, however, significantly more difficult than the linear problem obtained for a temperature or concentration field.

Numerical schemes such as Eq. (36) for computing the field values at the control volume faces from the corresponding values at the surrounding nodes are of key importance within the FVM. Such so-called *differencing schemes* have a major influence on the accuracy and boundedness of the numerical solution as well as on the stability of iterative equation solvers. Especially the discretization of the convective term was found to be central to the FVM and is has been investigated in much detail in the past decades (see, e.g., [79]). In some sense the simplest differencing scheme is given by Eq. (36) and known as *central differencing*. Assuming that h is the extension of a computational cell of an equidistant grid, with Eq. (36) the convective flux at the cell faces is approximated up to corrections of order h^2. A major disadvantage of the central differencing scheme is the fact that when the flow velocity exceeds a certain value, some of the coupling coefficients a_i in Eq. (37) assume a negative value [81]. Negative coupling coefficients can induce unphysical oscillations in the solution and may be a cause of potential stability problems of iterative equation solvers. Hence the central differencing scheme is rarely used in practice. Another very simple interpolation method is the *upwind scheme* which, when applied to the east face of a control volume, results in an expression of the form

$$\Phi_e = \begin{cases} \Phi_P & \text{for} \quad u_x \geq 0 \\ \Phi_E & \text{for} \quad u_x < 0 \end{cases} \tag{39}$$

It can easily be shown that for the upwind scheme all coefficients a_i appearing in Eq. (37) are positive [81]. Thus, no unphysical oscillatory solutions are found and stability problems with iterative equation solvers are usually avoided. The disadvantage of the upwind scheme is its low approximation order. The convective fluxes at the cell faces are only approximated up to corrections of order h, which leaves room for large errors on course grids.

The *hybrid scheme* (see, e.g., [81]) represents a compromise between central and upwind differencing. It is based on the idea of using central differencing at low velocities, where the coupling coefficients are all positive. At higher velocities, where at least one of the coupling coefficients changes sign, the upwind scheme is used. The switching between central and upwind differencing is done locally, i.e. the differencing scheme for a specific control volume depends on the local flow velocity. Hence, for low-velocity flows, the central differencing scheme is used at almost all control volumes and the discretization errors are approximately of order h^2. At high velocities, the upwind scheme dominates and error terms approximately of order h are found. However, stability is maintained and unphysical oscillations are avoided.

In order to increase the accuracy of the approximation to the convective term, not only the nearest-neighbor nodes, but also more distant nodes can be included in the sum appearing in Eq. (37). An example of such a higher order differencing scheme is the QUICK scheme, which was introduced by Leonard [82]. Within the QUICK scheme, an interpolation parabola is fitted through two downstream and one upstream nodes in order to determine Φ on the control volume face. The unknown field value is taken to be the value of the interpolation curve on the control volume face. In Figure 2.9 this is exemplified for the east face of a control volume. The value of Φ on the east face is determined from the corresponding values at nodes W, P and E. On an equidistant grid, Φ_e is given by

$$\Phi_e = -\frac{1}{8}\Phi_W + \frac{3}{4}\Phi_P + \frac{3}{8}\Phi_E . \tag{40}$$

The QUICK scheme has a truncation error of order h^3. However, similarly as in the case of the central differencing scheme, at high flow velocities some of the coupling coefficients of Eq. (37) become negative.

A few years ago, a division of the European Research Community on Flow, Turbulence and Combustion assessed the status of available CFD packages and issued a set of best practice guidelines for the use of CFD methods [83]. In these guidelines, it is recommended not to use differencing schemes with a low approximation order (such as upwind) in industrial applications. However, as low-order schemes are usually numerically more robust and less prone to convergence problems, they might be used to initialize a calculation. As convergence is approached, a higher order scheme should be used.

One should bear in mind that not only the differencing scheme, but also the structure of the computational grid determines the approximation order of a numerical model. On an equidistant grid, the central differencing scheme produces a second-order truncation error. However, if the size of neighboring control volumes is different, a truncation error of order h is found. The effects of grid non-uniformity can be compensated by a modification of the coupling coefficients related to the grid nodes [84], which, however, introduces a considerable additional bookkeeping effort.

Besides the convective fluxes, the diffusive fluxes on the control volume faces have to be determined. As apparent from Eq. (33), an expression for $\partial\Phi/\partial x_i$ containing the nodal values of Φ is needed. In the case of an orthogonal grid aligned with the axes of a Cartesian coordinate frame, the expression

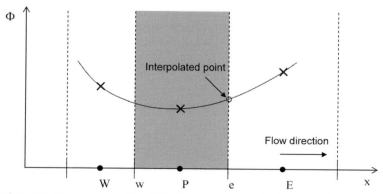

Figure 2.9 Determination of the field values on the control volume faces by interpolation within the QUICK scheme.

$$\left(\frac{\partial \Phi}{\partial x}\right)_e = \frac{\Phi_E - \Phi_P}{x_E - x_P} \tag{41}$$

for the diffusive flux in the x-direction suggests itself. Such a simple linear interpolation of the derivative terms is used as a standard within the FVM, and other differencing schemes are of minor importance.

When attempting to solve a partial differential equation such as Eq. (32) with a numerical approximation scheme, the accuracy of the approximation is a key issue for many practical applications. In chemical engineering technology, an increase in yield of a certain reaction of only 1% may be associated with a significant cost reduction. Even if in some cases numerical models with an order-of-magnitude accuracy may be useful (for example in cases where elaborate models and experimental data are not available), deviations of a few percent at the maximum between model predictions and the real world are usually demanded. Specifically in micro process technology and in microfluidics in general, there is one predominant discretization artefact often limiting the accuracy of numerical predictions: *numerical diffusion*.

The term numerical diffusion describes the effect of artificial diffusive fluxes which are induced by discretization errors. This effect becomes visible when the transport of quantities with small diffusivities [with the exact meaning of 'small' yet to be specified in Eq. (42)] is considered. In macroscopic systems such small diffusivities are rarely found, at least when being looked at from a phenomenological point of view. The reason for the reduced importance of numerical diffusion in many macroscopic systems lies in the turbulent nature of most macro flows. The turbulent velocity fluctuations induce an effective diffusivity of comparatively large magnitude which includes transport effects due to turbulent eddies [1]. The effective diffusivity often dominates the numerical diffusivity. In contrast, micro flows are often laminar, and especially for liquid flows numerical diffusion can become the major effect limiting the accuracy of the model predictions.

In order to be more explicit, the upwind differencing scheme will be considered. Again, an orthogonal computational grid aligned with the axes of a Cartesian coordinate system is assumed. When inserting the expression from the upwind scheme into the transport equation Eq. (33) and re-expressing the field values at the grid nodes by the corresponding values at the control-volume faces via a Taylor expansion, error terms of the same structure and the same sign as the original diffusive terms are found [81]. Hence the discretization error has the effect of artificially increasing the diffusive fluxes. The effective diffusion constant in the x-direction is given as

$$\Gamma_{\text{eff},x} = \Gamma + \frac{u_x h}{2}. \tag{42}$$

The second term on the right-hand side is the numerical diffusivity, depending on the flow velocity in the x-direction, u_x, and the length of the control volume in the x-direction, h. Similar expressions hold for the other coordinate directions. Apparently, the numerical diffusion increases with increasing flow velocity and

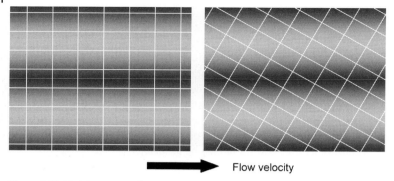

Flow velocity

Figure 2.10 Multilamination flow oriented along the computational grid (left) and forming a tilted angle with the grid cells (right).

increasing size of the grid cells. Furthermore, it is direction dependent and depends on the relative orientation of the local flow velocity vector and the control volume.

A problem often encountered in microfluidic systems is multilaminated flow with the flow direction equal to the orientation of the fluid lamellae. An example could be a multilamination micro mixer in which thin layers with varying concentration of a solvent are created. Again assuming a computational grid aligned with the axes of a Cartesian coordinate frame, Eq. (42) shows that the numerical diffusivity perpendicular to the fluid lamellae vanishes when the flow direction coincides with the orientation of the grid. On the other hand, when the flow is not aligned with the grid, an artificial diffusive flux between different fluid lamellae is induced. This situation is depicted in Figure 2.10 where the fluid lamellae are shaded with different gray tones. On the left-hand side, the flow is aligned with the grid and numerical diffusion between different lamellae is suppressed. On the right-hand side, the velocity vector stands at some finite angle with respect to the grid cell orientation. In such a case an artificial diffusive flux between different lamellae is induced.

In practice, situations where the computational grid can be fully aligned with the local flow velocity vectors are very rare. In order to reduce discretization artefacts, a differencing scheme of higher order can be chosen. In a strict sense, higher order differencing schemes are free of numerical diffusion, as, unlike for the upwind scheme, the error terms are no longer of the same structure as the diffusive transport terms. However, discretization errors of higher order are encountered having more subtle effects than purely enhancing the diffusivity. A comparison of typical results obtained with the upwind and the QUICK scheme for a hypothetical layered flow of a solute with zero diffusivity is shown in Figure 2.11. Owing to the vanishing diffusivity, in the exact solution of the transport equations the concentration profile should be given by a step function throughout the whole flow domain. This is indicated by a dashed line. In case of misalignment of the flow with the computational grid, the upwind scheme results in a broadening of the concen-

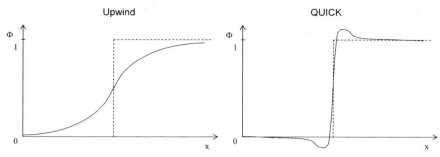

Figure 2.11 Step-function concentration profile and typical numerical results obtained with the upwind (left) and the QUICK scheme (right).

tration profile, as shown on the left. A significantly reduced broadening is found with the QUICK scheme (right). However, overshooting of the physical limits of the scalar field, which are assumed $\Phi = 1$ and $\Phi = 0$ for illustration, may occur. Even with higher order differencing schemes the modeling of convection–diffusion processes of solutes in liquid micro flows remains a challenge, and often satisfactory results are obtained only on very fine grids.

When transient problems are considered, the time derivative appearing in Eq. (32) also has to be approximated numerically. Thus, besides a spatial discretization, which has been discussed in the previous paragraphs, transient problems require a temporal discretization. Similar to the discretization of the convective terms, the temporal discretization has a major influence on the accuracy of the numerical results and numerical stability. When Eq. (32) is integrated over the control volumes and source terms are neglected, an equation of the following form results:

$$\frac{\partial \Phi}{\partial t} = A' \, \Phi \, , \tag{43}$$

with Φ being the vector of unknowns at the grid nodes and A' a coefficient matrix which is slightly modified compared to Eq. (38). For the sake of simplicity, a constant density was assumed and the source terms were set to zero. Often the time derivative is approximated as

$$\frac{\partial \Phi}{\partial t} \approx \frac{\Phi^{n+1} - \Phi^{n}}{\Delta t} \, , \tag{44}$$

where the superscripts denote the time steps n and $n + 1$. There are two major classes of time discretization schemes, *explicit* and *implicit* schemes. The explicit Euler scheme for the solution of Eq. (43) is given as

$$\frac{\Phi^{n+1} - \Phi^{n}}{\Delta t} = A' \, \Phi^{n} \, . \tag{45}$$

In the implicit Euler scheme, the unknown function at the new time step appears on the right-hand side.

$$\frac{\Phi^{n+1} - \Phi^n}{\Delta t} = A' \, \Phi^{n+1} . \tag{46}$$

Within explicit schemes the computational effort to obtain the solution at the new time step is very small: the main effort lies in a multiplication of the old solution vector with the coefficient matrix. In contrast, implicit schemes require the solution of an algebraic system of equations to obtain the new solution vector. However, the major disadvantage of explicit schemes is their instability [84]. The term stability is defined via the behavior of the numerical solution for $t \to \infty$. A numerical method is regarded as stable if the approximate solution remains bounded for $t \to \infty$, given that the exact solution is also bounded. Explicit time-step schemes tend to become unstable when the time step size exceeds a certain value (an example of a stability limit for PDE solvers is the von-Neumann criterion [85]). In contrast, implicit methods are usually stable.

For practical purposes, implicit schemes are the methods of choice when the solution is smooth and well behaved as a function of time. In that case much larger time steps can be taken than with explicit schemes, thus allowing a reduction in the computational effort. When large temporal gradients and rapid variations are expected, accuracy constraints set severe limits to the time-step size. In that case explicit schemes might be favorable, as they come with a reduced numerical effort per time step.

A method which represents a compromise between the explicit and implicit Euler scheme is the Crank–Nicolson method:

$$\frac{\Phi^{n+1} - \Phi^n}{\Delta t} = \frac{1}{2} A' \, \Phi^n + \frac{1}{2} A' \, \Phi^{n+1} . \tag{47}$$

The Crank–Nicolson method is popular as a time-step scheme for CFD problems, as it is stable and computationally less expensive than the implicit Euler scheme.

2.3.2
Solution of the Navier–Stokes Equation

The strategies discussed in the previous chapter are generally applicable to convection–diffusion equations such as Eq. (32). If the function Φ is a component of the velocity field, the incompressible Navier–Stokes equation, a non-linear partial differential equation, is obtained. This stands in contrast to Φ representing a temperature or concentration field. In these cases the velocity field is assumed as given, and only a linear partial differential equation has to be solved. The non-linear nature of the Navier–Stokes equation introduces some additional problems, for which special solution strategies exist. Corresponding numerical techniques are the subject of this section.

The flow phenomena described by the Navier–Stokes equation fall into two classes discriminated by the nature of the compressibility effects to be taken into account. For compressible flow, the Navier–Stokes equation [Eq. (1)] has to be solved in combination with the mass conservation equation [Eq. (2)]. In such a case an equation of motion for the pressure field results from the mass conservation equation when an equation of state:

$$\rho = \rho(p,T) \tag{48}$$

defining a relationship between density, pressure and temperature, is utilized. As an example, Eq. (48) could represent the ideal gas law.

The situation is different for incompressible flow. In that case, no equation of motion for the pressure field exists and via the mass conservation equation Eq. (17) a dynamic constraint on the velocity field is defined. The pressure field entering the incompressible Navier–Stokes equation can be regarded as a parameter field to be adjusted such that the divergence of the velocity field vanishes.

As will be outlined below, the computation of compressible flow is significantly more challenging than the corresponding problem for incompressible flow. In order to reduce the computational effort, within a CFD model a fluid medium should be treated as incompressible whenever possible. A 'rule of thumb' often found in the literature and used as a criterion for the incompressibility assumption to be valid is based on the Mach number of the flow. The Mach number is defined as the ratio of the local flow velocity and the speed of sound. The rule states that if the Mach number is below 0.3 in the whole flow domain, the flow may be treated as incompressible [84]. In practice, this rule has to be supplemented by a few additional criteria [3]. Especially for micro flows it is important to consider also the total pressure drop as a criterion for incompressibility. In a long micro channel the Mach number may be well below 0.3, but owing to the small hydraulic diameter of the channel a large pressure drop may be obtained. A pressure drop of a few atmospheres for a gas flow clearly indicates that compressibility effects should be taken into account.

For the important case of incompressible flow, a variety of numerical schemes for computing the velocity and pressure fields have been developed. A thorough discussion of even the few most important methods would be beyond the scope of this book. The discussion will be limited to so-called pressure-correction schemes which have found widespread applications in commercial FVM solvers. The essence of the methods lies in the coupling between velocity and pressure. On computational grids with coinciding nodes for the velocity and the pressure field, unphysical oscillations of the pressure field have been observed [81]. Such small-wavelength oscillations can be avoided when a staggered arrangement of grids for velocity and pressure is chosen [86]. A simple one-dimensional example of a staggered grid is displayed in Figure 2.12, together with the standard arrangement with colocated variables. In the colocated arrangement the nodes at which the velocity and the pressure variables are defined coincide. In the case of staggered grids, the pressure nodes are located on the control volume faces of the velocity grid. As staggered grids require an additional bookkeeping effort, especially for complex geometries, interpolation schemes for the velocity field have been developed which allow one to suppress the unphysical pressure oscillations [87]. These methods have to some degree rendered the staggered arrangement superfluous and have found widespread applications in commercial FVM solvers.

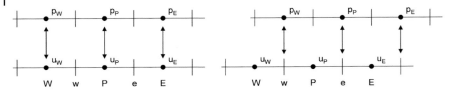

Figure 2.12 Computational grids with colocated (left) and staggered (right) arrangement of the velocity and pressure nodes.

A discretized version of the steady-state incompressible Navier–Stokes equation derived from Eq. (16) can be written as

$$a_P^{u_i} u_{i,P} + \sum_l a_l^{u_i} u_{i,l} = b_{i,P} - \left(\frac{\partial p}{\partial x_i}\right)_P . \tag{49}$$

In this expression, a node P is coupled to neighboring nodes l via coupling coefficients $a_l^{u_i}$ which themselves depend on the unknown velocity values u_i. The source term $b_{i,P}$ represents gravity or other volumetric forces. The spatial derivative of pressure taken at node P stands for any discrete version of this expression, either evaluated on a colocated or a staggered grid. The goal is to solve Eq. (49) in combination with the mass conservation equation Eq. (17). The basic idea of a pressure correction scheme is to begin an iterative procedure with a guess for the pressure field, denoted as p^{m-1}, and a corresponding guess for the velocity field u_i^{m-1}. Subsequently, a first approximation $u_i^{(1)\,m-1}$ for the new velocity field is computed by solving the equation

$$a_P^{u_i^{m-1}} u_{i,P}^{(1)m} + \sum_l a_l^{u_i^{m-1}} u_{i,l}^{(1)m} = b_{i,P} - \left(\frac{\partial p^{m-1}}{\partial x_i}\right)_P . \tag{50}$$

At that stage, the approximation obtained for the velocity field does generally not fulfil the mass conservation equation. In order to ensure mass conservation, corrections to the velocity and pressure field are introduced via

$$u_i^m = u_i^{(1)m} + u_i', \quad p^m = p^{m-1} + p' . \tag{51}$$

By demanding that the new velocity u_i^m field fulfils both the momentum and the mass conservation equation, the following equations for the velocity and pressure correction are derived:

$$a_P^{u_i^{m-1}} u_{i,P}' + \sum_l a_l^{u_i^{m-1}} u_{i,l}' = -\left(\frac{\partial p'}{\partial x_i}\right)_P , \tag{52}$$

$$\frac{\partial}{\partial x_i}\left[\frac{1}{a_P^{u_i^{m-1}}}\left(\frac{\partial p'}{\partial x_i}\right)\right]_P = \left(\frac{\partial u_i^{(1)m}}{\partial x_i}\right)_P - \frac{\partial}{\partial x_i}\left(\frac{1}{a_P^{u_i^{m-1}}} \sum_l a_l^{u_i^{m-1}} u_{i,l}'\right)_P . \tag{53}$$

By solving the pressure and velocity correction equations, the corresponding correction terms and thus the new velocity field u_i^m, which by construction fulfils the mass conservation equation, can be determined. However, in general a single correction step does not yield a solution of the Navier–Stokes equation, the reason for this fact being traced back to Eq. (52) and Eq. (50). The coupling coefficients $a_P^{u_i^{m-1}}$ appearing in these equations are evaluated with the help of the old velocity values u_i^{m-1}. Hence u_i^m does not represent a self-consistent solution of the Navier–Stokes equation and further iterations are required to obtain the final solution. For this purpose p^m and u_i^m are chosen as new initializations and the procedure described above is repeated until convergence is achieved.

In practice, the full versions of the pressure and velocity correction equation are rarely considered. In order to determine the velocity correction, the sum over neighboring nodes:

$$\sum_l a_l^{u_i^{m-1}} u_{i,l}' \tag{54}$$

appearing in Eqs. (52) and (53) would require a linear system of equations to be solved at each iteration. As this would introduce a considerable numerical effort, some simplifications are usually made. The most drastic approximation is to disregard completely the sum over neighboring nodes. Such a modification leaves the final values of the pressure and the velocity field unaffected, as for a fully converged solution the corrections are zero and the neglect of a vanishing term does not alter the determining equations. The corresponding method is known as the SIMPLE algorithm [88]. The SIMPLE algorithm is prone to convergence problems and usually requires under relaxation. For this reason, a slightly more complicated but yet more robust method has been devised. Instead of neglecting the sum over neighboring nodes it is approximated as

$$\sum_l a_l^{u_i^{m-1}} u_{i,l}' \approx u_{i,P}' \sum_l a_l^{u_i^{m-1}} . \tag{55}$$

A pressure correction scheme based on such an approximation is known as the SIMPLEC algorithm [89]. The SIMPLEC algorithm does not require under relaxation of the pressure correction, is efficient and has found widespread applications.

Compared with incompressible flow, the computation of compressible flow is significantly more involved. In a compressible medium, sound waves may propagate and may induce complicated spatio-temporal patterns superposed on the flow. For transsonic and supersonic flows, non-linear wave phenomena, so-called shock waves, are observed. Shock waves are sharp wave fronts where flow quantities such as density or pressure change in a discontinuous manner. In order to capture shock fronts numerically, very fine grids in the close vicinity of the shock are needed. As the location of the shock fronts is *a priori* unknown, meshes with adaptive grid refinement are required to describe corresponding flow phenomena with a reasonable degree of accuracy while keeping the numerical effort within tolerable limits [90, 91]. Even without shock waves compressible flows bear some additional numerical challenges. For incompressible flow the scale of the numerical time-step size is governed by the time it takes the fluid to convect through a single computational cell. When compressible flow is considered, a new time-scale is introduced via the speed of sound and the corresponding time a sound wave takes to propagate through a computational cell. If the speed of sound is much larger than the flow velocity, the allowable time-step size in a numerical computation of compressible flow phenomena might be much smaller than the corresponding quantity for incompressible flow.

Another principle difference of compressible flow computations as compared with incompressible flow is an enhanced coupling of the governing equations. Via the equation of state Eq. (48) a change in density is in general related to a temperature change. Vice versa, via convective transport of heat the velocity field has a substantial impact on the spatio-temporal temperature distribution. On the other hand, a change in temperature may induce a flow via the induced change in density. Hence there is a two-way coupling between the Navier–Stokes equation and the enthalpy equation. Another effect adding to the coupling is viscous heating. The viscosity term in the Navier–Stokes equation:

$$
\frac{\partial}{\partial x_k}\left[\mu\left(\frac{\partial u_i}{\partial x_k}+\frac{\partial u_k}{\partial x_i}-\frac{2}{3}\delta_{ik}\frac{\partial u_j}{\partial x_j}\right)\right] \tag{56}
$$

describes the dissipation of kinetic energy into heat. Consequently, from this term a source term for the enthalpy equation can be derived (see, e.g., [29]). The coupling of the momentum and the enthalpy equation induced is not only restricted to compressible flow, but in principle has to be considered in the case of incompressible flow also. However, the corresponding effect, known as viscous heating, is of only minor importance in many practical cases. A phenomenon of notable importance in practice is the temperature dependence of the dynamic viscosity. For non-isothermal, compressible and incompressible flow it should be taken into account that the viscosity depends on temperature and that the corresponding relationship allows the temperature field to have an effect on the flow field.

For the computation of compressible flow, the pressure–velocity coupling schemes previously described can be extended to pressure–velocity–density coupling schemes. Again, a solution of the linearized, compressible momentum equation obtained with the pressure and density values taken from a previous solver iteration in general does not satisfy the mass balance equation. In order to balance the mass fluxes into each volume element, a pressure, density and velocity correction on top of the 'old' values is computed. Typically, the detailed algorithms for performing this task rely on the same approximations such as the SIMPLE or SIMPLEC schemes outlined in the previous paragraph.

2.3.3
Computational Grids

The quality of the numerical grid is essential for the accuracy of the numerical results obtained in a CFD calculation. Not only the size but also the geometric shape of the grid cells determines the discretization errors of the numerical approximation. Whereas 20 years ago only simple geometries with a comparatively small number of degrees of freedom could be considered, nowadays even standard workstations allow the study of flow phenomena in complex geometries on grids with several million cells. This necessitates highly automated methods for the generation of high-quality grids.

Given a geometric model of the flow domain to be considered, the problem is to create a grid which is ideally adjusted to the geometry and produces comparatively small numerical errors. The latter requirement can be translated to an adjusted spatial variation of the grid resolution. Ideally, the grid should be fine in regions where large gradients of the computed field quantities are found, whereas it can be comparatively coarse in the rest of the computational domain. The decision on where to provide a high-mesh density is often left to the user. An experienced user is usually able to predict some of the regions where high numerical accuracy is needed, but even an expert will experience occasional surprises. In contrast, in the past few years some powerful automated methods for local grid refinement on the basis of *a posteriori* error estimates and target functions have been developed [92, 93]. However, the discussion of such methods is beyond the scope of this book, which can only scratch the surface of the most common techniques of computational fluid dynamics. In the following paragraphs an overview of the different types of grids, including a brief discussion of their advantages and disadvantages, is given.

There are at least two different ways to adjust a computational grid to the geometry of the flow domain. One strategy is to overlay a Cartesian grid on the domain and to approximate those parts of the domain boundary which are not parallel to the coordinate directions by a stepwise borderline. Sample geometry with a corresponding Cartesian grid is shown on the left side of Figure 2.13. Another strategy is to give up the requirement for orthogonality and to adjust the grid geometry in such a way that the domain boundary coincides with the boundary of the grid. Such a body-fitted grid is shown on the right side of Figure 2.13.

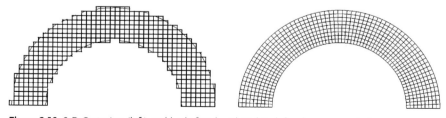

Figure 2.13 2-D Cartesian (left) and body-fitted grid (right) defined on a curved pipe.

The advantage of a Cartesian grid is the fact that discretization schemes can be easily implemented on a rectangular geometry. However, the approximation of the domain boundary by a step function is often poor and, correspondingly, boundary conditions are implemented only in a very crude fashion. In contrast, body-fitted grids allow for an unproblematic implementation of boundary conditions. Furthermore, with body-fitted grids, numerically optimal configurations such as grid cells aligned with the flow direction are often easier to achieve. As an example, the grid cells of the body fitted grid in Figure 2.13 are to a large degree aligned with the flow. According to the discussion in Section 2.3.1, an alignment of the velocity vectors with the grid cells guarantees a minimum degree of numerical diffusion. The disadvantage of body-fitted grids lies in the numerical complexity and a somehow increased computational effort related to solving the transport equations on a non-orthogonal grid. Nevertheless, body-fitted grids have found widespread applications in commercial CFD solvers, for which they have become almost a standard.

The standard way of implementing the transport equations for momentum, heat and matter on a body-fitted grid is based on mapping the complex flow domain in physical space to a simple (for example rectangular) flow domain in computational space. This is achieved via a curvilinear coordinate transformation. As an example, for the problem of computing the flow field in a curved pipe the physical domain shown on the left hand side of Figure 2.14 is mapped to a simple rectangular domain in computational space. The simple grid geometry in computational space has the advantage that discretization of the transport equations is comparatively straightforward. However, through the curvilinear coordinate transformation a number of extra terms in the transport equations are introduced, examples of which are the centrifugal or the Coriolis force in the transformed momentum equation.

For the example in Figure 2.14 it would be possible to perform the coordinate transformation analytically by introducing cylindrical coordinates. However, in general, geometries are too complex to be described by a simple analytical transformation. There are a variety of methods related to numerical curvilinear coordinate transformations relying on ideas of tensor calculus and differential geometry [94]. The fundamental idea is to establish a numerical relationship between the physical space coordinates x_i and the computational space curvilinear coordinates ξ_i. The local basis vectors of the curvilinear system are then given as

$$(\mathbf{e})_i = \frac{\partial \mathbf{x}}{\partial \xi_i} \tag{57}$$

Physical space   Computational space

Coordinate transformation

Figure 2.14 Example of a grid structure in physical space (left) and in computational space (right).

and the transport equation for a field quantity Φ can be transformed by means of the relationship

$$\frac{\partial \Phi}{\partial \xi_i} = \frac{\partial \Phi}{\partial x_j} \frac{\partial x_j}{\partial \xi_i}. \tag{58}$$

When facing the task of creating a computational grid for solving a fluid flow problem, a decision has to be made whether a structured or an unstructured grid is more suited for the problem under consideration. An example of 2-D structured and unstructured grids was given in Figure 2.7. A *structured grid* typically consists of quadrilateral or hexahedral cells and is defined according to the specifications of the user. In order to define a structured grid, the flow domain is divided into several blocks which are topologically equivalent to quadrilaterals (in 2-D) or hexahedra (in 3-D). Subsequently, the user has to specify subdivisions on the edges of the blocks which are directly used to define a grid of quadrilateral or hexahedral cells within the blocks. Inside each block a grid cell can be uniquely identified by a set of indices. For example, in two dimensions a label (i,j) denotes a cell in the ith column and jth row, where rows and columns are aligned with the local coordinate system of the block. The advantage of a structured grid is the fact that owing to the regular geometry of the computational cells, discretization errors related to specific differencing schemes for the various terms in the transport equations are minimized. A disadvantage of structured grids is related to local adaptation to steep gradients or sudden changes of the field quantity to be determined. Owing to the simple construction scheme of structured grids, it is very difficult to enhance the grid density in localized regions. Typically, in order to refine the grid at a specific point in space, the grid density has to be enhanced also in other regions where owing to the lack of steep gradients a lower grid density would be sufficient. As a further disadvantage, in complex geometries it can be very difficult or time consuming to define a structured grid. The standard approach to overcome such problems is the automatic generation of an unstructured grid.

In many fields of engineering, *unstructured grids* represent the only way to study flow phenomena in realistic geometries. Nowadays, CDF simulations based on unstructured grids of the scale of multi-million grid cells can be run even on standard workstations. In Figure 2.15 an example of a fluidic connector meshed with an unstructured grid is displayed. Typically, an unstructured grid consists of triangular (in 2-D) or tetrahedral (in 3-D) cells. The cell connectivity of such a grid is more complex than for an unstructured grid and a simple labeling of the cells with row and column indices is no longer possible. An advantage of unstructured grids is the fact that they can be created in a highly automated way, with minimum user input required. A further important aspect of automated grid generation is the possibility of automatic grid refinement according to local properties of the solution field. The only efficient way to adjust the grid density locally in response to a current approximation to the computed field quantities is an automatic subdivision of the grid cells. Even when no automatic refinement based on a sequence of grids is necessary and a single grid is sufficient, unstructured grids offer ways to

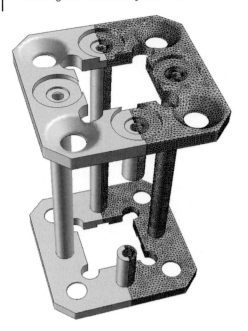

Figure 2.15 Fluidic connector
with an unstructured tetrahedral grid.

adjust the grid density in localized regions in space unparalleled by structured grids. The disadvantage of unstructured grids is a reduced accuracy due to the lower approximation order of many differencing and interpolation schemes when implemented on triangular or tetrahedral geometries.

Algorithms for the generation of unstructured meshes usually start on the boundary of the domain, for example in the three-dimensional case first a surface grid is created. After that, tetrahedra with a base on the surface are formed and the algorithm subsequently works its way towards the center of the domain. Frequently used algorithms for unstructured mesh generation are the advancing front method [95] and the method of Delaunay triangulation [96].

In addition to the grid density, the quality of a grid depends on a number of other factors such as the aspect ratio of the cells and the relative size of neighboring cells. A comprehensive listing of the quality criteria for computational grids is given elsewhere [83]. Some of the important points to be remembered are:

- Highly skewed cells should be avoided. The angles between the grid lines of a hexahedral mesh should be ~ 90°. Angles < 40° or > 140° often imply a reduced accuracy or numerical instabilities.
- The aspect ratio of hexahedral cells should be not too high, typically below 20–100. If high-aspect ratio cells are used, the accuracy and possible convergence problems depend greatly on the flow direction.
- The mesh expansion ratio, i.e. the size ratio of neighboring cells, should be kept small. Particularly in regions of large gradients, mesh size discontinuities should be avoided.

2.3.4
Solution Methods for Linear Algebraic Systems

The efficiency of modern CFD solvers relies heavily on the availability of fast solution methods for linear systems of algebraic equations. In the course of a CFD simulation, linear equation systems such as Eq. (38) have to be solved successively. The problem to be solved might be a linear one in itself, as would be the case for heat transfer phenomena. Then, given a steady-state problem, a single solution of Eq. (38) suffices to determine the unknown field quantity Φ in the whole computational domain. In the case of time-dependent linear problems, a linear algebraic system usually has to be solved for each time step. Finally, an inherently non-linear problem such as the solution of the Navier–Stokes equation requires successive solution of a system such as Eq. (38) in the course of the iteration process set up to solve the resulting non-linear algebraic system. If, in that context, a time-dependent problem is considered, a nested set of computational tasks consisting of successive time steps, successive non-linear iterations and successive solutions of linear algebraic systems is obtained. Thus, fast solution methods for linear equation systems are of paramount importance for any CFD solver.

In principle, the task of solving a linear algebraic systems seems trivial, as with Gauss elimination a solution method exists which allows one to solve a problem of dimension N (i.e. N equations with N unknowns) at a cost of $O(N^3)$ elementary operations [85]. Such solution methods which, apart from roundoff errors and machine accuracy, produce an exact solution of an equation system after a predetermined number of operations, are called *direct solvers*. However, for problems related to the solution of partial differential equations, direct solvers are usually very inefficient. Methods such as Gauss elimination do not exploit a special feature of the coefficient matrices of the corresponding linear systems, namely that most of the entries are zero. Such sparse matrices are characteristic of problems originating from the discretization of partial or ordinary differential equations. As an example, consider the discretization of the one-dimensional Poisson equation

$$\frac{d^2\Phi}{dx^2} = f(x) \tag{59}$$

by means of a simple finite-difference approach on a uniform grid. The resulting system of linear algebraic equations is given as

$$\frac{1}{(\Delta x)^2}(\Phi_{i-1} - 2\,\Phi_i + \Phi_{i+1}) = f_i, \tag{60}$$

where Δx is the grid spacing. From Eq. (60), it is obvious that a node i is coupled only to the neighboring nodes $i - 1$ and $i + 1$. The corresponding coefficient matrix has entries only along its diagonal and along two bands above and below the diagonal.

In order to exploit the sparseness of the matrix, *iterative solvers* can be applied. The iterative procedure is initialized with a guess for the solution vector Φ. In subsequent iterations this guess is successively improved until the required accuracy is reached. In general, iterative solvers produce an exact solution of a linear algebraic system only after an infinite number of iterations. However, as even direct solvers never produce exact results owing to roundoff errors, iterative solvers might yield approximations of comparable accuracy after a comparatively small number of iterations. In addition, there are inaccuracies related to the spatial and temporal discretization, and it is in general not useful to require a higher accuracy from the equation solvers than inherent in the discretization scheme. For large linear systems, iterative solvers are usually much more efficient than direct solvers. In general, an analytical expression for the operation count of iterative solvers as a function of the problem dimension cannot be derived. However, for the case of the numerical solution of the Poisson equation sketched above, the operation count of iterative solvers required for obtaining an approximate solution with relative accuracy ε can be computed analytically. Compared with Gauss elimination with its $O(N^3)$ effort, iterative solvers show a much better performance. Examples are the conjugate gradient method with its $O(N^{3/2} \ln\varepsilon)$ and the full multigrid method with its $O(N)$ operation count [97].

The general structure of an iterative solution method for the linear system of Eq. (38) is given as

$$M \, \Phi_{n+1} = N \, \Phi_n + b \,, \tag{61}$$

with

$$A = M - N \,. \tag{62}$$

The iteration scheme allows to compute an improved approximation Φ_{n+1} given a previous approximation Φ_n as an input. If after a number of iterations $\Phi_{n+1} \approx \Phi_n$ (i.e. the iteration scheme has converged), an approximate solution to the linear algebraic system has been found. The choice of the matrices M and N determines the specific form of the iterative solution scheme. After n iterations, the approximate solution will in general not satisfy the linear equation system exactly, but a non-zero residual ρ_n will remain, defined via

$$A \, \Phi_n = b - \rho_n \,. \tag{63}$$

In CFD, a number of different iterative solvers for linear algebraic systems have been applied. Two of the most successful and most widely used methods are *conjugate gradient* and *multigrid methods*. The basic idea of the conjugate gradient method is to transform the linear equation system Eq. (38) into a minimization problem

$$F = \Phi^T \, A \, \Phi - \Phi^T \, b \overset{!}{=} \min. \tag{64}$$

If the matrix A is positive definite, i.e. it is symmetric and has positive eigenvalues, the solution of the linear equation system is equivalent to the minimization of the bilinear form given in Eq. (64). One of the best established methods for the solution of minimization problems is the method of steepest descent. The term 'steepest descent' alludes to a picture where the cost function F is visualized as a landscape of hills and valleys in an *N*-dimensional space. In order to find a minimum of *F* starting from a specific point in that *N*-dimensional space, one simply walks 'downhill', i.e. in the direction of the negative gradient of F. A simple strategy would be to minimize *F* along a line defined by the negative gradient, compute the gradient at the minimum point determined in such a way, do another line minimization, etc. Often, in practice this strategy is not very efficient, as the search direction may oscillate rapidly without substantially lowering the value of the cost function. It is in fact not advisable always to take the steepest descent, rather the search should be performed along directions which are *conjugate* to each other. Two unit vectors e_1 and e_2 are conjugate with respect to a matrix *A* if they satisfy

$$e_1^T A e_2 = 0. \tag{65}$$

If a line minimization along e_1 is followed by a minimization along a conjugate direction e_2, the second minimization will avoid undoing what has been achieved in the first minimization as far as possible. In fact, it can be proven that the minimum of the cost function can be found by means of *N* subsequent line minimizations along conjugate directions [98]. However, the number of iterations required to find the exact solution will be very large for a typical CFD problem. For this reason, it is advisable to view the conjugate gradient method not as an algorithm to find exact solutions to systems of linear equations, but rather as an iterative scheme to determine approximate solutions. One of the essential ingredients of the conjugate gradient method is an algorithm for finding conjugate unit vectors with respect to the coefficient matrix under study. There exist methods which allow one to determine a unit vector conjugate to an already existing set of unit vectors [98]. In this way it is possible to start with one search direction and to subsequently generate new search directions which are optimal in the sense described above.

In practice, it is necessary to supplement and modify the basic principles described here. The coefficient matrix *A* will in general not be positive definite, so first the problem has to be transformed into one to which a minimization procedure is applicable [170]. In addition, without further provisions, convergence of the conjugate gradient method will be slow. The convergence speed of the method depends on the condition number of the matrix *A*, which is the ratio of the largest and the smallest eigenvalue. It is advisable first to do a transformation which improves the condition number before the conjugate gradient method is started. Hence the conjugate gradient method is applied to a transformed linear system of the form

$$\tilde{A}^{-1} A \Phi = \tilde{A}^{-1} b. \tag{66}$$

The matrix $\tilde{A}$ is known as the preconditioner and has to be chosen such that the condition number of the transformed linear system is smaller than that of the original system.

In contrast to the conjugate gradient method, the multigrid method is rather a general framework for iterative solvers than a specific method. The multigrid method exploits the fact that the iteration error

$$\varepsilon_n = \Phi - \Phi_n \tag{67}$$

often behaves in a very well-defined way as a function of the iteration number n. The iteration error is, similarly to Φ, a field quantity and depends on the spatial coordinates. It can be decomposed into Fourier components which show a different behavior as a function of n. Many iterative solvers reduce the short-wave components of ε_n fairly fast, whereas they are very slow in eliminating the long-wave components. The multigrid method makes use of the fact that in order to eliminate the long-wave components, a coarse grid is sufficient. On a coarse grid the effort for obtaining the updated field Φ_{n+1} from Eq. (61) is reduced. In addition, the whole iteration scheme usually converges faster, thus further saving computational effort. Within a multigrid method a number of grids with different resolution are employed to reduce the different Fourier components of ε_n. In order to perform that task, a minimum number of numerical building blocks is necessary:

- An iterative solution method for linear algebraic systems which damps the short-wave components of the iteration error very fast and, after a few iterations, leaves predominantly long-wave components. The Gauss–Seidel method [85] could be chosen as a suitable solver in this context.
- A method for smoothing the residual obtained on the fine grid in order to compute the corresponding residual on the coarse grid. In the terminology of the multigrid method, this step is called *restriction*.
- A method for interpolating the update to the solution obtained on the coarse grid to the fine one. The interpolation step is denoted *prolongation*.

In one cycle of the multigrid method, first a few iterations are performed on the fine grid in order to obtain a comparatively smooth iteration error. After that the obtained residual is restricted to the coarse grid, where further iterations are performed in order to damp out the long-wave components of the solution error. Subsequently the coarse-grid solution is interpolated to the fine grid and the solution on the fine grid is updated.

In order to damp effectively the solution error on all length scales, it is advisable to work not just with two, but with a multitude of different grids. There are in principle two different strategies for directing the information flow between grids of different resolution [99]. One can take the point of view that one moves to a coarser grid as soon as the short-wavelength Fourier components have been damped on the finer grid and any further iteration would only effect a very slow reduction of long-wave components. Damping the long-wave components of the iteration error on a coarser grid is preferable owing to the reduced computational cost. A recursive implementation of such a strategy is depicted on the left side of Figure 2.16. The solution procedure starts with a few iterations on a grid with spacing h, then the solution is restricted to a grid with spacing $2h$ and the steps are repeated until the coarsest grid is reached. After that, the coarsest-grid solution is prolongated to a finer grid ($8h$), and the corrected solution obtained from the prolongation is taken as initialization for a few further iterations on the finer grid. The procedure is repeated until the finest grid is reached, on which the approximative solution of the problem is obtained.

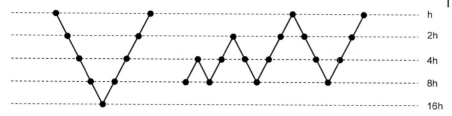

h

2h

4h

8h

16h

Figure 2.16 Two different schemes for the information flow between grids of different scale.

The other point of view concerning the information flow between different grids assumes that the iterations on a coarse grid provide an initial guess for the solution procedure on a fine grid. Correspondingly, one starts at the coarsest grid and successively moves on to finer grids. When implementing this idea and allowing for intermediate restrictions to coarser grids in order to damp the long-wave components of the iteration error, a scheme as depicted on the right side of Figure 2.16 is obtained.

Multigrid methods have proven to be powerful algorithms for the solution of linear algebraic equations. They are to be considered as a combination of different techniques allowing specific weaknesses of iterative solvers to be overcome. For this reason, most state-of-the-art commercial CFD solvers offer the multigrid capability.

2.4
Flow Distributions

In this and the following sections, the application of modeling and simulation techniques in the field of chemical micro process technology will be discussed. Several examples highlight the work that has already been done in this area and also the specific problems arising in the context of micro reactor and process design. Since the basic framework of numerical techniques was outlined in the previous section, in the following only specific models and special techniques allowing the corresponding problems to be solved efficiently will be discussed. In this context, not everything will rest on the fundaments of the FVM, but also other approaches, based on, e.g., finite-element or macro model approaches, play a role. As a discussion of the details of these numerical methods would be beyond the scope of this book, the reader is referred to the references given in the previous section for further information on these topics.

The focus of the examples given in this chapter is clearly on micro reactors for chemical processing in contrast to µ-TAS or Lab-Chip systems for bioanalytical applications. In the latter microfluidic systems, the fluidic requirements are somehow different from those in micro reactors. Typically, throughput plays only a minor role in µ-TAS systems, in contrast to micro reactors, where often the goal is to achieve a maximum molar flux per unit volume of a specific product. Moreover, flow control plays a much greater role in µ-TAS systems than in micro reactors. In μ-TAS systems it is often required to transport fluid volumes in a very exact and highly synchronized way, and dosing and metering operations are of major importance. Whereas μ-TAS systems are mostly based on liquid flows, many micro reactors rely on gas flows or multiphase flows. According to this distinction, some of the effects which have been utilized to meet the specific requirements of μ-TAS systems will not be discussed here. Examples of such effects are electrophoresis, electroosmosis or liquid transport by electrowetting.

Owing to the specific microstructuring technology employed to build up micro reactors, the geometric shape of the flow domain is often different from that in macroscopic equipment. Whereas usually the elements for fluid transport such as pipes are of circular cross-section, the channels in micro reactors have a rectangular or trapezoidal cross-section. Depending on the specific microstructuring technology, tub-like grooves or close-to triangular shapes might also be obtained, especially when after microfabrication a catalyst layer is deposited on the channel walls. Furthermore, a characteristic feature that many micro reactors have in common is a comparatively wide flow distributor followed by a large number of parallel micro channels. In the following, the flow distributions in these characteristic geometries and methods to obtain approximate flow distributions in highly parallel flow domains will be discussed.

2.4.1
Flow in Rectangular Channels

For laminar flow in channels of rectangular cross-section, the velocity profile can be determined analytically. For this purpose, incompressible flow as described by Eq. (16) is assumed. The flow profile can be expressed in form of a series expansion (see [100] and references therein), which, however, is not always useful for practical applications where often only a fair approximation of the velocity field over the channel cross-section is needed. Purday [101] suggested an approximate solution of the form

$$u(x,y) = u_{max} \left[1 - \left(\frac{x}{a} \right)^s \right] \left[1 - \left(\frac{y}{b} \right)^r \right] \tag{68}$$

for a rectangular channel oriented along the z-axis with a width 2 a in the x-direction and a depth 2b in the y-direction, where u is the local flow velocity and u_{max} the maximum velocity. The exponents s and r depend on the aspect ratio b/a of the channel; the most common correlations can be looked up elsewhere [100]. Typically, Eq. (68) approximates the exact velocity profile with an accuracy of a few percent. Often, not the detailed velocity profile is of interest, but only the friction factor f which determines the pressure drop over a channel of a given length. The fanning friction factor is defined as

$$f = -\frac{dp}{dz} \frac{D_h}{2 \rho U^2} , \tag{69}$$

with dp/dz being the pressure gradient along the channel, ρ the density, D_h the hydraulic diameter and U the mean flow velocity. The hydraulic diameter is given by $4A/P$, with A being the cross-sectional area of the channel and P its perimeter. Shah and London [102] derived a comparatively simple expression for the friction factor in rectangular channels which deviates from the analytical solution by less than 0.05%:

$$f = \frac{24}{Re}(1 - 1.3553\,\alpha + 1.9467\,\alpha^2 - 1.7012\,\alpha^3 + 0.9564\,\alpha^4 - 0.2537\,\alpha^5), \quad (70)$$

where α is the aspect ratio of the channel ($\alpha \leq 1$ by definition) and the length scale entering the Reynolds number Re is the hydraulic diameter.

Most expressions for the flow profile in rectangular channels assume that the flow is fully developed, i.e. that the flow velocity is oriented along the z-axis and does not change in streamwise direction. Close to the entrance of a channel this assumption is not valid, and the flow undergoes a development from an entrance distribution to a fully developed profile. Correspondingly, the pressure distribution deviates from that observed in a fully developed flow and expressions for the friction factor such as Eq. (70) are not valid in the entrance region. In order to determine the developing velocity distribution and friction factor, various approaches have been employed, such as analytical calculations based on a linearized inertia term, numerical solutions of the Navier–Stokes equation or experimental velocity and pressure measurements. An overview of the results was given by Hartnett and Kostic [100]. In general, the pressure drop per unit length of an entrance flow is higher than that of a fully developed flow. In order to compare entrance flow effects, a hydrodynamic entrance length L_{hy} can be defined which is the length necessary to achieve a centerline velocity equal to 99% of the fully developed value. Usually, a non-dimensional quantity is used, defined as

$$L_{hy}^+ = \frac{L_{hy}}{D_h\,Re}. \quad (71)$$

Depending on the aspect ratio of the channel, values between 0.01 and 0.1 are found for the non-dimensional entrance length [100]. From Eq. (71) it can be deduced, with L_{hy}^+ being Re independent, that L_{hy} increases linearly with the hydraulic diameter and the Reynolds number.

2.4.2
Generalized Channel Cross-Sections

A number of authors have considered channel cross-sections other than rectangular [102–104]. Figure 2.17 shows some examples of cross-sections for which friction factors and Nusselt numbers were computed. In general, an analytical solution of the Navier–Stokes and the enthalpy equations in such channel geometries would be involved owing to the implementation of the wall boundary condition. For this reason, usually numerical methods are employed to study laminar flow and heat transfer in channels with arbitrary cross-sectional geometry.

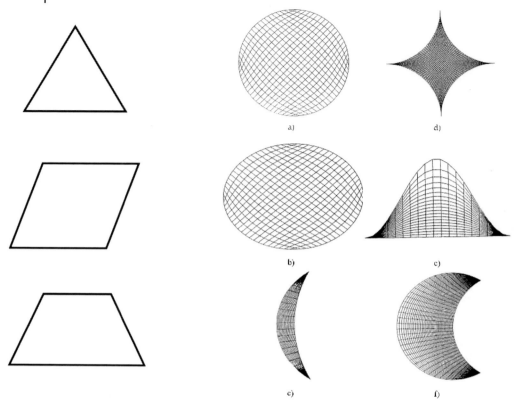

Figure 2.17 Channel cross-sections considered by Shah [103] (left) and by Richardson et al. [104] (right) in their numerical studies of laminar flows.

Shah [103] used a spectral method to compute the friction factors in ducts of triangular, rhombic and trapezoidal cross-section, among others. The results were given as tabulated values for different geometric parameters. Richardson et al. [104] considered the cross-sectional geometries which are shown on the right side of Figure 2.17 together with the computational grid. They used a finite-volume method to compute the velocity and the temperature field inside such channels and displayed their results in the form of an entropy generation rate going along with viscous friction.

2.4.3
Periodic and Curved Channel Geometries

Besides channels with different cross-sections, channels with specific periodic shapes were considered by several authors, mostly in the framework of 2-D simulations. Three different channel shapes having been studied are shown in

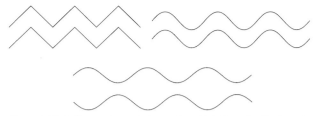

Figure 2.18 Different channel shapes for which flow distributions have been computed, a zigzag (upper left), a sinusoidally curved (upper right) and a converging–diverging channel (bottom).

Figure 2.18, a zigzag (upper left), a sinusoidally curved (upper right) and a converging–diverging channel (bottom). Asako and Faghri [105] computed velocity and temperature fields in zigzag channels based on the FVM for Reynolds numbers up to 1500. They found that above a specific Reynolds number depending on the geometry parameters, flow separation occurs, i.e. in the corners of the channel recirculation zones are being formed. Whereas the friction factor for straight channels displays a linear decrease as a function of Reynolds number, as is apparent from Eq. (70), the friction factor in zigzag channels becomes nearly independent of the Reynolds number for Re > 1000. Flow in sinusoidally curved channels for Reynolds numbers up to 500 was studied by Garg and Maji [106]. Again, a finite-volume discretization was used. For most of the parameter space explored, flow separation was not observed. However, for Re = 500 and a large enough sine-wave amplitude as compared with the period, there were some indications of recirculation zones forming in the recesses of the channel. Unfortunately, no attempt was made to compute friction factors.

Guzmán and Amon [107] studied flow in converging–diverging channels and paid special attention to the transition between laminar and turbulent flow. They used a spectral-element method where the flow domain is divided into macro-elements over which a set of polynomial test functions is defined. Their method allowed damping of small-scale fluctuations due to numerical viscosity to be suppressed and is thus well suited to study the transition from stationary to oscillatory and chaotic flow. The computed streamline patterns indicate a transition from a non-separated flow at low Reynolds numbers to a flow with recirculation zones within the recesses of the channel, occurring at comparatively small Reynolds numbers between 10 and 20. Streamline patterns for various Reynolds numbers are depicted in Figure 2.19. The phenomenology of flow distributions in such channel domains was found to be diverse. At low Reynolds numbers the flow is stationary, but at a Reynolds number of 150 oscillations begin to develop, with the vortices still being confined in the recesses. At Re = 400 the viscous forces are no longer strong enough to confine the vortices in the recesses and vortex ejection is observed. At even higher Reynolds numbers the flow becomes aperiodic and chaotic. Guzmán and Amon further analyzed the flow patterns using methods established in chaos theory based on, e.g., Poincaré sections, pseudo-phase space representations and Lyapunov exponents, and thereby provided a detailed flow map of the laminar-to-turbulent transition in converging–diverging channels.

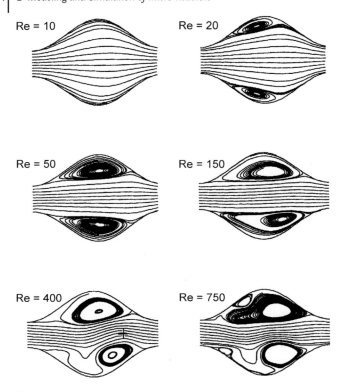

Figure 2.19 Streamline patterns in a converging–diverging channel for various Reynolds numbers, taken from [107].

In some cases, for example when large residence times are desired, comparatively long micro channels have to be integrated into a compact micro reaction device, a task which can only be achieved with curved channels. In addition, straight channels are often not suited for connecting a micro flow domain to the external world. One of the first theoretical studies on flow in curved channels was done by Dean [108, 109], who investigated the secondary flow perpendicular to the main flow direction induced by inertial forces using a perturbative analysis. The dimensionless group characterizing such transversal flows is the Dean number, defined as

$$K = \text{Re}\sqrt{\frac{D_h}{R}}\,, \tag{72}$$

where D_h is the hydraulic diameter and R the mean radius of curvature of the channel. The typical secondary flow pattern in curved channels is given by two counter-rotating vortices separated by the plane of curvature. The strength of these vortices increases with increasing Dean number. Originally, the studies of secondary flow in curved channels were performed for pipes of circular cross-section.

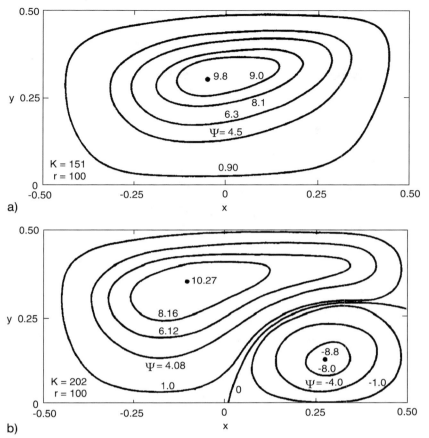

Figure 2.20 Streamline patterns of secondary flow in quadratic channels for $K = 151$ (above) and $K = 202$ (below), taken from [110]. Only the upper half of the channel cross-section is shown.

Of much greater relevance in micro reactors are rectangular channels, which were the subject of a study by Cheng et al. [110], among others. They solved the Navier–Stokes equation for channel cross-sections with an aspect ratio between 0.5 and 5 and Dean numbers between 5 and 715 using a finite-difference method. The vortex patterns obtained as a result of their computations are depicted in Figure 2.20 for two different Dean numbers.

As a result of the inertial forces exerted on the fluid elements due to the curvature of the channel, a pair of counter-rotating vortices transporting fluid perpendicular to the main flow direction is formed. When the Dean number is increased to values larger than 200, a second pair of vortices emerges close to the outer channel wall. In addition to the secondary flow patterns, Cheng et al. computed the axial velocity distribution and friction factors for different Reynolds and Dean numbers. Not surprisingly, the friction factor in curved channels was found to be larger than that in straight channels.

2.4.4
Multichannel Flow Domains

In chemical micro process technology, on the one hand it is important to guarantee well-defined and reproducible reaction conditions in micro channels, and on the other a high throughput should be achieved. For this purpose, the process fluid is guided through a large number of parallel micro channels, where heat exchange and/or chemical reactions occur. One of the characteristic problems in micro process technology is to distribute the incoming fluid equally over the micro channels. A fluid maldistribution would induce unequal residence times in different channels, with undesirable consequences for the product distribution of a chemical reaction being conducted inside the reactor. When a process with various competing side reactions and by-products is considered, the contact time of the process fluid in the reaction region should be as well defined as possible in order to maximize the selectivity of the process. However, it should be pointed out that not only the maldistribution of the fluid over a multitude of micro channels induces variations of the contact time, but also hydrodynamic dispersion of concentration tracers in the channels themselves. The corresponding effects will be discussed in a subsequent section devoted to mass transfer.

Various concepts for the equipartition of fluid over a multitude of micro channels have been developed. One concept relies on guiding the incoming fluid through a flow splitter with subsequent bifurcations [111]. The most widely used design is based on a comparatively wide inlet region leading into a multitude of narrow micro channels. A corresponding design, developed for methanol reforming, is shown in Figure 2.21. The figure shows the computational domain of the CFD model of the reactor, where use of reflection symmetry was made and only one quarter of the portion of the reactor to be considered was modeled. An inlet pipe leads to a flow distribution chamber connected with a multitude of micro channels. The essence of such designs is to be seen in the pressure barrier of the micro channels. The narrower the micro channels are, the higher will be the pressure

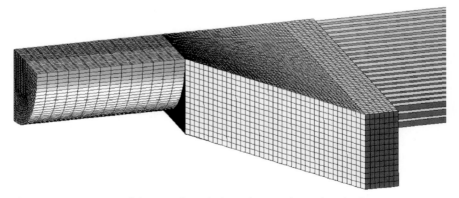

Figure 2.21 Computational domain of a multichannel reactor for methanol reforming.

drop in the channels themselves as compared with the pressure drop in the flow distribution chamber, and the more uniform the flow distribution will be.

Walter et al. studied the flow distribution in simple multichannel geometries by means of the finite-element method [112]. In order to reduce the computational effort, a 2-D model was set up to mimic the 3-D multichannel geometry. Even at a comparatively small Reynolds number of 30 they found recirculation zones in the flow distribution chamber and corresponding deviations from the mean flow rate inside the channels of about 20%. They also investigated the influence of contact time variation on a simple two-step reaction.

Also a simulation of the flow field in the methanol-reforming reactor of Figure 2.21 by means of the finite-volume method shows that recirculation zones are formed in the flow distribution chamber (see Figure 2.22). One of the goals of the work focused on the development of a micro reformer was to design the flow manifold in such a way that the volume flows in the different reaction channels are approximately the same [113]. In spite of the recirculation zones found, for the chosen design a flow variation of about 2% between different channels was predicted from the CFD simulations. In the application under study a washcoat cata-

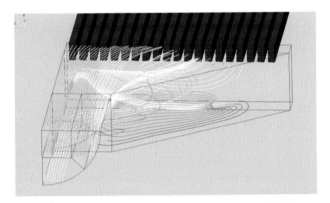

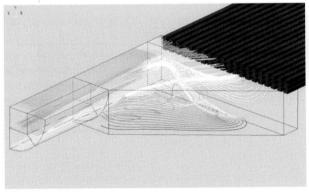

Figure 2.22 Streamlines in the flow distribution chamber of a multichannel reactor.

lyst layer is applied to the micro channels. The thickness variations of the catalyst layer are likely to play the leading role in the non-uniformities in flow distribution to be expected. Hence a flow manifold with intrinsic volume flow variations of a few percent over different reaction channels is usually satisfactory.

For the computation of the flow distribution in the methanol-reforming reactor, a reduced-order flow model for the micro channels was used. In such a model a fully developed flow profile as given by Eq. (68) is assumed, which means that entrance flow effects are neglected. This approximation is justified when the entrance length is small compared with the total length of the channel. A major advantage of a reduced-order flow model is the significant reduction of the degrees of freedom entering the simulation. Each micro channel is then only represented by a single degree of freedom which is the total volume flow, and a resolution of the flow domain by a computational grid is no longer required. In this way the simulation of microfluidic devices comprising a large number of channels becomes possible.

Another special feature allowing efficient simulation of flows entering from a flow-distribution chamber into a multitude of channels is the possibility to combine non-matching grids. Figure 2.23 displays an interface between a rectangular and a body-fitted grid. The center of the figure shows a projection of the two grids on to each other. A face of a rectangular grid cell overlaps with several faces of body-fitted grid cells, as depicted on the right. Domain decomposition methods with non-matching grids are an active research field of applied mathematics. In a specific class of methods the communication between different domains (i.e. different non-matching grids) is done via interfaces on which a space of shape functions is defined [114]. The solution of the full problem is then equivalent to the solution of an adjoint problem defined on the lower dimensional interface between the domains and the original problem in the different sub-domains. The solution of the adjoint equation defined on the interface provides the boundary conditions for the solution of the original transport equations within the sub-domains. Such methods were first developed in the framework of the finite-element approximation and were then extended to finite volumes [115]. The possibility of combining

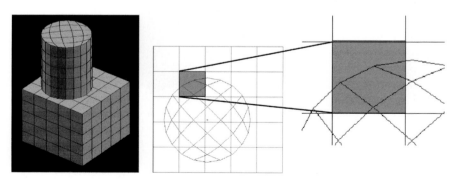

Figure 2.23 Arbitrary interface between a body-fitted and a rectangular grid. A face of a rectangular-grid cell overlaps with several faces of body-fitted grid cells.

non-matching grids allows one to grid each sub-domain with the most appropriate mesh while paying little attention to the transition between different meshes. In this way it is possible to set up a computational domain of, e.g., a flow distribution chamber interfacing with a number of channels, without the need to define an unstructured grid.

Finite-volume grids with arbitrary interfaces in combination with reduced-order flow models for micro channels allow one to solve flow distribution problems in 3-D for which the standard approach of interface-matched grids and discretized channel cross-sections would be computationally too expensive. However, when the goal is to find designs with an optimum flow equipartition by tuning specific geometric parameters, even an approach based on arbitrary grid interfaces might lead to intolerable CPU times. An optimization usually requires doing a large number of evaluations of a corresponding cost function, which means that re-peated computations of the flow distribution for specific geometries would be nec-essary. Apart from a few special cases, such optimizations are only possible with simplified models. Commenge et al. studied the flow distribution in multichannel micro reactors with the help of macro models [116]. The geometry they considered was a microstructured plate of a heat-exchanger stack developed at the Institute of Microtechnology Mainz. The plate together with arrows indicating the flow direc-tion is shown on the left of Figure 2.24, and the right the model set up by Commenge et al. together with the geometric parameters is displayed. The model is based on the idea of subdividing the flow domain into a number of channel segments with rectangular cross-section over which the flow is distributed. Through the channel segments in the inlet zone the volume flows V_1, V_2, ..., V_{nc} are transported, where a part of the flow branches off into the actual micro channels of width W_c and depth e. For each of the channel segments a relationship between the pressure drop Δp and the average flow velocity u of the form

$$\Delta p = 32\,\lambda_{nc}\,\frac{\mu\,L\,u}{D_h^2} \tag{73}$$

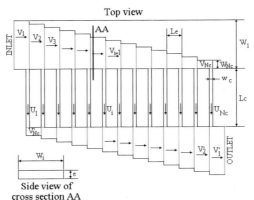

Figure 2.24 Geometry of the microstructured plates (left) and subdivision of the flow domain into channel segments (right) as considered in [116].

was used, where the hydraulic diameter is defined as

$$D_h = \frac{2\,w\,e}{w+e}.$$

(74)

In these expressions, μ is the dynamic viscosity, L the length of the channel segment, w and e their width and depth and λ_{nc} a correction factor accounting for the non-circularity of the channels. Clearly, the above equations rely on the assumption of a hydrodynamically developed flow.

By means of the model depicted in Figure 2.24, Eqs. (73) and (74), it is possible to compute the flow distribution just by solving a comparatively low-dimensional system of linear algebraic equations. The problem resembles the task of computing the current distribution of an electric circuit. *A priori* it is not clear if the approximation to subdivide the flow domain in the described way is justified. In order to assess the quality of the chosen approximation, Commenge et al. [116] computed the flow field by means of the finite volume method. The results obtained suggest that, owing to the orientation of the isobars of the flow and the absence of recirculation loops, the chosen subdivision into channel segments is a reasonable simplification, at least for the geometry and the flow regime considered. The model allows one to study the flow distribution for a variety of different geometries with a minimum computational effort. In Figure 2.25 the results for different geometries of the inlet chamber are displayed in the form of volume flow as a function of channel number. Depending on the geometry of the inlet chamber, the flow distribution over the channels assumes a concave (left), flat (middle) or convex (right) shape. In order to determine optimized chamber geometries, a numerical optimization algorithm was used. The velocity differences between different channels obtained with such optimized geometries are $< 0.1\%$.

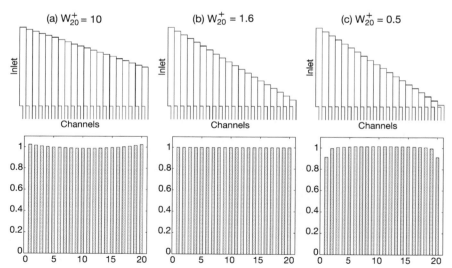

Figure 2.25 Normalized velocity distributions over 20 micro channels obtained by Commenge et al. [116] for a specific class of flow chamber geometries.

Although the strategy described above is a way to obtain effective, low-dimensional models for multichannel reactors in various cases, it clearly has its limitations. *A priori* the applicability of such models has to be carefully justified, and the simple approximations lose their validity when recirculation loops emerge. In geometries with hundreds of micro channels, another method can be utilized to compute the flow field. In a situation where the flow is distributed over a large number of parallel micro channels, the multitude of channels can be regarded as a porous medium. For porous media, effective, volume-averaged transport equations have been known for a long time (for an overview, see, e.g., [117]). In order to solve a flow distribution problem for a multichannel geometry, the flow domain can be split into several regions, one of them being the flow distribution chamber and another the region comprising the multitude of channels. In the flow distribution chamber, the ordinary transport equations are solved, whereas in the multichannel domain, the effective, volume-averaged description of the transport processes is used. A corresponding geometry is shown in Figure 2.26. The incoming flow distributes over a large number of channels with a width w_c separated by walls with a width $w - w_c$.

The most straightforward porous media model which can be used to describe the flow in the multichannel domain is the Darcy equation [117]. The Darcy equation represents a simple model used to relate the pressure drop and the flow velocity inside a porous medium. Applied to the geometry of Figure 2.26 it is written as

$$\frac{d}{dy} \langle p \rangle_f + \varepsilon \frac{\mu}{K} \langle u \rangle_f = 0, \tag{75}$$

with the porosity ε and the permeability K given as

$$\varepsilon = \frac{w_c}{w}, \quad K = \frac{\varepsilon \, w_c^2}{12}. \tag{76}$$

The flow velocity, pressure and dynamic viscosity are denoted u, p and μ and the symbol $\langle ... \rangle_f$ represents an average over the fluid phase. Kim et al. used an extended Darcy equation to model the flow distribution in a micro channel cooling device [118]. In general, the permeability K has to be regarded as a tensor quantity accounting for the anisotropy of the medium. Furthermore, the description can be generalized to include heat transfer effects in porous media. More details on transport processes in porous media will be presented in Section 2.9.

Compared with the use of arbitrary grid interfaces in combination with reduced-order flow models, the porous medium approach allows one to deal with an even larger multitude of micro channels. Furthermore, for comparatively simple geometries with only a limited number of channels, it represents a simple way to provide qualitative estimates of the flow distribution. However, as a coarse-grained description it does not reach the level of accuracy as reduced-order models. Compared with the macromodel approach as propagated by Commenge et al., the porous medium approach has a broader scope of applicability and can also be applied when recirculation zones appear in the flow distribution chamber. However, the macromodel approach is computationally less expensive and can ideally be used for optimization studies.

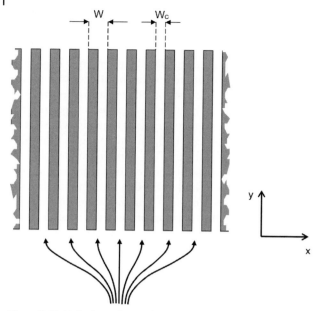

Figure 2.26 Multichannel geometry with channels and separating walls of uniform width.

2.5
Heat Transfer

Heat transfer phenomena belong to the key issues to be studied in micro reactors. Owing to the small thermal diffusion paths, micro reactors bear the potential to allow fast heat transfer and control of temperature distributions with very high accuracy. The excellent thermal performance of micro reactors is a key issue on which many processes of micro chemical engineering are based. Correspondingly, modeling and simulation of heat transfer phenomena and their reliable prediction are of paramount importance for process design. The simulation of temperature distributions in micro reactors often requires the solution of a conjugate heat transfer problem, i.e. the temperature fields in the fluid phase and in the solid wall material have to be computed. Owing to the microstructuring technologies used to fabricate micro reactors, the share of wall material in the total reactor volume is often higher than in conventional equipment. Hence solid heat conduction effects become important and usually have to be taken into account when the temperature field inside a reactor is to be computed.

2.5.1
Fundamental Equations of Heat Transport

The governing equation for the transport of heat contains convective and diffusive contributions. Inside the fluid phase, convective transport often dominates. Within the solid walls the convective contribution is zero and heat is transported solely by conduction. The transport equation for heat, sometimes termed the energy or the enthalpy equation, can be written as (see, e.g., [119])

$$\rho \left(\frac{\partial e}{\partial t} + u_i \frac{\partial e}{\partial x_i} \right) + p \frac{\partial u_i}{\partial x_i} = \frac{\partial}{\partial x_i} \left(\lambda \frac{\partial T}{\partial x_i} \right) + \dot{q}_v + \dot{q} , \tag{77}$$

where e denotes the thermal energy density, ρ the mass density, u_i the flow velocity, p pressure, T temperature, λ the thermal conductivity, $\dot{q}_v$ the heat source due to viscous dissipation and $\dot{q}$ all other heat sources. In incompressible media the enthalpy equation simplifies to

$$\frac{\partial T}{\partial t} + u_i \frac{\partial T}{\partial x_i} = \frac{1}{\rho\, c_p} \left[\frac{\partial}{\partial x_i} \left(\lambda \frac{\partial T}{\partial x_i} \right) + \dot{q}_v + \dot{q} \right] , \tag{78}$$

where c_p denotes the specific heat at constant pressure. The viscous dissipation term describes the transformation of kinetic energy of the flow into heat and is given by

$$\dot{q}_v = \frac{\mu}{2} \left(\frac{\partial u_i}{\partial x_k} + \frac{\partial u_k}{\partial x_i} \right)^2 , \tag{79}$$

where μ denotes the dynamic viscosity. From the structure of the term it is obvious that viscous dissipation is primarily important in regions with large velocity gradients. Eq. (77) contains a number of contributions describing effects of different nature. The first term on the left-hand side describes the transport of heat via convection with the flow velocity u_i. The derivative $\partial u_i / \partial x_i$ in the second term on the left-hand side indicates compression or expansion of fluid volumes. Via such processes internal energy is transformed into potential energy and vice versa at a pressure level p. The first term on the right-hand side describes transport of heat via conduction. Finally, $\dot{q}$ contains all remaining source terms, such as heat generation via the absorbance of radiation. It should be emphasized that Eq. (77) is clearly not the most general form of an enthalpy equation. As an example, the thermal conductivity of a medium might be anisotropic, such that the scalar thermal conductivity λ would have to be replaced by a thermal conductivity tensor. Furthermore, inclusion of the cross-diffusion effects discussed in Section 2.2.1 would require adding terms proportional the gradient of chemical species concentration on the right-hand side.

The Navier–Stokes equation and the enthalpy equation are coupled in a complex way even in the case of incompressible fluids, since in general the viscosity is a function of temperature. There are, however, many situations in which such interdependencies can be neglected. As an example, the temperature variation in a microfluidic system might be so small that the viscosity can be assumed to be constant. In such cases the velocity field can be determined independently from the temperature field. When inserting the computed velocity field into Eq. (77) and expressing the energy density e by the temperature T, a linear equation in T is obtained if the specific heat can be assumed to be temperature independent. The corresponding equation is much easier to solve than the non-linear Navier–Stokes equation.

2.5.2
Heat Transfer in Rectangular Channels

As stated in the previous section, a standard geometry of channels contained in microfluidic systems is a rectangular or close-to-rectangular cross-section. With a given velocity profile, the temperature field inside rectangular channels can often be determined analytically. However, it should be pointed out that the problem of determining a temperature profile in a channel geometry is much more multi-faceted than the computation of a flow distribution. Whereas in most cases a zero-velocity boundary condition at the channel walls is prescribed for the flow, the wall-boundary conditions for the temperature field can be diverse. On the one hand, either a heat flux or a temperature can be prescribed. On the other hand, the thermal boundary conditions on the four walls of a rectangular channel might be different. Owing to this complexity, the solution for the temperature field itself is usually not reported. Rather, the Nusselt number, defined as

$$\mathrm{Nu} = \frac{k\, D_{\mathrm{h}}}{\lambda} \tag{80}$$

is determined, where D_{h} is the hydraulic diameter, λ the thermal conductivity and k the heat transfer coefficient measuring the transmitted thermal power per unit area divided by a characteristic temperature difference. The Nusselt number is a dimensionless quantity characterizing the efficiency of heat transfer. Similarly to the velocity field, the temperature field will assume an invariant profile far enough downstream from the channel entrance. However, owing to the continuous heat transfer from or to the channel walls, only the shape of the temperature distribution stays invariant, and the normalization changes. Close to the entrance of the channel, a thermally developing flow may be observed. A thermal entrance flow is *a priori* not related to a hydrodynamic entrance flow, i.e. a thermally developing flow might be observed even in regions where the flow is hydrodynamically developed and vice versa.

Hartnett and Kostic give an overview of the Nusselt numbers obtained for rectangular channels with various aspect ratios and various thermal boundary conditions [100]. Depending on the thermal boundary condition, the Nusselt number for thermally fully developed flow either increases or decreases when the aspect ratio is increased. When the Nusselt number for thermally developing and hydrodynamically developed flow is plotted as a function of position along the channel, a divergence is observed when the channel entrance is approached. This means that with decreasing distance to the channel entrance, increasing heat transfer efficiency is found. The same observation is made when a simultaneously (i.e. hydrodynamically and thermally) developing flow is considered. Analogous to hydrodynamically developing flows, a thermal entrance length L_{th} can be defined.

It is given as the duct length required for the Nusselt number to fall within a 5% interval of the fully developed value. Again, a dimensionless quantity

$$L_{th}^+ = \frac{L_{th}}{D_h \, \text{Re} \, \text{Pr}} \qquad (81)$$

is used, where the Prandtl number Pr is the ratio of the momentum diffusivity (i.e. the kinematic viscosity) and the thermal diffusivity. The dimensionless thermal entrance length is a quantity depending only on the aspect ratio of the channel and the thermal boundary conditions. Hence L_{th} is a linear function of the hydraulic diameter, the Reynolds number and the Prandtl number.

2.5.3
Generalized Channel Cross-sections

In their studies of friction factors in channels with a number of different cross-sectional geometries, Shah [103] and Shah and London [102] also computed heat transfer properties. A few characteristic cross-sections for which Nusselt numbers were obtained are displayed on the left side of Figure 2.17. Their results include both the Nusselt numbers for fixed temperature and fixed heat flux wall boundary conditions and are given as tabulated values for different geometric parameters.

In addition to those more or less regular channel cross-sections, Richardson et al. [104] studied heat transfer in some more exotic channels as displayed on the right side of Figure 2.17. They numerically computed Nusselt numbers and expressed them as a dimensionless entropy generation rate, defined as

$$N_S = \frac{\dot{S}}{\dot{m} \, c_p}, \qquad (82)$$

where $\dot{S}$ is the entropy generation per unit time and $\dot{m}$ the mass flow rate. In addition to the contribution from heat transfer between the fluid and the channel walls, $\dot{S}$ contains a contribution from viscous heating, which is discussed in Section 2.5.5.

2.5.4
Periodic Channel Geometries

Similarly as with the computation of friction factors, heat transfer has not only been studied in straight channels, but also in channels with specific periodic shapes. Asako and Faghri [105] computed Nusselt numbers in the zigzag channels shown in Figure 2.18 using a finite-volume approach in two dimensions. They found a significant increase in the Nusselt number as compared with a straight channel, i.e. a parallel-plates arrangement. For suitable geometry parameters, a Reynolds number in the range of 1000 and a Prandtl number of 8, a heat transfer enhancement factor of about 13 was computed. For a Prandtl number of 0.7 (corresponding to air) they compared the ratio of the heat fluxes achieved with the zigzag and the straight channels at identical values of pumping power and pressure drop. It was found that the zigzag channel outperforms the straight channel by a factor of up to 6, i.e. for the same pumping power or pressure drop a considerably higher heat flux is achieved. This is an interesting result in view of the design of micro heat exchangers, where the goal is often to increase the heat flux while limiting the pressure drop.

Heat transfer in sinusoidally curved channels was studied by Garg and Maji [106]. Surprisingly, at a Reynolds number of 100 they obtained a value for the Nusselt number which is lower than the corresponding value in a parallel-plates channel. Such a suppression of heat transfer stands in contradiction to results recently obtained by Hardt et al. [120]. They considered sine-wave channel walls of comparatively large amplitude and solved the Navier–Stokes and the enthalpy equations using an FVM solver. The streamline patterns obtained from the simulations showed flow separation with recirculation zones developing in the recesses of the channel, as displayed on the left side of Figure 2.27. On the right side of the figure, the computed Nusselt number, being evaluated close to the exit of the computational domain, is plotted as a function of Reynolds number (note that in this case, in contrast to many other studies of 2-D channel geometries, the Reynolds number was based on the hydraulic diameter, which is twice the wall distance). The triangles represent the corresponding values obtained in parallel-plates geometry. The slight increase in the Nusselt number in the parallel-plates channel as a function of Reynolds number is due to the fact that in the entrance region of a channel, which increases with increasing Reynolds number, the Nusselt number is higher than for fully developed flow. For the sine-wave channel, a heat transfer enhancement increasing as a function of Reynolds number is observed. For the highest Reynolds number considered, heat transfer is enhanced by a factor of about 4.

In the articles cited above, the studies were restricted to steady-state flows, and steady-state solutions could be determined for the range of Reynolds numbers considered. Experimental work on flow and heat transfer in sinusoidally curved channels was conducted by Rush et al. [121]. Their results indicate heat-transfer enhancement and do not show evidence of a Nusselt number reduction in any range

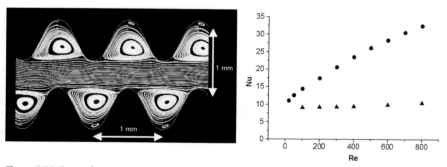

Figure 2.27 Streamline patterns in a channel with sinusoidal walls (left) and Nusselt number as a function of Reynolds number for the same channel (right), taken from [120]. For comparison, the triangles represent the Nusselt number obtained in parallel-plates geometry.

of Reynolds numbers. However, the flow patterns begin to exhibit some unsteady behavior for Reynolds numbers greater than about 200 in the geometry considered. Interestingly, oscillations were predominantly observed close to the exit of the channel and were found to move upstream when the Reynolds number was increased. Rush et al. attributed much of the heat-transfer enhancement found to the unsteady character of the flow.

The converging–diverging channel geometry for which Guzmán and Amon [107] computed flow fields was also considered by Wang and Vanka [122]. They computed the flow and temperature field in an elementary cell of the channel based on the finite-volume method using periodic boundary conditions. As far as the transition between a steady-state and an unsteady flow is concerned, they obtained similar results as Guzmán and Amon. At Reynolds numbers beyond the transition to the unsteady flow regime, the temperature field becomes complex. This is indicated in Figure 2.28, which shows the evolution of (normalized) temperature contour lines at a Reynolds number of 328. In order to quantify heat-transfer enhancement, the time-averaged Nusselt number was computed. The increase in Nusselt number compared with a straight channel was found to be significant and reached values as high as 7.54 at a Reynolds number of 520. However, the experimental results of Rush et al. [121], obtained for a comparable channel geometry, suggest that the assumption of periodic flow boundary conditions over one elementary cell of the channel might not be justified. The experiments show clear evidence of the flow pattern downstream in the channel being different from that one close to the channel entrance. Heat-transfer enhancement was also found in the experiments; however, the Nusselt numbers determined are significantly smaller than those predicted by Wang and Vanka.

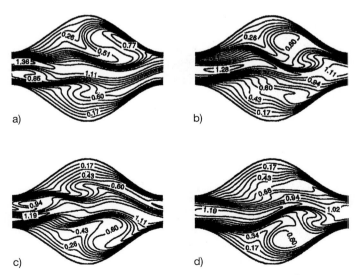

Figure 2.28 Evolution of the temperature field in the recesses of a converging–diverging channel, as obtained by Wang and Vanka [122].

2.5.5
Viscous Heating

In flow and heat transfer simulations in micro channels, the effect of viscous heating, described by Eq. (79), is often neglected. At fixed flow velocity, heat generation due to viscous heating scales as a power of $1/d^2$, where d is the diameter of the channel. Hence, when the channel diameter decreases, the importance of viscous heating grows rapidly and it might no longer be justified to neglect the corresponding effects. Sekulic et al. [123] and Richardson et al. [104] computed the entropy generation due to viscous heating for a variety of different channel cross-sections. As a second source of entropy besides viscous heating, the heat transfer from the fluid to the channel walls was considered. The transport equations for heat and momentum were solved using a finite-volume method on a body-fitted grid, with a fluid inlet temperature which was allowed to differ from the temperature of the channel walls. Typical results of these calculations for water as a fluid and a circular duct of 610 µm diameter are shown in Figure 2.29 as a function of Reynolds number. The labels on the different curves represent the different ratios of the inlet temperature and the temperature of the channel walls; for $\tau^* = 1.0$ both temperatures are equal. When τ^* increases, the heat transfer from the fluid to the channel walls begins to dominate the entropy generation. Heat transfer and viscous heating show a different behavior as a function of Reynolds number. When the Reynolds number increases, the residence time of the fluid in the channel the

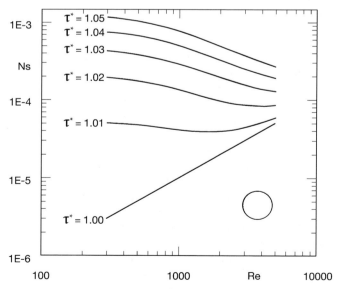

Figure 2.29 Dimensionless entropy generation rate as a function of Reynolds number for a circular channel of 610 µm diameter and different ratios of the fluid inlet temperature and the wall temperature, taken from [104].

entropy generation caused by the exchange of heat decreases. In contrast, an increasing Reynolds number is translated to an increasing flow velocity, which enhances the entropy generation due to viscous heating linearly. At a value of $\tau^* = 1.01$, the two effects add up in such a way that a minimum of the entropy generation curve is created within the range of Reynolds numbers considered. For Reynolds numbers in the region of 3000 the viscous heating effects become comparable in size to the entropy generation caused by the heat transfer from a fluid to channel walls with a temperature which is 1% lower.

Xu et al. [124] numerically computed the adiabatic temperature rise in a micro channel due to viscous heating and expressed their results by a correlation based on dimensionless groups. They introduced a dimensionless temperature rise $\Delta T^* = \Delta T / T_{\text{ref}}$ with a reference temperature of 1 K. The correlation they found is given by

$$\Delta T^* = \frac{93.419 \, \text{Vi} \, \text{Pr}^{-0.1}}{5.2086 + \text{Vi} \, \text{Pr}^{-0.1}}, \tag{83}$$

where Pr is the Prandtl number and Vi the viscous number, given by

$$\text{Vi} = \frac{\mu \, \bar{u} \, L}{\rho \, c_{\text{p}} \, T_{\text{ref}} \, D^2}, \tag{84}$$

where $\bar{u}$ denotes the average velocity, L the length of the channel and D its diameter. Using this correlation, it can easily be judged when viscous heating has to be taken into consideration. The calculations of Sekulic et al., Richardson et al. and Xu et al. give an indication of how fast viscous heating effects become important when the channel dimension is reduced and may serve as a guideline for the design of microfluidic systems.

2.5.6
Micro Heat Exchangers

Micro heat exchangers represent a standard example of heat transfer in rectangular micro channels. Micro channel heat exchangers are used for rapid heat transfer between a hot and a cold fluid or, alternatively, as heat sinks for cooling devices. A schematic drawing of a counter-current heat exchanger is shown in Figure 2.30. A characteristic feature of such devices is the fact that the thickness of the wall material separating the channels is of the same order of magnitude as the channel depth itself. A micro heat exchanger might as well be operated in co-current mode, where the hot and the cold fluid flow in the same direction, or an arrangement with flow directions perpendicular to each other, the so-called cross-current mode, might be chosen.

Stief et al. [125] investigated the performance of counter-current micro heat exchangers, especially with respect to the wall conduction effects and the choice of

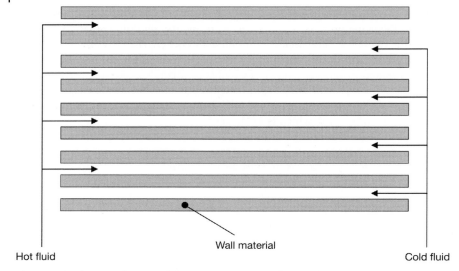

Hot fluid   Wall material   Cold fluid

Figure 2.30 Schematic drawing of a counter-current micro heat exchanger.

the wall material. They considered a flow of nitrogen in channels with a length of 10 mm and a depth of 50 µm and separating walls of thickness between 125 and 500 µm. In the plane perpendicular to the flow direction, periodicity was assumed. In order to reduce the computational effort, the heat exchanger geometry was only discretized in the axial direction, i.e. in the direction of the flow. In the direction perpendicular to the flow a fixed heat transfer coefficient was used to describe the exchange of heat between the fluid and the channel walls. The resulting set of ordinary differential equations was solved by a special algorithm with adaptive step size control. When the thermal conductivity of the wall material is varied, characteristic temperature profiles for the channels with the hot and cold fluid and the channel wall are obtained, as displayed in Figure 2.31. For very small values of the

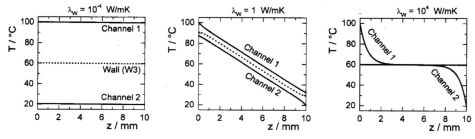

Figure 2.31 Characteristic temperature profiles in a counter-current micro heat exchanger for a very low (left), intermediate (middle) and very high (right) thermal conductivity of the wall material and equal volume flows inside the two channels, reproduced from [125].

thermal conductivity, no heat is exchanged and the temperatures stay approximately constant. At intermediate conductivities, almost linear temperature profiles are obtained, and at very high thermal conductivities the wall assumes a constant temperature and the temperature of the fluids changes rapidly in the entrance region of the channels.

A quantity characterizing the performance of a heat exchanger is the efficiency, which is the ratio of the transmitted heat and the maximum transmittable heat, given as

$$\varepsilon = \frac{T_{in}^{hot} - T_{out}^{hot}}{T_{in}^{hot} - T_{in}^{cold}} \tag{85}$$

in the case of equal volume flows two incompressible fluids. T_{in} and T_{out} denote the temperatures at the channel inlet and outlet, respectively. Stief et al. computed the heat exchanger efficiency for two characteristic geometries as a function of the wall thermal conductivity. A typical result for equal volume flows is displayed in Figure 2.32. At low conductivities almost no heat is transferred, as expected. At high conductivities the wall assumes a uniform temperature due to axial heat conduction and the device becomes indistinguishable from a co-current heat exchanger, which is limited to an efficiency of 50% at equal volume flows. At values around 1 W mK^{-1} the curve assumes a maximum, as axial heat conduction inside the walls is suppressed to a sufficient degree and transversal conduction to the opposite channels is still efficient. The position of the maximum is insensitive to the geometric parameters of the heat exchangers, Hence, in typical micro heat exchangers the use of low-conductivity materials such as glass or polymers is preferable, and the common stainless-steel materials are expected to reduce the efficiency.

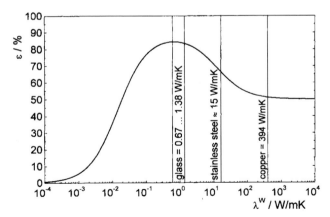

Figure 2.32 Heat exchanger efficiency as a function of the thermal conductivity of the wall, taken from [125].

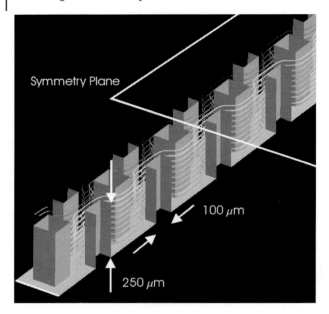

Figure 2.33 Heat exchanger micro channel with a checkerboard arrangement of micro fins, as considered by Hardt et al. [120].

Conventional heat exchangers rarely rely on simple flow geometries such as parallel channels [126]. Their design is usually optimized with regard to compactness, i.e. the transmitted heat flux per unit volume should be maximized. In general, flow in micro heat exchangers is laminar, in contrast to their macroscopic counterparts. Thus, owing to these different physical scenarios, the design concepts of conventional heat exchangers are not necessarily transferable to the world of micro process technology. Generic design features of heat exchangers are so-called fins, which are solid barriers introduced into the flow to remove heat from or introduce heat to the fluid efficiently. The problem of heat transfer enhancement in micro reactors by channels containing fins was studied by Hardt et al. [120]. They considered a channel with a checkerboard arrangement of micro fins of a cross-section of 100 μm × 100 μm, which is shown in Figure 2.33 together with the computed streamlines of the flow. Transversal to the main flow direction, the micro fin pattern is assumed to be repeated periodically *ad infinitum*.

The simulations of fluid flow and heat transfer in such microstructured geometries were carried out with an FVM solver. Air with an inlet temperature of 100 °C was considered as a fluid, and the channel walls were modeled as isothermal with a temperature of 0 °C. The streamline pattern is characterized by recirculation zones which develop behind the fins at comparatively high Reynolds numbers. The results of the heat transfer simulations are summarized in Figure 2.34, which shows the Nusselt number as a function of Reynolds number. For

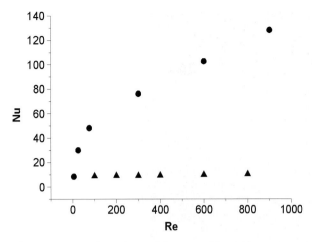

Figure 2.34 Nusselt number as a function of Reynolds number for a channel with a checkerboard arrangement of micro fins, taken from [120]. The triangles represent the Nusselt number obtained in a parallel-plates geometry.

comparison, the triangles represent the Nusselt number computed for channels of the same length as in the computational model, but with simple parallel-plates channel walls. The checkerboard arrangement of micro fins shows a significant increase in the Nusselt number as a function of Reynolds number.

The checkerboard micro fin arrangement with Nusselt numbers above 100 promises to allow a very high heat flux per unit volume. As the Nusselt number is a dimensionless measure of the heat flux per unit area, there is an additional increase in the total heat flux by the increase in surface area as compared with simple parallel-plates channel geometry. The increase in the Nusselt number in the micro fin geometry can be explained as a continuous entrance flow effect. As outlined previously, in the entrance region of a channel significantly higher Nusselt numbers are found than in downstream regions where the flow is fully developed. By the staggered arrangement of the micro fins shown in Figure 2.33 the flow is continuously forced to enter into barriers formed by two opposite micro fins and never reaches the equivalent of a developed state. However, the price to pay for the very fast heat transfer in the micro fin arrangement is a significant increase in the pressure drop in comparison with Poiseuille flow.

In order to examine the trade-off between heat exchange and pressure drop more closely, Drese and Hardt [127] studied micro fins of varying geometry. They chose a diamond-shaped fin cross-section as displayed on the left of Figure 2.35 and varied the fin length and the total fin length, while the fin width and the channel width were fixed to 50 μm. The goal was to identify geometries allowing for a maximum heat exchange for a given pressure drop. As a local dimensionless quantity which may serve as an indicator for such an efficiency figure, they identified the ratio of the Nusselt and the Poiseuille number. The Poiseuille number is the ratio of the friction factor as defined in Eq. (69) and the Reynolds number. In order to determine optimized fin geometries, CFD simulations for flow of air and constant-temperature wall boundary conditions were performed for the staggered fin arrangement shown in Figure 2.35. The results as a function of mass flow per unit area are displayed below. The curves are distinguished by the different total fin length considered. Apparently, the ratio of Nusselt and Poiseuille numbers assumes distinctively different values for different fin geometries. For relatively elongated fins (200 µm total fin length), the ratio of Nusselt and Poiseuille numbers assumes favorable values above 0.4, thus indicating high heat transfer per unit pressure drop. For short fins, dead water areas or recirculation zones behind the fins give rise to a less favorable ratio of heat transfer to pressure drop. Hence efficient micro heat exchangers with low pressure drop may be designed by utilizing channel geometries with a staggered arrangement of comparatively long, diamond-shaped fins which suppress the formation of dead water areas and recirculation zones.

When a large number of parallel heat exchanger channels are considered, the problem of computing the temperature distribution inside the channels and the channel walls becomes involved, similar to the flow distribution problem discussed

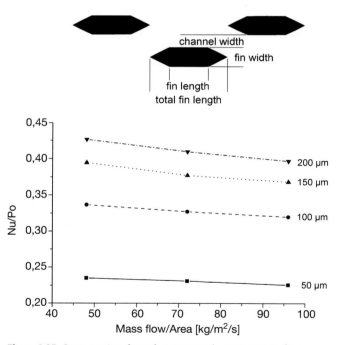

Figure 2.35 Cross-section through a staggered arrangement of micro fins designed for heat transfer enhancement in a micro channel (above) and ratio of Nusselt and Poiseuille numbers as a function of air flow per unit area for different total fin lengths (below), taken from [127].

in Section 2.4.4, and simplified models are needed. In such a case the use of the method of reduced-order flow models is not as straightforward as in the case of a pure flow distribution problem, owing to thermal cross talk between the different channels. Similarly, porous media models describing heat transfer in the multi-channel domain are involved, as a temperature gradient orthogonal to the direction of flow induces a heat flux with components in the flow direction as well as orthogonal to it.

For simplicity, consider an incompressible medium flowing through the multi-channel domain depicted in Figure 2.36. In a number of practical applications the heat flux in the x-direction will be very small compared with that in the y-direction. Then the volume-averaged enthalpy equation for the solid walls can be written as

$$\lambda_{s,eff} \frac{\partial^2 \langle T \rangle_s}{\partial y^2} = \alpha \, a_{spec} \left(\langle T \rangle_s - \langle T \rangle_f \right), \tag{86}$$

where $\langle ... \rangle_s$ represents a volume average over solid regions and $\lambda_{s,eff}$ denotes the effective thermal conductivity of the solid matrix, T the temperature, α the heat transfer coefficient between the fluid and the solid and a_{spec} the internal surface area per unit volume of the multichannel domain. A corresponding averaging volume, embracing a number of channel and wall segments, is shown in Figure 2.36. The volume-averaged enthalpy equation for the fluid domain can be expressed as

$$\lambda_{f,eff} \frac{\partial^2 \langle T \rangle_f}{\partial y^2} - \varepsilon \, \rho_f \, c_f \, \langle u \rangle_f \frac{\partial \langle T \rangle_f}{\partial y} = \alpha \, a_{spec} \left(\langle T \rangle_f - \langle T \rangle_s \right), \tag{87}$$

where $\lambda_{f,eff}$ denotes the effective thermal conductivity of the fluid domain, ε the porosity and ρ_f, c_f the fluid density and heat capacity, respectively. The volume-averaged fluid velocity $\langle u \rangle_f$ can be determined by solving Eq. (70) and, optionally, the momentum equation for the region upstream of the multichannel domain. The effective thermal conductivities in both regions are expressed as

$$\lambda_{s,eff} = (1 - \varepsilon) \, \lambda_s, \quad \lambda_{f,eff} = \varepsilon \, \lambda_f, \tag{88}$$

where λ_s and λ_f are the solid and fluid thermal conductivity, respectively.

Effectively, Eqs. (86) and (87) describe two interpenetrating continua which are thermally coupled. The value of the heat transfer coefficient α depends on the specific shape of the channels considered; suitable correlations have been determined for circular or for rectangular channels [100]. In general, the temperature fields obtained from Eqs. (86) and (87) for the solid and the fluid phases are different, in contrast to the assumptions made in most other models for heat transfer in porous media [117]. Kim et al. [118] have used a model similar to that described here to compute the temperature distribution in a micro channel heat sink. They considered various values of the channel width (expressed in dimensionless form as the Darcy number) and various ratios of the solid and fluid thermal conductivity and determined the regimes where major deviations of the fluid temperature from the solid temperature are found.

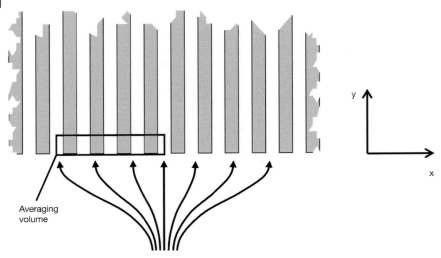

Averaging
volume

Figure 2.36 Multichannel flow domain with typical averaging volume
for obtaining volume-averaged enthalpy equations.

2.5.7
Thermal Optimization of Micro reactors

One of the most prominent application areas of modeling and simulation techniques in chemical micro process technology is the design optimization of micro
reactors. In this context, thermal problems are often comparatively easy to solve, as
the transport equation for heat is linear in temperature and in various cases the
flow field is influenced by the temperature field to only a very small degree and can
be assumed as given. An optimization based solely on variations of the temperature distribution was demonstrated by Quiram et al. [128]. They investigated thermal issues of the T micro reactor developed at MIT for gas-phase reactions. The
name 'T micro reactor' stems from the design with two feed channels and a reaction channel forming a T-shaped geometry. The temperature in the reaction channel can be controlled by a platinum heating meander integrated on the membrane
sealing the channel.

For the computation of the velocity and the temperature field inside the reactor,
a stabilized FEM method was used. Within the stabilized FEM method, numerical
parameters are introduced to overcome instability problems frequently occurring
for convection-dominated problems [129]. The same method, in combination with
a special procedure to deal with highly non-linear sets of algebraic equations, was
employed subsequently to compute the species concentration obtained via chemical reactions occurring in the reaction channel. By computing the temperature
distribution in the reaction channel, Quiram et al. [130] were able to show that the
initially proposed heater configuration produces a non-uniform temperature profile. By varying the geometry of the platinum heaters, a significantly flatter tem-

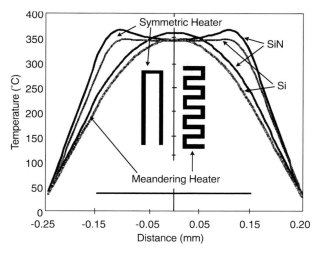

Figure 2.37 Temperature profiles across the membrane covering the reaction channel of the T micro reactor for a silicon and a silicon nitride membrane and two different heater designs, as discussed by Quiram et al. [128].

perature profile could be obtained. The obtained temperature profiles over the width of the reaction channel are displayed in Figure 2.37 for two different materials covering the channel. Attached to the cover membrane is a platinum heater, where one version thereof spans a considerable section of the channel width and the other version consists of strips located in both halves of the channel. The version with two heating strips apparently produces temperature profiles which qualify better to carry out reactions at a uniform temperature level.

2.6
Mass Transfer and Mixing

Similarly to heat transfer, fast mass transfer is one of the key aspects of micro reactors. Again, owing to the short diffusion paths, micro reactors permit a rapid mass transfer and a uniform solute concentration within the flow domain. Good control of reactant concentration throughout the whole reactor volume is a prerequisite for highly selective chemical reactions and helps to avoid hazardous operation regimes. In addition, overcoming mass transfer limitations by rapid mixing allows the exploitation of the rapid intrinsic kinetics of chemical reactions and allows a higher yield and conversion. However, when dealing with liquid-phase reactions, fast mixing remains a challenge even at length scales of 100 µm or less owing to the small diffusion constants in liquids.

2.6.1
Transport Equation for Species Concentration

The governing equation for mass transport in the case of an incompressible flow field is easily derived from the general convection–diffusion equation Eq. (32) with $\Phi = c$ and is given by

$$\frac{\partial c}{\partial t} + u_i \frac{\partial c}{\partial x_i} = \frac{\partial}{\partial x_i}\left(D \frac{\partial c}{\partial x_i}\right) - r \,, \tag{89}$$

where c is the concentration of a solute in units of mol/volume, u_i the flow velocity, D the molecular diffusivity and r a source term due to chemical reactions. For a given velocity field and a vanishing source term, the solution of Eq. (89) is governed by a single dimensionless group, which is the Peclet number:

$$\text{Pe} = \frac{u\,d}{D} \,. \tag{90}$$

In this expression, u is a typical velocity scale and d a typical length scale, for example the diameter of a micro channel. The Peclet number represents the ratio of the diffusive and the convective time-scales, i.e. flows with large Peclet numbers are dominated by convection.

2.6.2
Special Numerical Methods for Convection-Dominated Problems

As already indicated in Section 2.3.1, the numerical solution of convection–diffusion equations such as Eq. (89) is often impaired by numerical diffusion. From Eq. (42), it is obvious that for the upwind scheme the dimensionless group determining the strength of numerical diffusion is the cell-based Peclet number, i.e. the Peclet number to which the dimension of a grid cell enters as a length scale. When considering mass transfer in gases, the cell size can often be chosen small enough to suppress numerical diffusion artefacts. However, for mass transfer in liquids this is often not possible owing to the molecular diffusion constant, which is about three orders of magnitude smaller.

The solution of a pure convection equation for a scalar field Φ:

$$\frac{\partial \Phi}{\partial t} + \frac{\partial(u_i\,\Phi)}{\partial x_i} = 0 \tag{91}$$

can be regarded as a benchmark problem for the validation of corresponding numerical schemes. Considering a constant velocity field u_i and an initial distribution $\Phi(t=0)$, Eq. (91) describes the displacement of this distribution along the streamlines of the flow. Apart from such a displacement, the distribution of the scalar field remains undisturbed. A numerical solution usually cannot reproduce this behavior, rather some distortion and a dispersion of the field distribution will be induced. In order to shed more light on the numerical structure of the problem,

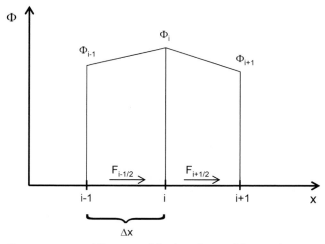

Figure 2.38 Finite-difference grid for the solution of the one-dimensional convection equation. The $F_{i \pm 1/2}$ denote the fluxes from node i to the neighboring nodes.

consider a one-dimensional version of Eq. (91) discretized on a finite difference grid as shown in Figure 2.38 Based on first-order explicit time differencing, the discretized version of the convection equation can be written in the general form

$$\Phi_i^{n+1} = \Phi_i^n - \frac{F_{i+1/2} - F_{i-1/2}}{\Delta x} , \tag{92}$$

where the superscripts refer to the temporal and subscripts to the spatial coordinates. The $F_{i \pm 1/2}$ are the fluxes between node i and its neighboring nodes which effect an increase or decrease of Φ_i in the next time step. They are in general a function of the values of Φ at different node locations, depending on the specific differencing scheme used. In stable discretization schemes, the fluxes usually have a diffusive component which damps off fluctuations. An undesired side effect of such numerical diffusion is that it also disperses the real, physical concentration peaks. In contrast, diffusion-free schemes such as central differencing may lead to instabilities, since unphysical over- and undershoots and new maxima and minima of the solution field are produced.

In order to minimize numerical diffusion, Boris and Book [131] formulated the idea of blending a low-order stable differencing scheme with a higher order, potentially unstable, scheme in such a way that steep concentration gradients are maintained as well as possible. The algorithm they proposed consists of the following steps:

- Compute an approximation of the solution at the next time step using a low-order scheme (indicated by superscript L):

$$\tilde{\Phi}_i^{n+1} = \Phi_i^n - \frac{F_{i+1/2}^{L} - F_{i-1/2}^{L}}{\Delta x} . \tag{93}$$

- Based on a higher order scheme (indicated by superscript H), compute 'anti-diffusive' fluxes:

$$A_{i\pm1/2} = F_{i\pm1/2}^{H} - F_{i\pm1/2}^{L} . \tag{94}$$

- Multiply $A_{i\pm1/2}$ with a flux limiter:

$$A_{i\pm1/2}^{C} = C_{i\pm1/2} \, A_{i\pm1/2} . \tag{95}$$

- Add the limited antidiffusive fluxes to the approximate solution

$$\Phi_i^{n+1} = \tilde{\Phi}_i^{n+1} - \frac{A_{i+1/2}^{C} - A_{i-1/2}^{C}}{\Delta x} . \tag{96}$$

Without the third step, the multiplication with the flux limiter, addition of the antidiffusive fluxes would just remove the terms causing diffusion from the approximate solution obtained with the low-order scheme. However, such a simple correction would usually result in an unstable numerical scheme. For this reason, the flux limiter $C_{i\pm1/2}$ is chosen such that the antidiffusion stage generates no new maxima or minima of Φ and does not accentuate already existing extrema. The choice of the flux limiter is crucial for the quality of the numerical approximation. Effectively, the procedure described above can be regarded as switching between high- and low-order approximations adaptively depending on the smoothness of the solution. The potentially unstable high-order approximation is used in regions where owing to the structure of Φ, an instability cannot be created.

Methods based on the addition of antidiffusive fluxes to a low-order differencing scheme are termed flux-corrected transport (FCT) algorithms. A multidimensional version of a FCT method was first developed by Zalesak [132]. The quality of corresponding numerical schemes is usually assessed by solving benchmark problems such as the convection of a rectangular pulse. For a considerable number of such examples, it has been shown that FCT methods are able to preserve steep gradients and stepwise changes of the solution field much better than standard differencing schemes [131, 132].

2.6.3
Mixing Channels

In the literature, a variety of different micro mixing devices have been described [133], most of which operate in the laminar flow regime. In order to understand the dynamics of micro mixers and to determine optimized designs, modeling and simulation techniques were applied in some cases. The simplest micro mixer is the so-called mixing tee which is displayed on the top of Figure 2.39. Two inlet channels merge into a common mixing channel where mixing of the two co-flowing fluid streams occurs. The mixing characteristics of T-type mixers was investigated by Gobby et al. [134] using CFD methods. Mixing of gases in a channel of 500 μm width was considered and certain geometric parameters such as the aspect

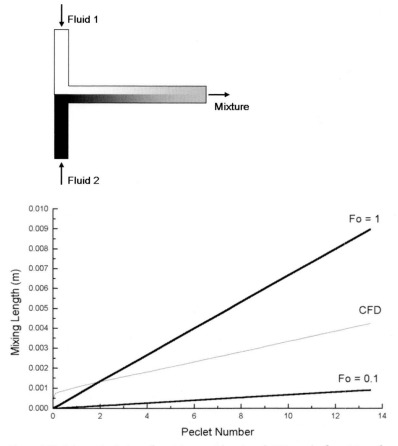

Figure 2.39 Schematic design of a mixing tee (above) and CFD results for mixing of gases in a channel of 500 µm width and 300 µm depth, taken from [134] (below).

ratio of the mixing channel or the angle at which the two inlet channels meet were varied. In order to quantify mixing, a mixing length was defined as the length in flow direction after which the gas composition over all positions of a channel cross-section deviates by no more than 1% of the equilibrium composition. The CFD results for the mixing length as a function of Peclet number are displayed on the right side of Figure 2.39.

In addition to the CFD results, estimates of the mixing length based on the Fourier number:

$$\text{Fo} = \frac{D\,t}{l^2} \tag{97}$$

are shown in Figure 2.39. The Fourier number relates a residence time t in the mixing channel to the binary diffusion constant and a characteristic length scale l, which is the width of the channel. For a given value of Fo and a given flow rate, the length along the mixing channel necessary to achieve the corresponding Fourier number was determined. As is apparent from Figure 2.39, the Fourier number is a reasonable indicator for mixing which occurs for Fourier numbers between 0.1 and 1.0. The linear increase of the CFD-based mixing length as a function of Peclet number points to a very simple mixing mechanism via diffusion between co-flowing fluid lamellae. Obviously, complex convection-dominated mixing mechanisms (for example driven by swirls or recirculating flows) are absent in the simple mixing-tee configuration for the range of Peclet numbers studied.

Microfluidic mixing tees are useful for the processing of species with high diffusion constants at comparatively low volume flows. However, for mixing of liquids at only moderate volume flows, more elaborate micro mixing concepts are needed. The next level of complexity in micro mixer design is to go from a straight mixing channel to a curved channel or a channel with structured walls. In that context, a design that has been studied in some detail is a zigzag micro channel, the geometry of which is shown in Figure 2.40. Mengeaud et al. [135] studied mixing in zigzag micro channels in a 2-D finite-element model and conducted experiments using fluorescein to visualize the mixing process. In order to suppress artefacts from numerical diffusion, the Peclet number of the flow was fixed at moderate values (Pe = 2600). When the flow velocity was varied, the diffusion constant was varied along with it in order to keep the Peclet number fixed. Owing to this fact, diffusion constants of up to 10^{-6} m^2 s^{-1} were used, three orders of magnitude larger than typical liquid diffusion constants. Several channel geometries with varying widths of the order of 100 μm and varying periods of the zigzag structures were considered. The configuration with maximum mixing efficiency was studied in more detail and exhibits an interesting behavior when the Reynolds number is increased. For Reynolds numbers up to 80 the mixing efficiency is equal to an equivalent straight mixing channel, indicating that convective mixing effects are negligible. For larger Reynolds numbers an increase of the mixing efficiency sets in, indicating that the flow pattern changes and convective mass transfer promotes mixing. Specifically, recirculation zones as shown on the right side of Figure 2.40 begin to form. Hence, for comparatively large Reynolds numbers, zigzag mixing channels achieve a distinctively higher mixing efficiency than straight channels.

Figure 2.40 Zigzag micro mixer with concentration field (left) and flow stream lines (right) obtained from a CFD simulation for a Reynolds number of 38. In [135] a sawtooth geometry of larger amplitude was considered and distinctive recirculation zones were found only at Reynolds numbers larger than 80.

This observation is supported by experiments carried out with a phosphate buffer and a fluorescein solution for visualization of the mixing process. However, in the experiments there are indications that the critical Reynolds number where the mass transfer enhancement sets in is lower (~7) than predicted by the simulations, a fact which is not well understood.

Another strategy to promote the mixing of two fluid streams merging in a channel is to induce a helical flow which redistributes the fluids in the mixing channel. Stroock et al. [136, 137] described a method to induce helical flows in straight channels. As a fundamental principle to induce recirculating flow patterns they considered the flow over grooved surfaces. Their theoretical studies were done for a slab geometry where one boundary is a sinusoidally modulated surface [136]. By writing down a perturbation series in the amplitude of the sinusoidal grooves they were able to solve the Stokes equation for flow parallel and perpendicular to the grooves. By a linear superposition of the two specific solutions, the general solution for an arbitrary flow direction is obtained. It turns out that on length scales large compared with the amplitude and the inverse wavenumber of the sinusoidal modulations, the grooved surface can be regarded as inducing a slip flow the direction of which in general does not coincide with the direction of the pressure gradient. The slip velocity u_i is obtained from a boundary condition of the form

$$ u_i\big|_{z=0} + (Z_{\text{eff}})_{ij} \frac{\partial u_j}{\partial z}\bigg|_{z=0} = 0 \tag{98} $$

on the grooved surface. The 2×2 tensor Z_{eff} depends on the applied pressure gradient and the orientation of the grooves in the surface located at $z = 0$. When at least one of the walls of a micro channel contains grooves standing at an angle θ to the main flow direction, a situation as depicted in Figure 2.41 is encountered. Owing

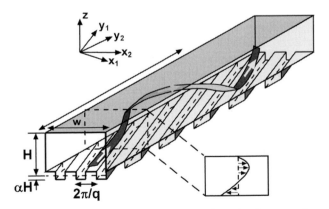

Figure 2.41 Schematic diagram of a micro channel with grooves in the bottom wall, taken from [136]. The ribbon indicates a typical streamline in the channel. In the box at the bottom the velocity profile over a channel cross-section is drawn schematically.

to the slip flow not being aligned with the pressure gradient along the channel, a helical flow is induced with helices extending over the entire cross-section of the channel.

For micro mixing applications, a helical flow would allow one to increase the interfacial area between the two fluid lamellae to be mixed, but it would usually not effect a dramatic improvement of mixing efficiency. Aref [138] showed that two vortex structures need to be superposed in an alternating fashion in order to create a chaotic flow pattern ('blinking vortex flow'). The corresponding flow can be de-scribed mathematically as a non-integrable Hamiltonian system. Aref computed the distribution of tracer particles by the blinking vortices in the chaotic regime and was able to show that the particles spread over the entire flow domain after some cycles.

In order to realize the blinking vortex flow principle in a mixing channel, Stroock et al. proposed to use a periodic, staggered arrangement of grooves to induce a chaotic flow pattern even at low Reynolds numbers [137]. The corresponding mixer geometry is shown in Figure 2.42. The bottom of the channel of height h and width w contains a staggered arrangement of grooves, where a fraction p of the lower channel wall has grooves inducing a left-handed recirculation and the remaining wall fraction induces a right-handed recirculation. A schematic view of the channel cross-section is shown on top of Figure 2.42. If $p \neq 1/2$, the vortex pattern is asym-metric, and a superposition of two patterns exhibiting the larger of the two vortices on the left and right sides, respectively, could result in chaotic flow. The superposition is achieved by the alternating, staggered groove patterns shown in Figure 2.42. The lower part of the figure shows channel cross-sections with two streams of fluores-cent and clear solutions after 0, 0.5 and 1 cycle. The images were recorded using a confocal microscope. The two different vortices are clearly visible and the third

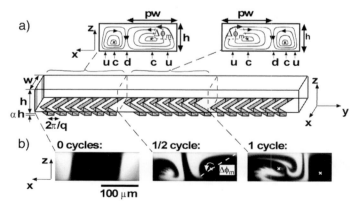

Figure 2.42 Micro mixer geometry with staggered groove structures on the bottom wall, as considered in [137]. The top of the figure shows a schematic view of the channel cross-section with the vortices induced by the grooves. At the bottom, confocal micrographs showing the distribution of two liquids over the cross-section are displayed.

frame shows first indications of a chaotic disturbance of the flow. By analyzing the gray-scale distribution of their confocal micrographs, Stroock et al. were able to show that chaotic mixing occurs in their mixer comprising a staggered arrangement of grooves. An indication of the chaoticity of the mixing process is the mixing length which was found to scale as ln Pe.

The theoretical results obtained by Stroock et al. are based on the Stokes flow regime and rely on surface modulations of comparatively small amplitude. Schönfeld and Hardt [139] studied helical flows in micro channels by numerically solving the full Navier–Stokes equation for different channel geometries including channels with corrugated walls. They compared the transversal velocities due to the helical flow to the analytical results and found good agreement even for grooves almost as deep as the channel itself.

Furthermore, as an alternative principle to induce helical flows, they studied the Dean effect as described in Section 2.4.3. In contrast to the single vortex induced by simple grooved channel walls, in curved channels at least two counter-rotating vortices are formed. Schönfeld and Hardt showed that in the curved micro channels considered, the transverse velocities due to the vortex-like structures are of comparable order of magnitude to those obtained with grooved channel walls. Typical results obtained from the numerical solution of the convection–diffusion equation for the concentration field are displayed in Figure 2.43. The left side of the figure shows the channel geometry, where typical dimensions are $a = 200$ μm and $R = 1000$ μm. On the right side the evolution of two initially vertical fluid lamellae

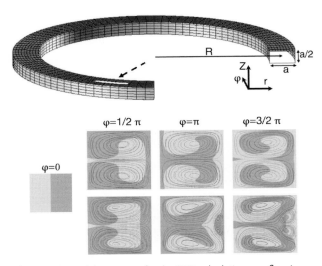

Figure 2.43 Model geometry for the CFD calculations on flow in curved micro channels (above) and time evolution of two initially vertical fluid lamellae over a cross-section of the channel (below), taken from [139]. The secondary flow is visualized by streamlines projected on to the cross-sectional area of the channel. The upper row shows results for $K = 150$ and the lower row for $K = 300$.

for a vanishing diffusion constant is displayed, in the upper row for a Dean number of $K = 150$ and in the lower row for $K = 300$. Depending on the Dean number, a flow pattern with either two or four counter-rotating vortices is found.

In order to permit efficient mixing, a chaotic flow pattern would be desirable. Schönfeld and Hardt suggested a method to realize a version of the blinking vortex principle in curved channels [139]. By a periodic arrangement of channel sections of different hydraulic diameters, the two- and the four-vortex pattern could be generated in an alternating sequence, thus effecting a stretching and folding of the fluid lamellae in a chaotic manner. Alternatively, at Dean numbers high enough for the four-vortex pattern to emerge, a periodic sequence of channel sections of alternating curvature could allow switching between flow patterns with the two small vortices either on the left or on the right side of the channel cross-section.

2.6.4
Estimation of Mixing Efficiency by Flow-field Mapping

For flows created in channels which exhibit a certain periodicity, due either to alternating groove patterns in the walls or to alternating curved sections, there is an elegant method to compute the growth of interfacial area per spatial period and to estimate the mixing efficiency. By the fluid streamlines, each point of a channel cross-section (x_k, y_k) is mapped to another point (x_{k+1}, y_{k+1}) of the corresponding cross-section after one period of the fluidic structures, as shown in Figure 2.44. Such a so-called Poincaré map can be written formally as

$$(x_{k+1}, y_{k+1}) = P[(x_k, y_k)].\tag{99}$$

The function P can be computed from either an analytical or a numerical representation of the flow field. In such a way, a 3-D convection problem is essentially reduced to a mapping between two-dimensional Poincaré sections. In order to analyze the growth of interfacial area in a spatially periodic mixer, the initial distri-

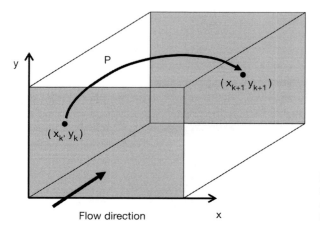

Figure 2.44 Section of a mixing channel with a map P connecting the points of two cross-sectional planes.

bution of fluids may be represented as a set of grid cells or points distributed over the inlet patch and tagged with different colors according to binary concentration values. By successively applying the mapping of Eq. (99) the points are redistributed or the grid cells are tagged with different colors. In such a way the growth of interfacial area and the stretching and folding of fluid lamellae can be analyzed. The Poincaré map method was applied by Linxiang et al. [140] to study mixing in curved channels.

2.6.5
Multilamination Mixers

Usually, even with zigzag mixing channels or chaotic mixers, liquid micro mixing can only be completed at moderate volume flows. In chemical process technology, throughput is often an important issue, and for this reason micro mixer designs going beyond the concept of two streams merging in a single mixing channel are needed. When abandoning mixer architectures where the fluid streams to be mixed are guided through only a single layer and going to multilayer architectures, the principle of multilamination becomes accessible. In multilamination mixers the two fluid streams are split into a multitude of sub-streams which are subsequently merged to form an interdigital arrangement of fluid lamellae. A design of a multilamination mixer is shown on the left side of Figure 2.45. From the flow distribution zone an interdigital arrangement of fluid lamellae enters a constriction with a width of 500 μm where the width of the lamellae is decreased, a principle known as hydrodynamic focusing. Subsequently the fluid streams enter a constriction which opens up to a wider channel.

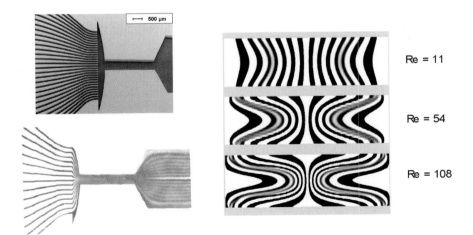

Figure 2.45 Design of a multilamination mixer with hydrodynamic focusing (upper left) and flow pattern in such a mixer for a total volume flow of 10 ml h^{-1} of water (lower left), taken from [141]. The right side of the figure shows the orientation of liquid lamellae over a cross-section of the constriction for different Reynolds numbers [142].

Hessel et al. [141] studied mixing in interdigital micro mixers experimentally, where different degrees of hydrodynamic focusing and mixers with and without opening of the constriction to a wider channel were considered. A result of such an experiment at a total volume flow of 10 ml h^{-1} is displayed at the lower left of Figure 2.45. Pure water and water colored with a dye were used as liquids. Apparently, it can be deduced that mixing has not proceeded to any sufficient degree, since the different colors are still visible in the reopening section. When the volume flow is increased substantially, downstream from the center of the constriction an average color becomes visible. Such a result could possibly be interpreted as complete mixing, but CFD simulations performed for this mixer geometry [142] offer a different explanation.

The CFD simulations were done in a 3-D model of the mixer based on the finite-volume method. Owing to artefacts from numerical diffusion occurring at high Peclet numbers, it was not possible to obtain satisfactory solutions of the convection–diffusion equation for typical liquid diffusion constants. In order to study the flow patterns in the mixer, a streamline-tracking technique was employed. From the velocity field obtained as solution of the Navier–Stokes equation, the flow streamlines were determined by numerical integration starting at the interface from the flow distribution zone to the actual mixer. The streamlines corresponding to pure and dyed water were tagged with a different color. By this means it is possible to compute flow patterns such as the one displayed on the right in Figure 2.45, which however, do not incorporate effects of diffusive mass transport. The diagram shows a cross-section through the constriction of the mixer for different Reynolds numbers. For small Reynolds numbers, the fluid lamellae are mainly oriented vertically. When the Reynolds number increases, the lamellae become deformed and assume a U-shape. This explains why at high flow rates a uniform color is observed in the mixing zone: a light beam cutting vertically through the flow domain intersects lamellae of different color, thus averaging out the coloring of the two different liquids. Hence it follows that experimental techniques for characterizing multilamination mixing processes based on the diffusion of a dye have to be viewed with care.

The streamline-tracking technique allows the extraction of qualitative information on the flow patterns in micro mixers, but does not permit a quantitative prediction of mixing efficiency. When hydrodynamic focusing does not proceed as rapidly as in the mixer displayed in Figure 2.45, the distortion of liquid lamellae is less pronounced and a semi-analytical method can be used to predict the efficiency of multilamination micro mixers, as proposed by Hardt and Schönfeld [142]. Approximately, a mixing process can be viewed as a pure diffusion process in a frame of reference co-moving with the average velocity of the flow. The solution of the diffusion equation for an initial condition being defined as a parallel, interdigital arrangement of liquid lamellae of width L and concentration values of 0 and 1 can be written as

$$c(x,t) = \frac{1}{2} + \sum_{n=1,3,\ldots} \frac{2}{n\,\pi} (-1)^{\frac{n-1}{2}} \exp(-D\,k_n^2\,t) \cos(k_n\,x), \tag{100}$$

where D is the diffusion constant and $k_n = n\pi/L$. On the basis of Eq. (100), the concentration field at different positions in the mixing channel of a multilamination mixer can be computed approximately. Hardt and Schönfeld used this approach to compare the mixing efficiency of different multilamination mixers and to design a new mixer allowing for a very high throughput and short mixing times [142].

2.6.6
Active Micro Mixing

In chemical micro process technology there is a clear dominance of pressure-driven flows over alternative mechanisms for fluid transport. However, any kind of supplementary mechanism allowing promotion of mixing is a useful addition to the toolbox of chemical engineering. Also in conventional process technology, actuation of the fluids by external sources has proven successful for process intensification. An example is mass transfer enhancement by ultrasonic fields which is utilized in sonochemical reactors [143]. There exist a number of microfluidic principles to promote mixing which rely on input of various forms of energy into the fluid.

The large surface-to-volume ratio of micro flows suggests the use of an actuation mechanism based on surface forces. One suitable mechanism is electroosmotic flow which is induced due to the force that an external electric field exerts on the electric double layer (EDL) building up at a solid/liquid interface. The extension of the EDL is usually very small compared with the width of a micro channel and the electroosmotic flow can be modeled by the Helmholtz–Smoluchowski slip-flow boundary condition Eq. (21). The zeta potential ζ determines the magnitude and direction of the electroosmotic flow close to the solid surface. Among other factors, ζ depends on the electric potential of the surface. An electric field normal to the solid/liquid interface can be imposed with the aid of electrodes embedded in the channel walls. In such a way, it is possible to reverse the direction of the electroosmotic flow building up in a micro channel. For this purpose, an electric field along the axis of the channel is created. A field perpendicular to the channel walls due to the embedded electrodes is used to tune the zeta potential. The direction of the flow induced by the driving electric field then depends on the potential of the electrodes. When using arrays of electrodes embedded in the upper and lower walls of a micro channel, periodic recirculating flow patterns can be generated. A possible electrode arrangement (represented by black rectangles) together with streamlines of flow patterns emerging in such channels is depicted on the left side of Figure 2.46.

Qian and Bau [144] have analyzed such electroosmotic flow cells with embedded electrodes on the basis of the Stokes equation with Helmholtz–Smoluchowski boundary conditions on the channel walls. They considered electrode arrays with a certain periodicity, i.e. after k electrodes the imposed pattern of electric potentials repeats itself. An analytic solution of the Stokes equation was obtained in the form of a Fourier series. Specifically, they analyzed the electroosmotic flow patterns with regard to mixing applications. A simple recirculating flow pattern such as the one

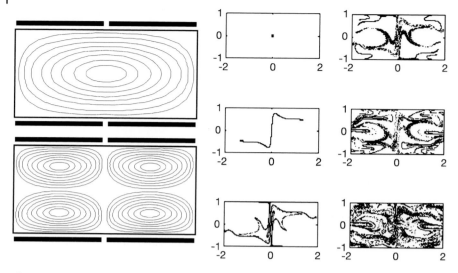

Figure 2.46 Section of a micro channel with electrodes embedded in the channel walls (left). When an electric field is applied along the channel, different flow patterns may be created depending on the potential of the individual electrodes. The right side shows the time evolution of an ensemble of tracer particles initially positioned in the center of the channel for a flow field alternating between the single- and the four-vortex pattern shown on the left [144].

shown on left side of Figure 2.46 will not promote mixing substantially, since tracer particles with a small diffusivity will follow the streamlines, while transport transverse to the streamlines will be negligible. In order to speed up mixing, an alternating flow created by switching between the single-vortex and the four-vortex pattern shown on the left side of Figure 2.46 was considered. Inertial effects were neglected and it was assumed that for the first half of the switching cycle T the flow is given by the single-vortex pattern, whereas for the second half of the cycle the four-vortex pattern prevails. With the flow pattern given, the time evolution of an ensemble of massless particles can be computed by integration along the streamlines. Qian and Bau positioned a small blob of massless particles in the center of the flow cell and computed their trajectories by solving the kinematic equations numerically. Their numerical results for times between 0 and $15T$ are displayed on the right side of Figure 2.46 The tracer particles spread over the volume of the cell in a chaotic manner and filled almost the complete cell volume in the final frame. The results suggest that chaotic advection based on alternating electroosmotic flows is a powerful principle for mixing of chemical species with a small diffusion constant. The principle suggests itself especially for fluidic cells with a slowly flowing liquid or a liquid at rest.

An electroosmotic mixer allowing to enhance the efficiency of mixing tees was proposed by Meisel and Ehrhard [145]. The corresponding geometry with a cylinder in the center of a mixing channel is displayed on top of Figure 2.47. When mixing ionic liquids, an electric double layer will form above the solid/liquid inter-

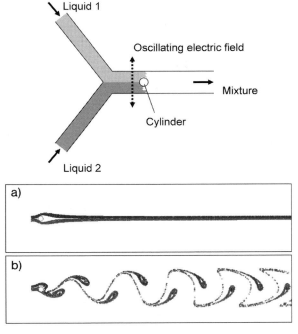

Figure 2.47 Micro mixer based on the excitation of an electro-osmotic flow around a cylinder by an oscillatory electric field (top). The bottom of the figure shows particle traces on both sides of the liquid/liquid interface with no electric field (above) and with the electric field switched on (below), as described in [145].

faces including the surface of the cylinder. An oscillatory electric field applied perpendicular to the channel in the region of the cylinder induces an electroosmotic flow around the cylinder perpendicular to the main flow along the mixing channel which may be pressure driven. Meisel and Ehrhard studied the performance of such cylindrical mixing structures by solving the Navier–Stokes equation in the channel numerically using a finite-element approach. The electroosmotic flow on the surface on the cylinder was modeled by imposing the Helmholtz–Smoluchowski slip-flow boundary condition. Characteristic results of these simulations are displayed on the bottom of Figure 2.47. The diagrams show the paths of massless particles on both sides of the interface separating the two liquids (which is not an interface in the strict sense since the liquids are miscible). In the upper diagram, the electric field is switched off and the particles follow a straight path. In the lower diagram a vortex street is created due to the oscillatory electroosmotic flow. Such vortex structures increase the interfacial area between the two liquids and promote mixing. A more quantitative analysis of the mixing efficiency and an optimization of the mixing device based on an arrangement of multiple cylinders requires further studies.

An alternative mechanism allowing promoting mass transfer between two liquids that has been studied in some detail is magneto-hydrodynamic mixing. The idea is based on flow patterns originating from the force an external magnetic field exerts on ions moving in a liquid. In order to induce a motion of the ions, electrodes are integrated in a micro channel which create an electrophoretic current when a voltage is applied. When an external magnetic field acts on the moving ions, they experience a Lorentz force driving them to a direction orthogonal to their momentary velocity and orthogonal to the magnetic field and drag the surrounding liquid molecules along with them. In short, the magneto-hydrodynamic forces are implemented by adding a source term

$$(\mathbf{J} \times \mathbf{B})_i \tag{101}$$

which is the cross product of the current density of the ions and the magnetic field to the Navier–Stokes equation Eq. (16). The current density is given by

$$J_i = \sigma \left[E_i + (\mathbf{u} \times \mathbf{B})_i \right], \tag{102}$$

where σ is the conductivity of the liquid, E_i the electric field strength and u_i the fluid velocity.

On top of Figure 2.48 a schematic design of a magneto-hydrodynamic mixer as proposed by Bau et al. [146] is displayed. The mixer consists of a micro channel of width w and depth $2h$ and contains electrodes separated by a distance L. A DC voltage is applied in such a way that the potential alternates between + and − for neighboring electrodes. Orthogonal to the channel a magnetic field is applied. In their theoretical treatment, Bau et al. exploited the symmetry of the problem and considered only the section of the channel enclosed by the gray faces. The rectangular patches orthogonal to the channel walls intersect the centerline between two electrodes. Based on the assumptions $h \ll w$ and $h \ll L$, they were able to obtain an approximate solution of the Navier–Stokes equation in the form of a series expansion. Their solution describes the flow in a static electric field along the channel and a static magnetic field perpendicular to the channel. The corresponding evolution of the interface between two liquid lamellae initially oriented in the x direction and each filling half of the channel is displayed on the bottom of Figure 2.48. Each of the diagrams shows a x–y section through the computational domain and the different diagrams are labeled with a time coordinate non-dimensionalized by a characteristic time-scale of the system. By the magneto-hydrodynamic forces a vortex flow is induced which causes an entanglement of the liquid lamellae. In a pure diffusion process without convection, mass transfer between the lamellae is governed by a $t^{1/2}$ law. In the magnetohydrodynamic mixer of Bau et al., the interfacial area is found to increase approximately linearly with time, an effect which superposes diffusion and effects a substantial reduction in mixing times. However, the linear interfacial stretching falls short of chaotic advection which exhibits an exponential growth as a function of time. In experiments conducted with a prototype system, the theoretical results were confirmed qualitatively [146].

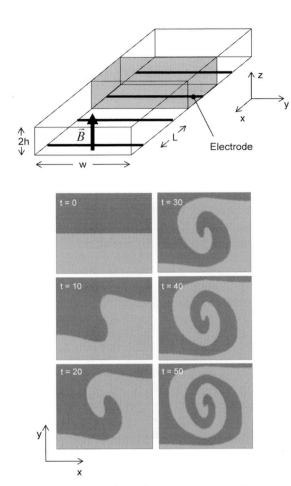

Figure 2.48 The top shows the schematic design of a magneto-hydrodynamic mixer with equally spaced electrodes arranged in a micro channel and an external magnetic field oriented along the z-axis. On the bottom theoretical results for the evolution of two parallel liquid lamellae as a function of dimensionless time are shown [146].

Gleeson and West [147] proposed another type of magneto-hydrodynamic micro mixer based on an annular geometry. The device, sketched on the left of Figure 2.49, consists of an annular micro channel, where the curved inner and outer walls of the channel are plated and act as electrodes. Perpendicular to the plane of the channel a magnetic field is applied. Both the electric and the magnetic field are alternating and in-phase. It can be shown that a magneto-hydrodynamic force is induced which acts in the azimuthal direction and drives an azimuthal flow. Gleeson and West obtained a solution of the equations of magneto-hydrodynamics in the form

Figure 2.49 The left side shows the schematic design of an annular magneto-hydrodynamic mixer. On the right, the evolution of the interface between two liquids, as described in [147], is depicted.

of a Fourier–Bessel series. Based on this solution, the growth of the interfacial area of two liquids initially positioned in two 180° azimuthal regions of the mixer was computed. The results of these analyzes are displayed on the right of Figure 2.49. By the azimuthal motion of the fluid, the interfacial area is increased substantially. A calculation of the length of the interface [147] shows that the long-time asymptotics are characterized by a linear increase, a result similar to the findings of Bau et al. [146] for their magneto-hydrodynamic mixer.

Whereas the previous studies were restricted to liquids with dissolved species of zero diffusivity, Gleeson et al. [148] extended their analysis to include finite diffusion constants. They considered the regimes of pure diffusion (zero Peclet number) and intermediate and high Peclet numbers and found analytical and numerical solutions for the mixing time in all three regimes. The predictions derived from the analytical expressions agreed fairly well with the numerical results. It was shown that convection speeds up mixing considerably, for example at Peclet numbers around 10^2 the mixing times are a factor of 10^2–10^3 shorter than the mixing times obtained from pure diffusion.

2.6.7
Hydrodynamic Dispersion

Mass transfer of a solute dissolved in a fluid is not only the fundamental mechanism of mixing processes, it also determines the residence-time distribution in microfluidic systems. As mentioned in Section 1.4, in many applications it is desirable to have a narrow residence-time distribution of concentration tracers being transported through a microfluidic system. An initially narrow concentration tracer will suffer a broadening (i.e. a dispersion) due to two different effects. First, in some regions of the flow domain of a system the fluid velocity will be smaller than in others, thus leading to a longer residence time of molecules being transported preferably through these parts of the domain. However, owing to Brownian motion the molecules will also sample some of the other regions with higher flow velocity. Hence molecular diffusion might reduce the dispersion of a concentration tracer. On the other hand, by diffusion an initially localized concentration tracer in a fluid at rest will become dispersed. From these arguments it becomes clear that hydrodynamic dispersion depends on a subtle interplay of convective and diffusive mass transfer, and the evolution of a concentration tracer as it is transported through the flow domain depends on various factors such as the flow profile, and the magnitudes of the flow velocity and the diffusion constant.

The key analysis of hydrodynamic dispersion of a solute flowing through a tube was performed by Taylor [149] and Aris [150]. They assumed a Poiseuille flow profile in a tube of circular cross-section and were able to show that for long enough times the dispersion of a solute is governed by a one-dimensional convection–diffusion equation:

$$\frac{\partial \bar{c}}{\partial t} + \bar{u}\,\frac{\partial \bar{c}}{\partial x} = D_e\,\frac{\partial^2 \bar{c}}{\partial x^2}\,, \tag{103}$$

where $\bar{c}$ denotes the concentration averaged over the cross-section of the tube, $\bar{u}$ the average velocity and D_e an effective diffusivity, also denoted dispersion coefficient, which is given by

$$D_e = D + \frac{\bar{u}^2\,R^2}{48\,D}\,, \tag{104}$$

where D is the molecular diffusivity and R the radius of the tube. The factor $1/48$ multiplying the velocity-dependent term is generic for tubes of circular cross-section and is modified when other geometries are considered. In many cases the second term, which can be rewritten as $D\,Pe^2/48$, dominates over the first, which is a purely diffusive contribution. Hence, due to convection a concentration tracer is usually dispersed much more strongly than it would have been by diffusion alone. A notable feature of Eqs. (103) and (104) is their independence of any initial condition. Independent of how the solute is distributed over the channel cross-section and along the channel initially, the description given by Taylor and Aris will be valid in the limit of long times ($t \to \infty$). When exactly this limit is reached with a given accuracy depends on the initialization of the concentration field. A rough guideline is provided by the Fourier number of Eq. (97) evaluated with the tube radius as length scale. The Fourier number can be regarded as a dimensionless time coordinate which compares the actual time with the time a molecule needs to sample the cross-sectional area of the tube. The validity of the Taylor–Aris description should be related to the condition that the Fourier number assumes values of order 1 or larger.

The analysis of Taylor and Aris was extended to arbitrary time values by Gill and Sankarasubramanian [151] for the dispersion of an initially plug-like profile, i.e.

$$c(x,r,0) = \begin{cases} c_0 & (|x| \leq l/2) \\ 0 & (|x| > l/2) \end{cases},$$ (105)

where the radial coordinate of the tubular geometry is denoted by r and l is the length of the plug. They derived a generalized evolution equation for the area-averaged concentration of the form:

$$\frac{\partial \overline{c}}{\partial t} = \sum_{n=1}^{\infty} k_n(t) \frac{\partial^n \overline{c}}{\partial x^n}$$ (106)

which is valid without any restriction on t. The derivatives in the infinite series appearing on the right side are multiplied by time-dependent dispersion coefficients k_n. In the Taylor–Aris limit, all of the dispersion coefficients except k_1, which describes the convection of the tracer with the flow, and k_2, which determines the spreading of the tracer, are negligible. When moving to smaller times, the time-dependence of k_2 needs to be taken into account, while all higher dispersion coefficients are still negligible [151]. Only at very small times do the higher dispersion coefficients become important. For the case they considered, Gill and Sankarasubramanian found that k_2 can be regarded as time independent for Fourier numbers greater than about 0.5.

Although the results discussed above highlight some of the most important aspects of hydrodynamic dispersion, they were based on cylindrical ducts which are not the generic geometry used in the field of microfluidics. In chemical micro process technology, tubular sections are used to connect different units; however, the channels contained in micro reactors typically have a rectangular or tub-like cross-section. Dispersion in rectangular channels was studied in detail by Doshi et al. [152] and Dutta and Leighton [153]. The evolution equation Eq. (106) is still valid in this case, but the expression for the dispersion coefficient Eq. (104) needs to be modified. While Aris [154] was still able to obtain a simple analytical expression for the dispersion coefficient related to flow between parallel plates, the corresponding expression for rectangular channels is a complicated series expansion. This is not very surprising, since the exact form of the flow profile in a rectangular channel is given in the form of an infinite series as well. Dutta and Leighton [153] found a simple functional dependence which approximates the exact expression for the dispersion coefficient in rectangular channels within an error of 10%. In addition, they considered tub-like channel cross-sections which are typically obtained by isotropic etching processes. For the latter they employed a numerical scheme allowing the computation of the dispersion coefficient. On this basis, they compared different channel geometries and identified favorable and less favorable designs.

While the previous studies refer to straight channels exceptionally, microfluidic devices often comprise channels with a curvature. It is therefore helpful to know how hydrodynamic dispersion is modified in a curved channel geometry. This aspect was investigated by Daskopoulos and Lenhoff [155] for ducts of circular cross-section. They assumed the diameter of the duct to be small compared with the radius of curvature and solved the convection–diffusion equation for the concentration field numerically. More specifically, a two-dimensional problem defined on the cross-sectional plane of the duct was solved based on a combination of a Fourier series expansion and an expansion in Chebyshev polynomials. The solution is of the general form

$$D_e^{cur} = D\left[1 + Pe^2 f(K,Sc)\right], \tag{107}$$

where D_e^{cur} is the dispersion coefficient in curved ducts, D the molecular diffusivity, Pe the Peclet number of the flow, and the function f depends on the Dean number K defined in Eq. (72) and on the Schmidt number Sc, which is the ratio of the kinematic viscosity and the diffusivity. Daskopoulos and Lenhoff found the following asymptotic behavior for the ratio of the dispersion coefficients in curved and straight ducts:

$$\frac{D_e^{cur}}{D_e} \quad \begin{cases} = 1 & K \to 0 \\ \propto K^{-1} & K \to \infty \end{cases}. \tag{108}$$

As mentioned earlier, in curved channels a secondary flow pattern of two counter-rotating vortices is formed. Similarly to the situation depicted in Figure 2.43, these vortices redistribute fluid volumes in a plane perpendicular to the main flow direction. Such a transversal mass transfer reduces the dispersion, a fact reflected in the K^{-1} dependence in Eq. (108) at large Dean numbers. For small Dean numbers, the secondary flow is negligible, and the dispersion in curved ducts equals the Taylor–Aris dispersion of straight ducts.

Mass transfer in micro channels can exhibit complex dynamics going far beyond the usual dispersion phenomena when adsorption on the channel walls is taken into account. The reason for this complex behavior lies in the kinetics of the adsorption process, which depend on the concentration of the adsporbed species and on temperature [156]. Fedorov and Viskanta [157] set up a model of micro channels with coated walls, where the coating layer adsorbs certain species dissolved in the gas flowing through the channel. They solved the momentum equation in combination with the enthalpy equation and the mass transport equation for the adsorbable species using a finite-difference method. The problem studied was the transport of a step-function concentration and temperature profile through the channel. By virtue of alternating adsorption and desorption processes, complex oscillatory temperature and concentration patterns appeared. Such examples illustrate that when mass transfer dynamics are coupled to adsporption/desorption dynamics at solid surfaces, a behavior qualitatively different from dispersion phenomena in micro channels might emerge.

2.7
Chemical Kinetics

Most plants or reactors in chemical micro process technology inevitably contain a unit where chemical conversion takes place. The goal might be to produce fine chemicals with a high yield and selectivity or to screen a large number of reactions in parallel. Hence a thorough understanding of chemical kinetics is a key requirement for the successful design and optimization of micro reaction devices. For this purpose, reliable models of reaction kinetics coupled to the transport equations of momentum, heat and matter are needed.

2.7.1
Kinetic Models

The effect of chemical kinetics on mass transport in incompressible flows is summarized by the reaction term r in Eq. (89). Applied to a chemical species a, it describes the rate of disappearance of this species per unit volume:

$$r_a = -\frac{1}{V}\frac{dn_a}{dt}, \tag{109}$$

where n_a denotes the molar amount of a. Especially in gas-phase reactions a complicated coupling between chemical kinetics on the one hand and momentum, heat and mass transfer on the other might occur. An exothermic or endothermic reaction releases or consumes energy, an effect which has to be included as a source term in the enthalpy equation Eq. (77). Furthermore, certain chemical species are consumed or produced, which is expressed by the source term in Eq. (89).

The reaction rate r_a determines how fast the concentration of a chemical species a increases or decreases due to chemical reactions. It depends on temperature and on the concentrations of other chemical species involved in the reaction. Consider the case of a simple reaction:

$$\nu_a\, A + \nu_b\, B \rightarrow \nu_c\, C, \tag{110}$$

where ν_a, ν_b, ν_c are stoichiometric coefficients and it is assumed that the stoichiometric equation truly represents the mechanism of the reaction. The law of mass action then states that the reaction rate of species a is given by

$$r_a = k\, c_a^\alpha\, c_b^\beta, \tag{111}$$

where k is a rate constant, c_a and c_b denote the species concentrations and $\alpha = \nu_a$, $\beta = \nu_b$. In general, the reaction mechanism will be more complex than suggested by Eq. (110), and the exponents α and β may take non-integer values. The rate constant of many reactions is given by the Arrhenius equation (see, e.g., [126]):

$$k = k_0 \exp\left(\frac{-E}{R\,T}\right), \tag{112}$$

where k_0 is the pre-exponential factor, E the activation energy, R the gas constant and T temperature. The exponential dependence on temperature and the occurrence of an activation energy indicate that the reaction proceeds via an intermediate state which is accessible to the molecules in the high-energy tail of the thermal ensemble.

Power-law kinetic models such as Eq. (111) in combination with the Arrhenius equation and their obvious generalizations to a larger number of reacting species find widespread applications in the simulation of reacting flows. However, strictly the validity of such models is questionable when solid catalytic reactions are considered. Solid catalysis is of major importance in chemical micro process technology, and prominent examples of reactions being conducted in micro reactors are partial oxidations or steam reforming reactions. Such heterogeneously catalyzed reactions are described by more complex models which take into account the adsorption kinetics on the solid surface. In this context, a simple picture of a reaction as described by Eq. (110) could be the following: first, both species are adsorbed to the catalyst surface, where the surface coverage depends on the gas-phase concentration in the vicinity of the surface and on temperature. After having been adsorbed, the species react and the products are released to the gas phase. Assuming first-order adsorption kinetics, the adsorption rate of a chemical species to a catalyst surface can be written as

$$r_{abs} = F \, s_{abs} (1 - \theta) , \tag{113}$$

where F is the number of molecules per unit area and time hitting the surface, s_{abs} is the adsorption probability at an active site of the surface and θ is the surface coverage, i.e. the percentage of active sites occupied by molecules. The desorption rate is given by

$$r_{des} = s_{des} \, \theta , \tag{114}$$

with a site-specific desorption probability s_{des}. The adsorption equilibrium is determined by equating the adsorption and the desorption rate. Taking into account that the flux F is proportional to the partial pressure p of the chemical species, the surface coverage can then be written as

$$\theta = \frac{b \, p}{1 + b \, p} , \tag{115}$$

where b can be determined from the parameters appearing in Eqs. (113) and (114). Assuming that the species react while being adsorbed at the catalyst surface, the rate of a reaction $A + B \rightarrow C$ is obtained as

$$r = k \, \theta_a \, \theta_b , \tag{116}$$

with a rate constant k. The mechanism just described is known as the *Langmuir–Hinshelwood* mechanism and is the most prominent model describing catalysis on solid surfaces. Depending on the specific adsorption and desorption mechanisms, $1 - \theta$ and θ in Eqs. (113) and (114) may have to be replaced by more general expressions $f_{ads}(\theta)$ and $f_{des}(\theta)$ [156]. Furthermore, the rate equation has to be modified when dissociation reactions have to be taken into account or when an adsorbed species reacts with molecules in the gas phase [126]. For practical applications such as industrial processes it is often very difficult to determine uniquely all the parameters of a Langmuir–Hinshelwood model experimentally. For this reason, often power-law kinetic models such as Eq. (111) are employed to describe solid-catalytic reactions. Such models are usually not justified from a first-principles standpoint, but they may provide a reasonable parametrization of the kinetics in a limited temperature and partial-pressure range.

2.7.2
Numerical Methods for Reacting Flows

Numerical computations of reacting flows are often difficult owing to the different time-scales involved and the highly non-linear dependence of the reaction rate on concentrations and temperature. The solution of the species concentration equations in combination with the momentum and the enthalpy equation generally requires an iterative procedure such as the one outlined in Section 1.3.4. A rough sketch of the numerical structure of a stationary reacting-flow problem is given as

$$
\begin{bmatrix} A_{cc} & A_{cu} & A_{cT} \\ A_{uc} & A_{uu} & A_{uT} \\ A_{Tc} & A_{Tu} & A_{TT} \end{bmatrix} \begin{bmatrix} c \\ u \\ T \end{bmatrix} = \begin{bmatrix} b_c \\ b_u \\ b_T \end{bmatrix},
\tag{117}
$$

where c, u and T denote the vector of concentration, velocity and temperature fields, respectively. Owing to the non-linear nature of the problem, the coefficients of the different matrices $A_{\alpha\beta}$ still depend on the unknowns c, u and T. The cross-coupling between different field quantities is provided by those matrices $A_{\alpha\beta}$ with $\alpha \neq \beta$. The set of non-linear algebraic equations is solved iteratively, i.e. starting with an initial guess the approximation is successively improved until convergence is reached. Depending on the nature of the chemical reaction term entering the species-concentration equation, different strategies may be applied to solve Eq. (117). For intrinsic kinetics characterized by a much shorter time-scale than transport of momentum, heat and matter, it is often preferable to set up an iteration scheme where a number of iterations of the species-concentration equation are performed during one iteration cycle of the remaining equations. However, for a fast reaction which is heat and mass transfer limited (for example, in a situation where the reactants are not premixed), comparable iteration cycles of the species-concentration equation and the remaining equations might be sufficient.

Apart from the coupling of chemical kinetics with the transport equations, the chemical reaction dynamics itself may pose numerical challenges when a number of different reactions are superposed. In such a case the rate of disappearance of a chemical species i can be written as

$$
-\frac{\dot{n}_i}{V} = R_{ij}\, r_j,
\tag{118}
$$

where r_j is the rate of the jth reaction and R_{ij} is a matrix defining how a specific reaction contributes to a change in concentration of the chemical species involved. Frequently it occurs that the time-scales characterizing the different reactions vary by orders of magnitude, such that the fast reactions are already completed while the slow reactions have not yet progressed to any appreciable degree. The corresponding stiff differential equations are usually solved using an implicit time-integration scheme which allows comparatively large time steps without suffering from numerical instabilities or predicting unrealistic asymptotic states [85]. However, implicit time integrators involve the solution of a (generally non-linear) algebraic system of equation for each time step which is done by some iterative scheme such as Newton's method. For reaction systems with a broad spectrum of time-scales, these iteration schemes can fail to converge, with the consequence that very small time steps have to be chosen. Such a situation is related to high computational costs, and methods are needed to simulate extremely stiff reaction systems more efficiently.

Methods based on the partitioning of a reaction system into fast and slow components have been proposed by several authors [158–160]. A key assumption made in this context is the separation of the space of concentration variables into two orthogonal subspaces Q_s and Q_f spanned by the slow and fast reactions. With this assumption the time variation of the species concentrations is given as

$$-\frac{\dot{n}_i}{V} = (Q_{s,j})_i \, (\dot{y}_s)_j + (Q_{f,j})_i \, (\dot{y}_f)_j \, . \tag{119}$$

The notation is such that $(Q_{s,j})_i$, $(Q_{f,j})_i$ denotes the ith component of the jth basis vector in the subspace of slow and fast reactions, respectively. The corresponding expansion coefficients are $(\dot{y}_s)_j$ and $(\dot{y}_f)_j$, respectively, and are expressed by the reaction rates via

$$(\dot{y}_s)_i = \left(Q_{s,j}^T \right)_i R_{jk} \, r_k \, , \tag{120}$$

$$(\dot{y}_f)_i = \left(Q_{f,j}^T \right)_i R_{jk} \, r_k \, . \tag{121}$$

If the time-scale of the fast reactions is much shorter than that of the slow reactions, it can be assumed that the former are completed at an initial stage of the latter. Mathematically, this assumption reads

$$\left(Q_{f,j}^T \right)_i R_{jk} \, r_k = 0 \, . \tag{122}$$

Eq. (122) represents a set of algebraic constraints for the vector of species concentrations expressing the fact that the fast reactions are in equilibrium. The introduction of constraints reduces the number of degrees of freedom of the problem, which now exclusively lie in the subspace of slow reactions. In such a way the fast degrees of freedom have been eliminated, and the problem is now much better suited for numerical solution methods. It has been shown that, depending on the specific problem to be solved, the use of simplified kinetic models allows one to reduce the computational time by two to three orders of magnitude [161].

2.7.3
Reacting Channel Flows

In chemical micro process technology there exists one class of reactor designs which deserves the term 'generic', since many of the micro reactors reported in the literature are based on this design concept. The design comprises at least one rectangular micro channel, often a multitude thereof, with a solid catalyst attached on the channel walls. The reacting fluid flows through the channel, while the reagents diffuse to the channel wall where they undergo chemical reactions. There exist two versions of this design concept, as displayed in Figure 2.50. Either a smooth surface, often a metal layer, acts as catalyst, or the reaction occurs in a catalytically active porous medium. Clearly, the advantage of the porous catalyst layer is the higher specific surface area offering more reaction sites to the reagents. However, all of the studies reported in this section are based on the concept of wall-catalyzed reactions.

For the reasons described above, reaction–convection–diffusion problems tend to be difficult to solve numerically. Hence the simulation of reacting flows in three dimensions or parameter studies of micro reaction devices may be very time consuming. In order to permit rapid prototyping of micro reactors, efficient modeling strategies with a minimum expenditure of computational resources are needed. The modeling approach developed by Gobby et al. [162] allows the assessment of a limited class of reacting micro channel flows very quickly. They assumed a micro channel of length L and depth h with a first-order reaction occurring at one of the channel walls, as depicted in Figure 2.51. In cases where the flow profile is independent of the axial position in the channel and the problem can be approximated by a two-dimensional model, the mass transport equation for a chemical species a can be written in dimensionless form as

$$\frac{\text{Pe } h}{L} u(\eta) \frac{\partial c_a}{\partial \zeta} = \frac{\partial^2 c_a}{\partial \eta^2} , \tag{123}$$

where the axial and transverse coordinates ζ and η were non-dimensionalized by the channel length L and the channel depth h. The reactant concentration is denoted by c_a, the velocity by u, and the Peclet number is expressed by the average

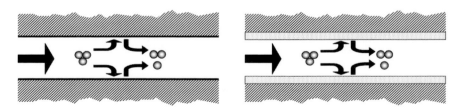

Figure 2.50 Reaction channels with a smooth surface (left) and a porous medium (right) as catalyst.

velocity $\bar{u}$ and the diffusion constant D as $\text{Pe} = \bar{u}\,h/D$. Eq. (123) is solved subject to the boundary conditions of an impermeable upper channel wall and a first-order reaction with rate constant k occurring at the lower channel wall. Such a first-order reaction term surely does not adequately capture the mechanism of heterogeneous catalysis; it might, however, be a reasonable approximation to the kinetics in a limited parameter or operation range. An important dimensionless group characterizing the reactive flow is the Damköhler number, defined as

$$\text{Da} = \frac{k\,h}{D}\,, \tag{124}$$

which characterizes the ratio of the diffusive and the reactive time-scale. The mass transport equation has a separable solution of the form

$$c_a(\zeta,\eta) = \bar{c}_a(\zeta)\,f_a(\eta)\,, \tag{125}$$

where

$$\bar{c}_a(\zeta) = \int_0^1 c_a(\zeta,\eta)\,\mathrm{d}\eta\,. \tag{126}$$

On inserting this ansatz into Eq. (123), the solution can be determined in the form of an eigenfunction expansion, as shown by Walker [163]. The parameter controlling the number of terms of this expansion having to be taken into account is $\text{Pe}\,h/L$, which is usually of the order of $O(0.01 - 1)$ in micro reactors. For this reason, often only the first term contributes. With the entrance condition $\bar{c}_a(\zeta) = 1$, the axial dependence can then be written as

$$\bar{c}_a(\zeta) = \exp(-\lambda_a\,\zeta)\,, \tag{127}$$

where the eigenvalue λ_a is given as the solution of a non-linear algebraic equation. Gobby et al. compared their analytical results with full numerical simulations and found good agreement. In addition to isothermal flows, they also determined analytical solutions for non-isothermal reacting flows and extended their model to second-order kinetics. Hence they developed a class of models which may provide a simple characterization of reacting flows in micro channels without the need to do a full numerical simulation.

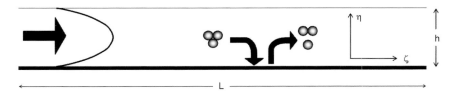

Figure 2.51 Two-dimensional model geometry of a micro channel with a reaction occurring at the lower channel wall.

Commenge et al. [164] used a similar analytical model for reacting flows in micro channels to assess the quality of simple plug-flow models which may be used to estimate reaction-rate constants. Micro reactors lend themselves to measure intrinsic rate constants of chemical reactions, as owing to the short diffusion paths, heat and mass transfer limitations can be eliminated. The simplest way to deduce the rate constant k of a first-order heterogeneously catalyzed reaction at the walls of a tube is by assuming the reaction to occur in the volume of a plug-flow reactor. In this way the wall reaction is replaced by a pseudo-homogeneous reaction and the velocity profile of the flow is ignored, which means that effectively a one-dimensional model is used. By measuring the inlet and outlet concentrations of the reacting component, the rate constant is then obtained as

$$k = \frac{\bar{u}\,R}{2\,L}\ln\left[\frac{c_a(\zeta = 1)}{c_a(\zeta = 0)}\right].\tag{128}$$

where the notation is chosen similarly to the previous paragraph and L and R measure the length and radius of the tube, respectively. Two effects are not taken into account by this expression. First, radial concentration gradients are ignored. Second, dispersion in the tube, as discussed in Section 2.6.7, is neglected.

Commenge et al. extended the one-dimensional model of reacting flows to include Taylor–Aris dispersion, i.e. they considered an equation of the form

$$\frac{d^2 c_a}{d\zeta^2} - \frac{Pe^*}{2}\frac{dc_a}{d\zeta} - \beta c_a = 0\,,\tag{129}$$

where Pe^* is a modified Peclet number containing the Taylor–Aris dispersion constant instead of the diffusivity and β is a dimensionless parameter representing the pseudo-homogeneous reaction. In order to study the influence of dispersion on chemical conversion, the solution of Eq. (129) was compared with the solution of the corresponding two-dimensional problem, obtained in a similar way as sketched in the previous paragraph. It turned out that for a Damköhler number of 1, no satisfactory agreement between the one- and the two-dimensional models was achieved. The inclusion of Taylor–Aris dispersion improved the concentration profiles to a certain degree with respect to a plug-flow model; however, the main reason for the deviations are the radial concentration gradients which are not accounted for in the one-dimensional models. Hence, when attempting to extract intrinsic reaction-rate constants from comparisons of experimental results with results of one-dimensional reactor models, care should be taken to work in a regime of Damköhler numbers significantly smaller than 1.

2.7.4
Heat-exchanger Reactors

The design of multichannel micro reactors for gas-phase reactions is typically based on a stack of micro structured platelets. For strongly endothermic or exothermic reactions, it lends itself to alternate between layers of reaction channels and heating or cooling gas channels which supply energy to or withdraw it from the reaction. Such a set up is similar to the heat exchanger design depicted in Figure 2.30. Within this class of micro reactor designs a choice can be made between different flow schemes of the gas streams in adjacent layers (co-, counter- or cross-current). The counter-current coupling of an endothermic reaction to a heating gas stream in a multi-layer architecture was studied by Hardt et al. [120]. The 2-D geometry their model was based on is displayed in Figure 2.52.

The dynamics of a heterogeneously catalyzed gas-phase reaction occurring in a nanoporous medium in combination with heat and mass transfer was simulated using a finite-volume approach. In contrast to other studies of similar nature, heat and mass transfer in the nanoporous medium were explicitly accounted for by solving volume-averaged transport equations in the porous medium (for a discussion of transport processes in porous media, see Section 1.9). Such an approach made it possible to compare the transport resistances in the gas phase and in the porous medium and to study the trade-off between maximization of catalyst mass and minimization of mass transfer resistance due to pore diffusion. A typical concentration profile of a reacting chemical species which is converted by the catalyst is displayed in Figure 2.53. Owing to the small pore size with an average diameter of 40 nm, the effective diffusivity in the porous medium is small and large concentration gradients build up, whereas in the micro channel the gradients are negligible. Typical catalyst effectiveness factors for a 100 µm catalyst layer were found to be of the order of 0.4. One of the outstanding potential features of micro reactors is an efficient utilization of the catalyst material. In conventional fixed-bed technology, catalyst pellets for liquid reactions are usually of a size of 2–5 mm [126]. Owing to diffusive limitations in such comparatively large pellets, reactions often occur in a region close to the surface.

A main objective of the work of Hardt et al. was to study the influence of heat transfer on the achievable molar flux per unit reactor volume of the product species. They compared unstructured channels to channels containing micro fins such as shown in Figure 2.31. Heat transfer enhancement due to micro fins resulted in a different axial temperature profile with a higher outlet temperature in the reaction gas channel. Owing to this effect and by virtue of the temperature dependence

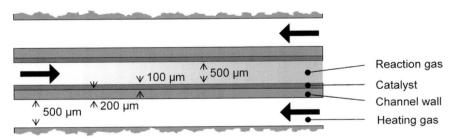

Figure 2.52 2-D model of a counter-current heat-exchanger reactor with a nanoporous catalyst layer deposited on the channel wall.

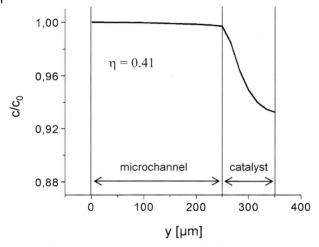

Figure 2.53 Normalized concentration profile of a reacting species across a micro channel of 500 µm width with a 100 µm catalyst layer deposited on the wall.

of the reaction rate, an improvement of heat transfer resulted in a significantly higher specific product molar flux. For the system under study, the heat transfer enhancement achievable with micro fins was found to increase the specific molar flux by about a factor of two. Such model studies show that a complex interplay between flow, heat and mass transfer may occur in micro reactors and underline the need for fully coupled simulations incorporating conjugate heat transfer and transport in porous media.

The optimization of heat transfer in a heat-exchanger reactor was also the objective of the work of TeGrotenhuis et al. [165]. Specifically, the exothermic water–gas shift (WGS) reaction:

$$CO + H_2O \rightleftharpoons H_2 + CO_2 \tag{130}$$

which is utilized in fuel reformers to reduce the level of carbon monoxide was considered. When the temperature level of an exothermic, reversible reaction such as the WGS reaction increases, the kinetics are accelerated but the equilibrium is shifted more towards the feed components. As a result, neither very low nor very high temperatures are optimal when the goal is to maximize the space–time yield for a given conversion. Rather, there is a specific temperature trajectory, i.e. a specific functional dependence of the reaction temperature on time, which allows the space–time yield to be maximized. Owing to their short thermal diffusion paths, micro reactors allow the temperature profile in a reaction channel to be controlled much better than conventional equipment.

TeGrotenhuis et al. studied a counter-current heat-exchanger reactor for the WGS reaction with integrated cooling gas channels for removal of the reaction heat. The computational domain of their 2-D model on the basis of the finite-element method

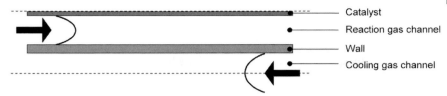

Figure 2.54 Model of a counter-current heat-exchanger reactor for exothermic reactions. The dashed lines indicate symmetry planes.

is sketched in Figure 2.54. The reactor design does not allow for a detailed adjustment of the temperature profile in the reaction gas channel; however, by varying the cooling gas inlet temperature and the ratio of cooling gas and reaction gas flow rates, different temperature profiles can be imposed.

Simulation results for the CO conversion as a function of the ratio of cooling and reaction gas flux are displayed in Figure 2.55. All of the results shown are based on a fixed reaction gas inlet temperature of 350 °C and a fixed inlet composition of the reaction gas. All of the curves obtained with different cooling gas inlet temperatures start at a conversion of about 70%. This is due to the fact that by release of reaction heat close to the inlet, the reaction gas temperature rises to above 400 °C, a region where the conversion is limited to 70% by the chemical equilibrium. The maximum achievable conversion increases by 2.7% when the cooling gas temperature is raised from 125 to 225 °C. The reaction dynamics are such that, owing to the fast kinetics at high temperatures, a large degree of conversion is obtained in the inlet region of the channel. For the remaining few percent conversion, a comparatively large reactor volume is needed. Hence even minor differences in conversion

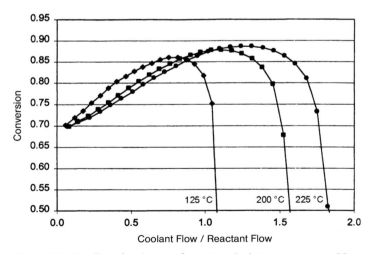

Figure 2.55 The effect of cooling gas flow rate and inlet temperature on CO conversion in the WGS reactor, as described in [165]. The cooling gas flow rate was varied for a fixed reaction gas flow rate and three different inlet temperatures were considered.

achievable with improved temperature control may result in a considerable reduction in reactor size and the required amount of catalyst. When considering more advanced reactor designs allowing for fine-tuning of the temperature trajectory, CFD simulations are indispensable for performance optimization.

2.7.5
Periodic Processing

As compared with macroscopic reactors, micro reactors not only permit fast heat and mass transfer, they also allow improved process control. In this context, one aspect that has been studied in some detail is periodic process control. It is well known that in some cases periodic variations of the process parameters permit improved reactor performance [166–168]. In macroscopic reactors, short cycle times are often not accessible owing to limitations of heat and mass transfer. In micro reactors, the large heat transfer coefficients achievable allow rapid thermal cycling. Furthermore, low-Peclet number flow in narrow channels allows concentration plugs to be transported without axial mixing. The potential of periodic process control for enzymatic reactions in micro reactors was investigated by Stepanek et al. [169]. They studied a reaction–adsorption process in a micro channel with alternating wall segments, as displayed on top of Figure 2.56. The channel comprises alternating segments with reaction and adsorption zones. In the reaction zones an

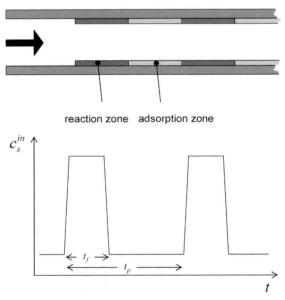

Figure 2.56 Micro channel with alternating segments for reaction and adsorption (above) and operating cycle with pulsed substrate inlet concentration (below).

immobilized enzyme catalyzes the synthesis of a biomolecule, which can be removed from the flow by wall adsorption in the adsorption zones. The reason for such different channel segments is the kinetics of the reaction which show both product and substrate inhibition, i.e. the reaction rate decreases with increasing substrate and product concentration. Hence, by removing the product from the flow in the adsorption zones, the reaction rate can be increased.

Stepanek et al. [169] used a one-dimensional plug flow model to describe the reaction dynamics in a micro channel for a periodic processing regime. In such a model the reaction is regarded as pseudo-homogeneous. An additional reaction term was included for the product to describe removal from the flow by adsorption. The partial differential equations for the two chemical species, substrate and product, were solved for periodic inlet conditions which are depicted on the bottom of Figure 2.56. For a fraction t_f/t_p of the temporal period t_p, the substrate concentration at the inlet is adjusted to a specific value and the pH is chosen in such a way as to promote product adsorption. For the remaining fraction of the period, the substrate concentration is zero and the pH is shifted to values favoring elution.

Typical results for the outlet concentration of substrate and product are shown in Figure 2.57 for two different values of t_f/t_p. When $t_f/t_p = 0.9$, for most of a cycle the

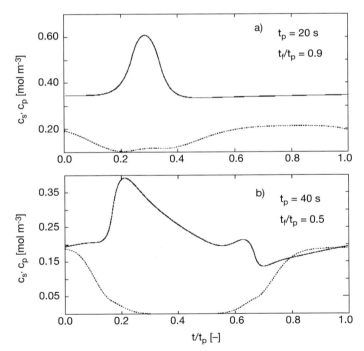

Figure 2.57 Substrate (dashed line) and product (full line) concentrations at the channel outlet for two different operation modes, taken from Stepanek et al. [169].

substrate is fed to the channel, whereas the substrate inlet concentration is zero for half of the cycle when $t_f/t_p = 0.5$. In the former case, the average product concentration is fairly high, but substrate and product are not easily separated, since for most of the cycle a mixture of substrate and product is found at the channel outlet. In the latter case, the average product concentration is lower, but the separability of substrate and product is considerably higher. Stepanek et al. [169] made a systematic parameter study of the reactor varying the cycle period, the switch-on time of the substrate and the inlet flow velocity. They found a trade-off between product yield and separability of the different chemical components. A high product yield usually resulted in a mixture of product and substrate (low separability) at the outlet and vice versa. Furthermore, the reactor performance increased with decreasing cycle time. Hence, in cases where the capital and/or the process costs for product separation are considerable, a micro reactor with periodic process control may offer a solution, since an adjustment of the process parameters allows a product separation *in situ* and the reactor can be operated with favorable short cycle times.

2.8
Free Surface Flow

Multi-phase flows of immiscible fluids are omnipresent in chemical process technology. In various processes, gases or liquids are dispersed in a surrounding liquid phase, or aerosols of liquid droplets in a gas phase are formed. A standard method to model such types of flow is the Euler–Euler description of interpenetrating continua [171]. In this approach, both the disperse phase, i.e. the droplets or the bubbles, and the surrounding continuous phase are treated as continua which occupy a certain fraction of each volume element of the fluid domain. Via exchange terms in the transport equations, the two phases are coupled to each other and may exchange momentum, heat and matter. In micro reactors the Euler–Euler description is often inappropriate. When the size of droplets or bubbles becomes comparable to the channel dimensions, the continuum assumption breaks down and the dynamics of multi-phase flow confined in narrow spaces deviates considerably from that in macroscopic vessels. Hence, in general the usual modeling approaches for multi-phase flow cannot be applied in micro reactors and special techniques are needed to predict the flow patterns.

In most cases the only appropriate approach to model multi-phase flows in micro reactors is to compute explicitly the time evolution of the gas/liquid or liquid/liquid interface. For the motion of, e.g., a gas bubble in a surrounding liquid, this means that the position of the interface has to be determined as a function of time, including such effects as oscillations of the bubble. The corresponding transport phenomena are known as *free surface flow* and various numerical techniques for the computation of such flows have been developed in the past decades. Free surface flow simulations are computationally challenging and require special solution techniques which go beyond the standard CFD approaches discussed in Section 2.3. For this reason, the most common of these techniques will be briefly introduced in this section, followed by a number of examples highlighting their application in the field of micro process technology.

2.8.1
Computational Modeling of Free Surface Flows

In order to describe correctly the dynamic evolution of a fluid/fluid interface, a number of boundary conditions have to be implemented into the computational models.

The kinematic condition requires that no fluid can transverse the interface, i.e. the local flow velocity u_i relative to the velocity of the interface u_i^{int} should be zero

$$(u_i - u_i^{int}) \, n_i \big|_{int} = 0 \,, \tag{131}$$

where n_i are the components of a unit vector normal to the interface and the whole expression is to be evaluated at an interfacial position.

The dynamic condition requires that the net force on any portion of the interface has to vanish. In a local coordinate frame attached to an interfacial position, three constraints are derived expressing the force balance for each of the three coordinate directions:

$$n_i \, \tau_{ij} \, n_j \big|_1 + \sigma \left(\frac{1}{R_t} + \frac{1}{R_s} \right) = -n_i \, \tau_{ij} \, n_j \big|_2 \,, \tag{132}$$

$$n_i \, \tau_{ij} \, t_j \big|_1 - \frac{\partial \sigma}{\partial t} = n_i \, \tau_{ij} \, t_j \big|_2 \,, \tag{133}$$

$$n_i \, \tau_{ij} \, s_j \big|_1 - \frac{\partial \sigma}{\partial s} = n_i \, \tau_{ij} \, s_j \big|_2 \,, \tag{134}$$

where τ_{ij} represents the stress tensor, σ the interfacial tension and R_t and R_s, the radii of curvature of the interface along the two orthogonal coordinate directions. The subscripts 1 and 2 refer to the two different phases. The position vector is expressed by two unit vectors t_i and s_i in the tangent plane of the interface and one orthogonal unit vector n_i as

$$\mathbf{r} = t \cdot \mathbf{t} + s \cdot \mathbf{s} + n \cdot \mathbf{n} \,. \tag{135}$$

Eq. (132) states that the interfacial tension has to be balanced by a pressure difference between the two phases. The terms containing derivatives of σ in Eqs. (133) and (134) are non-zero only if there are local variations of the interfacial tension, which might be due to differences in concentration or temperature. The flow induced by such an effect is known as Marangoni convection.

The constraint to be implemented at the three-phase contact line between the two fluids and a solid surface requires that the contact angle θ (compare Figure 2.58) assumes a prescribed value. As discussed in Section 2.2.3, the contact angle might also be allowed to vary with the velocity of the contact line. Especially in microfluidic

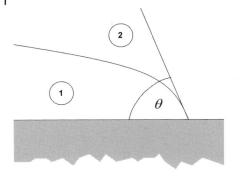

Figure 2.58 Two fluids in contact with a solid surface at a contact angle θ.

systems surface effects often dominate over volumetric effects. Therefore, a correct implementation of the boundary conditions outlined above is of major importance for the modeling of free surface flows in micro reactors.

There are two major classes of numerical methods for free surface flow simulations, *interface tracking* and *interface capturing* methods. In the interface tracking method (see, e.g., [172] or [173]), the interface coincides with a specific grid line, i.e. each cell either belongs to phase 1 or phase 2. When the interface becomes deformed, the grid follows its motion and the grid cells are adjusted in such a way that their identification with the fluid phases is maintained. An example of a corresponding grid deformation for a gas/liquid free surface flow is shown in Figure 2.59. In that case, only the liquid is modeled, and the stresses exerted on the surface by the gas phase above the liquid are so small that they can be neglected. The part of the grid shown in the figure follows the motion of a surface wave travelling from right to left. An advantage of the interface tracking method is the fact that by construction a sharp, well-localized interface is maintained throughout the simulation. Furthermore, in the case of gas/liquid flows the gas can often be neglected and only the liquid domain needs to be modeled, as indicated in Figure 2.59. However, a disadvantage is the fact that changes in topology of the flow domain, for example the decay of a liquid volume into droplets or droplet coalescence, are difficult to take into account. As such processes play an important role in chemical process technology, interface tracking methods will not be further considered here.

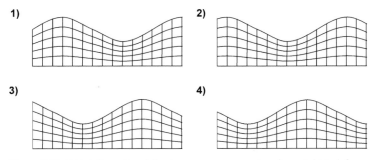

Figure 2.59 Grid deformation following a wave propagating from right to left on a liquid surface.

In contrast to the interface tracking method, in interface capturing techniques a fixed grid is used. Relative to that grid the two fluid phases move and the location of the interface has to be reconstructed. In general, different fluid phases cannot be uniquely assigned to different cells, there may be cells containing fractions of both fluids. The great advantage of interface capturing methods lies in their ability to model in principle all kinds of topological changes fluid volumes surrounded by a second immiscible phase might undergo. A disadvantage of some interface capturing methods is an artificial smearing of the interface due to numerical diffusion, as discussed in Section 2.3.1. Furthermore, local or even global mass conservation is sometimes not guaranteed, and the grid always needs to be larger than the fluid volume to be considered, since it is not co-moving with the fluid.

Among interface capturing methods, one of the most popular and most successful schemes is the volume-of-fluid (VOF) method dating back to the work of Hirt and Nichols [174]. The VOF method is based on a volume-fraction field c, assuming values between 0 and 1. A value of $c = 1$ indicates cells that are filled with phase 1, and phase 2 corresponds to $c = 0$. Intermediate values of c indicate the position of the interface between the phases; however, the goal is to maintain a sharp interface in order to identify the different fluid phases uniquely. Volumes assigned to the different phases are moving with the local flow velocity u_i, and therefore the evolution of c is determined by a convection equation:

$$\frac{\partial c}{\partial t} + \frac{\partial}{\partial x_i}(u_i\, c) = 0\,. \tag{136}$$

Via Eq. (136) the kinematic condition Eq. (131) is fulfilled automatically. Furthermore, a conservative discretization of the transport equation such as achieved with the FVM method guarantees local mass conservation for the two phases separately. With a description based on the volume fraction function, the two fluids can be regarded as a single fluid with spatially varying density and viscosity, according to

$$\rho = \rho_1\, c + \rho_2\, (1 - c), \quad \mu = \mu_1\, c + \mu_2\, (1 - c)\,, \tag{137}$$

where the subscripts 1 and 2 refer to phase 1 and 2, respectively.

When the transport equation for c is solved with a discretization scheme such as upwind, artificial diffusive fluxes are induced, effecting a smearing of the interface. When these diffusive fluxes are significant on the time-scale of the simulation, the information on the location of different fluid volumes is lost. The use of higher order discretization schemes is usually not sufficient to reduce the artificial smearing of the interface to a tolerable level. Hence special methods are used to guarantee that a physically reasonable distribution of the volume fraction field is maintained.

A simple method used to maintain a sharp interface is based on a correction algorithm [175]. After each time step, the amount of fluid of a specific phase having penetrated the interface at $c = 0.5$ is determined. Then the fluid is redistributed such that the 'voids' on the other side of the interface are filled up. The redistribution is done globally, as information on the origin of fluid volumes having penetrated the interface is not available. Owing to the global nature of the correction step, mass conservation is only fulfilled globally and not locally. Especially when a number of disconnected fluid volumes exist (for example, various droplets), artefacts might be induced by the correction algorithm. The artefacts show up as an exchange of mass between the disconnected volumes, i.e. one volume might increase at the expense of another.

Typically, the interface obtained with the versions of the VOF method described above is smeared over a few grid cells, which, on sufficiently fine grids, allows one to identify uniquely the simply connected volumes belonging to the different phases. Instead of regarding the dynamic conditions of Eqs. (132)–(134) as boundary conditions, surface tension can be implemented as a volume force in those cells where c lies between 0 and 1. In the method developed by Brackbill et al. [176], a momentum source term of the form

$$(f_{st})_i \, (x_j) = \sigma \, \kappa(x_j) \frac{\partial c(x_j)}{\partial x_i} \tag{138}$$

is added to the Navier–Stokes equation, where σ is the interfacial tension and κ the local curvature of the interface. Owing to the spatial derivative appearing in Eq. (138), the interfacial tension acts only in those regions where steep gradients of the volume-fraction field exist, which are the regions around the interface. Even for a sharp interface, the description given by Brackbill et al. can still be applied when c is replaced by a smoothed volume-fraction function $\tilde{c}$, which is the result of the convolution of c with, e.g., a Gaussian kernel.

In the methods described above, *a priori* no information on the position of the interface within computational cells or on the radii of curvature of the interface is available. This information has to be obtained from the values of the volume-fraction field in the neighborhood of the point of interest. However, in some versions of the VOF method, an algorithm allowing tracking the position of the interface within single computational cells is available. Two important examples of such algorithms are the Single-Line Interface Construction (SLIC) [177] and the Piecewise-Linear Interface Construction (PLIC) (see [178] and references therein) schemes. In the SLIC scheme for rectangular grids the interface in each computational cell is represented as a line (or a plane in 3-D) which is parallel to the faces of the cell. The more exact PLIC scheme still assumes a linear interface within single computational cells, but allows for an arbitrary interface orientation. A basic form of the SLIC algorithm is represented graphically in Figure 2.60. The interface orientation within a cell in a specific time step depends on the status of the cell under consideration and its neighbors in the previous time step. The figure shows an 'upwind' cell and a cell located downstream with respect to the local flow velocity (the 'downwind' cell). Depending on the status of both of these cells, the interface orientation in the next time step is computed according to the diagram on the right. In this context, a cell shaded in gray contains fluid 1 (for example, a liquid), a white cell fluid 2 (for example a gas) and a hatched area represents a cell containing a mixture of both fluids.

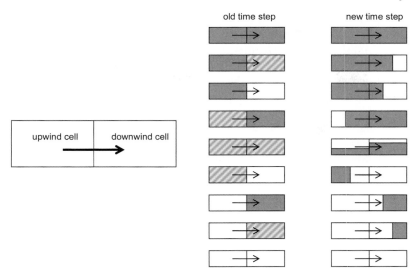

Figure 2.60 Pictorial representation of the SLIC scheme showing the updating scheme for an upwind and a downwind cell. Cells filled with fluid 1 are indicated in gray, those with fluid 2 in white. Cells containing a mixture of both fluids are represented by hatched areas. In the right column the configuration at the new time step is shown, with interface positions depicted explicitly.

Within the PLIC scheme, an interface is allowed locally to stand at any tilted angle with respect to the grid cell. Such a more realistic description goes along with increased accuracy and computational complexity. By reconstructing the position of the interface within single computational cells, both the SLIC and PLIC schemes add information on sub-grid scales to the VOF model. In such a way, a sharp interface is maintained throughout the complete simulation and volumes belonging to different immiscible fluids may be tracked accurately over long time-scales.

Another popular class of schemes for interface capturing is constituted by level-set methods, introduced by Osher and Sethian [179]. Level-set methods are based on the same type of equation as Eq. (136), but the field c is no longer interpreted as a volume fraction field. Instead, c is regarded as a level-set function, assuming positive values in the regions occupied by fluid 1 and negative values in the regions of fluid 2. The interface is interpreted as the surface for which c is equal to zero. The absolute values of c measure the distance to the interface, which, by construction, is always maintained as a well-defined surface, since it is simply identified with the value $c = 0$. However, the price to pay for this somewhat artificial construction is a violation of mass conservation. In order to enforce mass conservation, algorithms have been developed which re-initialize c after a number of time steps based on a special evolution equation [180].

In addition to the methods described above, there exist a number of other methods for the computation of free-surface flows which allow a sharp interface to be maintained. The approach which resembles computational methods for single-phase flows most closely is the front-tracking scheme. This technique essentially assumes a single fluid with local variations of density and viscosity, thus allowing modeling of different phases. The interface is marked by tracer particles moving with the local flow velocity. Surface tension is included as a volumetric force being applied to the computational cells in the vicinity of the interface.

A current version of the front-tracking method was developed by Tryggvason et al. [181]. As the tracer point density may become too small or too large in specific regions during the evolution of the interface, their algorithm creates additional points or deletes points wherever necessary. For 3-D problems, the tracer points are connected by triangular elements which may then have to be subdivided or merged. Since the front is represented by a line or a surface over which the points are distributed, a scheme for smoothing the surface tension on to the higher dimensional computational grid is needed. The advancement of the tracer particles is done with the local flow velocity, thus ensuring that there is no flow across the interface. Changes in the front topology are difficult to model in the framework of front-tracking schemes. Essentially it is necessary to equip the method with a search algorithm which determines interfacial sections that are close to each other and decides to change the connectivity of the interface whenever needed.

2.8.2
Micro Flows of Droplets and Bubbles

Owing to the transient nature of free-surface flows, the simulation of corresponding flow phenomena in multi-phase micro reactors remains a challenge, especially when 3-D models are needed. Nevertheless, in a few examples the use of surface capturing methods to study free-surface flows for applications in micro process technology has been demonstrated. So far the work has been focused mostly on pure flow phenomena and seems to have been restricted to non-reacting flows. One of the best established applications of micro process technology in the context of multi-phase flows is the generation of emulsions and dispersions. Although micro mixers showed some advantages for creating disperse systems compared with, e.g., conventional stirred tanks, it remained unclear by which physical mechanism droplet or bubble formation occurs and how the formation mechanism can be varied and regulated to create tailor-made emulsions and dispersions. In order to understand the physical phenomena better, multilamination mixers were fabricated from glass, thus allowing optical images to be taken showing the formation of droplets or bubbles. In parallel, simulations based on the VOF method were carried out, the results of which have been summarized [182, 183]. As an example of the experimental results, the formation of water droplets in silicone oil in a multilamination mixer with a rectangular mixing chamber is shown in Figure 2.61. From the figure it is apparent that in the mixing chamber liquid 'lamellae' are formed which subsequently decay into droplets.

The decay of liquid lamellae of circular cross-section at rest was studied in a 2-D model using the VOF method without subcellular tracking of the interface (in the following denoted 'basic VOF method') in combination with a correction algorithm

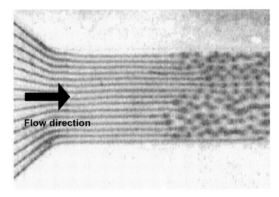

Figure 2.61 Formation of water droplets in silicone oil in a multilamination micro mixer.

to compensate numerical diffusion [175]. In order to initiate the decay, a sinusoidal fluctuation was imposed on the surface of the cylinder. The corresponding time evolution of a water cylinder is shown in Figure 2.62. The initial fluctuation imposed on the surface becomes damped, but initiates a decay of the cylinder into droplets with about twice the wavelength of the initial perturbation. The CFD results are in agreement with linear hydrodynamic stability theory pioneered by Rayleigh (see, e.g., [184]). Rayleigh stability theory predicts that fluctuation wavelengths below a certain cut-off are damped, and that a liquid cylinder decays with a

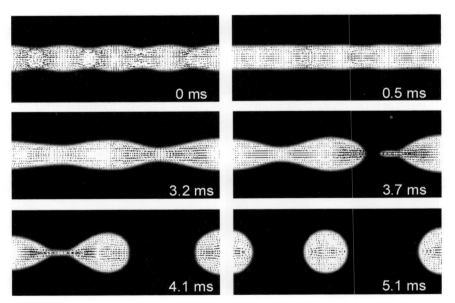

Figure 2.62 VOF-based simulation of a water cylinder decaying into droplets by a hydrodynamic instability [182].

preferred wavelength. In the free-surface flow simulation the correct decay wavelength is selected [182].

While the 2-D simulation has to be regarded as a test case rather than a serious attempt to describe droplet formation in micro mixers, a more elaborate 3-D model incorporating the actual geometry of the mixing chamber and a non-zero flow velocity was set up, again using the basic VOF method. Specifically, the system silicone oil/water was considered at various flow rates. Also with the 3-D model liquid lamellae were found to decay into droplets, while the decay mechanism depends on the contact angle between the glass surface and the two fluids. At contact angles of about 40° (referring to a drop of silicone oil resting on a glass surface and surrounded by water), almost cylindrical rods of water detaching from the walls of the mixing chamber are formed; at contact angles close to 90° the liquid lamellae of approximately rectangular cross-section are still in contact with the channel walls when they decay into droplets [183]. Snapshots of these two different droplet formation mechanisms are shown in Figure 2.63.

Experimentally, contact angles around 40° were determined for the system under study, which means that droplet formation should proceed as shown on the left of Figure 2.63. For a quantitative comparison of experiments and simulations, the droplet diameter, the decay wavelength (i.e. the distance between successive droplets) and the diameter of the water lamellae were determined [183]. In general, the agreement between the simulation results and the experimental data was fairly good. Especially the decay wavelength was found to agree reasonably with Rayleigh's linear stability theory, thus indicating that droplet formation occurs through the well-known Rayleigh plateau instability. Obviously, the dynamics of droplet formation are influenced by shear forces due to the non-zero flow velocity inside the mixer only to a very small extent. Based on this work, it may be claimed that one significant dynamic aspect of emulsion formation in micro mixers was revealed.

Another numerical study of free-surface flow patterns in narrow channels was conducted by Yang et al. [185]. They considered the flow of bubbles of different size driven by body forces, for example the rising of bubbles in a narrow capillary due to buoyancy. The lattice Boltzmann method [186] was used as a numerical scheme

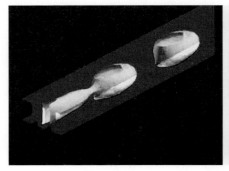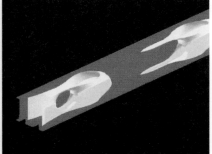

Figure 2.63 Droplet formation in a micro mixer for a wall contact angle of 40° (left) and 90° (right), with silicone oil being the continuous and water the disperse phase.

which recovers a numerical solution of the Navier–Stokes equation from a solution of the Boltzmann equation. Correspondingly, in addition to the computational grid in position space, a grid in velocity space is introduced and a collision term is used which rapidly drives the phase-space distribution into thermal equilibrium.

An important quantity determining the nature of the bubble flow considered by Yang et al. is surface tension, which often plays a dominant role in free-surface micro flows. However, viscous forces are also important in many cases. Hence the ratio of the viscous force and the surface tension force:

$$Ca = \frac{\mu u}{\sigma} \qquad (139)$$

termed 'capillary number', is an important dimensionless group characterizing the flow. When a bubble of diameter d enters a capillary of smaller diameter, a thin liquid film is formed between the bubble and the capillary wall. A quantity which has been determined experimentally is the thickness of this liquid film. Yang et al. compared their simulation results with corresponding experimental results for a range of capillary numbers between 0 and 0.117 and found good agreement. An interesting result of their studies is the fact that the rise velocity depends strongly on the bubble size. Owing to this effect, coalescence occurs in a rising column of bubbles of different sizes. The evolution of such a multi-bubble arrangement is shown in Figure 2.64. The smaller bubbles approach the bigger ones in front of them and undergo coalescence. For an ensemble with a certain size distribution this is a multi-step process, as shown in the figure.

t = 0   400   600   800   6400 6600 6800   8000 16000

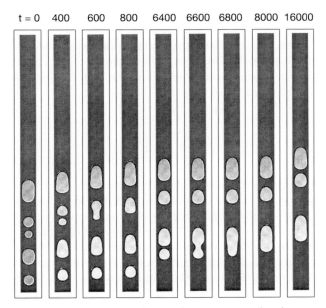

Figure 2.64 Time sequence showing the movement and coalescence of an ensemble of bubbles in a narrow capillary, taken from [185].

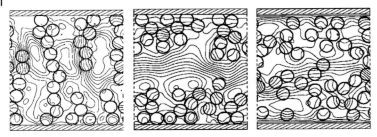

Figure 2.65 Droplet distribution in a channel due to an electric field perpendicular to the channel walls, as described in [187]. At vanishing flow rate the droplets line up in columns (left) and accumulate at the walls as the flow rate is increased (middle). At comparatively high flow rates the droplets get resuspended over the channel (right).

In another study of free-surface flow in confined geometries, Tryggvason et al. [187] used their front-tracking method to study the movement of droplets in channels under the influence of an electric field. When droplets are suspended in a phase of different conductivity and dielectric permittivity, the dielectric mismatch between the two phases induces a stress at the fluid interface. Furthermore, in an external electric field the droplets attain a dipole moment. The dipoles attract each other and the droplets tend to line up in columns. Tryggvason et al. studied the conformation of a droplet suspension in a channel limited by two parallel plates between which an electric field is applied. At vanishing flow rate the droplets line up in columns between the plates, as depicted on the left of Figure 2.65. The lines in the figure represent the streamlines of the flow. When the flow is turned on, the columns break up and the droplets accumulate at the walls. At comparatively high flow rates, the droplets are resuspended all over the channel, as shown on the right of Figure 2.65. Although giving valuable insight into the dynamics of, e.g., micro emulsions inside electric fields, these simulations still suffer from the fact that droplet coalescence has not been accounted for.

2.9
Transport in Porous Media

In the last few years, an increasing number of micro reactors containing micro channels with porous catalyst layers have been reported [13, 14]. In addition to such devices providing a large specific surface area, also the smooth metal surface of the channel walls itself was used as catalyst. The latter approach is suited for proof-of-principle studies rather than the development of systems meeting certain performance requirements such as a specific molar flux of the desired product. Nowadays, micro reactors are starting to be utilized in commercial processes. Hence demands for reaching certain performance benchmarks are raised and the use of smooth metal surfaces as catalysts often is no longer sufficient owing to the disadvantageous surface-to-volume ratio as compared with porous catalyst media.

Porous media have a long-standing history in chemical process technology. One important application is the use of fixed beds of catalyst pellets for heterogeneously catalyzed gas-phase reactions, where a gas mixture is being chemically converted in a tubular section filled with catalyst. In such processes gas is forced through the voids between the pellets by a pressure gradient and the mass transfer characteristics are determined by the transport of chemical species from the void space to the pellet surface and by diffusion within the porous pellets themselves. Analogously, porous media find their application as catalyst carriers in micro reactors. However, in contrast to the fixed-bed technology, porous layers are attached to the micro channel walls with a gas flow passing by.

Owing to the importance of transport in porous media not only in chemical process technology but also in other areas such as geology, more than a century of intense research has resulted in a variety of theoretical and experimental results. Depending on whether flow, heat or mass transfer is considered, a number of different transport mechanisms have to be taken into account. Flow through a porous medium is equivalent to convection of a fluid through the voids or the pore throats. Mass transfer is often dominated by diffusive transport and, similar to flow, occurs exceptionally in the void space of the porous medium. In contrast, heat transfer may occur in the fluid occupying the voids and also in the surrounding solid phase. In chemical micro process technology with porous catalyst layers attached to the channel walls, convection through the porous medium can often be neglected. When the reactor geometry allows the flow to bypass the porous medium it will follow the path of smaller hydrodynamic resistance and will not penetrate the pore space. Thus, in micro reactors with channels coated with a catalyst medium, the flow velocity inside the medium is usually zero and heat and mass transfer occur by diffusion alone.

2.9.1
Morphology of Porous Media

A porous medium consists of a pore space and a solid matrix. Depending on the chemical composition and the preparation technique used to deposit the porous catalyst layer on the micro channel walls, a variety of pore space geometries and topologies may be found. Figure 2.66 shows an image of a catalyst layer deposited in a micro channel using a wash-coat technique and a scanning electron microscopy (SEM) image of the surface morphology of such catalyst carriers. The SEM image gives an impression of the pore structure of the catalyst layer. One of the most important quantities characterizing a porous medium is the porosity ε, defined as the volume fraction of the pore space. The porosity of some materials, e.g. foams, may assume values in the proximity of 1 [117]. An important topological feature of a porous medium is the coordination number Z, defining the number of pore throats meeting at a node of the pore space. Another essential topological aspect is the existence of closed loops in the pore network, i.e. alternative pathways connecting two given points. On a larger scale, the connectivity of the pore space is an aspect which influences the transport properties of the medium. For a given mate-

Figure 2.66 Cross-section of a micro channel coated with a catalyst layer (left) (source: INM, Saarbrücken, Germany) and typical surface morphology of wash-coat catalyst carriers (right).

rial sample there might not exist a sample-spanning cluster of pores connecting two opposite boundaries. In such a case, flow or mass transfer between these boundaries cannot occur. Besides these topological features, the geometry of the pore space characterizes a porous medium. Roughly speaking, there might be comparatively small nodes connected by long pore throats, or there might exist a pore space which is dominated by the nodes and does not possess distinct pore throats.

2.9.2
Volume-averaged Transport Equations

As far as modeling of transport phenomena in porous media is concerned, the task is to provide a generic description which is applicable to as broad a class of materials as possible. The models should to some extent be idealized, allowing them to capture a broad class of phenomena without the need to model all geometric details of the pore space and allowing for a fundamental understanding of transport processes in porous media.

A popular semi-empirical approach is to assign effective transport coefficients to the porous medium. In the context of such models, taking heat transfer as an example, the porous medium would act as if it was equipped with an effective thermal conductivity λ_e being a function of the fluid and solid thermal conductivity λ_f and λ_s. The formal justification of this approach relies on a volume-averaging procedure for the local transport equations. In Figure 2.67 a schematic representation of a porous medium together with the volume over which the averaging is performed is displayed. The averaging volume should be much larger than a typical pore dimension and much smaller than the size of the material sample taken into consideration in order to allow for meaningful variations of the field quantity being looked at. In the case of heat conduction in a porous medium, the fundamental equation to be solved is

$$(\rho\, c_p)_i\, \frac{\partial T_i}{\partial t} = \frac{\partial}{\partial x_k}\left(\lambda_i\, \frac{\partial}{\partial x_k}\, T_i \right), \tag{140}$$

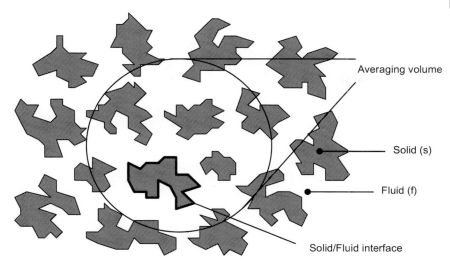

Averaging volume

Solid (s)

Fluid (f)

Solid/Fluid interface

Figure 2.67 Microstructure of a porous medium together with sample volume over which field quantities are averaged.

where ρ, c_p and T denote density, specific heat and temperature and the subscript i either indicates the fluid (f) or the solid (s) phase. In order to obtain an exact solution of the heat conduction problem, Eq. (140) would have to be solved together with appropriate continuity conditions on the solid/fluid interface. Clearly, this is an unmanageable task for macroscopic material samples with a complex pore structure.

Following the line of arguments developed by Carbonell and Whitaker [188] and Nozad et al. [189], a volume-averaged temperature is introduced as

$$\langle T \rangle_i = \frac{1}{V_i} \int_{V_i} T \, dV , \qquad (141)$$

where the subscript i indicates either fluid or solid and the integral is either over the fluid or the solid volume embraced by the averaging volume shown in Figure 2.67. One proceeds by averaging Eq. (140) over the fluid or solid volume, depending on which phase is considered. In addition, the temperature field is split into a mean value (defined by the local averaging procedure) and a fluctuating component, according to

$$T_i = \langle T \rangle_i + T_i' . \qquad (142)$$

Assuming local thermal equilibrium, i.e. the equality of the averaged fluid and solid temperature, a transport equation for the average temperature results which still contains and integral over the fluctuating component. In order to close the equation, a relationship between the fluctuating component and the spatial derivatives of the average temperature of the form

$$T_i' = (b_i)_k \frac{\partial \langle T \rangle}{\partial x_k} \tag{143}$$

is assumed, where the transformation vector $(b_i)_k$ might depend on the spatial coordinates. With that assumption a diffusion equation for the average temperature can be derived, given as

$$\left[\varepsilon (\rho \, c_p)_f + (1 - \varepsilon) (\rho \, c_p)_s \right] \frac{\partial \langle T \rangle}{\partial t} = \frac{\partial}{\partial x_k} \left[(\Lambda_e)_{kl} \frac{\partial \langle T \rangle}{\partial x_l} \right]. \tag{144}$$

The effective thermal conductivity tensor Λ_e depends on the transformation vector $(b_i)_k$ introduced above and on the geometry of the pore space. Hence, under the assumptions made, heat conduction in a porous medium is described by an effective transport coefficient matrix Λ_e. Similarly, for mass transfer and flow effective transport equations may be derived using the volume-averaging approach. In such a way, the well-known Darcy equation for flow in porous media may be obtained [190]:

$$\langle u_k \rangle = -\frac{K}{\mu} \frac{\partial p}{\partial x_k}, \tag{145}$$

where $\langle u_k \rangle$ denotes the average velocity, K the permeability of the medium, μ the fluid viscosity and p pressure.

2.9.3
Computation of Transport Coefficients

In order to be useful in practice, the effective transport coefficients have to be determined for a porous medium of given morphology. For this purpose, a broad class of methods is available (for an overview, see [191]). A very straightforward approach is to assume a periodic structure of the porous medium and to compute numerically the flow, concentration or temperature field in a unit cell [117]. Two very general and powerful methods are the *effective-medium approximation* (EMA) and the *position-space renormalization group method*.

The EMA method is similar to the volume-averaging technique in the sense that an effective transport coefficient is determined. However, it is less empirical and more general, an assessment that will become clear in a moment. Taking mass diffusion as an example, the fundamental equation to solve is

$$\frac{\partial c}{\partial t} = \frac{\partial}{\partial x_k} \left(D \frac{\partial c}{\partial x_k} \right), \tag{146}$$

where c is the concentration field and D the diffusivity. Following the derivation given in [191], a grid is introduced with nodes i, j connected by bonds which are characterized by transport coefficients W_{ij} being proportional to the local diffusivity between i and j. The bonds represent the pore throats and the transport coefficients W_{ij} reflect the geometry of the pore throats, for example a very narrow pore will have assigned a very small value of W_{ij}. Implemented on such a grid, Eq. (146) translates to

$$\frac{\partial c_i}{\partial t} = \sum_j W_{ij} \left[c_i(t) - c_j(t) \right], \tag{147}$$

where c_i is the concentration at node i. The initial concentration is assumed to be of the form $c_i(t = 0) = c_0 \, \delta_{i0}$. Taking the Laplace transform of Eq. (147), one obtains

$$\omega \, \tilde{c}_i(\omega) - \delta_{i0} = \sum_j W_{ij} \left[\tilde{c}_j(\omega) - \tilde{c}_i(\omega) \right], \tag{148}$$

where ω is the variable conjugate to t and the tilde indicates a transformed function. The essential idea of the EMA method is to introduce a node independent, but generally ω-dependent, transport coefficient which represents the average properties of the medium, according to

$$\omega \, \tilde{c}_i^e(\omega) - \delta_{i0} = \sum_j \tilde{W}_e(\omega) \left[\tilde{c}_j^e(\omega) - \tilde{c}_i^e(\omega) \right], \tag{149}$$

where $\tilde{c}_i^e(\omega)$ is the Laplace transform of the effective concentration field obtained in the medium described by the effective transport coefficient. By subtracting Eq. (148) from Eq. (149), an equation involving both $\tilde{c}_i(\omega)$ and $\tilde{c}_i^e(\omega)$ is obtained. When solving this equation approximately and demanding

$$\langle \tilde{c}_i(\omega) \rangle = \tilde{c}_i^e(\omega), \tag{150}$$

where the average is defined with respect to a specific probability distribution for W_{ij}, the function $\tilde{W}_e(\omega)$ can be determined. When computing an effective diffusivity from the transport coefficient and translating the results obtained on a grid to a continuum model, a diffusion equation with a memory term is obtained:

$$\frac{\partial c(x_k, t)}{\partial t} = \int_0^\infty D_e(t - \tau) \, \nabla^2 \, c(x_k, \tau) \, d\tau. \tag{151}$$

The effective diffusivity depends on the statistical distribution of the pore transport coefficients W_{ij}. The derivation shows that the semi-empirical volume-averaging method can only be regarded as an approximation to a more complex dynamic behavior which depends non-locally on the history of the system. Under certain circumstances the long-time ($t \to \infty$) diffusivity will not depend on t (for further details, see [191]). In such a case, the usual Fick diffusion scenario applies. The derivation presented above can, with minor revisions, be applied to the problem of flow in porous media. When considering the heat conduction problem, however, some new aspects have to be taken into account, as heat is transported not only inside the pore space, but also inside the solid phase.

As a second method to determine effective transport coefficients in porous media, the position-space renormalization group method will be briefly discussed.

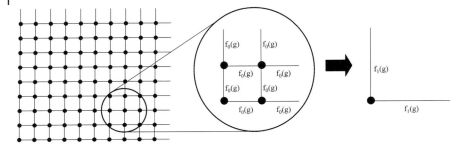

Figure 2.68 Grid model of a porous medium (left) and renormalization group transformation replacing a cluster of grid cells by a unit cell of larger scale (right).

Consider, as an example, a porous medium represented by the two-dimensional grid shown in Figure 2.68, where each bond stands for a pore throat and the coordination number is 4. To each bond a transport coefficient ('conductance') is assigned, similar to the coefficients W_{ij} appearing in the previous paragraph. The transport coefficients are not equal, but are sampled from a probability distribution $f_0(g)$ characteristic of the porous medium under consideration. The idea is to successively replace the fine grid shown on the left of Figure 2.68 by a series of coarser grids which mimic the fine grid as far as the transport properties are concerned. This replacement is indicated on the right of Figure 2.68, going along with the task of determining a new probability distribution $f_1(g)$. In case the relationship between $f_0(g)$ and $f_1(g)$ is simple enough this mapping can be iterated until a grid with cells on a macroscopic scale is reached or, alternatively, until the probability distribution for the transport coefficients remains invariant under a further scale transformation.

Formally, the expression one allowing to compute the probability distribution for a bond of the renormalized cell shown on the right of Figure 2.68 is

$$f_1(g) = \int f_0(g_1)\, dg_1\, f_0(g_2)\, dg_2 \ldots f_0(g_n)\, dg_n\, \delta(g - g'). \tag{152}$$

In this expression, the integral is over the conductances of the bonds of the original cell, which, when assigned values $g_1, g_2, \ldots, g_n$, result in a conductivity g' of a bond in the renormalized cell. The problem of finding g' can be regarded as the problem of determining the conductivity of an electric circuit, where, taking the example of Figure 2.68, the clamps are arranged either horizontally (for the horizontal bond of the renormalized cell) or vertically (for the vertical bond). Usually, for not too large unit cells, g' is a comparatively simple function of g_i ($i = 1, \ldots, n$). Clearly, for reasons of symmetry the horizontal bonds in Figure 2.68 are equivalent to the vertical ones and the same probability distribution is found for each type of bond in the renormalized cell. Since the renormalized cell is of the same structure as the cells of the original grid, the mapping defined in Eq. (152) can be applied over and over again until, e.g., an invariant probability distribution $f_i(g)$ is obtained.

Although in principle a powerful and elegant method, the position-space renormalization group method yields very complex expressions for the renormalized

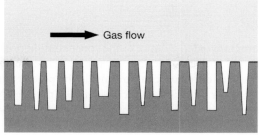

Figure 2.69 Structure of a porous micro channel surface (left) and schematic diagram of the flow scheme and pore arrangement (right).

probability distribution $f_i(g)$ when iterated a few times. Even when using symbolic algebra packages, the number of terms generated soon starts to exceed the capabilities of most computers. For this reason, the probability distribution is often replaced by a distribution of predetermined structure which approximates the true distribution [191]. With such so-called optimized distributions an initial distribution $f_0(g)$ usually converges rapidly to a stable distribution whose shape does not change under further rescaling.

In addition to the approaches for modeling transport phenomena in porous media briefly discussed here, a large number of alternative methods with specific advantages for specific applications exist. An appropriate discussion of only the most important methods lies beyond the scope of this book. The reader is referred to the books by Sahimi [191] and Kaviany [117] for a more detailed presentation of models for flow, heat and mass transfer in porous media. Nevertheless, one further class of modeling approaches which is of special importance for micro reactors should be discussed here. A special class of catalytic surface described in the literature is a monolith with more or less regular, straight and parallel pores [192, 193]. The walls of micro channels may be equipped with such monoliths in order to increase the specific surface area and the number of active sites catalyzing chemical reactions. A microscopic image of a corresponding channel surface is displayed in Figure 2.69, together with the typical flow scheme in corresponding reactors. Corresponding pore structures can be created by anodic oxidation, an electrochemical process in the course of which spots of Al_2O_3 are created on an aluminum substrate.

2.9.4
Reaction-diffusion Dynamics Inside Pores

The simple pore structure shown in Figure 2.69 allows the use of some simplified models for mass transfer in the porous medium coupled with chemical reaction kinetics. An overview of corresponding modeling approaches is given in [194]. The reaction-diffusion dynamics inside a pore can be approximated by a one-dimensional equation

$$\frac{d^2 c_i}{dz^2} = \frac{2}{D_i \, \bar{r} \, S_{cat}} \frac{dn_i}{dt},$$ (153)

where time independence of the concentration profile was assumed. The concentration of species i is denoted by c_i, z is the coordinate along the axis of the pore, D_i the species diffusion constant inside the pore, $\bar{r}$ the mean radius of the pore, n_i the number of moles of species i and S_{cat} the surface area of the catalyst material. The second factor on the right-hand side is the reaction rate per unit surface area of the catalyst material. In order for Eq. (153) to be applicable, the radial concentration gradients inside the pore should be negligible. This condition can be expressed mathematically as [194]

$$\sum_i \frac{\bar{r}}{3 \, v_i \, D_i} \frac{1}{S_{cat}} \frac{\partial(dn_i/dt)}{\partial c_i} \ll 1,$$ (154)

where v_i is the stoichiometric coefficient of species i in the chemical reaction under consideration. Assuming a reaction of the type A → B, Eq. (153) can be solved for all reaction orders. The functional forms of the solutions are given in [194] and allow the diffusional flux of species i at the pore mouth to be expressed as a function of concentration.

When modeling the reaction dynamics in micro channels equipped with such porous walls, the arguments outlined above allow the computations to be simplified considerably. Given the applicability of the one-dimensional approximation for the reaction-diffusion dynamics inside the pores, the domain of the catalyst monolith does no longer have to be included explicitly in the computational model. Consider, for example, a CFD model of a reactor with gas flow past a catalyst monolith with regular, parallel pores, such as depicted in Figure 2.69. It would be very challenging in terms of CPU time and memory requirements to include the pore space into the computational domain and define a grid inside the pores which would have to be matched with the grid of the flow channel. Rather than that, the techniques discussed above allow one to define the interface between the catalyst monolith and the channel as the boundary of the computational domain, similar to a smooth, unstructured surface acting as catalyst. The flow, heat and mass transfer problem would then only be solved in a computational domain defined by the flow channel. In such a model, the diffusional fluxes of chemical species at the interface to the monolith would be related to the concentrations via the relations for various reaction orders derived from the one-dimensional pore transport model, Eq. (153). However, when employing such an approach care has to be taken that all conditions allowing neglect of the catalyst domain in an explicit model are fulfilled. As an example, the temperature boundary condition at the interface to the monolith may *a priori* not be clear and heat conduction effects inside the porous medium might have to be taken into account.

References

1 Pope, S. B., *Turbulent Flows*, Cambridge University Press, Cambridge (**2000**).

2 Karniadakis, G. E., Beskok, A., *Micro Flows – Fundamentals and Simulation*, Springer-Verlag, New York (**2002**).

3 Gad-el-Hak, M., *The fluid mechanics of microdevices*, J. Fluids Eng. **121** (**1999**) 5–33.

4 Bird, G., *Molecular Gas Dynamics and the Direct Simulation of Gas Flows*, Oxford University Press, New York (**1994**).

5 Schaaf, S., Chambre, P., *Flow of Rarified Gases*, Princeton University Press, Princeton, NJ (**1961**).

6 Seidl, M., Steinheil, E., *Measurement of momentum accomodation coefficients on surfaces characterized by Auger spectroscopy, SIMS and LEED*, in Proceedings of the 9th Int. Symposium on Rarified Gas Dynamics, pp. E9.1–E9.2 (**1974**).

7 Lord, R., *Tangential momentum coefficients of rare gases on polycrystalline surfaces*, in Proceedings of the 10th Int. Symposium on Rarified Gas Dynamics, pp. 531–538 (**1976**).

8 Beskok, A., Karniadakis, G. E., *Simulation of heat and momentum transfer in complex micro-geometries*, J. Thermophys. Heat Transfer **8** (**1994**) 355–370.

9 Tu, J. K., Huen, T., Szema, R., Ferrari, M., *Filtration of sub-100 nm particles using a bulk-micromachined, direct-bonded silicon filter*, J. Biomed. Microdevices **1** (**1999**) 113–119.

10 Han, J., Craighead, H. G., Separation of long DNA molecules in a micro-fabricated trap array, *Science* **288** (**2000**) 1026–1029.

11 Han, J., Craighead, H. G., *Characterization and optimization of an entropic trap for DNA separation*, Anal. Chem. **74** (**2002**) 394–401.

12 Foquet, M., Korlach, J., Zipfel, W., Webb, W. W., Craighead, H. G., *DNA fragment sizing by single molecule detection in submicrometer-sized closed fluidic channels*, Anal. Chem. **74** (**2002**) 1415–1422.

13 Zapf, R., Berresheim, K., Cominos, V., Gnaser, H., Hessel, V., Kolb, G., Löb, P., *Detailed characterization of various porous alumina based catalyst coatings within microchannels and their testing for methanol steam reforming*, Chem. Eng. Res. Des., special issue on Chemical Reaction Engineering (**2003**) submitted for publication.

14 Haas-Santo, K., Fichtner, M., Schubert, K., *Preparation of micro-structure compatible porous supports by sol–gel synthesis for catalyst coatings*, Appl. Catal. A **220** (**2001**) 79–92.

15 Reif, F., *Fundamentals of Statistical and Thermal Physics*, McGraw-Hill, New York (**1965**).

16 Bhatnager, P., Gross, E., Krook, K., *A model for collision processes in gases*, Phys. Rev. **94** (**1954**) 511–524.

17 Cercignani, C., *The Boltzmann Equation and its Applications*, Springer-Verlag, New York (**1988**).

18 Cercignani, C., *Rarified Gas Dynamics: From Basic Concepts to Actual Calculations*, Cambridge University Press, New York (**2000**).

19 Chapman, S., Cowling, T., *The Mathematical Theory of Non-uniform Gases: An account of the Kinetic Theory of Viscosity, Thermal Conduction and Diffusion in Gases*, Cambridge University Press, Cambridge (**1970**).

20 Sharipov, F., Sleznev, V., *Data on internal rarified gas flows*, J. Phys. Chem. Ref. Data **27** (**1998**) 657–706.

21 Sharipov, F., Kalempa, D., *Gaseous mixture flow through a long tube at arbitrary Knudsen numbers*, J. Vac. Sci. Technol. A **20** (**2002**) 814–822.

22 Oh, C. K., Oran, E. S., Sinkovits, R. S., *Computations of high-speed, high Knudsen number microchannel flow*, J. Thermophys. Heat Transfer **11**, 4 (**1997**) 497–505.

23 Cai, C., Boyd, I. D., Fan, J., Candler, G. V., *Direct simulation methods for low-speed microchannel flows*, J. Thermophys. Heat Transfer **14**, 3 (**2000**) 368–378.

24 Kaplan, C. R., Oran, E. S., *Nonlinear filtering for low-velocity gaseous microflows*, AIAA J. **40** (**2002**) 82–90.

25 Fan, J., Chen, C., *Statistical simulation of low-speed unidirectional flows in transitional*

regime, in Brun, R. C. E. (Ed.), *Rarified Gas Dynamics*, Cepadues Editions, Toulouse, Vol. 2, pp. 245–252 **(1999)**.

26 Sun, H., Faghri, M., *Effect of surface roughness on nitrogen flow in a microchannel using the Direct Simulation Monte Carlo Method (DSMC)*, Numerical Heat Transfer A **43** (2003) 1–8.

27 Roveda, R., Goldstein, D., Varghese, P., *Hybrid Euler/particle approach for continuum/rarified flows*. J. Spacecraft Rockets **35** (1998) 258–265.

28 Atkas, O., Aluru, N. R., *A combined continuum/DSMC technique for multiscale analysis of microfluidic filters*, J. Comput. Phys. **178**, 2 **(2000)** 342–372.

29 Landau, L. D., Lifshitz, E. M., *Hydrodynamik*, Akademie-Verlag, Berlin **(1991)**.

30 Wrobel, L. C., *The Boundary Element Method*, Wiley, New York **(2002)**.

31 Pfahler, J., Harley, J., Bau, H., Zemel, J., *Liquid transport in micron and submicron channels*, Sens. Actuators A **21–23** (1990) 431–434.

32 Peng, X. F., Peterson, G. P., Wang, B. X., *Frictional flow characteristics of water flowing through rectangular microchannels*, Exp. Heat Transfer **7** (1994) 249–264.

33 Peng, X. F., Peterson, G. P., *The effect of thermofluid and geometrical parameters on convection of liquids through rectangular microchannels*, Int. J. Heat Mass Transfer **38**, 4 (1995) 755–758.

34 Flockhart, S. M., Dhariwal, R. S., *Experimental and numerical investigation into the flow characteristics of channels etched in (100) silicon*, J. Fluids Eng. **120** (1998) 291–295.

35 Harms, T. M., J., K. M., Gerner, F. M., *Developing convective heat transfer in deep rectangular microchannels*, Int. J. Heat Fluid Flow **20** (1999) 149–157.

36 Mala, G. M., Li, D., *Flow characteristics of water in microtubes*, Int. J. Heat Fluid Flow **20** (1999) 142–148.

37 Qu, W., Mala, G. M., Li, D., *Pressure-driven water flows in trapezoidal silicon microchannels*, Int. J. Heat Mass Transfer. **43** (2000) 353–364.

38 Brochard, F., de Gennes, P. G., *Shear-dependent slippage at a polymer/solid interface*, Langmuir **8** (1992) 3033–3037.

39 Migler, K. B., Hervet, H., Leger, L., *Slip transition of a polymer melt under shear stress*, Phys. Rev. Lett. **70** (1993) 287–290.

40 Baudry, J., Charlaix, E., Tonck, A., Mazuyer, D., *Experimental evidence for a large slip effect at a nonwetting fluid-solid interface*, Langmuir **17** (2001) 5232–5236.

41 Craig, V. S. J., Neto, C., Williams, D. R. M., *Shear-dependent boundary slip in an aqueous Newtonian liquid*, Phys. Rev. Lett. **87** (2001) 54504–54507.

42 Bonaccurso, E., Kappl, M., Butt, H. J., *Hydrodynamic force measurements: boundary slip of water on hydrophilic surfaces and electrokinetic effects*, Phys. Rev. Lett. **88** (2002) 76103–76106.

43 Thompson, P. A., Troian, S. M., *A general boundary condition for liquid flow at solid surfaces*, Nature **389** (1997) 360–362.

44 Ashcroft, N. W., Mermin, N. D., *Solid State Physics*, Harcourt Brace College Publishers, Fort Worth, TX **(1976)**.

45 Probstein, R. F., *Physicochemical Hydrodynamics*, Wiley, New York **(1994)**.

46 Rice, C., Whitehead, R., *Electro kinetic flow in a narrow cylindrical capillary*, J. Phys. Chem. **69** (1965) 4017–4023.

47 Adamson, A. W., Gast, A. P., *Physical Chemistry of Surfaces*, 6th ed., Wiley, New York **(1997)**.

48 Zeng, S., Chen, C. H., Mikkelsen, J. C., Santiago, J. G., *Fabrication and characterization of electroosmotic micropumps*, Sens. Actuators B **79** (2001) 107–114.

49 Chen, C. H., Santiago, J. G., *A planar electroosmotic micropump*, J. Microelectromech. Syst. **11** (2002) 672–683.

50 Lazar, I. M., Karger, B. M., *Multiple open-channel electroosmotic pumping system for microfluidic sample handling*, Anal. Chem. **74** (2002) 6259–6268.

51 Burgreen, D., Nakache, F., *Electrokinetic flow in ultrafine capillary slits*, J. Phys. Chem. **68** (1964) 1084–1091.

52 Hunter, R. J., *Zeta Potential in Colloid Science: Principles and Applications*, Academic Press, New York **(1981)**.

53 Yang, C., Li, D., Masliyah, J. H., *Modeling forced liquid convection in rectangular microchannels with*

electrokinetic effects, Int. J. Heat Mass Transfer **41 (1998)** 4229–4249.

54 CHENG, L., FENTER, P., NAGY, K. L., SCHLEGEL, M. L., STURCHIO, N. C., *Molecular-scale density oscillations in water adjacent to a mica surface*, Phys. Rev. Lett. **87**, 15 **(2001)** 6103–6104.

55 TRAVIS, K. P., GUBBINS, K. E., *Poiseuille flow of Lennard–Jones fluids in narrow slit pores*, J. Chem. Phys. **112**, 4 **(2000)** 1984–1994.

56 QIAO, R., ALURU, N. R., *Ion concentrations and velocity profiles in nanochannel electroosmotic flows*, J. Chem. Phys. **118**, 10 **(2003)** 4692–4701.

57 KOPLIK, J., BANAVAR, J. R., *Continuum deductions from molecular hydrodynamics*, Ann. Rev. Fluid Mech. **27 (1995)** 257–292.

58 ALLEN, M., TILDESLEY, D., *Computer Simulations of Liquids*, Clarendon Press, Oxford **(1994)**.

59 NAKANO, A., KALIA, R. K., VASHISHTA, P., CAMPBELL, T. J., OGATA, S., SHIMOJO, F., SAINI, S., *Scalable atomistic simulation algorithms for materials research*, Sci. Programming **10**, 4 **(2002)** 263–270.

60 PALM, B., *Heat transfer in microchannels*, Microscale Therm. Eng. **5 (2001)** 155–175.

61 PENG, C. F., WANG, B. X., *Forced convection and flow boiling heat transfer for liquid flowing through microchannels*, Int. J. Heat Mass Transfer **36 (1993)** 3421–3427.

62 PENG, X. F., WANG, B. X., *Cooling characteristics with microchanneled structures*, J. Enhanced Heat Transfer **1 (1994)** 315–326.

63 PENG, X. F., WANG, B. X., *Experimental investigation of heat transfer in flat plates with rectangular microchannels*, Int. J. Heat Mass Transfer **38 (1995)** 127–137.

64 GOLDENFELD, N., *Lectures on Phase Transitions and the Renormalization Group*, Addison Wesley, Boston **(1992)**.

65 TODA, M., KUBO, R., SAITÔ, N., *Statistical Physics I: Equilibrium Statistical Mechanics*, Springer-Verlag, Berlin **(1992)**.

66 FISHER, M. E., NAKANISHI, H., *Scaling theory for the criticality of fluids between plates*, J. Chem. Phys. **75 (1981)** 5857–5863.

67 EVANS, P., MARCONI, U., TARAZONA, P., *Fluids in narrow pores: adsorption, capillary condensation and critical points*, J. Chem. Phys. **84 (1986)** 2376–2399.

68 KRECH, M., *Casimir forces in binary liquid mixtures*, Phys. Rev. E **56 (1997)** 1642–1659.

69 JACOBS, D. T., MOCKLER, R. C., O'SULLIVAN, W. J., *Critical-temperature and coexistence-curve measurements in thick films*, Phys. Rev. Lett. **37 (1976)** 1471–1474.

70 DE GENNES, P. G., *Wetting: statics and dynamics*, Rev. Mod. Phys. **57 (1985)** 827–863.

71 JOANNY, J. F., DE GENNES, P. G., *A model for contact angle hysteresis*, J. Chem. Phys. **81 (1984)** 552–562.

72 DE GENNES, P. G., HUA, X., LEVINSON, P., *Dynamics of wetting: local contact angles*, J. Fluid. Mech. **212 (1990)** 55–63.

73 DE GENNES, P. G., *Deposition of Langmuir-Blodgett layers*, J. Colloid Polym. Sci. **264**, 463–465.

74 COX, R. G., *The dynamics of the spreading of liquids on a solid surface. Part 1. Viscous flow*, J. Fluid. Mech. **168 (1986)** 169–194.

75 HLAVACEK, V., PUSZYNSKI, J. A., VILJOEN, H. J., GATICA, J. E., *Model reactors and their design equations*, in ARPE, H. J. (Ed.), *Ullmann's Encyclopedia of Industrial Chemistry*, Wiley, New York, Vol. B4, pp. 121–165 **(1992)**.

76 FLETCHER, C. A. J., *Computational Techniques for Fluid Dynamics*, 2nd ed., Springer-Verlag, Berlin **(1991)**.

77 GRESHO, P. M., SANI, R. L., *Incompressible Flow and the Finite Element Method, Volume 1: Advection-Diffusion and Isothermal Laminar Flow*, Wiley, New York **(2000)**.

78 GRESHO, P. M., SANI, R. L., ENGELMAN, M. S., *Incompressible Flow and the Finite Element Method, Volume 2, Isothermal Laminar Flow*, Wiley, New York **(2000)**.

79 VERSTEEG, H. K., MALALASEKRA, M., *An Introduction to Computational Fluid Dynamics*, Longman Scientific & Technical, Harlow **(1995)**.

80 COURANT, R., HILBERT, D., *Methoden der Mathematischen Physik*, Springer-Verlag, Berlin **(1993)**.

81 NOLL, B., *Numerische Strömungsmechanik*, Springer-Verlag, Berlin **(1993)**.

82 LEONARD, B. P., *A stable and accurate convection modeling procedure based on quadratic upstream interpolation,* Comput. Methods Appl. Mech. Eng. **19** (1979) 59–98.

83 CASEY, M., WINTERGERSTE, T., *Special Interest Group on Quality and Trust in Industrial CFD: Best Practice Guidelines,* European Research Community on Flow, Turbulence and Combustion (2000).

84 FERZIGER, J. H., PERIC, M., *Computational Methods for Fluid Dynamics,* 3rd ed., Springer-Verlag, Berlin (2002).

85 PRESS, W. H., TEUKOLSKY, S. A., VETTERLING, W. T., FLANNERY, B. P., *Numerical Recipes in Fortran 77,* Cambridge University Press, Cambridge (1992).

86 HARLOW, F. H., WELSH, J. E., *Numerical calculation of time dependent viscous incompressible flow with free surface,* Phys. Fluids **8** (1965) 2182–2189.

87 RHIE, C. M., CHOW, W. L., *Numerical study of the turbulent flow past an airfoil with trailing edge separation,* AIAA J. **21** (1983) 1527–1532.

88 CARETTO, L. S., GOSMAN, A. D., PATANKAR, S. V., SPALDING, D. B., *Two calculation procedures for steady, three-dimensional flows with recirculation,* in Proceedings of the 3rd Int. Conf. Numer. Methods Fluid Dyn., Paris (1972).

89 VAN DOORMAL, J. P., RAITHBY, G. D., *Enhancement of the SIMPLE method for predicting incompressible fluid flows,* Numerical Heat Transfer **7** (1984) 147–163.

90 GERRITSEN, M., OLSSON, P., *Designing an efficient solution strategy for fluid flows. II. Stable high-order central finite difference schemes on composite adaptive grids with sharp shock resolution,* J. Comput. Phys. **147** (1998) 293–317.

91 KINOSHITA, T., INOUE, O., *A parallel adaptive mesh approach for flowfields with shock waves,* Shock Waves **12** (2002) 167–175.

92 BECKER, R., RANNACHER, R., *A feedback approach to error control in finite element methods: basic analysis and examples,* East-West J. Numer. Math. **4** (1996) 237–264.

93 HEUVELINE, V., RANNACHER, R., *A posteriori error control for finite element approximations of elliptic eigenvalue problems,* Adv. Comp. Math. **15** (2001) 107–138.

94 THOMPSON, J. F., WARSI, Z. U., MASTIN, C. W., *Boundary-fitted coordinate systems for numerical solution of partial differential equations,* J. Comput. Phys. **47** (1982) 1–108.

95 LO, S. H., *A new mesh generation scheme for arbitrary planar domains,* Int. J. Num. Methods Eng. **21** (1985) 1403–1426.

96 THOMPSON, J. F., SONI, B. K., WEATHERILL, N. P., *Handbook of Grid Generation,* CRC Press, Boca Raton, FL (1998).

97 TROTTENBERG, U., OOSTERLEE, C. W., SCHÜLLER, A., *Multigrid,* Academic Press, New York (2001).

98 STOER, J., BULIRSCH, R., *Introduction to Numerical Analysis,* Springer-Verlag, New York (1980).

99 BRIGGS, W. L., HENSON, V., McCROMICK, S. F., *A Multigrid Tutorial,* 2nd ed., Society for Industrial and Applied Mathematics, Philadelphia (2000).

100 HARTNETT, J. P., KOSTIC, M., *Heat transfer to Newtonian and non-Newtonian fluids in rectangular ducts,* Adv. Heat Transfer **19** (1989) 247–356.

101 PURDAY, H. F. P., *An Introduction to the Mechanics of Viscous Flow,* Dover, New York (1949).

102 SHAH, R. K., LONDON, A. L., *Laminar flow forced convection in ducts,* Adv. Heat Transfer Suppl. 1 (1978).

103 SHAH, R. K., *Laminar flow friction and forced convection heat transfer in ducts of arbitrary geometry,* Int. J. Heat Mass Transfer **18** (1975) 849–842.

104 RICHARDSON, D. H., SEKULIC, D. P., CAMPO, A., *Low Reynolds number flow inside straight micro channels with irregular cross sections,* Heat Mass Transfer **36** (2000) 187–193.

105 ASAKO, Y., FAGHRI, M., *Finite-volume solutions for laminar flow and heat transfer in a corrugated duct,* J. Heat Transfer **109** (1987) 627–634.

106 GARG, V. K., MAJI, P. K., *Flow and heat transfer in a sinusoidally curved channel,* Int. J. Eng. Fluid Mech. **1** (1988) 293–319.

107 GUZMÁN, A. M., AMON, C. H., *Dynamical flow characterization of*

transitional and chaotic regimes in converging–diverging channels, J. Fluid. Mech. **321** (**1996**) 25–57.

108 DEAN, W. R., *Note on the motion of a fluid in a curved pipe*, Philos. Mag. **4** (**1927**) 208–223.

109 DEAN, W. R., *The stream-line motion of fluid in a curved pipe*, Philos. Mag. **5** (**1928**) 673–695.

110 CHENG, K. C., LIN, R.-C., OU, J.-W., *Fully developed laminar flow in curved rectangular channels*, J. Fluids Eng. **3** (**1976**) 42–48.

111 CHEN, Y., CHEN, P., *Heat transfer and pressure drop in fractal tree-like microchannel nets*, Int. J. Heat Mass Transfer **45** (**2002**) 2643–2648.

112 WALTER, S., FRISCHMANN, G., BROUCEK, R., BERGFELD, M., LIAUW, M., *Fluiddynamische Aspekte in Mikro-reaktoren*, Chem. Ing. Tech. **71**, 5 (**1999**) 447–455.

113 COMINOS, V., HARDT, S., HESSEL, V., KOLB, G., LÖWE, H., WICHERT, M., ZAPF, R., *A methanol steam micro-reformer for low power fuel cell applications*, Chem. Eng. Commun. (**2003**) accepted for publication.

114 VERFÜRTH, R., *Non-overlapping domain decomposition methods*, in BRISTEAU, M. O., ETGEN, G. J., FITZGIBBON, W., LIONS, J. L., PERIAUX, J., WHEELER, M. F. (Eds.), *Computational Science for the 21st Century*, Wiley, New York (**1997**).

115 EWING, R., LAZAROV, R., LIN, T., LIN, Y., *Mortar finite volume element approximations of second order elliptic problems*, East-West J. Numer. Math. **8**, 2 (**2000**) 93–110.

116 COMMENGE, J. M., FALK, L., CORRIOU, J. P., MATLOSZ, M., *Optimal design for flow uniformity in microchannel reactors*, AIChE J. **48**, 2 (**2000**) 345–358.

117 KAVIANY, M., *Principles of Heat Transfer in Porous Media*, Springer-Verlag, New York (**1995**).

118 KIM, S. J., KIM, D., LEE, D. Y., *On the local thermal equilibrium in microchannel heat sinks*, Int. J. Heat Mass Transfer **43** (**2000**) 1735–1748.

119 WHITE, F. M., *Fluid Mechanics*, McGraw-Hill, New York (**1994**).

120 HARDT, S., EHRFELD, W., HESSEL, V., VANDEN BUSSCHE, K. M., *Strategies for size reduction of microreactors by heat transfer enhancement effects*, Chem. Eng. Commun. **190**, 4 (**2003**) 540–559.

121 RUSH, T. A., NEWELL, T. A., JACOBI, A. M., *An experimental study of flow and heat transfer in sinusoidal wavy passages*, Int. J. Heat Mass Transfer **42** (**1999**) 1541–1553.

122 WANG, G., VANKA, S. P., *Convective heat transfer in periodic wavy passages*, Int. J. Heat Mass Transfer **38** (**1995**) 3219–3230.

123 SEKULIC, D. P., CAMPO, A., MORALES, J. C., *Irreversibility phenomena associated with heat transfer and fluid friction in laminar flows through singly connected ducts*, Int. J. Heat Mass Transfer **40** (**1997**) 905–914.

124 XU, B., OOI, K. T., MAVRIPLIS, C., ZAGHLOUL, M. E., *Evaluation of viscous dissipation in liquid flow in microchannels*, J. Micromech. Microeng. **13** (**2003**) 53–57.

125 STIEF, T., LANGER, O.-U., SCHUBERT, K., *Numerical investigations of opimal heat conductivity in micro heat exchangers*, Chem. Eng. Technol. **21**, 4 (**1999**) 297–302.

126 PERRY, R. H., GREEN, D. W., *Perry's Chemical Engineers' Handbook*, 7th ed., McGraw-Hill, New York (**1997**).

127 DRESE, K.-S., HARDT, S., *Characterization of micro heat exchangers*, in Proceedings of the VDE World Microtechnologies Congress, MICRO.tec 2000 (25–27 September **2000**), VDE Verlag, Berlin, EXPO Hannover, pp. 371–374.

128 QUIRAM, D. J., HSING, I.-M., FRANZ, A. J., JENSEN, K. F., SCHMIDT, M. A., *Design issues for membrane-based, gas phase microchemical systems*, Chem. Eng. Sci. **55** (**2000**) 3065–3075.

129 BROOKS, A. N., HUGHES, T. J. R., *Streamline upwind Petrov/Galerkin formulation for convection dominated flows with particular emphasis on the incompressible Navier–Stokes equations*, Comput. Methods Appl. Mech. Eng. **32** (**1982**) 99–259.

130 QUIRAM, D. J., HSING, I.-M., FRANZ, A. J., SRINIVASAN, R., JENSEN, K. F., SCHMIDT, M. A., *Characterization of microchemical systems using simulations*, in EHRFELD, W., RINARD, I. H.,

WEGENG, R. S. (Eds.), *Process Miniaturization: 2nd International Conference on Microreaction Technology, IMRET 2, Topical Conf. Preprints,* AIChE, New Orleans, pp. 205–211 (1998).

131 BORIS, J. P., BOOK, D. L., *Flux-corrected transport – I. SHASTA, a fluid transport algorithm that works,* J. Comput. Phys. 11 (1973) 38–69.

132 ZALESAK, S. T., *Fully multidimensional flux-corrected transport algorithm for fluids,* J. Comput. Phys. 31 (1979) 335–362.

133 EHRFELD, W., HESSEL, V., LÖWE, H., *Microreactors,* Wiley-VCH, Weinheim (2000).

134 GOBBY, D., ANGELI, P., GAVRIILIDIS, A., *Mixing characteristics of T-type microfluidic mixers,* J. Micromech. Microeng. 11 (2001) 126–132.

135 MENGEAUD, V., JOSSERAND, J., GIRAULT, H. H., *Mixing processes in a zigzag microchannel: finite element simulations and optical study,* Anal. Chem. 74 (2002) 4279–4286.

136 STROOCK, A. D., DERTINGER, S. K. W., WHITESIDES, G. M., AJDARI, A., *Patterning flows using grooved surfaces,* Anal. Chem. 74, 20 (2002) 5306–5312.

137 STROOCK, A. D., DERTINGER, S. K. W., AJDARI, A., MEZIC, I., STONE, H. A., WHITESIDES, G. M., *Chaotic mixer for microchannels,* Science 295, 1 (2002) 647–651.

138 AREF, H., *Stirring by chaotic advection,* J. Fluid. Mech. 143 (1984) 1–21.

139 SCHÖNFELD, F., HARDT, S., *Simulation of helical flows in microchannels,* AIChE J. (2003) accepted for publication.

140 LINXIANG, W., YURUN, F., YING, C., *Animation of chaotic mixing by a backward Poincaré cell-map method,* Int. J. Bifurcation Chaos 11, 7 (2001) 1953–1960.

141 HESSEL, V., HARDT, S., LÖWE, H., SCHÖNFELD, F., *Laminar mixing in different interdigital micromixers – Part I: experimental characterization,* AIChE J. 49, 3 (2003) 566–577.

142 HARDT, S., SCHÖNFELD, F., *Laminar mixing in different interdigital micromixers – Part 2: numerical simulations,* AIChE J. 49, 3 (2003) 578–584.

143 THOMPSON, L. H., DORAISWAMY, L. K., *Sonochemistry: science and engineering,* Ind. Eng. Chem. Res. 38 (1999) 1215–1249.

144 QIAN, S., BAU, H. H., *A chaotic electroosmotic stirrer,* Anal. Chem. 74, 15 (2002) 3616–3625.

145 MEISEL, I., EHRHARD, P., *Simulation of electrically-excited flows in microchannels for mixing applications,* in Proceedings of the 5th International Conference on Modeling and Simulation of Microsystems, San Juan, Puerto Rico, 22–25 April 2002, Computational Publications, Boston, pp. 62–65 (2002).

146 BAU, H. H., ZHONG, J., YI, M., *A minute magneto hydro dynamic (MHD) mixer,* Sens. Actuators B 79 (2001) 207–215.

147 GLEESON, J. P., WEST, J., *Magnetohydrodynamic mixing,* in Proceedings of the 5th International Conference on Modeling and Simulation of Microsystems, San Juan, Puerto Rico, 22–25 April 2002, Computational Publications, Boston, pp. 318–321 (2002).

148 GLEESON, J. P., ROCHE, O. M., WEST, J., *Modeling annular micromixers,* in Proceedings of the Nanotechnology Conference and Trade Show, Nanotech 2003, San Francisco, 23–27 February 2003, Computational Publications, Boston, pp. 206–209 (2003).

149 TAYLOR, G. I., *Dispersion of soluble matter in solvent flowing slowly through a tube,* Proc. R. Soc. London, A 219 (1953) 186–203.

150 ARIS, R., *On the dispersion of a solute in a fluid flowing through a tube,* Proc. R. Soc. London, A 235 (1956) 67–77.

151 GILL, W. N., SANKARASUBRAMANIAN, R., *Exact analysis of unsteady convective diffusion,* Proc. R. Soc. London, A 316 (1970) 341–350.

152 DOSHI, M. R., DAIYA, P. M., GILL, W. N., *Three dimensional laminar dispersion in open and closed rectangular conduits,* Chem. Eng. Sci. 33 (1978) 795–804.

153 DUTTA, D., LEIGHTON, D. T., *Dispersion reduction in pressure-driven flow through microetched channels,* Anal. Chem. 75, 1 (2001) 57–70.

154 ARIS, R., *On the dispersion of a solute by diffusion, convection and exchange between phases,* Proc. R. Soc. London, A 252 (1959) 538–550.

155 DASKOPOULOS, P., LENHOFF, A. M., *Dispersion coefficient for laminar flow in curved tubes*, AIChE J. **34**, 12 (**1988**) 2052–2058.

156 THOMAS, J. M., THOMAS, W. J., *Principles and Practice of Heterogeneous Catalysis*, VCH, Weinheim (**1997**).

157 FEDOROV, A. G., VISKANTA, R., *Heat and mass transfer dynamics in the microchannel adsorption reactor*, Microscale Therm. Eng. **3** (**1999**) 111–139.

158 MAAS, U., *Efficient calculation of intrinsic low-dimensional manifolds for the simplification of chemical kinetics*, Comput. Visualization Sci. **1** (**1998**) 69–82.

159 MAAS, U., POPE, S. B., *Simplifying chemical kinetics: intrinsic low-dimensional manifolds in composition space*, Combust. Flame **88** (**1992**) 239–264.

160 KIEHL, M., *Partitioning methods for the simulation of fast reactions*, Zentralbl. Angew. Math. Mech. **78** (**1998**) 967–970.

161 POPE, S. B., *Computationally efficient implementation of combustion chemistry using in situ adaptive tabulation*, Combust. Theory Model. **1** (**1997**) 41–63.

162 GOBBY, D., EAMES, I., GAVRIILIDIS, A., *A vertically-averaged formulation of catalytic reactions in microchannel flows*, in MATLOSZ, M., EHRFELD, W., BASELT, J. P. (Eds.), *Microreaction Technology – IMRET 5: Proc. 5th International Conference on Microreaction Technology*, Springer-Verlag, Berlin (**2001**), pp. 141–149.

163 WALKER, R. E., *Chemical reaction and diffusion in a catalytic tubular reactor*, Phys. Fluids **4** (**1961**) 1211–1216.

164 COMMENGE, J. M., FALK, L., CORRIOU, J. P., MATLOSZ, M., *Microchannel reactors for kinetic measurement: influence of diffusion and dispersion on experimental accuracy*, in MATLOSZ, M., EHRFELD, W., BASELT, J. P. (Eds.), *Microreaction Technology – IMRET 5: Proc. 5th International Conference on Microreaction Technology*, Springer-Verlag, Berlin (**2001**), pp. 131–140.

165 TEGROTENHUIS, W. E., KING, D. L., BROOKS, K. P., J., G. B., WEGENG, R. S., *Optimizing microchannel reactors by trading-off equilibrium and reaction kinetics through temperature management*, in *Proceedings of the 6th International Conference on Microreaction Technology, IMRET 6*, 11–14 March 2002, AIChE Pub. No. 164, New Orleans (**2002**), pp. 18– 28.

166 ROUGE, A., SPOETZL, B., GEBAUER, K., SCHENK, R., RENKEN, A., *Microchannel reactors for fast periodic operation: the catalytic dehydration of isopropanol*, Chem. Eng. Sci. **56** (**2001**) 1419–1427.

167 WALTER, S., LIAUW, M., *Fast concentration cycling in microchannel reactors*, in *Proceedings of the 4th International Conference on Microreaction Technology, IMRET 4*, 5–9 March 2000, AIChE Topical Conf. Proc., Atlanta, GA (**2000**), pp. 209–214.

168 LIAUW, M., BAERNS, M., BROUCEK, R., BUYEVSKAYA, O. V., COMMENGE, J.-M., CORRIOU, J.-P., FALK, L., GEBAUER, K., HEFTER, H. J., LANGER, O.-U., LÖWE, H., MATLOSZ, M., RENKEN, A., ROUGE, A., SCHENK, R., STEINFELD, N., WALTER, S., *Periodic operation in microchannel reactors*, in EHRFELD, W. (Ed.), *Microreaction Technology: 3rd International Conference on Microreaction Technology, Proc. of IMRET 3*, Springer-Verlag, Berlin (**2000**), pp. 224–234.

169 STEPANEK, F., KUBICEK, M., MAREK, M., ADLER, P. M., *Optimal design and operation of a separating microreactor*, Chem. Eng. Sci. **54** (**1999**) 1494–1498.

170 SAAD, Y., SCHULTZ, M. H., *GMRES: a generalized residual algorithm for solving nonsymmetric linear systems*, SIAM J. Sci. Stat. Comput. **7** (**1989**) pp. 856–869.

171 SOMMERFELD, M., TSUJI, Y., CROWE, C. T., *Multiphase Flows with Droplets and Particles*, CRC Press, Boca Raton, FL (**1997**).

172 THÉ, J. L., RAITHBY, G. D., STUBLEY, G. D., *Surface-adaptive finite-volume method for solving free-surface flows*, Numer. Heat Transfer **B26** (**1994**) 367–380.

173 RAITHBY, G. D., XU, W. X., STUBLEY, G. D., *Prediction of incompressible free Surface flows with an element-based finite volume method*, Comput. Fluid Dyn. J. **4** (**1995**) 353–371.

174 HIRT, C. W., NICHOLS, B. D., *Volume of fluid (VOF) method for dynamics of free boundaries*, J. Comput. Phys. **39** (**1981**) 201–221.

175 CFX4 User Manual, *Surface sharpening in free surface flows*, Ansys CFX.

176 Brackbill, J. U., Kothe, D. B., Zemach, C., *A continuum method for modeling surface tension*, J. Comput. Phys. **100** (1992) 335–354.

177 Noh, W. F., Woodward, P. R., *SLIC simple line interface method*, Lect. Notes Phys. **59** (1976), pp. 330–340.

178 Rider, W. J., Kothe, D. B., *Reconstructing volume tracking*, J. Comput. Phys. **141** (1998) 112–152.

179 Osher, S., Sethian, J. A., *Fronts propagating with curvature-dependent speed: algorithms based on Hamilton–Jacobi formulations*, J. Comput. Phys. **79** (1988) 12–49.

180 Zhang, H., Zheng, L. L., Prasad, V., Hou, T. Y., *A curvilinear level set formulation for highly deformable free surface problems with application to solidification*, Numer. Heat Transfer **34** (1998) 1–20.

181 Tryggvason, G., Bunner, B., Esmaeeli, A., Juric, D., Al-Rawahdi, N., Tauber, W., Han, J., Nas, S., Jan, Y. J., *A front-tacking method for the computations of multiphase flow*, J. Comput. Phys. **169** (2001) 708–759.

182 Hardt, S., Schönfeld, F., Weise, F., Hofmann, C., Ehrfeld, W., *Simulation of droplet formation in micromixers*, in *Proceedings of the 4th International Conference on Modeling and Simulations of Microsystems*, 19–21 March 2001), Hilton Head Island, SC (2001) pp. 223–226.

183 Hardt, S., Schönfeld, F., *Simulation of hydrodynamics in multi-phase micro-reactors*, in *Proceedings of the 5th World Congress on Computational Mechanics, WCCM*, 7–12 July 2002, Vienna (http://wccm.tuwien.ac.at).

184 Eggers, J., *Nonlinear dynamics and breakup of free-surface flows*, Rev. Mod. Phys. **69**, 3 (1997) 865–929.

185 Yang, Z. L., Palm, B., Sehgal, B. R., *Numerical simulation of bubbly two-phase flow in a narrow channel*, Int. J. Heat Mass Transfer **45** (2002) 631–639.

186 Higuera, F. J., Succi, S., Benzi, R., *Lattice gas dynamics with enhanced collisions*, Europhys. Lett. **9** (1989) 345–349.

187 Tryggvason, G., Fernández, A., Lu, J., *The effect of electrostatic forces on droplet suspensions*, in *Proceedings of the Second M.I.T. Conference on Computational Fluid and Solid Mechanics*, June 17–20 2003, M.I.T., Cambridge, MA, Elsevier, Oxford, UK (**2003**), pp. 1166–1168.

188 Carbonell, R. G., Whitaker, S., *Heat and mass transfer in porous media*, in *Fundamentals of Transport Phenomena in Porous Media*, Martinus Nijhoff, Dordrecht (**1984**) pp. 121–198.

189 Nozad, I., Carbonell, R. G., Whitaker, S., *Heat conduction in multiphase systems I: theory and experiments for two-phase systems*, Chem. Eng. Sci. **40** (1985) 843–855.

190 Whitaker, S., *Flow in porous media I: A theoretical derivation of Darcy's Law*, Trans. Porous Media **1** (1986) 3–25.

191 Sahimi, M., *Flow and Transport in Porous Media and Fractured Rock*, VCH, Weinheim (**1995**).

192 Wiessmeier, G., Hönicke, D., *Microfabricated components for heterogeneously catalyzed reactions*, J. Micromech. Microeng. **6** (1996) 285–289.

193 Wunsch, R., Fichtner, M., Schubert, K., *Anodic oxidation inside completely manufactured microchannel reactors made of aluminum*, in *Proceedings of the 4th International Conference on Microreaction Technology, IMRET 4*, 5–9 March 2000, AIChE Topical Conf. Proc., Atlanta, GA (**2000**), pp. 481–487.

194 Xu, Y., Platzer, B., *Concepts for the simulation of wall-catalyzed reactions in microchannel reactors with mesopores in the wall region*, Chem. Eng. Technol. **24**, 8 (2001) 773–783.

3
Gas-phase Reactions

Gas-phase reactions, together with single-liquid-phase organic reactions, nowadays belong to the most frequently investigated processes in microstructured reactors. This is not only due to the economic and scientific impact of the corresponding investigations, it is also due to the good suitability of gas-phase reactions, from a process engineering point of view, to be carried out in micro reactors.

Wörz et al. [1] give a definition of processes suitable for micro reactors. Investigation of such processes is supposed to be advantageous that combine high reaction rates with large heat release, in particular when involving multiple phases. The first aspect holds for many gas-phase reactions, e.g., for some total or partial oxidations, proceeding in the millisecond regime at elevated temperatures. Concerning the second issue raised by Wörz et al., this also applies to most gas-phase reactions; usually they are carried out with catalyst contact, and hence involve two phases (gas/solid). Accordingly, since all criteria are fulfilled, gas-phase reactions seem to be particularly suited for chemical micro processing.

There is an additional point to be made about this type of processing. Many gas-phase processes are carried out in a continuous-flow manner on the macro scale, as industrial or laboratory-scale processes. Hence already the conventional processes resemble the flow sheets of micro-reactor processing, i.e. there is similarity between macro and micro processing. This is a fundamental difference from most liquid-phase reactions that are performed typically batch-wise, e.g. using stirred glass vessels in the laboratory or stirred steel tanks in industrial pilot or production plants.

It is worth while to go into the latter discussion a bit further. While micro reactors and technical reactors have similar ways of processing, they have distinctly different sorts of catalyst implementation. Conventional reactors typically rely on catalyst beds, either of fixed type, e.g. formed by pellets, or of fluidized type, e.g. formed by fine, agitated powders. Thin film coatings on reactor walls, which are widely used in chemical micro processing, are seldom found with conventional processes (although a few exceptions are known). In micro reactors, washcoats, which have wide use in automobile exhaust gas treatment, are deposited on the micro channel wall and serve as carriers for catalysts which are inserted by wet-chemical post-treatment, e.g. by impregnation via precursor solutions. In turn, powder and grain beds in micro channels are not very favorable as high pressure drops result, the laminar flow pattern is changed in a way more difficult to describe by simulation, and it is difficult to generate these materials in a reliable manner. Nevertheless, the indisputable advantage of powders and beds is that conventional catalysts, without the need for post-processing, may be analysed in that way on the micro scale. Indeed, micro processing with powders inserted in micro channels has been described a few times in the literature.

Considering the major importance of catalysts, especially for gas-phase reactions, a separate section was allocated to the description of techniques for catalyst layer formation in micro channels and the respective analytical characterization (see Section 3.1).

3.1
Catalyst Coating in Micro Channels: Techniques and Analytical Characterization

With the increasing quest for chemical micro processing research on catalyzed gas-phase reactions, both the catalysts themselves and their carriers have become the focus of scientific investigations – on their preparation, morphology, porosity, composition, etc.

Catalysts and their carriers are provided in micro channels by various means and in various geometric forms. In a simple variant, the catalyst itself constitutes the micro-reactor construction material without need for any carrier [2–4]. In this case, however, the catalyst surface area equals that of the reactor wall and hence is comparatively low. Accordingly, applications are typically restricted to either fast reactions or processing at low flow rates for slow reactions (to enhance the residence time).

Methods specifically made for or adapted to the needs of a micro channel's coating are available. For this reason, more and more recent reports are concerned with catalyst/carrier layers with typical depths of 1–50 µm. There are objective arguments in favor of catalyst coatings; a very large variety with regard to porosity, material, composition, mass, shape and (crystal) structure is thereby possible. A disadvantage of such coatings was, however, their non-uniform geometric cross-sectional and longitudinal shape. This is mainly the consequence of the dominance of surface forces on the micro scale which can lead to a significant difference in profiles before and after coating. For instance, coatings in rectangular micro channels tend to have a U-shape; those in semi-circular channels often have a V-shape (Figure 3.1) [5]. In the latter case, meanwhile, advanced compositions of the slurry solution have been identified by which U-shaped catalyst layers in semi-circular micro channels result, which is desired here owing to their uniform layer thickness [6].

The placement of catalysts/carriers in micro channels can be done by various means. In a conventionally oriented variant, catalyst powders or small grains are inserted as mini fixed beds [7]. However, more specific catalyst arrangements are also known, originally designed for novel ways of processing at the macro scale, such as catalyst filaments [8], wires [9] and membranes (Figure 3.2) [10, 11].

Among the non-traditional routes for formation of catalyst and catalyst/carrier coatings, the most prominent way is the washcoat route followed by wet impregna-

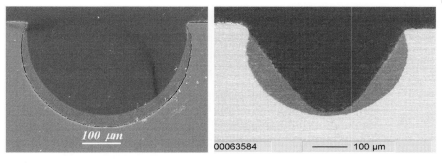

Figure 3.1 Cross section of micro channels coated with alumina washcoat exhibiting U- or V-shapes [6].

tion (see, e.g., [12–14]). Washcoats constitute an industrially accepted class of catalysts. Another class of industrially employed catalysts/carriers are created in micro reactors via the zeolite growth method [15–17]. Anodic oxidation is also widely used (see, e.g., [18]) to generate nano-porous oxide carrier layers, when aluminum reactors can be employed. Besides this, thin-film techniques such as CVD (see, e.g., [4]) and PVD, namely sputtering [11], serve for generating thin catalyst films. To complete this list of approaches towards catalyst and catalyst/carrier coatings, various other techniques have been tested, such as aerosol techniques [19], sol–gel techniques [20], an advanced plasma electrochemical process belonging to anodic spark deposition [21] and electrolysis [22].

Detailed descriptions of the procedures for the preparation of catalysts/carriers, for instance, have been given explicitly for washcoats (see, e.g., [5, 12, 14]) and for zeolite growth (see, e.g., [15–17]). For washcoats, a typical sequence of catalyst preparation steps is as follows (Figure 3.3) [6, 23]. The inlet and outlet chambers, encompassing the micro channel arrays on micro reactor plates, are protected with a thin polymer film. The suspension is deposited on the micro channel plate and the excess suspension is wiped off. Such a derived coating is dried at room temperature accompanied by shrinkage of the washcoat coating. Having cleaned the top parts of the micro channel fins, the dried washcoats are calcined typically at 500–600 °C to burn out the binder and to remove the protecting polymer film [12, 13]. Thereafter, the catalyst is brought into the washcoat by means of impregnation.

At present, very few comprehensive reports are devoted solely to catalyst/carrier coatings in micro channels, providing a deep insight in the subject and a detailed characterization of the coatings in terms of preparation, morphology, porosity, composition, etc. (see, e.g., [5]). However, an increasing number of reports concern the preparation of one sort of catalyst, e.g. for zeolites [15, 16, 24, 25]. The majority of these shorter reports provide also some basic characterization of the catalysts, with a focus on direct surface imaging, determining porosity and sometimes on cross-sectional profiles. In many further publications, SEM or other types of images of catalyst surfaces are found, most often before and from time to time after use [4,

SiO

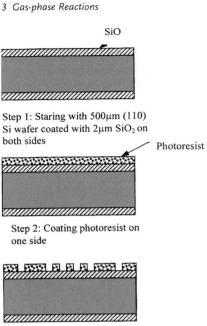

Step 1: Staring with 500μm (110)
Si wafer coated with 2μm SiO₂ on
both sides

Photoresist

Step 2: Coating photoresist on
one side

Step 3: Pattern frontside on
photoresist layer

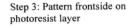

Step 4: Coating the backside
with photoresist

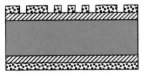

Step 5: pattern backside
on photoresist layer

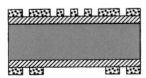

Step 6: Transfer patterns on
SiO₂ layer by etching with
BOF and remove photoresist

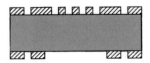

Step 7: Anisotropic etching in 30%
KOH solution for 10 minutes to form
15-20μm deep chamber on backside

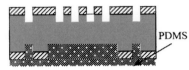

PDMS

Step 8: Protect wafer backside by
covering it with PDMS

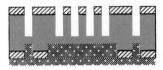

Step 9: KOH anisotropic etching for 3hr and
40 minutes until only 60μm silicon left

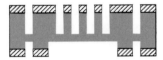

Step 10: Peel off silicone elastomer protection

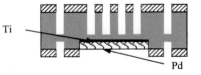

Ti

Pd

Step 11: Sputter 6 μm Pd with 10-nm
Ti as an adhesion layer

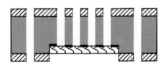

Step 12: Etch through Si wafer using
RIE until the metal film is exposed

Figure 3.2 Flow chart for palladium membrane fabrication process [10].

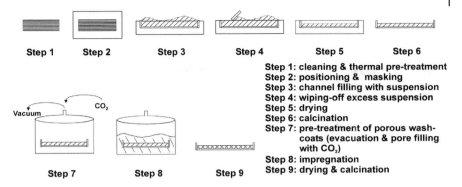

Step 1: cleaning & thermal pre-treatment
Step 2: positioning & masking
Step 3: channel filling with suspension
Step 4: wiping-off excess suspension
Step 5: drying
Step 6: calcination
Step 7: pre-treatment of porous wash-
coats (evacuation & pore filling
with CO$_2$)
Step 8: impregnation
Step 9: drying & calcination

Figure 3.3 Typical preparation steps needed for preparing a washcoat layer in a micro channel [23].

26]. The few reports on surface and cross-sectional imaging often indicate that today's micro channel coatings have uneven profiles, in both semicircular and rectangular channels (see the discussion above). Further, the coating depth varies according to its position in the channel [19, 20, 27]. Recently, considerable progress has been made here [5]. Details on temperature profiles inside a catalyst under operation, another important feature, are rarely found, an exception being, e.g., the simulation of such profiles for an alumina layer [28].

3.2
Micro Reactors for Gas-phase Reactions

3.2.1
Housing-encased Single-platelet and Multi-platelet Stack Micro Reactors

One of the most frequently used micro reactor types relies on the use of micro-structured platelets with multiple parallel channels, typically manufactured by methods other than routinely used for chip processing, encased in a housing [3, 4, 12, 13, 18, 28–39]. If more than one platelet is used, which is usually done to increase throughput, a stack-like arrangement is preferred for parallel feed. Such stacks are either welded directly from the outside [29, 30], are encompassed by a cover [3, 18, 31, 32, 37–39], have end caps with fluidic connectors [12, 13, 33] or are inserted into a recess of a housing, which is typically composed of two parts [4, 28, 34–36, 40–41].

Reversible sealing is achieved by seals, e.g. made of graphite or other materials, to achieve tightness between the plates and also between different functional fluidic elements on one plate [12, 13]. For achieving tightness in this way, compression of the housing is needed, usually done by screwing. In rare cases, specially polished surfaces are directly compressed without any seal. More often, the latter is used to seal stacks within a housing, omitting the individual seal between the platelets [4].

As an alternative to seals, irreversible bonding can be applied, e.g. by laser welding the surface of a microstructured stack [29, 30] or by diffusion bonding via vacuum compression of a microstructured stack [18, 37–39]. For better handling and fluid interconnection, diffusion-bonded stacks may be surrounded by a shell [18, 37–39]. Diffusion-bonded stacks typically are more compact. In addition, this interconnection technique is principally amenable to small-series production. Accordingly, it is seen as a proper way to realize future commercial, off-the-shelf micro reactors.

3.2.1.1 Reactor 1 [R 1]: Reactor Module with Different Multi-channel Micro Reactors

A complete reactor module was built, consisting of the actual micro reactor and an encasement that serves for temperature setting [28]. The latter consists of two parts, a furnace for setting the high temperature in the reactor inlet collection zone and in the reaction zone and a cooler for the outlet collection zone. The micro reactor has a housing with standard tube connections. An electric furnace serves for heating. Temperatures can be measured in the furnace, at the furnace/micro reactor border and in the outlet collection zone. For thermal insulation, a 2 mm ceramic

Reactor type	Multi-plate-stack micro reactor with outer module	Catalyst No. 1 material; formation	Pt, impregnated
Furnace material	Copper	Reactor No. 2: monolith channel diameter; length	500 μm; 9.0 mm
Cooler material	–	Cube No. 2 dimensions; material	10 mm × 10 mm × 9 mm; Pt
Micro reactor housing material	Nickel	Reactor No. 3: total number of reaction channels	49
Platelet material	Aluminum; Pt	Reaction channel No. 3 width; depth; length	280 μm; 140 μm; 9.0 mm
Temperature of cooler/furnace	–20 °C; max. 430 °C	Platelet No. 3 material; width; length	Aluminum, anodically oxidized 9.0 mm; 9.0 mm
Power input furnace	max. 185 W	Catalyst No. 3 material; formation	Pt, impregnated
Operating temperature	370 °C	Reactor No. 4: total number of reaction channels	20
Operating pressure	–	Reaction channel No. 4 diameter; length	145 μm; 6.5 mm
Reactor No. 1: platelet width; depth; length	4.34 mm; 0.3 mm; 7 mm	Platelet No. 4 width; length	6.5 mm; 6.5 mm
Platelet No. 1 material	Aluminum, anodically oxidized	Cooling channel No. 4 diameter; length	300 μm; 6.5 mm

ring is placed between furnace and cooler. The cooler removes heat from the reaction zone and quenches the reaction gas mixture. The reactor has three versions, each with a set of reaction channels of various dimensions. The first is a parallel-plate reactor, the second a monolith and the third a stack of microstructured platelets. A fourth reactor type is a combined reactor/heat exchanger which so far is only a concept study.

3.2.1.2 Reactor 2 [R 2]: Steel Multi-plate-stack Reactor with Micro Mixer

The reactor is a two-piece housing which is sealed by flat sealing by screws using graphite seals [4, 26, 40, 41]. The bottom piece contains two closely positioned square recesses made by die sinking, a μEDM process. In the recesses, stacks of platelets of micro channels, not connected and without seals, are inserted. The recesses are connected via a breakthrough that functions as diffusion zone to guarantee mixing of the reactant gas before entering the reaction zone (Figure 3.4). The stacks of platelets are compressed when screwing the top and bottom housing piece together. The inlet and outlet tubes are welded to the bottom piece of the housing.

The first recess contains a stack of mixer platelets made by a combination of laser-LIGA and electroforming. These platelets have curved micro channels that make a 90° fluidic turn. In order to have equal flows in each curved channels, different channel widths had to be used to compensate the differences in channel length. CFD simulation was used to determine the mixer channel width values and confirmed 99% mixing for all flow rates investigated. Two types of mirror-imaged platelet designs allow the creation of gas multi-lamellae in an alternating stack arrangement. The second recess originally contained a stack of silver reaction platelets made by LIGA and electroforming [4, 26, 40]. In a later version, also chemically etched steel and milled Aluchrom steel platelets were used [4]. Aluchrom is an

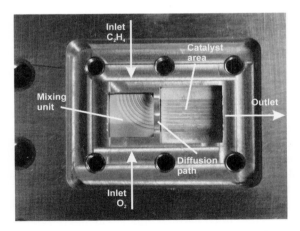

Figure 3.4 Magnified view inside a steel multi-plate-stack reactor.
Mixing unit (left), diffusion zone (middle) and stacked silver catalyst platelets (right).

Reactor type	Multi-plate-stack with mixer-reactor sections	Diffusion zone length	1 mm
Mixer + reaction platelet/housing material	nickel-gold plated + silver/stainless steel	Diffusion zone volume (= explosive volume)	0.042 cm^3
Heating	Forced convection flow	Reaction channel (laser-LIGA) width; depth; length	500 µm; 50 µm; 9.5 mm
Operating temperature	200–350 °C	Reaction channel (etched) width; depth; length	500 µm; 80 µm; 9.5 mm
Operating pressure	1–25 bar	Reaction channel (Aluchrom) width; depth; length	500 µm; 90 µm; 9.5 mm
Mixing channel width; depth	148–469 µm; 200 µm	Reaction plate thickness	300 µm
Mixing plate thickness	300 µm	Total number of reaction plates	14
Total number of mixing plates	14	Total number of reaction channels	126

aluminum containing stainless steel normally used metallic support for automobile exhaust catalysts (Krupp VDM e-20Cr-5Al). Heating the Aluchrom material to 1100 °C with air oxygen creates an α-Al$_2$O$_3$ surface.

The surface of the silver reaction channels was enhanced by means of the oxidation and outgassing reduction (OAOR) process, which relies on oxidation at 250 °C using pure oxygen and subsequent reduction. An increase in surface area by a factor of 2–3 was reached as indicated by chemisorption data.

3.2.1.3 Reactor 3 [R 3]: Modular Multi-plate-stack Reactor

A modular concept was developed to fit to the typical demands of laboratory reactors, flexibility, ease of handling, and fast change of parameters (Figure 3.5). It is based on five different assembly groups, namely microstructured platelets, a cylindrical inner housing, two diffusers and a cylindrical outer shell with a flange [42, 43]. The microstructured platelets are inserted in a recess of the bottom part of the inner housing, which is a rectangular mill cut (10 × 10 × 50 mm). Cylindrical 1/8 in tube connectors guide the flow from the reactor inlet via the diffuser to the platelet stack in the mill cut. The flange and cylindrical outer housing are held by six 5 mm screws and tightened via insertion of a copper gasket. The platelets are fabricated by means of thin-wire µEDM in the alloy AlMg3. Each platelet has 14 parallel micro channels. The surface of the micro channels was rough.

The micro reactor can be operated at temperatures up to 480 °C [44]. Platelet exchange can be performed in a short time, needing only 15–30 min of cooling from operational to ambient temperature. Heat production rates of about 30 W can be achieved without the need for external cooling [43].

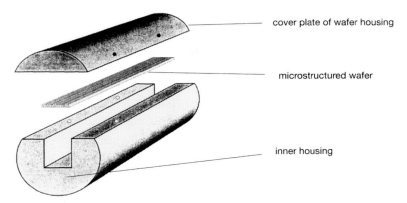

cover plate of wafer housing

microstructured wafer

inner housing

Figure 3.5 Schematic of the modular multi-plate-stack reactor [43].

The micro reactor was made in two versions:

(a)

Reactor type	Multi-plate-stack in cylindrical housing	Reaction channel width; depth; length	700 µm; 300 µm; 50 mm
Platelet material	AlMg3	Sputtered catalyst layer thickness	50–1400 nm
Operating Temperature	480 °C	Sol–gel catalyst layer thickness	1 µm
Operating pressure	3 bar	Total number of micro channels	14
Outer platelet dimensions: width; depth; length	10 mm; 1 mm; 50 mm		

For this micro reactor version, the catalyst was coated on the AlMg3 platelet as a thin silver layer by sputtering [43, 44]. A further set of platelets was covered with an α-alumina layer by sol–gel technique and impregnated by a three-step procedure with silver lactate.

(b)

Reactor type	Multi-plate-stack in cylindrical housing	Outer platelet dimensions (length No. I; No. II; No. III) = reaction channel length	12.5 mm; 37.5 mm; 50 mm
Platelet material	AlMg3	Reaction channel No. 1 width; depth	400 μm; 400 μm
Heating	external	Reaction channel No. 2 width; depth	200 μm; 200 μm
Operating Temperature	400 °C	Reaction channel No. 3 width; depth	80 μm; 80 μm
Operating pressure	0.1 MPa	Total number of micro platelets: No. 1; No. 2; No. 3	4; 8; 8

For this micro reactor version, the microstructured platelets were treated by anodic oxidation to obtain a nano-porous layer and impregnated with precursor solutions in organic solvents to obtain a $V_2O_5/P_2O_5/TiO_2$ catalyst.

3.2.1.4 Reactor 4 [R 4]: Multi-plate-stack Micro Reactor with Diffusers

This reactor is a professional tool made by small-series manufacturing that is nearly ready for commercialization [18, 43, 44]. A cylindrical reactor core is connected to two diffusers at each end that serve for gas-stream equipartition and collection. The reactor core contains a stack of microstructured metallic platelets (Figure 3.6). The platelets are made by mechanical micromachining using micro milling with special cutting tools. A thicker plate with borings for thermocouples can be inserted in the center of the stack so that temperature monitoring can be applied.

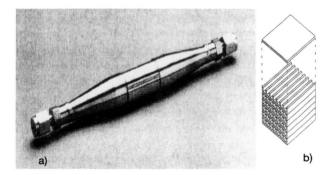

Figure 3.6 Multi-plate-stack reactor with diffusers.
Photograph (left) and schematic of the plate stack (right) [42].

The multi-plate stack micro reactor was made in two versions:

(a)

Reactor type	Multi-plate-stack with two diffusers	Outer platelet dimensions: width; depth; length	10 mm; 0.3 mm; 50 mm
Platelet material	AlMg3	Reaction channel width; depth; length	200 µm; 200 µm; 50 mm
Heating	Electrical	Sputtered catalyst layer thickness	1200 nm
Operating temperature	210 °C	Total number of micro channels	33
Operating pressure	3 bar	Total number of reaction plates	26
Increase in surface-to-volume ratio by anodic oxidation	10^4–10^5	Typical length of reactor (with diffusers)	17 mm

For this version, the micro structured AlMg3 platelets were coated with silver by CVD in [43]. In [44], the platelets were either totally made of silver (as construction material) or of AlMg3 and then coated by PVD with silver. In the latter version, two sub-versions were made with and without anodic oxidation to a generate nano-porous surface structure.

(b)

Reactor type	Multi-plate-stack with two diffusers	Reaction channel No. 1 width; depth; length	400 µm; 400 µm; 50 mm
Platelet material	AlMg3	Reaction channel No. 2 width; depth; length	200 µm; 200 µm; 50 mm
Heating	Electrical	Reaction channel No. 3 width; depth; length	80 µm; 80 µm; 50 mm
Operating temperature	400 °C	Alumina catalyst layer thickness	40 µm
Operating pressure	0.1 MPa	Total number of micro channels: No. 1; No. 2; No. 3	255; 550; 1165
Outer platelet dimensions: width; length	10 mm; 50 mm	Total number of reaction plates	15; 25; 37

For this version, the microstructured platelets were treated by anodic oxidation to obtain a nano-porous layer and impregnated with precursor solutions in organic solvents to obtain a $V_2O_5/P_2O_5/TiO_2$ catalyst.

3.2.1.5 Reactor 5 [R 5]: Cross-flow Multi-Plate Stack Micro Reactor

This professional tool, available in small series, is a derivative of a micro heat exchanger [31, 45–47] made by FZK, which was developed earlier. By insertion of catalytically active material, the 'micro heat exchanger' functions as a reactor. Quadratic platelets with straight micro channels are assembled into a stack so that two adjacent platelets have a 90° turn (Figure 3.7 and Figure 3.8). By this means, a cross-flow configuration is created with two separated fluid passages for reacting and heat transferring fluids.

Microfabrication of the parallel channels was performed by mechanical surface cutting of metal tapes [31]. In the case of aluminum alloys, ground-in monocrystalline diamonds were used [45]. In the case of iron alloys, ceramic micro tools have to be used owing to the incompatibility of diamonds with that material. Such a microstructured platelet stack is provided with top and cover plates, diffusion bonded and connected to suitable fittings for the inlet and withdrawal ducts by electron beam welding (Figure 3.9).

Figure 3.7 Alternated 90° turned adjacent platelets forming a stacked cross-flow configuration. Image of thermally bonded devices (top). Complete mounted multi-plate stack micro reactors (bottom) [45].

Figure 3.8 Schematic of stacked platelets (right) and a micrograph showing micro channel openings from a corner (left) [115].

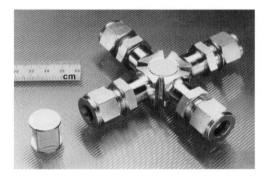

Figure 3.9 Image of a completely mounted cross-flow micro heat exchanger with 1 cm³ active volume [46].

Optimized microfabrication and advanced assembly led to the use of thin platelets, in an original version 100 μm thick with a 80 μm micro channel depth, so that very thin walls (20 μm in the case sketched) remain for separating the fluids. Therefore, also the total inner reaction volume with respect to the total construction volume or the 'active internal surface area' is very large. The latter surface amounts to 300 cm² (for both the heat transfer and reaction sides) at a cubic volume of 1 cm³. Indeed, the micro heat exchangers exhibited high heat transfer coefficients for gas [46] and liquid (Figure 3.10) [47, 48] flows.

The reactor can be obtained in many materials such as aluminum alloys, copper, silver, titanium and stainless steel. The number of stacked platelets, the dimensions of the micro channels on the platelets and the fluidic connectors were also varied. Pressure tightness up to 25 bar and He tightness were demonstrated, although this is certainly not the upper limit.

More details on the reactor are available [1, 49–51].

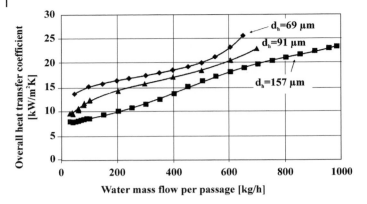

Figure 3.10 Calculated heat transfer coefficient depending on micro-channel dimensions and water flow rate. Experimental data are given in [47].

Reactor type	Multi-plate-stack, cross-flow	Operating pressure	–
Platelet/housing material	Silver/stainless steel	Reactor cube dimensions	10 mm × 10 mm × 10 mm
Heating	Cross-flowing gas	Reaction channel width; depth; length	400 µm; 300 µm; 10 mm
Operating Temperature	390 °C	Total number of channels per passage	200

3.2.1.6 Reactor 6 [R 6]: Counter-flow Multi-plate Stack Micro Reactor

This micro reactor consists of a stack of two sets of microstructured platelets which are arranged in such a way that two alternately positioned flow configurations are generated [13, 27]. Thereby, two fluids can be guided in a counter-flow mode, although by using one set of platelets only a parallel transport of one fluid can also be achieved. In the first case, a reactor/heat exchanger configuration is created, whereas only a reactor is set up in the second case (Figure 3.11).

The stack of platelets is encompassed by two end caps bearing the external fluidic connections. If desired, a third housing part can be introduced in between the end caps to shield the stack. As a further design modification, ceramic Macor™ insulating plates can be inserted between the end caps and platelet stack to prevent heat losses from the stack to the housing.

The platelets comprise an array of parallel micro channels with microstructured triangular-shaped headers at both ends. In the headers holes form conducts to the external feed fluid supply. Optimization of the shape of the header with respect to flow equalization was the topic of various simulation studies [52, 53].

The micro channels were made by isotropic wet chemical etching of metal plates. The plates were tightened by various means: they were either glued, stacked together with graphite sealing or fixed within a closed stainless steel housing (Figure 3.12). The micro reactor can be heated by various means, e.g. electrically with heating cartridges or frames, by setting it in an oven or by using fluid heat exchange in the counter-flow mode.

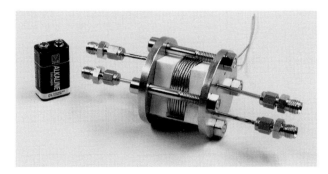

Figure 3.11 Photograph of the multi-plate stack reactor, originally designed as a counter-flow heat exchanger; this type of reactor was also used for periodic operation [13].

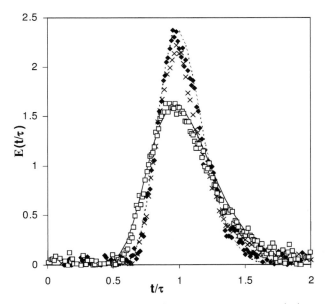

Figure 3.12 Residence time distribution in a micro reactor which is tightened by different means. (♦) Glued reactor without catalyst coating; (X) glued reactor with catalyst coating; (□) reactor with graphite joints. Calculated curves for tubular reactors with the Bodenstein number Bo = 33 (solid line) and Bo = 70 (dashed line).

Reactor type	Multi-plate-stack with counter-flow guidance	Total number of reaction platelets	6; 10
Housing (end cap) material	Stainless steel	Platelet material	Aluminum
Heating	Forced convection flow; oven; electrical heating	Number of reaction channels per platelet	34
Operating temperature	480 °C	Micro channel hydraulic diameter; length	230–280 µm; 20 mm
Operating pressure	1.2 bar	Catalyst material; formation	$V_{30}Ti_{70}O_x$-SiO_2; suspension + impregnation; finally anodic oxidation
Stack dimensions	40 mm × 40 mm × 0.5 mm		

3.2.1.7 Reactor 7 [R 7]: Multi-Plate Stack Micro Reactor in Heatable Holding Unit

The central part of this reaction system is an Rh construction material monolith made from a welded stack of platelets carrying parallel micro channels [3]. As initial material for this stack, hard-rolled Rh foils were first thermally treated and then micro structured. Micro milling can be applied for channel widths exceeding 200 µm. To generate smaller internal dimensions, thin-wire µEDM is applicable. Rectangular shapes were achieved by milling, whereas µEDM led to U-shapes. Laser or electron beam welding are the favored interconnection techniques, although diffusion bonding works is also satisfactory (Figure 3.13).

The welded Rh stack was welded to a cover plate and inserted into a ceramic holder, made of a special shrinkage-free material that can be machined as green compact [3]. The ceramic holder prevents heat losses from the metallic stack.

Figure 3.13 Photograph of welded stacks of Rh platelets and supporting parts [3].

The outer wall of this holder is provided with a spiral groove for insertion of a heating wire. The holder is installed into a pressure vessel via a few screw connections, giving a gas-tight system. Quartz glass inspection flanges at both ends of the vessel serve for pyrometric inspection to judge the catalyst temperatures at the stack inlet and outlet.

Compared with laboratory fixed-bed reactors or conventional extruded monoliths, such a microstructured monolith is smaller in characteristic dimensions, lower in pressure loss by optimized fluid guiding and constructed from the catalytic material solely [3]. The latter aspect also leads to enhanced heat distribution within the micro channels, giving more uniform temperature profiles.

The whole set-up for partial oxidation comprises a micro mixer for safe handling of explosive mixtures downstream (flame-arrestor effect), a micro heat exchanger for pre-heating reactant gases, the pressure vessel with the monolith reactor, a double-pipe heat exchanger for product gas cooling and a pneumatic pressure control valve to allow operation at elevated pressure [3].

The micro reactor was made in three versions, termed a–c, differing in internal dimensions [3].

Reactor type	Multi-plate-stack, welded and with external heating	Platelet width; length; thickness	(a) 5.5 mm; 5.0 mm; 200 μm (b) 5.5 mm; 5.0 mm; 200 μm (c) 5.5 mm; 20 mm; 200 μm
Housing	Special ceramic	Reaction channel width; depth; length	(a) 120 μm; 131 μm; 5.5 mm (b) 60 μm; 137 μm; 5.5 mm (c) 120 μm; 108 μm; 20.0 mm
Heating	Electrical heating by heating wire	Hydraulic diameter	(a) 125 μm; (b) 83.5 μm; (c) 114.1 μm
Operating temperature	1090–1190 °C	Total number of channels per stack	(a) 644; (b) 1152; (c) 675
Operating pressure	0.15–12 MPa	Residence time, $1\,l\,h_{STP}^{-1}$	(a) 3.0 ms; (b) 2.8 ms; (c) 10.6 ms
Stack dimensions (W × H × L)	(a) 5.5 × 5.5 × 5.0 mm³ (b) 5.5 × 5.5 × 5.0 mm³ (c) 5.5 × 5.5 × 20.0 mm³	Geometric surface per cm reactor length	(a) 32.4 cm²; (b) 45.4 cm²; (c) 30.9 cm²

3.2.1.8 Reactor 8 [R 8]: Ceramic Platelet Micro Reactor

This micro reactor contains an exchangeable platelet with a multi-channel reaction zone (Figure 3.14) [54, 55]. The housing is symmetric, i.e. inlet and outlet diffusers and tube connectors are the same and have mirror-imaged positions. Between the triangular-shaped diffusers a recess is placed where the reaction platelets are inserted. The reaction housing is covered with a top plate. Highly polished ceramic surfaces allow tight sealing. This version of the reactor is referred to as *Model A*.

Model B was especially designed for methane conversion to ethylene [54, 55]. This reaction needs pre-heating to a defined temperature before reaction. This is achieved by ceramic heaters in the housing. In addition, the gases do not enter as a

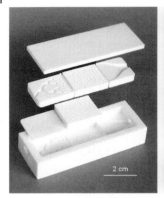 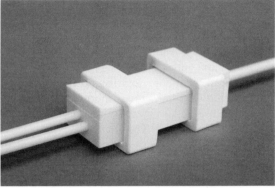

2 cm

Figure 3.14 Ceramic platelet micro reactor; bottom housing with recess for platelet stack, platelets and top plate [54].

mixture, but are separated and only mixed shortly before reaction. The model has an almost constant cross section over the reactor length to give an even flow. The catalyst-coated platelets are again exchangeable.

Fabrication was done by a combination of stereolithography and low-pressure injection molding [54, 55]. A 3D-CAD model was transferred in silicone polymer molds by stereolithography. These were filled with an alumina feedstock. After demolding, organic residues were removed from the green bodies by slow heating. Thereafter, sintering up to 1700 °C was applied. The tubes for gas feed were joined by a commercial glass–ceramic.

Reactor type	Ceramic platelet reactor	Micro channel width; length (type A + B)	500 µm; 25 mm
Catalyst material	$LiAlO_2$ (sol–gel)	Device inner volume	650 mm^3
Device length	70 mm	Operational temperature	1000 °C
Feed and withdrawal tubes	2 mm	Operational pressure	1.2 bar
Length of diffuser zone	16 mm		

3.2.1.9 Reactor 9 [R 9]: Micro Heat Transfer Module

The micro heat transfer module (Figure 3.15) comprises a stack of micro structured platelets which are irreversibly bonded [29, 30]. The module is heated by external sources, e.g. by placing it in an oven or by resistance heating. The single parallel flows are all guided in the same direction on the different levels provided by the platelets. Before and after, distribution and collection zones are found, connected to inlet and outlet connectors.

The micro structured plates are made by wet-chemical etching. The platelet stack is bonded by laser welding. The inlet and outlet connectors are also laser welded.

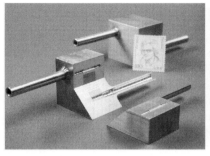

Figure 3.15 Micro heat transfer module [29].

Reactor type	Micro heat transfer module	Device inner volume	$2.6 \cdot 10^{-7}$ m^3
Module material	Stainless steel	Operational temperature	Up to 600 °C
Number of platelets	50	Operational pressure	0.4 bar
Number of channels on one platelet	13	Operational pressure	0.4 bar
Micro channel width; depth; length	500 μm; 100 μm; 8 mm	Outer device dimensions	32 mm × 32 mm × 26 mm
Total micro-channel surface area	$6.24 \cdot 10^{-3}$ m^2	Tube connectors: inner diameter; length	6 mm; 40 mm

3.2.2
Chip Micro Reactors

The design and fabrication of some gas-phase micro reactors are oriented on those developed for chip manufacture in the framework of microelectronics, relying deeply on silicon micromachining. There are obvious arguments in favor: the infrastructure exists at many sites world-wide, the processes are reliable, have excellent standards (e.g. regarding precision) and have proven mass-manufacturing capability. In addition, sensing and control elements as well as the connections for the whole data transfer (e.g. electric buses) can be made in this way.

Accordingly, chip micro reaction systems are frequently described in the literature. Most of them are made of silicon (see, e.g., [19, 56–62]); Glass can be manufactured by similar routes as for silicon and could hence constitute gas-phase micro reactors; however, the glass chip micro reactors described so far were made for liquid-phase applications (see, e.g., [63–70]).

Today's chips are often simple two-wafer bonded microstructured devices. However, exceptions are known. Complex multi-wafer arrangements, having a separte function on each level, have been described already in the pioneering phase of chemical micro processing and investigated by the chemical industry [71]. Another pioneering chip comprised a multitude of heating and sensing functions [19, 56–62].

3.2.2.1 Reactor 10 [R 10]: Catalyst Membrane Si-chip Micro Reactor with Sensing and Heating Functions

An Si-chip micro reactor (Figure 3.16) contains an etched micro channel of T-shape which was covered by a thin membrane [19, 56–62]. The membrane bears on its bottom side, i.e. facing the micro channel, a thin layer of catalyst material and on its top side heating elements and temperature sensors. Besides these carrier functions, the main task of the membrane is to transfer heat from the reaction zone to the outside by means of convective mechanisms. Accordingly, by variation of the material, i.e. change of thermal conductivity, control over heat removal can be exerted and, therefore, over the reaction temperature. For a given reactor configuration with a given membrane, heat generation and reaction temperature are determined by the exothermic release of the reaction and the power input of the heaters.

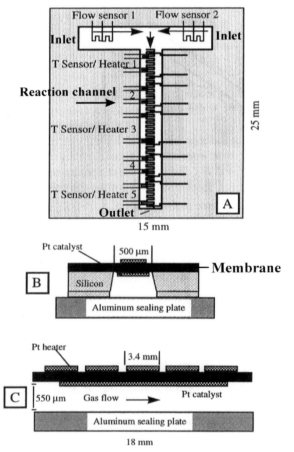

Figure 3.16 Schematic of Si-chip catalyst membrane micro reactor. Top view (A), end-on cross section of reaction channel (B); side-view cross section of reaction channel (C) [60].

An optimized design with an Y-shaped configuration (see also [57]) mainly served for reducing the mechanical stress in the chip device which can lead to rupture, especially under high temperature and pressure operation [19]. By changing the configuration, the large area of the free-standing membrane in the mixing intersection is reduced. The improved design withstands much higher temperatures and pressures (see table below) [19].

The micro reactor was made from a double-side, polished (100 mm diameter) Si wafer coated with 1 µm thick low-stress SiN_x that comprises the membrane [57]. SiN_x/Si reactors were fabricated in an Si-on-insulator (SOI) wafer with a 2.6 µm Si device layer covered by 150 nm of low-stress SiN_x. The SiN_x on the back side of the wafer was patterned by photolithography and plasma etched to expose the underlying silicon. Pt heaters and sensors were defined by patterning lift-off photo resist on the front side of the wafer, followed by e-beam deposition (with Ti as adhesion layer). The gas flow channels were formed from the back side by either etching the exposed Si in KOH solution or using deep reactive ion etching. The latter process allowed more freedom concerning structural design. The wafer was thereafter diced into the individual micro reactors and each die was affixed to the Al-base plate. Inlet and outlet holes matched the fluidic connections of the experimental rig. Electrical connections to bond pads of the heaters and sensors were made using a probe card (Figure 3.17).

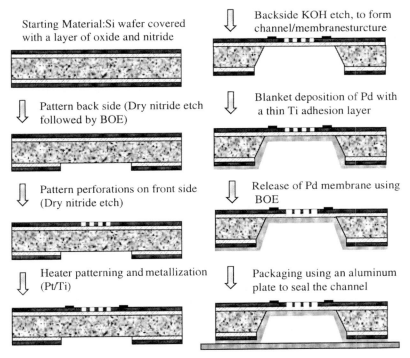

Figure 3.17 Microfabrication sequence for the silicon component of the catalyst membrane micro reactor [57].

Reactor type	Chip reactor with thin-film sensors and membrane	Catalyst material	Pt
Chip material	Silicon, aluminum	Catalyst layer thickness	0.1 µm
Membrane material	SiN_x; Si	Reaction channel width; depth; length	500 µm; 550 µm; 18 mm
Membrane thickness SiN/Si	1; 1.5, 2.6	Overall chip dimensions	15 mm × 25 mm
Maximal temperature	650 °C original design, 800 °C optimized design	Reaction channel (Aluchrom) width; depth; length	500 µm; 90 µm; 9.5 mm
Maximal pressure	0.5 bar original design, 2.7 bar optimized design	Reaction plate thickness	300 µm
Operating temperature	570 °C original design	Total number of reaction plates	14
Operating pressure	–	Total number of reaction channels	126
Chip material	Silicon, aluminum	Catalyst layer thickness	0.1 µm

3.2.2.2 Reactor 11 [R 11]: Single-channel Chip Reactor

This reactor comprises a single-channel reaction zone followed by a quenching (cooling) zone [72]. Gases, pre-heated in a separate zone, are contacted in an T-junction and mixed in a short passage thereafter. Such mixed gases enter the above-mentioned reaction zone.

Fabrication was done by photolithography and deep reactive ion etching (DRIE). The catalyst was inserted by sputtering. Such a prepared microstructure was sealed with a Pyrex cover. The bonded micro device was placed on a heating block containing four cartridge heaters. Five thermocouples monitored temperature on the back side. A stainless-steel clamp compressed the device with graphite sheets.

Reactor type	Single-channel reactor	Operating temperature	530 °C
Catalyst material	Sputtered silver	Operating pressure	1 atm
Reaction channel width; depth	600 µm; 130 or 70 µm		

3.2.2.3 Reactor 12 [R 12]: Multi-channel–One-plate Chip Reactor

This simple reactor concept is based on a microstructured silicon chip (Figure 3.18) covered by a Pyrex-glass plate by anodic bonding [73, 74]. The silicon microstructure comprises, in addition to inlet and outlet structures, a multi-channel array. Only the Pyrex-glass plate acts as cover and inlet and outlet streams interface the silicon chip from the rear.

Standard silicon micromachining processes were applied. The starting material was a 100 mm silicon (110) orientation wafer covered with thermal oxide. Stand-

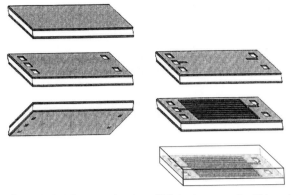

Figure 3.18 Schematic drawing of fabrication steps for Si-based micro reactors. The fabrication is carried out batch-wise on an 100 mm Si-wafer [73].

ard photolithography and KOH etching processes were applied for micro structuring. Separation into individual pieces was done using a diamond wafer saw.

Catalysts can be incorporated by the various known methods. So far, sputtered platinum was used. Such films are dense so that the catalyst surface area equals the channel surface.

Reactor type	Multi-channel chip reactor	Pt film thickness	10–40 nm
Catalyst material	Platinum	Catalyst surface area	2.2×10^{-4} m^2 (100 µm) 2.8×10^{-3} m^2 (5 µm)
Chip materials	Silicon; Pyrex glass	Operating temperature	200 °C
Channel width, depth, length	100 µm; 100 µm; 18/19 mm or 5 µm; 100 µm; 18 mm	Operating pressure	1 bar
Number of channels	39 (100 µm); 780 (5 µm)		

3.2.2.4 Reactor 13 [R 13]: Micro-strip Electrode Reactor

A test reactor was made of stainless steel which contains a so-called micro-strip electrode array [75]. This array is composed of thin strips surrounded by larger objects. The anodes are thin gold strips evaporated on glass bulk. The cathodes have a more complex bulky pattern similar to an oval.

Reactor type	Micro-strip electrode reactor	Cathode width	500 µm
Catalyst material	Gold emitting electrodes	Period (from anode to anode)	1000 µm
Anode width	12 µm	Size of electrode array	30 mm × 30 mm

3.2.2.5 Reactor 14 [R 14]: Self-heating Chip Micro Reactor

Thermally oxidized silicon wafers were structured by photolithography and wet-chemical etching [76]. In this way, a meandering pre-heating channel structure followed by a meandering reactor channel structure was prepared (Figure 3.19). Next to the channel connecting these two units a thermocouple well was placed. On the back side of the wafer a meandering Pt heating element was patterned. Glass plates covered the double-sided Si structure on both sides via anodic bonding. Holes were drilled for gas inlet and withdrawal and stainless-steel tubes attached.

A Pt catalyst was applied by dry and wet techniques. By means of sputtering using a mask process protecting parts of the microstructure, the micro channel bottom was coated selectively. In addition, an γ-alumina layer was applied by the sol–gel technique. Initially, the whole micro structure was covered by such a layer. Then, photoresist was applied and patterned so that only the channel part remained covered. After removal of the exposed photoresist and unprotected γ-alumina, only the channel bottom was coated with γ-alumina.

Reactor type	Self-heating chip reactor	Pt heating element: diameter; length	100 μm; 200 mm
Catalyst material	Pt (sputtered); Pt/γ-alumina (sol–gel) 2.5 μm	Glass plate thickness	1 mm
Si wafer	10 mm × 40 mm	Steel tube diameter	300 μm
Reactor channel: upper width, lower width; depth; length	600 μm; 515 μm; 60 μm; 78 mm	Thermocouple element wire diameter	100 μm
Reactor channel: width; depth; length	200 μm; 60 μm; 95 mm	Operational temperature	150 °C

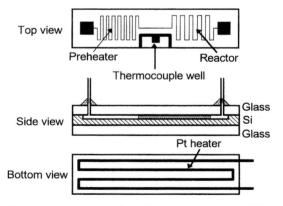

Figure 3.19 Schematic of a self-heating micro reactor [73].

3.2.2.6 Reactor 15 [R 15]: Modular Multi-functional Chip Reaction System

This micro-chip reaction system was developed for pioneering industrial investigations at the beginning of micro-reactor developments [71].

It comprises several wafers with different functional elements such as micro heat exchanger, mini fixed-bed catalyst chambers, manifolding structures and other components not disclosed. The flow partitioning was performed in a hierachic manner. A patent gives deeper insight into the generic construction architecture of the micro-reaction system. The wafers were bonded to a stack and equipped with fluid connections [137].

Heat transfer was accomplished by guiding flows throught different wafer levels, some acting as energy source, others as heat sink. For cooling, circulating liquids were applied.

The microstructured plates were made by wet-chemical etching. The platelet stack was bonded by standard welding. The inlet and outlet connectors were also welded.

Reactor type	Modular multi-functional chip reaction system	More details are given in [137]
Module material	Silicon	

3.2.3
Mini Fixed-bed Micro Reactors

Fixed-bed technology is a very common approach in conventional gas-phase processing, on a laboratory and industrial scale. The difficulty in transferring this concept to the micro scale stems from the restrictions on the availability of fixed-bed particles much smaller than the internal dimensions of the micro device. If they are available, handling of the particles certainly causes expenditure, as the tiny particles have to be filled into a micro channel, packed properly, should not dislocate on operation, and may have to be removed completely afterwards. In addition, the smaller the size of the bed particles, the more care has to be taken to achieve a proper packing, e.g. avoiding large conduits. For this reason, regular-sized particles were applied recently which result in a tight packing having interstices of the order of the internals of typical micro channels [7, 77–80]. For these beds, first considerations on mass and heat transfer based on the Weisz modulus and the Anderson criterion were made.

The undoubted advantage of mini fixed-bed micro reactors is that they follow a widely accepted processing path and in principle can use all of the commercial catalysts, if they can be crushed to a size much below the micro-channel diameter. Hence catalyst material flexibility is a major driver.

Today's massive efforts in nanotechnology will certainly provide more well-defined, regular-shaped particles in the submicron range, and mini fixed-bed technology will profit from that.

3.2.3.1 Reactor 16 [R 16]: Wide Fixed-bed Reactor with Retainer Structures, Pressure-drop Channels and Bifurcation-cascade Feed/Withdrawal

This concept, also termed cross-flow reactor, integrates multiple, short packed beds to one entity, a continuous, wide packed bed [7, 77, 78]. This is done for reasons of increasing the effective catalyst area and enlarging the throughput. The same residence time as an axial-flow reactor is claimed, but at larger throughput or smaller pressure drop, respectively. Owing to the short contact time, the reactor is amenable to differential operation (low conversions), which is helpful for investigating kinetics.

A bifurcation cascade with micro channels feeds a wide fixed bed (channel void space for particle insertion), followed by a multitude of catalyst retainers, which act like frits, i.e. support the catalyst particles and prevent their loss [7, 77, 78]. Besides supporting the particles, these parts have a size-exclusion function to the lower size limit of about 35–40 µm. The retainers are followed by an array of elongated channels that serve to build up a uniform pressure drop along the wide retainer bed. Finally, the streams are collected in a bifurcation cascade of identical shape as the feeding cascade, but mirror-imaged in position.

The pressure drop in the bifurcation channels is much larger than any other contribution of the whole device, i.e. exceeds the fixed-bed share by far [7, 77, 78]. Hence small deviations therein, e.g. due to different packing or particles of varying size, do not contribute to changing the residence time at one location.

Four side wells for thermocouples are designed to achieve easily thermal equilibrium in the reactor, made of the highly conductive material silicon [7, 77, 78].

One way to fabricate such a reactor is by deep reactive ion etching (DRIE) with a time-multiplexed inductively coupled plasma etcher (most details on fabrication are given in [77]) [7, 77, 78]. Regions of major importance such as the retainers are etched through to avoid differences in structural depth which may cause uneven flow. To generate various channel depths in one design, both front-side and back-

Reactor type	Wide fxed-bed reactor	Pressure-drop channels: number, width, depth	256; 40 µm; 20–25 µm
Catalyst material	Pd/Al$_2$O$_3$; Pt/Al$_2$O$_3$; Rh/Al$_2$O$_3$	Meander net centerline length	~2.2 mm
Catalyst particle diameter	70–100 µm	Silicon wafer: thickness; diameter	500 µm; 100 mm
Feed (bifurcate) channels: width, depth; number	350 µm; 370 µm; 64	Outer device dimensions	15 mm × 40 mm × 1.5 mm
Catalyst bed: width; depth; length; volume	25.55 mm; 500 µm; 400 µm; 5.1 µl	Operating temperature	550 °C (with Pyrex cover); 1000 °C (with Si cover)
Retainer posts: width	50 µm	Operating pressure	0.14 MPa

side etching are performed. Such prepared channels are capped on top and bottom with Pyrex wafers by anodic bonding. Holes are drilled for inlet and outlet flow guiding. A metal cover plate and thin elastomer gaskets are used for housing. Catalyst particles of 50–70 μm diameter are fed through an inlet port using a vacuum applied at the outlet. By applying high pressure, all particles are blown out from the device.

A number of experiments and finite-element simulations were done to confirm even flow distribution, uniform pressure drop and isobaric properties and also to analyse quantitatively mass and thermal transfer for the wide packed-bed reactor [78].

3.2.3.2 Reactor 17 [R 17]: Mini Packed-bed Reactor

The central element of this reactor is an elongated channel in which small catalyst particles can be filled to give a mini-packed bed (Figure 3.20) [79, 80]. Gas streams enter this reaction zone as a mixture via an interleaved channel section, which also prevents the small particles penetrating the gas-feed channels. A similar type of microstructured 'frit' is placed at the end of the packed bed for the same function. Next to the inlet channels on the right and the left catalyst-loading channels are placed to insert suspensions with catalyst particles (by applying a vacuum at the exit). Thermocouple wells serve for temperature monitoring.

The structures were etched using a time-multiplexed inductively coupled plasma etch (Figure 3.21). On the back side of such a structured silicon wafer holes were

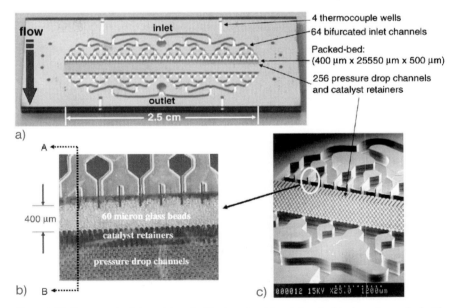

Figure 3.20 Photograph of the mini packed-bed reactor (a); mini packed-bed reactor packed with 50 μm glass beads (b); detailed SEM image (c) [80].

a. Wafer with deposited oxide

b. Pattern oxide

c.DRIE- packed bed & ports

d. Pattern 10μm photoresist

e. DRIE- p. drop channels masked

f. Strip resist and DRIE ~ 20μm

g. Strip oxide and anodic bond Pyrex

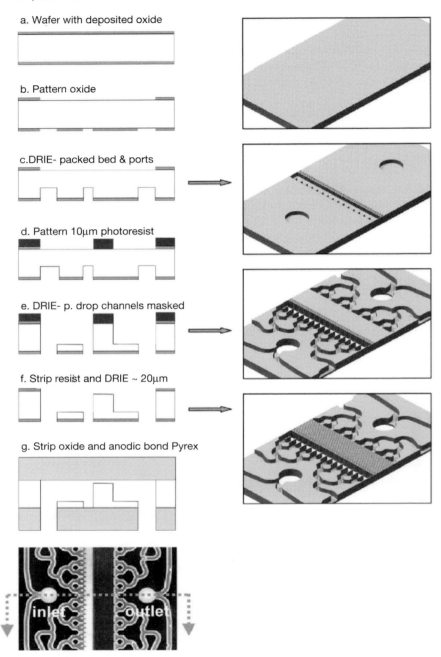

inlet   outlet

Figure 3.21 Schematic of the fabrication sequence for the mini packed-bed reactor [80].

etched for inlet and outlet flow guiding. A 500 nm silicon dioxide layer, made in a wet-etching furnace, protected the silicon from being etched by chlorine. By anodic bonding, the device was capped with a Pyrex cover. Such bonded wafer stacks were cut with a die saw to obtain individual pieces.

Reactor type	Mini packed bed-reactor	Catalyst-loading channels: width	400 µm
Catalyst material	Activated carbon	Post channels at outlet: width	25 µm
Catalyst particle diameter	53–73 µm	Silicon wafer: thickness, diameter	500 µm; 100 mm
Catalyst surface area	850 m² g⁻¹	Outer device dimensions	10 mm × 40 mm × 1.0 mm
Reaction channel: width; depth; length; volume	625 µm; 300 µm; 20 mm; 3.75 µl	Operating temperature	25–200 °C
Interleaved channels at inlet: width	25 µm	Operating pressure	1.40 atm

3.2.4
Thin-wire and µGauze Micro Reactors

These micro reactors do not have real micro channels, but rather have holes, in the case of the µgauze, or create microflow conduits, by placing a thin wire in a micro channel. Especially the first concept is derived from laboratory and industrial-scale processing of extremely fast reactions.

3.2.4.1 Reactor 18 [R 18]: Modular Integrated 3D System with Electrically Heated µGauze
This reactor has a rather complex construction (Figure 3.22) concerning the number of microstructured parts, their integration, tight arrangement, the variety of materials and the numerous surfaces that had to be tightened [2, 41, 81]. Since it was built from reversibly sealed parts, the system is modular, e.g. easily allowing the insertion of a different catalyst material or another micro heat exchange unit. The reactor is not a component, but an integrated system as it contains three micro-structured components in one assembly, namely a pre-heater with microstructured outlets, a microstructured gauze and a micro heat exchanger.

The pre-heater is a massive stainless-steel block with three bores that guide and heat the gases. This block is heated via three heating cartridges. At the end of the block three outlet bores were made by µEDM drilling. A small chamber serves for mixing the gases and the outlet holes prevent explosions or flames from moving upstream (flame-arrestor effect). Below the mixing chamber, a metallic strip with a micro hole array in the center is positioned. This strip is completely made from the catalytic material via shaping a foil by thin-wire µEDM. The micro holes are drilled

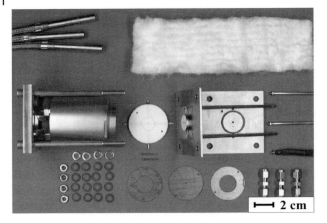

Figure 3.22 Photograph of the dismantled reactor with all parts [2].

by μEDM. The strip has electrical connectors at both ends so that resistance heating can be applied. The center of the strip, i.e. the hole array, has the highest resistance, and hence becomes hottest until glowing is achieved at very high temperatures. At the center of the strip, two smaller lines are attached that serve for temperature monitoring via measuring the thermal change of electrical resistance.

The micro hole array is an arrangement similar to monoliths and particularly to gauzes employed for the same purposes, and hence is termed μgauze in the following. The μgauze strip is inserted in a structured ceramic frame that contains a recess for the strip. Embedded silver and metal solder rods serve for electrical connection via the ceramic material (Figure 3.23).

After leaving the hot catalyst zone, the product gas enters a small chamber made in a plate and is distributed into four micro heat exchange channels of cross-flow type arrangement. Each micro channel is surrounded by two other channels that guide the cooling gas. In addition, the micro heat exchanger plate contains guidance for cooling liquid flow.

Figure 3.23 Photograph of the catalyst structure inserted in the ceramic support. Close-up of the holes in the platinum catalyst strip [63].

Reactor type	Modular integrated system with µgauze catalyst	Operating temperature	800–970 °C
Pre-heater (with micro mixer) material	Stainless steel	Operating pressure	1 bar
µGauze material; µGauze holder material	Platinum; Macor™ ceramic	Micro hole in mixer: diameter	60 µm
Micro heat exchanger material	Stainless steel	Micro hole in µgauze: diameter; depth	70 µm; 250 µm
Heating	Electrical resistance heating within the µgauze	Product gas channel width; cooling gas channel width	60 µm; 90 µm

3.2.4.2 Reactor 19 [R 19]: Catalyst-wire-in-channel Micro Reactor

A simple, but efficient reactor concept was developed based on the insertion of metallic wires that serve as a catalyst into a micro channel. The wire extends over the channel length and can thus be contacted electrically for heating purposes. It is sealed by graphite seals at both reactor ends. In this way, an easy, flexible and cheap concept for catalyst exchange and reactor assembly is provided.

The catalyst wires are inserted either in silicon micro channels, obtained by wet-chemical etching (Figure 3.24) [82], or in commercially available quartz glass tubes with internal dimensions in the sub-millimeter range [9]. The silicon plates are contained in a stainless-steel housing made by µEDM. The quartz glass tubes are mounted in a hollow cylindrical housing made of a machine-workable ceramic. Two separate feed holes at the reactor inlet serve for perpendicular introduction of two gas streams (like a mixing tee) which mix by impinging. An inspection window in the ceramic housing permits optical monitoring, e.g. for documenting glowing of the gas mixture.

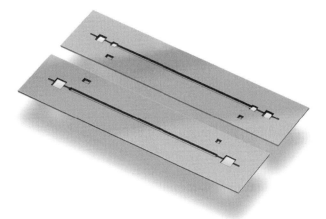

Figure 3.24 Image of the catalyst-wire-in-channel micro reactor [82].

Reactor type	Catalyst-wire-in-channel	Housing a: diameter; thickness	50 mm; 7.5 mm
Catalyst material	Platinum	Housing b: diameter; length	22 mm; 40 mm
Channel carrier a, b	Microstructured silicon platelet; quartz glass tube	Reaction channel a: cross-sectional area; depth; length	0.167 mm²; 525 μm; 20 mm
Housing material a, b	Stainless-steel plates; ceramic hollow cylinder	Tube b: internal diameter; external diameter; length	600 μm; 6 mm; 20 mm
Operating temperature	1150 °C	Wire diameter	150 μm
Operating pressure	1 bar		

3.2.5
Thin-membrane Micro Reactors

Membrane reactors are known on the macro scale for combining reaction and separation, with additional profits for the whole process as compared with the same separate functions. Microstructured reactors with permeable membranes are used in the same way, e.g. to increase conversion above the equilibrium limit of sole reaction [8, 10, 11, 83]. One way to achieve this is by preparing thin membranes over the pores of a mesh, e.g. by thin-film deposition techniques, separating reactant and product streams [11].

3.2.5.1 Reactor 20 [R 20]: Permeable-separation Membrane Chip Reactor

The membrane in this device serves to control the flux of one gas penetrating another gas [11]. The membrane separates two gas streams while being permeable for one of the gases. By adjusting membrane thickness, gas pressure drop over the membrane and temperature, the gas flow through the membrane can be adjusted. This serves to remove the product from the reaction mixture to enhance the yield, prevents undesired complete mixing of gases, e.g. when operating in the explosive regime, allows the use of gas mixtures when only one component thereof is needed (e.g. as for syngas) or can enhance selectivity.

A special version of the membrane reactor using Pd was made for separating hydrogen and oxygen and their controlled reaction.

The design of the Pd-membrane reactor was based on the chip design of reactor *[R 10]*. The membrane is a composite of three layers, silicon nitride, silicon oxide and palladium. The first two layers are perforated and function as structural support for the latter. They serve also for electrical insulation of the Pd film from the integrated temperature-sensing and heater element. The latter is needed to set the temperature as one parameter that determines the hydrogen flow.

Microfabrication of the silicon part of the device is done by processing a silicon wafer with LPCVD and other thin-film techniques, standard photolithography, dry and wet etching. Temperature resistors are obtained by e-beam deposition and a standard lift-off process. By special etching, very thin Pd membranes can be made. One channel is made by KOH etching (in the framework of the above-mentioned technology steps), the other by a molding process using an PDMS mold and an epoxy cast.

Reactor type	Membrane chip reactor	Membrane width	700 μm
Catalyst material	Palladium	Operating temperature	500 °C
Chip materials	Silicon; PDMS	Operating pressure	5 bar
Channel length	12 mm	Membrane width	700 μm
Pd film thickness	A few tens of μm		

3.2.6
Micro Reactors without Micro Channel Guidance – Alternative Concepts

Micro-flow processing is not an exclusive domain of micro-channel devices made by microfabrication. This approach can be applied to any packing of regular-shaped objects which results in interstices of the same internal dimensions and the same precision as given for micro channels. Obviously, interstices made from extended, but thin objects resemble best the nature of micro channels. Hence the use of filaments for constituting a micro-flow assembly was recently described [8].

It is to be expected that in the near future more of such concepts will find application, simply for cost reasons. Laboratory-scale investigations with precisely microfabricated reactors in advance of the use of such devices can give valuable information, providing a best-case scenario. From then, one can look for alternative micro-flow solutions of lower cost, higher reliability, higher flexibility and so on.

3.2.6.1 Reactor 21 [R 21]: Filamentous Catalytic-bed Membrane Reactor
This is the first reactor reported where the aim was to form micro-channel-like conduits not by employing microfabrication, but rather using the void space of structured packing from smart, precise-sized conventional materials such as filaments (Figure 3.25). In this way, a structured catalytic packing was made from filaments of 3–10 μm size [8]. The inner diameter of the void space between such filaments lies in the range of typical micro channels, so ensuring laminar flow, a narrow residence time distribution and efficient mass transfer.

In addition, the filament reactor can contain a membrane-separation function by grouping threads of filaments around an inner empty reactor core, that guides the permeate and may also increase permeation by reaction. Thus, the tube reactor constructed in such a way comprises two concentric zones, separated by a permeable Pd/Ag alloy membrane in the form of a tube. The reaction takes place in the filament zone. One product such as hydrogen is removed via the membrane and

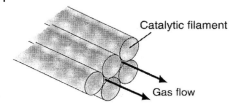

Catalytic filament

Gas flow

Figure 3.25 Schematic of flow guidance in a packed filament reactor [8].

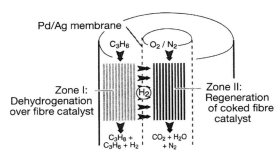

Pd/Ag membrane

C_3H_8 O_2 / N_2

Zone I:
Dehydrogenation
over fibre catalyst

H_2

Zone II:
Regeneration
of coked fibre
catalyst

$C_3H_8 +$ $CO_2 + H_2O$
$C_3H_6 + H_2$ $+ N_2$

Figure 3.26 Schematic of a membrane reactor packed with filamentous catalyst [8].

reacted in the inner zone. The heat generated in such a way is needed when conducting endothermic processes in the outer zone. Also, a shift from equilibrium can result from this. The coke formed at the catalyst by the reaction can be removed by burning with air.

Reactor type	Filament-bed membrane reactor	Porosity	0.8
Catalyst material	Pt/Sn on alumina/silica filament	Specific surface area	$108 \ m^2 \ m^{-3}$
Filament material	Aluminum borosilicate glass fibers	Membrane material	Pd/Ag (23 wt.-%)
Specific surface area of fibers	$2 \ m^2 \ g^{-1}$ initially; $290 \ m^2 \ g^{-1}$ after etching	Membrane tube diameter, thickness	6 mm; 70 µm
Thread	100 bundles of 0.5 mm diameter	Quartz tube diameter, length	8.6 mm; 140 mm
Catalyst carrier	γ-alumina ($100–230 \ m^2 \ g^{-1}$)	Operating temperature	550 °C
Catalyst	Pt/Sn	Operating pressure	0.14 MPa

3.2.6.2 Reactor 22 [R 22]: Various Other Reactor Designs

All other reactions designs are included in this category. Most of them comprise one multi-channel platelet embedded in a housing.

3.3
Oxidations

Oxidations and partial oxidations constitute one of the most important classes of gas-phase reactions. For this reason, the drivers for and benefits when undergoing micro-channel investigations of these reactions are discussed here in advance of the class as whole. In addition, the specific drivers and benefits are discussed for the individual oxidation reactions as throughout all chapters of this book with regard to reactions.

3.3.1
Drivers for Performing Oxidations in Micro Reactors

The target or value product of many oxidations is only partially oxidized, whereas complete oxidation leads to the useless carbon dioxide. In terms of elementary reactions, partial and total oxidation can be a series of reactions, rendering the latter a consecutive process. In other cases, total oxidation is a side reaction to the partial oxidation. Cases are even known where total oxidations are both side and consecutive reactions concurrently. As a consequence, controls over residence time and reaction temperature are key points to enhance the selectivity of partial oxidation. Concerning the latter, one has to be aware of the large reaction enthalpies that result from total oxidation. Conventionally, owing to insufficient heat transfer, this can lead to considerable hot spots in the reaction zone, which in turn favor total oxidation as the unselective route. In ethylene oxide synthesis, for example, the reaction enthalpy of the total oxidation to carbon dioxide is more than 10 times larger than that of the partial oxidation which locally induces very hot temperatures (hot spots) with corresponding negative consequences for the reaction course.

Micro heat exchangers and also any kind of micro channel devices, heated or cooled externally, offer considerably improved heat transfer owing to their large internal specific surface areas. Hence they offer unique possibilities to steer oxidations to increased selectivity of the partial-oxidation products.

A further driver for performing oxidations in micro reactors is the hazardous nature of oxidation reactions when using high oxygen contents or even pure oxygen. Flame formation or even explosions may result. As a result, much effort in industry is aimed at the finding of explosive regimes and the installation of security measures to prevent accidents. Usually, this also implies declaring explosive regimes to be 'forbidden zones' for further investigations. This means restricting chemistry to certain process windows and the impossibility of exploiting the full range of process parameters. For some gas-phase processes, such as the oxihydrogen reaction, this basically leads to a more or less total neglect of this process route, at least when facing practical issues concerning throughput, etc.

It has been shown, particularly for the latter reaction and for the ethylene oxide process, that micro reactors allow safe processing of otherwise hazardous oxidations [4, 26, 40, 42, 43, 84]. This is first due to the fact that the inner volume of micro reactors is small so that explosions also 'happen only on a micro scale'. The associated damage, if can be named such at all, is most often negligible. In a number of cases, it is simply the overcompensation of the large heat release by the good heat transfer properties which make micro reactors safe. Even more referring to an intrinsic 'micro reactor property', the large internal specific surface areas are said to actively change the reaction mechanism in the case of radical-chain-type explosions. Here, chain propagation is 'quenched' by wall collision and oxidations can be performed somehow in a completely new chemical way, with results different from those that were obtained before – with risk to equipment and sometimes even life! Veser [9] refers to this aspect of chemical micro processing as 'intrinsic safety'.

A general further driver for performing oxidations in micro reactors is the possibility of having fast dynamic changes of process parameters such as concentration, temperature, pressure, partial pressure, etc. [12, 13, 27, 85]. Micro reactors are continuously processed tools with an extremely small internal volume that favors fast dynamic changes, at least when dealing with single-channel or single-plate devices, which are perfect measuring tools. Hence the gathering of precise process information in a fast mode is a further benefit of micro-reactor studies. Such detailed investigations have been reported, e.g., for ethylene oxide formation, oxidation with ammonia or syngas generation.

As far as extinction/ignition behavior is concerned, oxidations in micro reactors can exhibit varied temperature profiles [19, 56, 57, 59–61]. As a consequence of their very distinct heat transfer characteristics, micro reactors can allow autothermal operation at a different temperature level compared with processing in conventional reactors. As an example, this may raise the selectivity of value products.

3.3.2
Beneficial Micro Reactor Properties for Oxidations

As indicated above, it has been demonstrated many times that the small reaction volumes in micro reactors and the large specific surface areas created allow one to cope with the release of the large amounts of heat. Knowledge of heat transfer characteristics seems to be a top priority when designing a micro reactor for oxidations.

The precise and, where needed, short setting of the residence time allows one to process oxidations at the kinetic limits. The residence time distributions are identical within various parallel micro channels in an array, at least in an ideal case. A further aspect relates to the flow profile within one micro channel. So far, work has only been aimed at the interplay between axial and radial dispersion and its consequences on the flow profile, i.e. changing from parabolic to more plug type. This effect waits to be further exploited.

Particularly valuable for the viable nature of oxidations is the flame-arrestor effect of micro reactors affecting radical-chain propagation.

3.3.3
Oxidation of Ammonia

Peer-reviewed journals: [28, 61]; proceedings: [19, 56, 57, 59, 60, 98]; sections in reviews: [58, 86–97].

3.3.3.1 Drivers for Performing the Oxidation of Ammonia

The oxidation of ammonia, first of all, is a thoroughly investigated gas-phase process and hence an excellent reference reaction for benchmarking chemical micro processing results [19, 56–62, 75, 86–98]. All the reactions for most conditions are known at atmospheric or low pressure (see original citations in [98]). A detailed mechanism based on a large number of elementary reactions was proposed. For many of these reactions, the rate constants are also known, although for some still large uncertainties exist. After adsorption, ammonia and oxygen form several active adsorbed species on the Pt catalyst surface at different adsorption sites and react via them to give the products.

The short contact times required, typically being considerably below, also favor operation in micro channels. Hence high reaction rates and high exothermicity are characteristic of the oxidation of ammonia.

Typical products of the oxidation of ammonia in micro reactors are dinitrogen oxide (N_2O), nitrogen oxide (NO) and nitrogen (N_2), depending on the process conditions. Dinitrogen oxide is a selective oxidizing agent for attractive syntheses e.g. to phenol from benzene via zeolite catalysts [28, 98]. Known conventional syntheses to NO_2, such as the thermal decomposition of ammonium nitrate or by biochemical means via nitrite reduction, are economically not acceptable. Instead, the autothermal oxidation of ammonia at very short contact times and low temperatures, necessitating chemical micro processing, is regarded as a promising alternative [28, 98].

Oxidations of ammonia display ignition/extinction characteristics and autothermal reaction behavior. At low heat supply, only low conversion is observed and temperature remains nearly constant. With increasing heat supply and approaching a certain temperature, the reaction heat generated can no longer be transferred completely totally to the reactor construction material. At this stage, the reaction 'starts up'. Suddenly, the temperature is raised by increased heat production until heat generation and removal are in balance. The reaction can now be carried out without a need for external heat supply, namely in autothermal mode.

3.3.3.2 Beneficial Micro Reactor Properties for the Oxidation of Ammonia

The investigations refer to the general capability of micro reactors to perform short-time processing with highy intensified mass and heat transfer. A special focus of most investigations on the oxidation of ammonia was the heat management. The use of new concepts for heat supply and removal opens the door to operation in new process regimes with very different product spectra.

Gas-phase reaction 1 [GP 1]: Oxidation of ammonia

$$NH_3 \xrightarrow[-H_2O]{O_2/Cat.} N_2O + (NO + N_2)$$

3.3.3.3 Typical Results

Conversion/selectivity/yield

[GP 1] [R 1] With an optimized diverse micro reactor design (for *[R 1]*) an N_2O selectivity up to 50% was reached at complete conversion in the temperature range from 300 to 380 °C (0.022–0.543 mg Pt on Al_2O_3; 6 vol.-% NH_3, 88 vol.-% O_2, balance He; 600–4430 cm^3 min^{-1} (STP); 260–380 °C) [98].

Conversion rates

[GP 1] [R 10] Conversion rates over platinum catalyst were determined in a chip-based reactor [56]. In the temperature range from 180 to 310 °C rates from 0 to 0.5 mol m^{-2} s^{-1} were found.

Residence time

[GP 1] [R 1] N_2O selectivity increased with increase in oxygen content, even in the presence of a large oxygen surplus (Pt auf Al_2O_3; 6 vol.-% NH_3, 40 or 88 vol.-% O_2, balance He; 260–380 °C) [98]. Normally, one would have expected a decrease due to follow-up oxidation of N_2O owing to the large excess of oxygen. It is assumed that the special modular design with closely linked furnace and cooler results in efficient cooling. In this way, the residence time at elevated temperature can be reduced to the kinetic needs and especially thermal follow-up reactions can be reduced, as there are no hot parts after the reaction zone any more.

Catalyst thermal behavior/reactor material

[GP 1] [R 1] A change from aluminum to platinum as construction material results in reduced micro-reactor performance concerning oxidation of ammonia, decreasing N_2O selectivity by 20% [28]. This is explained by the lower thermal conductivity of platinum, which causes larger temperature differences (hot spots) within the micro channels, i.e. at the catalyst site, e.g. due to insufficient heat removal from the channels or also by non-uniform temperature spread of the furnace heating.

Pt loading

[GP 1] [R 1] A comparison of four micro reactors (see Table 3.1) with different Pt loadings (Pt impregnated on an anodically oxidized alumina support) confirmed that higher conversions were obtained at higher Pt loadings (6 vol.-% NH_3, 88 vol.-% O_2, balance He; 600–4430 cm^3 min^{-1} (STP); 260–380 °C) [28, 98]. At near complete conversion, 48% N_2O selectivity was found (Figure 3.27).

Table 3.1 Characteristics of platinum catalysts used for experimental investigations [28].

Micro reactor	Pt mass (mg)	Al_2O_3 mass (mg)	Pt loading (wt.-%)	Pt dispersion (%)	Mean Pt cluster size (Å)
A1	0.022	44	0.05	100	< 9
A2	0.086	44	0.2	96	9
A3	0.560	44	1.3	48	19
B	microstructured solid Pt				
C	0.543	16	3.5	40	23

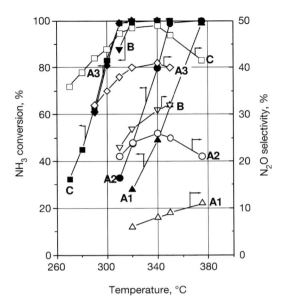

Figure 3.27 Conversion of NH$_3$ (open symbols) and selectivity to N$_2$O (closed symbols) for the ammonia oxidation process on Pt catalyst. Micro reactors A1 (▲), A2 (●), A3 (◆), B (▼) and C (■) were used (see Table 3.1) [98].

Pt cluster size

[GP 1] [R 1] A comparison of four micro reactors with different Pt loadings (Pt impregnated on anodically oxidized alumina support) and different Pt structures confirmed that cluster size has an impact on the single Pt-atom activity (6 vol.-% NH$_3$, 88 vol.-% O$_2$, balance He; 0.51 ms; 260–380 °C) [28, 98]. At low Pt loadings, isolated atoms are formed. Calculated ammonia consumption rates amount to 20 s^{-1} at 300 °C. At high Pt loadings, clusters are formed. Turn over frequencies (TOF) of about 40 s^{-1} are determined.

Steerable catalysis

[GP 1] [R 13] So-called micro-strip electrodes (MSE) can act as electrically steerable catalysts when used to switch on and off the conversion of ammonia at moderate voltages, several hundred volts (6 vol.-% NH$_3$, 88 vol.-% O$_2$, balance He; 0.51 ms; 260–380 °C) [75]. Thereby, NO formation was observed. By emitting and accelerating electrons in the range of mA cm^{-2} current density from the solid to the gas phase, radicals were formed, typically much more than the number of released electrons, e.g. 10 radicals per electron. This efficient use of energy is referred to as *dynamic catalysis*. The gas phase near the electrodes contains hot and cold radicals, thus providing a two-temperature system.

Time steps of about 60 s between switching on and off were realized. The performance of the MSE structures depends on their geometry, the electric field strength and the gas pressure.

Ignition/extinction behavior

[GP 1] [R 10] The ignition/extinction behavior of the oxidation of ammonia at three different membranes – 1 µm SiN; 1.5 µm Si and 2.6 µm Si – in an Si-chip micro reactor was compared [19]. Whereas the first membrane exhibits ignition already at low power input leading to a temperature rise of up to 570 °C and from there autothermal operation up to 500 °C, the latter membrane shows the opposite behavior. For a large power input range of the heating elements only a smooth increase in temperature was observed, followed by only a further small increase by ignition (from ca 270 °C to ca 330 °C) [19, 56]. An ignition/extinction hysteresis is hardly developed, which is very different from conventional reactors. The 1.5 µm Si membrane gives intermediate behavior, namely a fast increase up to only ca 520 °C, but with pronounced hysteresis formation.

Deformation of SiN membranes during ignition/extinction cycles up to membrane rupture have been reported [57]; images of membrane shapes during the ignition sequence were given. For the SiN membrane, ignition quickly travels upstream and stabilizes at the entrance region of the heater. When the power was decreased, the ignited area was reduced, first downstream.

Process temperature

[GP 1] [R 10] While temperature is more or less fixed in conventional reactors, a set of differently constructed micro reactors now enables temperature to be changed for the ammonia oxidation [19]. As a consequence of this variation of processing temperature by a change of micro-reactor construction, the selectivity of exothermic high-temperature reactions can be changed. For the oxidation of ammonia, the NO/N_2 selectivity was chosen as a sensitive parameter which is increased on raising the reaction temperature. Operation in the chip micro reactor in the range

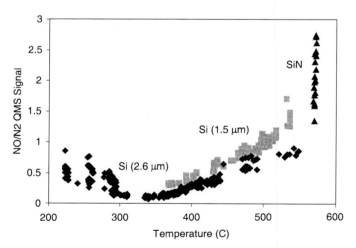

Figure 3.28 Unexpected increase in NO/N_2 selectivity for ammonia oxidation reaction in a micro membrane reactor [19].

from 340 to 570 °C resulted in an increase of the NO/N$_2$ selectivity from 0.5 to approximately 2.5. Maybe even more remarkable, the micro reactor allowed investigations even below 340 °C, which was not possible so far in conventional reactors. Hence a formerly 'hidden process window' was opened. Indeed, the information gained was surprising. In the low-temperature range an unexpected increase in NO/N$_2$ selectivity was found (Figure 3.28). This is an indication of a change in reaction mechanism [19].

Temperature distribution in reactor wall

[GP 1] [R 10] By proper heater design, membrane-based reactors with internal heaters allowed one to reach quasi-uniform temperatures at the membrane, which determines the catalyst temperature [19]. This thermal uniformity was checked during reaction, i.e. when large heats were released in the oxidation of ammonia and needed to be transferred out of the reaction zone (Figure 3.29). Thin-film-coated temperature sensors in the center and at the edges of the membrane served to monitor the lateral temperature difference.

When carrying out the reaction at an SiN membrane (1 µm thick) by increasing the power input, the temperature difference follows the ignition/extinction loop, i.e. relatively large temperature differences (maximum 14 K) between the membrane center and edge occur. In the case of the better conducting Si (2.6 µm) membrane, the reaction heat is transferred to the outside more effectively and hence hardly exhibits an ignition/extinction loop; the corresponding maximal temperature difference amounts to 4 K. Uniform catalyst temperatures can thus be achieved even for highly exothermic reactions (Figure 3.30).

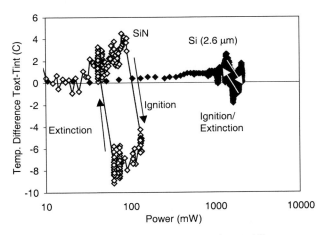

Figure 3.29 Ammonia oxidation over a Pt catalyst in different membrane micro reactors. Experimental results show good temperature uniformity across the catalyst regions [19].

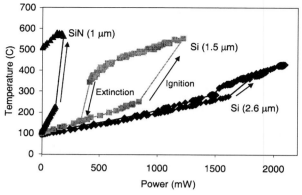

Figure 3.30 Ignition/extinction loops for ammonia oxidation over platinum performed in micro reactors with different membranes [19].

Simulation and modeling

[GP 1] [R 1] A kinetic model for the oxidation of ammonia was coupled to a hydrodynamic description and analysis of heat evolution [98]. Via regression analysis and adjustment to experimental data, reaction parameters were derived which allow a quantitative description of reaction rates and selectivity for all products under equilibrium conditions. The predictions of the model fit experimentally derived data well.

[GP 1] [R 1] Numerical simulations prove that isothermal processing is possible in micro reactors even under severe reaction conditions which correspond to an adiabatic temperature rise up to 1400 °C [98].

Proposals for exploitation of the results on the model reaction

Investigations with the modular multi-channel [28, 98] and silicon chip [19, 56–62] micro reactors demonstrate that by exact temperature control the oxidation of ammonia can be run with increased and deliberately steered selectivity. A major application is provided by carrying out former high-temperature reactions in the low-temperature regime. In the case of ammonia oxidation in the chip micro reactor, the yield of the value product NO was actually lower in that regime. In the case of the multi-plate-stack micro reactor, higher yields of the value product NO_2 were achieved.

However, ammonia oxidation and its products are only seen as models for other high-temperature reactions and substances with potentially more industrial impact when being carried out in a micro reactor. In this context, explicitly mentioned are syntheses of important value products such as anhydrides, ketones or alcohols (see especially [57], but also [19, 56]). Moreover, the work on oxidation of ammonia showed that it is possible to expand the operational regime for laboratory-scale investigations to gather information, e.g., on new product compositions and new process parameters. The feasibility being demonstrated, we are now waiting for more detailed investigations and for transfer to other, hopefully even more important, reactions. In this way, micro reactors may become a widely accepted tool in the field of kinetic analysis of heterogeneous catalysis.

3.3.4
Oxidation of Ethylene – Ethylene Oxide Formation

Peer-reviewed journals: [4, 40]; proceedings: [26, 42, 43, 84]; reactor description: [44, 99]; sections in reviews: [87, 90, 95–97, 100].

3.3.4.1 Drivers for Performing Ethylene Oxide Formation

Heat management is of crucial importance for ethylene oxide synthesis (see original citations in [4]). The reaction enthalpy of the total oxidation to carbon dioxide ($\Delta H = -1327$ kJ/mol) is more than 10 times larger than that of the partial oxidation ($\Delta H = -105$ kJ/mol), which induces locally very hot temperatures (hot spots) with corresponding negative consequences on the reaction course.

Industrial reactors work below or above the explosion regime, hence operation cannot be carried out at any ethylene-to-oxygen ratio (see original citations in [43]). The best industrial processes are based on ethylene and oxygen contents of 20–40 and 8%, respectively, using methane as inert gas. It is known that high oxygen contents improve selectivity.

Without promoters, a selectivity of up to about 50–70% is reported (see original citations in [43]); 90% of heat production is due to total oxidation. Advanced industrial reactors have a selectivity of up to 90% and at the lowest a 50% contribution of the total oxidation to the heat generation.

For pure silver catalysts, a selectivity of only 40% was reported (see original citations in [4]).

3.3.4.2 Beneficial Micro Reactor Properties for Ethylene Oxide Formation

Beneficial micro reactor properties mainly refer to improving heat management as a key for obtaining a partial reaction in a consecutive sequence, when large heats are released by reaction steps other than the partial one (see also Section 3.3.1).

Special attention was also paid to the search for operation in the explosive regime as micro reactors are said to have much greater safety here. In this way, improvements in terms of space–time yield were expected.

Gas-phase reaction 2 [GP 2]: Oxidation of ethylene to ethylene oxide

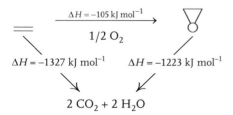

$\Delta H = -105$ kJ mol^{-1}

$1/2\ O_2$

$\Delta H = -1327$ kJ mol^{-1} $\Delta H = -1223$ kJ mol^{-1}

$2\ CO_2 + 2\ H_2O$

Ethylene oxide synthesis is one of the largest-volume industrial processes with a production rate of some plants of several 100 000 t a^{-1} (see original citations in [4]). In 1995, the world capacity for ethylene oxide was approximately 11 200 000 t a^{-1}. As industrial catalyst silver on alumina is employed. In addition to large production, small-throughput and mobile applications are known. This includes the on-site generation of ethylene oxide for sterilizing purposes, e.g. for cleaning in hospitals. Corresponding small-scale production units are on the market. Ethylene oxide synthesis is, besides its industrial application, a well-studied process. Kinetic data are available in the literature.

3.3.4.3 Typical Results
Start-up behavior
[GP 2] [R 2] New microstructured silver platelets have no initial activity for ethylene oxide formation [26, 40]. After treatment with the OAOR process, a small increase in activity was observed. After 1000 h of operation under oxygen conditions, larger amounts of ethylene oxide were produced.

[GP 2] [R 3a] Initially sputtered silver on aluminum platelets displays high selectivity at low activity. This ratio decreases until a constant performance is reached, typically after 24–48 h (20 vol.-% ethylene, 80 vol.-% oxygen; 3 bar; 0.23–2 s; 250 °C) [43].

[GP 2] [R 3a] Catalysts need to be initially activated on-stream with a mixture of 20% ethylene and 20% oxygen in methane as balance [44]. The temperature was raised until first formation of carbon dioxide became notable. The initial selectivity is close to 70% and after time-in-stream for 1 day at 250 °C decreases to 62% at 1.3% conversion. This loss in selectivity at the expense of conversion is a general phenomenon during all investigations conducted in [44]. Non-promoted catalysts show a certain decrease in selectivity within a few days, particularly at high temperature and conversion.

Conversion/selectivity/yield
[GP 2] [R 2] Initial work with a steel multi-plate-stack micro reactor with microstructured platelets made of polycrystalline silver (OAOR modified) confirmed a selectivity of up to 49% at conversions of about 12% (3 vol.-% ethylene, 50 vol.-% oxygen, balance nitrogen; 5 bar; 4 l h^{-1}; 277 °C) [4, 26, 40].

In a more comprehensive follow-up work, the selectivity on OAOR-modified silver could be raised to 65%, still without the presence of promoters such as 1,2-dichloroethane [4]. This value is by far better than most values known in the literature for the same catalyst. The best value finally obtained was 69% and approaches the 'industrial limit' of 80% that was obtained with promoters and a different, better catalyst, Ag/Al$_2$O$_3$. A similar catalyst type (Aluchrom catalyst) was also tested in the micro reactor, but so far yielding lower results, the best selectivity measured being 58%.

Variation of gas composition affects conversion on OAOR-modified silver. At a selectivity of 47%, a conversion exceeding 24% was observed (0.75 vol.-% ethylene, 33 vol.-% oxygen, balance nitrogen; 5 bar; 2 l h^{-1}; 277 °C) [4, 26, 40].

[GP 2] [R 3a] A nearly constant selectivity of up of about 60% at conversions ranging from 20 to 70% was determined for sputtered silver on anodically oxidized (porous) aluminum alloy (AlMg3) with two different ethylene loads (4 or 20 vol.-% ethylene, 80 or 96 vol.-% oxygen; 0.3 MPa; 230 °C) [44]. The highest yield

of 39% was obtained for the same catalyst at 4.5 s residence time (4 vol.-% ethyl-ene, 96 vol.-% oxygen; 0.3 MPa; 230 °C).

[GP 2] [R 3a] The selectivity–conversion behavior was determined for the com-mercial Shell Series 800 catalyst, in a fixed bed and electrophoretically deposited in micro channels (20 vol.-% ethylene, 80 vol.-% oxygen; 0.3 MPa; 230 °C) [101]; 54% selectivity at 17% conversion was found at the maximum, when processing with-out promoters.

Temperature and pressure
[GP 2] [R 2] On increasing temperature from 240 to 290 °C, reaction rates on OAOR-modified silver increase from $4.5 \cdot 10^{-5}$ mol s^{-1} m^{-2} to $1.2 \cdot 10^{-4}$ mol s^{-1} m^{-2} (3 vol.-% ethylene, 16.5 vol.-% oxygen, balance nitrogen; 5 bar; 0.124 s; 5 l h^{-1}) [4]. The selectivity decreases from 64 to 45%.

An increase from 2 to 5 bar total pressure increases the space–time yield by about 20% (15 vol.-% ethylene, 85 vol.-% oxygen, 2–20 bar; 0.235–3.350 s; 1 l h^{-1}) [4]. At higher pressures, 10 and 20 bar, a decrease activity is observed. Since industrial processes occur at up to 30 bar, at first sight this result is surprising. The decreas-ing activity with pressure was partially explained by catalyst deactivation, probably as a consequence of the longer residence times applied.

Isothermicity/radial temperature distribution
[GP 2] [R 2] The radial temperature distribution was determined by modeling, us-ing a worst-case scenario (5 Nl h^{-1}; stoichiometric mixture without inert; 100% conversion; 80% selectivity) [102]. The maximum radial temperature difference amounts to approximately 0.5 K. Thus, isothermal behavior in the radial direction can be diagnosed.

Periodic processing – cycle times and reactor performance
[GP 2] [R 2] A survey on the possibilities of periodic processing regarding to ethyl-ene oxide synthesis is given in [102]. The cycle times of periodic temperature and concentration control differed by 3–4 powers of ten. Cycle times of 0.05 s showed concentration effects, while times longer than 5 s can have temperature effects. The reason for this difference stems from the relatively high thermal inertia of the wall material influencing temperature changes.

The impact of such periodic processing on reactor performance was, however, slightly negative (assuming: 3 mol% ethylene; 20 mol% oxygen; balance nitrogen; 5 Nl h^{-1}; 10.63% conversion; 59.13% selectivity) [102]. In absolute terms conver-sion decreased by about 0.2% from 10.63%; hence in relative terms a decrease of 2% is given. Conversion decreased with increasing cycle time to level off at about 0.3 s. A small increase in conversion, but not reaching stationary performance, was observed above 100 s, due to temperature effects. For short cycle times, a small increase in selectivity of about 0.05% was found. For longer cycles, a performance worse than stationary was detected (Figure 3.31).

These mostly negative effects are explained as being due to the mechanistic char-acteristics of ethylene oxide formation, having a reaction order lower than one [102].

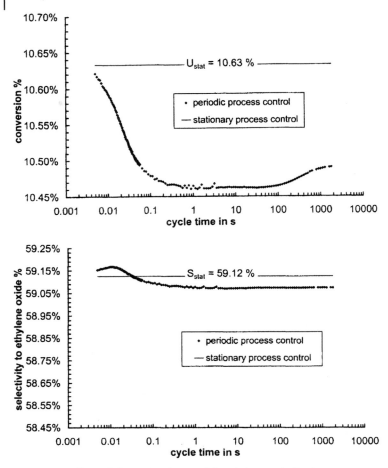

Figure 3.31 Effects of alternating change of the ethylene mole fraction. Conversion (left) and selectivity (right) as a function of cycle time [102].

It could be demonstrated that for any reaction with a reaction order higher than one, positive effects such as a conversion increase can be expected. Also, adaptation of the reactor geometry can improve the positive impact of periodic processing.

Residence time
[GP 2] [R 2] A residence time variation was performed under constant gas composition, temperature and pressure, but with varying flow rate. On increasing the residence time from 0.5 to 8 s, reaction rates on OAOR-modified silver decreased notably from $9.5 \cdot 10^{-5}$ mol s^{-1} m^{-2} to about $1 \cdot 10^{-4}$ mol s^{-1} m^{-2} (5 vol.-% ethylene, 50 vol.-% oxygen, balance nitrogen; 20 bar; 0.5–8 s) [4]. The selectivity decreases from 43 to 21%.

[GP 2] [R 3a] [R 4a] An investigation of sputtered silver (dense) on aluminum alloy (AlMg3) platelets and of sol–gel-Al_2O_3/silver (porous) on aluminum alloy (AlMg3) platelets served to show the impact of residence time [43]. At low conversion, the highest selectivity was found in both cases, which remained constant until 20% conversion. The dense silver had a maximum selectivity of 61%, exceeding the performance of porous sol–gel-Al_2O_3/silver (51%) (20 vol.-% ethylene, 80 vol.-% oxygen; 3 bar; 230 °C). For conversion exceeding 20%, a steep drop in selectivity was found.

[GP 2] [R 3a] The dependences of conversion, yield and selectivity on residence time were investigated in a comparative study with silver, sputtered silver (dense) on aluminum alloy (AlMg3) and sputtered silver on anodically oxidized (porous) aluminum alloy (AlMg3) as construction materials [44]. The ethylene content was varied, using 4% or 20%. Generally, conversion and yield increased with increasing residence time, whereas selectivity slightly decreased (after a short initial increase) or even remained the same. Proper operation using the construction material silver needed the longest residence times, e.g. up to 14.0 s to achieve 22% at selectivity of 48% (20 vol.-% ethylene, 80 vol.-% oxygen; 0.3 MPa; 250 °C). On lowering the ethylene content to 4%, conversion increases to 27% at a selectivity of 54% and 5 s residence time (4 vol.-% ethylene, 96 vol.-% oxygen; 0.3 MPa; 250 °C). Silver (dense) on aluminum alloy (AlMg3) gives a conversion of 43% at 58% selectivity and 5.5 s residence time (20 vol.-% ethylene, 80 vol.-% oxygen; 0.3 MPa; 250 °C). Sputtered silver on anodically oxidized (porous) aluminum alloy (AlMg3) gives a conversion of 33% at 60% selectivity and 7.0 s residence time (20 vol.-% ethylene, 80 vol.-% oxygen; 0.3 MPa; 230 °C). The performance is similar to that of the dense silver catalyst but at temperatures 20 °C lower. Both sputtered silver catalysts show nearly constant selectivity for the range of conversions investigated. The performance of both sputtered silver catalysts at 4% ethylene follows the trend depicted for the construction material silver. The highest yield of 39% was obtained for the sputtered silver on anodically oxidized (porous) aluminum alloy (AlMg3) at 4.5 s residence time (4 vol.-% ethylene, 96 vol.-% oxygen; 0.3 MPa; 230 °C).

Oxygen partial pressure

[GP 2] [R 2] Studies were performed at constant ethylene content and residence time and with varying ethylene content by exchange versus nitrogen. Selectivity increases from 44.7 to 49.9% at conversions of 10.0 and 11.6%, respectively, on OAOR-modified silver on increasing the oxygen partial pressure by varying the volume content from 10–50% (3 vol.-% ethylene, balance nitrogen; 5 bar; 4 l h^{-1}; 277 °C) [4, 26, 40].

The reaction rate increases with increasing oxygen partial pressure on OAOR-modified silver with an order of 0.78 with respect to oxygen (15 vol.-% ethylene, 0.2–3.5 hPa oxygen partial pressure, balance nitrogen; 4 bar; 0.469 s; 1 l/h; 290 °C) [4]. The selectivity increases from 30 to 37%. The investigations at such high oxygen partial pressures are within the explosion regime.

[GP 2] [R 3a] Studies were performed at constant ethylene content and residence time and by varying the oxygen content by exchange versus methane. An increase

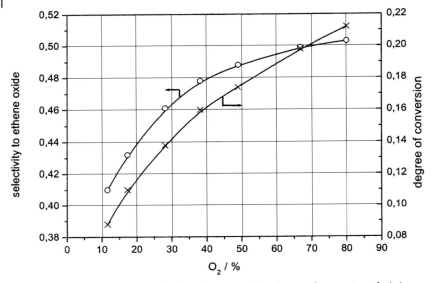

Figure 3.32 Experimental results for the selectivity and the degree of conversion of ethylene as a function of oxygen concentration [43].

in selectivity on sol–gel-Al_2O_3/silver on aluminum (AlMg3) from 41 to 51% is observed on increasing the oxygen content from 12 to 80% (20 vol.-% ethylene, 12–80 vol.-% oxygen, balance methane; 3 bar; 0.30 s; 210 °C) [43]. Conversion increases slightly from 9 to 21%. Consequently, at 80% oxygen content, the highest yield of about 21% was obtained (Figure 3.32).

Ethylene partial pressure
[GP 2] [R 2] Studies were performed at constant oxygen content and residence time and by varying the ethylene content by exchange versus nitrogen. An increase of the reaction rate on laser-LIGA OAOR silver with ethylene partial pressure at nearly constant selectivity of about 41% is observed, but does not follow first order (1.2–6.5 hPa ethylene partial pressure, 50 vol.-% oxygen, balance nitrogen; 4 bar; 0.235 s; 2 l h^{-1}; 290 °C) [4]. An averaged formal order of 0.53 with respect to ethylene was calculated.

[GP 2] [R 3a] Studies were performed at constant residence time and by varying the ethylene and oxygen content concurrently or individually. An increase of selectivity on sol–gel-Al_2O_3/silver on aluminum (AlMg3) from 52 to 55% is observed on increasing the ethylene content from 13 to 60% at constant residence time (13–60 vol.-% ethylene, balance oxygen; 3 bar; 0.53 s; 210 °C) [43]. Conversion drops steeply from 21% to 2%. At 30% ethylene content, the highest yield was obtained.

Addition of promotor
[GP 2] [R 2] The addition of the promoter 1,2-dichloroethane improves selectivity from 52 to 69%, but at the expense of reducing the space–time yield from 0.78 to

$0.42 \, t \, h^{-1} \, m^{-3}$ for laser-LIGA OAOR silver (6 vol.-% ethylene, 30 vol.-% oxygen, balance nitrogen; 5 bar; 0.124 s; $5 \, l \, h^{-1}$; 290 °C) [4].

Influence of construction material and micro fabrication

[GP 2] [R 2] A comparative study with laser-LIGA OAOR silver, etched OAOR silver and sawn Aluchrom catalyst was reported (Table 3.2) [4]. The selectivity was 44–69% (laser-LIGA OAOR silver), 38–69% (etched OAOR silver) and 42–58% (sawn Aluchrom catalyst); for details of the experimental protocols, see [4]. The conversions were 2–15% (laser-LIGA OAOR silver) 5–20% (etched OAOR silver), and 2–6% (sawn Aluchrom catalyst). The space–time yields were $0.01–0.07 \, t \, h^{-1} \, m^{-3}$ (laser-LIGA OAOR silver), $0.03–0.13 \, t \, h^{-1} \, m^{-3}$ (etched OAOR silver), and $0.01–0.06 \, t \, h^{-1} \, m^{-3}$ (sawn Aluchrom catalyst).

Table 3.2 Process parameters for an industrial process compared with parameters achieved for micro reactors machined by different techniques.

	Oxygen-based industrial process	Micro reactor-based process (laser-LIGA)	Micro reactor-based process (etched)	Micro reactor-based process (Aluchrom)
C_2H_4 (vol.-%)	15–40	1.5–6	3–15	15
O_2 (vol.-%)	5–9	10–41	5–85	85
CH_4 (vol.-%)	1–60			
Temperature (°C)	220–275	240–290	240–290	270
Pressure (bar)	10–22	5	2–20	5
Residence time (s)	0.9–1.8	0.1–0.2	0.1–1.5	1.2
C_2H_4 conversion (%)	7–15	2–15	5–20	2–6
Selectivity (%)	80	44–69	38–69	42–58
Space–time yield ($t \, h^{-1} \, m^{-3}$	0.13–0.26 (reactor)	0.01–0.07 (foils) 0.14–0.78 (channels)	0.03–0.13 (foils) 0.18–0.67 (channels)	0.01–0.06 (foils) 0.08–0.36 (channels)

Silver catalyst morphology

[GP 2] [R 2] The smooth surface of the laser-LIGA OAOR silver is roughened and pitted after OAOR treatment and used during more than 1000 h of ethylene oxide synthesis [4]. The etched OAOR silver shows pronounced sintering and agglomeration for the same type of treatment.

[GP 2] [R 3a] The sputtered silver on aluminum alloy (AlMg3) platelets, machined by thin-wire µEDM, were smooth and dense. On prolonged operation under reaction conditions, small silver particles are generated by surface diffusion so that also the blank aluminum platelet surface is exposed (20 vol.-% ethylene, 80 vol.-% oxygen; 3 bar; 0.23–2 s; 250 °C) [43].

Dense and porous catalysts

[GP 2] [R 3a] [R 4a] Sputtered silver (dense) on aluminum alloy (AlMg3) platelets was compared with sol–gel-Al_2O_3/silver (porous) on aluminum alloy (AlMg3) platelets [43]. However, investigations were not performed in the same, but rather in two different micro-reaction devices which differ in the specific micro channel surface. The aim was to compensate for the difference in catalyst surface area by varying the micro channel surface area. Hence the intention was not to compare porosity, but catalyst material instead. Investigations were made by varying residence time, thereby affecting conversion. The dense silver had a maximum selectivity of 61%, exceeding the performance of porous sol–gel-Al_2O_3/silver (51%) (20 vol.-% ethylene, 80 vol.-% oxygen; 3 bar; 230 °C). This performance is constant in both cases up to 20% conversion and then drops steeply.

Silver catalyst particle size

[GP 2] [R 3a] To vary the silver particle size, which is known to have a distinct influence on catalyst activity and selectivity, silver coatings with varying thickness from 50 to 1400 nm were applied on microstructured platelet by multiple-step sputtering (20 vol.-% ethylene, 80 vol.-% oxygen; 3 bar; 0.23–2 s; 250 °C) [43]. For three platelets silver coatings of different thickness, a constant selectivity at a low degree of conversion was found (Figure 3.33). At a certain degree of conversion, the selectivity drops. For the 1400 nm layer this occurs at 30% conversion and for the 400 nm layer at 23%.

Catalysts with very low thicknesses such as 50 nm display a different behavior. A high initial selectivity of 56% at about 2% conversion decreases steeply with a change of conversion to 1%. This is explained, however, as insufficient initial activation of the catalyst (20 vol.-% ethylene, 80 vol.-% oxygen; 3 bar; 0.23–2 s; 250 °C) [43].

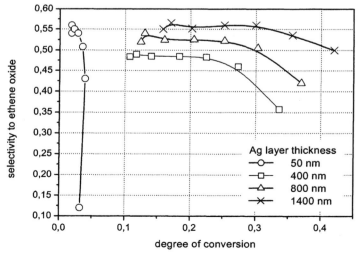

Figure 3.33 Experimental results for degree of selectivity and conversion for ethylene when using micro-channel reactors with different thicknesses of silver coatings [42].

Commercial catalyst as fixed bed and in micro channels
[GP 2] [R 3a] A Shell Series catalyst was measured in a fixed-bed configuration and deposited in micro channels electrophoretically (20 vol.-% ethylene, 80 vol.-% oxygen; 0.3 MPa; 230 °C) [101]. The selectivity was lower in the micro channels (51%) than in the fixed bed (57%) at a conversion of 17%. In a further investigation, a sputtered silver catalyst (cesium promoted) was better than both systems (68%) at higher conversion (25%).

Space–time yield
[GP 2] [R 2] The definition of space–time yield in a micro reactor depends on the definition of the 'reactor volume'. Owing to the large amount of construction material relative to the reaction channels and the neglect of some reactor parts ('abstraction to the real reaction zone'), several more or less useful definitions can be made. In the following, two definitions concerning the time yield divided by the 'pure' reaction channel volume and the platelet volume were used.

Following the first definition, a space–time yield of 0.78 t h^{-1} m^{-3} using a OAOR-modified silver is obtained, which exceeds the industrial performance considerably (0.13–0.26 t h^{-1} m^{-3}) [4]. Following the second definition and hence orienting more on outer than on inner dimensions, a space–time yield of 0.13 t h^{-1} m^{-3} is obtained, still within the industrial window.

Benchmarking to industrial reactor performance
[GP 2] [R 2] The selectivity of 49% on OAOR-modified silver (3 vol.-% ethylene, balance nitrogen; 5 bar; 4 l h^{-1}; 277 °C) is lower than the industrial performance [4, 26, 40]. The latter, however, relies on the presence of chlorine compounds such as 1,2-dichloroethane. At a temperature of 237 °C, a selectivity of 65% is reached, which exceeds values reported for polycrystalline silver catalysts.

For a further comparison of details of the experimental protocols and of conversion, selectivity and space–time yields, see [4].

For a discussion of space–time yields, see the section above.

Benchmarking to fixed-bed reactor performance
[GP 2] [R 3a] The performance of one micro reactor with three kinds of catalyst – construction material silver, sputtered silver (dense) on aluminum alloy (AlMg3), and sputtered silver on anodically oxidized (porous) aluminum alloy (AlMg3) – was compared with three fixed beds with the same catalysts [44]. The fixed beds were built up by hackled silver foils, aluminum wires (silver sputtered) and hackled aluminum foils (anodically oxidized and silver sputtered), all having the same catalytic surface area as the micro channels. Results were compared at the same flow rate per unit surface area.

All three comparisons reveal that the 'micro reactor catalysts' have higher conversion and yield, owing to enhanced mass transfer (20 vol.-% ethylene, 80 vol.-% oxygen; 0.3 MPa; 3.17 l h^{-1}; 230/250 °C) [44]. The selectivity is slightly better in the fixed beds, but, only by about 1–5%. The best fixed-bed and micro reactor selectivity is 65 and 63%, respectively. The best fixed-bed and micro reactor conversion is 37 and 66%, respectively.

A comparison at the same conversion was made for one fixed-bed reactor/micro reactor pair. The micro reactor gave better selectivity at the same conversion, e.g. 66 versus 57% at 62% conversion (4 vol.-% ethylene, 96 vol.-% oxygen; 0.3 MPa; 230 °C).

Activation energy – reaction rate

[GP 2] [R 2] An activation energy of 48 kJ mol^{-1} was observed for ethylene oxide synthesis on OAOR-modified silver [4, 26, 40].

A reaction rate of 4.8 · 10^{-5} mol s^{-1} m^{-2} was observed for OAOR-modified silver (3 vol.-% ethylene, 16.5% vol.-% oxygen; balance nitrogen; 5 bar; 239 °C) [4]. This value is higher than a literature value (1.7 · 10^{-6} mol s^{-1} m^{-2} at 2 vol.-% ethylene, 7% vol.-% oxygen; 1 bar; 230 °C). Correcting the micro reactor value for the different experimental conditions, a value of 1.9 · 10^{-6} mol s^{-1} m^{-2}, close to the literature value, was obtained. Doubling the content of reactants leads to an increase in reaction rate by a factor of nearly two for OAOR-modified silver (6 vol.-% ethylene, 33% vol.-% oxygen; balance nitrogen; 5 bar; 4 l h^{-1}; 267 °C) [4, 26, 40].

An increase in reaction rate with ethylene partial pressure was observed, but does not follow a first-order law. An averaged formal order of 0.53 was calculated [4]. The reaction rate increases with increasing oxygen partial pressure on OAOR-modified silver with an order of 0.78 with respect to oxygen [4].

[GP 2] [R 3a] A reaction rate of 6.0 · 10^{-6} mol s^{-1} m^{-2} was observed for sputtered silver on aluminum (20 vol.-% ethylene, 20% vol.-% oxygen; balance methane; 3 bar; 250 °C) (see [4, 26, 40] reporting results from [43]).

Operation in explosive regimes

[GP 2] [R 2] Safe operation under explosive conditions using pure oxygen was demonstrated in a steel multi-plate-stack micro reactor (e.g. 3 vol.-% ethylene, 50 vol.-% oxygen, balance nitrogen; 5 bar; 4 l h^{-1}; 277 °C) [4, 26, 40]. The cross-sections of the various micro channels employed were from 500 × 50 μm^2 to 500 × 90 μm^2.

By operating at high oxygen partial pressures on OAOR-modified silver, data on reaction rates could be gathered which previously could be obtained in a similar way only under low-pressure conditions (15 vol.-% ethylene, 0.2–3.5 hPa oxygen partial pressure, balance nitrogen; 4 bar; 290 °C) [4].

[GP 2] [R 3a] [R 4a] No explosion or flame formation was observed when working under 20 vol.-% ethylene, 80 vol.-% oxygen conditions on sputtered silver (dense), on aluminum alloy (AlMg3) platelets and sol–gel-Al$_2$O$_3$/silver (porous) on aluminum alloy (AlMg3) platelets (20 vol.-% ethylene, 80 vol.-% oxygen; 3 bar; 230 °C) [43]. The cross-sections of the micro channels of the two micro reactors employed amounted to 200 × 200 μm^2 and 700 × 300 μm^2, respectively.

Exploitation of the results

The very promising results for ethylene oxide synthesis by micro-channel processing given above still await industrial implementation. Selectivity needs to be further improved above 80%, as the costs for ethylene contribute 80% to the overall process costs [4]. In addition to the costs argument, the usual requirements for transfer from laboratory to industrial scale will face the micro reactor: reliability, proper process control and much more.

3.3.5
Oxidation of 1-Butene – Maleic Anhydride Formation

Proceedings: [84, 103]; sections in reviews: [90, 95, 97, 100].

3.3.5.1 Drivers for Performing Maleic Anhydride Formation in Micro Reactors

Maleic anhydride is an important industrial fine chemical (see original citations in [43]). The oxidation of C_4-hydrocarbons in air is a highly exothermic process, therefore carried out at low hydrocarbon concentration (about 1.5%) and high conversion. The selectivity of 1-butene to maleic anhydride so far is low. The reaction is composed of a series of elementary reactions via intermediates such as furan and can proceed to carbon dioxide with even larger heat release. As a consequence, hot spots form in conventional fixed-bed reactors, decrease selectivity and favor other parallel reactions.

3.3.5.2 Beneficial Micro Reactor Properties for Maleic Anhydride Formation

Beneficial micro reactor properties mainly refer to improving heat management as a key for obtaining a partial reaction in a consecutive sequence, when large heats are released by reaction steps other than the partial one (see also Section 3.3.1).

Special attention was also drawn to the search for operation in the explosive regime as micro reactors are said to have much greater safety here. In this way, improvements in terms of space–time yield were expected.

Gas-phase reaction 3 [GP 3]: Oxidation of 1-butene to maleic anhydride

$$\Delta_R H^0_{400} = -1315 \text{ kJ mol}^{-1}$$

Catalyst: $V_2O_5/P_2O_5/TiO_2$

$$CO_2 + H_2O \qquad \Delta_R H^0_{400} = -2542 \text{ kJ mol}^{-1}$$

3.3.5.3 Typical Results
Conversion/selectivity/yield

[GP 3] [R 3b] At conversions from 73 to 85%, selectivity for maleic anhydride of about 33% was achieved with a $V_2O_5/P_2O_5/TiO_2$ catalyst (0.4 vol.-% 1-butene in air; 0.1 MPa; 400 °C) [103]. The residence time was varied in these investigations. A 90% conversion in the 80×80 μm^2 micro channels corresponds to a 25 ms residence time.

Micro-channel diameter

[GP 3] [R 3b] At similar conversion and residence time, smaller micro channels (80 μm) have better selectivity than larger ones (200 μm) with a $V_2O_5/P_2O_5/TiO_2$ catalyst (0.4 vol.-% 1-butene in air; 0.1 MPa; 400 °C) [103]. The residence time was varied in these investigations.

The ratio of space–time yields referring to various micro channels equals the ratio of their cross sections.

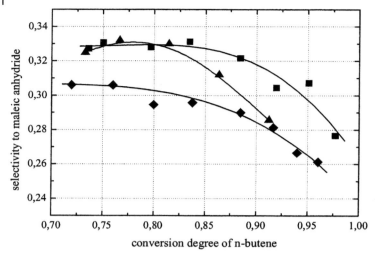

Figure 3.34 Experimental data on selectivity to maleic anhydride vs degree of conversion of 1-butene for different reactor types: (◆) channel width 0.08 mm and (▲) channel width 0.2 mm micro reactor; (■) fixed-bed reactor [103].

Benchmarking to fixed bed

[GP 3] [R 3b] The maximum selectivity of about 33% was the same for the best micro reactor and the fixed bed (Figure 3.34) at the same conversions from 73 to 85% with a $V_2O_5/P_2O_5/TiO_2$ catalyst [103]. At still higher conversion, the fixed bed has a better performance. However, the residence times needed for comparable conversion are one order of magnitude shorter than in the fixed-bed reactor.

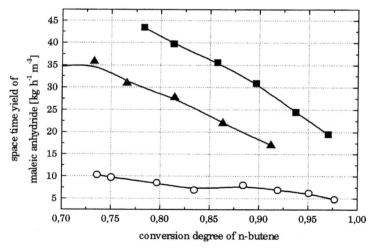

Figure 3.35 Space–time yield of maleic anhydride vs degree of conversion for 1-butene when using different reactor types: (■) channel width 0.4 mm and (▲) channel width 0.08 mm micro reactor; (O) fixed-bed reactor [103].

Space–time yield

[GP 3] [R 3b] The space–time yield of chemical micro processing was a factor of five larger than that of a conventional fixed-bed reactor (Figure 3.35) (0.4% 1-butene in air; 0.1 MPa; 400 °C) [103]. This is due to the shorter residence time needed in the micro reactors for the same conversion as in the fixed bed. Differences from the fixed bed become smaller when operating at very high conversion, up to 95%. The space–time yield of the 400 μm × 400 μm micro channels exceed that of the 80 μm × 80 μm micro channels.

The ratio of space–time yields referring to various micro channels equals the ratio of cross sections [103].

Post-catalytic effects

[GP 3] [R 3b] [R 4a] The presence of a metallic diffuser can contribute significantly and in a negative manner to the reaction, e.g. at long residence times by decomposing 21% of the product formed [103]. An exchange against a glass diffuser can notably reduce the product degradation.

Safe operation

[GP 3] [R 3b] [R 4a] Safe operation in the explosive regime was demonstrated [103]. Catalytic runs with 1-butene concentrations up to 10 times higher than the explosion limit were performed (5–15% 1-butene in air; 0.1 MPa; 400 °C). A slight catalyst deactivation, possibly due to catalyst active center blockage by adsorption, was observed under these conditions and not found for lower 1-butene concentrations. Regeneration of the catalyst is possible by oxidation.

An ignition experiment at 1-butene concentrations as high as 5% was performed to test instability in reaction behavior as an indication of unsafe operation (5% 1-butene in air; 0.1 MPa; 400 °C) [103]. The degree of conversion increased linearly and converged without any sign of instability. The power input corresponded to 6.5 W with an adiabatic temperature rise of more than 2000 °C. Plugging, however, was the major concern under these severe conditions.

Additionally, potential strong thermal changes in the micro reactors could be deliberately induced by strong changes in conversion (5% 1-butene in air; 0.1 MPa; 400 °C) [103]. For this reason, pulses of high 1-butene concentration were inserted in the micro reactor. Remarkably low axial temperature gradients within the explosion regime at high thermal power were found. The zone of the highest reaction rate shifts with respect to the micro channel length.

3.3.6
Oxidation of Methanol – Formaldehyde Formation

Proceedings: [72]; sections in reviews: [95, 97, 100].

3.3.6.1   Drivers for Performing Formaldehyde Synthesis in Micro Reactors

The oxidative dehydrogenation of methanol to formaldehyde is a model reaction for performance evaluation of micro reactors (see description in [72]). In the corresponding industrial process, a methanol–air mixture of equimolecular ratio of methanol

and oxygen is guided through a shallow catalyst bed of silver at 150 °C feed temperature, 600–650 °C exit temperature, atmospheric pressure and a contact time of 10 ms
or less. Conversion amounts to 60–70% at a selectivity of about 90%.

Both oxidative and non-oxidative routes with similar share are followed, yielding
hydrogen or water as additional products. As by-products, carbon dioxide and carbon monoxide, methyl formate and formic acid are generated. It is advised to quench
the exit stream as formaldehyde decomposition can occur.

3.3.6.2 Beneficial Micro Reactor Properties for Formaldehyde Synthesis

The reaction is a model reaction, hence the respective investigations evaluate micro reactor properties in a more general way.

Beneficial micro reactor properties mainly refer to improving heat management
as a key for obtaining a partial reaction in a consecutive sequence, when large heats
are released by reaction steps other than the partial one (see also Section 3.3.1).

Special attention has to be drawn to having good mass transfer at millisecond
contact times.

Gas-phase reaction 4 [GP 4]: Oxidation of methanol to formaldehyde

$$H_3C–OH + 1/2\ O_2 \xrightarrow{\text{Ag/510 °C}} \underset{H}{\overset{H}{>}}{=}O + H_2O$$

3.3.6.3 Typical Results
Conversion/selectivity/yield
[GP 4] [R 11] For methanol conversion over sputtered silver catalyst, conversions
up to 75% at selectivities of about 90% were found (8.5 vol.-% methanol; 10–90%
oxygen; balance helium; 510 °C; 10 ms; slightly > 1 atm) [72].

Oxygen content
[GP 4] [R 11] For methanol conversion over sputtered silver catalyst, variation in
oxygen content from 10% to more than 90% of the gas mixture results in a slight
decrease of conversion from 75 to 70% and of selectivity from 91 to 89% (8.5 vol.-%
methanol; balance helium; 510 °C; 10 ms; slightly > 1 atm) [72].

Oxygen reaction order
[GP 4] [R 11] For methanol conversion over sputtered silver catalyst, selectivity is
hardly affected by changing the oxygen concentration (8.5 vol.-% methanol; 10–90%
oxygen; balance helium; 510 °C; 10 ms; slightly > 1 atm) [72]. This is in line with
assuming a zero-order dependence on oxygen concentration and agrees with literature findings.

Catalyst deactivation
[GP 4] [R 11] For methanol conversion over sputtered silver catalyst, no catalyst deactivation at high oxygen : methanol ratios (e.g. over 0.2 : 1) was observed, different from findings in the literature with conventional catalysts and reactors (8.5 vol.-%
methanol; 10–90% oxygen; balance helium; 510 °C; 10 ms; slightly > 1 atm) [72].

Since in the literature such deactivation was attributed to a sudden temperature rise and decline under the same process conditions, it was assumed that the improved thermal control of micro channels is responsible for the higher stability.

Temperature
[GP 4] [R 11] For methanol conversion over sputtered silver catalyst, conversion increases almost linearly from 430 to 530 °C (8.6 vol.-% methanol; balance oxygen; 10 ms; slightly > 1 atm) [72]. A slight increase in selectivity towards carbon dioxide at the expense of formaldehyde was observed. At 530 °C, about 75% conversion at 90% formaldehyde selectivity was achieved.

Residence time
[GP 4] [R 11] For methanol conversion over sputtered silver catalyst, conversion increases from 57 to 73% on applying six times longer residence times (8.5 vol.-% methanol; balance oxygen; 510 °C; 4–27 ms; slightly > 1 atm) [72]. Selectivity decreases slightly from more than 90% initially.

Micro-channel diameter
[GP 4] [R 11] For methanol oxidation over sputtered silver catalyst, conversion is higher when using micro channels of smaller diameter (8.5 vol.-% methanol; balance oxygen; 510 °C; 4–27 ms; slightly > 1 atm) [72]. For two channels of the same width, but different depths (70 μm, 130 μm), concentration differences of nearly 10% at the same residence time were detected, all other parameters being equal. The increase in conversion in the 70 μm channel was partly at the expense of selectivity (3–6% in the range investigated).

Kinetics: reaction rate and activation energy
[GP 4] [R 11] For methanol conversion over sputtered silver catalyst, reaction rates and an activation energy (Figure 3.36) of 14.3 kcal mol^{-1} were reported (8.5 vol.-% methanol; balance oxygen; 10 ms; slightly > 1 atm) [72]. Since the latter is much lower than literature values (about 22.5–27 kcal mol^{-1}), different kinetics may occur or limitations of the reactor model may become evident.

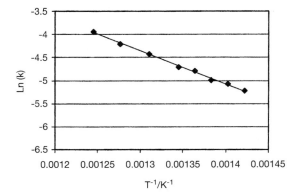

Figure 3.36 Arrhenius plot for the oxidative dehydrogenation of methanol to formaldehyde performed in a micro reactor [72].

A comparison of experimental findings and theoretical predictions is given in [72]. Although qualitatively consistent, experiments confirm a weaker dependence on parameters such as residence time and micro-channel diameter.

3.3.7
Oxidation of Derivatized Alcohols – Derivatized Aldehyde Formation

Peer-reviewed journals: [1, 49]; journals: [50]; proceedings: [51, 72, 104, 105]; reactor description: [31, 32, 46, 47, 106]; sections in reviews: [87, 90, 95–97, 100, 107].

3.3.7.1 Drivers for Performing Derivatized Aldehyde Synthesis in Micro Reactors

The oxidative dehydrogenation of methanol to formaldehyde was choosen as model reaction by BASF for performance evaluation of micro reactors [1, 49–51, 108]. In the industrial process a methanol–air mixture of equimolecular ratio of methanol and oxygen is guided through a shallow catalyst bed of silver at 150 °C feed temperature, 600–650 °C exit temperature, atmospheric pressure and a contact time of 10 ms or less. Conversion amounts to 60–70% at a selectivity of about 90%.

Both oxidative and non-oxidative routes with similar shares are followed, yielding hydrogen or water as additional products. As by-products, carbon dioxide and carbon monoxide, methyl formate and formic acid are generated. It is advised to quench the exit stream as formaldehyde decomposition can happen.

The oxidation of an undisclosed methanol derivative to the corresponding formaldehyde compound is a large-scale BASF process which was established in recent years, whereas the similar methanol-to-formaldehyde process, performed on a much larger scale, has been practised at BASF for more than 100 years [1, 49–51, 108]. The exact nature of the substituent(s) was not disclosed by BASF for reasons of confidentiality, although many publications on that topic appeared. The nature of the substituent makes the derivative, as the results of the investigations show, more labile to temperature.

Both processes – referring to the non-substituted and substituted methanol reactant – utilize elemental silver catalyst by means of oxidative dehydrogenation. Production is carried out in a pan-like reactor with a 2 cm thick catalyst layer placed on a gas-permeable plate. A selectivity of 95% is obtained at nearly complete conversion. This performance is achieved independent of the size of the reactor, so both at laboratory and production scale, with diameters of 5 cm and 7 m respectively.

3.3.7.2 Beneficial Micro Reactor Properties for Derivatized Aldehyde Synthesis

The reaction is an industrial process; the micro reactor was taken as a precise instrument giving analytical information which was transferred to the large-scale process in a non-disclosed way.

Beneficial micro reactor properties mainly refer to improving heat management as a key for obtaining a partial reaction which is part of a consecutive sequence, when large heats are released by reaction steps other than the partial one (see also Sectoin 3.3.1). An even more import selectivity issue refers to the suppression of side reactions, which relate to the other functionality present in the reactant. This, again, profits from improved heat transfer.

Gas-phase reaction 5 [GP 5]: oxidation of derivatized methanol to derivatized formaldehyde

$$R\diagdown OH + 1/2\ O_2 \xrightarrow[\text{Ag}]{550\,°C} R\diagdown O + H_2O$$

3.3.7.3 Typical Results

Conversion/selectivity/yield

[GP 4] [R 5] For an undisclosed methanol derivative, a selectivity of 96% at 55% conversion was found for the micro reactor with silver as construction material (390 °C), which exceeds the performance of laboratory pan-like (40%; 50%; 550 °C) and short shell-and-tube (85%; 50%; 450 °C) reactors (Figure 3.37) using elemental silver [1, 49–51, 108]. At slightly higher conversion, selectivity of the chemical micro processing decreases, and 89% selectivity at 59% conversion is found.

Hot spots

[GP 4] [R 5] For an undisclosed methanol derivative, no hot spot (close to 0 K rise) was found for the construction-material silver micro reactor (operational temperature: 390 °C); hot spots of 160 and 60 K were found for laboratory pan-like (40%; 50%; 550 °C) and short shell-and-tube reactors, respectively, using elemental silver [1, 49–51, 108].

Relevance for industrial process development

[GP 4] [R 5] Formaldehyde synthesis has been known at BASF for more than 100 years [1, 49–51, 108]. Hence it was expected to be able to handle the synthesis of substituted analogue, an undisclosed methanol derivative, with the same processsing concepts, major problems not being anticipated. This expectation was still supported by first attempts with the tried and tested pan-like reactor concept (5 cm diameter), which were promising. At 50% conversion, a selectivity of 90% was achieved. However, transfer to production scale using a 3 m production reac-

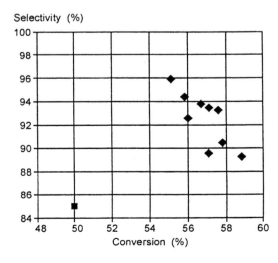

Figure 3.37 Selectivity-conversion diagram of the oxidative dehydrogenation reaction [1].

tor was difficult. At 50% conversion, the selectivity was only 40%. Considering the process again, the solution to prevent this large drop in selectivity became apparent. Hot spot formation due to the large heat release led to an increase in temperature in the reaction zone up to 160 °C. The reactant and product were judged to be more labile than methanol and formaldehyde. Consequently, thermally induced side reactions reduced the selectivity.

Based on this assumption, the needed measures were evident: to reduce the hot spots and, possibly concurrently, the operating temperature and, at best, to reduce the residence time substantially. An ideal combination of all these parameters is provided by operation in the micro reactor. As a result, a much better performance of the micro reactor than the conventional laboratory reactors was found (see the previous section for more details); no hot spot was found.

All conventional reactors, tested before using the micro reactor (simply since micro reactors were hardly available at that time), only fulfilled the demands of one measure, at the expense of the other measures. For instance, a single-tube reactor can be operated nearly isothermally, but the performance of the oxidative dehydrogenation suffers from a too long residence time. A short shell-and-tube reactor provides much shorter residence times at improved heat transfer, which however is still not as good as in the micro reactor.

3.3.8
Oxidation of Propene to Acrolein

Proceedings: [37]; master thesis: [109]; sections in reviews: [90].

3.3.8.1 Drivers for Performing the Oxidation of Propene to Acrolein
Propene is an intermediate utilized in the chemical and pharmaceutical industries. The partial oxidation of propene on cuprous oxide (Cu_2O) yields acrolein as a thermodynamically unstable intermediate, and hence has to be performed under kinetically controlled conditions [37]. Thus in principle it is a good test reaction for micro reactors. The aim is to maximize acrolein selectivity while reducing the other by-products CO, CO_2 and H_2O. Propene may also react directly to give these products. The key to promoting the partial oxidation at the expense of the total oxidation is to use the Cu_2O phase and avoid having the CuO phase.

3.3.8.2 Beneficial Micro Reactor Properties for the Oxidation of Propene to Acrolein
The reaction was of a scouting nature, actually one of the most common investigations done concerning gas-phase reactions in micro reactors. Hence it addresses in a general way the investigation of general micro reactor properties such as mass and heat transport and residence time.

Of particular concern was the finding of a suitable catalyst. Owing to the scouting nature, virtually no know-how base was available that time. The investigation gave highly valuable hints for later catalyst development. Actually, they motivated a search for catalysts of higher porosity and better defined composition. As a result, anodically oxidized alumina supports for catalysts were developed (see Sections 3.1 and 3.4.2).

Gas-phase reaction 6 [GP 6]: oxidation of propene to acrolein

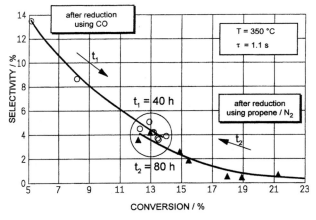

propene   acrolein

3.3.8.3 Typical Results
Conversion/selectivity/yield
[GP 6] [R 5] With a stabilized Cu$_2$O catalyst layer, by addition of bromomethane (ppm level), 20% selectivity at 5% conversion was found (0.5 vol.-% propene; 0.1 vol.-% oxygen; 2.25 ppm promoter; 350 °C) [37]. This is far better than with non-conditioned copper oxide catalysts which contain CuO besides Cu$_2$O. It is expected that the first species promotes more total oxidation, whereas the latter steers partial oxidation. In the above experiment, selectivity rises from 7 to 30% at slightly reduced conversion after 3 h of promoter conditioning.

Catalyst conditioning
[GP 6] [R 5] For Cu$_2$O catalyst layers, a non-uniform temperature dependence of the selectivity–conversion behavior was found (0.1 vol.-% propene; 0.02 vol.-% oxygen; 350, 363, 375 °C) [37]. A drop in selectivity with the first temperature variation was followed by a rise. This was attributed to oxidation of the initial Cu$_2$O catalyst layer during the course of the reaction (see the previous section) so that non-optimized and non-stationary catalyst compositions occur during and between the experiments. This was further evidenced by pre-conditioning of the copper oxide layer by reduction with either CO or propene/nitrogen, giving either Cu$_2$O- or CuO-rich layers. After 40 h time-on-stream, nearly the same selectivity was achieved in both cases (Figure 3.38).

Figure 3.38 Comparison of selectivity to acrolein vs conversion of propene depending on time on-stream. Pre-treated in CO (t_1); pre-treated in propene/nitrogen (t_2) [37].

3.3.9
Oxidation of Isoprene – Citraconic Anhydride Formation

Proceedings: [13, 27, 85]; sections in review: [90, 95, 97, 100, 110].

3.3.9.1 Drivers for Performing Citraconic Anhydride Formation
Citraconic anhydride is, among other applications, a starting material in the synthesis of pharmaceuticals (see original citation in [13, 27]). The currently applied synthesis from itaconic acid is complex and hence there is a quest for alternatives. One of these is the partial oxidation of isoprene in the gas phase. Owing to the high reaction enthalpy of 1000 kJ mol^{-1}, generation of large heats at high conversion is possible. In that case, local overheating (hot spots) can promote the total oxidation to carbon dioxide and water.

3.3.9.2 Beneficial Micro Reactor Properties for Citraconic Anhydride Formation
Beneficial micro reactor properties mainly refer to improving heat management as a key for obtaining a partial reaction which is part of a consecutive and parallel sequence, when large heats are released by reaction steps other than the partial one (see also Section 3.3.1).

Gas-phase reaction 7 [GP 7]: oxidation of isoprene to citraconic anhydride

$+ 5/2\ O_2 \rightarrow$ ... $+ 2\ H_2O$ $\Delta_R H \approx -1000$ kJ mol^{-1}

Catalyst: $V_{30}Ti_{70}O_x$

$+ 7\ O_2 \rightarrow$ $5\ CO_2 + 4\ H_2O$ $\Delta_R H \approx -3000$ kJ mol^{-1}

3.3.9.3 Typical Results
Conversion/selectivity/yield
[GP 7] [R 6] The selectivity (up to 28%) for micro-reactor processing using a $V_{30}Ti_{70}O_x$–SiO_2 catalyst at nearly complete conversion is similar to the performance of a fixed-bed reactor with a $V_{30}Ti_{70}O_x$ catalyst (0.6 vol.-% isoprene, 20 vol.-% oxygen; $p = 1.2$ bar) [27]; however, for the same performance, the processing in the micro reactor needs to be done at temperatures exceeding fixed-bed operation by 40 K. The micro reactor is operated at 400 °C whereas the fixed bed is processed at 360 °C. This difference is attributed to using a similar, but not identical, catalyst formulation; the 'micro reactor catalyst' $V_{30}Ti_{70}O_x$–SiO_2 is regarded as being less active (Figure 3.39). Furthermore, total oxidation of the citraconic acid product may occur in the hot outlet section of the micro reactor. For this reason, an optimized design with a cooler directly behind the reaction zone could be advantageous.

A maximum yield of 23% is found in the micro reactor [27].

[GP 7] [R 8] In a later study, a yield of about 40% is found for the same steel reactor [55]. For a ceramic reactor, an even higher yield of 45% is reported. This is explained by a reduction in blank activity (Figure 3.40) (see the sections Activity of

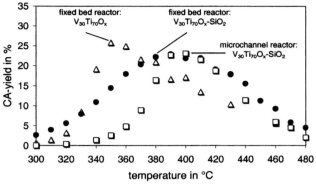

Figure 3.39 Partial oxidation of isoprene in a fixed-bed reactor and a micro reactor. The yield of citraconic anhydride is plotted as a function of reaction temperature and catalyst composition [27].

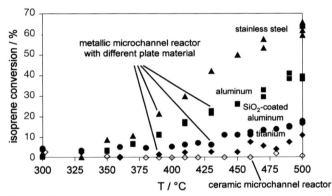

Figure 3.40 Experimental results for isoprene conversion in metallic and ceramic micro reactors. The metallic micro reactors were operated without catalyst to determine blank activity of the various construction materials. In addition, conversion data were calculated. ($\Diamond$) Calculated values for micro-channel reactor model; (full symbols) experimental values for different reactor materials [27].

the same catalyst coating on different microstructured materials and Blank activity of microstructured material below).

Process temperature
[GP 7] [R 6] For a $V_{30}Ti_{70}O_x$–SiO_2 catalyst, conversion increases with temperature until complete conversion is reached at 390–420 °C depending on the volume flow (0.6 vol.-% isoprene, 20 vol.-% oxygen; $p = 1.2$ bar) [27]; the selectivity passes a maximum at 390–420 °C depending on the volume flow. The yield correspondingly exhibits maxima with optimal performance at 390–420 °C.

Blank activity of microstructured material
[GP 7] [R 8] In absence of a catalyst, blank activities were determined for both metallic and ceramic micro-channel reactors [55]. The following sequence of materials with regard to undesired blank activity was determined: steel > aluminum >

titanium. Coating of aluminum with inert layers such as silica can also reduce blank activity. However, the best option is to use ceramic materials. Whereas ceramics have not more than 1–2% isoprene conversion at the maximum temperature investigated of 500 °C, steel has about 60% conversion, leading to total oxidation (0.5 vol.-% isoprene in air; 50 Nml min^{-1} and 200 Nml min^{-1}).

Similar findings were made by BASF in studies investigating an undisclosed gas-phase reaction in capillaries made of quartz, catalyst material and reactor-wall material [105]. The dimensions were chosen in such a way that they match the of surface-to-volume ratio of a fixed-bed reactor used previously for the same reaction. A quartz capillary shows no conversion, whereas reactor-wall material actually has a greater activity than the catalyst itself. Hence BASF came to the, at first sight, surprising conclusion that in their production process it was the reactor wall, and not the catalyst, which catalyzes the reaction. The reactor wall was 70 times more active than the catalyst; it needs a temperature increase of about 100 °C to have both at equal conversion.

Activity of the same catalyst coating on different microstructured materials

[GP 7] [R 8] For a suspended $V_{75}Ti_{25}O_x$–SiO_2 catalyst, conversion in a steel reactor is lower than when employing a ceramic reactor at a given temperature [55]. This was explained by more efficient heat removal of the steel construction material. This good isothermicity should in turn promote selectivity; however, this is outperformed by the higher blank activity. At 400 °C, complete conversion is given for the ceramic-supported catalyst, whereas about 40% is found for the steel-supported catalyst (0.6 vol.-% isoprene in air; 6 g_{cat} min mol$_{tot}^{-1}$). The corresponding yields are about 40 and 10%, respectively.

Catalyst formulation

[GP 7] [R 6] The performances of three different catalysts at constant flow per micro channel were compared (Figure 3.41): $V_{30}Ti_{70}O_x$–Al_2O_3 (by anodic oxidation and impregnation), $V_{30}Ti_{70}O_x$–SiO_2 (by a suspension method similar to washcoat treatment) and VO_x–SiO_2 (by a suspension method) [27]. The results for the suspension catalysts were superior. The maximum citraconic anhydride selectivity for the two suspension catalysts was 25 and 22%, whereas the selectivity of the anodic oxidation-impregnated catalyst was only 12% (0.6 vol.-% isoprene, 20 vol.-% oxygen; $p = 1.2$ bar). For the latter, conversion and selectivity curves are also shifted to higher temperatures.

Accordingly, the suspension catalysts were chosen for all subsequent investigations. However, the results were slightly worse compared with the $V_{30}Ti_{70}O_x$ catalyst in fixed bed (see Conversion/selectivity/yield section). The conventional formulation of the catalyst had to be slightly modified (from $V_{30}Ti_{70}O_x$ to $V_{30}Ti_{70}O_x$–SiO_2) for micro-reactor application to allow insertion in the micro channels via suspension [27]. Investigation of the new $V_{30}Ti_{70}O_x$–SiO_2 system, in both fixed-bed and micro reactors, shows the same activity, lower than $V_{30}Ti_{70}O_x$. This is an indication that the higher temperatures needed for comparable micro-reactor operation are due to lower catalyst activity and are not inherent to the chemical micro processing itself.

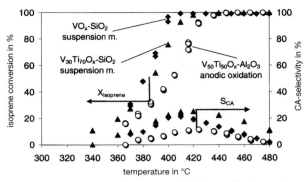

Figure 3.41 Conversion of isoprene and selectivity for citraconic anhydride as a function of temperature and catalyst used [27].

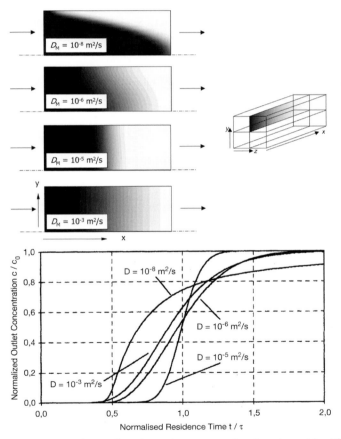

Figure 3.42 Evolution of a pulse at the entrance of a micro channel for different diffusion coefficients. Calculated concentration profile (left) and cumulative residence time distribution curve (channel: 300 μm × 300 μm × 20 mm; flow velocity: 1 m s^{-1}; t = 10 s) [27].

Simulations in periodic concentration processing

[GP 7] [R 6] The influence of the channel diameter on the cumulative residence time distribution (RTD) was simulated (Figure 3.42) [85] ($D = 3 \cdot 10^{-5}$ m^2 s^{-1}; channel length: 20 mm; channel diameter: 1200–75 µm; 1 m s^{-1}). The width of the RTD decreases initially, but remains finite for small diameters such as 75 µm.

The impact of the diffusion coefficient was also analysed; here values of isoprene and oxygen were taken [85]. It was found that the RTD of one component in a cycling experiment may be narrow, while the other component is not at optimum. By changing the channel diameter, such a sequence may even be reversed. It is advised [85] to operate in the diffusion-dominated branch rather than in the convection-dominated branch.

For a short description of the aims of experimental concentration cycling with respect to citraconic acid formation, see [13]. It is also demonstrated there that the response to rectangular concentration pulses for a given reactor configuration of a multi-channel-stack micro reactor can be improved by increasing the number of platelets and channels (while reducing their diameter). Such behavior was shown for an incompressible fluid, i.e. is not solely related to the above-mentioned reaction.

3.3.10
Partial Oxidation of Methane – Syngas Generation

Peer-reviewed journals: [3]; proceedings: [111, 112]; sections in reviews: [86, 87, 90, 96, 97, 113].

3.3.10.1 Drivers for Performing Syngas Generation

Syngas is a mixture of carbon monoxide and hydrogen made from oxidation of methane. The generation of syngas plays a role in some ideas on future exploitation of natural gas deposits (see original citations in [3]). At remote sites, many small- and medium-volume deposits containing natural gas or mixtures thereof with petroleum exist. Owing to the limited deposit volumes and the difficulty of installing transportation facilities in remote areas, it is usually not economical to exploit such resources. One solution to this problem could be to liquefy the gaseous products, in particular the most prominent species methane, so allowing more facile transportation. The transformation of methane into hydrogen-rich gases is part of this gas-liquefaction strategy, finally yielding liquid products such as methanol or liquid hydrocarbons. For this purpose, mobile, small-scale, efficient and flexible production systems are required. It is generally believed that micro reactors with their flexible numbering-up concept could provide a technical platform that may in future allow economical gas liquefaction at the remote deposit sites.

Partial methane oxidation comprises very high rates so that high space–time yields can be achieved (see original citations in [3]). Residence times are in the range of a few milliseconds. Based on this and other information, it is believed that syngas facilities can be far smaller and less costly in investment than reforming plants. Industrial partial oxidation plants are on the market, as e.g. provided by the Syntroleum Corporation (Tulsa, OK, USA). Requirements for such processes are operation at elevated pressure, to meet the downstream process requirements, and autothermal operation.

3.3.10.2 Beneficial Micro Reactor Properties for Syngas Formation

The efficient mass and heat transfer in general render micro reactors interesting for this very fast reaction. Given such performance, high throughputs or space–time yields are in prinicple achievable; therefore, the use for real production, of whatever capacity, is not out of discussion. Favorably, this will be applied for small, compact and light-weight systems for energy generation where the syngas formation is one step.

Gas-phase reaction 8 [GP 8]: Syngas generation by partial oxidation of methane

$$CH_4 + 2\,O_2 \;\rightleftharpoons\; CO_2 + 2\,H_2O$$

$$CH_4 + H_2O \;\rightleftharpoons\; CO + 3\,H_2$$

$$CO_2 + 2\,H_2 \;\rightleftharpoons\; CO + 2\,H_2O$$

3.3.10.3 Typical Results

Start-up behavior

[GP 8] [R 7] Ignition occurs at a rhodium catalyst at catalyst temperatures between 550 and 700 °C, depending on the process parameters [3]. Total oxidation to water and carbon dioxide is favored at low conversion (< 10%) prior to ignition. Once ignited, the methane conversion increases and hence the catalyst temperature increases abruptly.

Catalyst formation

[GP 8] [R 7] Rhodium catalysts generally show no pronounced activation phase as given for other catalysts in other reactions [3]. In the first 4 h of operation, methane conversion and hydrogen selectivity increases by only a few percent. After this short and non-pronounced formation phase, no significant changes in activity were determined in the experimental runs for more than 200 h.

Operational changes of the catalyst

[GP 8] [R 7] The structure of the rhodium catalyst changed during operation. Owing to the microfabrication process (thin-wire μEDM), the surface of the micro channels was rough before catalytic use [3]. After extended operational use, small crystallites are formed, especially in oxygen-rich zones such as the micro channels' inlet. Thereby, the surface area is enlarged by a factor of 1–1.5.

Noble metal loss of about 0.1 wt.-% of the honeycomb rhodium catalyst was observed during 200 h of operation; similar effects are also known in commercial ammonia combustion processes [3]. This did not lead to a decrease in catalyst activity as rhodium was the only construction material.

By inspection windows and use of a pyrometer, visual inspection of the catalyst and temperature monitoring on-site in a contactless manner were performed. It turned out that a glowing, homogeneous texture occurs at catalyst temperatures between 900 and 1200 °C, GHSV values up to $10^6\ h^{-1}$ and pressures less than 1 MPa [3]. This is an indication of the absence of soot deposits. At lower temperatures or higher pressures, separate soot nests were detected and can be completely removed by oxidation.

In-Site temperature monitoring

[GP 8] [R 7] By pyrometric analysis, local temperatures at the rhodium catalyst surface were determined, although the accuracy is limited [3]. In this way, temperatures at the inlet were found to be 50–120 K higher than at the outlet for the operational range investigated. Conditions favoring conversion such as a pressure increase lead to even steeper temperature profiles. This is consistent with predictions that the entire oxygen amount is spent at the inlet region and that endothermic reforming reactions, reducing temperature, should dominate thereafter.

Conversion/selectivity/yield

[GP 8] [R 7] In the temperature range between 1090 °C and 1190 °C, complete oxygen conversion was achieved at a rhodium catalyst [CH_4/O_2: 2.0; 0.12 MPa; $7.8 \cdot 10^5 \, h^{-1}$ (STP)] [3, 112]. At 1190 °C, CO and H_2 selectivities of about 92 and about 78%, respectively, are found at about 63% methane conversion (Figure 3.43).

At a different methane/oxygen ratio (Figure 3.44), the methane conversion increases to 96%, giving CO and H_2 yields of 85 and 80%, respectively [CH_4/O_2: 1.5; 0.15 MPa; $7.8 \cdot 10^5 \, h^{-1}$ (STP)] [3].

The equilibrium values are not reached at a rhodium catalyst on a microstructured reactor within the limits of the experimental conditions and the constructional constraints [3]. As possible explanations post-catalytic reactions at lower temperatures or, more likely, insufficient catalyst activity concerning the short residence times are seen.

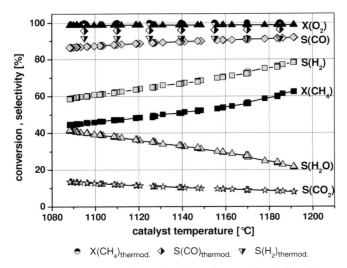

Figure 3.43 Conversion rates and product selectivity of partial methane oxidation as a function of the catalyst temperature. Experimental data (points) and calculated thermodynamic values (lines) [112].

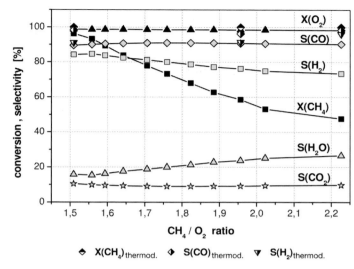

Figure 3.44 Conversion rates and product selectivity of partial methane oxidation performed under constant heating power as a function of the methane/oxygen ratio [112].

Process temperature
[GP 8] [R 7] Oxygen conversion at rhodium catalyst is complete within the temperature range 1090–1190 °C [CH_4/O_2: 2.0; 0.12 MPa; $7.8 \cdot 10^5$ h^{-1} (STP)] [3]. This shows that all oxygen is consumed by total oxidation of methane. With increasing temperature, CO and H$_2$ selectivities increase at increasing methane conversion owing to the increasing contribution of reforming reactions.

Process pressure
[GP 8] [R 7] Given constant catalyst temperature and GHSV, methane conversion and CO and H$_2$ selectivity decrease with increasing pressure at total oxygen consumption for a rhodium catalyst [CH_4/O_2: 2.0; 1–12 MPa; $1.17 \cdot 10^6$ h^{-1} (STP); 1200 °C] [3]. The decrease is larger than thermodynamically expected.

Methane/oxygen ratio
[GP 8] [R 7] Given process pressure, heating power and GHSV, the syngas yield decreases with increasing methane/oxygen ratio at a rhodium catalyst, as well as with the CO and H$_2$ selectivities, and particularly strong with the methane conversion [CH_4/O_2: 1.5–2.23; 0.15 MPa; 95 W; $7.8 \cdot 10^5$ h^{-1} (STP)] [3].

Residence time by changing flow velocity
[GP 8] [R 7] Residence times were varied at a rhodium catalyst by changing the flow velocity (GHSV). Methane conversion and CO and H$_2$ selectivity decreased by about 10% decreasing the flow velocity in the range investigated [CH_4/O_2: 1.66; 0.15 MPa; $2.0 \cdot 10^5 - 1.2 \cdot 10^6$ h^{-1} (STP); 1090 °C] [3]. The decline was attributed to

reaction kinetics, since it could be ruled out that mass transfer limitations exist for the operational conditions and the micro channel geometry investigated.

Residence time by changing micro channel length

[GP 8] [R 7] Residence times were varied at a rhodium catalyst by changing the micro channel length from 5 to 20 mm. Concerning different ways of comparison – in the autothermal state, for the same heating power and for the same GHSV – higher methane conversion and H_2 selectivity were observed for the longer micro channel (20 mm) reactor, whereas CO selectivity was slightly lower [CH_4/O_2: 2.0; 0.30 MPa; $2.9 \cdot 10^5 - 1.16 \cdot 10^6 \, h^{-1}$ (STP); autothermal or 130 W] [3]. The better performance, however, is at the expense of an increased pressure drop and four times more construction material. Lower temperatures at the micro channels' outlet were found for the 20 mm reactor compared with the 5 mm reactor owing to an increased contribution of reforming reactions and more heat losses due to the reactor volume increase.

Dilution of inert gas

[GP 8] [R 7] Dilution with the inert gas argon served to simulate the oxidation behavior when using air. Methane conversion and H_2 and CO selectivity remain constant for a long range of dilution until they finally drop at inert gas contents above 50% [CH_4/O_2: 2.0; 10 – 57 vol.-% Ar; 0.15 MPa; $7.8 \cdot 10^5 \, h^{-1}$ (STP); 105 W] [3]. Oxygen conversion is near-complete for all experiments. The micro channels' outlet temperatures drops on increasing the amount of inert gas.

Using air and a 20 mm long micro-channel reactor with rhodium catalyst results in high rates of syngas formation (CH_4/O_2: 1.75; oxygen at air level, balance nitrogen; 0.15–2 MPa; $1.95 \cdot 10^5 \, h^{-1}$ (STP); 1100 °C) [3]. However, to obtain high H_2 and CO selectivities, additional heating is required. Oxygen conversion was near complete for all experiments.

Benchmarking to laboratory scale reactor performance

[GP 8] [R 7] Syngas generation with commercial Pt–Rh gauzes, metal-coated foam monoliths and extruded monoliths has been reported. For similar process pressure, process temperature, and reaction mixture composition, methane conversions are considerably lower in the conventional reactors (CH_4/O_2: 2.0; 22 vol.-% methane, 11 vol.-% oxygen, 66 vol.-% inert species; 0.14–0.155 MPa; 1100 °C) [3]. They amount to about 60%, whereas 90% was reached with the rhodium micro reactor. A much higher H_2 selectivity is reached in the micro reactor; the CO selectivity was comparable. The micro channels' outlet temperatures dropped on increasing the amount of inert gas.

This is explained by a possible higher activity of pure rhodium than supported metal catalysts. However, two other reasons are also taken into account to explain the superior performance of the micro reactor: boundary-layer mass transfer limitations, which exist for the laboratory-scale monoliths with larger internal dimensions, are less significant for the micro reactor with order-of-magnitude smaller dimensions, and the use of the thermally highly conductive rhodium as construction material facilitates heat transfer from the oxidation to the reforming zone.

3.3.11
Oxidation of Carbon Monoxide to Carbon Dioxide

Peer-reviewed journals: [78]; proceedings: [7].

3.3.11.1 Drivers for Performing the Oxidation of Carbon Monoxide to Carbon Dioxide

This reaction serves for removal of carbon monoxide from gas mixtures and is usually carried out over supported metal catalysts. In reforming techniques, carbon monoxide, poisonous for the catalyst in fuel cells, is removed in such a way. It is also applied in automobiles for reducing the exhaust gas carbon monoxide to an environmentally acceptable level.

The carbon monoxide reaction is well studied and the observed kinetics are well understood. Of particular interest is the so-called 'CO-inhibiting regime', characterized by carbon dioxide covering and blocking the surface, so that the reaction rate is governed by CO desorption rate (see original citations in [78]).

3.3.11.2 Beneficial Micro Reactor Properties for the Oxidation of Carbon Monoxide to Carbon Dioxide

Owing to its nature as a test reaction, rather the reactor and its operational modes were tested, mainly to determine mass transfer limits (see Section 3.3.11.3). It was also used for kinetic studies on the performance of various catalysts.

Gas-phase reaction 9 [GP 9]: Oxidation of carbon monoxide to carbon dioxide

$$CO \xrightarrow[\;O_2\;]{\;Al_2O_3 / Pd\ (Rh,\ Pt)\;} CO_2$$

3.3.11.3 Typical Results
Kinetic data: turn-over frequency (TOF) and activation energies
[GP 9] [R 16] The reaction rate and activation energy of metal catalysts (Rh, Pt or Pd) supported on alumina particles (~3 mg; 53–71 μm) were determined for conversions of 10% or less at steady state (1% carbon monoxide; 1% oxygen, balance helium; 20–60 sccm; up to 260 °C) [7, 78]. The catalyst particles were inserted into a meso-channel as a mini fixed bed, fed by a bifurcation cascade of micro-channels. For 0.3% Pd/Al$_2$O$_3$ (35% dispersion), TOF (about 0.5–5 molecules per site

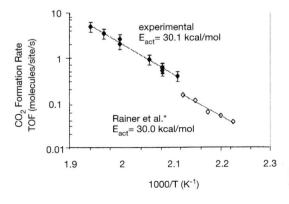

Figure 3.45 Experimental TOF for CO oxidation. Data given in [78].

and second) and apparent activation energy (30.1 kcal mol^{-1}) values compare well with literature values (Figure 3.45). The same was found for Pt/Al$_2$O$_3$ and Rh/Al$_2$O$_3$ catalysts.

Kinetic data: reaction orders

[GP 9] [R 16] When using metal catalysts (Rh, Pt or Pd) supported on alumina particles (~3 mg; 53–71 μm) in a wide mini fixed bed, experimental reaction orders of 0.79 for CO and +1.0 to +1.5 for O$_2$ were found (Figure 3.46) and compare to literature data [78].

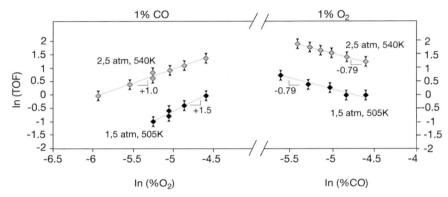

Figure 3.46 Determination of reaction orders for CO and O$_2$ for CO oxidation at different temperatures [78].

Thermal management/heat-sink properties

[GP 9] [R 16] When using metal catalysts (Rh, Pt, or Pd) supported on alumina particles (~3 mg; 53–71 μm) in a wide mini fixed bed, no temperature increase, as monitored by thermocouples, was observed [78]. This shows the high heat dissipating capacity of the silicon micro-reactor used, i.e. that it can be used efficiently as a heat sink. In the same way, the heat management is governed by the cartridge heaters, dominating the heat release by reaction.

Internal transport limits

[GP 9] [R 16] The extent of internal transport limits was analysed for the wide fixed-bed reactor, using experimental data on carbon monoxide conversion and matter and process parameter data for the reactants [78]. The analysis was based on the Weisz modulus and the Anderson criterion for judging possible differences between observed and actual reaction rates. As a result, it was found that the small particles eliminate internal transport limitations.

External transport limits

[GP 9] [R 16] The extent of external transport limits was made in an approximate manner as for the internal transport limits (see above), as literature data on heat and mass transfer coefficients at low Peclet numbers are lacking [78]. Using a Fick's law analysis, negligible concentration differences from the bulk to the catalyst sur-

face were found. Thus, mass transfer to the catalyst is not limiting. In an analogous manner, it was found that only a negligible temperature difference, such as 0.01 K, from the fluid bulk to the catalyst surface exists.

Diffusional mixing/differential behavior
[GP 9] [R 16] By finite-element reactor modeling, it was shown that for conversions as large as 34%, concentration differences within the mini wide fixed-bed reactor of only less than 10% are found [78]. Thus, the reactor approximates a continuous-stirred tank reactor (CSTR). This means that the mini wide fixed-bed reactor yields differential kinetics even at large conversions, larger than for reactors used so far (< 10% conversion).

Residence time
[GP 9] [R 16] When using metal catalysts (Rh, Pt or Pd) supported on alumina particles (~3 mg; 53–71 µm) in a wide mini fixed bed, residence times in the range 0.6–8.0 ms were applied (1% carbon monoxide, 1% oxygen, balance helium; 20–60 sccm; up to 260 °C) [78].

3.3.12
Andrussov Process

Proceedings: [2]; short mentioniong on the Degussa variant for hydrogen-cyanide synthesis (no details): [71]; sections in reviews: [87, 88, 90, 96, 97, 114]; trade press: [81].

3.3.12.1 Drivers for Performing the Andrussov Process
The Andrussov process is, in addition to the Degussa and BASF processes, the major way to generate hydrogen cyanide which is an important chemical product used for many syntheses such as poly(methyl methacrylate) production. The conventional Andrussov process is performed using several layered nets of platinum as catalyst material, so-called gauzes [2, 81]. Gauzes of various mesh numbers and wire diameters are assembled as a stack to adjust flexibly the overall fluid passage length and porosity. The reactant gases are pre-heated to about 600 °C, pass the gauze catalyst and enter a heat exchanger. Typical process parameters rely on temperatures of about 1050 °C and residence times of 0.1 ms. Methane is also converted completely with a selectivity of about 60%. The ammonia selectivity is still higher, of the order of 60–70%, but conversion is not complete and considerable amounts of ammonia remain unreacted.

3.3.12.2 Beneficial Micro Reactor Properties for the Andrussov Process
The investigations were done in the framework of contract research for six industrial companies. Since it was done in a very early stage of development, it can be considered as a pioneering effort. The motivation was industrially driven, however, in a more generic way, aimed at showing general advantages of micro reactors for commercial use. Not only was performance enhancement envisaged, but also it was intended to demonstrate the robustness and capability of the microfabrication techniques at that time. In this context, one selection criterion for the Andrussov process, besides its prominent nature as an industrial process for hydrogen cya-

nide, was the high temperature requirements (> 1000 °C) for performing this reaction. This posed considerable constraints on material choice and assembly. Furthermore, since hydrogen cyanide is very poisonous, the safe handling of small volumes of hazardous compounds was seen as a beneficial property. Although the investigations were only aimed at proving feasibility, on-site production in flexible quantities was a long-term goal behind such industrial investigations.

Gas-phase reaction 10 [GP 10]: Andrussov Process

$$CH_4 + NH_3 + 3/2\ O_2 \xrightarrow{\ Pt\ } HCN + 3\ H_2O \quad \Delta H = -474\ kJ/mol$$

3.3.12.3 Typical Results
Conversion/selectivity/yield
[GP 10] [R 18] Ammonia selectivity at a μ-gauze platinum catalyst is 62%, the methane selectivity amounts to 30% (70 ml h^{-1} methane; 70 ml h^{-1} ammonia; 500 ml h^{-1} air; 1 bar; 977 °C) [2, 81]. Ammonia and methane conversions are 45 and 98%, respectively. Significant amounts of carbon monoxide and dioxide are formed as total oxidation products. The best HCN yield determined is 31% (70 ml h^{-1} methane; 70 ml h^{-1} ammonia; 500 ml h^{-1} air; 1 bar; 963 °C).

Catalyst temperature
[GP 10] [R 18] Below 900 °C, no HCN formation at a μ-gauze platinum catalyst was observed, but a significant amount of total oxidation (70 ml h^{-1} methane; 70 ml h^{-1} ammonia; 500 ml h^{-1} air; 1 bar) [2]. At 900 °C, about 70% methane conversion is determined, giving about equal amounts of carbon dioxide and carbon monoxide. In the range 900–980 °C, an increasing amount of HCN is formed reaching finally a yield of about 31%. At this point, near complete methane conversion is achieved (Figure 3.47).

From 860 to 980 °C, the carbon monoxide yield increases from near zero to about 50%, while the carbon dioxide yield decreases from about 30 to 15% [2]. This shift is according to the Boudouard equilibrium.

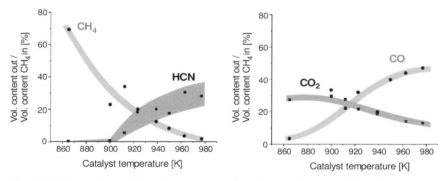

Figure 3.47 Dependence of the volume contents of methane, hydrogen cyanide, carbon monoxide and carbon dioxide, normalized to the methane content, on the reaction temperature (measurements made by BASF) [2].

Reactor cold spots/water condensation

[GP 10] [R 18] Cold spots in the reactor lead to water condensation [2]. This results in dissolution of water-soluble gases such as ammonia and hydrogen cyanide. As a consequence, the measurements of the gas levels of these compounds may differ considerably from the real values. After dissolution they are much too low; after re-evaporation of the aqueous solutions, unrealistically high values appear.

Benchmarking to industrial process

[GP 10] [R 18] The ammonia selectivity of 62% at a µ-gauze platinum catalyst is in the range of the technical process (Figure 3.48); the methane selectivity of 30% is only half the industrial performance (70 ml h^{-1} methane; 70 ml h^{-1} ammonia; 500 ml h^{-1} air; 1 bar; 977 °C) [2, 81]. The reason for this difference is due to much more total oxidation reactions in the micro reactor. The methane conversion of 98% is similar to the industrial performance (but at the expense of selectivity, as mentioned above). The HCN yield of 31% (70 ml h^{-1} methane; 70 ml h^{-1} ammonia; 500 ml h^{-1} air; 1 bar; 963 °C) is about half the value for the industrial process.

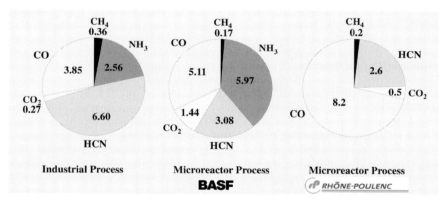

Figure 3.48 Product spectra obtained by measurements of BASF and Rhône-Poulenc compared with results of an industrial process. The data given are vol.-% of the product mixture [2].

Benchmarking to laboratory-scale monolith reactors

[GP 10] [R 18] The best HCN yield of 31% at a µ-gauze platinum catalyst (70 ml h^{-1} methane; 70 ml h^{-1} ammonia; 500 ml h^{-1} air; 1 bar; 963 °C) is much better than the performance of monoliths (Figure 3.49) having similar laminar flow conditions [2]. A coiled strip and a straight-channel monolith have yields of 4 and 16%, respectively. The micro-reactor performance is not much below the best yield gained in a monolith operated under turbulent-flow conditions (38%).

The difference in reactor performance is due to the difference in hydraulic diameters of the reaction channels, i.e. related to varying mass-transfer limitations. The micro channels of the µ-gauze platinum catalyst amount to 70 µm, whereas the monoliths have channel/pore diameters of 500–1200 µm.

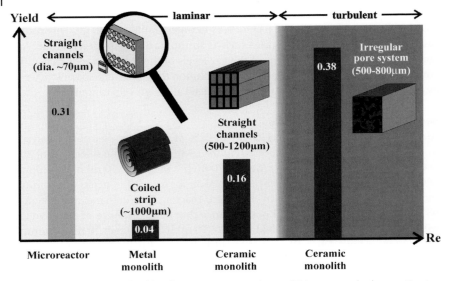

Figure 3.49 Comparison of yields of a micro reactor and monolithic reactors, both operating in the same flow regime [2].

3.3.13
Hydrogen/Oxygen Reaction

Peer-reviewed journals: [9]; proceedings: [82, 115–118]; sections in trade press: [92, 119]; sections in reviews: [90, 95–97, 100].

3.3.13.1 Drivers for Performing the Hydrogen/Oxygen Reaction

The reaction between hydrogen and oxygen leads to the formation of water. This reaction has extended explosive regimes with respect to the p, T, c-parameters. A mechanistic analysis of the elementary reactions is available and the explosion mechanisms are understood in detail. Accordingly, this reaction serves well as a model for other dangerous processes in the explosive regime such as many oxidations with pure oxygen.

Besides forming water as product, hydrogen peroxide can be generated by the oxidation of hydrogen by using special catalysts. Most of these processes involve the presence of a liquid phase in addition to the gas phase, e.g. relying on trickle-bed reactor design. The direct synthesis of hydrogen peroxide is attractive since it is believed to be a more economical and less waste-producing process than the established anthraquinone route. Industrial exploitation so far was hindered by the high safety risk of such processes. Indeed, for the handling of oxygen and hydrogen, a series of explosions, many even with lethal accidents, have been reported in the past.

3.3.13.2 Beneficial Micro Reactor Properties for the Hydrogen/Oxygen Reaction

Already the early pioneering reviews in the field have mentioned safe operation in micro reactors even in the explosive regime as one of the most relevant drivers [71,

120]. Later, much more detailed predictions of safe operation in micro reactors in the explosive regime with the example of the hydrogen/oxygen reaction could be given [9, 82, 117, 118]. It was predicted that micro-channel processing under given conditions can be considered as intrinsically safe.

Safe processing was found experimentally in parallel with this theoretical funda-ment [115, 116], both leading to further experimentation [9, 82, 117, 118]. Although the hydrogen/reaction is not of direct use itself, it stands as a prominent model reaction for other more valuable processes (see, e.g., [GP 2] and [GP 3]), for which benefits due to safe processing in novel explosive regimes are expected.

Gas-phase reaction 11 [GP 11]: hydrogen oxidation

$$2\ H_2 + O_2 \rightarrow 2\ H_2O \quad \Delta H = -241\ kJ\ mol^{-1}$$

3.3.13.3 Typical Results
Conversion/selectivity/yield
[GP 11] [R 5] Complete conversion is achievable, provided that proper process con-ditions are applied [115].

Process temperatures
[GP 11] [R 19] Temperatures close to 1200 °C, near the mechanical limit of the ce-ramic reactor material, have been achieved [9, 115]. With improved sealing and better material, processing could lead to 1300 °C [9].

Process safety
[GP 11] [R 19] Based on an analysis of the thermal and kinetic explosion limits, in-herent safety is ascribed to hydrogen/oxygen mixtures in the explosive regime when guided through channels of sub-millimeter dimensions under ambient-pressure conditions [9]. This was confirmed by experiments in a quartz micro reactor [9].

[GP 11] [R 19] Three kinetic explosion limits for a stoichiometric H_2/O_2 mixture and three different reactor diameters, d = 1 splm, 1 mm and 100 μm, are given in [9] (200–1450 °C; $10–10^8$ Pa).

[GP 11] [R 19] The third explosion limit is discussed in detail in [9] as it is impor-tant from both practical and mechanistic viewpoints (230–950 °C; $10–10^9$ Pa). This limit is normally responsible for the occurrence of explosions under ambient pres-sure conditions. In addition, these explosions are known to be kinetically induced by radical formation. The formation of these species is sensitive to size reduction of the processing volume owing to the impact of the wall specific surface area on radical chain termination. It turns out that the wall temperature has a noticeable, but not decisive influence on the position of the third limit.

The thermal explosion limit lies below the kinetic limit for all conditions speci-fied above (Figure 3.50) [9].

[GP 11] [R 19] The suppression of explosive homogeneous gas-phase reactions is not due simply to thermal quenching as a result of the heat losses from a micro reactor, but rather to radical quenching [9]. The micro reactor will therefore be safe even when heat losses from the reaction micro channel are reduced by design modi-fications.

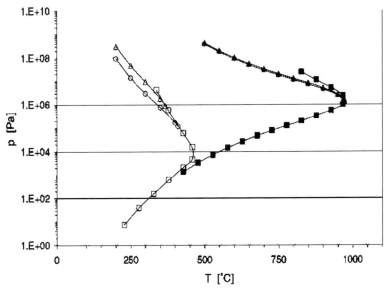

Figure 3.50 Extending kinetic explosion (squares) and thermal explosion limits by using a micro reactor with 300 μm channel diameter (filled symbols). Calculated values for $T_{wall} = T_{room}$ (circles) and $T_{wall} = T_{reaction}$ (triangles). Comparison with 1 m diameter (open symbols) [9].

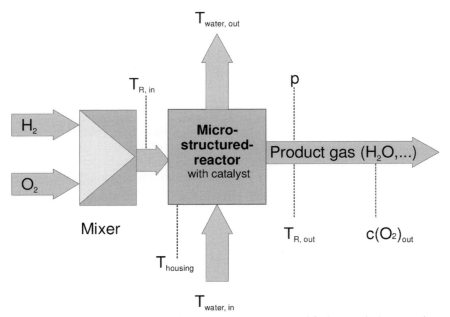

Figure 3.51 Schematic flow scheme of a micro-reactor system used for burning hydrogen and oxygen at large throuputs. A mixer is applied for reducing the volume of the explosive hydrogen/oxygen mixture in advance to the reaction [115].

[GP 11] [R 19] An impressive example of the impact of miniaturization on the explosion limit is given in [9]. For a conventional reactor of 1 m diameter, explosive behavior sets in at 420 °C at ambient pressure (10^5 Pa). In turn, an explosion occurs at about 750 °C when the reactor diameter is decreased to about 1 mm. A further reduction to 100 μm shifts the explosive regime further to higher pressures and temperatures. Even the first explosion limit is above ambient pressure. Now, explosive behavior can be excluded and so the reaction becomes inherently safe.

[GP 11] [R 5] Owing to the possibility of achieving complete conversion, process conditions can be found to yield off-gas, leaving the micro reactor, which is no longer in the explosion envelope [115].

Addition of water/kinetics
[GP 11] [R 5] The addition of up to 7 vol.-% of water has no detectable impact on the oxygen reaction rate [121]. This is a hint that desorption of water generated is not the rate-determining step (Figure 3.51).

Variation of oxygen concentration
[GP 11] [R 5] An increase in oxygen concentration at constant hydrogen concentration did not lead to a maximum of the oxygen reaction rate [121].

Variation of hydrogen concentration
[GP 11] [R 5] An increase in hydrogen concentration at constant oxygen concentration led to a maximum of the oxygen reaction rate (2.0–7.0 mmol l^{-1} hydrogen; 3.6 mmol l^{-1} oxygen; 48–70 °C) [121]. The maximum is found at a hydrogen/oxygen ratio of 1. This behavior could be described by Langmuir–Hinshelwood kinetics.

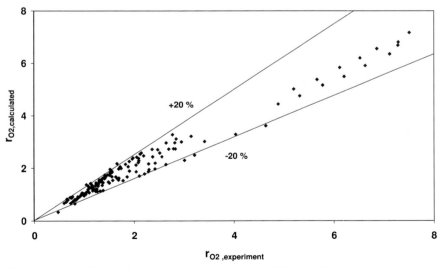

Figure 3.52 Parity diagram based on Langmuir–Hinshelwood kinetics of the oxygen reaction rate [121].

Applicability of Langmuir–Hinshelwood kinetics

[GP 11] [R 5] Langmuir–Hinshelwood kinetics adequately describe the observed results as a parity diagram (Figure 3.52), comparing experimental with theoretical values (2.0–7.0 mmol l^{-1} hydrogen; 3.6 mmol l^{-1} oxygen; 48–70 °C) [121].

Kinetic data

[GP 11] [R 5] Numerically iterated parameter values for the Langmuir–Hinshelwood kinetics were listed, including activation energy, oxygen reaction rate, and enthalpy (2.0–7.0 mmol l^{-1} hydrogen; 3.6 mmol l^{-1} oxygen; 48–70 °C) [121].

Mass-Transfer limitation/Mears criterion

[GP 11] [R 5] A judgement on mass-transfer limitations on the reaction rate according to the Mears criterion was made [121]. This inequality predicts no such limitations in the boundary layer when the Mears criterion is smaller than 0.15. Using process parameter data applied in a number of experiments, the highest value found is $6 \cdot 10^{-2}$ so that no mass-transfer limitations have to be assumed.

Heat-transfer limitation/Anderson criterion

[GP 11] [R 5] A judgement on heat-transfer limitations on the reaction rate according to the Anderson criterion was made [121]. This inequality predicts no such limitations in the boundary layer when the Anderson criterion is smaller than 1. Using process parameter data applied in a number of experiments, the highest value found was $2.2 \cdot 10^{-6}$ so that no heat-transfer limitations have to be assumed.

Temperature rise

[GP 11] [R 5] No temperature increase of the gas flow in micro channels was observed experimentally, being cooled by a guiding water mass flow of 4 g s^{-1} through perpendicularly arranged micro channels [121]. Referring to these experimental conditions, a maximum reaction power of 35 W was calculated for a maximum gas flow of 7.5 Nl min^{-1}.

Theoretical benchmarking to other reactor concepts

[GP 11] [R 5] Calculated heat-transfer coefficients, heat-exchange areas and estimated temperature gradients between the inner reactor zone and the outer wall were given, taking into account a reaction with 30 W heat generation [121]. In addition to considering a cross-flow micro reactor, the performances of a fixed-bed reactor and ceramic and metal honeycomb reactors were calculated (Table 3.3). The result was that the micro reactor gives the smallest estimated temperature increase (1.4 K) due to the largest heat exchange area and largest heat transfer coefficient.

Temperature distribution

[GP 11] [R 19] For an autothermal reactor, i.e. a device with neither internal nor external heat transfer, steep temperature profiles along the flow axis were found [9]. Via an inspection window, glowing of the front zone of the wire reactor was observed, indicating complete conversion within a few mm reaction passages. The

Table 3.3 Comparison of the calculated heat-transfer coefficients, heat-exchange areas and the estimated temperature gradients between the central reactor area and the outer wall of the reactor.

	k $(W\ m^2\ K^{-1})$	A (m^2)	ΔT (K)
Fixed-bed reactor (length 5 mm)	$\sim 7 \cdot 10^3$	$1.5 \cdot 10^{-4}$	27
Ceramic honeycomb reactor (number of honeycombs 32; diameter 0.63 mm; length 14 mm)	$\sim 2 \cdot 10^3$	$2.5 \cdot 10^{-4}$	58
Metal honeycomb reactor (number of honeycombs 17; diameter 1.17 mm; length 14 mm)	$\sim 1.1 \cdot 10^4$	$3.6 \cdot 10^{-4}$	8
Micro reactor (length 14 mm)	$1.539 \cdot 10^4$	$1.1 \cdot 10^{-3}$	1.4

rest of the reactor remained dark, which is explained by massive heat losses, since no special measures for insulation were taken.

[GP 11] [R 19] By increasing the flow rate, the above-mentioned hot spot can be moved downstream in the flow direction, as is evident from the moving zone of glow [9].

Residence time

[GP 11] [R 19] Using the above visual inspection technique, the minimum residence time can be estimated. Flow was increased and so the residence time decreased, until the whole reactor was glowing. This corresponds to a residence time as low as approximately 50 µs (0.1 slpm hydrogen, 0.14 slpm oxygen, 0.45 slpm nitrogen) [9].

Thermal management

[GS 11] [R 5] At high mass flows and with a large heat release, an undesired, substantial temperature increase even for highly efficient micro heat exchanger–reactors is observed (Figure 3.53) [115].

[GP 11] [R 19] Gas-exit temperatures ranging from ambient temperature to nearly 1200 °C were measured (Figure 3.54) as a function of the so-called equivalence ratio, a function of the H_2/O_2 ratio (0.5 slpm synthetic air, 0.1 slpm oxygen + 0.2 slpm nitrogen or 0.5 slpm oxygen + 0.75 slpm nitrogen; varying hydrogen content) [9]. A steep temperature increase was observed; a maximum was found at nearly stoichiometric conditions. For lower nitrogen dilutions, this maximum approaches nearly 1200 °C, whereas lower values are found when using air, less rich in oxygen. The maximum temperature also increases with increasing total flow rate. This is correlated with the heat losses, depending on flow rate and residence time.

[GP 11] [R 19] The gas-exit temperatures in a quartz micro reactor are higher than in a silicon micro reactor of similar design for reasons of strongly reduced heat losses, given the same process parameters (0.5 slpm synthetic air, 0.1 slpm oxygen + 0.2 slpm nitrogen or 0.5 slpm oxygen + 0.75 slpm nitrogen; varying hydrogen content) [9].

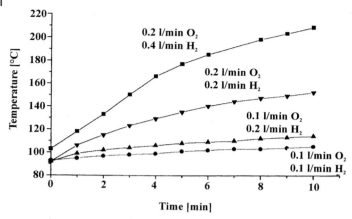

Figure 3.53 Temperature of the exiting reactant gas plotted against time on-stream for various operating conditions. N_2 as diluent, 1.0 l min^{-1}; N_2 as coolant, 3.0 l min^{-1} [115].

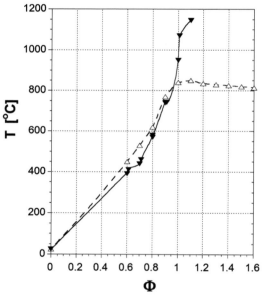

Figure 3.54 Measured gas exit temperatures for a catalytic H_2/O_2 reaction with varying H_2 content in 0.5 slpm synthetic air ($\triangle$) and oxygen ($\blacktriangledown$) enriched air (0.1 slpm oxygen + 0.2 slpm nitrogen) [115].

Reactor material/type of micro reactor

[GP 11] [R 19] The gas exit temperatures in a quartz micro reactor are higher than in a silicon micro reactor of similar design for reasons of strongly reduced heat losses, given the same process parameters (0.5 slpm synthetic air, 0.1 slpm oxygen + 0.2 slpm nitrogen or 0.5 slpm oxygen + 0.75 slpm nitrogen; varying hydrogen content) [9].

[GP 11] [R 19] No qualitative differences between the reaction performance of a wire-based silicon-chip and a quartz-shell micro reactor were observed [9].

Wall material

[GP 11] [R 19] The influence of different wall materials on the ignition behavior of the hydrogen/oxygen reaction using air was studied in numerical simulations with a two-dimensional boundary-layer model of a micro-channel reactor [117]. The model incorporates balance equations for mass, energy and momentum. Detailed elementary-step kinetics for both surface reactions and homogeneous gas-phase reactions were used in the simulations. Based on prior experimental evidence, parameters were chosen. The walls were divided into three categories: catalytic, so-called radical combination and inert. Platinum was taken for the first; the second represents materials such as quartz-glass and silicon, and the third stands for 'un-hampered' homogeneous gas-phase reaction.

Strong wall effects were observed for the catalytic and radical-combination walls [9]. In addition to the ignition-inhibiting effect found for both types, the catalytic wall displays a further effect, being ignition-promoting, which is explained by higher reaction rates on the catalytic surface as for the homogeneous reaction. This means that whereas flames and explosions may be suppressed in a micro channel, conversion at the heterogeneous surface may be high and so the range of reactive conditions is considerably increased.

Novel reactor concepts/membrane reactor

[GP 11] [R 20] Investigations with a Pd membrane reactor relied on reaction of streams separated via a membrane (to prevent complete mixing of reactants, not to enhance conversion) [11]. A hydrogen/nitrogen stream is guided parallel to an oxygen stream, both separated by the membrane and water is thereby formed. The membranes, made by thin-film processes, can sustain a pressure up to 5 bar.

3.3.14
Oxidation of Formamides – Synthesis of Methyl Isocyanate

Proceedings: [71]; short mention of cyclohexyl and butyl isocyanate syntheses (no details): [71]; sections in reviews; [95, 97, 100].

3.3.14.1 Drivers for Performing the Synthesis of Methyl Isocyanate

The corresponding industrial investigation had a truly pioneering character as it was the first industrial study on micro reactors to be published [71]. In the framework of this extended study, many catalytic gas-phase reactions were carried out in a micro structured reactor. The reactions were chosen according to selection criteria defined by the company owing to an extensive analysis of their suitability for micro flow processing. The following criteria were chosen: high temperature, dangerous, catalytic and photochemical. Concerning heterogeneous reactions, the catalyst was introduced in the form of solid particles as a mini fixed bed. For the synthesis of methyl isocyanate from methylformamide, similar conversions for the conventional synthesis could be determined, at low selectivities, however.

3.3.14.2 Beneficial Micro Reactor Properties for the Synthesis of Methyl Isocyanate
The main expectations of industrial researchers focused on improving heat management and increasing safety for hazardous process [71].

Gas-phase reaction 12 [GP 12]: synthesis of methyl isocyanate

$$\text{Me} - \underset{\underset{\text{H}}{|}}{\text{N}} \!\!\!\diagup\!\!\!\!\diagdown^{\text{O}}_{\text{H}} \quad \xrightarrow[\text{Ag}]{\text{O}_2} \quad \text{Me}-\text{N}=\text{C}=\text{O}$$

Methyl isocyanate is obtained by oxidation of methylformamide over a silver catalyst [71].

3.3.14.3 Typical Results
Conversion/selectivity/yield
[GP 12] [R 15] For the synthesis of methyl isocyanate from methylformamide similar conversions as for the conventional synthesis could be determined at low selectivities [71]. One reason for this is seen in the non-ideal temperature profiles within the reaction zone of the microstructured reactor packing.

Since then, numerous industrial laboratory investigations have been carried out [71].

3.4
Hydrogenations

3.4.1
Cyclohexene Hydrogenation and Dehydrogenation

Peer-reviewed journals: [74, 122]; proceedings: [20, 73, 123–126]; sections in reviews: [90, 94, 97].

3.4.1.1 Drivers for Performing the Cyclohexene Hydrogenation and Dehydrogenation
The reaction of cyclohexene in the presence of hydrogen at a Pt catalyst can lead to cyclohexane via hydrogenation and benzene via dehydrogenation. The hydrogenation and dehydrogenation of cyclohexene over a Pt catalyst are model reactions for important reaction classes in the petroleum industry and thus were studied extensively by many groups (see original citation in [74]). They serve to model hydrotreating, reforming and fuel processing.

3.4.1.2 Beneficial Micro Reactor Properties for Cyclohexene Hydrogenation and Dehydrogenation
The present investigations were largely motivated to show the serial-screening capabilities of the reactor concept used. The speed of process-parameter changes, consumption of small volumes only, preciseness of kinetic information, and robustness were major micro reactor properties utilized.

Gas-phase section 13 [GP 13]: hydrogenation and dehydrogenation of cyclohexene

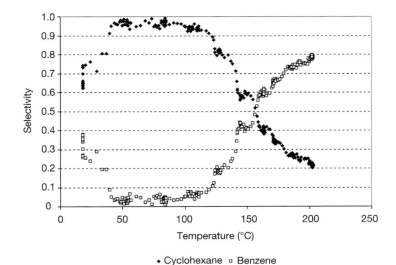

3.4.1.3   Typical Results
Conversion/selectivity/yield

[GP 13] [R 12] For the reaction of cyclohexene and hydrogen using a sputtered Pt catalyst on silicon, an initial conversion near 50% is found at room temperature (0.3 sccm hydrogen; 1.0 sccm argon saturated with cyclohexene vapor; 25–200 °C; 1 bar) [74]. An increase in temperature up to 55 °C leads to almost 100% conversion. At 120 °C and higher, conversion declines steeply, independent of temperature and only as a function of time (Figure 3.55). Conversion at 200 °C after several hours of operation was 55%, so nearly matching the initial room-temperature activity.

This room-temperature result followed by the steep increase in conversion up to 55 °C is correctly predicted by a reactor model [74]. The decline in conversion is not anticipated and is not due to equilibrium effects.

Product selectivities for cyclohexane (hydrogenation path) and benzene (dehydrogenation path) were monitored as a function of temperature (0.3 sccm hydrogen; 1.0 sccm argon saturated with cyclohexene vapor; 25–200 °C; 1 bar) [74]. Initial selectivities were about 60% for cyclohexene and 40% for benzene, but displayed transient behavior, i.e. shifted to larger cyclohexane contents with time. At 60 °C, selectivities are stable, as are the conversions. Now, 100% cyclohexane is formed.

Figure 3.55 Selectivity of cyclohexane and benzene depending on reaction temperature [74].

At temperatures exceeding 120 °C, selectivity shifts to the formation of benzene. This shift starts before the conversion decline, and hence is not related solely to it. At 200 °C, 80% benzene is generated.

Temperature dependence

[GP 13] [R 12] The results determined by increasing temperature via ramps are discussed in the previous section [74].

See also the section Characteristic inner diameter.

Transient catalyst behavior

[GP 13] [R 12] For the reaction of cyclohexene and hydrogen using a sputtered Pt catalyst on silicon, the initial transient shows that the catalyst is readily active for conversion of cyclohexene [74]. This shift is explained by establishing steady-state conditions on the catalyst surface. It needs time until the surface of the catalyst is covered by chemisorption of cyclohexene species. Hence local areas of the surface can promote dehydrogenation for a while until coverage is completed. The decline in catalyst activity is explained by an irreversible change of the catalyst surface. Irreversible species adsorption blocks the active catalyst sites and leads to poisoning.

See also the section Characteristic inner diameter.

Micro-channel dimensions

[GP 13] [R 12] A parametric study on the impact of channel depth (Figure 3.56) and length on conversion, based on a reaction probability model, was investigated for the dehydrogenation of cyclohexane to benzene, using literature-reported reaction probabilities for various Pt-carrier catalysts [10]. As expected, the channel dimensions have a large impact on conversion owing to the role of diffusion.

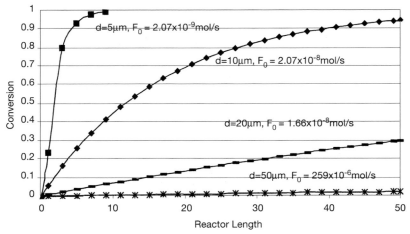

Figure 3.56 Results of a parametric study on the impact of channel depth on benzene conversion. Reaction probability $= 10^{-6}$; $T = 200$ °C; $p_{in} = 1.1$ atm; $p_{out} = 1$ atm [10].

For certain process parameters, complete conversion is achieved in a 5 μm channel, whereas zero conversion is given for a 50 μm channel. Similarly, the catalyst activity has a strong effect.

Hydrogen/cyclohexene partial pressure
[GP 13] [R 12] For the reaction of cyclohexene and hydrogen using a sputtered Pt catalyst on silicon, the benzene yield is the higher the lower is the ratio of partial pressures of the two reactants $p_{H_2}/p_{C_6H_{10}}$ ($p_{H_2}/p_{C_6H_{10}}$: 0.5–27; 0.3, 0.6, 1.0 sccm argon saturated with cyclohexene vapor; 200 °C; 1 bar) [74]. This is in line with the chemical equilibrium favoring dehydrogenation, at low hydrogen contents. In turn, high hydrogen amounts favor hydrogenation. Not evident at first sight, dehydrogenation does not occur unless small initial hydrogen contents are available. This is explained by the need to condition the catalyst surface with hydrogen.

The cyclohexane yield, different from benzene, depends only weakly on the $p_{H_2}/p_{C_6H_{10}}$ ratio [74].

See also the section Characteristic inner diameter.

Residence time
[GP 13] [R 12] For the reaction of cyclohexene and hydrogen using a sputtered Pt catalyst on silicon, the benzene yield is higher with increasing residence time (0.3–0.3 sccm hydrogen; 0.3, 0.6, 1.0 sccm argon saturated with cyclohexene vapor; 200 °C; 1 bar; 120–570 ms) [74]. The cyclohexane yield has no strong effect. Hence increasing the reactor length at constant flow will result in more benzene formation (Figure 3.57).

See also the section Characteristic inner diameter.

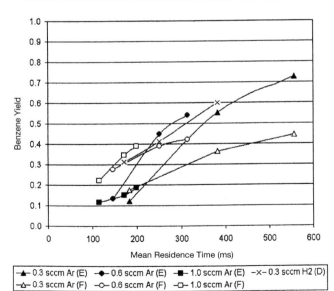

Figure 3.57 Strong effect of residence time on benzene yield [74].

Space–time yield

[GP 13] [R 12] For the reaction of cyclohexene and hydrogen using a sputtered Pt catalyst on silicon, the space–time yield strongly depends on the concentration of both reactants [74]. Increasing the hydrogen partial pressure reduces the benzene yield (p_{H_2}: 50–580 Torr; 0.3, 0.6, 1.0 sccm argon saturated with cyclohexene vapor; 200 °C; 1 bar); naturally, increasing the cyclohexene concentration has the opposite effect (0.3 sccm H_2; $p_{H_2}/p_{C_6H_{10}}$: 20–80 Torr; 0.3, 0.6, 1.0 sccm argon saturated with cyclohexene vapor; 200 °C; 1 bar). Cyclohexane formation has no distinct dependence on either reactant.

The maximum space–time yield observed at 200 °C is $8.0 \cdot 10^{-4}$ kg m$^{-2}_{(cat)}$ min^{-1} for benzene and $4.3 \cdot 10^{-4}$ kg m$^{-2}_{(cat)}$ min^{-1} for cyclohexane.

See also the section Characteristic inner diameter.

Reactant consumption/environmental

[GP 13] [R 12] For the reaction of cyclohexene and hydrogen using a sputtered Pt catalyst on silicon, only 3.1 g of cyclohexene were consumed in over 140 h of reaction [74]. Such designed silicon micro reactors in principle could be disposable, assuming economic mass fabrication and the installation costs of the fluidic peripherals to be relatively low. Overall, this implies an economical solution for process development and catalyst testing.

Only 32 mg of cyclohexene were needed for conducting an 18 h experiment [73].

Characteristic inner diameter

[GP 13] [R 12] For the reaction of cyclohexene and hydrogen using a sputtered Pt catalyst on silicon, the channel width was varied to study the impact of such enhanced mass transport on the conversion or selectivity as a function of various process parameters [74]. Two different reactor designs were employed, having micro channels of 100 and 5 µm width, respectively. The number of channels was adjusted to the ratio of widths, so that 39 and 780 channels were operated in parallel, respectively.

Concerning the dependence of conversion on mean residence time, the 5 µm channels show only a slight dependence, the 100 µm-channel whereas show a distinct increase in conversion with increasing residence time (0.1-, 0.3-, 1.0 sccm hydrogen; 0.3, 0.6, 1.0 sccm argon saturated with cyclohexene vapor; 200 °C; 1 bar; 150–650 ms) [74].

Concerning the dependence of conversion on the partial pressure ratio $p_{H_2}/p_{C_6H_{10}}$, the 5 µm channel show only a slight downwards slope, the 100 µm-channel whereas show a steep upwards slope with increasing ratio (0.1 sccm hydrogen; 0.3, 0.6 sccm argon saturated with cyclohexene vapor; 200 °C; 1 bar; $p_{H_2}/p_{C_6H_{10}}$: 1–27)[74]. At low ratios, both reactors have an initial upwards slope. Under the conditions applied, benzene formation dominates.

Concerning the dependence of selectivity towards cyclohexane and benzene on the partial pressure ratio $p_{H_2}/p_{C_6H_{10}}$, the 5 and 100 µm channels show qualitatively similar behavior. However, the 5 µm channels yielded similar contents of cyclohexane and benzene over the range of ratios studied, whereas the 100 µm-channels give about four times more benzene than cyclohexane (0.3, 1.0 sccm hydrogen; 0.3, 1.0 sccm argon saturated with cyclohexene vapor; 200 °C; 1 bar; $p_{H_2}/p_{C_6H_{10}}$: 1–27) [74].

Concerning the dependence of selectivity towards cyclohexane and benzene on temperature, the 5 and 100 µm channels show qualitatively similar behavior (0.3 sccm hydrogen; 1.0 sccm argon saturated with cyclohexene vapor; 200 °C; 1 bar; 25–210 °C) [74]. The slopes are shifted to higher temperatures for the 5 µm channels. In the region of constant selectivity (50–100 °C), a slightly higher benzene selectivity and slightly lower cyclohexane selectivity is found for the 5 µm channels.

Concerning the dependence of benzene space–time yield on cyclohexene partial pressure, the 5 µm channels show nearly constant behavior, whereas the 100 µm channels display a strongly increasing space–time yield with partial pressure (0.3, 1.0 sccm hydrogen; 0.3, 0.6, 1.0 sccm argon saturated with cyclohexene vapor; 200 °C; 1 bar; $p_{H_2}/p_{C_6H_{10}}$: 20–80) [74].

Mechanistic analysis
[GP 13] [R 12] For the reaction of cyclohexene and hydrogen using a sputtered Pt catalyst on silicon, there have been mechanistic discussions on intermediate species formation [74]. Also, it is supposed that carbonaceous coverage, which is larger in 5 than in 100 µm channels, reduces benzene selectivity with decreasing channel diameter.

Kinetic parameters
[GP 13] [R 12] For the reaction of cyclohexene and hydrogen using a sputtered Pt catalyst on silicon, the reaction probability ($1 \cdot 10^{-6.2}$) and turnover frequency (1.7 molecules per site pere second) were determined and found to be in good agreement with other hydrogenation reactions (0.1 sccm hydrogen; 0.1 sccm argon saturated with cyclohexene; 91% conversion; 200 °C; 1.01 bar) [73].

Conversion/selectivity/yield
[GP 13] [R 22] Using a sputtered platinum layer, studies on the impact of the flow regime in micro channels on conversion were done (8 ccm h^{-1} of condensed cyclohexane; 200 °C; feed and exit pressures 150 000 Pa and 1 hPa, respectively) [127]. During flow passage, different successive regimes take place: continuum flow, quickly changing to slip flow, followed by Knudsen flow. The extent of conversion depends on the regime due to the number of reactant–catalyst collisions. This was demonstrated by regionally sputtering the catalyst so that reaction is conducted mainly under one specific regime. Channels coated in their front part show no conversion of cyclohexane (slip conditions), whereas channels coated more downstream have 2–3% conversion (Knudsen conditions). This is consistent with the expected number of collisions of the molecules with the catalyst in each regime. Making the same experiment with a fully coated channel consequently results in 2–3% conversion.

3.4.2
Hydrogenation of *c,t,t*-1,5,9-Cyclododecatriene to Cyclododecene

Peer-reviewed journals: [38, 39]; proceedings: [18, 84, 128–130]; PhD thesis: [131]; sections in reviews: [87, 88, 90, 94–97, 100].

3.4.2.1 Drivers for Performing the Hydrogenation of *c,t,t*-1,5,9-Cyclododecatriene to Cyclododecene

Selective gas-phase reactions of unsaturated hydrocarbons such as *c,t,t*-1,5,9-cyclododecatriene, 1,5-cyclooctadiene and benzene are examples of reactions that are highly exothermic, mass-transfer limited and have high reaction rates [18, 130]. For bench-scale experiments, heat transfer problems can usually be neglected as only dilute mixtures with low reactant concentrations are used. Instead, mass transfer can be studied on the laboratory-scale. For example, the hydrogenation of *c,t,t*-1,5,9-cyclododecatriene to the corresponding monoene needs kinetically controlled conditions to diminish the formation of the most stable product cyclododecane [18].

The selective hydrogenation of the three compounds mentioned above to their corresponding cyclic monoalkenes is of industrial interest, as the latter can be selectively reacted to more valuable products. For example, cyclododecene can be finally converted to nylon or polyalkenamers.

3.4.2.2 Beneficial Micro Reactor Properties for the Hydrogenation of *c,t,t*-1,5,9-Cyclododecatriene to Cyclododecene

Beneficial micro reactor properties mainly refer to exerting process control over residence time as a key for improving a partial reaction in a consecutive sequence. Similar efforts have also been made for oxidations, but focused more on improving heat transfer (see also Section 3.3.1).

In addition, the studies on the hydrogenation of *c,t,t*-1,5,9-cyclododecatriene gave valuable insight into the way of preparing catalysts and in the impact of the catalyst shape and packing. Here, pioneering efforts were made, influencing later studies.

Together with the ammonia oxidation (see Section 3.3.3), this reaction was the first published, giving substantial details on how micro reactor properties affect the performance of gas-phase reactions.

Gas-phase reaction 14 [GP 14]: hydrogenation of *c,t,t*-1,5,9-cyclododecatriene to cyclododecene

cis,trans,trans-1,5,9- *cis,trans*-1,5 *trans,trans*-1,5-

cyclododecatriene (CDT)   cyclododecadiene (CDD)

cis-   *trans*-

cyclododecene (CDE)   cyclododecane (CDA)

Pd/Al$_2$O$_3$ catalyst.

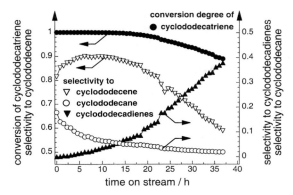

Figure 3.58 Experimental results for partial gas-phase hydrogenation of c,t,t-1,5,9-cyclododecatriene in a Pt micro-channel reactor [130].

3.4.2.3 Typical Results
Conversion/selectivity/yield
[GP 14] [R 4] On a Pd catalyst on nanoporous alumina (via anodic oxidation), time-on-stream measurements (40 h) gave a maximum selectivity of cyclododecene of 85–90% (Figure 3.58) ($p_{CDT}= 0.11$ kPa; $p_{H_2} = 0.33$ kPa; $p = 110$ kPa; 150 °C; 87 ms) [130]. The formation of cyclododecadiene increases from 0% initially to 40% after 37 h. In the same period, cyclododecane selectivity decreases from about 40 to 10%.

Time-on-stream
[GP 14] [R 4] On a Pd catalyst on nanoporous alumina (via anodic oxidation), time-on-stream measurements (80 h) showed constant complete c,t,t-1,5,9-cyclododeca-triene conversion over the whole period (see Table 3.4) [128]. Hydrogen conversion decreased in this period, whereas the selectivity to the cyclic mono-alkene cyclo-dodecene increased. Consequently, the selectivity of the fully hydrogenated prod-uct cyclododecane decreased.

Table 3.4 Reaction parameters for the performance of c,t,t-1,5,9-cyclododecatriene hydrogenation.

Temperature	393 K	Total pressure[a]	110 kPa
Total flow rate	7–26 l h^{-1}	Partial pressure of CDT	110 Pa
Molar flow rate	0.33–1.2 mmol/h	Partial pressure of H_2	330 Pa

a Equilibrated by nitrogen.

Pore length of anodically oxidized support
[GP 14] [R 4] On a Pd catalyst on nanoporous alumina (via anodic oxidation), the impact of the pore length on selectivity and conversion was determined (see also Figure 3.59) [128]. For this purpose, two pieces of activated aluminum wires which differed in pore length were compared. With increasing overall conversion, the

cyclododecene selectivity increased for both pore lengths. However, at very high concentrations exceeding 90%, a different behavior is found. For the short-pore-length pieces, selectivity becomes constant; for long pores, selectivity decreases.

A too long residence time in the catalyst zone, the nanopore, is said to be responsible for this difference. The species with more double bonds than cyclododecene displace cyclododecene readily from the active sites owing to the stronger interactions of their π-orbitals with the catalyst. This prevents further hydrogenation of cyclododecene to cyclododecane. However, if the concentrations of these species are too low, then total hydrogenation to cyclododecane will occur. This will happen when diffusion in the pores is no longer as fast as the reaction and adsorption/desorption processes, i.e. when the diffusion path is too long as a result of too long pores.

Flow pattern regularity – benchmarking to fixed beds

[GP 14] [R 4] On a Pd catalyst on nanoporous alumina (via anodic oxidation), various catalyst carriers other than microchannel based were tested [128] (see also [18] for a first description). The selection included conventionally coated granules, pieces of activated aluminum wires and fragments of activated aluminum foils (see Table 3.5). The total of four carriers differ in pore system, distribution of the catalytically active component and type of packing. Micro-channel platelets are regular with regard to all these features, whereas granules are irregular. The wire pieces and foil fragments have uniform pores and catalyst distribution, but may suffer from their irregular fixed-bed packing.

For the four catalyst carriers, cyclododecene yield was monitored versus selectivity (see Table 3.4) [128]. The yield when using the conventional granules decreases from 62 to 44% within the range of conversions investigated. For the foil fragments, a nearly constant selectivity of 73% was found, hence regular pores and uniform catalyst distribution have an impact. In addition, foils pack more densely than the granules probably give more uniform flow paths with less dead volume. Wire pieces and the micro-channel reactor both give notably better results than the

Table 3.5 Different types of catalysts used for the hydrogenation of *c,t,t*-1,5,9-cyclooctatriene and classification of these. Concerning the latter, each of the left rows is a "yes" function, the right rows serve as "no" function [128].

Catalyst	Type of the catalyst	Pore system (yes)	Pore system (no)	Cross section of catalyst	Distribution of catal. active component (yes)	Distribution of catal. active component (no)	Fixed bed (yes)	Fixed bed (no)	Cross section of fixed bed
CAT A	conventionally coated granules		+			+		+	
CAT B1	pieces of activated Al-wires	+			+			+	
CAT B2	pieces of activated Al-foils	+			+			+	
CAT C	stack of activated and microstructured Al-foils	+			+		+		

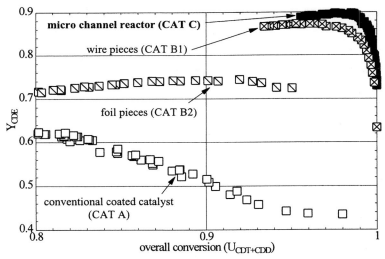

Figure 3.59 Cyclododecene yield vs conversion for different types of catalysts, respectively flow guidances consisting of three fixed beds and one micro-channel passage. Reaction parameters: see Table 3.4. Pore lengths: CAT A, 240 μm; CAT B1, 37 μm; CAT B2, 37 μm; CAT C, 37 μm [128].

foils; the micro-channel reactor is slightly better than the wires. Nearly constant selectivity for a small range is found, decreasing steeply at very high conversions. Wires give still tighter packing than foils which explains their superior performance. Compared with this, the impact of the more uniform flow pattern in the micro channels is relatively small, but exists.

Simulation of concentration profiles in the catalyst carrier's nanopores
[GP 14] [R 4] Using an effective kinetic approach, simulations on the concentration profiles in nanopores, which were models for the alumina support used in the above experiments, were carried out [132]. 'Representative' pores were used for the simulation based on average parameters of the real pores, such as radii and lengths. These virtual pores were grouped in a highly symmetric arrangement. Simulations and experiments matched only for data obtained at the reactor exit.

3.4.3
Hydrogenation of 1,5-Cyclooctadiene to Cyclooctene

Proceedings: [130]; sections in reviews: [90, 95, 97, 100].

3.4.3.1 Drivers for Performing the Hydrogenation of 1,5-Cyclooctadiene to Cyclooctene
See Section 3.4.2.1.

3.4.3.2 Beneficial Micro Reactor Properties for the Hydrogenation of 1,5-Cyclooctadiene to Cyclooctene
See Section 3.4.2.2.

Gas-phase reaction 15 [GP 15]: hydrogenation of 1,5-cyclooctadiene to cyclooctene

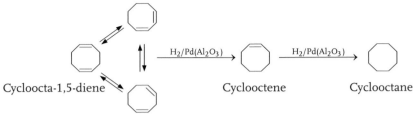

Cycloocta-1,3-diene

Cycloocta-1,5-diene

Cyclooctene

Cyclooctane

Cycloocta-1,4-diene

3.4.3.3 Typical Results

Deactivation

[GP 15] [R 4] The selective hydrogenation of 1,5-cyclooctadiene (COD) at a Pd catalyst is more facile than for *c,t,t*-1,5,9-cyclododecatriene, since no deactivation during a catalytic run is found [130].

Hydrogen partial pressure

[GP 15] [R 4] By increasing the partial pressure ratio of hydrogen to COD from 0.75 to 2, cyclooctene selectivity at a Pd catalyst decreases from nearly 100 to 88%, while conversion increases from 80 to nearly 100% (Figure 3.60) ($p_{COD} = 220$ Pa; $p = 110$ kPa; 150 °C; 87 ms; 12 l h^{-1}) [130].

Addition of carbon monoxide

[GP 15] [R 4] By increasing the concentration of carbon monoxide from 0 to 400 ppm, the cyclooctene yield at a Pd catalyst increases from 83 to 98% ($p_{COD} = 330$ Pa; $p_{H_2} = 660$ Pa; $p = 110$ kPa; 150 °C; 87 ms; 12 l h^{-1}) [130].

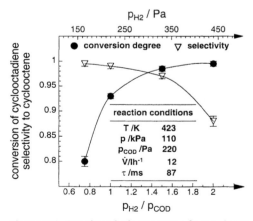

Figure 3.60 Gas-phase hydrogenation of 1,5-cyclooctadiene. Conversion and selectivity depending on hydrogen partial pressure [130].

Residence time

[GP 15] [R 4] By increasing the residence time from 35 to 115 ppm, the cyclooctene conversion at a Pd catalyst increases from 75 to nearly 100% (Figure 3.61), while selectivity decreases from 99.5 to 98% (p_{COD} = 110 Pa; p_{H_2} = 110 Pa; p = 110 kPa; 150 °C) [130].

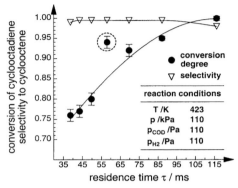

Figure 3.61 Increase of conversion of cyclooctadiene and nearly constant selectivity for cyclooctene with increasing residence time in a Pd micro-channel reactor [130].

Maximization of production

[GP 15] [R 4] By increasing the flux of 1,5-cyclooctadiene from 0.5 to 5.5 mmol h^{-1}, the cyclooctene yield at a Pd catalyst decreases from 98 to 84%, while production increases from 50 to nearly 500 mg h^{-1} (p_{H_2}/p_{COD} = 2; p = 110 kPa; 150 °C; 87 ms; 12 l h^{-1}) [130].

3.4.4
Hydrogenation of Benzene

Proceedings: [84]; sections in reviews: [90, 95, 97, 100].

3.4.4.1 Drivers for Performing the Hydrogenation of Benzene
See Section 3.4.2.1.

3.4.4.2 Beneficial Micro Reactor Properties for the the Hydrogenation of Benzene
See Section 3.4.2.2.

Gas-phase reaction 16 [GP 16]: hydrogenation of benzene

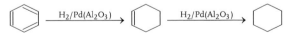

3.4.4.3 Typical Results
Catalyst variation

[GP 16] [R 4] An Ru–Zn catalyst was used for benzene hydrogenation, as a Pd-coated micro-channel reactor could not be applied successfully ($p_{benzene}$ = 110 Pa;

$p_{methanol}$ = 330 Pa; p_{H_2} = 44 kPa; p = 110 kPa; 80 °C; 235 ms) [130]. Methanol was added to the reaction mixture to act as diffusion barrier on the catalyst surface to prevent re-adsorption of the intermediate cyclohexene. The time-on-stream behavior over a period of 14 h was observed. The benzene conversion decreased rapidly and fell below steady-state conditions. Selectivity increased initially, until reaching a plateau of 36%.

[GS 16] [R 14] Sputtered and impregnated Pt catalysts were compared (Figure 3.62) regarding their reaction rates (< 1 ml min^{-1}; 100–150 °C; 100–600 ms) [76]. A sol–gel γ-alumina-based catalyst was notably more active than its sputtered counterpart. The temperature dependence of the reaction rate of both types of catalysts was also revealed.

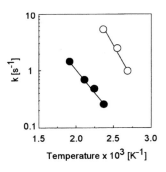

Figure 3.62 Comparison of reaction rates for hydrogenation of benzene in micro reactors: (○) sputtered Pt catalyst; (●) impregnated Pt catalyst [76].

Catalyst deactivation
[GP 16] [R 14] No catalyst deactivation was observed for sputtered and impregnated Pt catalysts in a Si chip reactor in the limits of the reaction conditions applied (< 1 ml min^{-1}; 100–150 °C; 100–600 ms) [76].

Kinetics
[GP 16] [R 14] First-order kinetics of the reaction rates were found for sputtered and impregnated Pt catalysts (< 1 ml min^{-1}; 100–150 °C; 100–600 ms) [76].

3.5
Dehydrogenations

3.5.1
Non-oxidative Dehydrogenation of Propane to Propene

Proceedings: [8]; sections in review: [90, 95, 97, 100, 110].

3.5.1.1 Drivers for Performing the Non-oxidative Dehydrogenation of Propane to Propene

The non-oxidative propane dehydrogenation is highly exothermic (129 kJ mol^{-1} at 550 °C and 0.14 MPa) and limited by the thermodynamic equilibrium (22% conversion under the same conditions) (see [8] and original citations therein). Coke formation results in rapid catalyst deactivation.

3.5.1.2 Beneficial Micro Reactor Properties for the Non-oxidative Dehydrogenation of Propane to Propene

The non-oxidative propane dehydrogenation served here as model reaction having specific limitations (see Section 3.5.1.1) to validate a completely novel reactor concept. The main benefit stems from combining reaction and separation. Using micro-scale flow serves for ensuring high mass transfer, providing low pressure drop and avoiding coke formation (see Section 3.5.1.3).

Gas-phase reaction 17 [GP 17]: dehydrogenation of propane to propene

$$\text{propane} \xrightarrow{\gamma\text{-Al}_2\text{O}_3/(\text{Pt, Sn})} \text{propene} + H_2$$

3.5.1.3 Typical Results
Hydrodynamics – residence time distribution
[GP 17] [R 21] The exit-age distribution for a filament-bed reactor (i.d.: 7 µm; porosity: 0.8), which was used for propane dehydrogenation, was compared with two random beds (Figure 3.63), differing in shape and size (from 100 µm to 1.5 mm) (response to switch of 30 Nml min^{-1} 10% argon/nitrogen to pure nitrogen; 25 °C; 0.1 MPa; tube-i.d.: 15 mm; tube length: 230 mm) [8]. Under identical conditions, the packed beds have a much broader residence time distribution than the structured filamentous packing, providing relatively well-defined flow conduits.

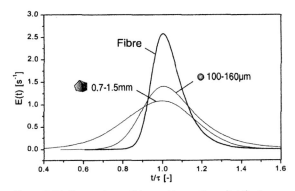

Figure 3.63 Comparison of the residence time distribution for packed-bed and filamentous-bed reactors [8].

Pressure drop
[GP 17] [R 21] The pressure drop in a filamentous-bed reactor (i.d.: 7 µm; porosity: 0.8), which was used for propane dehydrogenation, was compared with a randomly packed powder bed (sphere diameter from 100 to 160 µm) of similar hydraulic dimensions [8]. The filament-bed reactor has a pressure drop about five times lower [120 ml (STP) min^{-1} nitrogen; 25 °C; 1 bar]. The hydraulic diameter of the filament

bed of 70 μm is of the same order as the typical sizes of micro channels of micro reactors.

Benchmark to powder fixed-bed reactor: conversion/selectivity

[GP 17] [R 21] Similar conversions of 24–22% were found in time-on-stream (2 h) measurements for a quartz-tube reactor (i.d.: 6 mm) with powder fixed bed (sphere diameter from 100 to 160 μm) and a filamentous-bed reactor (i.d.: 7 μm; porosity: 0.8; 550 °C; 0.14 MPa; GHSV = 1161 h^{-1}; 3.1 s; m_{cat} = 0.375 g) [8]. The thermodynamic equilibrium (22%) was reached (Figure 3.64). Selectivity was slightly better for the filament-bed reactor. During time-on-stream, selectivity increased slightly to 95%, then being about 7% better than the powder bed. This was explained by a broader residence time distribution in the latter case, favoring cracking reactions of propane. Deactivation due to coke formation is slow for both types of reactors.

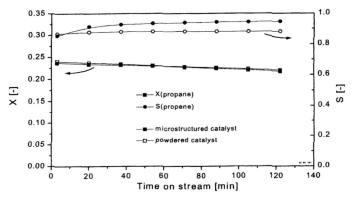

Figure 3.64 Similar conversion for propane was found for the powder fixed-bed and the filamentous-bed reactors, however, selectivity was better for the filamentous-bed reactor [8].

Removal of products by membrane separation

[GP 17] [R 21] If a membrane separation function is added to the filamentous-bed reactor (i.d.: 7 μm; porosity: 0.8), the performance can be further improved [8]. This is achieved by placing the filaments in an outer shell of a tube and separating them from air guidance, through the inner cylindrical conduit formed in that way, by a Pd/Ag membrane, permeable to hydrogen. In this way, hydrogen is removed from the reacting zone and burned to give water.

The measured conversion of 30% exceeds the equilibrium value (22%) for the first 30 min time-on-stream (550 °C; 0.14 MPa; GHSV = 1161 h^{-1}; 3.1 s; m_{cat} = 0.375 g) [8]. Owing to the removal of hydrogen, however, coke removal is more likely so that catalyst deactivation is more pronounced (Figure 3.65). After 140 h, conversion drops to about 12%, now being lower in performance than the powder bed. Propene selectivity is enhanced in the membrane reactor to 97%. Owing to the absence of hydrogen, hydroisomerization and hydrogenolysis reactions are reduced.

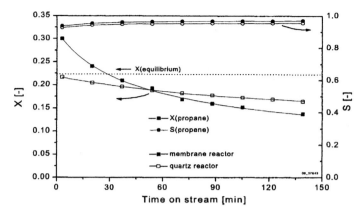

Figure 3.65 Measured enhancement of propane conversion and increased propene selectivity in a membrane reactor equipped with a filamentous catalyst [8].

3.5.2
Oxidative Dehydrogenation of Propane to Propene

Peer-reviewed journals: proceedings: [13]; sections in reviews: [90, 95, 97, 100, 110].

3.5.2.1 Drivers for Performing the Oxidative Dehydrogenation of Propane to Propene

The oxidative propane dehydrogenation is well investigated and also a highly exothermic reaction. In a fixed-bed reactor, steep temperature gradients are observable and the conversion of propane and selectivity of the reaction are strongly determined by temperature and total flow rate [133].

Potential benefits when performing combined oxidative and non-oxidative dehydrogenations by periodic operation have been briefly reviewed [13].

3.5.2.2 Beneficial Micro Reactor Properties for the Oxidative Dehydrogenation of Propane to Propene

The main aspect of the non-periodic investigations was the avoidance of hot-spot formation and the impact of hot spots on the reactor performance. This also allowed a detailed comparison of catalyst performance.

In addition, beneficial properties of micro reactors for periodic processing have been envisaged, details of which are given in [12, 13, 85].

Gas-phase reaction 18 [GP 18]: Oxidative dehydrogenation of propane

$$\xrightarrow[-CO,\,-CO_2,\,-H_2O]{+O_2\,/\,\gamma\text{-}Al_2O_3\,/\,VO_x}$$

3.5.2.3 Typical Results
Conversion/selectivity/yield

[GP 18] [R 6] The selectivity/conversion behavior over a VO_x/Al_2O_3 catalyst was compared for a multi-platelet stack micro-channel reactor and a conventional fixed bed

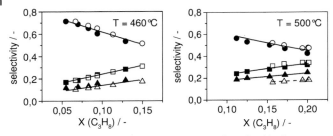

Figure 3.66 Selectivities of propane conversion for a fixed-bed (open symbols) and a micro-channel reactor (closed symbols) for two different inlet temperatures. $C_3H_8/O_2/Ne = 0.3/0.15/0.55$, $F_{tot} = 150$ ml min^{-1}. C_6H_6 (●,○); CO (■,□); CO_2 (▲,△) [133].

at two inlet temperatures (Figure 3.66) [133, 134]. Similar curves resulted, mainly showing that the catalyst was the determining factor for reactor performance. At a 460 °C inlet temperature, selectivity drops from more than 70 to about 50% with increasing conversion (from 5 to 15%). In turn, carbon monoxide and dioxide selectivities increase. To achieve similar conversion, the flow rate has to be varied.

Inlet temperature
[GP 18] [R 6] Experiments at fixed flow rate allowed a comparison of reactor performance over a VO_x/Al_2O_3 catalyst of a multi-platelet-stack micro-channel reactor and a conventional fixed bed as a function of the inlet temperature [133, 134]. The conversions rise steeply with increasing inlet temperature (Figure 3.67).

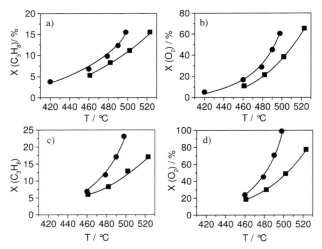

Figure 3.67 Steep increase of propane and oxygen conversions with increasing reactor inlet temperature. (●) Fixed-bed reactor; (■) micro-channel reactor. $C_3H_8/O_2/Ne = 0.3/0.15/0.55$, $F_{tot} = 150$ ml min^{-1} (a,b); $C_3H_8/O_2/Ne = 0.5/0.25/0.25$, $F_{tot} = 120$ ml min^{-1} (c,d) [133].

It turned out that in all cases investigated, higher propene and oxygen conversions result for the fixed-bed reactor. Propene conversions, for instance, differ by as much as about 10%. Measurements of local temperatures confirmed that hot spots are responsible for this difference; hence the real reaction temperatures may differ considerably from the inlet temperature.

Hot spots
[GP 18] [R 6] When using a VO_x/Al_2O_3 catalyst in a fixed bed, remarkable differences between inlet temperature and maximum temperature were found [133, 134]. Depending on the reaction conditions, hot spots ranging from 3 to 100 K were detected (Figure 3.68). Even when using diluted gases, the hot spots are as large as 20 K. Hence no isothermal operation could be established.

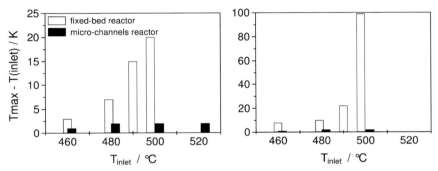

Figure 3.68 Comparison of rise in temperature between inlet and maximum in a micro-channel and fixed-bed reactor.
$C_3H_8/O_2/Ne = 0.3/0.15/0.55$, $F_{tot} = 150$ ml min^{-1} (left);
$C_3H_8/O_2/Ne = 0.5/0.25/0.25$, $F_{tot} = 120$ ml min^{-1} (right).

In turn, using the micro-channel reactor, isothermal processing was achievable for nearly all conditions applied, giving a maximum temperature increase of 2 K [133, 134].

Propene and oxygen partial pressure
[GP 18] [R 6] The influence of the propene and oxygen partial pressure on the space–time yield using a VO_x/Al_2O_3 catalyst was investigated [133, 134]. The fixed bed and the micro reactor give similar performance at an inlet temperature of 460 °C, whereas some differences were observed at higher temperature (fixed bed: 490 °C; micro reactor: 502 °C). At 460 °C, the space–time yield increased linearly (maximum: 0.3 mol kg^{-1} s^{-1}) on increasing the partial pressure of propene from 10 to 50 kPa. For the same temperature, the impact of oxygen partial pressure is less remarkable for both reactors. At higher temperatures (490, 502 °C), more carbon dioxide formation was noted for the micro reactor when increasing the propene partial pressure, which is explained by unspecific total oxidation due to the stainless-steel walls having a large specific area. For increasing oxygen partial pressure at the same temperatures, carbon monoxide formation in the micro reactor is high.

Periodic concentration cycling
[GP 17] [R 6] For a short description of potential process details and catalysts, see [13] (no experimental results are given there).

3.5.3
Dehydrogenation of Cyclohexane to Benzol

Proceedings: [127].

3.5.3.1 Drivers for Performing the Dehydrogenation of Cyclohexane
The reaction served as a model reaction for initial experimentation and modeling studies [127].

3.5.3.2 Beneficial Micro Reactor Properties for the Dehydrogenation of Cyclohexane
The studies referred mainly to determining conversion, exhibiting a relationship between pressure drop and internal dimensions, and analyzing collisions functions from kinetic theory under different flow regimes (slip flow; Knudsen) [127].

Gas-phase reaction 19 [GP 19]: Cyclohexane dehydrogenation to benzene

$$\text{C}_6\text{H}_{12} \xrightarrow{\ \text{Pt}\ } \text{C}_6\text{H}_6 + 3\,\text{H}_2$$

The dehydrogenation of cyclohexane to benzene is an endothermic process (206 kJ mol^{-1}), e.g., performed at 200 °C. The equilibrium conversion amounts to 18.9% [127].

3.5.3.3 Typical Results
Conversion/selectivity/yield
[GP 19] [R 22] Only 2–3% conversion was found at a sputtered catalyst (200 °C; 8 ml h^{-1} condensed liquid; slightly more than 1 s). The fluid enters the micro channel in continuous flow, passes through slip and transitional flow to Knudsen regime. Further details on the activity of the catalyst as a function of the channel length are also provided in [127].

3.6
Substitutions

3.6.1
Chlorination of Alkanes

Proceedings: [29]; sections in review: [95, 97, 100]; trade literature: [30].

3.6.1.1 Drivers for Performing the Chlorination of Alkanes
An industrial investigation studied the radical chlorination of alkanes in micro heat exchangers to analyse thermal effects on radical production [29, 30]. It was known from prior studies in a reactor consisting of two conventional tubes, one for

pre-heating and one for heating, each placed in an oven, that too slow approaching reaction temperature and over-heating are detrimental for reactor performance, i.e. decreasing conversion and space–time yield. Accordingly, a micro reactor with fast thermal ramping and without overshoots for the process fluids is required. In this way, the radicals formed can be most efficiently utilized for reaction, without losses, e.g. by recombination.

3.6.1.2 Beneficial Micro Reactor Properties for the Chlorination of Alkanes

The chlorination reaction benefited from the very efficient heat transfer provided by a micro heat exchanger [29, 30].

In terms of plant construction, the implementation of two such micro devices in a pilot-scale industrial chlorination plant is a good example of multi-scale processing, as only the pre-heating tube was replaced by a micro heat exchanger, while the conventional tube, surrounded by a large oven, was still used for reaction.

Gas-phase reaction 20 [GP 20]: Chlorination of alkanes

$$R_m C_n H_{2n+2-m} \xrightarrow{\text{Cl}_2} R_m C_n H_{2n+2-m-o} Cl_o$$

The exact nature of the reactant and the product was not disclosed for confidentiality reasons.

3.6.1.3 Typical Results

Conversion/selectivity/yield

[GP 20] [R 9] With a hybrid plant configuration, using two micro heat transfer modules and a conventional tube reactor attached, a significant increase in conversion by about 25% was achieved (500 °C; 0.3–2.3 s; 500 l (STP) h^{-1}; 0.4 bar) [29, 30].

Selectivity increases with decreasing residence time, both for the conventional two-tube reactor– and the micro module–tube reactor configurations [29, 30]. The equal performance of micro and macro processing is explained by running the process at thermodynamic equilibrium.

Size of equipment

[GP 20] [R 9] Externally heated micro heat transfer modules used for pre-heating in the framework of an alkane chlorination process are several times smaller than the previously used combination of a long tube reactor and a surrounding oven [29, 30].

Fast thermal ramping – thermal overshoots

[GP 20] [R 9] The micro heat transfer modules exhibited a steep increase in temperature of the processing fluid. Ten times faster heating of the reactants is achieved compared with the conventional pre-heating tube used formerly [29, 30].

In this way, the operating temperature is reached without thermal overshoots [29, 30].

Residence time

[GP 20] [R 9] The residence time in the micro heat transfer modules amounts to 10 ms; it is 3 ms when referred to the micro channels only [29, 30]. The residence

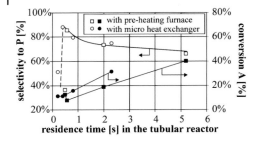

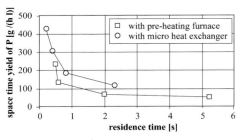

Figure 3.69 Selectivities and space–time yields for the hybrid-plant configuration, using two micro heat transfer modules and a conventional tube reactor attached, and for the conventional tube-reactor plant [29].

time in the conventional tube reactor is 400 ms. When using a conventional pre-heating tube, instead of the micro module, its residence time equals that of the reactor tube, i.e. by micro-flow heating the time could be reduced by a factor of 40.

Space–time yield

[GP 20] [R 9] With a hybrid-plant configuration, using two micro heat transfer modules and a conventional tube reactor attached, a significant increase in space–time yield from about 240 to about 430 g $h^{-1}l^{-1}$ was achieved (Figure 3.69) (500 °C; 0.3–2.3 s; 500 l (STP) h^{-1}; 0.4 bar) [29, 30].

Operational time

[GP 20] [R 9] A hybrid plant configuration, using two micro heat transfer modules and a conventional tube reactor attached, could be run for 24 h without shutdown (500 °C; 0.3–2.3 s; 500 l STP) h^{-1}; 0.4 bar) [29, 30]. Thereafter, incipient corrosion of the micro device was observed. However, this did not lead to immediate shutdown.

3.7
Eliminations

3.7.1
Dehydration of 2-Propanol to Propene

Peer-reviewed journals: [12]; proceedings: [13]; sections in reviews: [90, 95–97, 100, 110].

3.7.1.1 Drivers for Performing the Dehydration of 2-Propanol to Propene

Some current micro-reactor investigations on this reaction are concerned with the use of periodic concentration cycling. This is due to the possibility of achieving considerable increases in average reaction rate when switching the concentration from a certain level to zero (see original citations in [12]). This so-called stop-effect can be utilized for the catalytic dehydration of alcohols and deamination of amines on alumina or other amphoteric metal oxides. Different models to describe such periodic processing are available [12]. By this means, reactor performance can exceed the steady-state values. The correctness of such predictions was evidenced by experimental findings, e.g. on the dehydration of ethanol on γ-alumina [12]).

3.7.1.2 Beneficial Micro Reactor Properties for the Dehydration of 2-Propanol to Propene

Advantages of micro-reactor periodic processing have been summarized [12, 13, 85]. In particular, this refers to reducing signal dispersion when guiding the flow through reactors of small internal volumes, i.e. to achieve fast cycling times.

Gas-phase reaction 21 [GP 21]: dehydration of 2-propanol to propene

3.7.1.3 Typical Results

Hydrodynamics – exit-age distributions

[GP 21] [R 6] Exit-age distributions for an experimental set-up for 2-propanol dehydration with and without a multi-platelet-stack micro reactor were calculated, taking into account the measured outlet signal after switching from pure nitrogen to an argon-containing flow [12]. A mass spectrometer was used to determine the dispersion in the set-ups with and without the micro reactor; with the micro reactor attached, dispersion becomes more significant. From the micro-reactor data, dimensionless residence time distributions were derived (see next paragraph).

Hydrodynamics – influence of catalyst coating and type of sealing

[GP 21] [R 6] An analysis of the exit-age distributions for an experimental set-up for 2-propanol dehydration was made using multi-platelet-stack micro reactors, differing in the type of sealing and in the presence or absence of catalyst coatings [12]. The micro reactors were either glued or sealed by graphite sealing, which presumably leads to small variations of the geometry that adversely affects the repartition of flow between different channels. The presence of a coating reduces the cross-sectional area of the channels to an unknown extent.

By these variations, it was found that the micro reactor behaves almost like a plug-flow reactor with a Bodenstein number Bo = 70 for the uncoated and coated glued reactor and Bo = 33 for the reactor with graphite joints [12]. The catalytic coatings have hardly any influence on the residence-time distribution, whereas the variation in type of sealing has a distinct influence.

Hydrodynamics – calculation method for system response on arbitrary concentration variation

[GP 21] [R 6] By convolution of three functions, inlet function, reactor dispersion model and mass spectrometer response, the response of an experimental set-up with a micro reactor to arbitrary concentration variations at the reactor inlet can be estimated [12]. Measured and predicted curves show nearly perfect agreement.

By this method, the influence of the Bodenstein number and residence time on the response curves was derived [12]. For short mean residence times, the inlet square wave function is only slightly modified; sine functions are formed on applying longer residence times. The larger the dispersion, the more significant is this dependence.

Development of a kinetic model by fixed-bed measurements

[GP 21] [R 6] By making transient experiments (steps 2-PrOH/inert, inert/2-PrOH, and 2-PrOH/H$_2$O) a kinetic model was elaborated [12]. For this reason, a kinetic model from the literature was modified taking into account experimental findings obtained for 2-propanol dehydration performed in a fixed-bed reactor.

Periodic concentration cycling in the micro reactor

[GP 21] [R 6] Based on the kinetic model developed with the fixed-bed reactor and a reactor model, the dynamic behavior of the multi-platelet-stack micro reactor was simulated [12]. By this means, a quantitative description of experimental dynamic concentration changes was achieved. For this purpose, the concentrations of 2-propanol, propene and ether in a cycle period of 30 s were determined (see also [13]).

Periodic temperature cycling in the micro reactor

[GP 21] [R 6] The thermal behavior of a micro structured multi-channel-platelet reactor was investigated. A theoretical model was developed and compared with the measured transient temperature profiles (switches from 200 to 180 °C) [135]. The thermal response time in the center of the micro channel is of the order of 3 s. Temperature cycling times as short as about 20 s are achievable in the reactor (Figure 3.70).

When carrying out the dehydration of 2-propanol to propene over γ-alumina and imposing a temperature jump (Figure 3.71), the 2-propanol outlet concentration passes a peak as a function of time [135]. This is explained as being due to desorption and reaction phenomena, which gain impact on different time scales [temperature jumps: 190–210, 210–190 °C; 2-propanol: 0.92 mol m^{-3} (STP); 1.5 bar; 0.66 cm^3 s^{-1} (STP)]. The propene concentration adapts more readily to the new stationary value and remains constant after that.

The experimental results obtained at low frequencies can be described by a global kinetic model, whereas at high frequencies a more detailed approach has to be used comprising adsorption, desorption and surface reaction steps [135].

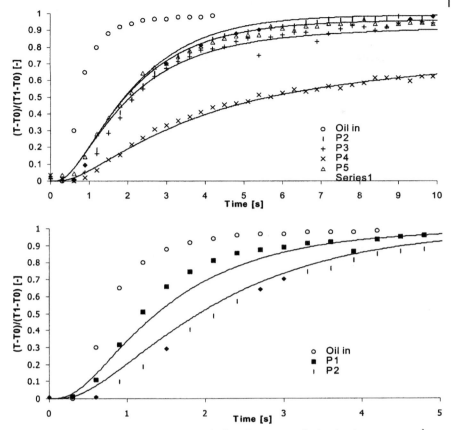

Figure 3.70 Temperature response measured after a heat carrier fluid (oil, velocity 0.35 m s^{-1}) has been switched from 200 to 180 °C at different locations on a micro heat exchanger platelet. Measured values (symbols); model predictions (solid lines) [135].

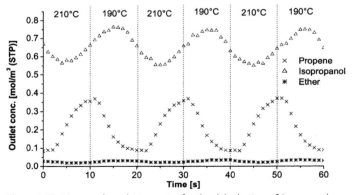

Figure 3.71 Measured product spectra for the dehydration of 2-propanol to propene with periodic temperature variation [135].

3.8
Additions and Coupling Reactions

3.8.1
Phosgene Formation

Peer-reviewed journals: [79]; short mention (no details): [71]; sections in reviews: [87, 88, 92, 114].

3.8.1.1 Drivers for Performing Phosgene Formation

Phosgene is an intermediate utilized in the chemical and pharmaceutical industries for the production of isocyanates for making polyurethane foams and for the synthesis of pharmaceuticals and pesticides (see original citations in [79]). Phosgene is extremely toxic and is an aggressive reactant. The reaction is moderately fast and exothermic (-26 kcal mol^{-1}). Phosgene formation demands specialized cylinder storage, environmental enclosures, and considerable preventive maintenance, just to mention a few differences in technical expenditure compared with other reactions. Hence most phosgene is consumed directly where it is produced.

For these reasons, phosgene formation is one example that potentially does not follow the economics of scale, when considering productivity. Safety and environmental constraints may pose a different view. Serious storage and shipping constraints could demand for production on-site and on-demand.

Micro reactors provide a flexible means to change phosgene productivity, simply by changing volume flow and by numbering-up for even larger flows. Owing to their small hold-up and small productivity per unit, failures may give only small releases. By multi-step processing, the need for transport may be further eliminated. Such an example is described in a patent application [136]. Multi-lamination micro mixers connected to a tube are used for the conversion of phosgene with amines. A flow sheet and an experimental protocol are given for one example, the synthesis of 1-methyl-2,4-diisocyanatocyclohexane from methyl-2,4-diaminocyclohexane by homogeneous gas-phase reaction at 350 °C.

3.8.1.2 Beneficial Micro Reactor Properties for Phosgene Formation

Phosgene formation profits from the small internal volumes of micro reactors due to the hazardous nature of this compound. The light-weight properties of micro reactors and the ability to group them as modules allow one to perform in principle on-site synthesis with flexible output. The latter relates directly to the numbering-up concept of micro reactors.

Gas-phase reaction 22 [GP 22]: phosgene formation

$$Cl_2 + CO \xrightarrow{\text{C}} COCl_2 \quad \Delta H = 26 \text{ kcal mol}^{-1}$$

3.8.1.3 **Typical Results**

Corrosion

[GP 22] [R 16] Silicon devices are strongly corroded by chlorine etching, as expected (50% chlorine; 50% carbon monoxide; 8 cm^3 (STP) min^{-1}; elevated temperature; 1.35–1.40 atm at inlet) [79]. Protection by thin oxide layers, e.g. 500 nm, made in a wet-oxidation furnace, totally prevents such corrosion (6 h exposure).

Thermal management/heat sink properties

[GP 22] [R 16] When using activated carbon catalysts (1.3 mg; 53–73 μm; surface area 850 m^2 g^{-1}) supported on alumina particles (~3 mg; 53–71 μm) in a wide mini fixed bed, no temperature increase, as monitored by thermocouples, was observed (33.3% chlorine; 66.7% carbon monoxide; 4.5 cm^3 (STP) min^{-1}; elevated temperature; 1.35–1.40 atm at inlet) [79]. This shows the high heat-dissipating capacity of the silicon micro-reactor used, i.e. that it can be used efficiently as a heat sink. In the same way, the heat management is governed by the cartridge heaters, dominating the heat release by reaction.

Conversion/selectivity

[GP 22] [R 16] When using activated carbon catalysts (1.3 mg; 53–73 μm; surface area 850 m^2 g^{-1}) supported on alumina particles (~3 mg; 53–71 μm) in a wide mini fixed bed (Figure 3.72), complete conversion was observed (33.3% chlorine; 66.7% carbon monoxide; 4.5 cm^3 (STP) min^{-1}; temperatures > 200 °C; 1.35–1.40 atm at inlet) [79]. At lower temperatures, a nearly linear increase with temperature was found.

No side products were observed under the above-mentioned conditions, most likely owing to avoidance of hot spots [79].

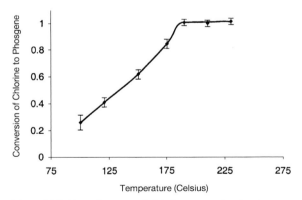

Figure 3.72 Linear increase of chlorine conversion to phosgene with increase in temperature. Experimental results are given for an alumina-supported catalyst in a mini fixed-bed reactor [79].

Catalyst deactivation

[GP 22] [R 16] When using activated carbon catalysts (1.3 mg; 53–73 μm; surface area 850 m^2 g^{-1}) supported on alumina particles (~3 mg; 53–71 μm) in a wide mini

fixed bed, no catalyst deactivation was observed within 6–10 h, possibly owing to the high purity of the gases employed (33.3% chlorine; 66.7% carbon monoxide; 4.5 cm^3 (STP) min^{-1}; 25–200 °C; 1.35–1.40 atm at inlet) [79].

Phosgene productivity
[GP 22] [R 16] When using activated carbon catalysts (1.3 mg; 53–73 μm; surface area 850 m^2 g^{-1}) supported on alumina particles (~3 mg; 53–71 μm) in a wide mini-fixed bed, a phosgene productivity of 3.5 kg a^{-1} (0.40 g h^{-1}) is achieved (33.3% chlorine; 66.7% carbon monoxide; 4.5 cm^3 (STP) min^{-1}; 25–200 °C; 1.35–1.40 atm at inlet) [79]. For a 50% chlorine/50% carbon monoxide mixture at 8.0 cm^3 (STP) min^{-1} flow, a productivity of 9.3 kg a^{-1} (1.1 g h^{-1}) results. Operational limits are not set by the pressure drop or by controlling issues, but rather by the thermal management. Higher flows would induce a thermal runaway of the system.

It is envisaged from prior experience with other reactions that numbering-up can further increase productivity [79]. Indeed, 10-fold units of the wide mini fixed bed exist. By extrapolation it is anticipated that these would probably give 93 kg a^{-1} (11 g h^{-1}), although the exact value certainly has to be validated experimentally.

Intraparticle mass and heat transfer limitations
[GP 22] [R 16] The extent of internal transport limits was analysed for the wide fixed-bed reactor, using experimental data on phosgene formation and matter and process parameter data for the reactants [79]. By applying the Anderson criterion and judging the Weisz modulus, it was found that transfer limitations are negligible.

An order of magnitude analysis was made for the mass transport to the catalyst particle [79]. It was found that concentration gradients are virtually suppressed.

Kinetics: activation energy and rate constants
[GP 22] [R 16] Rate constants for phosgene formation were extracted from wide mini fixed-bed data and were found to match literature data ('micro reactor' activation energy: 7.6 kcal mol^{-1}; literature data: 8.6 kcal mol^{-1}) [79].

3.8.2
Oxidative Coupling of Methane

Proceedings: [55].

3.8.2.1 Drivers for Performing the Oxidative Coupling of Methane
Since methane is available in large amounts, it is desired to convert methane to more valuable C$_2$ products that are important precursors for many chemicals. Thus, much research is dedicated to enhancing this reaction with acceptable performance. Oxidative methane coupling (OCM), usually performed at temperatures between 750 and 1000 °C, is a very fast reaction and is exothermic. In spite of many attempts to find suitable catalysts over many years, their performance still needs to be improved, e.g. maximum yields reached so far do not exceed 25% (see original citations in [55]).

3.8.2.2 Beneficial Micro Reactor Properties for the Oxidative Coupling of Methane

The oxidative coupling of methane is a reaction of industrial interest, which so far suffers from too low performance (see above). It is the general hope that the well-known beneficial properties of micro reactors as a new process-engineering tool can help to make the next step in the development of the reaction. Demonstration of feasibility is needed here to be able to provide more details on which property is the key to this.

Gas-phase reaction 23 [GP 23]: Oxidative coupling of methane

$$CH_4 + O_2 \xrightarrow{\text{LiAlO}_2} H_2, CO, CO_2, H_2O, \text{ethane, ethylene, acetylene, propane}$$

3.8.2.3 Typical Results
Benchmarking to monolith performance
[GP 23] [R 8] In a benchmarking study, the performance of two catalysts in micro-channel systems was compared with that of two catalysts in monoliths [55]. The 'micro-channel catalysts' were two LiAlO$_2$ materials made by a sol–gel method or by *in situ* decomposition of a metallic salt solvent. The 'monolithic' catalysts were of the same material, either a micromachined sintered compact with 17% porosity or as a ceramic slurry coated on a polymer foam with 75% porosity.

Both conversion and selectivity, and hence yield, for the micro-channel catalysts were much lower than for the monolithic (Figure 3.73). A yield of only 1.5% was found for the process parameters which correspond to a maximum monolith performance, 6.1% (total flow: 100 ml min^{-1}; methane/oxygen: 2; 950 °C).

Methane-to-oxygen ratio
[GP 23] [R 8] Using two types of LiAlO$_2$ catalysts coated in micro channels, it was shown that with increasing methane-to-oxygen ratio the conversion of methane rises from about 10 to 20% (total flow: 100 ml min^{-1}; methane/oxygen: 1, 2, 4; 950 °C) [55]. A qualitatively similar behavior was also exhibited by two monoliths having catalysts of the same material, but ranging from about 15 to 50%.

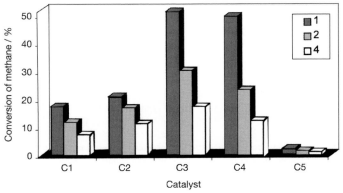

Figure 3.73 Comparison of methane conversion at various methane/oxygen ratios for micro-channel catalyst reactors (C1, C2) and monolithic catalyst reactors (C3, C4); non-coated catalyst (C5) [55].

References

1 Wörz, O., Jäckel, K.-P., Richter, T., Wolf, A., *Microreactors – a new efficient tool for reactor development*, Chem. Eng. Technol. **24**, 2 (**2001**) 138–143.

2 Hessel, V., Ehrfeld, W., Golbig, K., Hofmann, C., Jungwirth, S., Löwe, H., Richter, T., Storz, M., Wolf, A., Wörz, O., Breysse, J., *High temperature HCN generation in an integrated Microreaction system*, in Ehrfeld, W. (Ed.), *Microreaction Technology: 3rd International Conference on Microreaction Technology, Proc. of IMRET 3*, pp. 152–164, Springer-Verlag, Berlin (**2000**).

3 Fichtner, M., Mayer, J., Wolf, D., Schubert, K., *Microstructured rhodium catalysts for the partial oxidation of methane to syngas under pressure*, Ind. Eng. Chem. Res. **40**, 16 (**2001**) 3475–3483.

4 Kestenbaum, H., Lange de Olivera, A., Schmidt, W., Schüth, F., Ehrfeld, W., Gebauer, K., Löwe, H., Richter, T., *Silver-catalyzed oxidation of ethylene to ethylene oxide in a microreaction system*, Ind. Eng. Chem. Res. **41**, 4 (**2000**) 710–719.

5 Zapf, R., Becker-Willinger, C., Berresheim, K., Holz, H., Gnaser, H., Hessel, V., Kolb, G., Löb, P., Pannwitt, A.-K., Ziogas, A., *Detailed characterization of various porous alumina based catalyst coatings within microchannels and their testing for methanol steam reforming*, Trans IChemE **81**, A (**2003**) 721–729.

6 IMM, unpublished results.

7 Ajmera, S. K., Delattre, C., Schmidt, M. A., Jensen, K. F., *A novel cross-flow microreactor for kinetic studies of catalytic processes*, in Matlosz, M., Ehrfeld, W., Baselt, J. P. (Eds.), *Microreaction Technology – IMRET 5: Proc. 5th International Conference on Microreaction Technology*, pp. 414–423, Springer-Verlag, Berlin (**2001**).

8 Wolfrath, O., Kiwi-Minsker, L., Renken, A., *Filamentous catalytic beds for the design of membrane micro-reactor*, in Matlosz, M., Ehrfeld, W., Baselt, J. P. (Eds.), *Microreaction Technology – IMRET 5: Proc. 5th International Conference on Microreaction Technology*, pp. 192–201, Springer-Verlag, Berlin (**2001**).

9 Veser, G., *Experimental and theoretical investigation of H_2 oxidation in a high-temperature catalytic microreactor*, Chem. Eng. Sci. **56** (**2001**) 1265–1273.

10 Zheng, A., Jones, F., Fang, J., Cui, T., *Dehydrogenation of cyclohexane to benzene in a membrane reactor*, in *Proceedings of the 4th International Conference on Microreaction Technology, IMRET 4, 5–9 March 2000*, pp. 284–292, AIChE Topical Conf. Proc., Atlanta, GA (**2000**).

11 Franz, A. J., Jensen, K. J., Schmidt, M. A., *Palladium membrane micro-reactors*, in Ehrfeld, W. (Ed.), *Microreaction Technology: 3rd International Conference on Microreaction Technology, Proc. of IMRET 3*, pp. 267–276, Springer-Verlag, Berlin (**2000**).

12 Rouge, A., Spoetzl, B., Gebauer, K., Schenk, R., Renken, A., *Microchannel reactors for fast periodic operation: the catalytic dehydration of isopropanol*, Chem. Eng. Sci. **56** (**2001**) 1419–1427.

13 Liauw, M., Baerns, M., Broucek, R., Buyevskaya, O. V., Commenge, J.-M., Corriou, J.-P., Falk, L., Gebauer, K., Hefter, H. J., Langer, O.-U., Löwe, H., Matlosz, M., Renken, A., Rouge, A., Schenk, R., Steinfeld, N., Walter, S., *Periodic operation in microchannel reactors*, in Ehrfeld, W. (Ed.), *Microreaction Technology: 3rd International Conference on Microreaction Technology, Proc. of IMRET 3*, pp. 224–234, Springer-Verlag, Berlin (**2000**).

14 Pfeifer, P., Görke, O., Schubert, K., *Washcoats and electrophoresis with coated and uncoated nanoparticles on microstructured metal foils and micro-structured reactors*, in *Proceedings of the 6th International Conference on Microreaction Technology, IMRET 6, 11–14 March 2002*, pp. 281–285, AIChE Pub. No. 164, New Orleans (**2002**).

15 Wan, Y. S. S., Chau, J. L. H., Gavriilidis, A., Yeung, K. L., *TS-1 zeolite microengineered reactors for 1-pentene epoxidation*, Chem. Commun. (**2002**) 878–879.

16 Wan, Y. S. S., Chau, J. L. H., Gavriilidis, A., Yeung, K. L., *Design and fabrication of zeolite-based micro-*

reactors and membrane microseparators, Micropor. Mesopor. Mater. **42 (2001)** 157–175.

17 REBROV, E. V., SEIJGER, G. B. F., CALIS, H. P. A., DE CROON, M. H. J. M., VAN DEN BLEEK, C. M., SCHOUTEN, J. C., *The preparation of highly ordered single layer ZSM-5 coating on prefabricated stainless steel microchannels*, Appl. Catal. A **206 (2001)** 125–143.

18 WIESSMEIER, G., SCHUBERT, K., HÖNICKE, D., *Monolithic microstructure reactors possessing regular mesopore systems for the successful performance of heterogeneously catalyzed reactions*, in EHRFELD, W. (Ed.), *Microreaction Technology – Proc. of the 1st International Conference on Microreaction Technology, IMRET 1*, pp. 20–26, Springer-Verlag, Berlin **(1997)**.

19 FRANZ, A. J., AJMERA, S. K., FIREBAUGH, S. L., JENSEN, K. F., SCHMIDT, M. A., *Expansion of microreactor capabilities trough improved thermal management and catalyst deposition*, in EHRFELD, W. (Ed.), *Microreaction Technology: 3rd International Conference on Microreaction Technology, Proc. of IMRET 3*, pp. 197–206, Springer-Verlag, Berlin **(2000)**.

20 ZHAO, S., BESSER, R. S., *Selective deposition of supported platinum catalyst for hydrogenation in a micromachined reactor*, in *Proceedings of the 6th International Conference on Microreaction Technology, IMRET 6*, 11–14 March 2002, pp. 289–296, AIChE Pub. No. 164, New Orleans **(2002)**.

21 GORGES, R., KÄSSBOHRER, J., KREISEL, G., MEYER, S., *Surface-functionalization of microstructures by anodic spark deposition*, in *Proceedings of the 6th International Conference on Microreaction Technology, IMRET 6*, 11–14 March 2002, pp. 186–191, AIChE Pub. No. 164, New Orleans **(2002)**.

22 KUSAKABE, K., MIYAGAWA, D., GU, Y., MAEDA, H., MOROOKA, S., *Preparation of microchannel palladium membranes by electrolysis*, in MATLOSZ, M., EHRFELD, W., BASELT, J. P. (Eds.), *Microreaction Technology – IMRET 5: Proc. 5th International Conference on Microreaction Technology*, pp. 78–85, Springer-Verlag, Berlin **(2001)**.

23 KOLB, G., ZAPF, R., Personal communication, unpublished results **(2002)**.

24 REBROV, E. V., SEIJGER, G. B. G., CALIS, H. P. A., DE CROON, M. H. J. M., VAN DEN BLEEK, C. M., SCHOUTEN, J. C., *Synthesis and characterization of ZSM-5 zeolites on prefabricated stainless steel microchannels*, in *Proceedings of the 4th International Conference on Microreaction Technology, IMRET 4*, 5–9 March 2000, pp. 250–255, AIChE Topical Conf. Proc., Atlanta, GA **(2000)**.

25 WAN, Y. S. S., CHAU, J. L. H., GAVRIILIDIS, A., YEUNG, K. L., *Design and fabrication of zeolite-containing microstructures*, in MATLOSZ, M., EHRFELD, W., BASELT, J. P. (Eds.), *Microreaction Technology – IMRET 5: Proc. 5th International Conference on Microreaction Technology*, pp. 94–102, Springer-Verlag, Berlin **(2001)**.

26 KESTENBAUM, H., LANGE DE OLIVEIRA, A., SCHMIDT, W., SCHÜTH, F., GEBAUER, K., LÖWE, H., RICHTER, T., *Synthesis of ethylene oxide in a catalytic microreacton system*, in EHRFELD, W. (Ed.), *Microreaction Technology: 3rd International Conference on Microreaction Technology, Proc. of IMRET 3*, pp. 207–212, Springer-Verlag, Berlin **(2000)**.

27 WALTER, S., JOANNET, E., SCHIEL, M., BOULLET, I., PHILIPPS, R., LIAUW, M. A., *Microchannel reactor for the partial oxidation of isoprene*, in MATLOSZ, M., EHRFELD, W., BASELT, J. P. (Eds.), *Microreaction Technology – IMRET 5: Proc. 5th International Conference on Microreaction Technology*, pp. 387–396, Springer-Verlag, Berlin **(2001)**.

28 REBROV, E. V., DE CROON, M. H. J. M., SCHOUTEN, J. C., *Design of a microstructured reactor with integrated heat-exchanger for optimum performance of highly exothermic reaction*, Catal. Today **69 (2001)** 183–192.

29 BAYER, T., HEINICHEN, H., LEIPPRAND, I., *Using micro heat exchangers as diagnostic tool for the process optimization of a gas phase reaction*, in *Proceedings of the VDE World Microtechnologies Congress, MICRO.tec 2000*, 25–27 September 2000, pp. 493–497, VDE Verlag, Berlin **(2000)**.

30 HEINICHEN, H., *Kleiner Maßstab – große Wirkung: Mikrowärmeaustauscher zur Verfahrensoptimierung*, Chemie-Technik **30**, 3 (**2001**) 89–91.

31 BIER, W., KELLER, W., LINDER, G., SEIDEL, D., SCHUBERT, K., *Manufacturing and testing of compact micro heat exchangers with high volumetric heat transfer coefficients*, Sens. Actuators **19** (**1990**) 189–197.

32 SCHUBERT, K., BIER, W., LINDER, G., SEIDEL, D., *Herstellung und Test von kompakten Mikrowärmeüberträgern*, Chem. Ing. Tech. **61**, 2 (**1989**) 172–173.

33 MATSON, D. W., MARTIN, P. M., STEWARD, D. C., TONKOVICH, A. L. Y., WHITE, M., ZILKA, J. L., ROBERTS, G. L., *Fabrication of microchannel chemical reactors using a metal lamination process*, in EHRFELD, W. (Ed.), *Microreaction Technology: 3rd International Conference on Microreaction Technology, Proc. of IMRET 3*, pp. 62–71, Springer-Verlag, Berlin (**2000**).

34 TONKOVICH, A. L. Y., JIMENEZ, D. M., ZILKA, J. L., LaMONT, M. J., WANG, J., WEGENG, R. S., *Microchannel chemical reactors for fuel processing*, in EHRFELD, W., RINARD, I. H., WEGENG, R. S. (Eds.), *Process Miniaturization: 2nd International Conference on Microreaction Technology, IMRET 2, Topical Conf. Preprints*, pp. 186–195, AIChE, New Orleans (**1998**).

35 TONKOVICH, A. L., ZILKA, J. L., LaMONT, M. J., WANG, Y., WEGENG, R., *Microchannel chemical reactor for fuel processing applications – I. Water gas shift reactor*, Chem. Eng. Sci. **54** (**1999**) 2947–2951.

36 TONKOVICH, A. L., FITZGERALD, S. P., ZILKA, J. L., LaMONT, M. J., WANG, Y., VANDERWIEL, D. P., WEGENG, R., *Microchannel chemical reactor for fuel processing applications – II. Compact fuel vaporization*, in EHRFELD, W. (Ed.), *Microreaction Technology: 3rd International Conference on Microreaction Technology, Proc. of IMRET 3*, pp. 364–371, Springer-Verlag, Berlin (**2000**).

37 HÖNICKE, D., WIESSMEIER, G., *Heterogeneously catalyzed reactions in a microreactor*, in EHRFELD, W. (Ed.), *Microsystem Technology for Chemical and Biological Microreactors, DECHEMA Monographs*, Vol. 132, pp. 93–107, Verlag Chemie, Weinheim (**1996**).

38 WIESSMEIER, G., HÖNICKE, D., *Micro-fabricated components for heterogeneously catalyzed reactions*, J. Micromech. Microeng. **6** (**1996**) 285–289.

39 WIESSMEIER, G., HÖNICKE, D., *Heterogeneously catalyzed gas-phase hydrogenation of cis,trans,trans-1,5,9-cyclododecatriene on palladium catalysts having regular pore systems*, Ind. Eng. Chem. Res. **35** (**1996**) 4412–4416.

40 KESTENBAUM, H., LANGE DE OLIVERA, A., SCHMIDT, W., SCHÜTH, H., EHRFELD, W., GEBAUER, K., LÖWE, H., RICHTER, T., *Synthesis of ethylene oxide in a catalytic microreactor system*, Stud. Surf. Sci. Catal. **130** (**2000**) 2741–2746.

41 LÖWE, H., EHRFELD, W., GEBAUER, K., GOLBIG, K., HAUSNER, O., HAVERKAMP, V., HESSEL, V., RICHTER, T., *Microreactor concepts for heterogeneous gas phase reactions*, in EHRFELD, W., RINARD, I. H., WEGENG, R. S. (Eds.), *Process Miniaturization: 2nd International Conference on Microreaction Technology, IMRET 2, Topical Conf. Preprints*, pp. 63–74, AIChE, New Orleans (**1998**).

42 KURSAWE, A., HÖNICKE, D., *Epoxidation of ethene with pure oxygen as a model reaction for evaluating the performance of microchannel reactors*, in *Proceedings of the 4th International Conference on Microreaction Technology, IMRET 4*, 5–9 March 2000, pp. 153–166, AIChE Topical Conf. Proc., Atlanta, GA (**2000**).

43 KURSAWE, A., HÖNICKE, D., *Comparison of Ag/Al- and Ag/α-Al₂O₃ catalytic surfaces for the partial oxidation of ethene in microchannel reactors*, in MATLOSZ, M., EHRFELD, W., BASELT, J. P. (Eds.), *Microreaction Technology – IMRET 5: Proc. 5th International Conference on Microreaction Technology*, pp. 240–251, Springer-Verlag, Berlin (**2001**).

44 KURSAWE, A., PILZ, R., DÜRR, H., HÖNICKE, D., *Development and design of a modular microchannel reactor for laboratory use*, in *Proceedings of the 4th International Conference on Microreaction Technology, IMRET 4*, 5–9 March 2000, pp. 227–235, AIChE Topical Conf. Proc., Atlanta, GA (**2000**).

45 SCHUBERT, K., BIER, W., LINDER, G., SEIDEL, D., *Profiled microdiamonds for producing microstructures,* Ind. Diamond Rev. **50**, 5 **(1990)** 235–239.

46 BIER, W., KELLER, W., LINDER, G., SEIDEL, D., SCHUBERT, K., MARTIN, H., *Gas-to-gas heat transfer in micro heat exchangers,* Chem. Eng. Process. **32**, 1 **(1993)** 33–43.

47 SCHUBERT, K., BRANDNER, J., FICHTNER, M., LINDER, G., SCHYGULLA, U., WENKA, A., *Microstructure devices for applications in thermal and chemical process engineering,* Microscale Therm. Eng. **5** **(2001)** 17–39.

48 SCHUBERT, K., BIER, W., BRANDNER, J., FICHTNER, M., FRANZ, C., LINDER, G., *Realization and testing of microstructure reactors, micro heat exchangers and micromixers for industrial applications in chemical engineering,* in EHRFELD, W., RINARD, I. H., WEGENG, R. S. (Eds.), *Process Miniaturization: 2nd Internatio-nal Conference on Microreaction Techno-logy, IMRET 2, Topical Conf. Preprints,* pp. 88–95, AIChE, New Orleans **(1998)**.

49 WÖRZ, O., JÄCKEL, K.-P., RICHTER, T., WOLF, A., *Mikroreaktoren – Ein neues wirksames Werkzeug für die Reaktor-entwicklung,* Chem. Ing. Tech. **72**, 5 **(2000)** 460–463.

50 WÖRZ, O., JÄCKEL, K. P., *Winzlinge mit großer Zukunft – Mikroreaktoren für die Chemie,* Chem. Techn. **26**, 131 **(1997)** 130–134.

51 WÖRZ, O., JÄCKEL, K. P., RICHTER, T., WOLF, A., *Microreactors, a new efficient tool for optimum reactor design,* in EHRFELD, W., RINARD, I. H., WEGENG, R. S. (Eds.), *Process Miniaturization: 2nd International Conference on Microreaction Technology, IMRET 2, Topical Conf. Preprints,* pp. 183–185, AIChE, New Orleans **(1998)**.

52 COMMENGE, J. M., FALK, L., CORRIOU, J. P., MATLOSZ, M., *Optimal design for flow uniformity in microchannel reactors,* in *Proceedings of the 4th International Conference on Microreaction Technology, IMRET 4, 5–9 March 2000, pp. 24–30,* AIChE Topical Conf. Proc., Atlanta, GA **(2000)**.

53 COMMENGE, J. M., FALK, L., CORRIOU, J. P., MATLOSZ, M., *Optimal design for flow uniformity in microchannel reactors,* AIChE J. **48**, 2 **(2000)** 345–358.

54 KNITTER, R., GÖHRING, D., BRAHM, M., MECHNICH, P., BROUCEK, R., *Ceramic microreactor for high-temperature reactions,* in *Proceedings of the 4th International Conference on Microreaction Technology, IMRET 4, 5–9 March 2000, pp. 455–460,* AIChE Topical Conf. Proc., Atlanta, GA **(2000)**.

55 GÖHRING, D., KNITTER, R., RISTHAUS, P., WALTER, S., LIAUW, M. A., LEBENS, P., *Gas phase reactions in ceramic micro-reactors,* in *Proceedings of the 6th Inter-national Conference on Microreaction Technology, IMRET 6, 11–14 March 2002, pp. 55–60,* AIChE Pub. No. 164, New Orleans **(2002)**.

56 FRANZ, A. J., QUIRAM, D. J., SRINI-VASAN, R., HSING, I.-M., FIREBAUGH, S. L., JENSEN, K. F., SCHMIDT, M. A., *New operating regimes and applications feasible with microreactors,* in EHRFELD, W., RINARD, I. H., WEGENG, R. S. (Eds.), *Process Miniaturization: 2nd Inter-national Conference on Microreaction Technology, IMRET 2, Topical Conf. Preprints,* pp. 33–38, AIChE, New Orleans **(1998)**.

57 JENSEN, K. F., FIREBAUGH, S. L., FRANZ, A. J., QUIRAM, D., SRINIVASAN, R., SCHMIDT, M. A., *Integrated gas phase microreactors,* in HARRISON, J., VAN DEN BERG, A. (Eds.), *Micro Total Analysis Systems,* pp. 463–468, Kluwer Academic Publishers, Dordrecht **(1998)**.

58 JENSEN, K. F., *Microchemical systems: status, challenges, and oportunities,* AIChE J. **45**, 10 **(1999)** 2051–2054.

59 JENSEN, K. F., HSING, I.-M., SRINIVASAN, R., SCHMIDT, M. A., HAROLD, M. P., LEROU, J. J., RYLEY, J. F., *Reaction engineering for microreactor systems,* in EHRFELD, W. (Ed.), *Microreaction Technology – Proc. of the 1st International Conference on Microreaction Technology, IMRET 1, pp. 2–9,* Springer-Verlag, Berlin **(1997)**.

60 QUIRAM, D. J., HSING, I.-M., FRANZ, A. J., SRINIVASAN, R., JENSEN, K. F., SCHMIDT, M. A., *Characterization of microchemical systems using simulations,* in EHRFELD, W., RINARD, I. H., WEGENG, R. S. (Eds.), *Process*

Miniaturization: 2nd International Conference on Microreaction Technology, IMRET 2, Topical Conf. Preprints, pp. 205–211, AIChE, New Orleans **(1998)**.

61 SRINIVASAN, R., I-MING HSING, BERGER, P. E., JENSEN, K. F., FIREBAUGH, S. L., SCHMIDT, M. A., HAROLD, M. P., LEROU, J. J., RYLEY, J. F., *Micromachined reactors for catalytic partial oxidation reactions,* AIChE J. **43**, 11 **(1997)** 3059–3069.

62 QUIRAM, D. J., HSING, I.-M., FRANZ, A. J., JENSEN, K. F., SCHMIDT, M. A., *Design issues for membrane-based, gas phase microchemical systems,* Chem. Eng. Sci. **55 (2000)** 3065–3075.

63 BROADWELL, I., FLETCHER, P. D. I., HASWELL, S. J., MCCREEDY, T., ZHANG, N., *Quantitative 3-dimensional profiling of channel networks within transparent 'lab-on-a-chip' microreactors using a digital imaging method,* Lab. Chip **1 (2001)** 66–71.

64 FLETCHER, P. D. I., HASWELL, S. J., ZHANG, X., *Electrokinetic control of a chemical reaction in a lab-on-a-chip micro-reactor: measurement and qualitative modeling,* Lab. Chip **2 (2002)** 102–112.

65 FLETCHER, P. D. I., HASWELL, S. J., POMBO-VILLAR, E., WARRINGTON, B. H., WATTS, P., WONG, S. Y. F., ZHANG, X., *Micro reactors: principle and applications in organic synthesis,* Tetrahedron **58**, 24 **(2002)** 4735–4757.

66 GREENWAY, G. M., HASWELL, S. J., MORGAN, D. O., SKELTON, V., STYRING, P., *The use of a novel microreactor for high troughput continuous flow organic synthesis,* Sens. Actuators B **63**, 3 **(2000)** 153–158.

67 HASWELL, S. J., *Miniaturization – What's in it for chemistry,* in VAN DEN BERG, A., RAMSAY, J. M. (Eds.), *Micro Total Analysis System,* pp. 637–639, Kluwer Academic Publishers, Dordrecht **(2001)**.

68 SKELTON, V., HASWELL, S. J., STYRING, P., WARRINGTON, B., WONG, S., *A micro-reactor device for the Ugi four component condensation (4CC) reaction,* in RAMSEY, J. M., VAN DEN BERG, A. (Eds.), *Micro Total Analysis Systems,* pp. 589–590, Kluwer Academic Publishers, Dordrecht **(2001)**.

69 WATTS, P., WILES, C., HASWELL, S. J., POMBO-VILLAR, E., STYRING, P., *The synthesis of peptides using micro reactors,* Chem. Commun. **(2001)** 990–991.

70 WATTS, P., WILES, C., HASWELL, S. J., POMBO-VILLAR, E., *Solution phase synthesis of β-peptides using micro reactors,* Tetrahedron **58**, 27 **(2002)** 5427–5439.

71 LEROU, J. J., HAROLD, M. P., RYLEY, J., ASHMEAD, J., O'BRIEN, T. C., JOHNSON, M., PERROTTO, J., BLAISDELL, C. T., RENSI, T. A., NYQUIST, J., *Microfabricated mini-chemical systems: technical feasibility,* in EHRFELD, W. (Ed.), *Microsystem Technology for Chemical and Biological Microreactors, DECHEMA Monographs,* Vol. 132, pp. 51–69, Verlag Chemie, Weinheim **(1996)**.

72 CAO, E., YEONG, K. K., GAVRIILIDIS, A., CUI, Z., JENKINS, D. W. K., *Micro-chemical reactor for oxidative dehydro-genation of methanol,* in *Proceedings of the 6th International Conference on Micro-reaction Technology, IMRET 6, 11–14 March 2002,* pp. 76–84, AIChE Pub. No. 164, New Orleans **(2002)**.

73 BESSER, R. S., FORT, J., SURANGALIKAR, H., OUYANG, S., *Microdevice-based system for rapid catalyst development,* in MATLOSZ, M., EHRFELD, W., BASELT, J. P. (Eds.), *Microreaction Technology – IMRET 5: Proc. 5th International Conference on Microreaction Technology,* pp. 499–507, Springer-Verlag, Berlin **(2001)**.

74 BESSER, R. S., OUYANG, X., SURAN-GALIKAR, H., *Hydrocarbon hydrogenation and dehydrogenation reactions in microfabricated catalytic reactors,* Chem. Eng. Sci. **58 (2003)** 19–26.

75 ROTH, M., HAAS, T., LOCK, M., GERICKE, K. H., BRÄUNING-DEMIAN, A., SPIELBERGER, L., SCHMIDT-BÖCKING, H., *Micro-structure electrodes as electronic interface between solid and gas phase: electrically steerable catalysts for chemical reactions in the gas phase,* in EHRFELD, W. (Ed.), *Microreaction Technology – Proc. of the 1st International Conference on Microreaction Technology, IMRET 1,* pp. 62–69, Springer-Verlag, Berlin **(1997)**.

76 KUSAKABE, K., MIYAGAWA, D., GU, Y., MAEDA, H., MOROOKA, S., *Development of*

a *self-heating catalytic microreactor*, in Matlosz, M., Ehrfeld, W., Baselt, J. P. (Eds.), *Microreaction Technology – IMRET 5: Proc. 5th International Conference on Microreaction Technology*, pp. 70–77, Springer-Verlag, Berlin (2001).

77 Ajmera, S. K., Delattre, C., Schmidt, M. A., Jensen, K. F., *Microfabricated cross-flow chemical reactor for catalyst testing*, Sens. Actuators **82**, 2–3 (2002) 297–306.

78 Ajmera, S. K., Delattra, C., Schmidt, M. A., Jensen, K. F., *Microfabricated differential reactor for heterogeneous gas phase catalyst testing*, J. Catal. **209** (2002) 401–412.

79 Ajmera, S. K., Losey, M. W., Jensen, K. F., Schmidt, M. A., *Microfabricated packed-bed reactor for phosgene synthesis*, AIChE J. **47**, 7 (2001) 1639–1647.

80 Ajmera, S. K., Delattre, C., Schmidt, M. A., Jensen, K. F., *A novel cross-flow microreactor for kinetic studies of catalytic processes*, in Matlosz, M., Ehrfeld, W., Baselt, J. P. (Eds.), *Microreaction Technology – IMRET 5: Proc. of the 5th International Conference on Microreaction Technology*, pp. 414–423, Springer-Verlag, Berlin (2001).

81 Hessel, V., Ehrfeld, W., Golbig, K., Wörz, O., *Mikroreaktionssysteme für die Hochtemperatursynthese*, GIT **43**, 10 (1999) 1100.

82 Veser, G., Friedrich, G., Freygang, M., Zengerle, R., *A modular microreactor design for high-temperature catalytic oxidation reactions*, in Ehrfeld, W. (Ed.), *Microreaction Technology: 3rd International Conference on Microreaction Technology, Proc. of IMRET 3*, pp. 674–686, Springer-Verlag, Berlin (2000).

83 Karnik, S. V., Hatalis, M. K., Kothare, M. V., *Palladium based micro-membrane for water gas shift reaction and hydrogen gas separation*, in Matlosz, M., Ehrfeld, W., Baselt, J. P. (Eds.), *Microreaction Technology – IMRET 5: Proc. 5th International Conference on Microreaction Technology*, pp. 295–302, Springer-Verlag, Berlin (2001).

84 Kursawe, A., Dietzsch, E., Kah, S., Hönicke, D., Fichtner, M., Schubert, K., Wiessmeier, G., *Selective reactions in microchannel reactors*, in Ehrfeld, W. (Ed.), *Microreaction Technology: 3rd International Conference on Microreaction Technology, Proc. of IMRET 3*, pp. 213–223, Springer-Verlag, Berlin (2000).

85 Walter, S., Liauw, M., *Fast concentration cycling in microchannel reactors*, in *Proceedings of the 4th International Conference on Microreaction Technology, IMRET 4*, 5–9 March 2000, pp. 209–214, AIChE Topical Conf. Proc., Atlanta, GA (2000).

86 Brenchley, D. L., Wegeng, R. S., *Status of microchemical systems development in the United States of America*, in Ehrfeld, W., Rinard, I. H., Wegeng, R. S. (Eds.), *Process Miniaturization: 2nd International Conference on Microreaction Technology, IMRET 2, Topical Conf. Preprints*, pp. 18–23, AIChE, New Orleans (1998).

87 Ehrfeld, W., Hessel, V., Haverkamp, V., *Microreactors*, in *Ullmann's Encyclopedia of Industrial Chemistry*, Wiley-VCH, Weinheim (1999).

88 deWitt, S., *Microreactor for chemical synthesis*, Curr. Opin. Chem. Biol. **3** (1999) 350–356.

89 Jensen, K. F., Ajmera, S. K., Firebaugh, S. L., Floyd, T. M., Franz, A. J., Losey, M. W., Quiram, D., Schmidt, M. A., *Microfabricated chemical systems for product screening and synthesis*, in Hoyle, W. (Ed.), *Automated Synthetic Methods for Specialty Chemicals*, pp. 14–24, Royal Society of Chemistry, Cambridge (2000).

90 Ehrfeld, W., Hessel, V., Löwe, H., *Microreactors*, Wiley-VCH, Weinheim (2000).

91 Jensen, K. F., *Microchemical systems for synthesis of chemicals and information*, in *Proceedings of the Japan Chemical Innovation Institute (jap.) (JCII)*, Tokyo (2001).

92 Jensen, K. F., *Microreaction engineering – is small better?* Chem. Eng. Sci. **56** (2001) 293–303.

93 Jensen, K. F., *Microsystems for chemical synthesis, energy conversion, and bioprocess applications*, in Yoshida, J.-I. (Ed.), *Microreactors, Epoch-making Technology*

for Synthesis – High Technology
Information, pp. 63–74, cmc publisher
(2003).

94 HASWELL, S. T., WATTS, P., *Green
chemistry: synthesis in micro reactors,*
Green Chem. **5** (2003) 240–249.

95 HESSEL, V., LÖWE, H.,
*Mikroverfahrenstechnik: Komponenten –
Anlagenkonzeption – Anwenderakzeptanz
– Teil 2,* Chem. Ing. Tech. **74**, 3 (2002)
185–207.

96 GAVRIILIDIS, A., ANGELI, P., CAO, E.,
YEONG, K. K., WAN, Y. S. S., *Technology
and application of microengineered
reactors,* Trans. IChemE. **80**/A, 1 (2002)
3–30.

97 JÄHNISCH, K., HESSEL, V., LÖWE, H.,
BAERNS, M., *Chemie in Mikrostruktur-
reaktoren,* Angew. Chem. **44**, 3 or 4
(2003) in press.

98 REBROV, E. V., DE CROON, M. H. J. M.,
SCHOUTEN, J. C., *Development of a cooled
microreactor for platinum catalyzed
ammonia oxidation,* in MATLOSZ, M.,
EHRFELD, W., BASELT, J. P. (Eds.),
*Microreaction Technology – IMRET 5:
Proc. 5th International Conference on
Microreaction Technology,* pp. 49–59,
Springer-Verlag, Berlin (2001).

99 RICHTER, T., EHRFELD, W., GEBAUER, K.,
GOLBIG, K., HESSEL, V., LÖWE, H., WOLF,
A., *Metallic microreactors: components and
integrated systems,* in EHRFELD, W.,
RINARD, I. H., WEGENG, R. S. (Eds.),
*Process Miniaturization: 2nd Inter-
national Conference on Microreaction
Technology, IMRET 2, Topical Conf.
Preprints,* pp. 146–151, AIChE, New
Orleans (1998).

100 HESSEL, V., LÖWE, H., *Micro chemical
engineering: components – plant concepts –
user acceptance: Part II,* Chem. Eng.
Technol. **26**, 4 (2003) 391–408.

101 FÖDISCH, R., KURSAWE, A., HÖNICKE,
D., *Immobilizing heterogeneous catalysts
in microchannel reactors,* in *Proceedings of
the 6th International Conference on
Microreaction Technology, IMRET 6,* 11–
14 March 2002, pp. 140–146, AIChE
Pub. No. 164, New Orleans (2002).

102 STIEF, T., LANGER, O.-U., *Simulation
studies of periodic process control in
microreactors,* in *Proceedings of the 4th
International Conference on Microreaction*

Technology, IMRET 4, 5–9 March 2000,
pp. 215–226, AIChE Topical Conf. Proc.,
Atlanta, GA (2000).

103 KAH, S., HÖNICKE, D., *Selective oxidation
of 1-butene to maleic anhydride –
comparison of the performance between
microchannel reactors and fixed bed
reactor,* in MATLOSZ, M., EHRFELD, W.,
BASELT, J. P. (Eds.), *Microreaction
Technology – IMRET 5: Proc. 5th International Conference on Microreaction
Technology,* pp. 397–407, Springer-
Verlag, Berlin (2001).

104 WÖRZ, O., JÄCKEL, K. P., RICHTER, T.,
WOLF, A., *Microreactors, new efficient
tools for optimum reactor design,* in
*Microtechnologies and Miniaturization,
Tools, Techniques and Novel Applications
for the BioPharmaceutical Industry,* IBC
Global Conferences Limited, London
(1998).

105 WÖRZ, O., *Microreactors as tools in
chemical research,* in MATLOSZ, M.,
EHRFELD, W., BASELT, J. P. (Eds.),
*Microreaction Technology – IMRET 5:
Proc. 5th International Conference on
Microreaction Technology,* pp. 377–386,
Springer-Verlag, Berlin (2001).

106 SCHUBERT, K., *High Potentials –
Mikrostrukturapparate für die chemische
und thermische Verfahrenstechnik,* Chem.
Tech. **27**, 10 (1998) 124–127.

107 WÖRZ, O., *Wozu Mikroreaktoren?* Chem.
Unserer Zeit **34**, 1 (2000) 24–29.

108 JÄCKEL, K. P., *Microtechnology:
application opportunities in the chemical
industry,* in EHRFELD, W. (Ed.),
*Microsystem Technology for Chemical and
Biological Microreactors,* DECHEMA
Monographs, Vol. 132, pp. 29–50, Verlag
Chemie, Weinheim (1996).

109 WIESSMEIER, G., *Untersuchungen zur
heterogen katalysierten Oxidation von
Propen in einem Mikrostrukturreaktor,*
Master thesis, Karlsruhe (1992).

110 BASELT, J. P., FÖRSTER, A., HERMANN, J.,
*Microreaction technology: focusing the
German activities in this novel and
promising field of chemical process
engineering,* in EHRFELD, W., RINARD,
I. H., WEGENG, R. S. (Eds.), *Process
Miniaturization: 2nd International
Conference on Microreaction Technology,
IMRET 2, Topical Conf. Preprints,*

pp. 13–17, AIChE, New Orleans
(1998).

111 TONKOVICH, A. L. Y., ZILKA, J. L.,
POWELL, M. R., CALL, C. J., *The catalytic
partial oxidation of methane in a micro-
channel chemical reactor*, in EHRFELD, W.,
RINARD, I. H., WEGENG, R. S. (Eds.),
*Process Miniaturization: 2nd Inter-
national Conference on Microreaction
Technology, IMRET 2, Topical Conf.
Preprints*, pp. 45–53, AIChE, New
Orleans (1998).

112 MAYER, J., FICHTNER, M., WOLF, D.,
SCHUBERT, K., *A microstructured reactor
for the catalytic partial oxidation of
methane to syngas*, in EHRFELD, W. (Ed.),
*Microreaction Technology: 3rd Inter-
national Conference on Microreaction
Technology, Proc. of IMRET 3*,
pp. 187–196, Springer-Verlag, Berlin
(2000).

113 SCHOUTEN, J. C., REBROV, E., DE CROON,
M. H. J. M., *Challenging prospects for
microstructured reaction architectures in
high-throughput catalyst screening, small
scale fuel processing, and sustainable fine
chemical synthesis*, in *Proceedings of the
Micro Chemical Plant – International
Workshop*, 4 February 2003, pp. L5 and
25–32, Kyoto (2003).

114 WEGENG, R. W., CALL, C. J., DROST,
M. K., *Chemical system miniaturization*,
in *Proceedings of the AIChE Spring
National Meeting*, 25–29 February 1996,
pp. 1–13, New Orleans (1996).

115 HAGENDORF, U., JANICKE, M., SCHÜTH,
F., SCHUBERT, K., FICHTNER, M., *A Pt/
Al₂O₃ coated microstructured reactor/heat
exchanger for the controlled H₂/O₂-
reaction in the explosion regime*, in
EHRFELD, W., RINARD, I. H., WEGENG,
R. S. (Eds.), *Process Miniaturization:
2nd International Conference on Micro-
reaction Technology, IMRET 2, Topical
Conf. Preprints*, pp. 81–87, AIChE,
New Orleans (1998).

116 HAAS-SANTO, K., GÖRKE, O., SCHUBERT,
K., FIEDLER, J., FUNKE, H., *A micro-
structure reactor system for the controlled
oxidation of hydrogen for possible
application in space*, in MATLOSZ, M.,
EHRFELD, W., BASELT, J. P. (Eds.),
*Microreaction Technology – IMRET 5:
Proc. 5th International Conference on*

Microreaction Technology, pp. 313–320,
Springer-Verlag, Berlin (2001).

117 CHATTOPADHYAY, S., VESER, G., *Detailed
simulations of catalytic ond non-catalytic
ignition during H₂-oxidation in a micro-
channel reactor: isothermal case*, in
Proceedings of ChemConn-2001,
December 2001, pp. 1–6, Chennai,
India (2001).

118 VESER, G., FRIEDRICH, G., FREYGANG,
M., ZENGERLE, R., *A simple and flexible
micro reactor for investigations of
heterogeneous catalytic gas reactions*, in
FROMENT, G. F., WAUGH, K. C. (Eds.),
*Reaction Kinetics and the Development of
Catalytic Processes*, pp. 237–245, Elsevier
Science (1999).

119 DONNER, S., *Tausend Kanäle für eine
Reaktion*, Chem. Rundschau
(11.02.2003).

120 EHRFELD, W., HESSEL, V., MÖBIUS, H.,
RICHTER, T., RUSSOW, K., *Potentials and
realization of micro reactors*, in EHRFELD,
W. (Ed.), *Microsystem Technology for
Chemical and Biological Microreactors,
DECHEMA Monographs, Vol. 132*,
pp. 1–28, Verlag Chemie, Weinheim
(1996).

121 GÖRKE, O., PFEIFER, P., SCHUBERT, K.,
*Determination of kinetic data in the
isothermal microstructure reactor based on
the example of catalyzed oxidation of
hydrogen*, in *Proceedings of the 6th
International Conference on Microreaction
Technology, IMRET 6, 11–14 March
2002*, pp. 262–274, AIChE Pub. No. 164,
New Orleans (2002).

122 SURANGALIKAR, H., OUYANG, X., BESSER,
R. S., *Experimental study of hydrocarbon
hydrogenation and dehydrogenation
reactions in silicon microfabricated reactors
of two different geometries*, Chem. Eng.
J. 90, 4140 (2002) 1–8.

123 BESSER, R., PREVOT, M., *Linear scale-up of
micro reaction systems*, in *Proceedings of
the 4th International Conference on
Microreaction Technology, IMRET 4*,
5–9 March 2000, pp. 278–283, AIChE
Topical Conf. Proc., Atlanta, GA (2000).

124 CIU, T., FANG, J., MAXWELL, J., GARDNER,
J., BESSER, R., ELMORE, B., *Micromachining of microreactor for dehydrogenation of cyclohexane to benzene*, in
Proceedings of the 4th International

Conference on Microreaction Technology, *IMRET 4*, 5–9 March 2000, pp. 488, AIChE Topical Conf. Proc., Atlanta, GA (**2000**).

125 BESSER, R. S., OUYANG, X., SURANGALIKAR, H., *Fundamental characterization studies of a microreactor for gas-solid heterogeneous catalysis,* in *Proceedings of the 6th International Conference on Microreaction Technology, IMRET 6*, 11–14 March 2002, pp. 254–261, AIChE Pub. No. 164, New Orleans (**2002**).

126 SURANGALIKAR, H., BESSER, R. S., *Study of catalysis of cyclohexene hydrogenation and dehydrogenation in a microreactor,* in *Proceedings of the 6th International Conference on Microreaction Technology, IMRET 6*, 11–14 March 2002, pp. 248–253, AIChE Pub. No. 164, New Orleans (**2002**).

127 JONES, F., QING, D., FANG, J., CUI, T., *A fundamental study of gas-solid heterogeneous catalysis in microreactors,* in *Proceedings of the 4th International Conference on Microreaction Technology, IMRET 4*, 5–9 March 2000, pp. 400–409, AIChE Topical Conf. Proc., Atlanta, GA (**2000**).

128 WIESSMEIER, G., HÖNICKE, D., *Strategy for the development of micro channel reactors for heterogenously catalyzed reactions,* in EHRFELD, W., RINARD, I. H., WEGENG, R. S. (Eds.), *Process Miniaturization: 2nd International Conference on Microreaction Technology, IMRET 2, Topical Conf. Preprints,* pp. 24–32, AIChE, New Orleans (**1998**).

129 WIESSMEIER, G., HÖNICKE, D., *Microreaction technology: development of a micro channel reactor and its application in heterogenously catalyzed hydrogenations,* in EHRFELD, W., RINARD, I. H., WEGENG, R. S. (Eds.), *Process Miniaturization: 2nd International Conference on Microreaction Technology, IMRET 2, Topical Conf. Preprints,* pp. 152–153, AIChE, New Orleans (**1998**).

130 DIETZSCH, E., HÖNICKE, D., FICHTNER, M., SCHUBERT, K., WIESSMEIER, G., *The formation of cycloalkenes in the partial gas phase hydrogenation of c,t,t-1,5,9-cyclododecatriene, 1,5-cyclooctadiene and*

benzene in microchannel reactors, in *Proceedings of the 4th International Conference on Microreaction Technology, IMRET 4*, 5–9 March 2000, pp. 89–99, AIChE Topical Conf. Proc., Atlanta, GA (**2000**).

131 WIESSMEIER, G., *Monolithische Mikrostruktur-Reaktoren mit Mikroströmungskanälen und regelmäßigen Mesoporensystem für selektive, heterogen katalysierte Gasphasenreaktion,* Dissertation, Karlsruhe (**1996**).

132 XU, Y., PLATZER, B., *Concepts for the simulation of wall-catalyzed reactions in microchannel reactors with mesopores in the wall region,* Chem. Eng. Technol. **24**, 8 (**2001**) 773–783.

133 STEINFELDT, N., BUYEVSKAYA, O. V., WOLF, D., BAERNS, M., *Comparative studies of the oxidative dehydrogenation of propane in micro-channels reactor module and fixed-bed reactor,* in SPIVEY, J. J., IGLESIA, E., FLEISCH, T. H. (Eds.), *Stud. Surf. Sci. Catal.,* pp. 185–190, Elsevier Science, Amsterdam (**2001**).

134 STEINFELDT, N., DROPKA, N., WOLF, D., *Oxidative dehydrogenation of propane in a micro-channel reactor-kinetic measurements, modeling and reactor simulation,* in *Proceedings of the Microreaction Technology – IMRET 5: Proceedings of the 5th International Conference on Microreaction Technology,* 27–30 May 2001, Strasbourg, France (**2001**).

135 ROUGE, A., RENKEN, A., *Forced periodic temperature oscillations in microchannel reactors,* in MATLOSZ, M., EHRFELD, W., BASELT, J. P. (Eds.), *Microreaction Technology – IMRET 5: Proc. 5th International Conference on Microreaction Technology,* pp. 230–239, Springer-Verlag, Berlin (**2001**).

136 BECKER, G., FISCHER, K., FLINK, A., HERRMANN, E., WEISMANTEL, L., WIESSMEIER, G., SCHUBERT, K., FICHTNER, M., *Verfahren zur Phosgenierung von Aminen in der Gasphase unter Einsatz von Mikrostrukturmischern,* Ger. Pat., DE 19800529, Bayer AG, Leverkusen (**1998**).

137 ASHMEAD, J. W., BLAISDELL, C. T., JOHNSON, M. H., NYQUIST, J. K., PEROTTO, J. A., RYLEY, J. F., *Integrated chemical Processing Apparatus and*

Processes for the Preparation Thereof,
US 5690763, E.I. Du Pont de Nemours
and Company, Wilmington (USA)
(1993).

4
Liquid- and Liquid/Liquid-phase Reactions

4.1
Micro Reactors for Liquid-phase and Liquid/Liquid-phase Reactions

4.1.1
Tube Micro Reactors

This class is the simplest of all micro reactors and certainly the most convenient one to purchase, but not necessarily one with compromises or reduced function. HPLC or other tubing of small internal dimensions is used for performing reactions. There are many proofs in the literature for process intensification by this simple concept. As a micro mixer is missing, mixing either has to be carried out externally by conventional mini-equipment or may not be needed at all. The latter holds for reactions with one reactant only or with a pre-mixed reactant solution, which does not react before entering the tube.

4.1.1.1 Reactor 1 [R 1]: Electrothermal Tubing-based Micro Reactor
The central part of this device (Figure 4.1) is a liquid chromatography (LC)-type steel tubing suitable for high-pressure operation which is resistance-heated using electrical current from an external power supply [1]. On the tube, along the flow passage, a series of voltage taps monitor real-time temperature profiles. Thereby, the course of the reaction can be followed stage-wise provided that heat is released or consumed. The typical volume of a stage involves about 200 μl.

Reactor type	Electrothermal tubing-based micro reactor	Tubing length	1.5 m
Tubing material	300-series HPLC grade steel	Maximum working pressure for tubing	70 MPa
Tubing outer diameter	1.59 mm	Injection of further streams by	HPLC-type sample injection valves
Tubing inner diameter	1.27 mm		

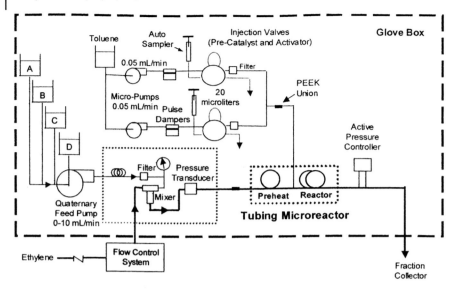

Figure 4.1 Flow scheme of a plant comprising an electrothermal tubing-based micro reactor configured for ethylene polymerization [1].

Before the actual reaction tube, a pre-heating tube is attached to bring the fluids close to reaction temperature [1]. Injection of further streams is performed by HPLC-type sample injection valves.

4.1.1.2 Packed-bed Tube or Capillary Micro Reactors

When performing catalytic reactions or reactions with immobilized reactants, a bed or support has to be filled into a tube or capillary. The filling may be a bed of powder, a bed of granules or a three-dimensional material network (e.g. a polymerized foam). By special choice of the filling, e.g. very regularly sized particles, it is attempted to improve the flow characteristics.

These filled tubes or capillaries hence are mini fixed beds with internal micro flow regions. Such approach is a link between real micro structured units and conventional equipment. The type of processing may be analogous to conventional processing, e.g. fixed-bed or trickle-bed operation.

4.1.2.1 Reactor 2 [R 2]: Packed-bed Capillary Micro fFow Reactor

This reactor concept, termed micro flow reactor [2], simply relies on the use of conventional polymer or glass tubing. The tubing has millimeter internal diameter. Functionalized Merrifield resin polymer beads with catalyst moieties attached are filled into the tubing and are held in place by plugs of glass wool. Standard HPLC or OmniFit connectors are attached to the tubing on one end for connection to syringe pumps. The other end of the tubing is connected to a syringe needle to enter into a vessel to quench the experiment.

Reactor type	Packed-bead capillary micro flow reactor	Packed-bead bed length	2–10 mm
Capillary material	Polypropylene; glass	Merrifield resin	200–400 mesh
Capillary inner diameter	2 mm (polypropylene); 1 mm (glass)	Nickel catalyst loadings	2–6 wt.-%

4.1.2.2 Reactor 3 [R 3]: Porous-polymer Rod in Tube Micro Reactor

A porous glass rod serves as holder for a polymer block. This material is introduced as monomer in the carrier and polymerized therein [3]. Such a glass rod was encapsulated within a pressure-resistant fiber-reinforced housing (Figure 4.2).

The polymerization applied produces spherical polymer particles (1–10 µm diameter) connected by polymer bridges [3]. Thus, a one-piece polymer phase is obtained. The interstices between the particles have a characteristic length of a few micrometers. Overall, the polymer structure can be ascribed as lose.

This polymer resin is in a first step chemically functionalized, yielding initial reactive moieties [3]. In a second step, other groups can be introduced by chemical reaction of these moieties, e.g. creating ammonium groups. By ion exchange, reactive anions are bound which serve as reactants for the reactions to be carried out later on [P 56].

Reactor type	Porous-polymer rod in tube micro reactor	Polymer load	10%
Polymer rod carrier material	Porous glass	Polymer type	Polystyrene-divinylbenzene
Polymer rod carrier inner diameter	50–300 µm	Polymer mass	250 mg
Material for housing for carrier	PTFE	Reactive group	Quaternary amine
Housing for carrier: diameter; length	5.3 mm; 110 mm		

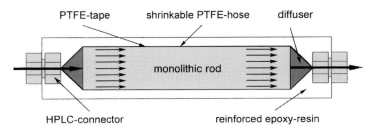

Figure 4.2 Schematic of the porous-polymer rod in a tube micro reactor [3].

4.1.3
Chip Micro-reactor devices

This section describes micro-channel reactors that are made by silicon micro machining, thin-film techniques and other related techniques, originally developed for microelectronic fabrication. In the following, only components are considered that are more or less completely made via this route. Typically, these reactors have a mixing function, e.g. as a mixing tee or Y-piece and a subsequent channel section for reaction. Additional functions, especially concerning sensing, controlling and heating, are easily implemented owing to the microelectronic fabrication route chosen. Typically, the way of assembly is to have a layered structure which is bonded or clamped. The overall size of the device is small, similar to a cheque card.

Liquid transport is achieved by hydrostatic action, pumping or electroosmotic flow (EOF). So far, chip reactors have been employed at low to very low flow rates, e.g. from 1 ml min^{-1} to 1 µl min^{-1}. Applications consequently were restricted to the laboratory-scale or even solely to analytics. However, this is not intrinsic. By choosing larger internal dimensions, similar throughputs as for the other classes of liquid or liquid/liquid micro reactors are in principle achievable.

4.1.3.1 Reactor 4 [R 4]: Chip Reactor with Micro-channel Mixing Tee(s)

A micro reactor comprising one or several (up to three) micro-channel mixing tees was made by photolithography and etching (Figure 4.3) [4–13]. Species transport is achieved by means of EOF and electrophoretic mobility (a detailed description on EOF is given in [14]). The number of mixing tees mainly derives from the number of reactants to be added. For instance, when using four reactants, such as for the Ugi reaction, three micro-channel mixing tees result. The mixing tees usually are fed by counter-flow (180°) or cross-flow (90°) arrangements of the two streams to be mixed.

Microfabrication is achieved by photolithography and isotropic etching of glass using HF [4–13]. Thermal bonding serves for interconnection. Holes are drilled in the top plate for connection to the fluidic peripherals.

Some of the chip reactors are equipped with micro porous frits (see, e.g., [13]).

Version (a) comprises one mixing tee only (Figure 4.4) [11]. Version (a2) can be heated up to 70 °C [9]. A T-shaped Peltier heater attached to the lower plate was used to adjust temperature.

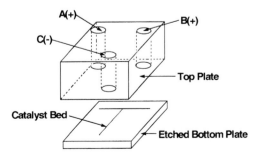

Figure 4.3 Schematic of the chip reactor with one micro-channel mixing tee [11].

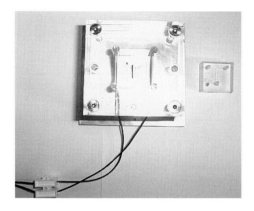

Figure 4.4 Mounted chip reactor with (right) and single micro-channel mixing tee chip (left) [8].

Reactor type	Chip micro reactor with one micro-channel mixing tee	Channel width; depth	200 μm (at top) × 100 μm
Micro reactor material	Borosilicate glass	Thickness of top plate	17 mm
Outer dimensions	20 × 20 × 25 mm³	Diameter of inlet and outlet holes	3 mm

Version (b) has a four-channel flow guidance that encompasses two mixing tees in two simple mixing tees (Figure 4.5) [8]. An example of this function is the flow guidance for the Michael addition. In a first step, the base and 1,3-dicarbonyl compound streams merge. The enolate stream thus formed is then mixed with the Michael acceptor. Microporous silica frits are set into the channels to minimize

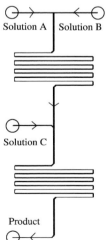

Figure 4.5 Schematic of the chip reactor with two micro-channel mixing tees.

hydrodynamic effects. Electroosmotic flow was applied to transport the reactants. Voltage setting and monitoring of the platinum electrodes, which are placed in the reservoirs, was achieved by the Lab-View program.

Reactor type	Chip micro reactor with two micro-channel mixing tees	Micro reactor material	Borosilicate glass
Channel dimensions	$150 \times 50 \ \mu m^2$	Outer dimensions	$20 \times 20 \times 25 \ mm^3$

In version (c), three mixing tees were realized in another version of the above-mentioned chip reactor concept [16].

Reactor type	Chip micro reactor with three micro-channel mixing tees	Micro reactor material	Borosilicate glass
Channel dimensions	$200 \times 100 \ \mu m^2$	Outer dimensions	$20 \times 20 \times 25 \ mm^3$

4.1.3.2 Reactor 5 [R 5]: Chip Micro Reactor with Multiple Vertical Injections in a Main Channel

This reactor is generically and according to fabrication closely related to the reactor concept [R 4]. Although the micro channel geometry is, on a first sight, only slightly different from [R 4], the functional principle and details on the liquid feeding (by EOF and electrophoresis) are not. A detailed description of EOF is given in [14].

The chip micro reactor [R 5] comprises a long micro channel connected to two vertically positioned shorter channels at each end which lead to two reservoirs [17]. These shorter channels are oriented in opposite directions so that a Z-type flow configuration results. In the downstream section of the long channel, two other vertically oriented shorter channels are also attached. These channels are each connected to a liquid reservoir. A total of five reservoirs are made in this way (Figure 4.6).

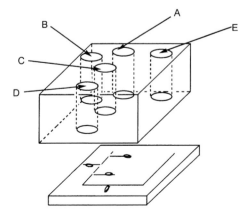

Figure 4.6 Schematic of the chip reactor with multiple vertical injections in a main channel [17].

Besides continuous operation, the flow of one or the other solution from the two feed reservoirs, or compounds, can be triggered by proper setting of voltages so that a slug of one solution (compound) in another can be formed [17]. By this means, migration of one component into a front of another, leading to mixing and subsequent reaction, can be performed as the mobilities of species are different when applying an electrical field.

In a special version of the device [17], all four shorter channels were discontinuous in the sense that they each comprised an array of very small channels which act as flow resistors. This ensures suppression of pressure-driven flow resulting from differences in the reservoir heights.

Reactor type	Chip micro reactor with multiple vertical injections in a main channel	Micro-channels etch mask width; etch depth	146 μm; 38 μm
Micro reactor material	Borosilicate glass	Flow resistor channels: etch mask width; etch depth	183 μm; 3 μm
Outer dimensions	25 × 25 × 20 mm³		

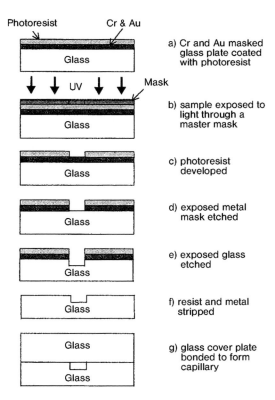

a) Cr and Au masked glass plate coated with photoresist

b) sample exposed to light through a master mask

c) photoresist developed

d) exposed metal mask etched

e) exposed glass etched

f) resist and metal stripped

g) glass cover plate bonded to form capillary

Figure 4.7 Steps in microfabrication of the chip reactor with multiple vertical injections [17].

Microfabrication was achieved by isotropic etching of glass using HF (Figure 4.7) [17]. Thermal bonding served for interconnection. Holes were drilled in the top plate for connection to the fluidic peripherals.

4.1.3.3 Reactor 6 [R 6]: Chip Micro Reactor with Multiple Micro Channel–Mixing Tees

This reactor concept is generically similar to [R 4], but relies on a more complex network of micro channels having more reservoirs (seven) [18]. The corresponding chip is a commercial product of Caliper Technologies Company (110 Caliper chip™), originally designated for µTAS applications. The version actually used for performing organic chemistry was optimized for hydrodynamic flow control (Figure 4.8). The chips were constructed from two glass plates by means of standard photolithography.

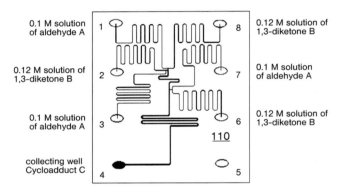

Figure 4.8 Schematic of the commercial 110 Caliper chip™ micro reactor for performing organic chemistry [18].

The etched micro channels have different widths for more stable flow, e.g. to avoid a dependence on capillary forces in the reservoirs [18]. The glass chip is glued to a polymer caddy for interfacing with a multiport control device, the Caliper 42™ Workstation.

Reactor type	Chip micro reactor with multiple micro-channel mixing tees	Caddy material	Polypropylene
Chip material	Soda lime glass	Widths of two sets of micro channels	29 µm; 74 µm

4.1.3.4 Reactor 7 [R 7]: Chip Micro Reactor with Z-type Flow Configuration

This reactor concept is generically similar to [R 4] and uses the same fabrication and assembly techniques, so the reader is referred to the corresponding description [19]. However, it has only one micro channel as it serves for transporting one liquid (or solution) only, i.e. was constructed for reactions with one reactant only such as eliminations [19].

The micro reactor contained a heating function (unlike [R 4] and the other versions of this reactor concept [R 5] and [R 6], decribed below) via a heating wire connected to a potentiostat [19]. This wire was integrated into the micro reactor by placing it in the mold before pouring the liquid PDMS.

For electroosmotic flow transport, a tube was inserted into the base plate, connected to the micro channel [19].

Reactor type	Chip micro reactor with Z-type flow configuration	Heater type/material	Nichrome wire
Micro reactor material	Borosilicate glass	Wire outer diameter	250 μm
Top plate material	Polydimethyl-siloxane (PDMS)	Tube for electro-osmotic flow: inner diameter; length	500 μm; 20 mm
Micro channel: width; depth; length	200 μm; 80 μm; 30 mm		

4.1.3.5 Reactor [R 8]: Chip Micro Reactor with Extended Serpentine Path and Ports for Two-step Processing

This is also (see [R 6]) a commercial chip ('Radiator'), provided by MCS, Micro Chemical Systems Ltd., The Deep Business Center [20]. A bottom plate contains an extensively wound serpentine channel. A top plate covers this microstructure. The two reactant solutions enter via capillary tubing through holes in the top plate. The first reactant is fed at the start of the serpentine path and the second enters this path in a short distance. Shortly before the end of the serpentine, a third stream can enter which may serve, e.g., for dilution and thus quenching of the reaction. After a very short passage, the diluted streams enter via a fourth port analytics. Commercially available capillary connectors were employed.

Microfabrication was made by wet-chemical glass etching [20]. Sealing was achieved by thermal bonding.

Reactor type	Chip micro reactor with extended serpentine path and ports for two-step processing	Serpentine micro channel: width; depth	100 μm; 25 μm
Chip material	Borosilicate glass	Type of commercial connectors	Standard fused-silica capillary connectors

4.1.3.6 Reactor 9 [R 9]: Chip System with Triangular Interdigital Micro Mixer–Reaction Channel

This system is a chip version of three dimensional micro mixer–tube reactor set-ups [21]. It comprises a triangular interdigital micro mixer with a focusing zone that thins the multi-lamellae and a subsequent reaction channel that is surrounded by two heat-transfer fluid channels (Figure 4.9). These channels themselves are surrounded by two insulation gaps to prevent heat losses. At the end of the reaction channel a temperature sensor may be placed.

The whole system is constructed from two silicon wafers, fabricated using photoresist by deep reactive ion etching (DRIE) [21]. The wafers were thermally bonded. Thereafter, inlet and outlet ports were machined and the single reactors isolated by DRIE.

Owing to the transparency of silicon in most parts of the IR spectrum, such systems can be used also for in-line chemical analysis, utilizing them as flow-through analysis cells (see Section 4.2.1.4, In-line IR monitoring) [21].

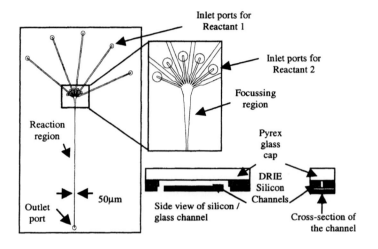

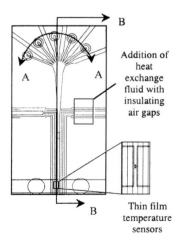

Figure 4.9 Chip system with triangular interdigital micro mixer–reaction channel. First- (top) and second- (bottom) generation reactor designs [22].

Reactor type	Chip system with triangular interdigital micro mixer–reaction channel	Mixing channel width	50 µm
Outer dimensions	24×29 mm² (1st generation); 21×27 mm² (2nd generation)	Heat transfer channel width; depth	50 µm; 350–500 µm
Feed channel width	60 µm		

4.1.3.7 Reactor 10 [R 10]: 2 × 2 Parallel Channel Chip Reactor

On this chip, two sets of two channels each branch in a T-configuration yielding eight channels [23, 24]. The latter are guided so that they merge to four channels in which the reactions are carried out (Figure 4.10). These channels comprise the full set of all 2×2 combinations of the reactants.

The idea of the chip is that all four permutated compounds from two reactants of one sort and from two of the other sort are created [23, 24]. To avoid crossing of reactant streams, a multi-layered architecture has to be used for construction, separating one sort reactant streams in one layer from the other sort in another layer. Extension of this principle to $n \times n$ parallel prepared permutations could be termed combinatorial.

Two microstructured layers of the 2×2 chip were fabricated by photolithography and wet etching in glass (Figure 4.11) [23, 24]. These top and bottom layers and a third thinner layer containing holes as conduits were thermally bonded to yield the chip. The way of guiding the micro channels, as described above, is referred to as two-level crossing.

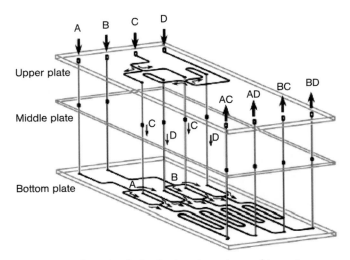

Figure 4.10 Schematic of a 2 × 2 micro-channel-array chip reactor; a generic design for performing combinatorial chemistry [23].

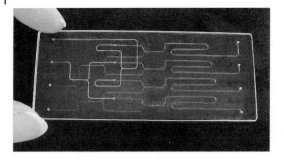

Figure 4.11 2 × 2 micro channel-array chip reactor realized in glass [23].

Reactor type	2 × 2 parallel channel chip reactor	Dimensions middle plate	30 × 70 × 0.2 mm³
Material	Glass	Micro-channel width; depth	240 μm; 60 μm
Dimensions top and bottom plate	30 × 70 × 0.7 mm³		

4.1.3.8 Reactor 11 [R 11]: Bifurcation-distributive Chip Micro Mixer

This device is based on multiple parallel bi-lamination using bifurcation cascade for generating multiple thin fluid olamellae [25]. The first feed stream is split into multiple sub-streams via a bifurcation cascade; in a similar way this is done for the second feed stream in another level. The corresponding sub-streams enter via nozzles into the first level. Here, the end of the channels of the bifurcation cascade and the nozzles lie next to each other. Thereby, bi-laminated sub-streams are formed and enter many parallel channels of an inverse-bifurcation cascade. These are recombined to multilayered stream in one main channel which has a serpentine shape, i.e. comprises extended length.

Reactor type	Bifurcation-distributive chip micro mixer	
Material	Glass/silicon	

Figure 4.12 On-line coupling of the bifurcation-distributive chip micro mixer to a Perseptive Biosystems Mariner™ TOF-MS [25].

4.1.3.9 Reactor 12 [R 12]: Micro Y-Piece Micro-channel Chip Reactor

This device is a generically simple reactor comprising a micro-channel Y-piece section and an elongated reaction micro channel attached [26, 27]. The microstructures were mechanically fabricated in PMMA using a flat end mill. A top plate was joined with the microstructured plate by baking under vacuum. The reaction temperature was fixed using a hot-plate.

Reactor type	Micro Y-piece micro-channel chip reactor	Micro-channel width; depth; length	200 µm; 200 µm; 400 mm
Material	PMMA		

Chip reactor of similar design was also proposed by [91].

4.1.3.10 Reactor 13 [R 13]: Triple Feed Continuous Multi-phase Chip Reactor

This device contains one main reaction and processing path to which several channels are attached, feeding the main stream with additional phases, miscible or immiscible, or withdrawing such phases [28]. As a consequence, phases are either mixed or move as continuous streams side-by-side without intermixing. By this means, bi- or even tri-layered phase arrangement are created. If needed, one phase leaves the other by moving into a side channel.

The chip comprises five inlet ports and two outlet ports [28]. Three streams, two aqueous phases and one organic phase, are contacted initially; a layered system is created in this way (Figure 4.13). The aqueous phase is removed from the main stream; two other aqueous streams encompass the remaining organic phase. The two streams fulfil different functions. In total, several (up to 10) chemical unit operations may be conducted in series in this way.

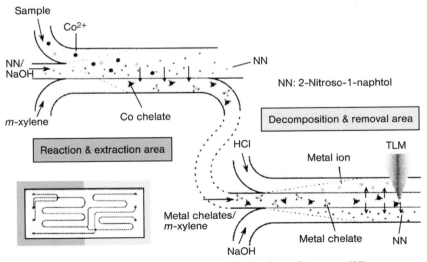

Figure 4.13 Schematic of the triple-feed continuous multi-phase chip reactor [28].

The chip is made from glass by photolithography and wet etching followed by thermal bonding.

Reactor type	Triple-feed continuous multi-phase chip reactor system	Micro-channel width; depth; length	Not given in [28]
Material	Glass		

4.1.3.11 **Reactor 14 [R 14]: Chip with Bi-/Tri-layer Flow Configuration Using Y-type Contact**

The polymer chips comprise two feed micro channels which are connected in a Y-type flow configuration to one elongated micro channel which is folded multiple times for compact design [29]. At the end of this channel a split into two channels is made, again by a Y-type flow configuration. The two-channel flow configuration serves for preparing water–oil parallel (bi-laminated) streams.

In another version, a three-channel flow configuration is created for continuous water–oil-water or oil–water–oil parallel streams [29].

The micro-channel chips were fabricated by an imprinting method. The templates for imprinting were made in silicon by conventional photolithography/dry etching [29]. The structural design was prepared on a glass plate serving as a photo mask by a photographic technique. A resist was coated on a silicon substrate, exposed and dry etched. This silicon master was then imprinted on a polymer substrate at elevated temperature. The structured polymer plate was sealed with a flat polymer plate by clamping.

Reactor type	Chip with bi/tri-layer flow configuration using Y-type contact	Outer dimensions	$30 \times 15 \times 1.7$ mm^3
Material	Polystyrene	Micro-channel width; depth; length	100 μm; 20 μm; 350 mm

4.1.3.12 **Reactor 15 [R 15]: Single-channel Chip Micro Reactor**

A single channel is fabricated in a wafer by etching using traditional silicon micromachining [30]. The catalyst is introduced using selective seeding, monolayer self-assembly and hydrothermal synthesis (see [P 47] for more details). The microstructured wafer is bonded to a glass cover. SU8-resist as sort of glue was spin coated on the glass and UV-exposed after joining to the wafer.

Reactor type	Single-channel chip micro reactor	Zeolite catalyst layer thickness	3 μm
Reaction flow-through chamber: width; depth; length	500 μm × 250 μm × 33 mm		

4.1.4
Chip–Tube Micro Reactors

This class of hybrid components comprises chip micro-reactor devices, as described in Section 4.1.3, connected to conventional tubing. This may be HPLC tubing which sometimes has as small internals as micro channels themselves. The main function of the tubing is to provide longer residence times. Sometimes, flow through the tube produces characteristic flow patterns such as in slug-flow tube reactors. Chip–tube micro reactors are typical examples of multi-scale architecture (assembly of components of hybrid origin).

4.1.4.1 Reactor 16 [R 16]: Liquid-Liquid Micro Chip Distributor–Tube Reactor
(a) This mixer–tube reactor configuration is composed of a liquid/liquid distributor and a tube [31]. By this means, capillary flow, alternating aqueous and organic slugs, can be created. A set of three distributors, of three-way type, was applied; one commercial and two specially manufactured ones made from the commercial (Figure 4.14). Two syringe pumps were used for liquid feed (acid mixture and organic compound). A collection bottle at the reactor output was used for quenching and separating the reaction mixture by diluting and coalescing the dispersed slug mixture to give two separate phases with a much smaller specific interface.

Reactor type	Liquid–liquid micro chip distributor–tube reactor	Tube diameters (PTFE)	150 µm; 178 µm
Mixing tee channel diameters in connector	150 µm; 500 µm; 800 µm for all (three) channels	Tube length (stainless steel)	Not reported
Tube material	Stainless steel; PTFE	Tube lengths (PTFE)	450 mm; 900 mm; 1350 mm
Tube diameters (stainless steel)	127 µm; 178 µm; 254 µm		

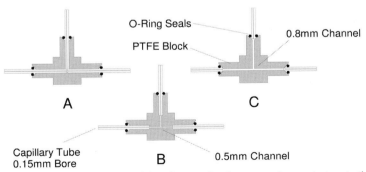

Figure 4.14 Schematic of liquid/liquid micro distributors used as contactors in the reaction system [31].

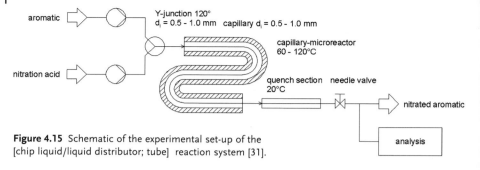

Figure 4.15 Schematic of the experimental set-up of the [chip liquid/liquid distributor; tube] reaction system [31].

(b) In another version of this reactor concept, a Y-piece was employed and attached to a PTFE capillary embedded in a thermostatically controlled jacket (Figure 4.15) [31]. The jacket maintains a uniform temperature by counter-flow with silicone oil. Owing to the high specific surface area of the capillary, it is assumed that isothermal processing is achieved in the reactor.

Reactor type	Liquid-liquid micro chip distributor–tube reactor	Tube diameter (PTFE)	150; 178 µm
Y-piece channel diameter	500–1000 µm	Tube length (PTFE)	1–8 m
Tube material	PTFE		

(c) In a further version of this reactor concept, a T-piece was employed and attached to both steel and PTFE capillaries embedded in a temperature bath [31]. The T-piece was drilled out, allowing a tight connection of all three capillaries. The organic and aqueous phases were fed through two incoming tubes into the T-piece, the merged phase leaving through a third tube. Flow rates were controlled by a single syringe driver loaded with 1 ml and 100 µl syringes. Isothermal behavior was assumed for the set-up.

Reactor type	Liquid/liquid micro chip distributor–tube reactor	Capillary tube length (reactor)	500–1800 mm
Capillary tube material	Stainless steel; PTFE	T-piece inner diameter	~1.59 mm, fits to capillary tubes
Capillary tube outer diameter	1.59 mm	Flow velocities	20–200 mm s^{-1}
Capillary tube inner diameter	127–254 µm	Flow ratio: acid : organic phase	10.5 : 1
Capillary tube length (feed)	300 mm		

4.1.4.5 Reactor 17 [R 17]: Fork-like Chip Micro Mixer–Tube Reactor

Central part of this reaction unit is a split–recombine chip micro mixer made of silicon based on a series of fork-like channel segments [32–36]. Standard silicon micro machining was applied to machine these segments into a silicon plate which was irreversibly joined to a silicon top plate by anodic bonding (Figure 4.16).

The fork-type chip mixer was used in connection with conventional tubes. PTFE tubing was applied in some studies [37, 38].

Reactor type	Fork-like chip micro mixer–tube reactor	Characteristic structure of the mixing stage	'G structure'
Brand name	accoMix (µRea-4 formerly)	Inner device volume	70 µl
Micro mixer material	Silicon	Outer device dimensions	$40 \times 25 \times 1.3$ mm^3
Inlet channel width	1000 µm	Tube material	PTFE
Number of parallel mixing channels to which the inlet flow is distributed	9	Tube diameter	Not reported
Number of mixing stages within one mixing channel	6; 3 in top plate, 3 in bottom plate	Tube length	Not reported
Channel width; depth of micro mixer	360 µm; 250 µm (triangular-shaped)		

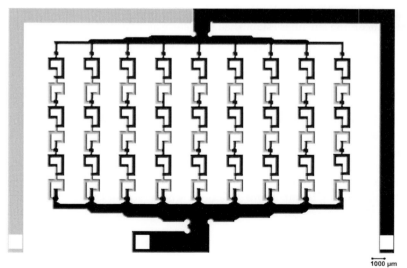

1000 µm

Figure 4.16 Schematic of fork-like chip micro mixer–tube reactor (Accoris GmbH).

4.1.5
3-D Microfab Reactor Devices

3-D microfabricated reactor devices are typically made by fabrication techniques other than stemming from microelectronics, e.g. by modern precision engineering techniques, laser ablation, wet-chemical steel etching or μEDM techniques. Besides having this origin only, these devices may also be of hybrid nature, containing parts made by the above-mentioned techniques and by microelectronic methods. Typical materials are metals, stainless steel, ceramics and polymers or, in the hybrid case, combinations of these materials.

3-D devices have complex requirements on assembly and, in particular, sealing. Integration of sensing is not as facile as for chip micro reactors. In turn, 3-D devices are robust and comprise the classical materials for chemical engineering apparatus.

4.1.5.1   Reactor 18 [R 18]: Interdigital Micro Mixers

Interdigital micro mixers (see [R 20]) are devices with an alternate feed micro-channel arrangement, connected to two feed reservoirs [39–42]. On introducing two liquids, a set of multi-laminated lamellae is created. This set may flow without any change in cross-section as long as needed for completion of mixing by diffusion. The corresponding micro mixer is then termed *rectangular interdigital mixer* (Figure 4.17). The multi-lamellae flow may be geometrically compressed by constantly reducing the micro-channel cross-section. By compressing the overall lamellae set, the individual lamella becomes thinner, speeding mixing by diffusion. The corresponding micro mixer is then termed a *triangular interdigital mixer*. When such geometric focusing is put at its extreme and a subsequent wider flow-through chamber (so-called focusing–expansion approach [43]) is added, secondary flow patterns rise to give eddies which speed up mixing. By this means, jet mixing of the multi-laminated incoming stream is established. The corresponding micro mixer is then termed a *slit-type interdigital mixer*, owing to the first such realized geometry in a stainless steel micro mixer.

The interdigital micro mixers are typically realized as 3-D microfab reactors having a micro structured inlay with the interdigital feed structure and the subsequent flow-through chamber in the top part of a housing [39, 41, 42]. The interdigital feed is realized by guiding multiple flows within overlapping micro channels in the counter-flow direction. The inlay is placed in a recess of the bottom housing part. A characteristic section of the flow-through chamber is the so-called slit, a shallow, but initially wide conduit which becomes much narrower in the direction of the flow (Figure 4.18). This slit, a section of an arc when given in 2-D projection, is connected to a tubing of small inner diameter which feds an outlet. Three such connectors of multiple origin (e.g. HPLC) are connected to the top part of the housing.

The interdigital feed can be fed in a counter-flow or co-flow orientation; the first principle is realized in metal/stainless steel or silicon/stainless steel devices [39, 41], the latter in glass chip devices [40, 44–46].

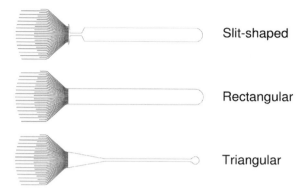

Slit-shaped

Rectangular

Triangular

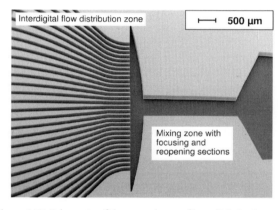

Interdigital flow distribution zone

⊢——⊣ 500 µm

Mixing zone with
focusing and
reopening sections

Figure 4.17 Schematic of the central part of interdigital micro mixers, the interdigital feed element and the flow-through chamber as mixing element. Different shapes of flow-through chambers allow one to perform micro mixing in different ways. Three different designs for interdigital 2-D micro mixers (top). SEM-image of the interdigital feed zone and part of the mxing chamber of the slit interdigital 2-D micro mixer [40].

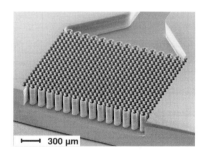

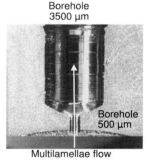

Borehole
3500 µm

Borehole
500 µm

⊢——⊣ 300 µm

Multilamellae flow

Figure 4.18 Left: interdigitated micro channels made by ASE (inlay for 3-D device). Right: cut through the top part of the housing of the slit interdigital 3-D micro mixer comprising the slit-type focusing zone and the subsequent small channel to the outlet [40, 41].

The inlay of the first sort of interdigital mixers is typically a metal part manufactured by the LIGA process, relying on synchrotron irradiation and electroforming [39, 41]. The slit is made by a die-sinking process, a variant of µEDM. The small bore attached to the slit is prepared by special drilling of the top housing part. Alternatively to LIGA, the inlay can be fabricated by ASE, µEDM and laser ablation. By this means, other materials can be used for the inlay such as stainless steel (including high-grade alloys such as Hastelloy), metals not consistent with LIGA (e.g. titanium), conductive ceramics, and polymers.

As a second sort of interdigital micro mixers, chip micro devices, based on thin microstructured plates, were developed [40, 44–46]. Here, the flow conduit has to be arranged in several plates to avoid cross-over of streams. Conduits in the plates serve for interconnection of streams where desired. Finally, all streams are directed to one plate comprising the interdigital feed structure (now in co-flow orientation) and the subsequent mixing chamber (Figure 4.19). The inlet and outlet feed is achieved directly via holes leading to the two feed reservoirs and the end of the mixing chamber, respectively.

This chip version is typically made in glass and has the great advantage that the flow can be directly visualized [40, 44–46]. Fabrication is achieved by photolithography and wet-chemical etching followed by thermal bonding of the plates covered with a thin layer of solder [47].

For the application referred to, the interdgital micro mixers were used on their own, without tubing attached, as reactors. Especially at low flow rates, the internal flow-through chamber acts as delay loop for providing a sufficient residence time.

Reactor type	Interdigital micro mixer	Triangular chamber: initial width; focused width; depth; focusing length; mixing length; focusing angle	3.25 mm; 500 µm; 150 µm; 8 mm; 19.4 mm; 20°
Mixer material	Metal/stainless steel; silicon/stainless steel; glass	Slit-type chamber: initial width; focused width; depth; focusing length; expansion width; expansion length; expansion angle	4.30 mm; 500 µm; 150 µm; 300 µm; 2.8 mm, 24 mm; 126.7°
Metal/silicon mixer feed channel width; depth	40 µm; 300 µm	Slit depth in steel housing	60 µm
Glass mixer feed channel width; depth	60 µm; 150 µm	Tubing attached to slit: diameter	500 µm
Type of flow-through chamber	Rectangular; triangular; slit-type	Device outer dimensions: diameter; height	20 mm; 16.5 mm
Rectangular chamber: width; depth; length	3250 µm; 150 µm; 27.4 mm		

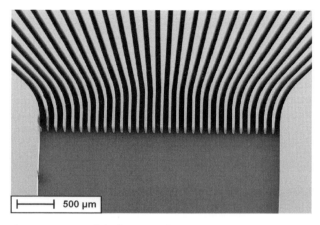

Figure 4.19 SEM of the feed part of a 2-D rectangular interdigital micro mixer [44].

4.1.6
3-D Microfab Mixer–Tube Reactors

3-D microfabricated micro mixers (see Section 4.1.5) may be connected to tubes for reasons of residence time prolongation, similar to chip–tube micro reactors (see Section 4.1.4). The tube may also have the function of creating distinct hydrodynamic features (see Section 4.1.4).

4.1.6.1   Reactor 19 [R 19]: Slit-Type Interdigital Micro Mixer–Tube Reactor

Interdigital micro mixers comprise feed channel arrays which lead to an alternating arrangement of feed streams generating multi-lamellae flows [39–42]. If processes have to be carried out with extended residence times (> 1 s) and/or at a temperature level different from the mixing step, tubes have to be attached to the interdigital micro mixers. Their internals comprise millimeter dimensions or below, if necessary.

One version of such a reactor concept is the combination of a slit-type interdigital metal/steel micro mixer (for a detailed description see [R 18]) with a conventional tube. In [37], PTFE tubing was applied (Figure 4.20).

Reactor type	Slit-type interdigial micro mixer–tube reactor	Mixer material	Stainless steel, nickel
		Tube material	PTFE
		Tube diameter	Not reported
Mixer channel: width, depth; slit width	40 μm; 300 μm; 60 μm	Tube length	Up to 150 cm

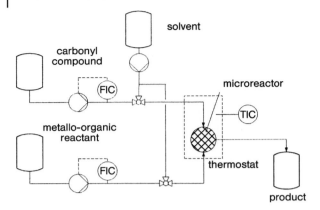

Figure 4.20 Schematic of laboratory-scale reaction system with a slit-type interdigital micro mixer as central element, used at Merck site [134].

4.1.6.2 Reactor 20 [R 20]: Triangular Interdigital Micro Mixer–Tube Reactor

One version of this reactor concept is the combination of the triangular interdigital metal/steel micro mixer (for a detailed description, see [R 18]) with conventional PTFE tubing (Figure 4.21) [46, 48].

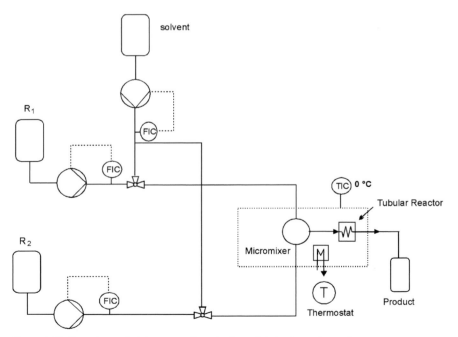

Figure 4.21 Flow sheet of a laboratory triangular interdigital micro mixer–tube reactor set-up, used for an industrial application, the so-called Clariant process [48].

Reactor type	Triangular interdigital micro mixer–tube reactor	Mixer material	Specialty glass (Foturan™)
		Tube material	PTFE
		Tube diameter	2 mm
Mixer channel: width, depth; slit width	60 μm; 150 μm	Tube length	6 m

4.1.6.3 Reactor 21 [R 21]: Caterpillar Mini Mixer–Tube Reactor

A caterpillar steel mini mixer is connected to conventional tubing, either stainless steel or polymeric. The caterpillar mixer acts by distributive mixing using the split–recombine approach which performs multiple splitting and recombination of liquid compartments [41, 42, 48, 49]. A ramp-like micro structure splits the incoming flow into two parts which are lifted up and down. Thereafter, the two new streams are reshaped separately in such a way that the two new cross-sections combined restore the original one. Then, the streams are recombined, again by the lifting up and down procedure using ramps. This procedure is repeated multiple times, yielding a multi-lamellae flow configuration – in the ideal case.

For a first-generation device, real-case deviations from this ideal pattern were found when experimentally visualizing the flow [50]. Splitting did not give identical compartments so that more stages may actually be needed than according to ideal-case simulations. At large volume flows, even no lamellae at all are formed, but rather twisted and intersegmented patterns. These induce secondary flows and turbulences which speed up mixing (however, not in the way the in which mixer actually is intended to do). For this reason, a second-generation device was made containing a splitting plate in the middle of the mixer (Figure 4.22).

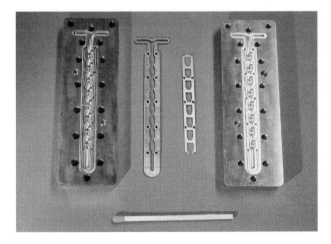

Figure 4.22 Second-generation caterpillar mini mixer with splitting plate and improved microstructure geometry [50].

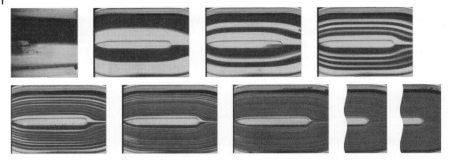

Figure 4.23 Near-ideal multi-lamination flow patterns in the second-generation caterpillar mini mixer as a result of introducing a splitting plate and improving micro structure geometry [50].

This plate cuts the flow into pieces which are better defined than the poorly defined ones obtained by the first-generation caterpillar mini mixer. In addition, the micro structure geometry was improved by means of simulation. As a result, near-ideal multi-lamination flow patterns were yielded (Figure 4.23), which showed excellent correspondence with simulation [50].

All types of split–recombine mixers generally have high volume flows (e.g. 100 l h^{-1} and more at moderate pressure drops) at favorable pressure drops (not exceeding 5 bar) as their internal micro structures can be held large [41, 42, 48, 49].

Reactor type	Caterpillar mini mixer–tube reactor, 1st generation	Micro structure in one plate: initial depth; maximum depth	600 μm; 850 μm
Mini mixer material	Stainless steel	Mini mixer stage: length	2400 μm
Number of plates needed to form mini mixer channel	2	Number of mixing stages	8
Mini mixer channel: initial width; maximum width	1200 μm; 2400 μm	Total length of caterpillar mini mixer	19.2 mm
Mini mixer channel (both plates): initial depth; maximum depth	1200 μm; 1700 μm	Device outer dimensions	50 × 50 × 10 mm³

4.1.6.4 Reactor 22 [R 22]: [Separation-layer Micro Mixer; Tube] – Reaction System

Separation layer mixers use either a miscible or non-miscible layer between the reacting solutions, in the first case most often identical with the solvent used [48]. By this measure, mixing is 'postponed' to a further stage of process equipment. Accordingly, reactants are only fed to the reaction device, but in a defined, e.g. multi-lamination-pattern like, fluid-compartment architecture. A separation layer technique inevitably demands micro mixers, as it is only feasible in a laminar flow regime, otherwise turbulent convective flow will result in plugging close to the entrance of the mixer chamber.

Both concentric and stacked fluid layer arrangements, corresponding to different versions of separation mixers, were developed, allowing either a drop- or stream-like injection of liquids in a reaction tube attached to the micro mixer [48].

Concentric separation layer micro mixer

This separation layer mixer is constructed as an assembly of stacked stainless-steel plates having three tubes [48]. These tubes are placed into each other and are inserted into a fit. The plates contain three feeding lines for reactants 1 and 2 as well as the separating fluid.

The stacked steel plates were manufactured by milling [48]. The PTFE tubes were home-made by means of turning and milling.

Reactor type	Concentric separation-layer micro mixer–tube reactor	Tube outer diameters	2.0 mm; 3.0 mm; 4.0 mm
Material	PEEK (tubes); stainless steel (housing)	Tube lengths	28.50 mm; 21.75 mm; 15.00 mm
Tube inner diameters	1.5 mm; 2.5 mm; 3.4 mm	Device outer dimensions	$41 \times 41 \times 24 \ mm^3$

Stacked separation layer micro mixer

This separation layer mixer was fabricated as an assembly of stacked glass plates which were irreversibly bonded by a thermal process [48]. An interdigital feeding structure generates alternately arranged lamellae of the three liquids. The mixing chamber is rectangular from the feed inlet until close to the outlet and becomes then tapered. In a later version, the same design concept was transferred into stainless steel. The steel device version was made by thin-wire μEDM.

Reactor type		Number of lamellae created	9
Material		Mixing chamber: width; depth; length	2.15 mm 2 mm 5 mm
Micro channel: width; depth; total length; Straight (i.e. uncurved) length	150 μm; 2 mm; 4 mm; 1 mm	Outer dimensions: length; width; depth	$58 \times 26 \times 8 \ mm^3$
Fin width	100 μm		

4.1.6.5   Reactor 23 [R 23]: [Impinging-jet Micro Mixer; Tube] – Reaction System

Impinging-jet micro mixers rely on a similar fluid guidance as given for impinging jet contactors for gas/liquid processes, namely on collision of two liquid streams [48]. Different from the latter, they generate a liquid jet rather than an extended liquid film. Since the collision of the streams can be performed in either a gaseous or a non-miscible medium, wall contact can be strongly diminished, thereby strongly reducing fouling.

The impinging-jet micro mixers are constructed as a cylindrical block comprising two feed tubes which become smaller towards the outlet [48]. Accordingly, the main characteristics of these mixers are the diameter of the outlet bore, the interspaces between the bores and the angle defined by the orientation of bores (relative to the normal). For one experimental study [48], nine impinging jet micro mixers were made which differ in these specifications.

Reactor type	Impinging-jet micro mixer–tube reactor	Bore angles	45°; 60°; 90°
Material	Stainless steel	Interspaces between bores	2 mm; 3 mm; 4 mm
Bore diameters	500 μm; 1000 μm		

4.1.7
3-D Microfab Micro Mixer–Micro Heat Exchangers

3-D microfabricated micro mixers (see Section 4.1.5) may be connected to 3-D microfabricated micro heat exchangers and also other combinations of micro devices performing unit operations and reactions are possible. These are combinations of components, rather than presenting total system approaches. Flexibility of component connection is generally higher compared with modifying a system, but integration and in some sense functionality are tentatively lower. Compared with the multi-scale micro mixer–tube reactor concept (see Section 4.1.6), combining micro components represents more a mono-scale solution.

4.1.7.1 Reactor 24 [R 24]: System with Series of Micro Mixers–Cross-Flow Reactor Modules

Various micro mixers and reaction modules with heat transfer function were connected (Figure 4.24). Details on the micro mixers were not given; the reaction modules comprise cross-flow configurations in a micro-channel platelet architecture [51]. The micro mixers are also connected in a serial manner to allow sequential mixing of up to three reactant solutions. The heat exchange modules are connected in series, by using commercial flange technology.

Figure 4.24 Modular micro-reactor system. Left: single reactor module with (length 60 mm). Right: mounted system of four single reactors (overall length 24 cm) [51].

Each of the reactor modules is fed by a heat transfer liquid, either water or a heat transfer fluid. The inlet temperatures of the reactor modules are set by these liquids; the outlet temperatures may be higher owing to release of reaction heat. By varying the inlet temperatures per module, a temperature profile along the reaction passage is created. Thereby, a series of operations, ignition/reaction/quenching is initiated, with the two inner modules performing a reaction. The cross-flow heat exchangers have so-called highly asymmetric passages, i.e. comprise much more (e.g. by an order of magnitude) micro channels on the heat transfer flow side than on the reaction side.

Reactor type	System with series of micro mixers–cross-flow reactor moldules	Number of reaction channels	169
Details on micro mixer	Not given in [51]	Number of heat transfer channels	1960
Reaction platelet material	Hastelloy C	Flange connection	Sandvik™ type L
Reaction channel width; depth; length	300 µm; 150 µm; 60 mm	Heat transfer fluid	Marlotherm™ SH

4.1.8
2-D Integrated Total Systems with Micro Mixing and Micro Heat Exchange Functions

Different from sole combinations of micro devices, this refers to a total system with many functional elements and flow-distribution and, recollecting zones, typically composed of 2-D plate-type architecture. Each of these plates usually has a separate function, comprising unit operations and reaction. Frequently, micro mixing and micro heat exchange functions and corresponding elements are employed. Often, the system can be composed of different elements resulting in different process flow combinations. Such an approach may be termed a construction kit.

Compared with the multi-scale micro mixer–tube concept (see Section 4.1.6), the total-system approach is a true mono-scale solution, and may be even termed monolithic. Integration of sensing and controlling is facile owing to the high order and repetition of construction units (plates).

4.1.8.1 Reactor 25 [R 25]: CPC Micro Reaction System CYTOS™
This is a commercial system comprising mixing and reaction/heat exchange functions [52] (see also [53, 54]). The system is constructed from various modules, typically caving stacked-plate units (Figure 4.25). The modules can be easily connected by a special interface. The plates are mounted together by metallic bonding. Numerous parallel channels fulfil functions of dividing reactants into substreams, performing reaction, maintaining the reaction temperature and collecting product substreams [55] The system is equipped with pumps and other fluidic peripherals so that it can be considered as a whole plant.

Figure 4.25 Specialty plates for laboratory and pilot plant micro reactor modules of modified CYTOS™ systems (left). Standard off-the-shelf CYTOS™ system (right) [55].

Details on microfabrication and on the internals in the stacked plates have not been substantially disclosed so far. Accordingly, no information on the mechanisms of mass and heat transfer was reported. In one version, geometrically focused multi-lamination is used for mixing liquid streams [55].

4.1.8.2 Reactor 26 [R 26]: Chip Micro Reaction System with Parallel Mixer–Reaction Channels

A chip-type micro reactor array comprises parallel mixer units composed of inverse mixing tees, each followed by a micro channel that it is surrounded by heat exchange micro channels (so called 'channel-by-channel' approach similar to the 'tube-in-tube' concept). Such an integrated device was developed as a stack of microstructured plates made of a special glass, termed Foturan™ (Figure 4.26). The integrated device was attached to PTFE tubes of various lengths.

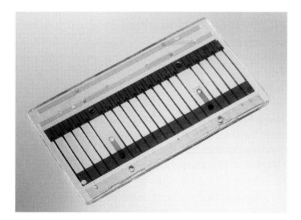

Figure 4.26 Chip reaction system with 20 parallel mixer–reaction channels, made of glass [with courtesy from mgt mikroglass AG, Germany].

Reactor type	Chip micro reaction system with parallel mixer–reaction channels	Plate thicknesses	2 × 0.2 mm; 5 × 0.7 mm; 1 × 1 mm
System material	Specialty glass (Foturan™)	System outer dimensions	90 × 50 × 4.9 mm^3
Mixer micro channels: width; depth	350 µm; 200 µm	Total number of plates	8
Reaction micro channels: width; depth; length	700 µm; 200 µm; 31.15 mm		

4.1.8.3 Reactor 27 [R 27]: [Bi-layer Contactor; High-aspect-ratio Heat Exchanger] – Reaction System

This micro reaction system is constructed as a stack of five machined plates [56] (see also [57–60] and further [61–63]). Between the first two plates, four parallel reaction platelets are laid. The first two plates serve for feeding one reactant each. Both feed streams are distributed to the four platelets which comprise an array of eight micro structured channels on both sides. First, the two streams, typically immiscible liquids, are contacted in a bi-layer configuration. This bi-layer may, under certain flow conditions, decompose to a dispersion, further enlarging the specific interface for mass transfer. Attached to the micro contactor is a micro heat exchanger formed by a reaction channel of large aspect ratio surrounded by cooling liquid channels of similar aspect ratio. The latter channels are structured on the back side of the platelet. Owing to the channels' large aspect ratio and the thinness of the platelet, the channels overlap, although they are not cross-linked. After passage to the heat transfer channel, the single flows are collected from the eight channels in an inverse bifurcation unit. A schematic of the bi-layer configuration platelet comprising reaction area, heat exchange area and collecting channels is shown in (Figure 4.27).

From there, the reaction flow either leaves the total system to be quenched or, more commonly, enters the next plate which contains a delay loop, a spiral channel [56]. Leaving that plate, the streams flow to the last structured plate containing a bifurcation–mini mixer unit. The streams are distributed in multiple streams and contacted with a likewise split water stream. This leads to fast dilution, e.g., of a concentrated sulfuric acid stream, and rapidly cools the reaction stream. The reaction is quenched more or less initially. The final plate is unstructured and acts as a cover plate with holes for liquid withdrawal (Figure 4.28).

The outer shape of the plates was made by thin-wire erosion [56]. Annealing processes served for improving flatness and eliminating stress. Thereafter, the microchannels were introduced by die sinking. A special variant was used here, based on rotating disk electrodes, enabling high-aspect-ratio micro structuring by facilitating withdrawal of material [60]. In a later development, a normal die sinking process was employed [56]. Tungsten carbide electrodes, having a negative shape of the microstructure to be manufactured, were made by wire erosion using a brass core wire-

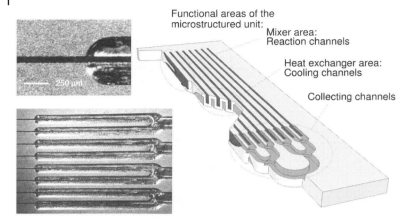

Figure 4.27 Schematic of the bi-layer contacting reaction platelet and photographs of details of the transfer region micro channel-collecting channel (left top) and of the array of the collecting channels (left bottom) [56].

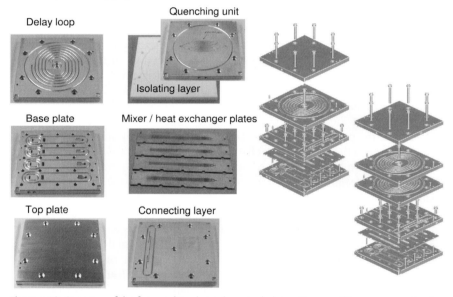

Figure 4.28 Overview of the five machined stainless steel plates. Two possible arrangements of all these system parts are given as well; many other are possible due to a flexible interconnection concept [56].

electrode 150 µm in diameter and coated with zinc. Using a special multi-clamping device, as many as five roughing and five finishing electrodes were employed in a single fabrication run without the need for any intervention by an operator. After electrode manufacturing, transfer to the die sinking machine was effected using the same clamping mechanism. The feeding plates, the delay-loop plate and the mini-mixer plate were made by conventional machining, drilling and milling.

Reactor type	[Bi-layer contactor; high-aspect-ratio heat exchanger]– reaction system	Residence times at 750 ml h^{-1}; 1 ml h^{-1} for different combinations of plates	1–10 s; 14 min – 2 h
Plate and platelet material	Stainless steel, grade 1.4539	Total number of plates; total number of platelets	5; 4
Sealing	Polymer O-rings	Reaction platelet thickness	1.2 mm
Reaction micro channel width; depth; length	70 µm; 900 µm; 61.4 mm	Total number of reaction channels; number on platelet	32; 8
Heat transfer micro channel width; depth; length	70 µm; 900 µm; 61.4 mm	Volume of total number of reaction channels	0.12 ml
Material thickness between reaction and heat transfer channels	100 µm	Volume of system without, with one and with two delay loops	< 1 ml; < 3 ml; < 5 ml
Volume flow at 1 bar pressure drop	~3 l h^{-1} (depending on plate combination)		

4.1.8.4 Reactor 28 [R 28]: Multi-channel Integrated Mixer-Heat Exchanger

This concept relies on integrated mixing and heat exchange functions, i.e. when mixing is initiated, cooling of the reaction mixture can be performed directly [64]. Both reactant streams are split into a multitude of reaction channels. One set of re-directed channels merges with another set of straight channels so that one channel of one type joins with another without having contact with a third channel. After a certain reaction passage, the parallel channels are combined to one channel. Thin liquid layers are achieved by having reaction and cooling micro channels of small depth at large width. The former have typically larger hydraulic diameters than the latter. The separating wall is kept thin to reduce resistance by heat conduction.

Reactor type	Multi-channel integrated mixer– heat exchanger	Hydraulic diameter of cooling channel	900 µm
Reactor material	Stainless steel	Surface-to-volume ratio of reaction micro channels	10 000 m^2 m^{-3}
Reaction micro channel width; depth; length	100 µm; 5000 µm	Heat transfer coefficient (cooling liquid not moving)	2000 W m^2 K^{-1}
Hydraulic diameter of reaction channel	200 µm	Thickness of wall separating reaction and cooling channels	1 mm
Cooling micro channel width; depth; length	500 µm; 5000 µm		

4.1.9
Electrochemical Micro Reactors

These devices have a special function which allows them to perform electro-organic synthesis. Typically, they contain electrode structures to generate electrons as tunable 'reactants'. Often, these electrodes are constructed as plate-type structures, sometimes also being the construction material for the channels themselves.

4.1.9.1 Reactor 29 [R 29]: Multi-sectioned Electrochemical Micro Reactor

The multi-sectioned electrochemical micro reactor comprises a multitude of alternating conducting and insulating layers [65]. Each conducting section is combined with independent current generators (Figure 4.29). This ensures that each section receives the planned average current density. In order to reach optimum reactor performance the layers have to be made sufficiently thin (micro-scale approach) and their number sufficiently large (parallel operation).

This new design is sought to overcome the limits of conventional porous fixed-bed reactors using an electrode phase flowing through the pores [65]. The latter systems suffer from the low conductivity of the electrolyte phase. This generates electrical resistance and leads to accumulation of the electrical current in certain reactor zones and hence results in a spatially inhomogeneous reaction. This means poor exploitation of the catalyst and possible reductions in selectivity.

Accordingly, the multi-sectioned reactor has advantages in terms of process refinement in the space domain by appropriate positioning of micro fabricated insulating and conducting surfaces [65]. However, it is said also to have advantages for process control in the time domain. By rapid and precise time-varying control of the electrical current in the microsecond range, steady-state and pulse operation can be improved. Concerning the latter, a concept of raster-pulse electrolysis was proposed [65]. Here, a steady-state baseline distribution of the current density, spatially uniform or non-uniform, is superimposed by a cleaning pulse. This pulse is applied periodically to each of the electrode sections in a programmed manner. By this means, both kinetics and catalyst poisoning can be taken into account.

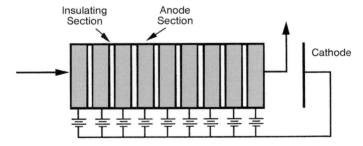

Figure 4.29 Electrochemical reactor with alternating conducting and insolating porous sections each connected to separate power supplies [65].

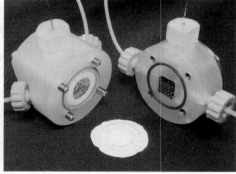

Figure 4.30 Electrochemical micro reactor, a diaphragm micro flow cell, applied to perform the 'cation flow' method. Assembled device (left). Disassembled device showing the two compartments of the cell within the housings and the diaphragm (right) [67].

4.1.9.2 Reactor 30 [R 30]: Electrochemical Diaphragm Micro Flow Cell

This type of electrochemical reactor is composed of two bodies by mechanical manufacturing [66, 67]. It contains a two-compartment cell with an anodic and cathodic chamber separated by a membrane as diaphragm. The anodic chamber is equipped with a carbon felt anode made of carbon fibers; a platinum wire is inserted in the cathodic chamber (Figure 4.30).

Reactor type	Electrochemical micro flow cell	Diaphragm material	PTFE
Reactor body materials	Diflone™; stainless steel	Carbon fibers: diameter	10 μm

4.1.9.3 Reactor 31 [R 31]: Electrochemical Capillary Micro Flow Reactor

This capillary micro flow reactor (Figure 4.31) is used for investigating spatially one-dimensional effects induced by electric fields which propagate parallel to the flow direction [68]. A rectangular cuvette with a rectangular capillary channel is fixed on both sides by filling chambers. These are attached by packing to larger electrode cells containing planar (plate) electrodes. The packing encompasses a microporous membrane. The whole system is tightened by screws and placed in a thermostatically controlled water bath. The cuvette is filled with a solution through openings in the filling chambers. The electrode cells are filled with the same solution. Thereby, electric current flows from electrode to electrode. The microporous membrane serves to separate the products of the electrode reactions from the reaction products in the capillary and thus to prevent intermixing.

By external stimulus at the plate electrode, migration of the ions in the solvent is induced, which changes the spatial concentration and so the local course of reactions [68]. By this means, weak electrical fields change the propagation velocity of the reaction zone through the capillary; strong electrical fields ('supercritical') further affect the global feature of the reaction in the capillary.

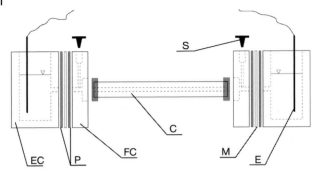

Figure 4.31 Electrochemical capillary micro flow reactor.
EC, electrode cells; FC, filling chambers; SP, silicon packings; C, rectangular capillary;
M, microporous PTFE membranes; E, platinum planar electrodes; S, stoppers [68].

Reactor type	Electrochemical capillary micro flow reactor	Electrode dimensions	$30 \times 30 \times 3$ mm³
Cuvette material	Optically clean glass	Electrode cell material	Organic glass
Cuvette outer dimensions	$9 \times 9 \times 84$ mm³	Electrode cell dimensions	$45 \times 85 \times 60$ mm³
Capillary material	Optically clean glass	Electrode cell volume	15 ml
Capillary cross-sections	1×1 mm; 0.7×0.7 mm; 0.5×0.5 mm;	Microporous membrane material	Teflon™
Filling chamber material	Organic glass	Packing material	Silicon

4.1.9.4 Reactor 32 [R 32]: Electrochemical Sheet Micro Flow Reactor

This sheet micro flow reactor (Figure 4.32) was used for investigating spatially two-dimensional effects in reaction media using agar gel induced by electric fields [68]. This device utilizes an adapted Petri dish which comprises a rectangular channel

Reactor type	Electrochemical sheet micro flow reactor	Reaction medium layer depth	600 μm
Polymer packing material	Lukopren™	Excess mixture depth	14 mm
Rectangular channel cross-section	4.5×2 mm²	Center hole diameter	300 μm
Shaped seal material	Optically clean glass	Wire electrode material	Plantinum

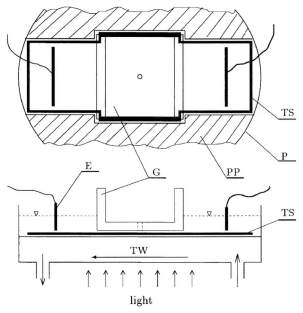

Figure 4.32 Electrochemical sheet micro flow reactor.
P, Petri dish; TS, PTFE supports; PP, polymerized packing (Lukopren™);
G, glass seal, E, platinum planar electrodes; TW, thermostated water [68].

after filling with polymerized packing. A specially shaped seal is placed on supports to create a thin layer of the reaction medium. The excess which is removed by this seal yields a layer in which the plate electrodes are inserted. In a small hole in the center of the glass seal a wire electrode is placed.

By external stimulus at the wire electrode, a reaction is initiated which propagates from the center in all directions evenly and so forms an expanding ring [68].

4.1.9.5 Reactor 33 [R 33]: Electrochemical Plate-to-Plate Micro Flow Reactor

This electrochemical micro reactor concept was based on the utilization of micro structuring techniques for electrochemical thin-layer cell technology, so far being applied in analytics [69]. The main component of the stacked-platelet-type micro reactor is a micro channel layer embedded between a working and a counter-electrode, referred to as a plate-to-plate configuration. Furthermore, an integrated cooling element serves for control of reaction temperature (Figure 4.33).

The micro structured platelets, hold in a non-conducting housing, were realized by etching of metal foils and laser cutting techniques [69]. Owing to the small Nernst diffusion layer thickness, fast mass transfer between the electrodes is achievable. The electrode surface area normalized by cell volume amounts to $40\,000\ \mathrm{m^2\ m^{-3}}$. This value clearly exceeds the specific surface areas of conventional mono- and bipolar cells of $10\text{--}100\ \mathrm{m^2\ m^{-3}}$.

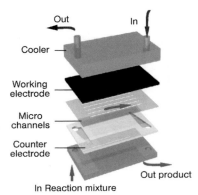

Figure 4.33 Electrochemical micro reactor with integrated electrode cooler. Left: overall view. Right: schematic of dismantled reactor [69].

Reactor type	Electrochemical plate-to-plate micro flow reactor	Reactor volume	35 µl
Micro channel width; depth; length	800 µm; 25 µm; 64 mm	Surface-to-volume ratio	400 cm^{-1}
Micro channel surface area	51.2 mm^2	Typical volume flow; Re number; pressure drop	6 ml h^{-1}; 0.2; 1.1 bar
Micro channel volume	1.3 µl	Typical residence time	21 s
Number of micro channels	27	Nernst diffusion layer	25 10^{-4} cm

4.1.9.6 Reactor 34 [R 34]: Ceramic Micro Reactor with Interdigitated Electrodes

This micro reactor consists of five ceramic layers [70, 71]. The top layer contains two bores for fluid feed and withdrawal. The second layer contains the flow distribution structures. The third comprises a micro channel array. The fourth carries the above-mentioned electrode structures. The last layer is an unstructured plate (Figure 4.34).

The ceramic micro reactor was fabricated in a three-step procedure, relying on screen printing, sintering and curing. Platinum interdigitated band electrodes were screen printed using a semi-automatic screen printer on a low-temperature co-fired ceramic soft tape [70, 71]. For gap widths exceeding 250 µm, the interdigital geometry was screen printed on the substrate via a patterned mask; direct writing with a laser beam on a printed platinum layer was applied when generating smaller gaps down to 30 µm. A dielectric layer was spread over the two connecting pads.

The micro channels were prepared with a cutter in a ceramic tape [70, 71]. Sealings served to ensure liquid tightness. Each end of the stack of five ceramic layers was clamped between two blocks, allowing a reversible interconnection.

Reactor type	Ceramic reactor with interdigitated electrodes	Number of electrodes	20 anodes + 20 cathodes
Ceramic tape material	Aluminum borosilicate and polymeric binders	Number of micro channels	7
Electrode material	Platinum	Micro channel width; depth; length	1000 μm; 254 μm; 50 mm
Anode–cathode gap	500 μm	Inlet and outlet hole diameter	2 mm
Pad width; length	5 mm; 80 mm	Seal material	PDMS
Working area	15 cm²	Frame material	Polycarbonate

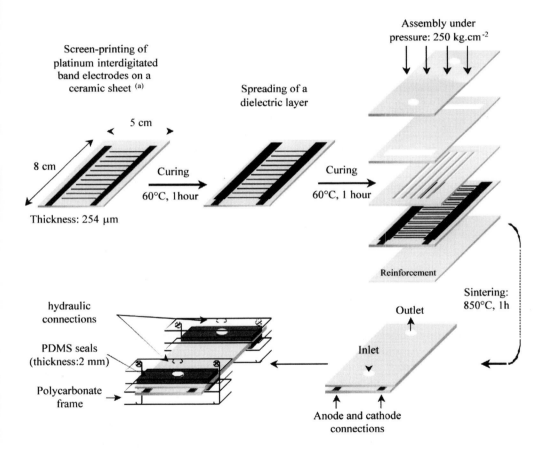

(a) Interdigitated band electrode characteristics: length = 29 mm, width = 1 m m and inter-electrode gap width =500 μm

Figure 4.34 Schematic of the assembly for the ceramic micro reactor with interdigitated electrodes [70].

4.1.10
Photochemical Micro Reactors

These devices have special function, namely to irradiate the liquid phase with light to induce a photoreaction or photoinduced reaction. Hence the characteristic feature is a transparent section within the reactor, often in the visible or commonly in the UV spectral region. The devices may have integrated photo energy sources or on-line analysis units. Otherwise, this is performed by external instruments.

Reaction 35 [R 35]: Photochemical Serpentine Chip Micro Reactor
This chip micro reactor was designed to have transparent reaction sections, to provide a large specific surface area of the flowing liquid, and to minimize fouling by crystallization of insoluble photoproducts [72–74].

(a) A first-generation micro device contained as main element a serpentine flow path, creating large surfaces at thin liquid layers [72–74]. Such a micro structured plate was covered by a transparent plate. One inlet and one outlet were used for solution feed and withdrawal.

A disadvantage of the first-generation device was that reaction and detection units were separated by HPLC tubing, which caused delay in analysis [72–74].

The device was realized by deep reactive ion etching (DRIE) using the SU-8™ technique, producing vertical side walls [72–74]. This fabrication route was chosen to avoid crystallization, which is known to occur at sharp channel edges. Using DRIE smooth, curved corners can be realized, unlike by conventional silicon wet etching.

Reactor type	Photochemical serpentine chip micro reactor, 1st generation	Reaction flow-through chamber: width; depth	500 μm; 50 μm
Channel material	Silicon	Device outer dimensions	20 × 25 mm²
Cover plate material	Pyrex™	On-line analysis	Delayed

(b) A second-generation device had integrated reaction and detection units and, in addition, allowed light of shorter wavelength (down to 254 nm) to pass [72–74]. For this purpose, part of the reactor has to be transparent from the top to the bottom. A sandwich structure of a silicon wafer encased between two quartz wafers fulfilled this criterion. The fabrication was as described above, but using a special thermal bonding technique with special bonding material in addition.

As the second-generation device contains integrated reaction and detection units, virtually real-time analysis could be achieved (compare to the delay in analysis for the first generation device described above) [72–74].

For both types of micro devices, a stainless steel frame was used as holder and Viton gaskets were applied for sealing [72–74]. A clear PMMA cover plate was also used and served for compression. This cover plate provided housing for a mini UV lamp used for irradiation. The detection unit was packaged in a similar fashion. Optical fibers were also integrated.

Reactor type	Photochemical serpentine chip micro reactor, 2nd generation	Reaction flow-through chamber: width; depth	500 µm; 500 µm
Channel material	Silicon	Device outer dimensions	20 × 20 mm²
Cover plate material	Pyrex™	On-line analysis	Not delayed

4.1.11
Complete Parallel-synthesis Apparatus

These are total systems or even plants made for parallel automated organic synthesis, typically in the liquid phase. In this section, no commercial devices (typically not relying on micro flow processing) are considered, but rather only specialty apparatus developed in the framework of chemical micro processing.

Whereas commercial systems usually employ stirred mini vessels, laboratory-developed apparatus may be operated in flow-through mode. For mini-vessels (vials, wells), the titer plate format is typical and widely accepted.

Reactor 36 [R 36]: Solid-phase Synthesis–Pneumatic Agitation–8-Reactor System

The design of a novel solid-phase parallel reaction system with 96 reactors in standard micro titer plate size, which is operated in semi-batch mode, is outlined in [75] (see also [76] and for microfabrication [77]). So far, a single-cell and an eight-cell prototype have been realized (Figure 4.35). Both comprise a stainless-steel housing with fluoropolymer plates containing the reactors and a membrane serving for mixing. The membrane is laser welded and the plates are compressed to tightness.

The reactors are cylindrical in shape and can carry up to 30 mg of resin. Polymer sieves at the top and bottom of the cylinders serve for liquid feed and withdrawal. The array of reactors is attached to a capillary system allowing feed to either columns or rows. This distribution system is said to provide uniform charges to the various reactors. A specific detail of the reaction system is that mixing is achieved by pneumatic actuation using a fluoropolymer membrane (Figure 4.36).

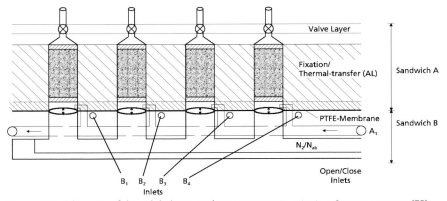

Figure 4.35 Schematic of the solid-phase synthesis–pneumatic agitation–8-reactor system [75].

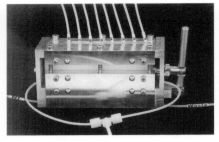

Figure 4.36 Solid-phase synthesis–pneumatic agitation–8-reactor system. Left: dismantled system. Right: same system, assembled [75].

4.2
Aliphatic Nucleophilic Substitution

4.2.1
Hydroxydehalogenation – Hydrolysis of Chlorides and Acid Chlorides

Proceedings: [46] (benzal chloride), [21] (acid chlorides).

4.2.1.1 Drivers for Performing Chloride Hydrolysis in Micro Reactors

The cleavage of two chloride groups on one C atom can be a very fast process. The hydrolysis of benzal chloride, for instance, is such a reaction between two immiscible media leading to huge heat release [46]. By uncontrolled mixing, the temperature can rise significantly accompanied by an increase in viscosity due to side reactions. For both reasons, the reaction can lead to a bursting of the whole processed sample. Therefore, the reaction is usually carried out by dropwise addition of one reactant and rigrous stirring. Accordingly, the driver for micro channel synthesis is to search for isothermal processing at high degrees of conversion and fast mixing.

Similar aggressive reaction conditions characterize the hydrolysis of acid chlorides, in particular when using short-chain alkyl-substituted acid chlorides such as propionic acid chloride. This fast reaction serves well as a model reaction for micro channel processing, especially for IR monitoring owing to the strong changes in the carbonyl peak absorption by reaction [21].

4.2.1.2 Beneficial Micro Reactor Properties for Chloride Hydrolysis

In the case of the above-mentioned dichloride hydrolysis, good mixing, i.e. emulsification, good heat transfer and restriction of residence time (to reduce side reactions) is demanded [46].

Concerning acid chloride hydrolysis, the advantage of micro chemical processing is that the micro reactor itself can be used as a flow-through cell for analysis, very unlike most large-scale conventional reactors [21]. For the case indicated, IR analysis is suitable and can be performed with silicon as encasing material, which is transparent for a wide range of the IR spectrum. In other cases, when reactions lead to color changes such as for the Wittig reaction [13], visible or UV detection may be required. Here, glass or quartz is the best micro-reactor construction material.

4.2.1.3 Chloride Hydrolysis Investigated in Micro Reactors
Organic synthesis 2 [OS 1]: Hydrolysis of benzal chloride

By reaction of sulfuric acid and benzal chloride benzaldehyde is generated. Both reactants are rather viscous and immiscible leading to the above mentioned reaction problems [46]. Due to temperature increase and too long reaction times, side reactions such as the oxidation to benzoic acid occur.

Organic synthesis 2 [OS 2]: Hydrolysis of 4-fluorobenzal chloride

This reaction is basically similar to [OS 1], but is expected to differ in reactivity owing to the I- and M-effects of the fluoro group.

Organic synthesis 3 [OS 3]: Hydrolysis of propionic acid chloride

$v = 1791 \ cm^{-1}$ $v = 1738 \ cm^{-1}$

This hydrolysis is accompanied by strong changes in IR absorption, particularly concerning a considerable shift of the carbonyl group peak (from 1791 to 1738 cm^{-1}) [21].

4.2.1.3 Experimental Protocols
[P 1] Through an interdigital micro mixer–tubular reactor set-up, sulfuric acid and benzal chloride (or 4-fluoro benzal chloride) are fed by piston pumps [46]. Sulfuric acid has to be used in excess; a ratio of acid to aromatic compound of 5 : 1 was applied. This also was chosen for reasons of emulsification, for the formation of small, relatively uniform droplets of the aromatic compound in the acid, the viscosity being not too high. Favorably high total flow rates, e.g. above 200 ml h^{-1}, had to be used to yield such patterns. At much lower flow rates, large segregated zones of the aromatic compound result, yielding worse reaction performance (Figure 4.37).

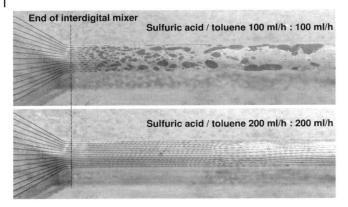

Figure 4.37 Images of contacting sulfuric acid with dyed (iodine) toluene in a rectangular-shaped interdigital micro mixer at two different flow rates. The system sulfuric acid/toluene (iodine) was taken as model for the reacting media sulfuric acid/benzal chloride [46].

4.2.1.4 Typical Results
Conversion/selectivity/yield
[OS 1] [R 20] [P 1] Yields range up to 69% (60 °C; 8 s). Nearly complete conversion is achieved for this parameter set [46].

Benchmarking to batch processing
[OS 1] [R 20] [P 1] The best yield obtained is 69% (60 °C; 8 s); batch synthesis is reported to result in a 65% yield [46].

Reaction temperature
[OS 1] [R 20] [P 1] The yield increases (Figure 4.38) from less than 10% at room temperature to 69% at 60 °C for a given residence time (8 s) [46].

Residence time
[OS 1] [R 20] [P 1] On increasing the residence time from 1 s to nearly 100 s the yield passed through a maximum, while conversion increased and reached a plateau. This was explained by the larger contribution of side reactions at longer residence times. Conversion, as to be expected, increased with residence time so that selectivity decreased. At about 10 s, nearly 100% conversion is achieved [46].

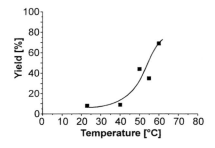

Figure 4.38 Increase in benzaldehyde yield with increase in reaction temperature by performing benzal chloride hydrolysis in a slit-shaped interdigital micro mixer [46].

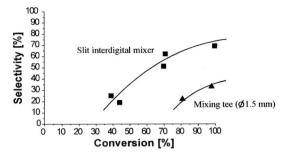

Figure 4.39 Comparison of benzal chloride hydrolysis
performed in a slit-shaped micro mixer and in a mixing tee [46].

Internal diameter/benchmarking to mixing tee
[OS 1] [R 20] [P 1] A comparison of the selectivity/conversion behavior of an
interdigital micro mixer–tube reactor with that of a mixing tee of about 1.5 mm
inner diameter (and thus of larger internal dimensions) was made (Figure 4.39).
For all data gathered, the performance of the micro mixer was much better, e.g.
about 30% more selectivity at a given conversion [46].

Synthesis of a substituted derivative
[OS 2] [R 20] [P 1]The feasibility of the micro channel hydrolysis of 4-fluorobenzal
chloride was shown [46]. A maximum yield of about 50% at 100% conversion was
reached (20–70 °C). Higher yields at low conversions were found compared with
benzal chloride hydrolysis.

In-line IR monitoring
[OS 3] [no protocol] The propionic acid chloride hydrolysis by water was characterized by in-line IR monitoring, using the micro reactor itself as a flow-through analysis cell that was placed in a commercial holder in a commercial IR spectrometer
[21]. By following the decrease of the 1791 cm^{-1} peak (acid chloride-substituted
carbonyl), and the corresponding increase of the 1738 cm^{-1}-peak (acid-substituted
carbonyl) the course of the reaction can be monitored [21]. Different flow rates
were investigated (0.002, 0.01 and 0.2 ml min^{-1}) showing the dependence of yield
on residence time. Details on how to achieve a maximum signal-to-noise ratio by
choosing proper silicon construction materials are given also [21].

4.2.2
Cyanodehalogenation – Preparation of Nitriles

Proceedings: [3].

4.2.2.1   Drivers for Performing Preparation of Nitriles in Micro Reactors
The carrying out of this reaction served to demonstrate organic synthesis on a
newly developed porous-polymer-rod micro reactor [3].

4.2.2.2 Beneficial Micro Reactor Properties for Preparation of Nitriles

A detailed study on velocity profiles, pressure drop and mass transport effects is given in [3]. This, in quantitative terms, precisely underlines the advantages (and limits) of the porous-polymer-rod micro reactor concept.

4.2.2.3 Preparation of Nitriles Investigated in Micro Reactors
Organic synthesis 4 [OS 4]: Nitrile synthesis from α-bromotoluene

α-Bromotoluene reacts with cyanide groups to give α-cyanotoluene [3].

4.2.2.4 Experimental Protocols

[P 2] α-Bromotoluene was nucleophilically substituted to give α-cyanotoluene in benzene at 70 °C (12 h) by cyanide coupled to a porous polymer resin [3]. The polystyrene polymer was cross-linked via linking divinylbenzene moieties in the main chain. Originally, the polymer contained benzylchloride groups, which were converted to quaternary ammonium groups. By means of ion exchange, functional anionic groups such as the reductive cyanide moiety can be introduced. Typical ion exchange capabilities of the micro reactor were about 0.1–1.0 mmol, depending on the polymer load.

4.2.2.5 Typical Results
Conversion/selectivity/yield
[OS 4] [R 3] [P 2] > 99% yield was obtained after 12 h of processing [3].

4.2.3
Thiocyanatodehydrogenation – Thiocyanation

Proceedings: [3].

4.2.3.1 Drivers for Performing Thiocyanation in Micro Reactors
See Section 4.2.2.1.

4.2.3.2 Beneficial Micro Reactor Properties for Thiocyanation
See Section 4.2.2.1.

4.2.3.3 Thiocyanation Investigated in Micro Reactors
Organic synthesis 5 [OS 5]: Rhodanide substitution of α-bromotoluene

α-Bromotoluene reacts with rhodanide groups to give α-thiocyanidetoluene [3].

4.2.3.4 Experimental Protocols

[P 3] This protocol was performed identically with [P 2], with the exception of the reactants and conditions. α-Bromotoluene was nucleophilically substituted in benzene to give α-thiocyanotoluene at 70 °C (12 h) using rhodanide coupled to a porous polymer resin [3].

4.2.3.5 Typical Results
Conversion/selectivity/yield

[OS 5] [R 3] [P 3] > 99% yield was obtained after 12 h of processing [3].

4.2.4
Azidodehalogenation – Formation of Azides

Proceedings: [3].

4.2.4.1 Drivers for Performing Azide Substitutions in Micro Reactors
See Section 4.2.2.1.

4.2.4.2 Beneficial Micro Reactor Properties for Azide Substitutions
See Section 4.2.2.1.

4.2.4.3 Azide Substitutions Investigated in Micro Reactors
Organis synthesis 6 [OS 6]: Azide substitution of α-bromotoluene

α-Bromotoluene reacts with azide groups to give α-azidotoluene [3].

4.2.4.4 Experimental Protocols

[P 4] This protocol was performed identically with [P 2], with the exception of the reactants and conditions. α-Bromotoluene was nucleophilically substituted in benzene to give α-azidotoluene at 70 °C (12 h) by azide coupled to a porous polymer resin [3].

4.2.4.5 Typical Results
Conversion/selectivity/yield

[OS 6] [R 3] [P 4] > 99% yield was obtained after 12 h of processing [3].

4.2.5
Aminodehalogenation – Menschutkin Reaction (Formation of Quaternary Amines)

Proceedings: [78]; sections in review: [79, 80].

4.2.5.1 Drivers for Performing the Menschutkin Reaction in Micro Reactors
Most of the known organic reaction yield sooner or later precipitates or involves directly the addition of solid reagents [78]. Hence fouling phenomena are rather

the rule than the exception when dealing with organic synthesis. By choosing an-other solvent, enhancing temperature, decreasing concentration, using an anti-fouling coating in micro channels or even changing the processing route, there are ways to cope with fouling. However, this often leads to compromises. Accordingly, one wants to have novel micro flow concepts that are virtually insensitive to foul-ing.

The Menschutkin reaction was carried out as a test reaction to show the feasibil-ity of such novel micro flow concepts that allow to process fouling-sensitive reac-tions (see also Section 4.2.6; here another test reaction is decribed for the same purpose) [78]. The reaction of alkyl bromide with ternary bases such as pyridine or triethylamine gives quaternary salts insoluble in most solvents. Often, fairly rapid precipitation of this salt occurs, hence ideally serving as a test reaction for fouling sensitivity of micro-channel devices. The reaction of 4,4'-bipyridyl and ethyl bromoacetate [78] belongs to the category of fast-precipitating Menschutkin reac-tions, as the halide function is activated by the carbonyl function.

4.2.5.2 Beneficial Micro Reactor Properties for the Menschutkin Reaction

The reduction of fouling sensitivity refers to new microfluidic concepts, based on delaying mixing or on free-flow guiding [81]. With regard to these criteria, the Menschutkin reaction is just one prominent example among a vast number of others, also chosen for reasons of having a very high reaction rate.

4.2.5.3 Menschutkin Reaction Investigated in Micro Reactors
Organic synthesis 7 [OS 7]: Formation of quaternary salts from 4,4'-bipyridyl and ethyl bromoacetate

This formation of a quaternary salt is performed in dichloromethane as solvent [78]. The halide function is a good leaving group owing to the presence of the neighboring carbonyl function, ensuring a fast reaction. Owing to the double-ionic character and the rigid core, hardly without any flexible chains, of the product, fast precipitation occurs when the reaction has proceeded to a certain conversion (see also Section 4.2.6; here another test reaction is decribed for monitoring fouling sensitivity).

4.2.5.4 Experimental Protocols

[P 5] Layers of 4,4'-bipyridyl (0.3 mol l^{-1} in dichloromethane), ethyl bromoacetate (0.3 mol l^{-1} in dichloromethane) and a separation layer of dichloromethane were fitted into each other by means of a concentric separation mixer (three-fluid nozzle with three tubes having diameters of 1.5, 3 and 4 mm, slotted into each other) [78]. Thereby, two circular liquid layers of a thickness of 200 μm and a center stream of 1.5 mm diameter were generated. The reaction temperature was 22 °C. The reac-tion solution was inserted as droplets or a continuous stream either directly or via

the tubular reactor in the beaker. The precipitate solution yielded was passed through a frit and the remaining solid was washed with dichloromethane and dried at elevated temperature and weighed.

Experiments with the following individual flow rates of the three liquids were performed [78]: 5 : 25 : 5; 5 : 100 : 5; 5 : 200 : 5; 5 : 250 : 5; 25 : 250 : 25; 50 : 150 : 50; 300 : 600 : 300; 300 : 1000 : 300 as bipyridyl in dichloromethane : dichloromethane : ethyl bromoacetate in dichloromethane in ml h^{-1}.

4.2.5.5 Typical Results
Conversion/selectivity/yield
[OS 7] [R 22] [P 5] Reactions were performed with wide variations of flow rates, ranging from 5 : 25 : 5 to 300 : 1000 : 300 (each value in ml h^{-1}) [78]. The corresponding yields were all about 75%. These yields are of the same order as for laboratory-batch operation [82].

Plug-free operation
[OS 7] [R 22] [P 5] For a set of flow rate variations, ranging from 5 : 25 : 5 to 300 : 1000 : 300 (each value in ml h^{-1}), stable operation for at least 2 h, sometimes ranging up to 8 h, could be achieved for the three streams of a separation-layer micro mixer [78].

Analysis and prevention of fouling in tubes attached to micro mixers
[OS 9] [R 23] [P 7] See discussion in Section 4.2.6; here another test reaction is decribed for monitoring fouling sensitivity) [78].

4.2.6
Aminodehalogenation – Acylation of Amines

Peer-reviewed publications: [23, 83]; proceedings: [78]; sections in reviews: [79, 80].

4.2.6.1 Drivers for Performing Acylations of Amines in Micro Reactors
In one example, the acylation of various types of amines (aliphatic/aromatic; branched/non-branched; varying alkyl chain length; straight chain/cyclic) was used to demonstrate the feasibility of a serial screening concept for liquid single-phase reactions [83]. In another investigation, this served to show novel micro flow concepts that allow one to process fouling-sensitive reactions [78]. Acylations with amines often use auxiliary bases that react with the hydrochloric acid released to give insoluble quaternary salts. In some cases, e.g. when using triethylamine, this is a rather fast precipitation, hence ideally serving as a test reaction for the fouling sensitivity of micro-channel devices.

For another investigation, amide formation was used as a model reaction to demonstrate the performance of parallel processing in micro-channel devices [23]. The target of such processing is combinatorial synthesis, the provision of multiple substances within one run.

4.2.6.2 Beneficial Micro Reactor Properties for Acylations of Amines

The first aim given above is methodologically oriented on the capability of performing serial screening with a small amount of back-mixing and consumption of low volumes. This is valid not only for amine acylation, but also for any other reaction. The second argument, reduction of fouling sensitivity, refers to new microfluidic concepts, based on delaying mixing or on free-flow guiding [81]. With regard to these criteria, the acylation reaction is just one prominent example among a vast number of others, also chosen for reasons of having a very high reaction rate.

Referring to highly parallel synthesis, the smallness of the micro-channel dimensions enables one to combine several micro unit operations on one chip [23]. By using multi-layered chip architecture complicated fluidic circuits with $n \times m$ combinations of fluid streams can be made. By this means, truly combinatorial parallel processing can be achieved.

4.2.6.3 Acylations of Amines Investigated in Micro Reactors

Organic synthesis 8 [OS 8]: Acetic anhydride acylation of diverse amines

This acylation reaction is performed in the presence of triethylamine using DMF or dioxane as solvent [83].

Organic synthesis 9 [OS 9]: Acetyl chloride acylation of n-butyl amine

This acylation reaction is performed in the presence of triethylamine using THF as solvent [78].

Organic synthesis 10 [OS 10]: 2 × 2 library from nitro- and dinitrobenzoyl chlorides and two amines

3-Nitrobenzoyl chloride and 3,5-dinitrobenzoyl chloride were each reacted with DL-1-phenylethylamine and 4-amino-1-benzylpiperidine using a phase-transfer reaction [23]. The amines were in the aqueous phase and the acid chlorides in the organic phase. By this means, a 2 × 2 library was created in one experimental run.

The reactions proceed via a phase-transfer mechanism [23]. The amine diffuses from the aqueous phase into the organic phase and reacts with the acid chloride. The amide formed remains in the organic phase, while the salt generated from the released HCl and the auxiliary base is transferred to the aqueous phase.

4.2.6.4 Experimental Protocols

[P 6] No protocol was given in [83].

[P 7] Layers of acetyl chloride (0.79 mol l^{-1}) in tetrahydrofuran (THF), *n*-butyl-amine (0.80 mol l^{-1}) and triethylamine (0.80 mol l^{-1}) in THF, and a separation layer of THF were fit into each other by means of the concentric separation mixer (three-fluid nozzle with three tubes having inner diameters of 1.5, 2.5 and 3.4 mm slotted into each other) [78]. Thereby, two circular liquid layers of a thickness of 200 µm and a center stream of 1.5 mm diameter were generated. The reaction temperature was 22 °C. The reaction solution was inserted as droplets or as continuous stream either directly or via the tubular reactor in a beaker containing water. By rigorous stirring, hydrolysis of the acid chloride and hence termination of the reaction were achieved. The phases were separated and the water phase was extracted with THF. The combined THF phases were dried over anhydrous Na$_2$SO$_4$. After filtration, the THF solvent was evaporated at 25 mbar. The remaining amide product was charac-terized by FTIR spectroscopy.

In a second experiment, higher concentrations were applied: acetyl chloride (0.198 mol l^{-1}), *n*-butylamine (0.200 mol l^{-1}) and triethylamine (0.200 mol l^{-1}) [78]. Experiments with the following individual flow rates of the three liquids were performed [78]: 5 : 25 : 5; 5 : 100 : 5; 5 : 200 : 5; 5 : 250 : 5; 25 : 250 : 25; 50 : 150 : 50; 300 : 600 : 300; 300 : 1000 : 300 as acetyl chloride in THF : THF : *n*-butylamine + triethylamine in THF in ml h^{-1}.

[P 8] A jet of acetyl chloride (0.197 mol l^{-1}) in THF at a flow rate of 1000 ml h^{-1} and a jet consisting of *n*-butylamine (0.200 mol l^{-1}) and triethylamine (0.200 mol l^{-1})

in THF at a flow rate of 1000 ml h^{-1} were generated by using an impinging-jet mixer (two 350 µm openings, separated by 3 mm, and inclined to each other by 45°) [78]. Both jets merged in a Y-type flow configuration. The rest of this experimental protocol is identical with [P 7].

[P 9] DL-1-Phenylethylamine and 4-amino-1-benzylpiperidine were dissolved in 0.1 M NaOH aqueous solution [23]. 3-Nitrobenzoyl chloride and 3,5-dinitrobenzoyl chloride were used as ethyl acetate solutions. The concentration of all reactants was set to 0.01 M. Syringe pumps served for liquid feed. The flow rate was 50 µl min^{-1} and room-temperature processing was applied. No further temperature control was exerted as the reaction is only mildly exothermic. After having passed the micro reactor, the phases were settled in test-tubes and the organic phase was withdrawn for analysis.

4.2.6.5 Typical Results
Conversion/selectivity/yield
[OS 9] [R 22] [P 7] Reactions were performed with wide variations of flow rates, ranging from 5 : 25 : 5 to 300 : 1000 : 300 (each value in ml h^{-1}) [78]. The corresponding yields were between 87 and 100%. The lower yields were obtained at high total flow rates.

[OS 8] [R 25] [P 6] The acylation of acetic anhydride with various amines was investigated; all yields ranged from 75 to 100% with the exception of one value of 46% [83]. The lowest yield was obtained for 1-naphthylamine and the highest for 1-hexylamine. Throughputs were from 3.8 to 68.3 g l^{-1}.

[OS 10] [R 10] [P 9] Yields of 82–93% were obtained for a set of two amines and two acid chlorides by means of a phase-transfer reaction (Figure 4.40) [23].

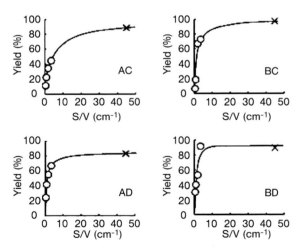

Figure 4.40 Dependence of yield on specific interfacial area between aqueous or organic phase. The four combinations AC–BD refer to the respective reactions between DL-1-phenyl-ethylamine (A), 4-amino-1-benzylpiperidine (B), 3-nitrobenzoyl chloride (C), and 3,5-dinitrobenzoyl chloride (D) [23].

Plugging-free operation

[OS 9] [R 22] [P 7] For a set of flow rate variations, ranging from 5 : 25 : 5 to 300 : 1000 : 300 (each value in ml h^{-1}), a stable operation of at least 1 h could be achieved for the three streams of a separation-layer micro mixer [78]. Particularly advantageous was the setting of the flow rates to 5 : 250 : 5. Here, the reaction could be operated for 3 h.

Analysis and prevention of fouling in tubes attached to micro mixers

[OS 9] [R 23] [P 7] Two set-ups comprising an impinging-jet micro mixer were tested, both being mixer–tubular reactor configurations [78]. These set-ups differ in the type of tube used. A first set-up was equipped with a spirally wound steel tube and the second comprised a straight glass tube. Both set-ups used a home-made housing in which the impinging-jet mixer was inserted. The housing chamber is tapered like a funnel in order to collect the liquid mixture and to introduce it directly into the tubular reactor attached, without any wake.

The first set-up comprised a 3.30 m long, spirally wound tube of 4 mm inner diameter, mounted vertical [78]. When performing the amide reaction (0.200 mol l^{-1}, total flow rate 1000 ml h^{-1}), this set-up was plugged after only a 42 s operation time. The second set-up using a 0.75 m straight glass tube of 0.3 mm inner diameter could be used for a much longer period of about 20 min. Regarding this value, it should be pointed out that, owing to the large amount of chemicals and solvents consumed, all operations described here were deliberately stopped instead of waiting for a stop caused by plugging.

The flow in the glass tube was relatively undisturbed [78]. Partly, bubble formation due to HCl gas evolution and passing of Et$_3$NHCl lumps was observed. These phases were moving with the liquid mixture and were always rinsed out of the tube, and hence were not obstacles causing a breakdown of the flow.

In a further run, the reactant concentration was doubled; 0.395 mol l^{-1} acetyl chloride in THF and 0.400 mol l^{-1} *n*-butylamine and 0.400 mol l^{-1} Et$_3$N in THF were processed in the second set-up with the straight tube at a total flow rate of 2000 ml h^{-1} [78]. Although extensive precipitation of Et$_3$NHCl was observed, these lumps are still carried out of the tube. For a 38 min operation, no plugging was observed.

The importance of the linear arrangement of mixer/funnel/tubular reactor is shown when processing in a set-up with a curved flow element (0.3 m long bent Teflon tube of 0.3 mm inner diameter) in between the funnel and tubular reactor [78]. If a straight tube of equal dimensions as given above is used, plugging occurs after 30 s. Hence even short curved flow passages are detrimental for micro-channel-based amidation studies.

Interfacial area

[OS 10] [R 10] [P 9] The specific interfacial area was varied for a phase-transfer reaction for four amide formations from two amines and two acid chlorides [23]. This was done by filling the solutions in normal test-tubes of varying diameter (1–5 × cm^{-1}) and using a micro reactor which had the largest specific interface (45 × cm^{-1}). The yields of all four reactions are highly and similarly dependent on

the specific interface, as to be expected for a phase-transfer reaction (Figure 4.40). The micro reactor yields approach 80–95%.

2 × 2 parallel synthesis – a first step towards combinatorial chemistry
[OS 10] [R 10] [P 9] The feasibility of 2 × 2 parallel synthesis using two amines and two acid chlorides for a phase-transfer reaction was demonstrated [23]. This paves the way for *n* × *m* parallel reaction combinations as a new micro flow approach for combinatorial chemistry.

The flow distribution was far from ideal [23]. Collected volumes at the individual outlets ranged from 2.15 to 3.65 ml, thus, differing by more than 50% at maximum. This was seen to be due to differences in pressure drops resulting from imperfections in microfabrication.

Despite the differences in the volumes collected, and hence in the concentrations for the various channel processing, the yields are (surprisingly) comparable to those from single micro reactors which do not suffer from flow deviations. The yields of the 2 × 2 processing (82–93%) were essentially the same as when the same products were obtained by single-micro-reactor processing (yields ranging from 83 to 98%) [23]. This good reactor performance, in lieu of the flow imperfections, can only be explained by having carried out the reaction at lower flow deviations as reported above. Indeed, the authors report strongly changing flow for the single channels when they disconnect one outlet port and reassemble it.

[OS 10] [R 10] [no protocol] In another study, the above-mentioned features were also investigated. Two amines, 0.01 M DL-1-phenylethylamine and 4-amino-1-benzylpiperidine as solutions in 0.1 M NaOH, were reacted with two acids, 0.01 M 3,5-dinitrobenzoyl chloride and 3-nitrobenzoyl chloride in ethyl acetate [24]. By using a 2 × 2 micro channel glass chip reactor, four separate flow passages were realized, each connected to two of the four feed streams (one amine and one acid chloride). Thereby, the formation of the four possible amides was demonstrated and hence the feasibility of micro-channel processing for parallel liquid screening in one chip. Especially, it was confirmed by thin layer chromatographic (TLC) analysis that from each channel mainly pure solutions were obtained, i.e. there was no detectable cross-over of reactant streams. The latter indicated proper distribution of reactant streams.

4.2.7
Aminodehalogenation – Acylating Cleavage with Acetyl Chloride

Proceedings: [75]; literature on micro reactor and microfabrication used: [76, 77].

4.2.7.1 Drivers for Performing Acylating Cleavage in Micro Reactors
The motivation for investigating the acylating cleavage stems from using it as a test reaction for showing the feasibility a newly developed miniaturized system (*[R 36]*) for performing ultra-high throughput screening [75]. Modern apparatus can test more than 100 000 compounds per day. The titer plate formate, consequently, has increased from 96 wells to 1536 wells. Conversely, the sample amount has decreased from several mg to < 1 mg. Solid-phase organic chemistry (SPOS) can reach this goal with only a few tens of milligrams of polymer resin. However, the purchase of

such apparatus is not state of the art, but needs self-developed, specialized solutions such as the micro reaction system [R 36].

4.2.7.2 Beneficial Micro Reactor Properties for Acylating Cleavage

SPOS has the benefit of employing an excess of reactant, which can be washed off afterwards, and of driving reactions in this way to completion [75]. Difficult purification steps are avoided. SPOS is facile as it needs only a few repetitive unit operations.

However, current solutions suffer from speed of the slowest step (e.g. filling with robots) or lack of automation [75].

4.2.7.3 Acylating Cleavage Investigated in Micro Reactors
Organic synthesis 11 [OS 11]: Synthesis of piperazine

AcCl, THF

purity 98% (GC)

4.2.7.4 Experimental Protocols

[P 10] The reaction was performed on 100 mg of Merrifield resin [75]. Absolute tetrahydrofuran was used as solvent and 4 h of agitation was employed. No other details are given in the reference. Generally, about 2 min were needed to perform a complete washing cycle.

4.2.7.5 Typical Results

[OS 11] [R 36] [P 10] The feasibility of performing the acylating cleavage of T2-triazene resin (100 mg) in absolute tetrahydrofuran to give piperazine was demonstrated [75]; 4 h of processing was required.

4.2.8
Alkoxydehydroxylation – Enzymatic Esterification of Acids with Alcohols

Proceedings: [84].

4.2.8.1 Drivers for Performing Enymatic Esterifications in Micro Reactors

By the enzymatic esterification of diglycerol with lauric acid, the corresponding monolaurate ester is obtained [84]. This is an important industrial reaction for the cosmetic, pharmaceutical and feed industries, since this ester is used as biodegradable non-ionic surfactant. In recent years, the synthesis of this and other polyglycerols with fatty acids has attracted growing interest in industry, leading also to a demand for enantiomerically and isomerically pure products.

4.2.8.2 Beneficial Micro Reactor Properties for Enymatic Esterifications

For the reasons mentioned above, the development of miniature test processes becomes increasingly important for such biotransformation reactions [84]. Micro reactors can handle small volumes and process them in a well-defined manner and have been shown to have high test throughput frequencies. Hence waste reduction, quality and robustness of information and speed of analysis qualify micro reactors as testing tools also for enzymatic esterifications.

4.2.8.3 Enymatic Esterifications Investigated in Micro Reactors
Organic synthesis 12 [OS 12]: Esterification of diglycerol with lauric acid with Novozym-435™

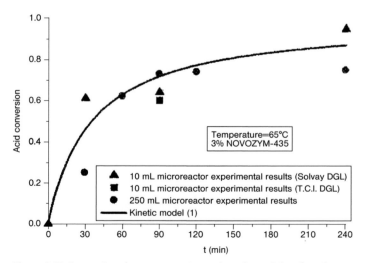

4.2.8.4 Experimental Protocols
[P 11] Reactions were performed in a completely stirred tank reactor of 10 ml volume [84]. The stirrer was set to 700 rpm. The reactor was immersed in a water bath. Owing to the small size of the reactor, special precautions had to be taken for stirring and for temperature control of the bath [84].

The catalysts was added after the reactants were fed in the tank reactor and pressure and temperature were set to the target values [84]. The study was performed using an immobilized lipase, Novozym-435™, as biocatalyst. The temperature was set to 65–75 °C and the pressure was reduced (60 mmHg). A catalyst concentration of 1–5% with an acid : alcohol ratio of 1 : 3, 1 : 1 or 3 : 1 was used.

4.2.8.5 Typical Results
Feasibility – comparison with kinetic model
[OS 12] [reactor given in [84]] [P 11] In [84], the scale-down from a 250 ml batch reactor to a 10 ml batch reactor is described. The validity of applying a pseudo-second-order kinetic model for the scaled-down processing was confirmed (Figure 4.41).

Figure 4.41 Comparison between experimental results and data from kinetic model [84].

Sample consumption

It was confirmed that much less of the reactants needs to be consumed for testing, which will be important if expensive compounds are used such as enantiomerically and isomerically pure products [84]. Hence the feasibility of mini-scale biotransformation processing was demonstrated using a model reaction and now allows this testing procedure to be extended to other more important reactions.

4.2.9
Amidodeamidation (trans-Amidation) – Desymmetrization of Thioureas

Proceedings: [85]; Reviews: [42]; sections in reviews: [42].

4.2.9.1 Drivers for Performing Desymmetrization of Thioureas in Micro Reactors

The investigations were carried out in the framework of an industrial study on the general applicability of micro reactors to organic synthesis for pharmaceutical applications, in the long run aiming at performing combinatorial chemistry by micro flow processing [85] (see a more detailed description in [42]). The scouting studies focused on determining suitable reaction parameters and monitoring yields as a function of time.

4.2.9.2 Beneficial Micro Reactor Properties for Desymmetrization of Thioureas

The studies mentioned above referred to the general advantages of micro flow processing in terms of enhanced heat and mass transfer [85] (see a more detailed description in [42]).

4.2.9.3 Desymmetrization of Thioureas Investigated in Micro Reactors
Organic synthesis 13 [OS 13]: Thiourea from phenyl isothiocyanate and cyclohexylamine

Organic synthesis 14 [OS 14]: Thiourea from diphenylthiourea and cyclohexylamine

4.2.9.4 Typical Results
Conversion/selectivity/yield

[OS 13] [R 17] [no protocol] Using a micro mixer/commercial tube reactor, the synthesis of a thiourea from phenyl isothiocyanate and cyclohexylamine at 0 °C was carried out [85] (see a more detailed description in [42]). A single mixing device connected to a stainless-steel tube of about 10 m length and 0.25 mm diameter was used. The feasibility of performing a nearly spontaneous reaction could be shown.

[OS 14] [R 17] [no protocol] Further studies related to the desymmetrization of thioureas showed that for the diphenylthiourea/cyclohexylamine system reasonable reaction rates and conversions were achieved [42, 85]. It is notable that the temperatures of up to 91°C applied slightly exceed the boiling point of the solvent acetonitrile.

4.2.10
Aminodehydroxylation – Acylation of Amines by Acids (Peptide Synthesis)

Peer-reviewed journals: [5, 86, 87]; proceedings: [88]; sections in reviews: [14, 89, 90].

4.2.10.1 Drivers for Performing Peptide Syntheses in Micro Reactors

Dipeptides and longer peptides are typically synthesized by solid-phase chemistry at polymer beads, a route discovered by and named after Merrifield [5, 88]. Disadvantages of this approach are that the polymer support is expensive and additional steps for linkage to and cleavage from the polymer are required. Hence solution chemistries are an alternative to the Merrifield approach.

Peptide synthesis from β-amino acids is particularly attractive for first feasibility micro-reactor tests as there are no chiral centers which may complicate analysis of the products [5, 88]. β-Peptides are also attractive owing to their structural and biological properties, especially concerning the stability versus degradation by peptidases as compared with their α-analogues (see original citations in [5]).

4.2.10.2 Beneficial Micro Reactor Properties for Peptide Syntheses

Micro reactors are continuous-flow devices consuming small reaction volumes and allowing defined setting of reaction parameters and fast changes. Hence they are ideal tools for process screening and optimization studies to develop solution-based chemistries.

4.2.10.3 Peptide Syntheses Investigated in Micro Reactors
Organic synthesis [OS 15]: β-Dipeptide synthesis by carbodiimide coupling using Dmab O-protection

Boc-β-alanine was *O*-protected (carboxylic moiety) by DMAP coupling (4-dimethyl-aminopyridine) yielding Dmab-β-alanine, whereas the Fmoc group was used for *N*-protection of β-alanine [88]. Thereby, orthogonal protecting groups were established. By carbodiimide coupling, Dmab-β-alanine and Fmoc-β-alanine reacted and the synthesis of the corresponding β-dipeptide was realized.

Organic synthesis 16 [OS 16]: β-Dipeptide synthesis using pentafluorophenyl *O*-activation

Fmoc-β-alanine was pre-activated by introducing the pentafluorophenyl function as an ester group [88]. Dmab-β-alanine and the pentafluorophenyl ester of Fmoc-β-alanine reacted and the synthesis of the corresponding β-dipeptide was realized.

Organic synthesis 17 [OS 17]: β-Dipeptide synthesis using pentafluorophenyl *O*-activation

Boc-β-alanine was pre-activated by introducing the pentafluorophenyl function as an ester group [88]. Dmab-β-alanine and the pentafluorophenyl ester of Boc-β-alanine reacted and the synthesis of the corresponding β-dipeptide was realized.

Organic synthesis 18 [OS 18]: Dipeptide from Fmoc-L-β-homophenylalanine

Dmab-β-alanine and Fmoc-L-β-homophenylalanine were reacted to give the dipeptide [5].

Organic synthesis 19 [OS 19]: Dipeptide from Fmoc-L-β-homo-p-chlorophenylalanine

Dmab-β-alanine and Fmoc-L-β-homo-p-chlorophenylalanine were reacted to give the dipeptide [5].

Organic synthesis 20 [OS 20]: Dipeptide from N-ε-Boc-L-lysine

N-ε-Boc-L-lysine is a more complex peptide owing to the additional amino function [5]. Dmab-β-alanine and N-ε-Boc-L-lysine were reacted to give the dipeptide.

Organic synthesis 21 [OS 21]: Dipeptide from N-α-Boc-L-lysine

This is a similar reaction to *[OS 20]* [5].

Organic synthesis 22 [OS 22]: Dipeptide from Dmab-Boc-glycine

Dmab-Boc-glycine and the pentafluorophenyl ester of Boc-β-alanine were reacted to give the dipeptide [5].

Organic synthesis 23 [OS 23]: Diverse deprotection and peptide bond-forming reactions

In order to extend the results obtained by undergoing *[OS 15]*, *[OS 16]* and *[OS 17]* reactions, routes for the preparation of longer chain peptides in micro reactors were searched for [88]. Therefore, deprotection and peptide bond forming reactions were needed.

Organic synthesis 24 [OS 24]: Tripeptide synthesis

Fmoc-β-alanine and two equivalents of Fmoc-β-alanine were reacted to give the corresponding tripeptide [5].

Organic synthesis 25 [OS 25]: α-Dipeptide synthesis from (R)-2-phenylbutyric acid

(*R*)-2-Phenylbutyric acid and (*S*)-α-methylbenzylamine react to the corresponding dipeptide via an EDCI coupling [86]. In a control experiment, (*R*)-2-phenylbutyric acid and (*R*)-α-methylbenzylamine were also reacted.

Organic synthesis 26 [OS 26]: α-Dipeptide synthesis from (S)-α-methylbenzylamine

Boc-D-alanine and (*S*)-α-methylbenzylamine react to give the corresponding dipeptide via an EDCI [3-ethyl-1-(3-dimethylaminopropyl)-carbodiimid] coupling [86]. In a control experiment, Boc-L-alanine and (*S*)-α-methylbenzylamine also reacted.

4.2.10.4 **Experimental Protocols**

[P 12] The micro channels were primed with anhydrous *N,N*-dimethylformamide (DMF) to remove air and moisture before carrying out the reaction [88]. A 50 µl volume of a solution of Fmoc-β-alanine (0.1 M) in anhydrous DMF was placed in one reservoir of a micro chip, driven by electroosmotic flow. A 50 µl volume of a solution of EDCI (0.1 M) in anhydrous DMF and 50 µl of a solution of Dmab-β-alanine (0.1 M) in anhydrous DMF were inserted in the other two reservoirs. Anhydrous DMF was placed in the fourth reservoir, for collection of the product. Room temperature was applied for reaction for a 20 min period. The voltage was set in the range 500–700 V.

[P 13] The micro channels were primed with anhydrous DMF to remove air and moisture before carrying out the reaction [5, 88]. A 50 µl volume of a solution of the pentafluorophenyl ester of Fmoc-β-alanine (0.1 M) in anhydrous DMF was placed in one reservoir of a micro chip, driven by electroosmotic flow (Figure 4.42). A 50 µl volume of a solution of Dmab-β-alanine (0.1 M) in anhydrous DMF was inserted in another reservoir. Anhydrous DMF was placed in the fourth reservoir, for collection of the product. Room temperature was applied for reaction for a 20 min period. The voltage was set in the range 600–700 V.

[P 14] For the reaction of the pentafluorophenyl ester of Fmoc-β-alanine with Dmab-β-alanine, a similar protocol to that for *[P 13]* was used [5, 88].

[P 15] Diverse protocols for routes for deprotection and peptide bond-forming reactions in micro reactors have been reported [5, 88]. These are needed for preparation of longer chain peptides.

[P 16] First reservoir 50 µl of a solution of the pentafluorophenyl ester of Fmoc-β-alanine (0.1 M; 1000 V) in anhydrous DMF; second reservoir, 50 µl of a solution of Dmab-β-alanine (0.1 M; 1000 V) in anhydrous DMF; third reservoir, 50 µl of a solution of 1,8-diazabicyclo[5.4.0]undec-7-ene (DBU) (0.1 M; 400 V) in anhydrous DMF; fourth reservoir, 50 µl of a solution of the pentafluorophenyl ester of Fmoc-β-alanine (0.1 M; 700 V) in anhydrous DMF; fifth reservoir, 40 µl of anhydrous DMF [86]. Room-temperature processing.

[P 17] First reservoir, 30 µl of a solution of the pentafluorophenyl ester of (*R*)-2-phenylbutyric acid (0.1 M; 600 V) in anhydrous DMF; second reservoir, 30 µl of a solution of (*S*)-α-methylbenzylamine (0.1 M; 1000 V) in anhydrous DMF; third reservoir, 30 µl of anhydrous DMF [86]. Room-temperature processing; 20 min reaction time.

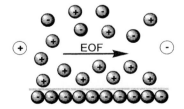

Figure 4.42 Principle of electroosmotic flow: movement of charged molecules under the action of an external electrical field [5].

[P 18] First reservoir, 30 µl of a solution of the pentafluorophenyl ester of Boc-D-alanine (0.1 M; 1000 V) in anhydrous DMF; second reservoir, 30 µl of a solution of (S)-α-methylbenzylamine (0.1 M; 1000 V) in anhydrous DMF; third reservoir, 30 µl of anhydrous DMF [86]. Room-temperature processing.

4.2.10.5 Typical Results
Conversion/selectivity/yield

[OS 15] [R 5] [P 12] Using stoichiometric ratios of reactants and adjusting the flow rate by setting the voltage (700 V) yielded only 10% conversion [88]. Increasing the residence time by decreasing the voltage to 500 V did not improve conversion. Using more coupling reagent EDCI (2.0 mol l^{-1} instead of 1.0 mol l^{-1}), however, increased the conversion to 20%. With stopped-flow techniques, periodically 'pushing' and so mixing the flow, an increase to 50% was reached (2.5 s injection; 10 s stop). A further increase in the concentration of the coupling reagent necessitated a change from EDCI to dicyclohexylcarbodiimide (DCC) for reasons of limited solubility in DMF of the first agent. Applying a 5.0 mol l^{-1} DCC solution resulted in a 93% yield of the dipeptide.

Benchmarking to Batch Processing

[OS 15] [R 5] [P 12] Batch synthesis of Dmab-β-alanine and Fmoc-β-alanine using EDCI to the corresponding β-dipeptide gave a yield of 50% [5]. The electroosmotic-driven micro-reactor processing resulted in 50% at best; using another coupling reagent (DCC), a 93% yield was obtained in the micro reactor [88] (see Conversion/selectivity/yield, above).

[OS 16] [R 5] [P 13] Using continuous flow in an electroosmotic-driven micro reactor gave a quantitative yield of the dipeptide in only 20 min (600 V for Dmab-β-alanine; 700 V for the Fmoc ester) [5, 88]. Batch synthesis under the same conditions gave only a 40–50% yield [5] (46% in [5]), needing 24 h.

[OS 17] [R 5] [P 14] Using continuous flow in an electroosmotic-driven micro reactor gave a quantitative yield of the dipeptide in only 20 min (700 V for both Dmab-β-alanine and the Boc ester) [88]. Batch synthesis under the same conditions gave only a 40–50% yield [5] (57% in [5]), needing 24 h.

Variation in amino acid composition

[OS 18] [R 5] [P 13] (adapted) Subsequent to making dipeptides from protected β-alanines, the dipeptide from Dmab-β-alanine and Fmoc-L-β-homophenylalanine was prepared [5]. Using continuous flow in an electroosmotic-driven micro reactor gave a quantitative yield of the dipeptide in only 20 min (600 V for Dmab-β-alanine; 900 V for the Fmoc ester) [88]. Batch synthesis under the same conditions gave only a 35% yield, needing 24 h [5].

[OS 18] [R 5] [P 13] (adapted) The same was done using Fmoc-L-β-homo-*p*-chlorophenylalanine [5]. Batch, 36%; micro reactor, quantitative in 20 min.

[OS 20] [R 5] [P 13] (adapted) Using more complex peptides such as *N*-ε-Boc-L-lysine is also feasible by micro-channel processing [5]. Batch, 9%; micro reactor, quantitative in 20 min.

[OS 21] [R 5] [P 13] (adapted) N-α-Boc-ʟ-lysine has been used [5]. Batch, 50%; micro reactor, quantitative in 20 min.

[OS 22] [R 5] [P 13] (adapted) The Dmab ester of Boc-glycine was also used [5]. Batch, 35%; micro reactor, quantitative in 20 min.

Reaction rate – comparison of batch with micro reactor
[OS 18] [R 5] [P 13] (adapted) Using 0.05 M solutions of Dmab-β-alanine and Fmoc-ʟ-β-homophenylalanine, a comparison between batch and micro-reactor processing was made [5]. Whereas the micro reactor gave a 100% yield in 20 min, only about 5% was reached by batch. Even after 400 h, only 70% conversion was achieved.

First steps to extended-peptide chain formation
[OS 23] [R 5] [P 15] Deprotection and peptide bond-forming reactions in a micro reactor and their yields have been described [5, 88]. Establishing protocols for these reactions paves the way to the preparation of longer chain peptides in micro reactors.

Formation of tripeptides
[OS 24] [R 5] [P 16] Dmab-β-alanine and Fmoc-β-alanine were reacted to give a dipeptide [5]. After cleavage of the Fmoc function, Fmoc-β-alanine was added to such a dipeptide resulting in tripeptide formation with 30% yield [5].

Racemization of α-dipeptides
[OS 25] [R 4] [P 17] For dipeptide formation from the pentafluorophenyl ester of (R)-2-phenylbutyric acid and (S)-α-methylbenzylamine an extent of racemization of 4.2% was found [86]. At higher concentration (0.5 instead of 0.1 M), a higher degree of racemization was found (7.8%). This experiment also served to demonstrate monitoring of the racemization of a simple carboxylic acid used in peptide synthesis.

[OS 26] [R 4] [P 18] For dipeptide formation from the pentafluorophenyl ester of Boc-ᴅ-alanine and (S)-α-methylbenzylamine an extent of racemization of 5.6% was found [86]. This experiment also served to demonstrate monitoring of the racemization of an α-amino acid used in peptide synthesis.

4.2.11
Hydroxydearyloxy Substitution + O-Aryl, O-Alkyl Substitution –
Hydrolysis and Transglycosylation

Peer-reviewed journals: [26, 27].

4.2.11.1 Drivers for Performing Hydrolysis and Transglycosylation in Micro Reactors
The synthesis of glycoconjugates opens the route to one of the most important class of biomolecules which play an active role in relevant biological reactions [26]. One way to do so is to use enzymes, which, however, suffer from instability and slow reaction rates.

4.2.11.2 Beneficial Micro Reactor Properties for Hydrolysis and Transglycosylation

Given the feasibility of micro-channel devices for enzyme-based oligosaccharide synthesis, enhancement of mass transfer therein could speed up this reaction. Also, it may be hoped that enzyme degradation may be reduced, for reasons that are not so straightforward.

4.2.11.3 Hydrolysis and Transglycosylation Investigated in Micro Reactors
Organic synthesis 27 [OS 27]: Hydrolysis of p-nitrophenyl-β-D-galactopyranoside

Organic synthesis 28 [OS 28]: Transgalactosylation of p-nitrophenyl-2-acetamide-2-deoxy-β-D-glucopyranoside

4.2.11.4 Experimental Protocols

[P 19] Two micro syringes were filled with 0.32 mM *p*-nitrophenyl-β-D-galacto-pyranoside in phosphate buffer (pH 8) and β-galactosidase (20 U) in 10 ml of the same buffer [26]. Both solutions were pumped into the micro channel at the same flow rate (a few μl min^{-1}). The reaction was carried out for 0–30 min at 37 °C using a hot-plate. The residence time was set by adjusting the flow rate. After passing the micro channel, the reaction mixture was dropped into hot water to inactivate the enzyme.

The micro channel may be pre-treated before processing in the following way: first, the channel is charged with β-galactosidase in phosphate buffer; then, only buffer solution is passed; finally, only *p*-nitrophenyl-β-D-galactopyranoside (without enzyme) is filled [26]. The respective flow rates orient on the protocol given above.

[P 20] Via two micro syringes, a mixture of 0.32 M *p*-nitrophenyl-β-D-galacto-pyranoside and β-galactosidase and 3.2 mM *p*-nitrophenyl-2-acetamide-2-deoxy-β-D-glucopyranoside in 50% phosphate buffer (pH 8)–acetonitrile and β-galactosi-dase (20 U) in 10 ml of the same solvent system were fed into the micro channel [26]. Identical flow rates were used and the reaction was carried out for 0–30 min at 37 °C.

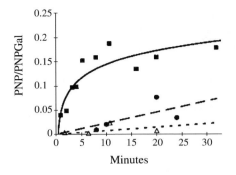

Figure 4.43 Benchmarking of untreated and pre-treated Y-piece micro reactors to a commercial micro test-tube for the hydrolysis of *p*-nitrophenyl-β-D-galactopyranoside. Y-piece micro reactor (■); commercial micro test-tube (●); pre-treated Y-piece micro-reactor (△) [26].

4.2.11.5 Typical Results
Conversion/selectivity/yield – benchmarking to batch processing

[OS 27] [R 12] [P 19] When monitoring the molar ratio of β-galactosidase and *p*-nitrophenyl-β-D-galactopyranoside (a measure for analyzing the product yield of D-galactose) as a function of time, hydrolysis in a micro-channel chip is about five times faster than in a commercial micro test-tube [26]. The ratios found were in the range 0.01–0.17. The test-tube was not stirred to avoid inactivation of the enzyme. As a possible explanation, faster diffusion in the micro channel compared with the bulk fluid is suggested. To underline this, experiments with a pre-treated micro channel were undertaken, leading to strongly reduced activity (Figure 4.43).

[OS 28] [R 12] [P 20] The rate of transgalactosylation, the reverse hydrolysis reaction (see above), was also increased by micro-channel processing [26]. The ratio of galactosylated *p*-nitrophenyl-2-acetamide-2-deoxy-β-D-glucopyranoside and *p*-nitrophenyl-2-acetamide-2-deoxy-β-D-glucopyranoside was taken as measure for yield determination as a function of time (Figure 4.44). The ratios were found to be in the range 0.005–0.05.

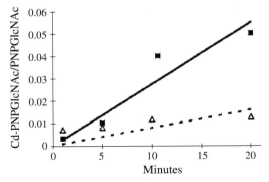

Figure 4.44 Benchmarking of the Y-piece micro reactor to a commercial micro test-tube for the transglycosylation of β-D-galactopyranoside. Y-piece micro reactor (■); commercial micro-test tube (△) [26].

4.2.12
Alkoxydehydroxylation – Esterification of Acids

Peer-reviewed publications: [91].

4.2.12.1 Drivers for Performing Esterifications in Micro Reactors

First investigations revealed a comparison of batch-scale performance with micro-channel processing [91]. Further investigations of the role of surface effects catalyzing the esterification concerned the deliberate enhancement in the number of silanol groups on the glass surface of such micro-channel reactors.

4.2.12.2 Beneficial Micro Reactor Properties for Esterifications

Concerning the impact of surface-induced reactions, the large specific surface area of the micro channels provides a correspondingly large number of reaction sites that are available without transport resistance [91]. By this means, a former bulk reaction may be changed to a heterogeneous reaction or combined heterogeneous–heterogeneous reaction. This may have a significant impact on the choice of reactants and auxiliary additives and also on reaction speed and process parameters such as temperature and concentration.

4.2.12.3 Esterifications Investigated in Micro Reactors
Organic synthesis 19 [OS 29]: Esterification of 4-(1-pyrenyl)butyric acid

4-(1-Pyrenyl)butyric acid reacts with ethanol in the presence of sulfuric acid [91].

4.2.12.4 Experimental Protocols

[P 21] Solutions of 10^{-4} M 4-(1-pyrenyl)butyric acid in ethanol and 10^{-4} M sulfuric acid in ethanol were contacted in a micro-mixing tee/micro channel flow configuration at room temperature and at 50 °C [91]. Pressure-driven feed was used. The glass surface of the micro channels was either tuned hydrophobic (by exposure to octadecyltrichlorosilane) or hydrophilic (by wetting with a sulfuric acid/hydrogen peroxide mixture).

4.2.12.5 Typical Results
Conversion/selectivity/yield – benchmarking to batch processing

[OS 29] [similar to R 12, details in [91]] [P 21] A 15–20% yield for a 4-(1-pyrenyl)butyric acid ester was obtained by micro-channel processing after 40 min of reaction at room temperature [91]. In turn, no reaction occurred at room temperature or at

50 °C by carrying out a batch experiment. At 50 °C, micro-channel processing resulted in a maximum yield of 83% (40 min; 0.1 µl min^{-1}).

Residence time

[OS 29] [similar to R 12, details in [91]] [P 21] On varying the residence time by changing the flow rate from 0.1–1 µl min^{-1} for micro-channel processing at 50 °C, the yields decrease from 83 to 17%. No figures of residence times are given in [91].

Absence of acid catalyst/role of silanol surface

[OS 29] [similar to R 12, details in [91]] [P 21] Without the presence of sulfuric acid no reaction to a 4-(1-pyrenyl)butyric acid ester was found in the micro reactor [91]. On activating the surface with a sulfuric acid/hydrogen peroxide mixture, however, a yield of 9% was achieved after 40 min at 50 °C. On making the surface hydrophobic by exposure to octadecyltrichlorosilane, no product formation was found. Using silica gel in a laboratory-scale batch experiment resulted in detectable conversion, however, being substantially lower than in the case of the micro reactor. The yield was no higher than 15% (40 min; 0.1 µl min^{-1}), whereas the best micro reactor result was 83% (40 min; 0.1 µl min^{-1}).

4.2.13
Allyldehydro-Substitution – C–C Bond Formation with Acyliminium Cations

Peer-reviewed journals: [66]; proceedings: [67, 92]; sections in reviews: [90].

4.2.13.1 Drivers for Performing Electrochemical C–C Bond Formation with Cations

Carbocations are highly reactive species that can be used for C–C bond formation. One driver for using continuous micro chemical processing is to employ also unstable cations, which are not amenable to batch synthesis because they decompose before they can actually be used [66, 67].

As C–C bond formation is an important step in organic synthesis, particularly for pharmaceutical applications, it is useful to look for operation modes of chemical micro processing that allow one to carry out combinatorial chemistry investigations. As such, the serial introduction of multiple reactant streams by flow switching was identified [66, 67]. The wide availability of precursors for acyliminium cations has led to the expression 'cation pool' [66, 67].

4.2.13.2 Beneficial Micro Reactor Properties for Electrochemical C–C Bond Formation with Cations

Micro reactors show, under certain conditions, low axial flow dispersion; reactions with unstable intermediates can be carried out in a fast, stepwise manner on millisecond time-scales. Today's micro mixers mix on a millisecond scale and below [40]. Hence in micro reactors reactions can be carried out in the manner of a quench-flow analysis, used for determination of fast kinetics [93].

Concerning combinatorial operation, micro reactors can be operated in both a highly parallel a and fast serial manner. The latter approach has been realized for the 'cation flow' method for C–C bond formation [66, 67].

4.2.13.3 Electrochemical C–C Bond Formation with Cations Investigated in Micro Reactors

Organic synthesis 30 [OS 30]: Electrooxidative C–C bond formation of carbamates

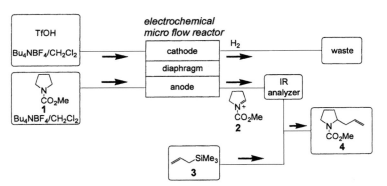

A substrate containing an amine carboxylate moiety is converted in an electrolyte solution in the presence of a strong acid to a cationic intermediate, an *N*-acyliminium cation, by electrooxidative reaction. This species is immediately reacted with an allylsilane [66, 67]. By nucleophilic reaction, C–C bond formation is achieved.

4.2.13.4 Experimental Protocols

[P 22] A 0.05 M solution of methyl pyrrolidinecarboxylate with 0.3 M Bu_4NBF_4 electrolyte in dichloromethane was fed by syringe pumping (2.1 ml h^{-1}) into the anodic chamber (carbon felt) of the electrochemical micro flow reactor at a temperature of –72 °C [66, 67]. Into the cationic chamber equipped with a platinum wire as electrode, a solution of the electrolyte and trifluoromethanesulfonic acid as a proton source was fed. The cationic intermediate was generated by low-temperature electrolysis (14 mA) and then immediately transferred to another reaction vessel where the final product was obtained by nucleophilic attack. Between the first and second reaction steps monitoring by a continuous-flow IR analyzer (ATR method) was undertaken, revealing the concentration of the carbocation (Figure 4.45).

Figure 4.45 Schematic of the 'cation flow' system [66].

4.2.13.5 Typical Results

Conversion/yield/selectivity

[OS 30] [R 30] [P 22] The synthesis of nine C–C bonded products was made from four carbamates and five silyl enol ethers [66, 67]. Conversions ranged from 49 to 69%; the corresponding selectivities ranged from 67 to 100%. Similar performance was achieved when serially processing the same reactions (see Serial combinatorial synthesis).

Cation pool – a variety of carbamates synthesized

[OS 30] [R 30] [P 22] The feasibility of generating a cation pool, i.e. of performing multiple reactions with various reactants, by means of electrooxidative micro flow processing was demonstrated [66, 67]. The micro reaction system was consequently termed 'cation flow'. By this means, various C–C bonded products were made from carbamates, having pyrrolidine, piperidine, diethylamine and trihydroisoquinoline moieties. These carbamates were combined with various silyl enol ethers, yielding nine products.

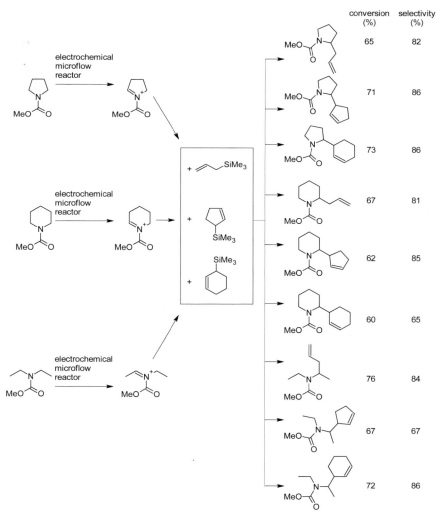

Figure 4.46 Schematic of serial combinatorial synthesis for creating a cation pool from diverse carbamates and silyl enol ethers [66].

Serial combinatorial synthesis

[OS 30] [R 30] [P 22] By simple flow switching, serial combinatorial synthesis for creating a cation pool from diverse carbamates and silyl enol ethers was accomplished (Figure 4.46) [66, 67]. The conversions and selectivities were comparable to continuous processing using three feed streams only (see Conversion/yield/selectivity, above).

4.3
Aromatic Electrophilic Substitution

4.3.1
Nitrodehydrogenation – Nitration of Aromatics

Peer-reviewed journal: [31, 94]; proceedings: [38, 95–98]; sections in reviews: [14, 83, 89, 99, 100]; additional information: [101, 102]. See also the information given on the nitration of aliphatics in Section 4.7.1.

4.3.1.1 Drivers for Performing Aromatic Nitrations in Micro Reactors

Most nitrations are highly exothermic and hence release a lot of reaction heat for most experimental protocols [37, 94]. This high exothermicity may even lead to explosions [37, 38]. Nitration agents frequently display acid corrosion [37]. For these reasons, nitrations generally are regarded as being hazardous [37, 38].

Owing to the heat release, nitrations often lack selectivity, i.e. many parallel, consecutive and decomposition processes are known to occur. As a result, product spectra are unusually wide and consequently yields and purity are low [37, 94].

The selectivity issue has been related to multi-phase processing [31]. Nitrations include both organic and aqueous phases. Oxidation to phenol as one side reaction takes places in the organic phase, whereas all other reactions occur in the aqueous phase and are limited by organic solubility. For this reason, enhancing mass transfer by large specific interfaces is a key to affecting product selectivity.

Having high mass transfer, in addition to good heat transfer, may change the product spectra, by increasing the conversion to product and decreasing the formation of some of the by-products [94]. Nitrations are well suited for two-phase capillary flow processing yielding uniform alternating slugs. In these slugs, internal circulation leads to high mass transfer. The defined setting of residence time can be achieved by establishing two-phase plug flow behavior in so-called capillary-flow reactors.

4.3.1.2 Beneficial Micro Reactor Properties for Aromatic Nitrations

The small reaction volumes in micro reactors and the large specific surface areas created are beneficial for coping with the problems caused by the release of the large heats, as mentioned above [37, 38]. Delicate temperature control is what is expected for micro-reactor operation; isothermal processing is said to be achiev-

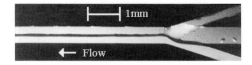

Figure 4.47 Image of parallel liquid/liquid flow through a micro channel [31].

able even when large reaction heats are released [94]. Small size should so increase process safety and suppress unwanted secondary reactions [37, 38].

In micro reactors, very unique flow patterns can be established using proper contactors/mixers. This holds particularly for two-phase flow by capillary-flow processing yielding uniform alternating slugs (see Figure 4.51) [94] or parallel-flowing streams (Figure 4.47) [31]. Besides advantages in terms of improved mass transfer, they uniformly set residence time by plug-flow motion. Apart from increasing reaction to the kinetic limits, the flow patterns may directly affect selectivity, which depends on interfacial area [31].

4.3.1.3 Aromatic Nitration Reactions Investigated in Micro Reactors

Nitration reactions are among the basic chemical pathways followed in organic synthesis. They are used for making pharmaceutical products, agricultural and pest control chemicals, pigments, precursors for polyurethane or polyamide production and explosives [37]. Typically, acidic nitration agents such as nitric acid are employed to insert the nitro function, using dehydrating agents such as sulfuric acid or anhydrous acetic acid. Dinitrogen pentoxide is usually more reactive and may be employed, if acid-sensitive substances have to be reacted.

Organic synthesis 31 [OS 31]: Mononitration of benzene

$$\text{benzene} \xrightarrow{\text{HNO}_3/(\text{H}_2\text{SO}_4)} \text{nitrobenzene (NO}_2\text{)}$$

The nitration of benzene is a standard organic synthesis process, well described in the scientific and patent literature [31, 97].

Organic synthesis 32 [OS 32]: Mononitration of toluene

$$\text{toluene} \xrightarrow{\text{HNO}_3/(\text{H}_2\text{SO}_4)} \text{nitrotoluene (NO}_2\text{)}$$

The nitration of toluene, similarly to the nitration of benzene, is a standard organic synthesis process, well described in the scientific and patent literature [31, 97].

Organic synthesis 33 [OS 33]: Nitration of substituted single-ring aromatics

The nitration of several single-ring aromatics with two substituents at various ring positions has been reported [94]. The exact nature of these species, however, was not disclosed, probably for intellectual property reasons, as the reactions were performed for an industrial company (BASF).

Organic synthesis 34 [OS 34]: Mononitration of naphthalene

The nitration of naphthalene was used as a test reaction [37]. As a consequence of having two aromatic rings, a particularly large variety of nitration products are in principle possible. This refers to multiplicity of nitration and to positional selectivity for each nitration step.

4.3.1.4 Experimental Nitration Protocols in Micro Reactors

[P 23] The nitration of two standard aromatic compounds, namely benzene and toluene, was performed utilizing nitric acid/sulfuric acid mixtures and the pure

organic reactant [31, 38, 97, 101, 103]. The sulfuric acid content in the mixture was set high, ranging from 70 to 85 mass-%, to achieve a high reaction rate, i.e. to test the mass transfer capabilities of the micro reactor when switching from kinetic to mass transfer control. Acid-to-organic ratios of flow rates were from 2 : 1 to 7 : 1; accordingly, the nitric acid : benzene molar ratios were about 0.3–2, depending on the afore mentioned flow ratios and sulfuric acid contents. The temperature was set from 60 to 90 °C. Residence times were several tens of seconds, typically 25 s.

[P 24] The nitration of naphthalene was carried out with dissolved or *in situ* generated N_2O_5 gas [37]. The temperature was set to –10 to 50 °C and residence times to 15–45 s. The reaction mixture processed in the micro reactor was quenched with water, extracted and analyzed by HPLC or GC with mass-selective detection.

[P 25] The nitration of various single-ring aromatics was performed in a capillary-flow micro reactor with capillary lengths of 1 to 8 m [94]. The aromatic and the pre-mixed nitration acid were fed by high-pressure piston pumps into a Y-piece contactor (0.5–1.0 mm inner diameter; 120°). The PTFE capillary of 0.5–1.0 mm inner diameter was inserted in a heating mantle, operated with silicone oil in counter-flow mode. By this means, isothermal conditions were established. The reactor was operated at 4 bar. After having passed the capillary, the reaction was quenched to room temperature to slow the reaction. Further addition of cooled water completely stopped the reaction by cooling and dilution. A needle valve served for releasing pressure.

4.3.1.5 Typical Results
Conversion/selectivity/yield
[OS 31] [R 4] [P 23] Under electroosmotic flow conditions, the reactant benzene was mobilized as a microemulsion using sodium dodecyl sulfate (SDS) as surfactant [103] (see also [14]). The nitronium ions, generated *in situ* from sulfuric and nitric acid, were moved by electrophoretic forces. By this means, a 65% yield of a nitrobenzene was obtained; consecutive nitration products such as 1,3-dinitrobenzene (8% yield) and 1,3,5-trinitrobenzene (5% yield) were also produced.

[OS 34] [R 17] [R 19] [R 26] [P 23] By performing naphthalene nitration with fuming HNO_3 in micro reactors, the selectivity for mononitronaphthalenes can be significantly enhanced (Figure 4.48) [98]. The selectivity could be raised to more than 95%, independent of the microreactor used.

In addition, the isomeric ratios are affected. Whereas in industrial processes the ratio of 1- to 2-mononitronaphthalene is about 20 : 1; this ratio is dramatically increased to more than 30 by using micro reactors [98].

[OS 34] [R 17] [R 19] [R 26] [P 24] During naphthalene nitration in micro reactors, 100% conversion was obtained for all experiments undertaken [37]. Mainly mono and dinitro products were obtained (Figure 4.49) [37]. For batch processing of naphthalene, a wider range of products are found containing many isomers of the above-mentioned species, but also tri- or tetranitrated products. In the micro reactor, even at 50 °C and using a excess of nitrating agent, high selectivity was maintained, as exhibited by the high degree of mononitronaphthalenes in the product mixture [37].

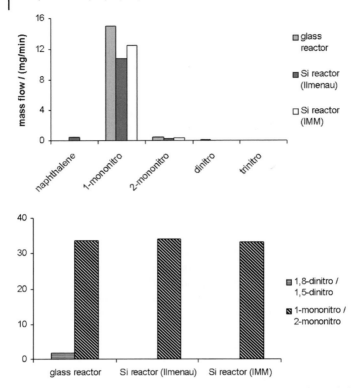

Figure 4.48 Product spectra and isomer ratios for the nitration of naphthalene with HNO_3 in micro reactors from different suppliers [98].

Ratio of regioisomers for aromatic substrates

[OS 34]] [R 17] [R 19] [R 26] [P 24] The isomeric ratio of two regioisomers, 1,5-dinitro-to 1,8-dinitronaphthalene, being constant at 1 : 3.5 for macroscopic batch reactors, changes to 1 : 2.8 in micro reactors [98]. The formation of the 1,5-dinitro product is, however, not favored (Figure 4.50, see also Figure 4.48 and Figure 4.49).

More detailed information concerning the characterization of micro mixers from different suppliers and of different mixing principles is given in [98] (this source may be difficult to obtain).

For conventional stirred tank processing of the nitration of benzene, see [102]. Here a description of regioisomer formation is given as a function of stirring intensity. Mechanistic analysis is also given there that may be applied to micro-reactor processing utilizing large specific interfaces.

Mixer/distributor impact on reaction

[OS 31] [R 16a] [R 26] [P 23] Reaction rates for the nitration of benzene increase strongly when the bore diameter is halved (from 127 to 254 μm) [31, 97]. In the temperature range investigated (60–90 °C), an increase in the reaction rate by a

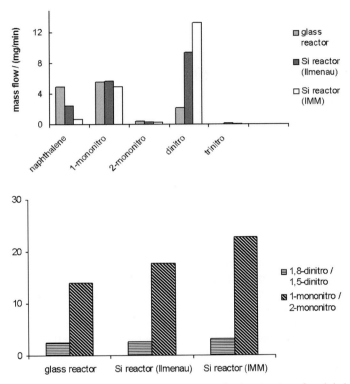

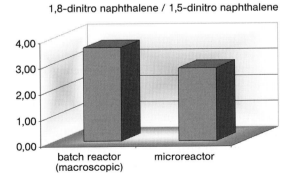

Figure 4.49 Product spectra and isomer ratios for the nitration of naphthalene with N_2O_5 in micro reactors of different types and from different suppliers [98].

Figure 4.50 Comparison of isomeric ratios for 1,5- and 1,8-dinitronaphthalene. Reaction was performed in a macroscopic batch reactor and micro reactors of different types and from different suppliers [37].

factor of 3–4 was observed. The reaction rates (based on a shell-reaction model and a 1.5th-order assumption) achieved were in the range from 1 to 5.5 min^{-1}.

Smaller bore diameters naturally produce slugs of smaller diameter [31, 97]. Typically, a smaller length can also be generated thereby. As a consequence, internal circulation in the slug and specific interface between the slugs are increased. It is assumed that the impact of the increase in internal circulation on mass transfer/ reaction processing is generally more dominant.

[OS 34] [R 17] [R 19] [R 26] [P 24] A detailed comparison of micro mixers from different suppliers and of different mixing principles, with regard to conversion, isomer formation and consecutive nitration, is given in [98].

For conventional stirred tank processing of the nitration of benzene, the dependence of conversion on impeller speed is given in [102].

Mixer/distributor impact on slug formation

[OS 33] [R 16b] [P 25] Uniform slugs of a single-ring aromatic/nitric acid were formed in a Y-piece having volumes according to the volumetric flows fed by piston pumps [94]. The volume of the individual slugs depends only on the internal diameter of the slugs. Thus, by decreasing the channel diameter, the interfacial area of the slugs can be increased. The capillary attached has a stabilizing effect on the slug flow and determines their length by setting the (fixed) internal diameter. The deviation of slug size distribution is very small and amounts only to about 5% from the average value. Hence the interfacial area is nearly constant for this type of capillary flow. A slug of the nitrating acid phase shown, for instance, has about a 0.75 mm diameter and 4 mm length; the slug of the organic phase is much smaller (Figure 4.51).

0.75 mm

Figure 4.51 Liquid/liquid two-phase plug-flow with plugs of nitrating agent (dark) and organic phase (white) in a capillary micro reactor [94].

Mixer/distributor impact on phase contamination

[OS 33] [R 16c] [P 25] The amount of phase contamination, i.e. the undesired dispersion of one phase into the other as a consequence of unbalanced surface energies, of a parallel liquid/liquid flow through a 500 μm wide channel using a kerosene/propylene glycol/water mixture was studied to monitor a similar flow with benzene/nitric acid/sulfuric acid [31]. The study was performed at different aqueous/organic viscosity ratios (0.56–22.1). The contamination was analyzed as a function of the kerosene flow proportion, including ideal and non-ideal splitting (Figure 4.52). The experimental findings were compared with CFD results assuming laminar-flow properties. The two data sets were in agreement.

For optimum splitting, a contamination of less than 2% was found. By operating at slightly higher flow rates, one stream could be made contamination free. Thus, a clean product phase can be obtained when performing the nitration in the T-piece/ tube reactor.

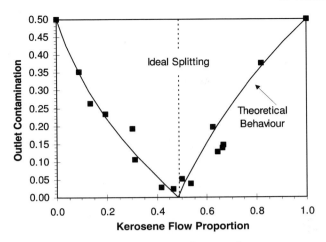

Figure 4.52 Observed flow splitting performance for a 1000 μm × 100 μm micro channel system [31].

Temperature impact

[OS 31] [R 16a] [P 23] On increasing the temperature, the reaction rate for nitration of benzene increases (Figure 4.53), as usually to be expected for most organic reactions [31]. For a capillary-flow micro reactor, more than doubling of the reaction rate was determined on increasing the temperature from 60 to 90 °C.

For toluene, a less significant impact of temperature (exceeding 70 °C) on reaction rates was observed than for benzene [31].

[OS 33] [R 16b] [P 25] For the nitration of a single-ring aromatic in a capillary-flow micro reactor, experiments were performed at two temperature levels, 60 and 120 °C [94]. Owing to the assumed increase in conversion rate with higher temperature, attempts were made to compensate for this by decreasing the capillary length at otherwise constant dimensions. For the 60 °C experiment, a very low

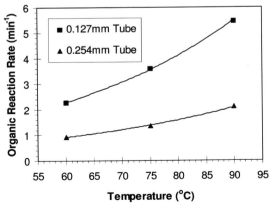

Figure 4.53 Increase of nitration performance as a function of temperature [31].

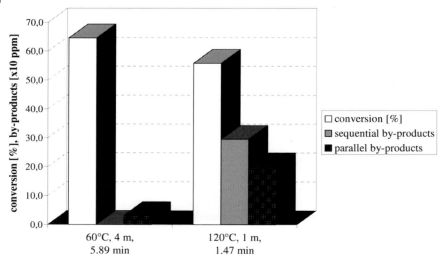

Figure 4.54 Comparison of conversion and by-product formation for two reaction temperatures [94].

level of parallel by-products (phenol derivatives) and consecutive by-products (dinitrated species) was found, not exceeding 50 ppm each (4 m capillary length; 0.75 mm diameter; 5.89 min). For the 120 °C experiment, high levels of parallel and consecutive by-products were found (1 m capillary length; 0.75 mm diameter; 1.47 min). Levels of 300 ppm dinitrated species and 200 ppm phenol derivative were detected (Figure 4.54). From 60 °C to 120 °C, the conversion decreased slightly from about 65 to about 56%. The increase in by-products with temperature shows that the activation energies of by-product formation are higher than for the product formation.

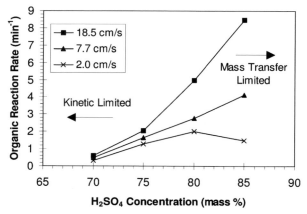

Figure 4.55 Varyiation of flow velocities to increase the nitration performance within the mass-transfer limited regime [31].

Flow velocity

[OS 31] [R 16c] [P 23] On increasing the flow velocity, the reaction rate for the nitration of benzene can be increased (Figure 4.55), provided that operation is performed in a mass-transfer limited regime [31, 97]. This was explained as being due to changing flow patterns which, in turn, affect the specific interface. Mass-transfer limited regimes were found at high sulfuric acid concentrations (> 75%). Reaction rates up to 8 min^{-1} were reached.

Sulfuric acid concentration/content of NO_2^+ ions

[OS 31] [R 16c] [P 23] By increasing the sulfuric acid content in the acid mixture, the acid strength is enhanced, i.e. more NO_2^+ ions are generated [31, 97]. This increases the reaction rate up to 8 min^{-1} (see Flow velocity, above).

Substrate reactivity

[OS 32] [R 16a] [P 23] Toluene nitration rates determined in the capillary-flow reactor were generally higher than benzene nitration rates [31, 97]. This is not surprising, as it stems from the higher reactivity of toluene towards electrophilic substitution owing to its more electron-rich aromatic core. For instance, at a reaction temperature of 60 °C, rates of 6 and 2 min^{-1} were found for toluene and benzene nitration, respectively. However, care has to be taken when quantitatively comparing these results, since experimental details and tube diameters vary to a certain extent or are not even listed completely.

Acid-to-organic flow ratio

[OS 32] [R 16a] [P 23] On increasing the ratio of flows in favor of the acid content, the toluene nitration reaction rate decreases, especially at low temperature (25 °C; 150 μm capillary tube) [31, 97]. The nitrotoluene concentration increases with increasing acid-to-organic flow ratio at a high temperature of 90 °C (178 μm capillary tube) [31, 97]. Mixtures having a ratio beyond 5 : 1 do not contribute any further to increasing nitrotoluene concentration as a consequence of the above mentioned decrease in the reaction rate. The concentration increases with increasing length of the capillary tube (450, 900, 1350 mm), which also shows that post-reaction in the collection vessel is of minor importance (Table 4.1).

Table 4.1 Influence of reaction temperature and acid-to-organic flow rate on the reaction rate of toluene nitration [31, 97].

to organic flow ratio	Reaction rate at 25 °C (min⁻¹)	Reaction rate at 60 °C (min⁻¹)
2 : 1	3.27	6.10
3 : 1	2.81	6.12
5 : 1	2.25	4.39
7 : 1	1.65	4.14

By-product formation

[OS 31] [R 16c] [P 23] The levels of dinitrobenzene, dinitrophenol and picric acid in the organic phase during benzene nitration in a micro reactor were monitored [31]. Picric acid levels were no higher than 100 ppm for all experiments conducted. Dinitrobenzene was the largest impurity fraction. A study revealed contents < 1000 ppm up to 34 mass-% on increasing the sulfuric acid content from 70 to 85%.

A more detailed and more accurate study showed the impact of temperature and tube inner diameter and took into account also the dinitrophenol impurity [31]. On increasing the temperature from 80 to 120 °C, the dinitrobenzene and dinitrophenol contents increase from 1640 and 340 ppm to 53 500 and 1233 ppm, respectively, for a 178 μm tube. For a 127 μm tube, the same temperature rise results in an increase in the dinitrobenzene and dinitrophenol contents from 970 and 90 ppm to 38 200 and 1281 ppm, respectively.

Tube diameter

[OS 31] [R 16c] [P 23] A decrease in tube diameter from 254 to 127 μm leads to an increased product yield for the nitration of benzene [31, 97]. At a temperature of 90 °C, the reaction rate increases by a factor of more than two.

Tube reactor length/residence time

[OS 32] [R 16a] [P 23] An increase in tube length from 45 to 135 cm leads to an increased product yield for the nitration of toluene [31, 97]. This is strongly dependent on the acid-to-organic flow ratio. The higher the latter, the larger is the difference in product yield (Figure 4.56).

The latter fact can be explained by the influence on the reaction time of varying the flow ratio [31, 97]. If the reaction time becomes longer than the residence time in any of the tubes investigated, differences in product yield result.

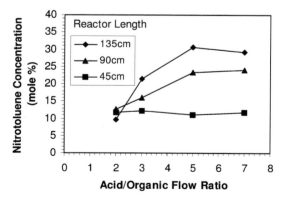

Figure 4.56 Nitrotoluene product formation at various acid-to-organic flow ratios using a reaction system with a liquid/liquid distributor and a tube at 90 °C [31].

Corrosion resistance

[R 16a] [R 19] [P 23] High sulfuric acid contents can lead to steel corrosion [37, 38, 97]. This may even lead to blockage owing to accumulation of corroded material in the tube. In [38] it is also claimed that steels are not suited for nitration; however, since the grade of the steel employed is not given, it cannot be excluded that high-alloy steels may behave better. Silicon, glass and titanium are recommended materials [38].

Benchmarking with patented literature

[OS 31] [R 16a] [P 23] For benzene nitration, the results achieved in the capillary-flow micro reactor were benchmarked against results claimed in the patent literature (see Table 4.2) [97]. An analysis of conversion, by-product level, reaction time and reaction rate showed that the results achieved in micro reactors and conventional equipment are competitive, i.e. were similar. As tendencies, it seemed that the micro reactor can lead to a lower by-product level owing to its better temperature guiding and that reaction times can be further shortened. However, the corresponding results are not absolutely comparable in terms of reaction conditions and hence further data are required here.

[OS 31] [R 16c] [P 23] In a further study, it is shown that about the same conversion (94%) can be achieved if the micro reactor is operated at lower temperature (90 °C instead of ~130 °C) and using higher sulfuric acid contents (78% instead of ~65%) [31]. The by-product level is, however, then much higher (4600 ppm instead of 1000–2000 ppm). The by-product level of the micro reactor can be decreased to commercial practice by decreasing the sulfuric acid content (to 72%), but at the expense of dramatically losing conversion (only 61%).

Table 4.2 Comparison of conversion and by-product formation for the nitration of benzene in conventional reactors and micro reactor set-ups [31].

Type of nitration process	Inlet (°C)	Outlet (°C)	H_2SO_4 (mass-%)	Conversion (%)	By-product (ppm)	Time (s)	Rate (min^{-1})
Conventional	80	128	60.6	89.5	1000	120	0.9
Conventional	80	134	65.2	99.9	2090	120	2.1
Conventional	95	120	69.5	90.0	1750	25.0	4.6
[97]	90	90	77.7	94.0 (178 µm capillary)	4600	24.4	5.9
[97]	90	90	72.2	60.7 (178 µm capillary)	< 1000	26.1	1.6

Mechanistic analysis of by-product formation

[OS 33] [R 16b] [P 25] For the nitration of a single-ring aromatic, the substituents of which were not disclosed, the mechanism of by-product formation was investigated in a capillary-flow micro reactor [94]. Dinitrated products were generated via consecutive nitration of the mononitrated product. It was concluded that phenolic by-products were formed directly from the aromatic, rather than from the mononitrated product. This proposed reaction mechanism could be confirmed by performing selective nitration of the mononitrated product. Here, even after relatively long residence times of about 20 min (at 120 °C), no phenolic moieties were detected in the product mixture. On the contrary, after less than 5 min phenols were formed using the single-ring aromatic (at 120 °C).

Interphase mass transfer between liquid–liquid slugs

[OS 33] [R 16b] [P 25] The influence of interphase mass transfer between liquid/liquid slugs was investigated for nitration of single-ring aromatics in a capillary-flow reactor [94]. This was achieved by changing flow velocity via volume flow setting. The residence time was kept constant by increasing the capillary length with respect to the flow change.

Conversion to the mono nitrated aromatic increased linearly with increasing flow velocity owing to enhanced mass transfer (120 °C; 2.95 min; 9.0–36.0 ml h^{-1}; 1.0–4.0 m capillary length; 0.75 mm capillary inner diameter) [94]. The formation of parallel by-products (phenols) increased in the same manner for similar reasons (Figure 4.57). In turn, consecutive by-products, dinitrated aromatics, were formed in a linear decreasing fashion. This was explained by a mass-transfer induced removal of the mononitrated product from the reacting slug.

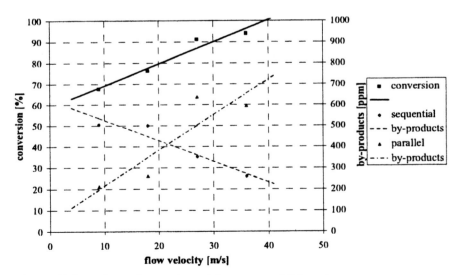

Figure 4.57 Dependence of conversion and the amounts of dinitrated and phenolic by-products on flow velocity (120 °C; 2.95 min) [94].

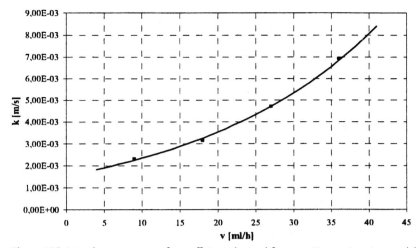

Figure 4.58 Interphase mass-transfer coefficient obtained for a reaction engineering model [94].

Reactor model

[OS 33] [R 16b] [P 25] For the nitration of single-ring aromatics in a capillary-flow reactor, a reactor model was developed, taking into account both mass transfer of organic components between the two phases and the homogeneous reaction within the aqueous phase, in the latter case relying on literature data [94]. For instance, results obtained at different flow velocities could be deduced in this way (Figure 4.58).

In a further step, an extended kinetic model was developed and applied, considering the kinetics of the homogeneous side reactions as well [94]. By this means, the activation energies of these processes could be derived.

Simulation of circulation zones

[OS 33] [R 16b] [P 25] For the nitration of single-ring aromatics in a capillary-flow reactor, internal circulation, exhibiting an inner downwards and an outer upwards flow, in the aqueous slug of a two-phase flow was simulated [94].

4.3.2
Diazotization + Arylazodehydrogenation – Diazonium Salt Formation + Diazo Coupling (Azo Chemistry)

Peer-reviewed journals: [4, 104] [107]; proceedings: [55]; sections in reviews: [14, 83, 89, 90, 105]; trade press: [106].

4.3.2.1 Drivers for Performing Azo Chemistry in Micro Reactors

Diazo coupling was used to demonstrate the feasibility of organic multi-step synthesis on a chip [4]. This involved showing the safe quenching of hazardous intermediates such as the diazonium salts needed as precursor for that synthesis. Especially, this involves the *in situ* generation and consumption of such species.

From an analytical point of view, azo coupling is easy to monitor as visual inspection of the color formation by reaction in transparent reactors can be performed.

As azo compounds are intensely colored, they find commercial use as pigments [55]. The motivation to use micro reactors here stems from the benefits of developing a continuous process – production of flexible quantities and eliminating the need for refining such as milling the pigment in the end of the production line, as typically given for batch processes. The new micro-reactor process with intensified mixing characteristics is also expected to provide pigments with improved features, as usually specified by color strength, transparency, brightness and purity.

Coupled with the latter investigations, it was shown that a diazo suspension can be fed through a micro device without leading to plugging [55]. This broadens the scope of micro processing applicability towards azo pigment formation.

4.3.2.2 Beneficial Micro Reactor Properties for Azo Chemistry

When considering the thermal management of very exothermic reactions and the avoidance of corresponding explosions, it is the minute volume of micro reactors that is addressed [4].

Micro mixers provide completely different and often improved mixing characteristics compared with conventional batch stirring. For this reason, it is very likely that the material properties of azo pigments produced by continuous micro chemical processing can be substantially changed [55]. By continuous processing using a simple micro mixer–tube reactor configuration, large throughputs can be realized, fulfilling the demands of pilot-scale processing.

4.3.2.3 Azo Chemistry Investigated in Micro Reactors

Organic synthesis 35 [OS 35]: Formation of 1-(phenylazo)-2-naphthol (Sudan I) [4]

Organic synthesis 36 [OS 36]: Formation of 1-(2-methylphenylazo)-2-naphthol [4]

Organic synthesis 37 [OS 37]: Formation of 1-(3-methylphenylazo)-2-naphthol [4]

In addition to pigment applications, azo dyes are also used in medicinal chemistry. Sulfanilamide, the metabolite of an azo dye, has antibacterial properties [4]. The Sudan series of azo dyes, which have also been synthesized in micro reactors, are commonly used as microbial stains. The thermally unstable nature of the diazonium precursors and reported explosions often demand extensive safety procedures when going to an industrial scale, which limits the commercial applicability of the azo reaction.

Organic synthesis 38 [OS 38]: Formation of an azobenzene derivative from N,N-dimethylaniline with 4-nitrobenzenediazonium tetrafluoroborate

The synthesis of N,N-dimethylaniline with 4-nitrobenzenediazonium tetrafluoroborate yielded the corresponding azobenzene derivative [107] (see also [14]).

Organic synthesis 39 [OS 39]: Formation of azobenzene derivatives for pigment production

Two commercial azo pigments, one yellow and one red, were prepared following the scheme given above; the nature of the substituents was not disclosed [55].

4.3.2.4 Experimental Protocols

[P 26] A chip micro reactor with two consecutive micro-mixing tees, each attached to a serpentine channel section, was used [4]. In the first mixing tee, a solution of the following composition was fed: 0.1 ml of aniline, 0.35 ml of concentrated hydrochloric acid, 2 ml of water and 12 ml of N,N-dimethylformamide DMF. The second mixing tee was fed with a solution comprising 0.75 g of sodium nitrite in 4 ml of water and 20 ml DMF. Both solutions were fed at a flow rate of 3.5 $\mu l \, min^{-1}$. Thereafter, the mixed flow passed a serpentine channel section which acted as a delay loop. The length of this loop was set to guarantee at least complete diffusion mixing. Thereafter, a third solution was fed by a second mixing tee into the reacted solution, comprising 0.15 g of β-naphthol, 9 ml of 10% sodium hydroxide solution and 20 ml of water in 290 ml DMF. This solution was fed at a flow of 7 $\mu l \, min^{-1}$ using a syringe pump. A second serpentine section was used to perform the azo coupling, which could be monitored by optical inspection.

[P 27] Diazotation was performed in batchwise manner [55]. A CPC micro reactor with multi-lamination mixer (lamellae < 100 μm) was embedded in a housing

and equipped with pumps, heating bath and vessels. No further details on the reactor were given. Into this micro reactor, the diazonium compound and the coupling agent were charged. The acid generated was buffered by an additional feed of 2–6% NaOH solution or by using internal buffer. No further details on the protocol were given. The pigment suspension formed was conventionally separated from the reaction solution, dried and milled.

4.3.2.5 Typical Results
Conversion/selectivity/yield

[OS 35] [R 4b] [P 26] A conversion of 52% was found [4].

[OS 36] [R 4b] [P 26] A conversion of 23% was found [4]. This result and that below, compared with the much higher value given above, show that the reaction is not optimized for azo chemistry. The large degrees in conversion do not stem from the different reactants applied solely.

[OS 37] [R 4b] [P 26] A conversion of 9% was found [4].

[OS 38] [reactor and protocol given in [107]] By reaction of N,N-dimethylaniline with 4-nitrobenzenediazonium tetrafluoroborate, the corresponding azobenzene derivative is obtained at a conversion of 37% using methanol (protic solvent) or acetonitrile (aprotic solvent) under electroosmotic flow conditions [107] (see also [14]).

Optical properties of pigments

[OS 39] [R 25] [P 27] The two azo pigments made in the micro reactor had a color strength of 119 and 139%, a 5 and 6 times glossier brightness and a 5 and 6 steps higher transparency compared with the same products made by batch-processing (Table 4.3) [55]. The beneficial product features were due to the formation of smaller particles with a narrower size distribution (micro reactor, $D_{50} = 250$ nm, $s = 1.5$; batch, $D_{50} = 600$ nm, $s = 2.0$).

Table 4.3 Color properties of pigments synthesized in two different micro reactors compared with the batch standard [55].

	Micro reactor pigment 1	*Micro reactor pigment 2*
Color strength (%)	119	139
Brightness	5 steps glossier	6 steps glossier
Transparency	5 steps more transparent	6 steps more transparent

The same features were found for pilot-size micro-reactor operation (Figure 4.59). Brightness and transparency were the same and color strength could be increased to 149% [55]. The mean particle size was even set to a lower value compared with the laboratory-scale processing (micro reactor, $D_{50} = 90$ nm, $s = 1.5$; batch, $D_{50} = 600$ nm, $s = 2.0$).

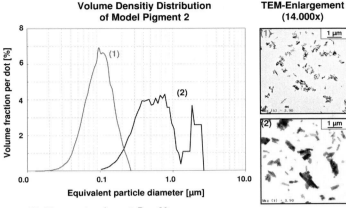

(1): Microreactor pigment, D_{50} = 90 nm

(2): Standard batch pigment, D_{50} = 598 nm

Figure 4.59 Properties of a model pigment (termed "2" in [55]) obtained in the pilot-plant micro reactor compared with the batch standard. Left: volume density distribution. Right: TEM enlargement [55].

Pilot-scale processing
[OS 39] [R 25] [P 27] Using a flow of 500 ml h^{-1}, a production of commercial azo pigments in the range of 10 t a^{-1} was estimated when accounting for 8000 h annual running time [55]. The increase in throughput compared to laboratory-scale micro reactors used previously (1 t a^{-1}; 20–80 ml h^{-1}) was achieved by both internal and external numbering-up accompanied by a slight scale-up of internal dimensions. More reaction plates were operated in parallel within one device; in addition, three such devices were operated in parallel. Furthermore, slightly larger micro channels were used, still ensuring laminar flow.

Fouling during suspension processing
[OS 39] [R 25] [P 27] A 24 h run of a pilot-scale micro reactor for azo pigment production was performed using a diazo suspension [55]. At the end of this period, the pressure loss of the micro reactor increased exponentially. Special means were developed to prevent clogging and unstable operation. By partial removal of the deposits, the pressure loss was brought back to normal.

4.3.3
Alkoxyborondehydro Substitution – Arylboron Formation

Proceedings: [48]; patents: [108]; sections in reviews: [79, 80, 90, 109].

4.3.3.1 Drivers for Performing Arylboron Formation in Micro Reactors
The main driver was to develop a laboratory-scale micro-channel process and transfer it to the pilot-scale, aiming at industrial fine-chemical production [48, 108]. This included fast mixing, efficient heat transfer in context with a fast exothermic reaction, prevention of fouling and scale-/numbering-up considerations. By this means, an industrial semi-batch process was transferred to continuous processing.

4.3.3.2 Beneficial Micro Reactor Properties for Arylboron Formation

The above motivation refers to the general advantages of improving mass and heat transfer at reduced residence times. This particularly refers to micro mixing features. Also, the more facile scalability of micro-channel processing is targeted when going from laboratory-scale to pilot-scale processing.

4.3.3.3 Arylboron Formation Investigated in Micro Reactors
Organic synthesis 40 [OS 40]: Formation of phenyl esters of boronic acid

The formation of phenyl esters from boronic acid and phenylmagnesium bromide in THF is a fast liquid reaction involving contacting of two reactants, R_1 (boronic acid) and R_2 (phenylmagnesium bromide), dissolved in the same solvent to yield a liquid mixture [48, 108]. This mixture is post-processed by a fast hydrolysis. This step was performed conventionally in a batch mode.

The two reactants R_1 and R_2 form an intermediate I_1 which reacts to a second intermediate I_2 via two reaction pathways, one needing the presence of R_2 [48, 108]. A number of additional reactions decrease the selectivity and yield. First, the reactants R_1 and R_2 are labile to moisture. Hydrolysis results in the formation of the side product reactants S_1 and S_2. By more complex pathways, three other known side products, S_3, S_4 and S_5, are generated. Second, a consecutive intermediate product I_3 is formed when I_2 reacts with R_1 (equivalent to R_2 reacting with two molecules of R_1). By addition of R_2 or by another process route, a further consecutive intermediate product I_4 is yielded. In the post-processing step, I_4 is converted to the final consecutive product C_1. A low content of this species is, in particular, important for a facilitated product separation. C_1 and all other side products remain in solution after finishing the process and cannot be separated simply by filtration.

The intermediates I_1 and I_3, however, precipitate in the course of the reaction [48, 108]. They are dissolved again by post-processing to the product P_1 and the consecutive product C_1.

4.3.3.4 Experimental Protocols

[P 28] A glass interdigital micro mixer was connected to PTFE tubes with an inner diameter of 2 mm which was immersed in an ice–water bath of appropriate temperature (laboratory set-up) [48, 108]. Operation was performed for 5–40 s at temperatures ranging from –12 to 50 °C.

Typically, 2 mol l^{-1} solutions of the reagents boronic acid and phenylmagnesium bromide were applied for the former semi-batch industrial process [48, 108]. However, operation with such highly concentrated solutions in a triangular interdigital mixer–tubular reactor resulted in intense fouling and even clogging. Therefore, it was attempted to dilute the solution to a concentration which can be properly handled. It turned out that a 0.5 mol l^{-1} reactant solution did fulfil this criterion. Thereby, it was possible to operate the set-up for about 15 min. Thereafter, fouling was clearly visible in the interdigital mixer by white stripes composed of precipitates along the interfaces between the liquid lamellae. A 15 min processing time turned out to be long enough to gather reliable data on process yield.

[P 29] A set-up comprising a steel caterpillar mini mixer and four steel tubes attached was used, being dipped into a cylinder completely filled with a cooling medium (scale-up set-up) [48, 108]. By means of a 5/2-way valve, it was possible to switch the reactants to either of the tubes acting as delay loops, differing in inner diameter and hence residence time.

Compared with the simple and flexible laboratory set-up, the scale-up set-up is said to be more robust and user-friendly [48, 108]. By selecting the proper delay tube via the 5/2-way valve, the residence time was varied 5, 10, 26 and 120 s at 10 and 20 °C, and 1, 5, 10 and 26 s at 30 and 40 °C. The stainless-steel tubes used had inner diameters of 0.7, 3.8, 4.8, 9.3 and 21.2 mm and a length of 1000 mm (four tubes were permanently mounted and one was exchanged, if needed).

4.3.3.5 Typical Results

Conversion/selectivity/yield

[OS 40] [R 20] [P 28] The best yield obtained in the interdigital micro-mixer laboratory-scale set-up was 83% (22 °C; 1000 ml h^{-1}; 8 s) [48]. Compared with the performance of the industrial production process (65%, batch), this is an improvement of nearly 20%. Most experiments done at widely different residence time and reaction temperature did not differ from this best yield by more than 5–10%.

Accordingly, for all these parameter sets, favorable mass and heat transfer could be achieved, which was explained as being mainly due to having the same flow patterns for all these sets and having large surface-to-volume ratios.

[OS 40] [R 21] [P 29] Using a caterpillar mini mixer inserted in the scale-up set-up, a maximum yield of 89% was determined [48]. This is even slightly higher than for the laboratory-scale setup (83%, see above).

[OS 40] [R 20] [P 28] The above-mentioned clogging in the interdigital micro mixer [R 20] naturally was an exclusion criterion for being applied also to pilot-scale and production operation [48]. Hence a mixer with larger internals, the caterpillar mini mixer, was used, following a similar choice already made for another industrial process (see Section 4.9.3.5). By this, fouling and clogging could be largely avoided.

Residence time

[OS 40] [R 20] [P 28] Using an interdigital micro mixer, a nearly constant yield could be obtained for the full range of residence times at 22 and 50 °C (5–40 s) [48]. At lower temperature (–12 °C), a maximum yield at a flow rate of 1000 ml h^{-1} (5 s) is obtained. Still higher flow rates lead to a too low residence time (Figure 4.60).

[OS 40] [R 21] [P 29] Using a caterpillar mini mixer, the yield differed strongly on changing the flow rate (Figure 4.61) [48]. This was explained by the much stronger dependence of micro mixing of this split–recombine mixer as compared with a multi-lamination mixer (see above). A CFD simulation is also given, showing non-ideal flow patterns in the caterpillar mixer, deviating from ideal multi-lamellae arrangements.

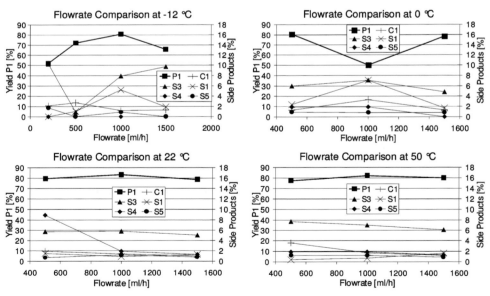

Figure 4.60 Yield of product and side/consecutive products as a function of flow rate, respectively residence time for the laboratory set-up with an interdigital micro mixer [48].

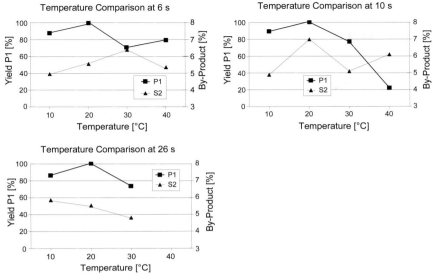

Figure 4.61 Yield of product and side/consecutive products as a function of temperature for the scale-up set-up with a caterpillar steel mini mixer [48].

This mixing-based explanation was confirmed by the finding that by changing the residence time at a constant flow rate, i.e. changing it by prolongation of delay-loop tubing, the yield was fairly constant [48]. Here, the same mixing quality is assured when the varying residence time.

Temperature
[OS 40] [R 20] [P 28] Similarly to residence time variation, fairly constant behavior was found on varying the temperature [48]. There is a slight increase in yield with a rise of temperature from –12 to 0 °C, then approaching a constant value. The fact that the yield does not decrease at high temperature is indicative of similar activation energies of the main and side reactions. It also confirms that the residence time was properly set, as a too long processing would enhance the contents of consecutive products.

[OS 40] [R 21] [P 29] A much larger dependence of yield on reaction temperature was observed for the caterpillar mixer–scale-up set-up than for the interdigital mixer–laboratory set-up [48]. At temperatures above ambient, significant decreases in yield of the order of several tens percent were determined.

Side and consecutive products
[OS 40] [R 20] [R 21] [P 28] Various side products were formed during the reactions, each typically not exceeding 5% [48]. The formation of side products decreases with increasing flow rate, i.e. providing a longer residence time than kinetically needed decreases selectivity (Table 4.4).

Table 4.4 Yields obtained for the target product and side/consecutive products using a laboratory-scale set-up with caterpillar steel micro mixer [48].

Temperature	−12 °C				0 °C			30 °C				40 °C		
Flow rate (ml h^{-1})	200	500	1000	1500	500	1000	1500	500	1000	1500	500	1000	1500	
P_1	51.9	72.0	80.7	65.5	80.6	50.1	78.4	79.6	83.2	78.7	77.4	82.1	80.1	
C_1	2.1	2.7	1.2	1.3	1.3	3.3	1.3	1.9	1.4	1.0	3.6	1.7	1.2	
S_3	10.2	0.9	7.9	9.7	6.0	7.1	4.8	5.7	5.8	5.0	7.7	7.0	6.1	
S_1	0.0	1.0	5.2	1.9	2.4	7.1	1.7	1.5	0.9	1.3	0.4	0.7	1.6	
S_4	0.0	0.0	0.0	0.0	1.9	1.9	0.0	8.9	1.9	1.3	1.9	1.9	1.3	
S_5	0.0	0.0	0.9	0.0	0.9	0.8	0.8	0.7	1.2	0.8	1.2	1.2	0.9	

[OS 40] [R 21] [P 28] The amount of the various side/consecutive products for the scale-up set-up was of the same order (Table 4.5) as for the laboratory-scale investigations [48].

Table 4.5 Yields obtained for the target product and side/consecutive products using a laboratory-scale set-up with caterpillar steel micro mixer [48].

Flow rate (ml h^{-1})	2000	10 000
Mass of isolated crystals (g)	26.29	43.83
Yield (%)	65	81
P_1 content (%)	98.9	98.2
C_1 content (%)	~1.0	~1.7
Total content of S_1, S_3, S_4, S_5 (%)	< 0.1	< 0.1

Inner diameter

[OS 40] [R 21] [P 29] With the exception of one experiment, the choice of internal dimensions, i.e. the inner diameter of tubing attached to the micro mixer, did not greatly affect the yield [48]. Accordingly, heat transfer was probably not a major issue for micro-channel processing. Tubes as large as 4.8 mm could be employed successfully, provided that the temperature was not set too high. Processing in tubes of 9.3 mm inner diameter, however, resulted in clogging, probably owing to the extremely slow velocities applied, hence providing a high degree of back-mixing.

Preparative isolation

[OS 40] [R 21] [P 29] To corroborate HPLC results, obtained prior to yield analysis, preparative isolation of the product was achieved by precipitation and filtration of the crystals [48]. A crystal mass up to about 44 g was so isolated at a yield of up to 81%. The product was of high purity (> 98% of the total crystal mass).

4.4
Aliphatic Electrophilic Substitution

4.4.1
Keto–Enol Tautomerism – Isomerization of Allyl Alcohols

Peer-reviewed journals: [110, 111]; proceedings: [112, 113]; sections in reviews: [90, 99, 114, 115].

4.4.1.1 Drivers for Performing Isomerization of Allyl Alcohols in Micro Reactors

Allyl alcohol isomerization is typically conducted as a single-phase reaction, needing efforts for separation of the catalyst [110, 113]. One driver was to exploit a catalyzed liquid/liquid route with aqueous (catalytic) and organic phases as commonly employed in the chemical industry.

To establish a multi-phase process for isomerization requires high mass transfer between the phases to be conducted in a kinetically controlled manner [110, 112]. Despite affecting conversion, mass transfer is known to impact enantio- and regio-selectivity for many reactions [110]. For this reason, also conventional micro-titration apparatus, typically employed in combinatorial chemistry of single-phase reactions, often suffers from insufficient mixing when dealing with multi-phases [110].

Furthermore, allyl alcohol isomerization is of interest for fine-chemical applications [112]. Owing to the large number of available allyl alcohol derivatives, by varying the two substituents, screening approaches are needed to exploit and investigate the full range of possible compounds.

4.4.1.2 Beneficial Micro Reactor Properties for Isomerization of Allyl Alcohols

Micro reactors are capable of generating large specific interfaces between two liquids, thereby enhancing mass transfer. For example, interdigital micro mixers are known to produce fine emulsions [116]. By using small samples, they also allow efficient screening. Besides parallel approaches, serial screening is increasingly recognized as an efficient way particularly for multi-phase micro-channel processing [110]. Detailed investigations concerning test-throughput frequency, sample consumption and significance of screening experiments have been reported, e.g. [111, 117].

4.4.1.3 Isomerization of Allyl Alcohols Investigated in Micro Reactors
Organic synthesis 41 [OS 41]: Isomerization of diverse substituted allyl alcohols

The substrates and products of this reaction remain mainly in the organic layer, whereas the catalyst is soluble only in the water layer

(a) In one major study, 1-hexen-3-ol was used as substrate and reacted with diverse catalysts (Table 4.6). A restricted library consisting of eight catalytic systems from four transition-metal precursors (Rh, Ru, Pd, Ni) and four sulfonated phosphane or disphosphane ligands was described [110].

Table 4.6 Yield of 1-hexen-3-ol isomerization products when using different catalysts [110].

Catalyst	Ligand : metal ratio	Product	Conversion (%)
RhCl$_3$/TPPTS	4.6 : 1		53
Rh$_2$SO$_4$/TPPTS	4.1 : 1		34
[Rh(cod)Cl]$_2$/DPPBTS	1.1 : 1		36
[Rh(cod)Cl]$_2$/BDPPTS	1.1 : 1		1.5
[Rh(cod)Cl]$_2$/CBDTS	1.3 : 1		1
RuCl$_3$/TPPS	4.0 : 1		61
PdCl$_2$/DPPBTS	2.6 : 1		3.5
Ni(cod)$_2$/TPPS	4.0 : 1		9
			3

(b) In a subsequent study, the substrate was also varied [112]. Five C$_4$–C$_8$ alcohols were employed.

4.4.1.4 Experimental Protocols

[P 30] The reaction was carried out at 40–80 °C using a catalyst in water/*n*-heptane. Organic pulses containing substrates (e.g. from a 1000 μl loop) were injected simultaneously with pulses containing catalyst solutions (e.g. from a 200 μl loop) and contacted in a slit-type interdigital micro mixer (Figure 4.62) [110]. Flow rates of the organic and aqueous phases were 60 and 300 ml h^{-1}, respectively, amounting to a residence time of 100 s. The mixer is part of a liquid/liquid HTS reaction system, comprising a reaction glass tube (4 mm inner diameter; 800 mm length),

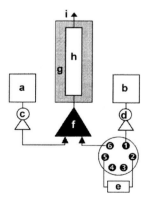

Figure 4.62 Experimental set-up for liquid/liquid experiments: (a) reservoir for the substrate in *n*-heptane; (b) water reservoir; (c, d) high-pressure liquid pumps; (e) HPLC injection valve with sample loop for catalyst injection; (f) micro mixer; (g) heating mantle; (h) tubular stainless-steel reactor; (i) outlet to analytics [112].

injection valves and high-pressure liquid pumps. By mixing, emulsification of the organic phase in water was achieved, creating a reacting segment (pulse) moving downstream the reactor tube. Such *n*-heptane-in-water emulsions were stable for more than 10 min.

4.4.1.5 Typical Results
Variation of transition metal used in the catalyst complex
[OS 41a] [R 19] [P 30] Palladium complexes were inactive with monophosphane ligands and only weakly active with diphosphane ligands (conversion 3.5%) [110]. Nickel complexes led to 1,3-transposition of the hydroxyl group. Ruthenium and rhodium complexes gave comparable conversions. Ruthenium had the highest activity of 61% conversion, but suffered from side reactions decreasing selectivity. Of various rhodium(I) complexes differing in the ligands, the ligand tris(*m*-sulfophenyl)phosphane gives the best result (conversion 53%). This catalyst also had activity for several other substrates than 1-hexen-3-ol.

Variation of alcohol substrate – benchmarking to conventional apparatus
[OS 41a] [R 19] [P 30] Ten different substrates (C_4–C_8 alcohols) were reacted with rhodium(I)–tris(*m*-sulfophenyl)phosphane [110]. The variance in conversions (ranging from about 1–62%) determined was explained by differences in the solubility of the alcohols in the aqueous catalytic layer and by their different intrinsic activities. Chain length and steric/electronic effects of the different alcohols affected their reactivity in a well-known pattern (Figure 4.63). The results obtained correspond to the conversions achieved in a well-mixed traditional batch reactor (40 cm^3). They further agreed with data from mono-phasic processing.

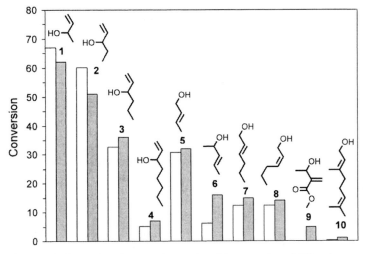

Figure 4.63 Comparison between the results of the screening of different substrates against one catalyst in a batch reactor (white columns) and micro reactor set-up (gray columns) [110].

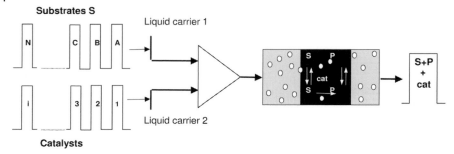

Figure 4.64 Schematic of a mixer–tube reactor set-up used for high-throughput sequential screening of (i) catalysts and (N) substrates. The substrate (S) is thus treated to form the product (P) [110].

Benchmarking to long-time small vial and short-time mini batch processing

[OS 41a] [R 19] [P 30] A study was undertaken to compare extended (1 h) processing in small vials (2 cm³) with short-time (100 s) continuous micro reactor and mini-batch (10 cm³) operation for 10 different substrates (C_4–C_8 alcohols) which were reacted with rhodium(I)–tris(*m*-sulfophenyl)phosphane [111]. The vials were either directly filled with the two phases yielding a bilayered fluid system with small specific interfaces or by interdigital micro mixer action yielding an emulsion with large specific interfaces.

Concerning the small-vial tests, it was evident that by introducing emulsions with small droplets, the conversion was generally increased [111]. An exception to this trend was a highly soluble substrate for which mass transfer was not an issue. The conversions found for the small vials, irrespective of whether a micro mixer was used or not, were generally high, reflecting the long reaction time provided. As a consequence, it was not possible to rank substrate reactivity, i.e. to perform kinetic studies. For this reason, a tool was needed providing shorter residence times such as continuous micro flow devices.

A micro mixer–tube reactor set-up, in which pulses of substrates and catalysts were injected (Figure 4.64), fulfilled that criterion [111]. By this means, a ranking of substrate reactivity was possible. It was shown that the different reactivities of the 10 substrates found were due to their varying solubility in the aqueous phase where the catalyst is provided.

By comparison with data from a vigorously stirred mini-batch reactor (10 cm³), it could be shown that this micro-reactor operation gave intrinsic kinetic data [111]. This is demonstrated, e.g., by the lower conversion of the branched iso-alcohols respective to the normal-chain ones.

Parity plot – flow modeling

[OS 41b] [R 19] [P 30] A Parity plot (Figure 4.65) for five different substrates (C_4–C_8 alcohols) was derived, showing a comparison of experimental data with a plug and laminar flow model [112]. In the plug-flow case, much higher conversions were theoretically expected than actually measured. Laminar-flow modeling describes the results much better and is in line with visual inspections of the liquid/liquid flow.

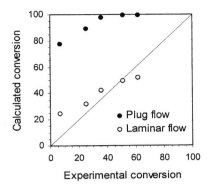

Figure 4.65 Parity plot comparing the conversion of a batch and laminar flow model for five different substrates made by allyl alcohol isomerization. For calculation, the same rate law was used for all substrates. The increase in conversion is due to increased substrate solubility [112].

4.5
Aromatic Nucleophilic Substitution

4.5.1
Aminodehalogenation – Alkylaminodefluorination in the Ciproflaxin® Multi-step Synthesis

Peer-reviewed journals: [83, 118].

4.5.1.1 Drivers for Performing Alkylaminodefluorinations in Micro Reactors
The synthesis of Ciproflaxin® was one among several syntheses being performed in contract research by a micro reactor developer for the pharmaceutical industry [83]. In this multi-step synthesis, alkylaminodefluorinations were an essential part of the chemistry.

The aims in performing the syntheses of functional molecules were manifold, but may be summarized by the term 'process intensification'. Syntheses should be carried out more rapidly and more reliably with more flexible quantities of better quality at higher yield [83]. Finally, improved economy was claimed.

The selection of functional molecules was oriented at targeting blockbuster syntheses for pharmaceutical companies [83]. In this way, an already marketed blockbuster synthesis was applied to micro flow processing, with the aim of finally achieving production in micro reactors for the same synthesis.

4.5.1.2 Beneficial Micro Reactor Properties for Alkylaminodefluorinations
The above-mentioned targets refer to general advantages of micro reactors [42, 80, 100, 114, 119]. Enhanced transfer and better controlled residence time improve conversion and selectivity. The tools have small internal volumes, allowing one to generate flexibly a multitude of samples in serial or parallel fashion. Synthesis can be combined with a multi-step procedure. The economy of micro-reactor processes has not really been analyzed so far; however, it is clear that as laboratory tools they allow in a number of cases technical expenditure, personnel and costs to be reduced.

4.5.1.3 Alkylaminodefluorinations Investigated in Micro Reactors
Organic synthesis 42 [OS 42]: Ciproflaxin® synthesis

This reaction scheme involves two substitutions of fluorine moieties at the aromatic ring by amines, yielding the final product for pharmaceutical applications [83, 118]. In total, five synthesis steps are actually required to obtain the target molecule.

4.5.1.3 Experimental Protocols
[P 31] No protocol is given in [83, 118].

4.5.1.4 Typical Results
Demonstration of feasibility
[OS 42] [R 25] [P 31] The multi-step synthesis of the pharmaceutical agent Ciproflaxin® was carried out in a CPC micro reactor [83, 118].

4.5.2
Photocyanation of Aromatic Hydrocarbons

Peer-reviewed journals: [29].

4.5.2.1 Drivers for Performing the Photocyanation of Aromatic Hydrocarbons
Photocyanations rely on photoinduced electron transfer [29]. This was demonstrated by monitoring cyanation yields as a function of the droplet size for oil-in-water emulsions. Hence increase in interfacial area is one driver for micro-channel processing. Typically, fluid systems with large specific interfacial areas tend to be difficult to separate and solutions for more facile separation are desired.

4.5.2.2 Beneficial Micro Reactor Properties for the Photocyanation of Aromatic Hydrocarbons
In micro reactors, large specific interfaces between immiscible phases can in general be achieved. In special laminating contactors, a stable two-phase flow with continuous phases can be obtained, while separation of the phases is facilitated when the two separate streams leave the micro reactor as there is no dispersion.

4.5.2.3 Photocyanation of Aromatic Hydrocarbons Investigated in Micro Reactors
Organic synthesis 43 [OS 43]: Photocyanation of pyrene

The pyrene molecule is transferred by irradiation to its cation radical [29]. This reacts at the oil/water interface by nucleophilic attack from the cyanide ion. Typically, the cyanated product remains in the organic phase.

4.5.2.4 Experimental Protocols
[P 32] Pyrene (20 mM), 1,4-dicyanobenzene (40 mM) and sodium cyanide (1 M) were reacted in propylene carbonate and water. A 100 μl solution of pyrene (20 mM), 1.4-dicyanobenzene (40 mM) in propylene carbonate and a 100 μl solution of sodium cyanide (1 M) in water were fed by programmable dual-syringe pumps via fused-silica capillary tubes into a micro-channel chip [29]. Both solutions were fed with equal flow velocity. A 300 W high-pressure mercury lamp was used as light source. After passing an optical filter made of a $CuSO_4$ solution, the whole chip was irradiated after formation of a stable oil/water interface inside. The oil phase was collected at the exit.

4.5.2.5 Typical Results
Conversion/Yield/Selectivity
[OS 43] [R 14] [P 32] A yield of 28% of the cyanated pyrene was obtained in a first run by two-liquid layer (oil/water) flow, using a residence time of 210 s and room-temperature processing (0.2 μl min^{-1}, 300 W, > 300 nm wavelength) [29]. Using a three-liquid layer (water/oil/water) flow resulted in a yield of 73%. The content of non-reacted pyrene was 8%. Thus, to close the balance, about 20% by-products had to be formed, i.e. conversion was high. The nature of these by-products was not identified (Figure 4.66).

Residence time
[OS 43] [R 14] [P 32] Changing the residence time from 70 to 210 s by varying the flow rates increases the yield of pyrene photocyanation from 7 to 28% [29].

Three-liquid layer processing – thinning of lamellae
[OS 43] [R 14] [P 32] Using a three-liquid layer (water/oil/water) flow instead of a two-liquid layer flow at constant channel dimensions decreases the liquid lamellae width and doubles the absolute value of the organic/aqueous interface. As a consequence, mass transport is facilitated compared with the two-flow configuration. Hence it was found that a much higher yield was obtained for the three-liquid layer flow when performing experiments of both flow configurations under the same experimental conditions (210 s, 0.2 μl min^{-1}, room temperature, 300 W, > 300 nm

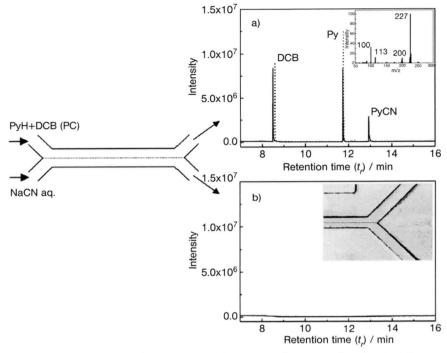

Figure 4.66 Photocyanation of pyrene (PyH) to the corresponding cyanated product (PyCN) in dicyanobenzene (DCB). Left: schematic of the flow inside the micro reactor and a microscope image of the chip micro channels. Right: GC and mass spectra of samples from micro flow processing [29].

wavelength) [29]. For the pyrene photocyanation, a yield of 73% was obtained (two-flow processing: 28%), hence an increase in reaction rate by a factor of about three.

Using a three-liquid layer (oil/water/oil) flow only gave a yield of about 35% (210 s, 0.2 μl min^{-1}, room temperature, 300 W, > 300 nm wavelength) [29]. This is explained by the fact that the specific interface here is not as large as for the water/oil/water flow. This is due to the larger volume of the oil phase compared with the water phase [29].

CN source
[OS 43] [R 14] [P 32] Experiments run without NaCN did not yield the cyanated photoproduct [29]. Therefore, NaCN and not 1,4-dicyanobenzene has to be considered as the source of the CN anion that is incorporated into the pyrene moiety.

Product extraction
[OS 43] [R 14] [P 32] The cyanated pyrene product leaves the reaction channel completely via the organic phase [29]. Since the reaction occurs with aqueous dissolved NaCN, one can draw the conclusion that complete extraction of the product was achieved.

4.6
Aromatic Substitution by Metal Catalysis or Other Complex Mechanisms

4.6.1
Aryldehalogenation – Suzuki Coupling using Pd(0)–Phosphine Catalysis

Peer-reviewed journals: [7]; proceedings: [6, 85]; sections in reviews: [14, 42, 83, 89, 90, 105].

4.6.1.1 Drivers for Performing Suzuki Couplings in Micro Reactors
One driver for performing Suzuki couplings in micro reactors refers to enhancement of yield by increased mass transfer and especially by in-line generation of hydroxide ions that promote reaction [6]. By this means, modified Suzuki couplings are possible that no longer need addition of a base. A second driver refers to minimize the loss of colloidal catalyst when conducting the reaction heterogeneously. This contaminates the product, needing further separation. Achievement of this goal would also be an advantage over the homogeneously catalyzed reaction, used predominantly so far.

4.6.1.2 Beneficial Micro Reactor Properties for Suzuki Couplings
In addition to the general improvement of transfer in micro reactors, there is evidence that the voltage of electroosmotic flow (for EOF see [14]) in combination with the large internal surface area in glass chips can induce hydroxide ion formation [6]. Concerning catalyst loss, there is no obvious direct correlation; rather, micro reactors can act as mini fixed beds fixing heterogeneous catalyst particles.

4.6.1.3 Suzuki Couplings Investigated in Micro Reactors
Organic synthesis 44 [OS 44]: C–C bond formation of 4-bromobenzonitrile and phenylboronic acid

A modified Suzuki coupling by Styring (see original citation in [6]) was applied in the micro reactor. Such Suzuki couplings usually have high selectivity for formation of carbon-to-carbon bonds; however, they suffer from the problem of catalyst degradation and loss into the product solution.

The Suzuki–Miyaura synthesis is one of the most commonly used methods for the formation of carbon-to-carbon bonds [7]. As a palladium catalyst typically tetrakis(triphenylphosphine)palladium(0) has been used, giving yields of 44–78%. Recently, Suzuki coupling between aryl halides and phenylboronic acid with efficient catalysis by palladacycles was reported to give yields of 83%.

Organic synthesis 45 [OS 45]: Suzuki coupling of 3-bromobenzaldehyde and 4-fluorophenylboronic acid

3-Bromobenzaldehyde and 4-fluorophenylboronic acid were coupled in DMF using tetrakis(triphenylphosphine)palladium(0) as catalyst [85] (see a more extended description in [42]).

The mechanism of the Suzuki coupling is known, although not in all details [6, 7]. The first step is an *oxidative addition* of the organohalide to Pd(0) (see a schematic of the mechanism below in Section 4.6.1.5, Addition of base). Pd(II) is generated and the organohalide splits into two moieties, serving as ligands for the central ion. The second step is a *transmetallation* with the organoboron compound. The aromatic moiety of the latter substitutes the halide group in the Pd(II) complex; boronic acid leaves as this step typically is assisted by base addition, usually as hydroxide ion. The exact nature of this hydroxide action is still under discussion. Not all proposed intermediates have been identified and different mechanisms have been proposed by various authors. The third step is a *reductive elimination* of Pd(II) to Pd(0); the starting catalytic material is obtained again. This is accompanied by cross coupling of the aromatic moieties, undergoing C–C bond formation at the position of the former halide and boron functionalities.

4.6.1.4 Experimental Protocols

[P 33] The catalyst bed was manually positioned in the micro channel (300 μm wide; 115 μm deep) at room temperature using a 10% (v/v) solution of formamide and potassium silicate [6, 7]. Micro reactor bottom and top plates were thermally bonded thereafter. Then, 75% THF (aqueous) solutions of 4-bromobenzonitrile and phenylboronic acid having equimolar concentrations were placed in the two reservoirs of a micro-mixing tee chip. In the collection reservoir, 30 μl of the THF solution was placed. Voltages ranging from 100 to 400 V were used, but kept constant only for one reservoir. The other one was switched on and off at 200 V for given time periods.

4.6.1.5 Typical Results

Conversion/selectivity/yield – benchmarking to batch processing

[OS 44] [R 4a] [P 33] A yield of 62% was found for micro flow processing, being about six times higher than for batch processing (10%) under comparable process conditions [6]. Furthermore, a base is required for the batch case, whereas the micro reactor gives a better performance without the need for any base.

[OS 45] [R 17] [no protocol] For the reaction of 3-bromobenzaldehyde and 4-fluorophenylboronic acid, improved conversion was found for a micro reactor, composed

of a mixer and reactor unit, compared with results obtained in a laboratory flask [42, 85]. This was explained by having temperature and concentration profiles within the flask, whereas both parameters are nearly uniform in the micro reactor.

Injection time/plug volume 4-bromobenzonitrile
[OS 44] [R 4a] [P 33] By switching on and off the electroosmotic-driven flow of one reactant, plugs can be inserted in the reaction channel [6]. 4-Bromobenzonitrile plugs (5 s) are inserted into a continuous phenylboronic acid stream. By this means, the effective concentration of the aryl halide in the micro channel can be increased. As expected, a maximum yield is found in this way.

An optimum injection length of 5 s was determined, corresponding to a yield of 14% [7].

Maximum frequency of 4-bromobenzonitrile injections/catalyst saturation
[OS 44] [R 4a] [P 33] 4-Bromobenzonitrile plugs (5 s) were inserted into a continuous phenylboronic acid stream for time intervals of 5, 15, 25, 30, 40 and 55 s (25 min reaction period) [6, 7]. For a 5–20 s delay, catalyst saturation occurs owing to the high injection rates rendering the phenylboronic acid concentration too low. In turn, for delay times of 30–55 s, the effective concentration of 4-bromobenzonitrile becomes too low. At 25 s, optimum performance with a yield of 62% was achieved (Figure 4.67).

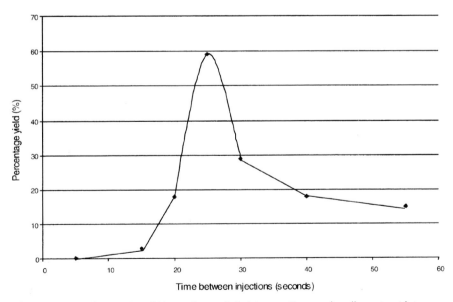

Figure 4.67 Periodic injection of 4-bromobenzonitrile into a continuous phenylboronic acid stream. Product yield is plotted as a function of injection interval [6, 7].

Flow rates/residence times/voltages

[OS 44] [R 4a] [P 33] On changing the voltages for electroosmotic flow and hence flow rates, the yield passes through a maximum (5 s injections with 25 s interval) [6, 7]. The best yields of 49–68% were obtained at 200 V (0.8 µl min^{-1}). At higher voltages (300–400 V), the flow rate (01.0–1.3 µl min^{-1}) is too high, thus residence time too low. At lower flow rate (0.65 µl min^{-1}), no effective transfer is observed.

Temperature

[OS 44] [R 4a] [P 33] At room temperature, no reaction is observed in the presence or absence of a base using a batch synthesis [7]. Using a base, sodium carbonate (aqueous, 0.2 M, 20 ml water), at 75–80 °C results in 10% conversion after 8 h. The best micro reactor conversion was 68%, using no base and operating at room temperature.

Addition of base

[OS 44] [R 4a] [P 33] Conventionally, base addition is needed for Suzuki couplings to activate the boronic acid group (Figure 4.68) [6, 7].

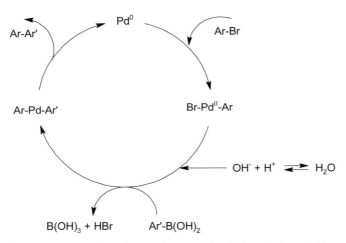

Figure 4.68 Proposed catalytic mechanism with addition of a base [6, 7].

Surprisingly, there is no need for addition of a base when performing Suzuki coupling in a glass chip micro reactor [6, 7]. This was explained as being due to local generation of a base at heterogeneous sites. Under the action of the voltage for electroosmotic micro flow processing, water can be converted to hydroxide ions. The high specific surface areas in the micro reactor probably accelerate this process. Although the corresponding hydroxide concentrations may be low in bulk, they potentially can be large at the catalyst surface where these species enrich. As a consequence, the Suzuki coupling can be performed without base in micro reactors. Testing with the same process parameters does not lead to any conversion in a batch reactor. Here, the addition of a base is essential.

Contamination of catalyst

[OS 44] [R 4a] [P 33] A residual content of the palladium catalyst of only 1.2–1.6 ppb in the outlet solution was found after micro-reactor processing using a heterogeneous catalyst [6, 7]. In turn, for batch processing with homogeneous catalysts, high catalyst impurities in the reacting solution are usually found.

Loss of catalyst activity

[OS 44] [R 4a] [P 33] For an operation of 35 h, performing many reactions, no loss of catalyst activity was detected [6, 7].

4.6.2
Alkinyldehydro Substitution – Sonogashira Coupling using Pd(II)–Phosphine Copper-free Catalysis

Peer-reviewed journals: [120].

4.6.2.1   Drivers for Performing Sonogashira Couplings in Micro Reactors

The performance of the Sonogashira reaction is claimed to be the first example of a homogeneously metal-catalyzed reaction conducted in a micro reactor [120]. Since the reaction involves multi-phase post-processing which is needed for the separation of products and catalysts, continuous recycling technology is of interest for an efficient production process. Micro flow systems with micro mixers are one way to realize such processing.

4.6.2.2   Beneficial Micro Reactor Properties for Sonogashira Couplings

Following the arguments given above, micro mixers are valuable tools for bench-scale continuous processing to fill the recycle loop. For first tests on feasibility, the total volume of the recycle loop has to be kept small, demanding also for small-scale processing units. Conventional devices would probably suffer here from too large internal volumes.

4.6.2.3   Sonogashira Couplings Investigated in Micro Reactors
Organic synthesis 46 [OS 46]: Copper-free Sonogashira reaction

$$Ar\text{-}X \quad + \quad H\text{—}\!\!\equiv\!\!\text{—}R \xrightarrow[\begin{array}{c} \left[Me\text{-}N\!\!\overset{+}{\frown}\!\!N\text{-}Bu \right] PF_6^- \\ \text{i-PrNH}_2 \text{ or Piperidine} \end{array}]{PdCl_2(PPh_3)_2} Ar\text{—}\!\!\equiv\!\!\text{—}R$$

The reaction between iodobenzene and phenylacetylene was carried out in the presence of a catalytic amount of $PdCl_2(PPh_3)_2$ in an ionic liquid [120].

The Sonogashira reaction is a transition metal-catalyzed coupling reaction which is widely used for the preparation of alkyl-, aryl- and diaryl-substituted acetylenes (Table 4.7) [120]. This reaction is a key step in natural product synthesis and is also applied in optical and electronic applications. Sonogashira reactions involve the use of an organic solvent with a stoichiometric portion of a base for capturing the acid produced in the reaction. Using ionic liquids facilitates catalyst separation since the organic products can be separated from the catalyst by extraction with organic solvents.

Table 4.7 Sonogashira coupling of aryl halides with terminal acetylenes in the presence of $PdCl_2(PPh_3)_2$ catalyst [120].

Ar-X	H≡≡—Ph	Ar≡≡—R	Yield (%)
(phenyl iodide)	H≡≡—Ph	Ph	95
MeO—(aryl)—I	H≡≡—Ph	MeO—(aryl)≡≡—Ph	91
(acetylphenyl)—I	H≡≡—Ph	(acetylphenyl)≡≡—Ph	91
O_2N—(aryl)—I	H≡≡—Ph	O_2N—(aryl)≡≡—Ph	97
(styryl bromide, E/Z = 86/14)^c	H≡≡—Ph	(styryl)≡≡—Ph	86 (E/Z = 93/7)^c
(phenyl)—I	H≡≡—C_6H_{13}	(phenyl)≡≡—C_6H_{13}	87
(phenyl)—I	H≡≡—CH_2OH	(phenyl)≡≡—CH_2OH	88
(naphthyl)—I	H≡≡—$C(CH_3)_2OH$	(naphthyl)≡≡—$C(CH_3)_2OH$	90
MeO—(aryl)—I	H≡≡—(cyclohexenyl)	MeO—(aryl)≡≡—(cyclohexenyl)	97
(thienyl)—I	H≡≡—Ph	(thienyl)≡≡—Ph	85

4.6.2.4 Experimental Protocols

[P 34] A mixture of phenylacetylene, iodobenzene and di-*n*-butylamine is fed into one inlet of the slit-shaped interdigital metal/steel micro mixer and the $PdCl_2(PPh_3)_2$ catalyst in the ionic liquid 1-butyl-3-methylimidazolium hexafluorophosphate is fed into the other inlet by means of a syringe pump [120]. The flow rate was set to 0.1 ml h^{-1} and the temperature to 119 °C. After being mixed in the micro mixer, the reactants pass through the flow-through chamber with about a 10 min residence time. From the outlet the product solution was sampled and the product was isolated by extraction with hexane/water (Figure 4.69).

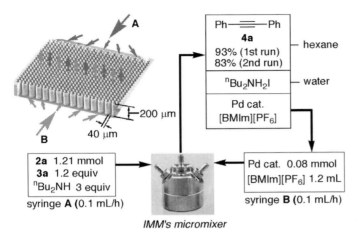

Figure 4.69 Flow scheme of a Sonogashira reaction performed in a micro flow system [120].

4.6.2.5 Typical Results
Conversion/selectivity/yield
[OS 46] [R 18] [P 34] In a recycling experiment, an initial yield of about 93% was found [120]. During the second recycle with recovered catalyst, a yield of 83% was determined.

Benchmarking to batch-recycle processing
[OS 46] [R 18] [P 34] The initial yield of batch-recycle processing was 96% (micro mixer: 93%) and the yield of the second recycle process was 80% (micro mixer: 83%) [120]. Hence the yields were comparable. For the batch recycling, the yields of the third and fourth runs were also determined, being 78 and 63% (no micro mixer data here).

4.6.3
Aryldehalogenation – Kumada–Corriu Reaction using Ni(II)–Phosphine Catalysis

Peer-reviewed journals: [2]; sections in reviews: [14, 90].

4.6.3.1 Drivers for Performing Kumada–Corriu Reactions in Micro Reactors
The Kumada–Corriu reaction is a metal-catalyzed reaction of great importance for the fine-chemical and pharmaceutical industries for making C–C bonds [2]. Owing to its mild conditions and clean conversion, this reaction is one of the most frequently applied processes for such chemistry. The motivation to use this reaction therefore was to demonstrate the feasibility of micro channel processing of quality comparable to standard processing [2]. Having done this, the 'normal' advantages of micro reactors, such as improved heat transfer and capability for numbering-up, were predicted to apply to the Kumada–Corriu reaction.

4.6.3.2 Beneficial Micro Reactor Properties for Kumada–Corriu Reactions

The above-mentioned research targets generally address the good mass and heat transfer properties achieved by micro channel processing, in particular referring to fast mixing and good heat transfer.

4.6.3.3 Kumada–Corriu Reactions Investigated in Micro Reactors
Organic synthesis 47 [OS 47]: Reaction between 4-bromoanisole and phenylmagnesium bromide [2]

In a Kumada–Corriu reaction, an aryl halide is oxidatively coupled with a homogeneous nickel(II)-phosphine catalyst [2]. This species reacts with a Grignard reagent to give biaryl or alkylaryl compounds. Later, palladium-phosphine complexes were also successfully applied. By this means, stereospecific transformations were achieved.

The Kumada–Corriu reaction is characterized by mild conditions and clean conversions [2]. A disadvantage of previous Kumada–Corriu reactions was due to the use of homogeneous catalysts, with more difficult product separation. Recently, an unsymmetrical salen-type nickel(II) complex was synthesized with a phenol functionality that allows this compound to be linked to Merrifield resin polymer beads (see original citation in [2]). By this means, heterogeneously catalyzed Kumada–Corriu reactions have become possible.

4.6.3.4 Experimental Protocols
[P 35] To test the feasibility of microfabricated reactor processing, first experiments were conducted in polypropylene or glass tubing of millimeter internal dimensions with flanged ends, so-called micro-flow reactors [2]. By use of syringe pumps, mixtures of equimolar (e.g. 1.0 M) solutions of 4-bromoanisole and phenylmagnesium bromide in THF were fed at room temperature through the tubing packed with catalyst beads. No reaction in the mixture occurred over 72 h in absence of catalyst. Flow rates of 13.3 μl min^{-1} or 33.3 μl min^{-1} were used, amounting to residence times of a few minutes. The output stream was passed directly into a flask with an aqueous solution containing sodium hydrogencarbonate to quench the reaction. This solution was extracted with diethyl ether.

See also [2] for a description of respective batch syntheses.

4.6.3.5 Typical Results
Conversion/selectivity/yield
[OS 47] [R 2] [P 35] A 60% conversion to the target compound was found, while 20% reacted to biphenyl as side product and 20% remained non-reacted [2].

Kinetics: rate constant/reaction time – benchmarking with batch processing
[OS 47] [R 2] [P 35] Observed rate constants have been determined [2]. For high-flow-rate processing (33.3 µl min^{-1}), an observed rate constant of 0.033 1 s^{-1} was obtained. This amounts to a rate enhancement of 3300-fold compared with the value for batch processing (Table 4.8). For a series of other reactants, less reactive iodoarenes, with Grignard reactants having two differing substituents, rate constants were also determined. The enhancement rates over batch processing are from about 16- up to 79-fold.

Table 4.8 Comparison of rate constants for a series of Kumada–Corriu coupling reactions in flow reactors with corresponding batch reaction [2].

R	X	R'	Flow rate (µl min^{-1})	Rate constant k (10^{-3} s^{-1})	Rate constant k (10^{-3} s^{-1})	Rate enhancement (-fold)
CH$_3$	I	Ph	25.0	2.67	3.50	76.3
OCH$_3$	I	Ph	25.0	1.59	3.50^a	45.4
OCH$_3$	I	CH$_3$	25.0	2.67	3.35	79.3
CH$_3$	I	CH$_3$	25.0	1.05	6.73	15.6
OCH$_3$	Br	Ph	33.3	33.00	1.00	3300.0

a Rate constant estimated, since reaction complete (all of the 4-iodoanisole consumed) before first aliquot removed.

This was explained by having only the contribution of surface reaction in the case of batch processing, whereas micro reactors profit, in addition, from processing inside the pores of the catalyst beads. The penetration of the reaction solution into the pores is achieved here by applying pressure [2]. By this means, the number of available catalyst sites is increased.

Accordingly, Kumada–Corriu batch processes that need hours can be conducted in micro reactors in a few minutes [2]. 116 mg of product (0.62 mmol) was so produced in only 2 h. Numbering-up to 10 reactor units would give gram quantities.

4.7
Free-radical Substitution

4.7.1
Nitrodehydrogenation – Nitration of Aliphatics

Proceedings: [38, 97, 98]; sections in reviews: [89, 90, 100]. See also the information given on the nitration of aromatics in Section 4.3.1.

4.7.1.1 Drivers for Performing Aliphatics Nitrations in Micro Reactors
Most nitrations are highly exothermic and hence release a lot of reaction heat for most experimental protocols [37, 94]. This high exothermicity may even lead to explosions [37, 38]. Nitration agents frequently display acid corrosion [37]. For these reasons, nitrations are generally regarded as hazardous [37] [38].

Due to the heat release, nitrations are often lacking selectivity, i.e. many parallel, consecutive and decomposition processes are known to occur. As a result, product spectra are unusually wide and consequently yields and purity are low [37, 94].

Having high mass transfer, in addition to good heat transfer, may change the product spectra, by increasing conversion to product and decreasing the formation of some of the by-products [94]. Nitrations are well suited for two-phase capillary-flow processing, yielding uniform alternating slugs. In these slugs, internal circulation leads to high mass transfer. The defined setting of residence time can be achieved by establishing two-phase plug flow behavior in so-called capillary-flow reactors.

Multi-phase processing of nitrated aromatics is also described in [31], including both organic and aqueous phases. Side reaction take place in the organic phase, whereas all other reactions occur in the aqueous phase and are limited by organic solubility. For this reason, enabling mass transfer by large interfaces is a key to affect product selectivity.

4.7.1.2 Beneficial Micro Reactor Properties for Aliphatics Nitrations
The small reaction volumes in micro reactors and the large specific surface areas created are seen as beneficial to cope with the problems caused by the release of the large amounts of heat, as mentioned above [37, 38]. Delicate temperature control is expected for micro-reactor operation; isothermal processing is said to be achievable even when high reaction heats are released [94]. Small size should increase process safety and suppress unwanted secondary reactions [37, 38].

In micro reactors, very unique flow patterns can be established using proper contactors/mixers. This holds particularly for two-phase flow by capillary-flow processing yielding uniform alternating slugs (Figure 4.51) [94] or parallel flowing streams (Figure 4.47) [31]. Besides demonstrating advantages in terms of improved mass transfer, the residence time was set in the latter two references uniformly by plug-flow motion. Apart from increasing reaction performance to the kinetic limits, the flow patterns may directly affect selectivity, which depends on interfacial area [31].

4.7.1.3 Aliphatic Nitration Reactions Investigated in Micro Reactors

Nitration reactions are among the basic chemical pathways followed in organic synthesis. They are used for making pharmaceutical products, agricultural and pest control chemicals, pigments, precursors for polyurethane and polyamide production and for explosives [37]. Typically, acidic nitration agents such as nitric acid are employed to insert the nitro function, using dehydrating agents such as sulfuric acid or anhydrous acetic acid. Dinitrogen pentoxide is usually more reactive and may be employed if acid-sensitive substances have to be reacted.

Organic synthesis 48 [OS 48]: Dinitration of *N,N'*-diethylurea

The nitration of *N,N'*-diethylurea gives nitrated products which are precursors for a new energetic plasticizer *N,N'*-dialkyl-*N,N'*-dinitrourea (DNDA). For macroscopic batch processing, this reaction is characterized by a lack of selectivity owing to mononitro derivative formation and thermal decomposition of the dinitro product due to increasing temperature during the course of reaction [37, 38].

Organic synthesis 49 [OS 49]: Mononitration/nitrosation of *N,N'*-diethylthiourea

The derivatized thiourea is a much cheaper starting material than its oxygen analogue [38]. Basically it leads so the same product DNDA as described in [OS 48]. Although nitrating agents are used, nitroso compounds results. This stems from oxidation of the sulfur of the thiourea moiety by the nitrating agent, giving also HNO_2 as nitrosation agent, which, in turn, reacts with the nitrogen moiety. In excess of HNO_3, HNO_2 forms N_2O_4 that is completely ionized to NO^+ and NO_3^-.

Organic synthesis 50 [OS 50]: Nitration of *N,N'*-diethyl-*N*-nitrosourea

By this reaction, the nitroso moiety is replaced by a nitro group, in addition to the nitration of the –NH-ethyl group [38].

Organic synthesis 51 [OS 51]: Nitration of alkanes such as methane and hexane with N$_2$O$_5$

These reactions also have high exothermicity so that safe and reliable processing in technical processes still demands high technical expenditure [37].

4.7.1.4 Experimental Nitration Protocols in Micro Reactors

[P 36] For the synthesis of *N,N'*-diethylurea and *N,N'*-diethylthiourea, the respective ureas dissolved in dichloromethane were exposed to pure liquid or dissolved nitrating agents [38]. As nitrating agents were used concentrated HNO$_3$ (65%, fuming), mixed HNO$_3$/H$_2$SO$_4$ in ratios from 1 : 1 to 6 : 1 or N$_2$O$_5$ (dissolved in dichloromethane or *in situ* generated gas).

Nitrating agents were applied in up to 10-fold excess [38]. Residence times were set to 0.6–82 s by using tubes of different lengths. Temperatures were between 0 and 20 °C. Reaction mixtures were quenched with ice–water and/or cold dichloromethane and extracted. Analysis was made by NMR, MS, IR, HPLC or GC/MS. Analytical devices and sensors were used for temperature and flow monitoring/control. By FTIR microscopy, on-line monitoring of intermediates and final products for urea nitration in silicon micro reactors was performed, relying on the high spatial resolution (> 10 μm) of this technique. Thereby, the progress of the reaction was accessible on-line.

[P 37] Nitrations were carried out using dissolved or gaseous N$_2$O$_5$ at temperatures between –10 and 50 °C and residence times of 15–45 s [37]. The reaction stream was quenched with water, extracted and passed to analysis.

As micro devices, mixers from various suppliers ([R 19], [R 17]) were used [37]. These devices were each connected to a PTFE tube of length up to 150 cm. In addition, a tailor-made micro reaction system with parallel channels and integrated cooling was used ([R 26]).

4.7.1.5 Typical Results
Conversion/selectivity/yield

[OS 48] [R 17] [R 19] [R 26] [P 36] Yields of up to 75% were achieved for the nitration of N,N'-diethylurea concerning the dinitro product [38]. The dinitro product was obtained in a significantly larger ratio (to the mono product) in the micro reactor as compared with batch [37]. Other side products, e.g. stemming from thermal decomposition, were also formed to a much lower extent.

[OS 49] [R 17] [R 26] [P 36] At almost quantitative conversion, yields of 90% of two (in a first run) unidentified products and of 10% N,N'-diethylurea were reported, accompanied by small amounts of the mono-product [38]. All products no longer contained any C=S moiety, hence were somehow attacked via a nucleophilic route. By subsequent MS and IR analysis, the two main products were identified as N,N'-diethyl-N-nitrosourea and, probably, N,N'-diethyl-N,N'-dinitrosourea. By optimization of the [P 23] procedure, 100% selectivity for the nitration of N,N'-diethylurea to N,N'-diethylurea was achieved.

[OS 50] [R 17] [R 26] [P 36] By nitration of N,N'-diethyl-N-nitroso-urea the target compound for DNDA synthesis N,N'-diethyl-N-dinitro-urea was achieved with 100% selectivity [38].

[OS 48] [R 17] [R 26] [P 36] The selectivity achieved for the nitration of N,N'-diethylurea concerning the dinitro and mono-nitro products were almost similar to macroscopic batch processing, giving, if at all, a slight tendency for the di-nitro product [37, 38]. Isothermal operation and shortening of residence time being the major differences between micro-conti- and conventional-batch, these results accordingly showed that for this combination of reaction/reactor/processing parameters, no benefits could be gained.

[OS 51] [R 17] [R 19] [R 26] [P 37] Methane and hexane nitration were successfully performed under safe conditions in micro reactors [37]. The very preliminary character of these experiments did not allow any detailed process information to be derived apart from demonstrating feasibility and the fact that selectivity was low in the first runs. No significant improvement over the batch processing was obtained here. The formation of many undesired products was particularly high for nitration of hexane [37].

Step-by-step synthesis

[OS 48] [R 17] [R 19] [R 26] [P 36] For the nitration of urea and thiourea derivatives, a selective introduction of nitro and nitroso groups one after the other was demonstrated. Via monosubstituted nitroso groups, dinitro products were achieved [38]. Still other steps (reduction of keto moiety) are required to establish a fully 'step-by-step mode' for micro reactors to give the target compound DNDA. This would be a multi-step, highly selective micro reactor route for the synthesis of dangerous explosives that probably would be a safe route for on-demand production reducing waste. From the starting material until the final hazardous product, all would be done in one line.

Corrosion resistance

[R 16a] [R 17] [R 19] [P 36] High sulfuric acid contents can lead to steel corrosion [37, 38, 97]. This may even lead to blockage by accumulation of corroded material in the tube. It is also claimed [38] that steels are not suited for nitration; however, since the grade of the steel employed is not given, it cannot be excluded that high-alloy steels may behave better. Silicon, glass and titanium are recommended materials [38].

4.8
Addition to Carbon–Carbon Multiple Bonds

4.8.1
Hydrobis(ethoxycarbonyl)methyl Addition – Michael Addition

Peer-reviewed journals: [8]; sections in reviews: [14, 89].

4.8.1.1 Drivers for Performing Michael Additions in Micro Reactors
The reduction of reaction times by changing from batch processing to micro-reactor technology is a major driver [8]. Typical Michael batch additions require hours to be completed.

4.8.1.2 Beneficial Micro Reactor Properties for Michael Additions
Although no special information is given in [8], it is obvious that efficient mass and heat transfer serve for reduction of processing times by establishing kinetic control.

4.8.1.3 Michael Additions Investigated in Micro Reactors
Organic synthesis 52 [OS 52]: Addition of 2,4-pentanedione enolate to ethyl propiolate [8]

The reactions [OS 52], [OS 53], [OS 54] and [OS 55] were chosen as test reactions among a wide class of reagents employed for Michael additions. 1,3-Dicarbonyl compounds were chosen because of their relatively high acidity since they enable one to use weak bases instead of strong bases such as sodium ethoxide. The latter is labile to moisture and can react with the Michael acceptor [8]. Diisopropylethylamine was chosen as a weak base.

Organic synthesis 53 [OS 53]: Addition of benzoyl acetone to ethyl propiolate [8]

Organic synthesis 54 [OS 54]: Addition of diethyl malonate to ethyl propiolate [8]

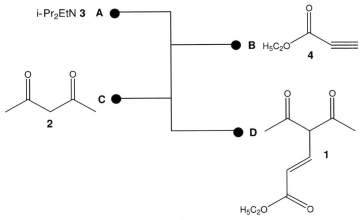

Organic synthesis [OS 55]: Addition of 2,4-pentanedione enolate to methyl vinyl ketone [8]

4.8.1.4 Experimental Protocols

[P 38] Ethanol solutions of ethyl propiolate and diisopropylethylamine were pumped via electroosmotic flow through the micro channels of the reactor [8]. By mixing thereof the enolate was obtained. By subsequent contacting with the 1,3-dicarbonyl compound, the product was obtained. The temperature was set to room temperature. In a period of 20 min a volume sufficiently large for analysis was sampled. The reaction product spectra was analyzed by GC/MS via comparison with synthetic standards. The remaining amount of diketone was used for calculating conversions.

Typically, 40 µl of 5.0 M alkyne compound and 40 µl of 5.0 M alkyne of the diketo compound were placed in the corresponding reservoirs (Figure 4.70) [8]. The same holds for diisopropylethylamine (40 µl; 5.0 M). The electrical fields applied were:

Figure 4.70 Flow configuration for the Michael addition using 2,4-pentanedione and ethyl propiolate in a two-mixing tee microreactor [8].

417, 318, 333 and 0 V cm^{-1}. The residence time was about 20 min and room temperature was applied.

The corresponding batch operation protocol refers to the following procedure [8]. The 1,3-dicarbonyl compound (0.25–0.50 g; 1.5–5.0 mmol) was added to a stirred solution of ethyl propiolate (0.15–0.49 g; 1.54–5.0 mmol) and diisopropylethylamine (0.40–1.29 g; 3.00–10.00 mmol) in ethanol (50 ml). After stirring overnight, the reaction mixture was concentrated and purified by silica gel chromatography.

4.8.1.5 Typical Results
Conversion/selectivity/yield
[OS 52 [R 4b] [P 38] Using a two-fold injection, the first for forming the enolate and the second for its addition to the triple bond, a conversion of 56% was achieved (batch synthesis: 89%) [8]. Using the stopped-flow technique (2.5 s with field applied; 5.0 s with field turned off) to enhance mixing, a conversion of 95% was determined.

Stopped-flow technique
[OS 52] [R 4b] [P 38] See discussions in the preceding and following sections [8].

Reactivity of different substrates with diketone moieties
[OS 52] [OS 53] [OS 54] [OS 55] [R 4b] [P 38] In a two-micro-mixing tee chip reactor, substrates with diketone moieties of known different reactivity, such as 2,4-pentanedione, benzoylacetone and diethyl malonate, were processed, each with the same acceptor ethyl propiolate [8]. Also, a reaction with the less alkynic Michael acceptor methyl vinyl ketone was carried out.

The conversions observed followed the sequence of reactivity known from batch experiments carried out in advance. For example, only 15% conversion was found for the less reactive reagent benzoylacetone in the micro reactor experiment, while 56% was determined when using the more reactive 2,4-pentanedione (batch syntheses: 78% and 89%, respectively) [8]. Using the stopped-flow technique (2.5 s with field applied; 5.0 s with field turned off) to enhance mixing, the conversions for both syntheses were increased to 34 and 95%, respectively. Using a further improved stopped-flow technique (5.0 s with field applied; 10.0 s with field turned off), the conversion could be further enhanced to 100% for the benzoylacetone case. For the other two substrates, diethyl malonate and methyl vinyl ketone, similar trends were observed.

Processing time
[OS 52] [R 4b] [P 38] To achieve comparable extents of conversion, 24 h operation was needed for batch synthesis, whereas micro-reactor operation needs only about 20 min [8].

Isomeric ratio
[OS 52] [R 4b] [P 38] The same isomeric ratio (99% trans; 1% cis) was observed for micro reactor and batch operations [8].

4.8.2
Cycloadditions – The Diels–Alder Reaction

Peer-reviewed journals: [18]; sections in reviews: [89, 90].

4.8.2.1 Drivers for Performing Cycloadditions in Micro Reactors
The aim of one study was to show that arrays of cycloadducts, from various precursors, can be made in a single run in one chip [18]. In addition, this study served more generally to demonstrate the feasibility and advantages of pressure-driven flow in micro chips exemplary of one prominent organic reaction. The advantages and drawbacks of pressure-driven flow as compared with electroosmotic flow (for EOF see [14]) were discussed [18].

4.8.2.2 Beneficial Micro Reactor Properties for Cycloadditions
The formation of arrays or libraries of compounds in microfluidic devices stems from the rapid changes of process parameters and reactants, mainly due to the small internal volumes allowing fast response. Furthermore, miniaturization offers a high degree of integration in a compact volume. Specifically, this allows one to introduce, mix and react many samples or reactants rapidly and in a confined area. As a further effect, the total processed volume is kept small. Especially when producing a large number of samples, this issue becomes relevant. Thus, micro channel processing saves time and resources. An estimation of test frequency and sample consumption concerning a library made by gas/liquid micro channel processing is given in [121].

4.8.2.3 Cycloadditions Investigated in Micro Reactors
[OS 56] Domino reactions

(A) (B) (C)

The domino reaction consists of a Knoevenagel condensation giving an intermediate which immediately undergoes an intramolecular hetero-Diels–Alder reaction with inverse electron demand [18].

As aldehydes, commercially available *rac*-citronellal and a synthesized aromatic aldehyde and also two commercially available 1,3-diketones, 1,3-dimethylbarbituric acid and Meldrum's acid, were selected [18]. By 2 × 2 combinations of these reactants, four different cycloadducts were generated ([OS 57]).

Organic synthesis [OS 57]: 2 × 2 combination of reactants to form four different cycloadducts

4.8.2.4 Experimental Protocols

[P 39] Protocol for single-run processing: reservoirs of the 110 Caliper chip™ were filled with the following solutions [18]: 0.10 M solutions of *rac*-citronellal and a synthesized aromatic aldehyde in methanol/water (80 : 20) and 0.12 M solutions of 1,3-dimethylbarbituric acid and Meldrum's acid in methanol/water (80 : 20) with 10% molar catalyst ethylenediamine diacetate (EDDA).

A 10 µl volume of the aldehyde was placed in three reservoirs and 10 µl of the 1,3-diketone in three other reservoirs [18]. The chip was inserted into the interfacing device and a script for defining pressures was set in the multiport control using a Caliper 42 Workstation™. A 30 min run was carried out. An eight-peristaltic pump system was used as pressure or vacuum source. No active quenching was performed, but it was assumed that by dilution in the collection reservoir reaction was stopped or at least notably slowed.

[P 40] A similar protocol was made for the 2 × 2 multi-reaction mode. Using a special script to guide flow over 700 or 1000 s, depending on the individual reaction, a residence time of 120 s was achieved for each reaction. The other features of the protocol are identical with [P 39].

See also [18] for references on corresponding batch syntheses.

4.8.2.5 Typical Results
Conversion/Selectivity/Yield
[OS 57] [R 6] [P 39] The conversion of the micro channel processing amounted typically to about 50–75%, depending on the nature of the cycloadduct and the residence time chosen [18].

Benchmarking to batch processing
[OS 57] [R 6] [P 39] The conversion using micro channel processing was comparable to that using batch syntheses [18].

Residence time
[OS 57] [R 6] [P 39] Increasing the residence time from 120 to 360 s results in an increase of about 6–10% in conversion for four chosen cycloadducts [18].

Parallel multi-reaction
[OS 57] [R 6] [P 40] One of the first examples of parallel multi-reaction was performed in a Caliper chip [18]. By 2 × 2 combinations of two aldehydes and two 1,3-diketones, four cycloadducts were generated simultaneously in one run on one chip (Figure 4.71). The conversions were comparable to those for the single runs, with one exception. Also, cross-contamination was observed. It ranged from a few percents to about 50%. It should be pointed out that despite these initial drawbacks the demonstration of multi-reaction feasibility is a further valuable step in micro channel processing.

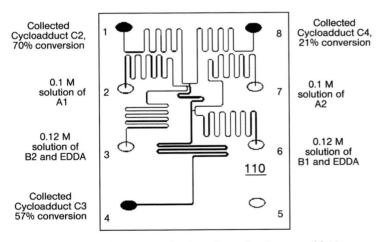

Figure 4.71 Design of a commercial Caliper chip and assignment of fluidic ports to reactant and product solutions used for carrying out (see [OS 57]) [18].

4.8.3
Addition of Oxygen – Epoxidations

Peer-reviewed journals: [30, 122]; sections in reviews: [90, 123]; literature for catalyst formation: [30, 123].

4.8.3.1 Drivers for Performing Epoxidations in Micro Reactors
1-Pentene oxidation over TS-1 catalyst is a fast reaction and hence fulfils a basic
requirement for being suited to micro-channel processing [30]. Thus, it can serve
as a model reaction to demonstrate the benefits of micro chemical engineering,
particularly for zeolite-catalyzed reactions. Apart from this, epoxidations are an
important class of organic reactions, also of industrial importance.

  As the zeolite catalyst preparation is not trivial, a catalyzed epoxidation reaction
serves to demonstrate the feasibility of gas/liquid/solid processing using this industrially well-applied catalyst class [30].

4.8.3.2 Beneficial Micro Reactor Properties for Epoxidations
The major attempts so far were to demonstrate the benefits of enhancing mass
transport in micro channels, especially by decreasing the hydraulic diameter [30].

4.8.3.3 Epoxidations Investigated in Micro Reactors
Organic synthesis 58 [OS 58]: Epoxidation of *n*-pentene by hydrogen peroxide

  The reaction is carried out using a titanium silicalite-1 (TS-1) zeolite catalyst [30,
122]. This type of catalyst is known to accelerate the selective oxidation of alcohols,
epoxidation of alkenes and hydroxylation of aromatics. These reactions have importance for fine-chemical production.

4.8.3.4 Experimental Protocols
[P 41] Catalyst preparation: selective seeding on the micro channel wall surfaces
was used to attach the TS-1catalyst [30]. Mercapto-3-propyltrimethoxysilane was
used to modify the micro channel's surface for this purpose by forming a monolayer which provides better adhesion for the colloidal catalyst. After introducing a
TS-1 seeding solution, the wafer containing the micro channels was calcined at
873 K. By hydrothermal synthesis using orthosilicates and titanates and an organic
templating agent, a layer of TS-1 catalyst was grown on the seeded layer. By thermal treatment in air, organic residues were removed. A very homogeneous layer of
3 μm thickness was generated.

  Reaction protocol: 0.9 M 1-pentene, 0.18 M hydrogen peroxide (30%) and 0.2 M
tert.-butyl methyl ether were dissolved in methanol [30]. This mixture was passed
over the catalyst at a rate of 30–120 μl h^{-1}. Reaction was carried out at room temperature. The product mixture was collected in an ice bath.

4.8.3.5   Typical Results
Conversion/selectivity/yield – benchmarking to conventional batch synthesis
[OS 58] [R 15] [P 41] The surface-to-volume ratio was varied by testing both 1000 and 500 µm wide micro channels [122]. By this comparably small increase in catalyst surface area, the yield of the epoxidation could be doubled.

Catalyst deactivation
[OS 58] [R 15] [P 41] When operating continuously for more than 100 h, deactivation of the catalyst occurs [122]. This seems not to be reversible, as calcinations by air do not bring back catalyst activity.

4.8.4
Dialkoxy Additions – Electrochemical Dialkoxylation of Heteroaromatics

Peer-reviewed journals: [70]; proceedings: [71, 124]; sections in reviews: [90].

4.8.4.1   Drivers for Performing Dialkoxylations
Electrochemical syntheses can be favorably performed at room temperature, thus saving economic resources [70, 71]. No additional reactants are required so that the reaction products are relatively pure. As the reaction occurs directly at the electrode surface, high regio- and stereoselectivity is achieved.

4.8.4.2   Beneficial Micro Reactor Properties for Dialkoxylations
Micro reactors are seen to have smaller inhomogeneities of the electrical field and less temperature rise in the reaction medium due to the Joule heating effect between the electrodes [70]. Submillimeter interelectrode gaps are expected to reduce the ohmic loss.

4.8.4.3   Dialkoxylations Investigated in Micro Reactors
Organic synthesis 59 [OS 59]: Di-methoxylation of furan

Furan was dimethoxylated to give 2,5-dihydro-2,5-dimethoxyfuran, using electro-generated bromine molecules generated from bromide salts in electrolyte solutions [71]. This reaction was characterized in classical electrochemical reactors such as pump cells, packed bipolar cells and solid polymer electrolyte cells. In the last type of reactor, no bromide salt or electrolyte was used; rather, the furan was oxidized directly at the anode. However, high consumption of the order of 5–9 kWh kg^{-1} (at 8–20 V cell voltage) was needed to reach a current efficiency of 75%.

Organic synthesis 60 [OS 60]: Dimethoxylation of methyl 2-furoate

Methyl 2-furoate was dimethoxylated using methanol in sulfuric acid to give methyl-2,5-dihydro-2,5dimethoxy-2-furan carboxylate [70]. The reaction mechanism at the electrodes is not completely known. However, the anodic reaction is said to be the oxidation of methanol. A two-electron process is assumed and hydrogen production is observed at the cathode.

4.8.4.4 Experimental Protocols

[P 42] Typically, a flow rate of 150 µl min^{-1} was used [71]. A 0.1 M furan solution containing 0.1 M NaClO$_4$ and 5 mM NaBr was used as reaction medium. Reaction was carried out at room temperature. Interdigitated band electrodes with 100 µm wide gaps and 500 µm band widths were tested in a 800 µm deep Perspex micro reactor.

[P 43] Typically, a flow rate of 250 µl min^{-1} was established by use of syringe or peristaltic pumps [70]. A 0.1 M methanol solution in sulfuric acid was used as reaction medium. Reaction was carried out at room temperature at a voltage of 4 V.

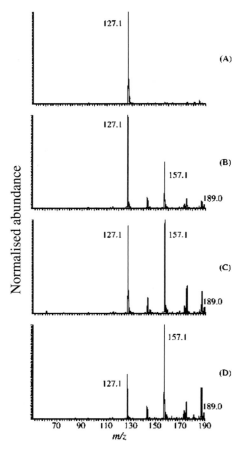

Figure 4.72 Mass spectrometric analysis of product formation of dimethoxylation of methyl 2-furoate for various flow rates. (A) Initial solution; (B) 250; (C) 150; (D) 50 µl min^{-1} [70].

4.8.4.5 Typical Results
Flow rate/residence time
[OS 60] [R 34] [P 43] Although no precise data on conversion and yield could be found, mass spectrometric analysis of the dimethoxylation reaction (Figure 4.72) allowed one to judge the impact when reducing the flow rate, respectively increasing residence time (250, 150, 50 µl min^{-1}) [70]. The peak assigned to the product increased relative to that of the reactant. Also, with increasing residence time, side reactions become evident, e.g. the formation of carboxylic moieties by saponification of the product.

Cell voltage/bromine production
[OS 59] [R 34] [P 42] The furan and dimethoxylated product concentrations were monitored as a function of the cell voltage [71]. The product concentration follows the sigmoidal shape of the bromide oxidation current. The furan concentration shows the inverse behavior. For a cell voltage of 3 V and a current of 30 mA, a concentration of 50% is approached. The product formation is limited by mass transfer of bromine generation.

Dimensions of inter-electrode gap
[OS 59] [R 34] [P 42] The furan dimethoxylation was conducted using interdigital electrodes with 100 and 500 µm inter-electrode gaps, for each case in the presence of electrolyte (0.1 M NaClO$_4$) and without [71]. Using 500 µm inter-electrode gaps, the conversion vs cell voltage curves differ considerably when using electrolyte and when not (150 µl min^{-1}). This is explained by the role of the electrolyte in maintaining the current for such a large electrode distance. Using a shorter distance (100 µm), this is obviously no longer needed. Here, the curves for operation with and without electrolyte are essentially the same (75 µl min^{-1}).

Bromide concentration/relation to energy consumption
[OS 59] [R 34] [P 42] Conversion was monitored versus cell voltage for four bromide concentrations, 5, 20, 100 and 200 mM NaBr (150 µl min^{-1}) [71]. Whereas a 50% conversion was achieved for the lowest salt content, the two highest concentrations led to 70% conversion. When the 5 and 100 mM operations were compared for the same conversion (70%), the energy consumption and the current efficiency were much worse when using the higher salt concentration (50 mM furan; 85 µl min^{-1}; 3 V).

Benchmarking to conventional reactors: current efficiency and energy consumption
[OS 59] [R 34] [P 42] Flow cells with interdigitated band electrode configurations provide a 50% energy saving when compared with conventional reactors such as packed-bed bipolar cells [71]. While 100 µm electrodes (no electrolyte) have an energy consumption of 1.5 kWh kg^{-1}, the bipolar cell needs double this amount. The current efficiency is 80 and 90% for the micro device and the bipolar cell, respectively. In view of the high selectivity of the micro flow process and the low contents of the electrolyte, successive purification steps are facilitated compared with conventional technology.

4.8.5
Polyalkenyl Addition – Polyacrylate Formation

Proceedings: [125]; patents: [126]; sections in reviews: [42, 90, 99, 100, 127, 128].

4.8.5.1 Drivers for Performing Polyacrylate Formation in Micro Reactors

Fouling is a major problem encountered during radical polymer formation [125]. High molecular-weight polymers, potentially of branched nature, form during polymerization and precipitate owing to their insolubility in the solvent.

Fouling is only one result stemming from the influence of mixing on establishing local concentration profiles and their impact on the course of reaction [125]. Hence, in a more comprehensive view, micro mixing can affect the local concentration of initiator, monomer and additives. This should have an impact on the molecular weight distribution of the polymer formed.

The mixing sensitivity of (fast) polymerizations is known and frequently described in literature. This is due to the fact that radical polymerizations typically take 1 s until chain termination [125]. However, typical mixing times of large-scale mixers, including conventional static mixers, are longer. Accordingly, the course of mixing has an effect on the product quality, i.e. the polymer-chain distribution.

4.8.5.2 Beneficial Micro Reactor Properties for Polyacrylate Formation

Micro mixers offer unique mixing physics by using different mixing mechanisms as compared with stirring and other conventional mixing means [125]. The range encompasses multi-lamination, distributive mixing (split–recombine) and chaotic mixing. By this means, more or less regular concentration profiles, with regard to time and space, are generated. If the reaction is faster than mixing (at least for conventional processing), mixing should have an influence on reaction.

Micro mixers permit mixing times much below 1 s. Some multi-lamination micro mixers even approach sub-millisecond mixing [40]. Therefore, mixing can performed faster than most known reactions, including fast radical polymerizations.

Potentially, the mixing physics in micro mixers may allow fouling-free processing, despite the use of tiny channels and nozzles, for cases which lead to plugging when using conventional equipment. Fast mixing in micro mixers may result in polymer molecular weight distributions having a ratio of the number average to the weight-average (M_n/M_w) of an ideally mixed radical polymerization [126].

4.8.5.3 Polyacrylate Formation Investigated in Micro Reactors

Organic synthesis 61 [OS 61]: Radical polymerization of acrylates

This reaction includes modified acrylates with or without addition of styrenes in combination with one or more initiators in a solvent [126]. In an example, tetrahydrofuran was used as solvent and the polymer concentrations amounted to about 5.6 g l^{-1}. Thus, the polymerization is carried out as solvent process.

4.8.5.4 Experimental Protocols

[P 44] The whole process was carried out in a pilot-scale plant (Figure 4.73) equipped with an interdigital micro mixer array and a tube reactor [126] (a shorter description of the set-up is also given in [125]). Monomer/solvent and initiator/solvent mixtures were taken from tanks and pre-heated in a heat exchanger, typically to a temperature below or approaching reaction temperature (50–180 °C). The pre-heated solutions entered the micro mixer array and left it either directly to a tube reactor or to a pre-mixer, set between the micro mixer and tube reactor. The tube reactor was actually composed of three tubes connected in series. The first two had an inner diameter of, e.g., 10 mm and contained commercial Kenics static mixers; the last one had an inner diameter of, e.g., 20 mm and contained a commercial Sulzer SMX static mixer. The tubes, however, can also be operated with the mixer internals. The diameters of the tubes were oriented on the throughput targeted. The total volume of all tubes was, for one reported type of equipment [126], about 0.5 l. The pre-mixer may be a Sulzer SMX with an internal diameter of 5 mm.

The ratio of monomer to initiator ranges from 1 : 1 to 10 : 1, and preferentially is set to 9 : 1 [126]. By setting initiator concentration, initiator type, residence time and temperature, the polymer molecular weight, conversion and solution viscosity are determined. The monomer is, e.g. acrylate-based with or without styrene. The pressure is regulated to be 2×10^5–5×10^6 Pa in order to avoid solvent and monomer boiling.

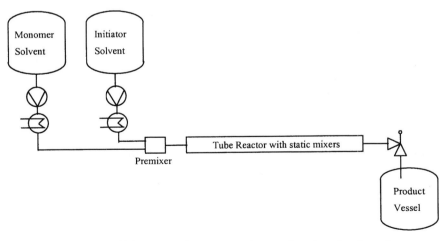

Figure 4.73 Schematic of laboratory-scale experimental set-up for polyacrylate formation. The Sulzer-type pre-mixer can be replaced by an interdigital micro mixer [125].

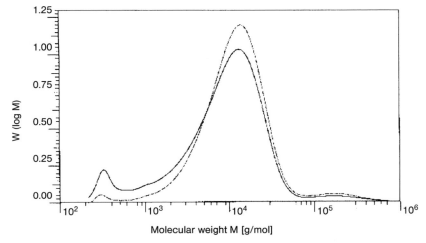

Molecular weight M [g/mol]

Figure 4.74 Radical polymerization of acrylates in a laboratory-scale experimental set-up with a Sulzer-type pre-mixer. Fouling at the feeding point of the static mixer (top) and molecular weight distribution (bottom) [125].

4.8.5.5 Typical Results
Molecular weight distribution – fouling – benchmarking to conventional processing
[OS 61] [R 20] [P 44] The polymer molecular-weight distribution of a micro-mixer based processing was given as the normalized frequency distribution $W(\log M)$, determined by both UV and refractive index analysis [126]. No high-molecular-weight contents above a mass of $> 10^5$ were found [125, 126]. As a result, no precipitates formed on the micro channels' surface, despite the large surface-to-volume ratio (Figure 4.74).

In contrast to this result, polymer samples taken from processing without a micro mixer displayed a small but significant fraction of high-molecular weight polymer with a mass $> 10^5$ [126]. Here, in some cases, heavy precipitation occurred, resulting even in plugging of the static-mixer internals of the tube reactors [125].

Micro mixing
[OS 61] [R 20] [P 44] The polymer molecular-weight distribution of a static mixer-based processing, which was determined both by UV and refractive index analysis,

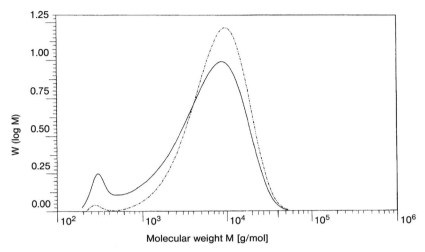

Figure 4.75 Radical polymerization of acrylates in a laboratory-scale experimental set-up with a micro mixer as premixer. Less fouling at the feeding point of the static mixer (compare with Figure 4.74) (top) and molecular weight distribution (bottom) [125].

showed a high-molecular-weight peak [125]. This was explained by the longer time needed for mixing than for reaction. Radical polymerizations are typically completed in about 1 s; mixing in static mixers may take longer. Hence reactions are conducted under segregated concentration profiles, partly promoting long-chain formation. Such high-molecular-weight polymers, especially when being branched, tend to be insoluble and cause fouling of the mixer (Figure 4.75).

In turn, the interdigital micro mixer mixes much faster, provides more unique concentration profiles before reaction takes place and consequently changes the course of the reaction [125]. As a result, no high-molecular-weight polymer fraction is observed by GPC measurement and no fouling occurs, although the specific wall surfaces of the micro device are expected to promote deposition.

Scale-up
[OS 61] [R 20] [P 44] A micro mixer-based laboratory plant would give 50 t a^{-1}, assuming an annual operation time of 8000 h. Based on these laboratory experiments, a pre-basic design comprising numbering-up of 28 micro mixers and four tube reactors was proposed [125]. Accordingly, the production of such plant was calculated to be in the order of 2000 t a^{-1}, assuming 8000 h operation.

4.8.6
Polyalkenyl Addition – Polyethylene Formation

Peer-reviewed journals: [1].

4.8.6.1 Drivers for Performing Polyethylene Formation in Micro Reactors

Screening of process conditions was one driver for performing polyethylene synthesis [1]. Thus, test-throughput frequency, the number of possible samples per day, is a target value. Also, flexibility with regard to temperature and pressure at low sample consumption is a major issue. In addition to fastness and flexibility, the quality of the information and the insight obtained is seen as a motivation for micro-channel studies.

4.8.6.2 Beneficial Micro Reactor Properties for Polyethylene Formation

The above-mentioned targets result from the build-up and operation of a complex apparatus rather than from the micro-channel device or the tubing itself [1]. It is the interplay between the use of advanced HPLC tools, delicate temperature monitoring, electronics (e.g. for pressure control), control system and modern software that allows high-level screening. The most striking micro-channel property here is the possibility of controlling temperature easily by ohmic self-heating and measuring the temperature evolution by voltage taps.

4.8.6.3 Polyethylene Formation Investigated in Micro Reactors
Organic synthesis 62 [OS 62]: Radical polymerization of ethylene

A stream of ethylene is fed into the reactor by use of quaternary LC pumps and subsequently dissolved in a 1.90 ml h^{-1} toluene stream [1]. Ethylene is handled at 60 °C, well above the critical temperature. Catalyst additions are fed via HPLC-type sample injection valves. Various combinations of precatalysts and activators were sampled and loaded by an autoinjector. Catalyst solutions typically were diluted 20-fold within the micro reactor.

4.8.6.4 Experimental Protocols

[P 45] The entire flow system was put in a glove-box in an oxygen-free environment (< 1 ppm, v/v) [1]. A quaternary feed system was used to provide different comonomer additives dissolved in toluene that are mixed with ethylene (25 h) in a mixer.

The total solvent flow rate was set to 2.0 ml min^{-1}. Separate streams of catalyst (0.02 M in toluene) and cocatalyst (0.02 M in toluene) formulations were fed to the tubing reactor in the microliter per minute range by use of special micro-pumps. The volume of the injection loops amounted to 20 μl; the catalyst stream was set to 50 μl min^{-1}. In an electrothermal pre-heat zone the ethylene/comonomer mixture in toluene was brought close to reaction temperature before adding catalyst. The reaction was carried out at 175 °C and a pressure of 2.8 MPa. In several subsequent zones each of 16 cm length (volume: 200 μl) along the reactor tubing, temperature was monitored by voltage pads.

4.8.6.5 Typical Results
Temperature profiles

[OS 62] [R 1] [P 45] Temperature profiles versus time for different positions along the reactor tubing were obtained for catalyst plug-induced ethylene polymerization [1]. The maximum rise in temperature was about 23 °C. With downstream flow passage, the profiles exhibit a decreasing maximum temperature and broaden in shape (Figure 4.76). This was explained by the laminar flow inducing increase in plug length by axial dispersion. Heat transfer effects are seen also to play a role.

Pressure monitoring – relationship to temperature increase

[OS 62] [R 1] [P 45] Improved pressure control was exerted for catalyst plug-induced ethylene polymerization by using advanced pressure control electronics [1]. In the regions of large temperature increase, the pressure fluctuated slightly; this effect diminished downstream.

By deliberately changing the pressure (in a loop), the temperature response followed immediately [1]. This proved that control of pressure is crucial for obtaining stable temperature baselines.

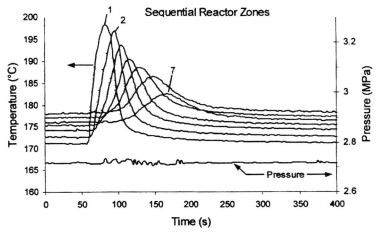

Figure 4.76 Temperature profiles during ethylene polymerization versus time for different positions along the reactor tubing [1].

Reliability of processing – statistics on temperature profiles

[OS 62] [R 1] [P 45] By monitoring temperature profiles, a high reproducibility for catalyst plug-induced ethylene polymerization could be demonstrated [1]. The standard deviation for the temperature peak area was found to be only 1.6%.

Variation in catalyst

[OS 62] [R 1] [P 45] The impact of the choice of catalyst on catalyst plug-induced ethylene polymerization was analyzed [1]. A constrained-geometry catalyst (CGC) with a cyclopentadienyl moiety was about 3.6 times more active than a CGC–indenyl catalyst.

Temperature

[OS 62] [R 1] [P 45] The impact of temperature on catalyst plug-induced ethylene polymerization was determined using a CGC–cyclopentadienyl catalyst [1]. Catalyst activity and polymer formation were higher at lower temperature (155 °C) than at higher temperature (175 °C). This was explained by thermally induced catalyst degradation (Figure 4.77).

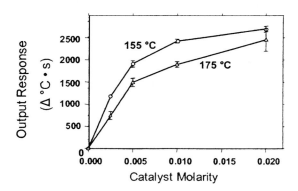

Figure 4.77 Effect of temperature and catalyst modularity on catalyst activity for ethylene polymerization [1].

Test-throughput frequency

[OS 62] [R 1] [P 45] Catalyst plug-induced micro-channel ethylene polymerization allows to process about 10 runs per hour [1]. This is considerably more than achievable with conventional equipment (Parr reactors) processing only 4 to 6 runs per day.

4.8.7
Dihydro Addition – Hydrogen-transfer Reduction

Peer-reviewed journals: [117].

4.8.7.1 Drivers for Performing H-transfer Reduction in Micro Reactors

Short residence time and high heat transfer capabilities are demanded for the transfer hydrogenation of dimethyl itaconate, as in general for comparable liquid/liquid reactions [117].

4.8.7.2 Beneficial Micro Reactor Properties for H-Transfer Reduction
The above-mentioned criteria can be met by suitable micro-channel devices.

4.8.7.3 H-transfer reduction Investigated in Micro Reactors
Organic synthesis 63 [OS 63]: H-transfer reduction of dimethyl itaconate

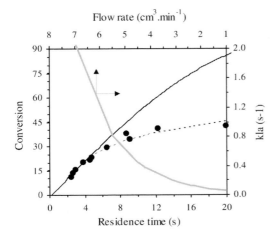

This H-transfer reduction with sodium formate and employing catalysis by a water-soluble rhodium-phosphine catalyst yields dimethyl methylsuccinate [117].

4.8.7.4 Experimental Protocols
[P 46] Prior to the liquid/liquid micro reaction system used, a microgrid serves for dispersing the phases [117]. No other details on performing the reaction are given.

4.8.7.5 Typical Results
Conversion/selectivity/yield – benchmarking to batch processing/kinetics
[OS 63] [R 27] [P 46] A 43% conversion was achieved by micro-channel processing, while both batch experiments and expectation from intrinsic kinetics indicated conversions close to 90% [117]; 80% was achieved by micro-channel processing with an additional micro mixer (see the Section Setting micro mixing prior to reaction, below).

Correction for liquid/liquid mass transfer resistance
[OS 63] [R 27] [P 46] Experimental results were compared with a kinetic model taking into account liquid/liquid mass transfer resistance [117]. Calculated and experimental conversions were plotted versus residence time; the corresponding dependence of the mass-transfer coefficient $k_l á$ is also given as well (Figure 4.78).

Figure 4.78 Liquid/liquid H-transfer reduction of dimethyl itaconate to dimethyl methylsuccinate: experimental results vs models. Main axis: experimental conversions (●); intrinsic kinetic model (solid line); kinetic model + mass transfer kinetics (dashed line). Secondary axis: variation of computed $k_l a$ with flow rate. Further details are given in [117].

The first two data sets were in good accordance, whereas a fit to the intrinsic kinetics without correction was only valid for short residence times and large flow rates. This is the consequence of achieving a low degree of liquid/liquid dispersion at low flow rates, i.e. the reaction becomes mass-transfer limited in this regime.

Setting micro mixing prior to reaction
[OS 63] [R 27] [R 18] [P 46] Using a slit-type interdigital micro mixer prior to a liquid/liquid reaction system improves the conversion to 80%, hence close to the kinetic limits [117]. This is an improvement over using a microgrid in front of the reactor (see the Section Conversion/selectivity/yield – benchmarking to batch processing/kinetics, above).

4.9
Addition to Carbon-Hetero Multiple Bonds

4.9.1
1/N-Hydro-2/C-(α-Acyloxyalkyl),2/C-Oxo Biaddition – Ugi Four component Condensation (4CC)

Proceedings: [16, 25]; theoretical analysis: [129–131]; reviews on classical Ugi chemistry: [130, 132].

4.9.1.1 Drivers for Performing Ugi reactions in Micro Reactors
The Ugi reaction is historically the first synthetic approach that truly can be termed combinatorial as in one step a multitude of reactants, typically four (but possibly up to seven, see Figure 4.79), react to give a complex compound [OS 64]. The Ugi reaction is carried out as a one-pot synthesis performing consecutive multiple reaction steps [16]. Accordingly, a general driver for performing Ugi reactions in micro reactors is oriented on the screening of pharmaceutical targets and on establishing a first methodology for using micro channel technologies for that purpose.

Organic synthesis 64 [OS 64]: General reaction scheme for Ugi reactions

Other drivers are low reagent consumption, the use of small volumes and rapid process optimization, especially with regard to combinatorial synthesis.

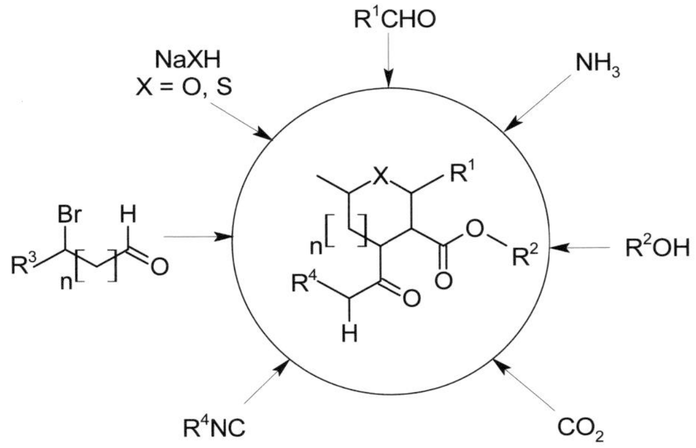

Figure 4.79 The so-called 7CRs, a new principle of designing chemical reactions [130].

In another concrete example of use, the Ugi reaction served as a prominent example to test the performance of one micro mixer and to test specifically the effect of micro mixing on such a multi-stage reaction [25].

Some Ugi reactions are highly exothermic and therefore have to be performed at reduced temperature (e.g. at 0 °C) [25].

Ugi reactions as multi-component reactions are also for interest for creating molecular libraries by parallel means [129–131]. Such approaches so far have only been postulated in a generic way, mostly describing the mathematic tools for automated optimization of the Ugi reaction by micro-channel processing. In this context, a multi-parameter on-line optimization by means of a genetic algorithm or other heuristics was reported to give better yield or selectivity [129]. Micro channel arrays are seen to be central for such an approach. The computer-aided syntheses of molecular libraries under optimized conditions result in an automaton, achieved by choosing the molecular subunits of a multi-component reaction.

In further work, the achievement of well-controlled reaction conditions in micro reactors is highlighted to provide 'chemical data' yielding a highly parallel system of problem-solving functions [131]. This is used to approach a class of problems in computer science that is called NP-complete, for which algorithms are very difficult to solve. In summary, this mathematical approach is used to describe chemical reactions which are highly parallel systems as the parameter space and the related dependencies are virtually infinite (Figure 4.80).

In another report, aspects for automating preparative chemistry are described [130]. A comprehensive description of the Ugi reaction is given in [132] and the vision of a 'micro multi-component reaction' as automated parallel micro-channel synthesis is sketched. An interesting point is to convert aldehydes, chiral primary amines, carboxylic acids and isocyanates into corresponding α-amino acids and peptides (U-4CR).

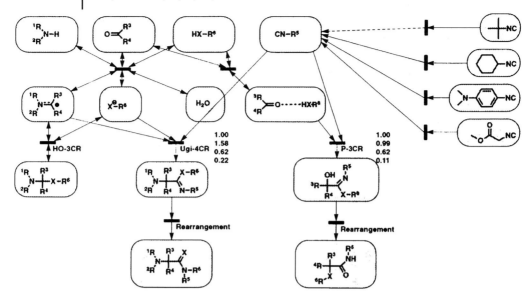

Figure 4.80 Multi-component synthesis for molecular libraries [131].

4.9.1.2 Beneficial Micro Reactor Properties for Ugi Reactions

Concerning the information given above, the defined setting of residence times is an aspect making the use of micro reactors attractive [16]. This refers in particular to the individual residence times between the multiple injections. This allows an individual development of the single reactions one after the other.

Micro reactors allow the sequential injection of the reaction partners, rather than introducing them in one step. By this means, the course of the reaction, in particular concerning the dynamic evolution of the concentration of reactants, may be changed. It is hoped that this may impact on selectivity.

Apart from such process optimization issues, it may be desired to conduct combinatorial screening in micro channels. Here, specific micro-channel geometries for both liquid serial [110] and parallel [23] screening in organic chemistry have already been described.

The possibility of isothermal processing in micro reactors may allow exothermic Ugi reactions to be conducted at room temperature rather than needing cooling much below ambient [25].

4.9.1.3 Ugi Reactions Investigated in Micro Reactors
Organic synthesis 65 [OS 65]: 4CC reaction between aldehyde, primary amine, acid and isocyanide

Methyl 4-formylbenzoate, benzylamine, 2-nitrobenzoic acid, and cyclohexyl iso-cyanide were converted to the corresponding 4CC Ugi adduct [16].

Organic synthesis [OS 66]: Ugi reaction between aldehyde, secondary amine salt and isocyanide

Formaldehyde was used; the exact nature of the secondary amine salt and isocya-nide were not disclosed. The corresponding α-dialkylacetamide was obtained [25].

A further summary of the many decades of achievement of the Ugi chemistry has been given [132].

4.9.1.4 Experimental Protocols
[P 47] Typically, a solution of 50 µl of 0.03 M cyclohexyl isocyanide in methanol and 50 µl of 0.075 M methyl 4-formylbenzoate in methanol were placed in correspond-ing reservoirs [16]. Benzylamine in methanol (50 µl; 0.03 M) and 2-nitrobenzoic

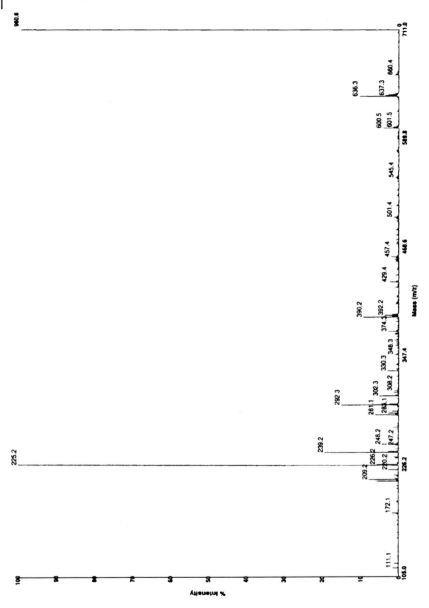

Figure 4.81 Proof of feasibility of performing Ugi MCR in a micro reactor by detecting the parent ion [25].

acid in methanol (50 µl; 0.03 M) were added to the other reservoir, methanol (30 µl) being placed in the collection reservoir. A field of 400 V cm^{-1}, a residence time of about 20 min and room temperature were applied.

[P 48] Initially, a solution of 2.0 mM formaldehyde in methanol (5 µl min^{-1}) and pure methanol (5 µl min^{-1}) were fed through a bifurcation-distributive chip micro mixer [25]. The methanol stream was replaced by a methanol solution of 2.0 mM isocyanide and 0.2 mM amine salt. The reaction was carried out at room temperature and no purification steps were applied.

4.9.1.5 Typical Results
Conversion/selectivity/yield
[OS 65] [R 4c] [P 47] The imine intermediate was isolated in 94% yield. The final product can be obtained at yields ranging from 15 to 60%, depending on the reaction conditions [16].

Increase of reaction temperature to ambient
[OS 66] [R 11] [P 48] Despite the exothermic nature of a specific Ugi reaction, which in conventional batch processing demands cooling to 0 °C, room-temperature processing was accomplished in a chip micro mixer [25]. Although no active cooling was involved, this was explained by enhanced heat dissipation via the large specific surface areas of the micro channels. The product formation was only semi-quantitatively analyzed by mass spectra (Figure 4.81). A large parent peak of the product shows that a considerable yield was achieved; the exact quantity, however, is not given.

4.9.2
Alkyliminodeoxo Bisubstitution – Hantzsch Synthesis

Peer-reviewed journals: [10]; proceedings: [9]; sections in reviews: [89, 90].

4.9.2.1 Drivers for Performing Hantzsch Syntheses in Micro Reactors
The Hantzsch synthesis was primarily chosen to evaluate the potential of a micro-mixing-tee chip reactor for carrying out reactions above room temperature (e.g. 70 °C) [9, 10]. It is said to be the first example of a heated organic reaction performed in a glass chip reactor under electroosmotic flow (EOF) control [10] (for EOF see [14]). The performance of this reactor for room-temperature reactions such as the Wittig reaction and Suzuki coupling was demonstrated before.

In a wider sense, the Hantzsch synthesis is a further example for evaluating the potential of microfluidic systems for high-throughput screening [9].

4.9.2.2 Beneficial Micro Reactor Properties for Hantzsch Syntheses
Following the above-mentioned motivation, precise control over temperature can be exerted in micro reactors [9, 10]. Also, parallel or fast serial screening, handling small volumes distributed over compactly arranged reaction flow-through chambers, can be achieved in micro reactors.

4.9.2.3 Hantzsch Syntheses Investigated in Micro Reactors
Organic synthesis 67 [OS 67]: Hantzsch syntheses using diverse ring-substituted 2-bromoacetophenones and 1-substituted-2-thioureas

From the diverse examples on Hantzsch syntheses reported, the reaction of 2-bromo-4'-methylacetophenone and 1-acetyl-2-thiourea was exemplarily chosen to be represented here [10].

Other reactions with different reactants have been described [9, 10].

4.9.2.4 Experimental Protocols
[P 49] Typically, a voltage of 400 V was applied for 30 min [9]; alternatively [10], voltages of 300–700 V were favorably applied. As solvent *N*-methylpyrrolidone was used. The temperature was set to 70 °C.

4.9.2.5 Typical Results
Conversion/selectivity/yield
[OS 67] [R 4a2] [P 49] Initial micro reactor yields ranged from 58% to almost complete conversion (99%), depending on the voltages applied and the reaction considered [9]. In a later series of experiments, yields from 42 to 99% were reported, depending on the voltages applied and the reaction considered.

Benchmarking to batch processing
[OS 67] [R 4a2] [P 49] Comparative and better yields were achieved when using a micro-mixing tee chip reactor as compared with conventional laboratory batch technology (Table 4.9). In the case of improvement, the increase in yield amounted to about 10–20% [9, 10].

Table 4.9 Benchmarking of mixing-tee chip micro-reactor operation to macro batch-scale synthesis for a series of 2-aminothiazoles [10].

Entry No.	R_1	R_2	R_3	Microreactor conversion (%)			Batch conversion (%)
				300 V	400 V	500 V	
1	Acetyl	H	H	42	63	14	44
2	Acetyl	H	OMe	53	58	14	53
3	Acetyl	H	Me	74	77	72	59
4	Acetyl	Br	H	91	95	99	83
5	Acetyl	NO_2	H	99	99	99	96
6	Phenylethyl	H	H	99	99	99	99

Electroosmotic parameters – voltage

[OS 67] [R 4a2] [P 49] For a given reaction, the dependence of yield on voltage, ranging from 100 to 700 V, was exemplarily given in [10] (see also [9]). Up to 300 V an
increase in yield was observed, and then a slight decrease up to 600 V was found,
followed by an increase again to 700 V.

4.9.3
O-Hydro, C-Alkyl Addition – Grignard Reaction (Mg Alkylation)

Proceedings: [118, 134]; sections in reviews: [42, 90, 99, 100, 127, 128].

4.9.3.1 Drivers for Performing Grignard Reactions in Micro Reactors
One investigation referred to using a Grignard reaction for industrial purposes
[134]. It served as model reaction meeting the relevant criteria: highly exothermic,
temperature sensitive, fast and difficult to handle in a stirred vessel. The reaction
chosen had a reaction enthalpy of 300 kJ mol^{-1} and occurred in a time frame of
about 10 s, as did most of the side reactions did. The avoidance of side reactions,
i.e. an increase in selectivity, was the main motivation for the development of a
micro-channel process.

4.9.3.2 Beneficial Micro Reactor Properties for Grignard Reactions
The above-mentioned research targets generally address the good mass and heat
transfer properties achieved by micro channel processing, in particular referring
to isothermal processing. Because the Grignard process is also very sensitive to the
concentration profiles of the reactants, i.e. the actual ratio of reactants, the capability to have good micro mixing was essential.

4.9.3.3 Grignard Reactions Investigated in Micro Reactors
**Organic synthesis 68 [OS 68]: Enolate formation by addition of the Grignard reactant
to a carbonyl**

The Grignard reactant having a long alkyl chain was added to a keto compound;
the substituents remain undisclosed [134]. Thereby, an enolate is formed which is
further reacted in the framework of a multi-stage fine-chemical industrial process.

**Organic synthesis 69 [OS 69]: Grignard reaction between cyclohex-2-enone and
diisopropylmagnesium chloride [118]**

(A)   (B)

4.9.3.4 Experimental Protocols

[P 50] The temperature of the micro channel processing was set to −10 °C at residence times < 10 s having volume flows e.g. in the range 0.4–2.0 l h⁻¹. The reaction was carried out as a laboratory-scale process with three pumps (two reactants, one solvent) [134]. A slit-type interdigital micro mixer was used which was placed into a thermostatic bath. From this mixer, the reacting solution was transferred via tubing into a product vessel. No further information on the type of solvent and concentration are given in [134].

4.9.3.5 Typical Results

Conversion/selectivity/yield – benchmarking to laboratory and industrial processes

[OS 68] [R 19] [P 50] A yield of 95% was obtained by a micro-mixer-based process (< 10 s, at −10 °C), while the industrial batch process (6 m³ stirred vessel) gave only 72% yield (5 h, at −20 °C) [134]. The laboratory-scale batch process (0.5 l flask; 0.5 h, at −40 °C) gave an 88% yield.

[OS 69] [R 25] [no protocol] Software-supported process optimization (factorial design) in a micro reactor was carried out for a Grignard reaction [118]. In this context, 14 different reaction conditions were investigated in 14 h. By this means, the initial yield of 49% could be improved to 78% with a simultaneous increase in regioisomer ratio (A : B) from 65 : 35 to 95 : 5.

Transfer from laboratory to industrial process – numbering-up concept

[OS 68] [R 19] [P 50] Having a yield of 95% by a micro-mixer-based process (< 10 s, at −10 °C), it was decided to perform pilot-scale studies with a self-built mini mixer for reasons of clogging, which was not decisive at the laboratory scale [134]. With one mini mixer at the pilot scale a yield of 92% was obtained (< 10 s, at −10 °C). The validity of the numbering-up concept was proven by operating also five mini mixers of the same type at a yield of 92% (< 10 s, at −10 °C). This was the central part of the actual production process, running for more than 3 years until the life-cycle of the commercial product of the corresponding multi-stage process ran out (Table 4.10).

Table 4.10 Comparison of residence time and yield for different reactor types [134].

Reactor type	Temperature (°C)	Residence time	Yield (%)
Flask, 0.5 l	−40	0.5 h	88
Production (stirred vessel, 6 m³)	−20	5 h	72
Microreactor (laboratory set-up)	−10	< 10 s	95
Mini reactor (pilot scale)	−10	< 10 s	92
Five Mini reactors (production)	−10	< 10 s	92

Statistical experimentation design – flow rate

[OS 68] [R 19] [P 50] The experiments on micro-channel processing followed a statistical design [134]. The dependence of the yield on temperature and on the ratio

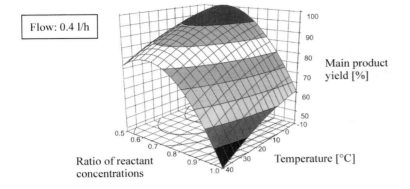

Flow: 0.4 l/h

Main product
yield [%]

Ratio of reactant
concentrations

Temperature [°C]

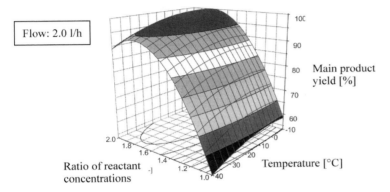

Flow: 2.0 l/h

Main product
yield [%]

Ratio of reactant
concentrations

Temperature [°C]

Figure 4.82 Results of trials according to the statistical design of experiments [134].

of the reactant concentrations for two flows $(0.4\,l\,h^{-1}; 2.0\,l\,h^{-1})$ was given. This dependence was qualitatively the same; however, the highest yield exceeding 90% was only found at the higher flow rate $(2.0\,l\,h^{-1})$, and also at higher temperatures as compared with the $0.4\,l\,h^{-1}$ processing. This is in accordance with the need for flow equilibration in the array of the slit-type interdigital micro mixer (Figure 4.82).

By the statistical approach, the complete parameter set for processing at 95% yield was discovered very quickly.

Thermal control/energy expenditure
[OS 68] [R 19] [P 50] The lower yield of the industrial batch process (6 m³ stirred vessel) of only 72% is due to limitations of the cooling system allowing processing only at –20 °C, and not at –40 °C as is possible for the better laboratory-scale batch process (0.5 l flask; 0.5 h; 88% yield) [134]. The industrial process had a surface-to-volume ratio of 4 m² m⁻³, the laboratory-scale process of 80 m² m⁻³ and the micro reactor of 10 000 m² m⁻³. Accordingly, the residence time had to be increased from 0.5 h to 5 h to allow less heat generation per unit time for the large-scale process.

As a consequence, the contribution of side and follow-up reactions is larger. In addition, micro-channel operation at −10 °C causes less energy expenditure and costs than the former batch processing at −20 °C.

Fouling/blockage

[OS 68] [R 19] [P 50] Fouling leading to blockage occurred by micro-mixer processing. Therefore, a new mixing element with wider channels was designed. The surface-to-volume ratio was 4000 m^2 m^{-3} (micro mixer: 10 000 m^2 m^{-3}). Using such a slight increase in internal dimensions and a more elaborate set-up, pilot-scale processing at 92% yield avoiding blockage could be achieved [134].

Production process – automation

[OS 68] [R 19] [P 50] A numbering-up of five mini mixers, tested at the pilot stage, was used [134]. Automation of the entire process was required; liquid-flow splitting to the single reactors was '... by no means trivial ...' [134]. The capacity of one mini reactor was 30 ml s^{-1}, i.e. 108 l h^{-1}. The complete setup hence should be operated close to 500 l h^{-1}. The micro-reactor plant was operated at intervals as the preceding step was carried out batchwise. The operation of the micro-reactor plant started in August 1998 after a period of only about 1.5 years for development.

4.9.4
O-Hydro, C-Alkyl Addition – Li Alkylation of Ketones

Peer-reviewed journals: [83].

4.9.4.1 Drivers for Performing Li Alkylations in Micro Reactors

One main driver was to increase the processing temperature for this class of metallation reactions more towards ambient. Typically, Li alkylations are conducted under cryogenic conditions (e.g. at −60 °C) [83]. Further motivation came from aiming at increasing the yield reducing investment and operating costs.

4.9.4.2 Beneficial Micro Reactor Properties for Li Alkylations

Improved control over heat and mass transfer as well as residence time by micro-channel processing often allows one to increase the reaction temperature of cryogenic processes without losing selectivity. It often leads to improved selectivity.

4.9.4.3 Li Alkylations Investigated in Micro Reactors
Organic synthesis 70 [OS 70]: Li alkylation of ketones yielding chiral alcohols

In a first step, an alkyl iodide is converted by lithium to an unstable Li intermediate [83]. This reacts with the ketone to give a chiral alcohol.

This lithium process is the analogue of the magnesium-based Barbier reaction [135].

4.9.4.4 Experimental Protocols
[P 51] No protocol is given in [83].

4.9.4.5 Typical Results
Conversion/selectivity/yield – benchmarking to batch processing
[OS 70] [R 25] [P 51] The total yield of a combined lithiation and alkylation of a ketone to a chiral alcohol was increased to 93% by micro-reactor processing, whereas the batch process gave only an 83% yield [83]. This increase in yield is related to reduced β-elimination of the unstable Li intermediate.

Reaction temperature
[OS 70] [R 25] [P 51] The temperature of a lithiation step was raised from –78 °C (batch) to –15 °C (micro reactor) without losing selectivity [83]. The temperature of the subsequent alkylation of a ketone to a chiral alcohol could also be increased from –60 °C (batch) to 0 °C (micro reactor).

4.9.5
Alkyliminodeoxo Bisubstitution – Formation of Imines (Schiff Bases)

Proceedings: [127].

4.9.5.1 Drivers for Performing Formation of Imines in Micro Reactors
The motivation for an industrial investigation was to gather kinetic and mechanistic information for a very fast and highly exothermic reaction [127]. In particular, by-product formation was analyzed.

4.9.5.2 Beneficial Micro Reactor Properties for Formation of Imines
Micro reactors can cope with situations as depicted above. They are excellent laboratory tools to gather such information.

4.9.5.3 Formation of Imines Investigated in Micro Reactors
Organic synthesis 71 [OS 71]: Reaction of an aldehyde with a primary amine

Neither the substituents of the aldehyde nor of the amine were disclosed, as the reaction probably is a proprietary industrial process [127]. The Schiff base formed reacts further by decomposition.

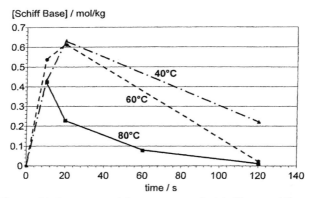

$$R_1\text{-}NH_2 \; + \; \genfrac{}{}{0pt}{}{R_2}{H}\!\!>\!\!C\!=\!O \quad \rightleftharpoons \quad [R_1\text{-}N\!=\!CHR_2] \; + \; H_2O \; \longrightarrow$$

Figure 4.83 Formation and decomposition of a Schiff base at different temperatures [127].

4.9.5.3   Experimental Protocols

[P 52] The reaction was carried out at 40–80 °C on a time-scale of 10–120 s. No further details are given in [127], as the reaction probably is a proprietary industrial process.

4.9.5.4   Typical Results
Conversion/selectivity/yield

[OS 71] [no details on reactor] [P 52] The Schiff base is formed (Figure 4.83) with maximum yield at about a 20 s residence time at 40 and 60 °C [127]. At 80 °C, this decreases to less than 10 s.

Mechanistic information

[OS 71] [no details on reactor] [P 52] An intermediate, previously unknown, was discovered during micro-channel Schiff base formation [127].

Residence time

[OS 71] [no details on reactor] [P 52] A shift in the residence time needed for achieving maximum yield at a given temperature was observed [127].

Reaction temperature

[OS 71] [no details on reactor] [P 52] By micro-channel investigations, it was found that the decomposition of the Schiff base was much more pronounced at higher temperature [127].

4.9.6
Dual Alkyliminodeoxo Bisubstitution and Ring Closure – Knorr Synthesis

Peer-reviewed journals: [20].

4.9.6.1 Drivers for Performing Knorr Synthesis in Micro Reactors

Drug synthesis and industrial applications thereof were in the focus of recent investigations [20]. To achieve a high structural diversity, today the creation of combinatorial libraries is ultimately needed. Of particular importance is to achieve a close combination of synthesis and analysis. Performing organic syntheses in micro channels is seen as one important future technology for that purpose. By this means, a miniaturized screening variant of ultra-high throughput chemical synthesis should be achievable.

The Knorr synthesis was chosen as an example to demonstrate that with an automated micro-reactor system a further step was made over existing parallel micro flow processing techniques, providing a 2×2 library in a chip micro reactor [20]. The new approach was designed for much higher diversity, aiming at a 7×32 library.

Hence the investigations made so far were aimed more at showing the capability of an automated micro-reactor system, using the Knorr synthesis as a model reaction; however, to a certain extent, information on this synthesis itself was also gained (see Section 4.9.6.5) [20]. The Knorr synthesis of pyrazoles chosen is of industrial interest since by this route compounds with a wide range of biological activity can be produced.

4.9.6.2 Beneficial Micro Reactor Properties for Knorr Synthesis

From the time when it was shown that micro flow reactors can provide valuable contributions to organic chemistry, it was obvious to develop them further and their workflow towards modern screening techniques [20]. It was especially the finding of high reaction rates, the capability to transport and transform minute sample volumes and the first integration of analytics that paved the way to a parallelization of micro flow processing. These benefits were combined with the ease of automation of a micro flow system. By this means, the potential of on-line analysis of the reactions can be fully exploited.

4.9.6.3 Knorr Synthesis Investigated in Micro Reactors
Organic synthesis 72 [OS 72]: Pyrazole ring closure from 1,3-dicarbonyl compounds and hydrazines

1,3-Dicarbonyl compounds react with the two nitrogen functionalities of hydrazines with ring closure to give pyrazoles [20]. The Knorr synthesis is of interest for drug applications as products with a wide range of biological activity can be generated in this way.

The reaction was also chosen for analytical reasons as the pyrazole ring is a good UV chromophore, facilitating analysis [20].

The 3×7 library in Table 4.11, consisting of diverse 1,3-dicarbonyl compounds and hydrazines, was synthesized.

Table 4.11 Synthesized 3 × 7 library consisting of diverse 1,3-dicarbonyl compounds and hydrazines [20].

4.9.6.4 Experimental Protocols

[P 53] Before operation, a start-up time of about 10 min was applied to stabilize pressure in the chip micro reactor ([R 6]) [20]. As a result, a stable flow pattern was achieved. The reactant solutions were filled into vials. Slugs from the reactant solutions were introduced sequentially into the micro chip reactor with the autosampler and propelled through the chip with methanol as driving solvent. The flow rates were set to 1 µl min^{-1}. The slug volume was reduced to 2.5 µl.

The pyrazole library was created sequentially using 10 mM solutions of the 1,3-dicarbonyl compound and 0.8 M solutions of the hydrazines, each introduced as a 2.5 µl slug [20]. This requires control of feeding of both reactant solutions so that the slugs enter the chip at the same time and mix thereafter. The residence time was 210 s. Thereafter, the reaction slugs were diluted on-chip by a 1 : 1 methanol–water stream at 8 µl min^{-1} and detected. Analysis of the nature of the products and the degree of conversion was done using standards of reactant and product materials.

The chip micro reactor ([R 6]) was only one part of a complex serial-screening apparatus [20]. This automated system consists of an autosampler (CTC-HTS Pal system) which introduces the reactant solutions in the chip via capillaries. A pumping system (µ-HPLC–CEC System) serves for fluid motion by hydrodynamic-driven flow. A dilution system [Jasco PU-15(5)] is used for slug dilution on-chip. The detection system was a Jasco UV-1575 and analysis was carried out by LC/MS (Agilent 1100 series capLC–Waters Micromass ZQ). All components were on-line and self-configured.

The UV intensity was recorded with the Lab-View program [20]. Such UV-decoded slug patterns were almost identical and near-rectangular with some axial dispersion at the slug edges (see the Section Slug dispersion, below, for more details on signal broadening).

4.9.6.5 Typical Results
Conversion/selectivity/yield

[OS 72] [R 6] [P 53] Quantitative conversion was obtained for 16 out of 21 reactions for a 3 × 7 library (Table 4.12) [20]. The other five reactions had conversions ranging from 35 to 85%.

Table 4.12 Conversions for a 3 × 7 library, the compounds in which (A1–A7; B1–B3) are depicted in Table 4.11 [20].

Reactant	Conversion (%)		
	B1	B2	B3
A1	99	99	99
A2	99	99	99
A3	99	99	99
A4	99	99	99
A5	99	99	99
A6	35	99	85
A7	57	56	49

In order to test the role of residence time for process performance, an experiment equal in process conditions to the one with the lowest performance (35 %) was conducted; the only exception being that residence time was doubled (420 instead of 210 s) [20]. A much improved conversion (52%) was obtained in this way (see also the Section Residence time, below).

Slug dispersion
[OS 72] [R 6] [P 53] Slug patterns, measured as UV absorption vs time, were almost identical for the various sequential slugs [20]. This proves that a regular and stable flow pattern was achieved for the micro flow in the chip. Although axial dispersion was evident from some broadening at the slug edges, over most of the slug length the same UV absorption was found, i.e. near-rectangular pulses were achieved. This shows that in the slug core-defined concentration profiles were maintained. Signal broadening increased with increasing residence time, as expected.

Residence time
[OS 72] [R 6] [P 53] For five reactions with non-quantitative conversions, processing was repeated at a doubled residence time (420 instead of 210 s). A control experiment with a reaction of quantitative conversion was also successfully made at the best residence times. Much higher conversions were obtained in this way for all five reactions, being about 20% in each case.

Automated sequential preparation 3 × 7 library – feasibility
[OS 72] [R 6] [P 53] A 3 × 7 library was made in a sequential and automated way with conversions of 35% (lowest) and 99% (quantitative; for 16 reactions) [20]. The results obtained were compared for consistency with single-reaction processing on the same chip.

Cross-contamination of 3 × 7 library
[OS 72] [R 6] [P 53] No cross-contamination was found during preparation of the 3 × 7 library [20]. Neither products, by-products nor the hydrazine in excess were intermixed.

4.9.7
Alkyliminodeoxo Bisubstitution – Formation of Enamines

Peer-Reviewed journals: [11]; sections in reviews: [14, 89].

4.9.7.1   Drivers for Performing Formation of Enamines in Micro Reactors
Research on enamine formation in micro reactors was focused on eliminating the need for using Lewis acid catalysts [11]. In addition, operation under mild conditions such as room-temperature processing was favored.

4.9.7.2   Beneficial Micro Reactor Properties for Formation of Enamines
The above-mentioned research targets generally address the good mass and heat transfer properties achieved by micro channel processing, in particular referring to isothermal processing.

4.9.7.3   Formation of Enamines Investigated in Micro Reactors
Organic synthesis 73 [OS 73]: Stork enamine formation from cyclohexanone and pyrrolidine

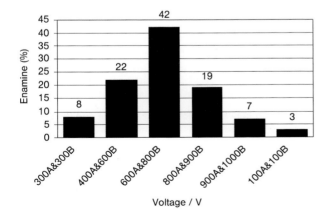

This carbon–carbon bond-generating reaction can be used extensively over a wide range of chemistries [11]. As the reaction is an equilibrium process, needing the removal of water to obtain high yields, chemical means have to be used to accomplish this task. 1,3-Dicyclohexylcarbodiimide (DCC) is a commonly used reagent for this purpose. Alternatively, molecular sieves find use for conventional processing, but are not so favorable for micro-reactor processing, because the sieve needs to be inserted into the micro channel (additional fabrication expenditure) and may disrupt the liquid transport if EOF is applied.

4.9.7.4   Experimental Protocols
[P 54] A 50 μl volume of a 0.3 M solution of cyclohexanone in anhydrous methanol with about 1 mg of DCC is placed in one reservoir of a micro-mixing tee chip reactor [11]; 50 μl of a 0.3 M solution of pyrrolidine is added to the other reservoir and anhydrous methanol is filled in the third, the collection reservoir. Voltages ranging from 300 to 1000 V are applied for a period of 40 min to transport the reaction species. The reaction is carried out at room temperature.

4.9.7.5   Typical Results
Conversion/selectivity/yield
[OS 73] [R 4a] [P 54] Depending on the voltages applied, enamine yields between 8 and 42% were achieved [11]. Using an external voltage combination of 600 V at one reservoir (cyclohexanone) and 800 V at the second reservoir (pyrrolidine) gives the highest yield without using a catalyst (Figure 4.84).

Figure 4.84 Conversion of enamine depending on the voltages applied to the reservoirs [11].

Benchmarking to batch synthesis/elimination of catalyst
[OS 73] [R 4a] [P 54] The micro reactor yield (up to 42%) is comparable to that for batch Stork-enamine reactions using *p*-toluenesulfonic acid in methanol under Dean and Stark conditions [11].

Reaction temperature
[OS 73] [R 4a] [P 54] Reasonable micro reactor yields (up to 42%) can be obtained at room temperature [11].

4.9.8
Bis(ethoxycarbonyl)methylenedeoxo Bisubstitution – Knoevenagel Condensation

Peer-reviewed journals: [18]; sections in review: [89]. A Knoevenagel condensation is described under 4.8.2 Cycloadditions – The Diels–Alder Reaction, since both reactions were performed combined in a domino-type process.

4.9.9
O-Hydro, C-(*a*-Acylalkyl) Addition – Aldol Reaction

Peer-reviewed journals: [15]; sections in reviews: [14, 89, 90].

4.9.9.1 Drivers for Performing Aldol Reactions in Micro Reactors
The aldol reaction is one of the best known means of C–C-bond formation in organic chemistry. The reaction needs the formation of enolates, which themselves are one of the most extensive species permitting C–C-bond formation [15].

Reducing the processing time is a driver for micro channel processing of aldol reactions [15]. Using reactive reactants such silyl enol ethers, this can be accomplished.

4.9.9.2 Beneficial Micro Reactor Properties for Aldol Reactions
The above-mentioned research targets generally address the good mass and heat transfer properties achieved by micro channel processing, in particular referring to fast mixing.

4.9.9.3 Aldol Reactions Investigated in Micro Reactors
Silyl enol ethers are an elegant means to 'protect' the reactive and hence labile enolate moiety [15]. At the time of reaction, the enolate group is generated as an intermediate and reacts with the carbonyl-carrying compound.

Organic synthesis 74 [OS 74]: Reaction between 4-bromobenzaldehyde and the silyl enol ether of cyclohexane [15]

Organic synthesis 75 [OS 75]: Reaction between 4-bromobenzaldehyde and the silyl enol ether of acetophenone [15]

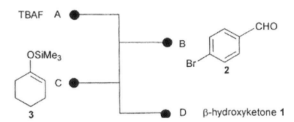

4.9.9.4 Experimental Protocols

[P 55] Before synthesis, a micro-mixing tee chip micro reactor (Figure 4.85) (with two mixing tees and four reservoirs) was primed with anhydrous tetrahydrofuran (THF). A 40 µl volume of a 0.1 M solution of tetrabutylammonium fluoride trihydrate in anhydrous THF is filled into one reservoir of a micro-mixing tee chip reactor [15], 40 µl of a 0.1 M solution of 4-bromobenzaldehyde in anhydrous THF is added to a second reservoir, 40 µl of a 0.1 M solution of the silyl enol ether (masking the enolate of a carbonyl compound such as cyclohexanone) in anhydrous THF is added to a third reservoir and anhydrous THF is filled into the fourth collection reservoir. Electrical fields of 417, 455, 476 and 0 V cm^{-1} are applied to transport the reaction species from the respective reservoirs. The reaction is carried out at room temperature.

See also [15] for a description of the corresponding batch syntheses.

TBAF A

OSiMe$_3$

B

CHO

Br

2

C

3

D β-hydroxyketone **1**

Figure 4.85 Flow configuration for the aldol reaction of silyl enol ethers in a mixing-tee chip micro reactor [15].

4.9.9.5 Typical Results

Conversion/selectivity/yield

[OS 74] [R 4b] [P 55] For the reaction of 4-bromobenzaldehyde with the silyl enol ether of cyclohexanone, only 1% conversion was achieved on applying protocol [P 55] [15]. Changing the set of electrical fields so that the concentration of tetrabutylammonium fluoride trihydrate was raised resulted in 100% conversion (417, 341, 333 and 0 V cm^{-1}).

Reaction time

[OS 75] [R 4b] [P 55] For the reaction of 4-bromobenzaldehyde with the silyl enol ether of acetophenone, 100% conversion with respect to the silyl enol ether was achieved in 20 min for a given set of electrical fields (375, 409, 381 and 0 V cm^{-1}) [15]. The corresponding batch synthesis time was about 1 day.

4.9.10
C,O-Dihydro Addition – BH₃–Carbonyl Hydrogenations

Proceedings: [3].

4.9.10.1 Drivers for Performing BH₃–Carbonyl Hydrogenations in Micro Reactors

This reaction served to demonstrate organic synthesis on a newly developed po-rous-polymer-rod micro reactor [3].

4.9.10.2 Beneficial Micro Reactor Properties for BH₃–Carbonyl Hydrogenations

A detailed study on velocity profiles, pressure drop and mass transport effects is given in [3]. This, in quantitative terms, precisely underlines the advantages (and limits) of the micro reactor concept.

4.9.10.3 BH₃–Carbonyl Hydrogenations Investigated in Micro Reactors

Organic synthesis 76 [OS 76]: Reduction of acetophenone

Acetophenone was reduced in methanol to 1-phenylethanol at 20 °C by boro-hydride moieties coupled to a porous polymer resin [3]. In principle, four hydro-gen atoms can be released from the borohydride; the reactivity, however, decreases with each hydrogen atom lost. Experimentally it was shown that the first two at-oms mainly contribute and to the reduction the other two remain on the polymer site.

4.9.10.4 Experimental Protocols

[P 56] A 1 mmol amount of acetophenone in methanol was reacted with borohydride moieties coupled to a porous polymer resin [3]. The reaction temperature was set to 40, 60 and 80 °C. The ratio of acetophenone to the solid borohydride was 0.90, 0.45 and 0.20. The polystyrene polymer was cross-linked by linking divinylbenzene moieties in the main chain. Originally, the polymer contained benzylchloride groups, which were converted to quaternary ammonium groups. By means of ion exchange, functional anionic groups such as the reductive cyanide moiety can be introduced. Typical ion-exchange capabilities of the micro reactor were about 0.1–1.0 mmol, depending on the polymer load.

4.9.10.5 Typical Results

Conversion/selectivity/yield – benchmarking to batch processing

[OS 76] [R 3] [P 56] After 500 h, total conversion was achieved for the micro reactor [3]. Polymer/glass rods and crushed rods were considerably less active, by a factor of 4–5 (Figure 4.86).

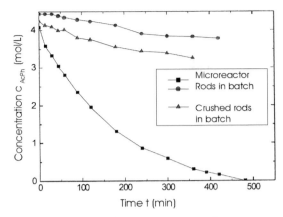

Figure 4.86 Dynamic changes in conversion for a micro reactor and batches comprising rods and crushed rods [3].

Kinetic investigations – rate constants and activation energy
[OS 76] [R 3] [P 56] Two reaction rate constants at different temperature and reactant ratios were determined applying a set of four kinetic equations [3]. This mathematical model described the experimental results very well. Using an Arrhenius plot, frequency factors and activation energies were determined as well. For stage-one reduction, an activation energy of 7.438 kcal mol^{-1} resulted. For stage two, an activation energy of 2.498 kcal mol^{-1} was derived.

4.9.11
Alkylidenedeoxo Bisubstitution – Wittig Reactions and Horner–Emmons Reactions

Peer-reviewed journals: [13]; proceedings: [12, 85, 136]; sections in reviews: [14, 42, 83, 89, 90].

4.9.11.1 Drivers for Performing Wittig Reactions in Micro Reactors
The Wittig synthesis is a two-step reaction comprising first the formation of the ylide and then the double-bond closure [12]. The possibility of precisely setting the reaction times of these two steps in miniature continuous flow systems allows one a high level of chemical control, which is not possible by batch systems to the same extent. Another driver for Wittig syntheses is to control stereoselectivity [12]. Cis and trans isomers of the double bond are formed. Their ratio can be influenced generally by setting the flow and particularly by manipulating it by means of electroosmotic flow (EOF) (for EOF see [14]).

Another driver is doing screening and combinatorial syntheses in micro reactors [13].

4.9.11.2 Beneficial Micro Reactor Properties for Wittig Reactions
The precise definition of residence times for various stages of reactions by introducing reactants in a spatially confined manner in micro flow devices allows new ways

of chemical control [12]. Even properties such as the ratio of stereoisomers, although intuitively not evident, seem to be influenced by flow properties in micro channels.

Mixing can be performed in a fast manner, so limiting residence time to the kinetic needs [12]. Particularly favorable is the application of EOF leading to plug-flow profiles, allowing one to set the residence time simply by changing voltages of the EOF, and also to perform mixing other than by diffusional means [12].

4.9.11.3 Wittig Reactions Investigated in Micro Reactors
Organic synthesis 77 [OS 77]: 2-Nitrobenzyltriphenylphosphonium bromide and methyl 4-formylbenzoate [12]

Organic synthesis 78 [OS 78]: 2-Nitrobenzyltriphenylphosphonium bromide and four aldehydes

The testing of 2-nitrobenzyltriphenylphosphonium bromide and methyl 4-formyl-benzoate and four other aldehydes, 3-benzyloxybenzaldehyde, 2-naphthaldehyde, 5-nitrothiophene-2-carboxaldehyde and 4-[3-dimethylamino)propoxy]benzaldehyde, has been reported [13].

Wittig reactions generally occur in a two-stage manner. First, the triphenyl-phosphonium bromide reacts fast with sodium methoxide to give the corresponding ylide intermediate [12]. The coloration by ylide formation allows one to follow the course of the reaction visually. Then, a second-order reaction with the aldehyde compound follows to give the product containing the double-bond moiety. Here, cis and trans isomers may be formed. Side products refer to triphenylphosphine oxide and the derivatives of the non-conjugated reactants.

Organic synthesis 79 [OS 79]: Methyl diethoxyphosphonoacetate and 4-methoxybenzaldehyde (Wittig–Horner–Emmons)

4-Methoxybenzaldehyde and methyl diethoxyphosphonoacetate were reacted via the Wittig–Horner–Emmons route to give the corresponding alkene product [85] (see a more detailed description in [42]).

4.9.11.4 Experimental Protocols

[P 57] One early investigation focused on plug production (Figure 4.87), flow being fed by electroosmotic means [12]. Therefore, the voltages for moving the different reactants in the three reservoirs were set to 594 V for both 2-nitrobenzyltriphenyl-phosphonium bromide and sodium methoxide and to 678 V for methyl 4-formyl-benzoate. The reaction was carried out at room temperature. No other details are given in [12].

[P 58] Another protocol focused on continuous contacting of the two reactant solutions. Again, flow was fed by electroosmotic means [13]. A 0.01 M methanol solution of 2-nitrobenzyltriphenylphosphonium bromide was used; a 0.02 M metha-nol solution for methyl 4-formylbenzoate with sodium methoxide (0.015 M) was used. Volumes of 80 μl of both solutions were set in the respective reservoirs on the chip and 40 μl of methanol in the collection reservoir. A voltage of 400 V was applied for both feed lines. The reactions were carried out at room temperature and run for 20 min.

[P 59] A third protocol differed from [P 57] and [P 58] by sequential insertion of all reactants instead of using a pre-mixed solution of nitrobenzyltriphenyl-phosphonium bromide and sodium methoxide. As above, flow was fed by electro-osmotic means [13]. A 40 μl volume of a methanol solution of 0.01 M 2-nitrobenzyl-triphenylphosphonium bromide was filled into the first reservoir, 40 μl of a methanol solution of 0.015 M sodium methoxide in the second, 40 μl of a methanol solution

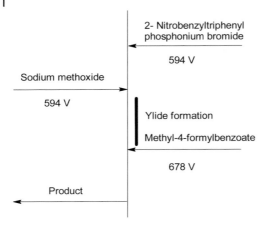

Figure 4.87 Flow configuration for a Wittig synthesis
in a chip micro reactor with three vertical injections [13].

of 0.01 M methyl 4-formylbenzoate in the third and 40 μl of methanol in the collection reservoir. Following the sketched sequence of feeding, the voltages applied were: 594, 660 and 678 V. The reactions were carried out at room temperature and run for 20 min.

4.9.11.5 **Typical Results**
Conversion/selectivity/yield – benchmarking to batch synthesis
[OS 77] [R 4b] [P 57] The migration of the base, sodium methoxide, was enhanced by electrophoretic force [12]. This led to faster mixing than by diffusion only. As a consequence, the conversion rate was increased.

[OS 79] [R 17] [no protocol] 4-Methoxybenzaldehyde and methyl diethoxyphosphonoacetate were reacted by means of the Wittig–Horner–Emmons reaction [85] (see a more detailed description in [42]). A modified micro reaction system consisting of two mixers, for deprotonation of the phosphonates and introduction of the aldehyde, connected to an HPLC capillary of 0.8 m length and 0.25 mm diameter was employed. The micro reactor showed higher yields than laboratory batch synthesis.

Testing of different aldehydes at 2 : 1 ratio – benchmarking to batch synthesis
[OS 78] [R 4a] [P 58] The Wittig reactions of five aldehydes with 2-nitrobenzyltriphenylphosphonium bromide were investigated (using a 2-to-1 excess of the aldehyde) [13]: methyl 4-formylbenzoate, 3-benzyloxybenzaldehyde, 2-naphthaldehyde, 5-nitrothiophene-2-carboxaldehyde and 4-[3-dimethylamino)propoxy]benzaldehyde. Whereas the first reaction gave an improvement in yield compared with batch synthesis, the other four reactions did not. The different chemical natures of these species was not considered as a possible explanation for this finding, but it was rather suggested that the flow conditions were not set correctly (the voltages used, i.e., as the protocol was optimized for the first reaction and only applied to the

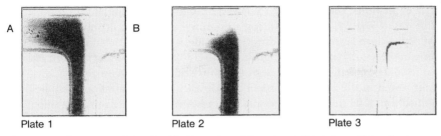

Plate 1   Plate 2   Plate 3

Figure 4.88 Flow patterns, visualized by formation of the colored ylide intermediate, allowing one to judge on mixing efficiency [13].

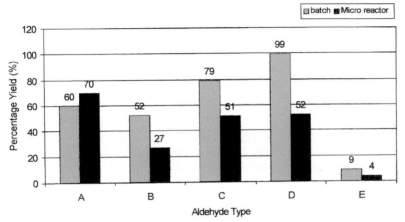

Figure 4.89 Comparison between yields obtained by one mixing tee micro reactor and batch operation using five different aldehydes at 2 : 1 stoichiometry [13].

other). Flow monitoring of the ylide formation from the inlet section of the micro-mixing tee to the reaction channel demonstrates this (Figure 4.88).

Using optimized reaction conditions, the Wittig reactions with four of the five aldehydes resulted in an improvement in their yields (see Figure 4.89 below) [13].

Testing of different aldehydes at a 1 : 1 ratio

[OS 78] [R 4a] [P 58] The Wittig reactions of five aldehydes with 2-nitrobenzyl-triphenylphosphonium bromide were investigated also at a 1 : 1 stoichiometric ratio [13]. Using optimized reaction conditions, improved yields were found for four of the five aldehydes.

For methyl 4-formylbenzoate, the yield decreased from 70% (2 : 1) to 59% (1 : 1) [13]. For batch processing, the yield decreased from 60% (2 : 1) to 48% (1 : 1), i.e. the reductions were similar. In both cases, micro channel processing was superior.

The 1 : 1 ratio was realized by plug insertion instead of continuous feed as in the 2 : 1 case. One injection per minute of 2-nitrobenzyltriphenylphosphonium bromide was made for a duration of 30 s into a continuous stream of the aldehyde solution at 400 V [13].

Use of pre-mixed solutions versus sequential single-reactant injection

[OS 78] [R 5] [P 59] Instead of using a premixed solution of nitrobenzyltriphenyl-phosphonium bromide and sodium methoxide, these two reactants were each feed into the reaction channel by single injection [136]. A total of three reactant streams were so inserted in a specially designed chip reactor for that purpose. Mixing was performed in the reaction channel and directly induced reaction. The yield obtained with fully single injection was 38%, whereas for the pre-mixed solution 59% was obtained. The reason for this is not really well understood; obviously the ylide formation was not completed under the conditions of protocol [P 59]. It was checked in advance that processing a pre-mixed solution in reactor [R 5] gave the same yield as for the originally used [R 4].

However, despite having a lower yield, sequential injection is seen to have the advantage of allowing more in-depth variation of process parameters, in particular concentration changes and performing individual stages of the reaction separately [136].

Cis/trans- (Z/E) isomer ratio

[OS 77] [R 4b] [P 57] By simply adjusting the voltages in an EOF driven chip, the ratio of cis and trans products (Z/E ratio) can be changed [12]. The Z/E ratio is also strongly influenced by moving the reactants separately or as a premixed solution. For a 1 : 1 ratio of the reactants, the Z/E ratio changed from 2.35–3.0 (premixed) to 0.82–1.09 (not pre-mixed, separate movement) [13].

These initial findings motivated a detailed analysis of the influence of flow properties, i.e. changing the voltage in an electroosmotic-driven experiment, on the Z/E ratio [13]. Voltages were varied between 300 and 700 V. It was shown that on decreasing the voltage, which is equivalent to increasing the residence time, the trans isomer is formed more predominately, but at the expense of a drastic reduction of the overall yield. This demonstrates that thermodynamic control is more effective here, as expected, for long residence times. For large changes in the Z/E ratio, relatively small voltage changes of the order of only 100 V were needed. For instance, changing the voltage from 694 to 494 V for one channel decreased the Z/E ratio from 2.3 to 0.57. The corresponding yields as outlined above, changed from 100 to 7%.

4.10
Eliminations

4.10.1
Hydro–Hydroxy Elimination – Dehydrations of Alcohols

Peer-reviewed journals: [19]; sections in reviews: [14, 89, 90]; microfabrication: [137, 138].

4.10.1.1 Drivers for Performing Dehydrations of Alcohols in Micro Reactors

Performing this reaction primarily served as a model to show the feasibility of micro flow processing for solid/liquid reactions [19]. In a similar way as for catalyzed gas-phase reactions, micro-reactor processing was expected to show benefits in terms of mass and heat transfer. Particularly this relates to transfer enhancement when using porous media.

4.10.1.2 Beneficial Micro Reactor Properties for Dehydrations of Alcohols

The benefits refer to the ability to achieve defined thin, highly porous coatings in micro reactors. In combination with the small length scales of the channels, diffusion to the active sites is facilitated. The residence time can be controlled, accurately minimizing consecutive reactions which may reduce selectivity.

4.10.1.3 Dehydrations of Alcohols Investigated in Micro Reactors

Organic synthesis 80 [OS 80]: Dehydration of 1-hexanol to hexene

$$\text{(1-hexanol)} \quad \xrightarrow[\text{-H}_2\text{O}]{\text{ZrO}_2,\ 155\text{-}160°\text{C}} \quad \text{(hexene)}$$

The dehydration of 1-hexanol to hexene was conducted over heterogeneous sulfated zirconium oxide catalyst [19, 138]. The zirconia was treated with sulfuric acid and is known as super acid catalyst, having well documented performance for many reactions [19]. The reaction conditions are notably milder as for other acid catalysts, such as silica–alumina.

Organic synthesis 81 [OS 81]: Dehydration of ethanol to ethene

$$\text{H}_3\text{C}\text{—OH} \quad \xrightarrow[\text{-H}_2\text{O}]{\text{ZrO}_2,\ 155\text{-}160°\text{C}} \quad \text{H}_2\text{C}=\text{CH}_2$$

4.10.1.4 Experimental Protocols

[P 60] The dehydration of 1-hexanol to hexane and of ethanol to ethane were conducted at 155 °C. Heating was accomplished by a heating wire inserted in the micro reactor's top plate. This wire was connected to a potentiostat (0–270 V); temperature was monitored by a digital thermometer with the probe close to the reaction channel. A syringe pump was applied for liquid transport [19]. A flow rate of 3 μl min^{-1} was applied. The alcohols were purged with nitrogen directly prior to reaction to minimize coke formation.

After activation by heating, the catalyst was dusted over the surface of a thin polydimethylsiloxane (PDMS) layer, being coated on the PDMS top plate of the micro reactor [19]. Such a modified plate was baked for 1 h at 100 °C. A high surface area and firm immobilization of the catalyst resulted. Then, the micro reactor was assembled from the top and another bottom plate, having at one micro-channel wall the catalyst layer. Stable operation with the PDMS micro reactor up to 175 °C could be confirmed.

[P 61] For the dehydration of ethanol to ethane, electroosmotic pumping was applied for liquid transport [19]. A flow rate of 0.9–1.1 µl min^{-1} was applied, giving longer residence times as in [P 60]. The other details of the protocol are identical with those for [P 60].

4.10.1.5 Typical Results

Conversion/selectivity/yield – benchmarking to large-scale reactors

[OS 80] [R 7] [P 60] The acid-catalyzed dehydration of of 1-hexanol to hexene was conducted in a micro reactor made of PDMS, which also contained a heating function [19, 138]. Sulfated zirconium oxide was coated as catalyst on the top plate of the micro reactor. A yield of 85–95% was obtained; by-products could not be detected. This performance exceeds those of conventional reactors (30%).

[OS 81] [R 7] [P 60] The acid-catalyzed dehydration of of ethanol to ethene over sulfated zirconium oxide led to a mixture containing 68% ethene, 16% ethane, and 15% methane [19, 138].

Operational time

[OS 80] [R 7] [P 60] The micro reactor was used constantly for 3 days without any hint of losing performance [19, 138].

Residence time

[OS 81] [R 7] [P 61] An increase in residence time by a factor of about 3 was accomplished by changing the flow rate from 3.0 to 0.9–1.1 µl min^{-1} [19]. By far the main reaction product detected was methane; otherwise only traces of methanol were present. Instead, at the shorter residence time a mixture containing 68% ethene, 16% ethane and 15% methane was obtained [19, 138]. Hence the presence of methane demonstrates that complete cracking occurred as a consecutive reaction to dehydration.

Electroosmotic pumping

[OS 81] [OS 80] [R 7] [P 60] [P 61] Whereas electroosmotic pumping can be achieved for ethanol, it is not feasible with hexanol, as the volume flow rate is inversely propoertional to the carbon chain length [19, 138].

4.11
Rearrangements

4.11.1
Rearrangements of Hydroperoxides

Patents: [64].

4.11.1.1 Drivers for Performing Rearrangements of Hydroperoxides in Micro Reactors

The only investigation so far refers to the second reaction of a two-step industrial process, the Hock process, which is used for phenol production world-wide [64].

Here, cumene hydroperoxide is rearranged to provide phenol and acetone via acid cleavage.

As this acid cleavage releases considerable heat, the apparatus used needs efficient heat exchange units [64]. Selectivity of the hydroperoxide cleavage is affected if temperature rises in an undesired manner. For this reason, not the technical 65–90 wt.-% solutions, but rather strongly diluted ones (e.g. below 10%) are employed. In the first case, a sudden loss of the heat exchange function could result in an temperature increase from 50 to 500 °C within seconds.

As a consequence, recycle processing is needed with high recycle ratios, e.g. 17 [64]. Depending on the acid/organic ratio, homogeneous or heterogeneous acid cleavage can be carried out. In the latter case, only a small portion of the acid is used. Accordingly, a mixture with very different ratios of the two solutions has to be prepared in a continuous-flow system which demands good mixing properties.

4.11.1.2 Beneficial Micro Reactor Properties for Rearrangement of Hydroperoxides

According to the above remarks, micro reactors have to provide efficient heat transfer and should be able to reduce the recycle ratios [64]. Further, they have to be able to mix streams at very different flow ratios.

The excellent heat transfer properties of micro reactors are a general feature that is not specifically related to the hydroperoxide rearrangement. Calculations on cumene hydroperoxide cleavage showed that even if the pumps stop working, i.e. the cooling fluid is no longer moving, no critical temperature increase has to be expected for micro-reactor processing [64]. This is a result of their extremely large surface-to-volume ratios. Thus, controllability of the hazardous substances used (peroxides) is notably improved.

The impact of micro reactor processing on the recycle ratio cannot be predicted in a straightforward manner. As this is somehow linked to mass transfer and mixing, improvements are possible; actually this was found experimentally, e.g. recycle ratios below 2 were used for micro flow processing (instead of 17 in the conventional industrial processing, see below) [64].

4.11.1.3 Rearrangements of Hydroperoxides Investigated in Micro Reactors
Organic synthesis 82 [OS 82]: Rearrangement of cumene hydroperoxide

The Hock process includes the oxidation of cumene by air to hydroperoxides using large bubble columns and the cleavage of the hydroperoxide via acid catalysis, which is reaction [OS 82]. This process is used for the majority of world-wide phenol production and, as a secondary product, also produces large quantities of acetone [64]. Phenol is used, e.g., for large-scale polymer production when reacted in a polycondensation with formaldehyde.

The reaction mechanism is based on protonation of the hydroxyl moiety, rearrangement of the phenyl group and simultaneous cleavage of water, creating a carbocation as intermediate [135]. This cation is hydroxylated by water. Thereby, an unstable hemiacetal is formed that splits into two molecules, phenol and water.

4.11.1.4 Experimental Protocols

[P 62] The acid cleavage was carried out at 45–75 °C at a pressure of 1–5 bar. Water may be added at levels of 0.3–1 wt.-% [64]. This addition was made upstream of the micro reactor or directly inside. The residence time was set in the range 0.5–5 min. Sulfuric acid was used as catalyst. By changing residence time and acid addition, the residual cumene hydroperoxide content was favorably reduced to 0.1–0.3 wt.-%. For this, an acid concentration of 50–500 ppm is typically required. Part of the so cleaved product stream may be recycled.

In a typical experiment in a micro reactor, 67 wt.-% technical cumene hydroperoxide was reduced to 1.0 wt.-% [64]. A recycle ratio of 2 was applied.

4.11.1.5 Typical Results

Conversion/selectivity/yield – benchmarking to conventional apparatus

[OS 82] [R 28] [P 62] High-boiling substances are typical by-products of acid cleavage which need to be reduced to facilitate product purification. Using 67 wt.-% technical cumene hydroperoxide and a recycle ratio of 2 yields a content of high-boiling substances of 0.12 wt.-% [64]. Conventional processing at a recycle ratio of 17 results in 0.21 wt.-% high-boiling substances.

An increase in cumene yield by 0.5% was thus achieved in the micro reactor [64]. Considering an annual production of 7 million tons, this is a considerable improvement regarding economics.

Recycle loop/recycle ratio

[OS 82] [R 28] [P 62] The recycle ratio of acid cleavage was considerably reduced in micro reactors [64]. At better selectivity, a ratio of 2 or below was applied, whereas conventional processing relies on much larger ratios, e.g. 17. Even operation without recycling was possible in the micro reactor. In this case, 0.1 wt.-% high-boiling substances was achieved. This is a reduction by a factor of two compared with the state-of-the-art industrial processing.

For reducing the recycle ratio, the acid content had to be increased. Although this was detrimental in terms of selectivity, this effect was more than counterbalanced by the reduction of loop passages [64].

Total system failure/cooling liquid not being pumped
[OS 82] [R 28] [P 62] Even when the cooling liquid is not being pumped (e.g. for reasons of pump failure), the high surface-to-volume ratio of the micro reactor still guarantees sufficient heat removal so that undesired heat generation or even explosions can be intrinsically avoided [64]. This is due to the high heat transfer coefficient (2000 W m^{-2} K^{-1}) resulting from the large specific surface area of the reaction channel (10 000 m^2 m^{-3}).

4.12
Oxidations and Reductions

4.12.1
C,O-Dihydro Elimination – Br(OAc)$_2^-$ Oxidations of Alcohols to Ketones

Proceedings: [3].

4.12.1.2 Drivers for Performing Br(OAc)$_2^-$ Oxidations of Alcohols in Micro Reactors
The carrying out of this reaction served to demonstrate organic synthesis on a newly developed porous-polymer-rod micro reactor [3].

4.12.1.2 Beneficial Micro Reactor Properties for Br(OAc)$_2^-$ Oxidations of Alcohols
A detailed study on velocity profiles, pressure drop and mass transport effects has been described [3]. This, in quantitative terms, precisely underlines the advantages (and limits) of the micro reactor concept.

4.12.1.3 Br(OAc)$_2^-$ Oxidations of Alcohols Investigated in Micro Reactors
Organic synthesis 83 [OS 83]: Br(OAc)$_2^-$ Oxidation of cyclohexanol

Cyclohexanol is reduced to cyclohexanone at 20 °C (6 h) by Br(OAc)$_2^-$ moieties [3].

4.12.1.4 Experimental Protocols
[P 63] Cyclohexanol was reduced in dichloromethane to cyclohexanone at 20 °C (6 h) by Br(OAc)$_2^-$ moieties coupled to a porous polymer resin [3]. The polystyrene polymer was cross-linked by reacting divinylbenzene moieties in the main chain with each other. Initially, the polymer contained benzylchloride groups, which were converted to quaternary ammonium groups. By means of ion-exchange, functional anionic groups such as the reductive Br(OAc)$_2^-$ moiety can be introduced. Typical ion exchange capabilities of the micro reactor were about 0.1–1.0 mmol, depending on the polymer load.

4.12.1.5 **Typical Results**
Conversion/selectivity/yield
[OS 83] [R 3] [P 63] A > 99% yield was obtained after 6 h of processing [3].

4.12.2
Oxodedihydro Bisubstitution – Catalyzed Oxidation of Ethanol with H_2O_2 to Acetic Acid

Proceedings: [51].

4.12.2.1 **Drivers for Performing Catalyzed Oxidations with H_2O_2 in Micro Reactors**
Catalyzed oxidations of organic compounds such as ethanol with hydrogen peroxide (H_2O_2) are highly exothermic processes releasing large heat power [51]. At high
conversion, there may be technical difficulties in removing the reaction heat so
that the temperature may rise in an uncontrolled manner. For this reason, these
reactions are typically carried out at low conversion, e.g. by dilution with solvent or
by operation in differential reactors.

The catalyzed oxidation of ethanol to acetic acid is a well-studied reaction, carried
out in continuous stirred tank reactors (CSTR) [51]. Hence it is a good test reaction
for benchmarking micro reactor results.

4.12.2.2 **Beneficial Micro Reactor Properties for Catalyzed Oxidations with H_2O_2**
Micro reaction channels can be operated in tight contact with cooling channels so
that large reaction heats can be removed. By this means, formerly uncontrollable
reaction conditions may be realized.

4.12.2.3 **Catalyzed Oxidations with H_2O_2 Investigated in Micro Reactors**
**Organic synthesis 84 [OS 84]: Catalyzed oxidations of ethanol with hydrogen peroxide
to acetic acid**

4.12.2.4 **Experimental Protocols**
[P 64] Flows of 0.2–0.9 kg h^{-1} ethanol (98%) and 0.015–0.050 kg h^{-1} of a 1 mol l^{-1}
aqueous solution of $Fe(NO_3)_3$ and acetic acid were fed by pumps and mixed in a
micro mixer; then this solution was mixed with 0.3–4.0 kg h^{-1} hydrogen peroxide
(35%) solution in a second micro mixer [51]. Such a mixed flow was passed through
a series of four reaction modules with cross-flow guided thermofluids (Figure 4.90).
The inlet temperatures of the modules were set to 70–115 °C. The modules had
different temperatures, creating a temperature profile along the flow axis. In terms
of process operations, this refers to the series ignition/reaction/quenching, with
modules performing reaction. The temperatures were monitored at the inlets and
outlets of each module. Pressure was held constant at 3–5 bar to avoid boiling.
Analysis was performed by in-line NIR flow-cell measurement.

The range of flow rates is not given for the heat transfer fluids in [51]. However,
for one typical experiment this was specified (0.71 kg h^{-1} ethanol; 3.025 kg h^{-1} hy-

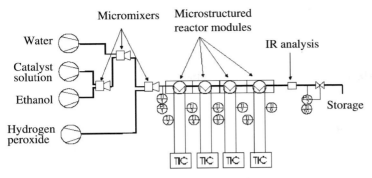

Figure 4.90 Scheme of test facility for ethanol oxidation by hydrogen peroxide [51].

drogen peroxide; 70 °C for ignition; 105 °C for reaction; 20 °C for quenching); 43 and 54 kg h^{-1} water were pumped through the ignition and quenching modules, respectively. The two inner reactor modules were each fed by 300 kg h^{-1} Marlotherm™ SH thermofluid.

Water could additionally be injected via a third micro mixer in the ethanol/catalyst solution mixture [51]. This served for heat transfer characterization, adjustment of temperature before reaction and most prominently dilution of the reaction mixture. By the last step, runaway situations occurring during the reaction can be managed.

4.12.2.5 Typical Results
Conversion/selectivity/yield – benchmarking to batch processing
[OS 84] [R 24] [P 64] Near-complete conversion (> 99%) at near-complete selectivity (> 99%) was found in a micro reaction system [0.71 kg h^{-1} ethanol, 98%; 3.025 kg h^{-1} hydrogen peroxide, 30%; 1 mol l^{-1} Fe(NO$_3$)$_3$ and 1 mol l^{-1} acetic acid; 70 °C for ignition; 105 °C for reaction; 20 °C for quenching] [51]. Processing in a continuous stirred tank reactor (CSTR) resulted in 30–95% conversion at > 99% selectivity (Table 4.13).

Table 4.13 Comparison of reaction conditions and results obtained in a micro reactor set-up with results obtained in a continuous stirred tank reactor [51].

	CSTR	*Micro structured plug flow reactor*
Residence time (s)	1760	3
Pressure (bar)	Atmospheric	3–5
Conversion (%)	30–99, oscillating	> 99
Selectivity (%)	> 99	> 99
Reaction volume (cm^3)^a	2900	3
Throughput (cm^3 h^{-1})	5930	4300
Space–time yield (h^{-1})	0.7–2.0, oscillating	500

a Reaction volume is defined as the volume in which the reaction took place, not the volume of the reaction system.

Reaction time

[OS 84] [R 24] [P 64] Micro-reactor operation was completed within 3 s, whereas processing in a CSTR takes nearly 30 min (exactly 1760 s) [0.71 kg h^{-1} ethanol, 98%; 3.025 kg h^{-1} hydrogen peroxide, 30%; 1 mol l^{-1} Fe(NO$_3$)$_3$ and 1 mol l^{-1} acetic acid; 70 °C for ignition; 105 °C for reaction; 20 °C for quenching] [51].

Space–Time yield

[OS 84] [R 24] [P 64] The space–time yield of micro-reactor operation amounted to 500 h^{-1}, while processing in a CSTR gives 0.7–2.0 h^{-1} [0.71 kg h^{-1} ethanol, 98%; 3.025 kg h^{-1} hydrogen peroxide, 30%; 1 mol l^{-1} Fe(NO$_3$)$_3$ and 1 mol l^{-1} acetic acid; 70 °C for ignition; 105 °C for reaction; 20 °C for quenching] [51]. As reaction volume, only the real processed volume was taken into account, i.e. not considering the complete outer volume. A comparison of the reaction volumes and throughputs of tank and micro reactors are also given [51].

Power release

[OS 84] [R 24] [P 64] Experimental power data ranged from 2.80 to 3.10 kW depending on reaction time, approaching the theoretical value of 3.13 kW [0.71 kg h^{-1} ethanol, 98%; 3.025 kg h^{-1} hydrogen peroxide, 30%; 1 mol l^{-1} Fe(NO$_3$)$_3$ and 1 mol l^{-1} acetic acid; 70 °C for ignition; 105 °C for reaction; 20 °C for quenching] [51]. These power data were derived from a number of temperature measurements along the flow passage through the four reactor modules (Figure 4.91). The temperature data are also given [51].

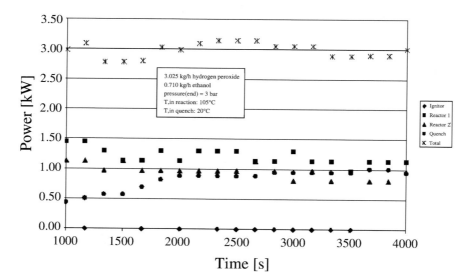

Figure 4.91 Power output calculated from measured temperature rise of the coolant fluid during oxidation reaction [51].

4.12.3

Oxidation of Arylmethanes – Electrochemical Alternative Routes to the Étard Reaction

Proceedings: [69]; patents: [139]; sections in reviews: [42, 90, 105].

4.12.3.1 Drivers for Performing the Electrochemical Oxidations of Arylmethanes in Micro Reactors

The first study on the oxidation of arylmethanes used this reaction as a model to show the general advantages of electrochemical micro processing and to prove the feasibility of an at this time newly developed reactor concept [69]. Several limits of current electrochemical process technology hindered its widespread use in synthetic chemistry [69]. As one major drawback, electrochemical cells still suffer from inhomogeneities of the electric field. In addition, heat is released and large contents of electrolyte are needed that have to be separated from the product.

4.12.3.2 Beneficial Micro Reactor Properties for Electrochemical Oxidations of Arylmethanes

Micro reaction systems may help to overcome or at least reduce some of the above-mentioned limitations [69]. Electrochemical micro reactors with miniature flow cells where electrodes approach to micrometer distances should have much improved field homogeneity. As a second result of confined space processing, the addition of a conducting salt may be substantially reduced. In addition, benefits from a uniform flow distribution and efficient heat transfer may be utilized.

4.12.3.3 Electrochemical Oxidations of Arylmethanes Investigated in Micro Reactors

Organic synthesis 85 [OS 85]: Oxidation of 4-methoxy toluene to 4-methoxy benzaldehyde

The electrosynthesis of 4-methoxybenzaldehyde (anisaldehyde) from 4-methoxy-toluene by means of direct anodic oxidation is performed on an industrial scale [69]. Via an intermediate methyl ether derivative, the corresponding diacetal is obtained, which can be hydrolyzed to the target product. The different types of products – ether, diacetal, aldehyde – correspond to three distinct single oxidation steps.

4.12.3.4 Experimental Protocols

[P 65] Carrying out the oxidation of 4-methoxytoluene (20%, 2.0 mol l^{-1}) in methanol, a constant current density of 79 mA cm^{-2} was employed using 0.1M KF as conducting salt [69]. A flow rate of 0.1 ml min^{-1} was applied, yielding a velocity of 0.046 mm s^{-1} (Re = 0.2), at a pressure drop of 1.1 bar. This corresponds to a residence time of 21 s. A potentiostat/galvanostat was used in combination with a charge counter. For temperature control, the micro reactor was immersed in a thermostat.

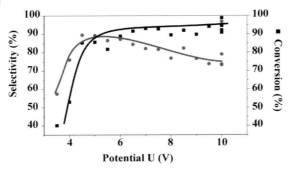

Figure 4.92 Conversion (■) and selectivity (●) diagram for the electrochemical reaction of 4-methoxytoluene with 0.1 M KF as conducting salt [69].

4.12.3.5 **Typical Results**
Conversion/yield/selectivity

[OS 85] [R 33] [P 65] High conversion (> 85%) and selectivity (> 85%) were found at a potential of 4.5 V (0.1 M 4-methoxytoluene) [69]. At lower potential, a strong increase in both quantities is observed (Figure 4.92). At still higher potential, conversion increases slightly to 95% (at 10 V), whereas selectivity decreases to 75% (at 10 V).

Composition of product mixture

[OS 85] [R 33] [P 65] By means of HPLC analysis, it could be shown that all three types of products were formed during reaction in the micro reactor [69]. For a typical experiment, the main fraction was composed of 4-methoxybenzaldehyde dimethylacetal, the second largest fraction consisted of the aldehyde and 4-methoxybenzyl methyl ether was generated in a smaller amount.

The change of content of these three products was monitored as a function of temperature, in the presence of conducting salt (0.1 M KF) or in its absence [69]. In the first case, a near linear increase for the ether and acetal was found, but not for the aldehyde, which remained constant. The contents of the first two compounds were more than doubled. On adding no salt, the ether and acetal increased only slightly and the aldehyde decreased to some extent, probably due to oxidation.

Addition of conducting salt/temperature

[OS 85] [R 33] [P 65] Compared to the industrial process using 0.1–0.2 M KF, low contents of conducting salt (0.01 M) were used [69]. In some cases, even no salt at all was needed (Figure 4.93).

Without conducting salt, the yield and conversion increased with rising temperature (10–60 °C) owing to increased diffusion [69]. The yield approaches conversion so that selectivity is high. With conducting salt (0.01 M KF), conversion is much higher, close to 100%, and decreases slightly at higher temperature. increases up to 25 °C and then reaches a plateau at about 75%. The relative difference between yield and conversion is larger than for the no-salt experiment, hence selectivity is lower. The best selectivity for micro-reactor operation was determined to be 96%.

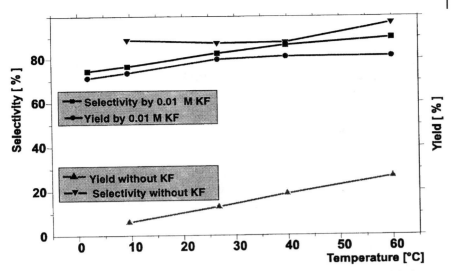

Figure 4.93 Comparison of selectivity and yield for the synthesis of 4-methoxybenzaldehyde depending on reaction temperature and amount of conducting salt [69].

Current efficiency

[OS 85] [R 33] [P 65] Using conducting salt (0.01 M) in a micro reactor yields a current efficiency of 60–65%, whereas operation without any salt has an efficiency of 96–98% [69]. For conventional electrochemical processing of 4-methoxybenzaldehyde, an efficiency of 49–54% is reported (Figure 4.94).

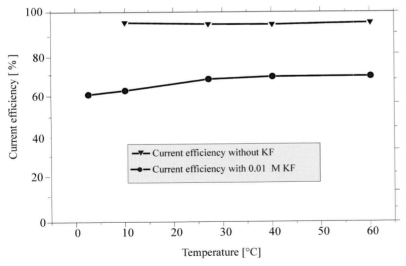

Figure 4.94 Comparison of current efficiency for the electrochemical process with 0.01 M KF and without conducting salt [69].

4.12.4
Oxidative CO₂ Elimination – Electrochemical Decarboxylations

Proceedings: [65].

4.12.4.1 Drivers for Performing Electrochemical Decarboxylations
The drivers for performing decarboxylations come so far mainly from the general advantages of electrochemical processing which apply also to other reactions. They stem from using electrons instead of chemicals directly for chemical synthesis and refer to the ease of creating a suitable reaction environment simply by providing the right electrical potential. As large-scale electrochemical reactors cannot exploit this potential fully owing to field inhomogeneities and release of heat, micro reactors overcoming these limits are seen as a suitable alternative.

4.12.4.2 Beneficial Micro Reactor Properties for Electrochemical Decarboxylations
The micro reactor properties concern process control in the time domain and process refinement in the space domain [65]. As a result, uniform electrical fields are generated and efficiency is thought to be high. Furthermore, electrical potential and currents can be directly measured without needing transducer elements. The reactor fabrication methods for electrical connectors employ the same methods as used for microelectronics which have proven to satisfy mass-fabrication demands.

4.12.4.3 Electrochemical Decarboxylations Investigated in Micro Reactors
Organic synthesis 86 [OS 86]: Electrochemical cecarboxylation of D-gluconic acid

D-Gluconic acid is decarboxylated to yield the five-carbon sugar D-arabinose, releasing carbon dioxide [65]. The reaction can be performed at a graphite electro-catalyst under conditions that do not imply further oxidation to low-order carbon sugars. Therefore, loss in selectivity does not come from consecutive reactions of arabinose, but rather from side reactions of the solvent or water giving oxygen. The side reactions should be reduced when having a more uniform electrical field, i.e. the potential distribution is equal.

4.12.4.4 Experimental Protocols
[P 66] No details of the electrical field are given in [65] other than mentioning the average charge densities. These amount for two experiments to 1 and 5 A m^{-2}.

4.12.4.5 Typical Results
Faradaic current efficiency
[OS 86] [R 29] [P 66] The Faradaic current efficiency, the electrical charge equivalent for conversion as a fraction of the total electrical charge, was measured for a

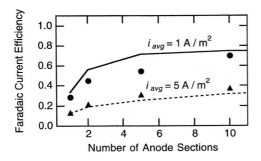

Figure 4.95 Faradaic current efficiency as electrical charge equivalent for conversion of D-gluconic acid given for two values of average current density. The symbols represent measured conversion and the solid and dashed lines calculated results [65].

multi-sectioned electrode design in a micro reactor as a function of the number of electrode sections when conducting the electrochemical decarboxylation of D-gluconic acid [65]. As desired by the multi-sectioned concept, this current efficiency increases with increasing number of anode sections and approaches a constant value after 8–10 anodes.

Average electrical current

[OS 86] [R 29] [P 66] On increasing the average electrical current from 1 to 5 A m^{-2} for the D-gluconic acid decarboxylation, the Faradaic current efficiency increases by a factor of 2–3 [65]. At best, a Faradaic current efficiency of about 75% is reached (Figure 4.95).

Modeling

[OS 86] [R 29] [P 66] A model is able to describe experimental results obtained for the electrochemical decarboxylation of D-gluconic acid (1 A m^{-2}) [65]. At an average electrical current of 5 A m^{-2} the model predicts better performance than is actually achieved.

4.12.5
Photochemical Reductive Biradical Coupling – Pinacol Formation

Peer-reviewed journals: [72, 73]; proceedings: [74]; sections in reviews: [14, 83, 89, 90, 115]; additional information: [73]; general theoretical analysis on optical photocatalytic dissociation: [140].

4.12.5.1 Drivers for Performing Photochemical Biradical Formation in Micro Reactors

The main driver here is not dedicated to the specific reaction, but rather stems from the general desire to establish photochemical paths for a wider range of reactions.

Photochemical reactions have the principal advantage of 'clean chemistry', as they use light of defined energy [72, 74]. Synthesis of vitamin D and photocleavage of protection groups, for example, are accepted organic synthesis routes. Nevertheless, no widespread use of photochemistry has been made so far as this technique suffers from lack of scaleability, efficiency and safety. As micro reactors are distinctively different here from other reactors, they may show completety different features with regard to such issues.

Large-scale reactors have low quantum yields as radiation does not penetrate deeply into the reaction vessel [72, 74]. As a consequence, high-power lamps have to be used causing a lot of excess heat and even posing safety constraints. These energy sources produce locally high quantities of radicals which may not mix thoroughly with the rest of the solution. Therefore, they may not find a second reaction partner, but instead react by themselves. This radical combination reduces selectivity and creates additional heat.

4.12.5.2 Beneficial Micro Reactor Properties for Photochemical Biradical Formation

Micro reactors may favorably use thin liquid films to have high quantum yields [72, 74]. The energy of the light should be usable completely as the light can penetrate the full layer, having small length scale, and is not absorbed to a large extent before inducing reaction.

Diffusion of the highly reactive radicals is facile in micro reactors so that recombination, as one source of selectivity reduction, should be diminished strongly [72]. Heat generation, even if it occurs at all, is under control since heat transfer can be set high.

See also the theoretical description of a micro reactor for optical photocatalytic dissociation of non-linear molecules in [140]. Here, a mathematical model for a novel type of micro reactor is given. Rotating non-linear molecules at excitation of valent vibrations are considered, having a magnetic moment. Resonance decay of molecules can be utilized with comparatively weak external energy sources only.

4.12.5.3 Photochemical Biradical Formation Investigated in Micro Reactors
Organic synthesis 87 [OS 87]: Benzopinacol formation by two-radical combination

The pinacol formation reaction follows a radical mechanism. Benzopinacol, benzophenone and the mixed pinacol are formed jointly with many radical species [72, 74]. In the course of the reaction, first a high-energy excited state is generated with the aid of photons. Thereafter, this excited-state species reacts with a solvent molecule 2-propanol to give two respective radicals. The 2-propanol radical reacts with one molecule of benzophenone (in the ground state, without photon aid) to lengthen the radical chain. By combination of radicals, adducts are formed, including the desired product benzopinacol. Chain termination reactions quench the radicals by other paths.

4.12.5.4 **Experimental Protocols**

[P 67] A stock standard solution of benzophenone in 2-propanol (0.5 M) with a drop of glacial acetic acid was diluted to give further standard solutions of 0.1–0.4 M [72, 74]. These solutions were stored at room temperature, protected from exposure to light and were employed for a device having an optical path of 50 µm. When using a second generation device with a 500 µm path, the stock solution concentration was reduced to 0.05 M.

The reactor was first primed with a cleaning solution, then with the reacting solution, and fed by pumping for a longer period [72, 74]. Then, the liquid flow was set to 1 ml min^{-1}. The samples were analyzed typically after 48 h to ensure completion of 'dark' follow-up reactions.

A miniaturized UV lamp (120 V AC; 0.20 A) and a deuterium detection lamp were used during the experiments [72, 74]. They were turned on for at least 20 min prior to use. UV spectra were obtained in a spectrometer with a 10 s integration time. Absorbance was determined at 280, 310, 330 and 360 nm. The power output of the lamp was measured by using a digital power meter and a reference silicon chip with channels etched through, which hence allowed the light to pass through and to be measured after such passage.

Besides on-line UV analysis, off-line HPLC analysis was performed after about 48 h 'dark reaction' [72, 74]. In this period, the photoinduced reaction proceeds by radical paths in the dark.

Owing to the high concentration applied (0.5 M), the Lambert–Beer law is no longer valid [72, 74]. Therefore, absorption coefficients were derived experimentally as described in detail in [72]. These measurements took advantage of having the thin layers in the micro reactor, i.e. they would not have been feasible in conventional cuvettes.

4.12.5.5 **Typical Results**
Conversion/selectivity/yield

[OS 87] [R 35] [P 67] Conversions of up to 20% of benzophenone were achieved [72, 74]. The conversion was measured by comparing the UV absorption spectra of reacted samples with those of standard solutions with defined degree of converted products.

Quality of first-generation on-line analysis

[OS 87] [R 35] [P 67] Generally, on-line analysis allows one to determine product concentrations, giving proper conversion [72, 74]. A limit is given by intermediates which also absorb in the same spectral range, also shifting the maximum of the absorption curve. It was assumed that the product pinacol does not contribute to the absorption, only the reactant and the intermediates. In addition, a delay arises between the end of irradiation and on-line analysis as the reaction mixture has to pass a conduit between the two locations. Hence the reaction proceeds further and this is dependent on the flow rate and residence time.

Quality of improved, second-generation on-line analysis
[OS 87] [R 35] [P 67] In a monolithic integrated device, the delay between the end of reaction and on-line analysis was reduced to only a few seconds [72, 74]. Consequently, higher absorbance, i.e. lower conversion, was found. In addition and surprisingly, this absorbance could be even higher than that of the unreacted solutions, indicating the identification of a short-living, strongly absorbing intermediate. This interference limited the quantitative judgement of the real performance of the second-generation analysis, but clearly its possibilities were improved.

Residence time/flow rate
[OS 87] [R 35] [P 67] The longer the residence time, the higher is the conversion, as expected [72, 74]. This trend is seen in the on-line UV- and off-line HPLC spectra. Whereas on-line UV absorption showed zero conversion at too short a residence time (flow rate 10 µl min^{-1}), a level of about 50% was found in the HPLC analysis. This clearly proves that the reaction proceeds by a radical path in the dark, if sufficient time is given.

Overall quantum efficiency – benchmarking to photochemical set-ups
[OS 87] [R 35] [P 67] Quantum efficiencies increase with flow rate and decrease with residence time [72, 74]. When the residence time is too long the light coming into the solution can no longer have a function. The quantum efficiencies, defined as moles reacted versus moles of photons, were of the order of 1–3.3, exceeding the performance of conventional photochemical set-ups at the higher end.

4.13
Organic Synthesis Reactions of Undisclosed Mechanism

4.13.1
Vitamin Precursor Synthesis

Peer-reviewed journals: [61, 62]; journals: [141]; proceedings: [127, 142, 143]; reactor description: [56, 58–60, 63]; sections in reviews: [42, 90, 105, 114, 115, 119, 144].

4.13.1.1 Drivers for Performing Vitamin Precursor Synthesis in Micro Reactors
The studies reported were mainly motivated by industrial investigations of process performance optimization [61, 62, 127, 142, 143]. For a known reaction which was part of a multi-step process finally yielding a vitamin, new process parameter data had to be analyzed, i.e. the micro reactor served to gather precise information. This concerned short-temperature processing (< 4 s), which simply was not possible using macroscopic bench-scale apparatus. In the latter case, almost 50% of the reaction heat was released already at the mixer unit, rather after entering the subsequent heat exchanger. The temperature rise led to side reactions, reducing the yield. Hence isothermal processing was the major issue of chosing micro-channel processing. Furthermore, the micro reactor has to allow short-time processing so

that residence time setting was another issue. The main reaction course and the side reactions were known to be very rapid. Therefore, avoiding back-mixing also had to be taken into account.

4.13.1.2 **Beneficial Micro Reactor Properties for Vitamin Precursor Synthesis**

Regarding the issue of isothermal processing, the excellent heat exchange properties of special micro-channel geometries are of major interest for the drivers outlined above [61, 62, 127, 142, 143]. Concerning this and the second issue, residence time setting and the integration of a micro mixing and heat exchange unit were inevitable. Concerning the device actually used, the mixer was directly followed by heat exchange micro channels having large heat exchange surfaces. This combination allowed processing in the second range and below. Owing to the laminar-flow properties, back-mixing can be largely excluded for micro-channel processing.

4.13.1.3 **Vitamin Precursor Synthesis Investigated in Micro Reactors**
Organic synthesis 88 [OS 88]: Synthesis of a vitamin precursor

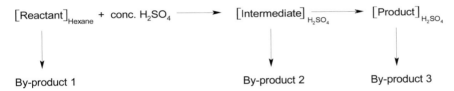

The reactants and the product were not disclosed in the open literature as the industrial process is proprietary [61, 62, 127, 142, 143]. The reactant is dissolved in hexane and the reaction is catalyzed by concentrated sulfuric acid which is present in quantitative amounts. Thus, the reaction is carried out as a liquid/liquid process. A reaction scheme is given in [61, 62]. The reactant quickly forms an intermediate which again quickly reacts to give the product. Thermally induced side reactions occur.

On an industrial level, the reaction is carried out in semi-batch mode with a yield of 70% [61, 62, 127]. First steps towards continuous production with a mixer and a cooler increased the yield to 80–85%; however, the reaction occurred partly in the mixer, thereby increasing the temperature in this unit. The reaction is quenched by dilution with water, also having a cooling function.

4.13.1.3 **Experimental Protocols**
[P 68] No detailed experimental protocol was given [61, 62, 142, 143]. Two reactant streams, the solution of the reactant in hexane and concentrated sulfuric acid, were fed separately in a specially designed micro reactor by pumping action. There, a bilayer was formed initially, potentially decomposed to a dispersion, and led to rapid mass transfer between the phases. From this point, temperature was controlled by counter-flow heat exchange between the reaction channel and surrounding heat-transfer channel. The reaction was typically carried out at temperatures from 0 to 50 °C and using residence times of only a few seconds. If needed, a delay loop of

larger internal dimensions served to prolong the reaction path and residence time. In a subsequent mini mixing unit, water was introduced and quenched the reaction.

4.13.1.4 **Typical Results**
Conversion/selectivity/yield – benchmarking to industrial process and laboratory-scale processing
[OS 88] [R 27] [P 68] A maximum yield of 80–85% was obtained at 4 s residence time and a temperature of 50 °C by micro reaction system processing [61, 62, 127, 142, 143]. Using ordinary laboratory-processing with standard laboratory glassware yielded only 25%. The continuous industrial process had a yield of 80–85%; the previously employed semi-batch industrial process gave a 70% yield. The temperature and the residence time of industrial and micro reactor continuous processing were identical.

Simultaneous residence time and temperature variation – benchmarking to industrial process
[OS 88] [R 27] [P 68] On increasing the residence time from 4 to 30 s and decreasing the reaction temperature from 50 of 20 °C, a yield of 90–95% was obtained [61, 62, 127, 142, 143]. This is a considerable improvement over the yield of the continuous industrial process of 80–85% (see also above).

Fouling – channel blockage
[OS 88] [R 27] [P 68] During extended laboratory studies of a micro reaction system, no blockage of the tiny micro channels (60 μm width) was found, although the use of reactants (viscous concentrated sulfuric acid) and the product mixtures known from industrial processing would have one led to such a conclusion [61, 62, 127, 142, 143]. As an explanation, the absence of temperature gradients and of averaged-velocity gradients between the individual flows was put forward.

4.13.2
Methylation of Aromatics

Proceedings: [127].

4.13.2.1 **Drivers for Performing Methylation of Aromatics in Micro Reactors**
To achieve excellent thermal control, to use short residence times and to have no back-mixing were the main drivers for an industrial investigation of the methylation of an aromatic. The target molecule yielded thereby, a precursor for a crop-protection product, is temperature sensitive [127]. Accordingly, cryo-processing has to be applied to avoid decomposition when the reaction is conventionally performed. A driver for micro reactor processing would be to enable room-temperature processing or, at least, to increase the reaction temperature closer to ambient.

4.13.2.2 **Beneficial Micro Reactor Properties for Methylation of Aromatics**
The above-mentioned research targets generally address the good heat transfer properties achieved by micro-channel processing, in particular referring to isother-

mal processing. Also, short average residence times can be realized largely avoiding back-mixing.

4.13.2.3 Methylation of Aromatics Investigated in Micro Reactors
Organic synthesus 89 [OS 89]: Methylation of a 1,3,4-substituted benzene

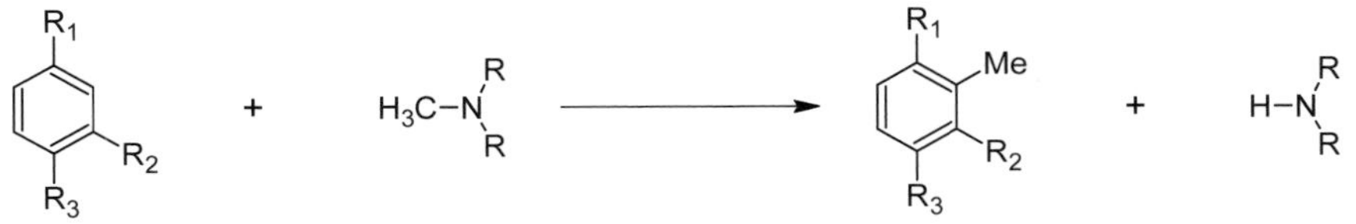

The 1,3,4-substituted benzene is reacted with an *N,N*-disubstituted-*N'*-methylamine resulting in transfer of the methyl group to the 2-position of the aromatic ring, thereby creating a crowded 1,2,3,4-substituted aromatic [127]. No details on the nature of the substituents and the presence of a solvent are given, as the process is proprietary. The temperature in the micro reactor was set to 0 °C.

4.13.2.4 Experimental Protocols
[P 69] No details on the solvent used and concentrations are given in [127], as the process most likely is proprietary (Figure 4.96). Probably the process is solvent-free as obviously one of the reactants has also the function of dissolving the other. The temperature for micro-channel processing was set to 0 °C. The residence time between the pre-reactor and micro mixer was 1 s and between the micro mixer and quench 5 s, totalling 6 s.

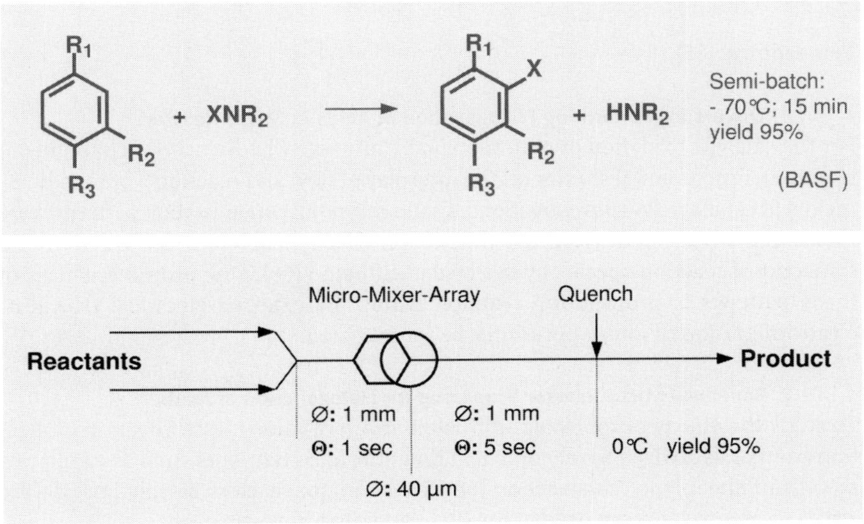

Figure 4.96 Flow sheet of the micro reactor set-up used for methylation of a 1,3,4-trisubstituted benzene [127].

4.13.2.5 **Typical Results**
Blockage – use of pre-reactor
[OS 89] [R 19] [P 69] Without a pre-reactor, blockage occurred owing to only partial solubility of one reactant in the other [127]. Processing was therefore not possible. With a tubular pre-reactor, having Y-shaped mixing guidance of the streams upstream the micro mixer, no blockage occurred for 0 °C processing. Probably, reaction was initiated, yielding some product which acted in a solution-enhancing manner.

Conversion/selectivity/yield – benchmarking to industrial process
[OS 89] [R 19] [P 69] Using a special reactor configuration with an interdigital micro-mixer array with pre-reactor, subsequent tubing and a quench, a yield of 95% at 0 °C was obtained [127]. The industrial semi-batch process had the same yield at – 70 °C.

Increase of reaction temperature close to ambient
[OS 89] [R 19] [P 69] Using the special reactor configuration outlined above, processing could be done at 0 °C, which is energetically much more favorable than the previous –70 °C industrial semi-batch processing [127].

4.14
Inorganic Reactions

4.14.1
Halogenation of Acids – The Belousov–Zhabotinskii Reaction

Proceedings: [68].

4.14.1.1 **Drivers for Performing Halogenation of Acids in Micro Reactors**
For the catalytic oxidation of malonic acid by bromate (the Belousov–Zhabotinskii reaction), fundamental studies on the interplay of flow and reaction were made. By means of capillary-flow investigations, spatio-temporal concentration patterns were monitored which stem from the interaction of a specific complex reaction and transport of reaction species by molecular diffusion [68]. One prominent class of these patterns is propagating reaction fronts. By external electrical stimulus, electromigration of ionic species can be investigated.

4.14.1.2 **Beneficial Micro Reactor Properties for Halogenation of Acids**
To study the effects of molecular diffusion and formation/destabilization of reaction fronts it is advised to rely on small flow-through chambers such as capillaries or cells of sheet-type cross-section [68]. These micro reactors simply provide the small-scale environment needed for such laboratory investigations.

4.14.1.3 Halogenation of Acids Investigated in Micro Reactors
Organic synthesis 90 [OS 90]: Catalytic oxidation of malonic acid by bromate

$$2\ Br^- + BrO_3^- + 3\ H^+ + 3\ H_2C \begin{smallmatrix} HO \\ =O \\ =O \\ HO \end{smallmatrix} \longrightarrow 3\ Br \begin{smallmatrix} HO \\ =O \\ =O \\ HO \end{smallmatrix} + 3\ H_2O$$

$$4\ Ce^{4+} + Br \begin{smallmatrix} HO \\ =O \\ =O \\ HO \end{smallmatrix} + 2\ H_2O \longrightarrow Br^- + 4\ Ce^{3+} + HCOOH + 2\ CO_2 + 5\ H^+$$

$$Br^- + BrO_3^- + 6\ H^+ \longrightarrow 3\ Br_2 + H_2O$$

$$Br- + HBrO_2 + H^+ \longrightarrow 2\ HOBr$$

$$BrO_3^- + HBrO_2 + H^+ \longrightarrow 2\ BrO_2 + H_2O$$

$$2\ BrO_2 + 2\ Ce^{3+} + 2\ H^+ \longrightarrow 2HBrO_2 + 2\ Ce^{4+}$$

$$BrO_3^- + HBrO_2 + Ce^{3+} + 2\ H^+ \longrightarrow Ce^{4+} + 2HBrO_2 + H_2O$$

This reaction, widely known as the Belousov–Zhabotinskii reaction, can proceed in an oscillatory fashion [68]. For overall slow conversion, the concentrations of intermediates and the catalyst undergo cyclic changes. By this means, many pulse-like reaction zones propagate in a spatially distributed system. Ferroin/ferriin can be applied as an optically detectable catalyst.

4.14.1.4 Experimental Protocols
[P 70] No details on the protocol are given in [68].

4.14.1.5 Typical Results
Weak electrical field stimulus
[OS 90] [R 31] [P 70] At weak electrical field, the propagation velocity of a reaction front in a capillary-flow reactor was increased or decreased depending on the mutual orientation of the electrical field and the reaction zone propagation [68]. The movement of two reaction fronts was given by optical images in [68].

Strong electrical field stimulus
[OS 90] [R 31] [P 70] At strong electrical field, global changes of the reaction in a capillary-flow reactor were induced which depend on the features of the reaction mechanism [68]. For fields of different strength, experimental findings showed that this involves the formation of several new reaction zones out off the original one, the reversal of the direction of the reaction zone propagation and termination of the reaction (Figure 4.97).

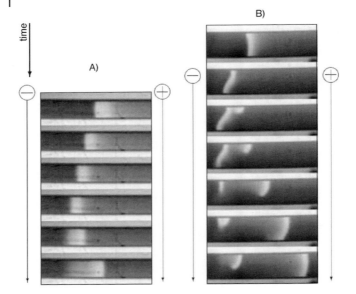

Figure 4.97 Global changes of the Belousov–Zhabotinskii reaction behavior in the capillary reactor under electrical field. (A) Reversal of the direction of the reaction zone (white stripe) propagation (0.5 mm capillary reactor); $E = 12$ V cm^{-1}; monitored length of the capillary $L = 3$ mm; Δt between snapshots is 20 s. (B) Break of the planar reaction zone in a 1 mm capillary reactor; $E = 6$ V cm^{-1}; $L = 4.4$ mm; $\Delta t = 40$ s [68].

Stability of reaction front

[OS 90] [R 31] [P 70] Capillary micro flow reactors of various inner cross-sections were used to study the stability of the reaction front and the homogeneity of the concentration profiles [68]. By concentration changes, hydrodynamic flow may be created which affects the reaction front stability. As expected, stronger influences were found for capillaries of larger cross-section. It was further found that the reaction front is perpendicular to the flow axis of the capillary for positive fields or when no field is applied (spatially one-dimensional case). For negative fields, bent or otherwise modified, reaction zones were found so that two- or three-dimensional propagation was induced. Splitting-off of zones and gradual spreading resulted from such geometry changes.

4.14.2
Redox Reaction Iodide/Iodate – Dushman Reaction

Proceedings: [22].

4.14.2.1 Drivers for Performing the Dushman Reaction

The Dushman reaction was chosen for its changes in UV–visible properties to demonstrate the capability of such in-line monitoring in a micro reactor [22].

4.14.2.2 Beneficial Micro Reactor Properties for the Dushman Reaction

The aim given above refers to the possibility of integration of multiple functions in a micro device; here, this refers to the combination of flow processing and sensing [22].

4.14.2.3 Dushman Reaction Investigated in Micro Reactors
Organic synthesis 91 [OS 91]: Redox reaction of iodate and iodide

$$6\,H^+ + IO_3^- + 5\,I^- \longrightarrow 3\,I_2 + 3\,H_2O$$

$$I_2 + I^- \longrightarrow I_3^-$$

This reaction is the reverse hydrolysis of iodine [22]. In a further reaction, the product iodine reacts with iodide to give triodide. Both iodine and triodide absorb in the visible region, hence the reaction can be monitored optically.

4.14.2.4 Experimental Protocols

[P 71] Only flow rates were given; no further details on process parameters were revealed [22]. The reactants were filtered and fed by syringe pumps.

Processing in the micro reactor was analyzed by a CCD camera with a long working distance magnifying lens [22]. Visible spectrometry was applied for in-line sensing. The change in product concentration was determined at 450 nm. The light was collected via an optical fiber and sent to the spectrometer.

4.14.2.5 Typical Results
Flow rate/residence time

[OS 91] [R 9] [P 71] By in-line spectrometry, the course of the reaction can be directly followed when changing the residence time (Figure 4.98). For decreasing reaction times, increasing flow rate, a decrease in absorption of the product was monitored, as expected [22].

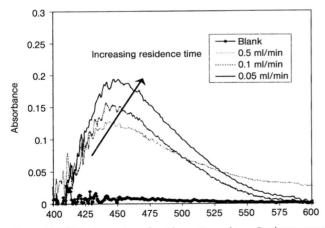

Figure 4.98 With decreasing of residence time a lower Dushman reaction product is observable, indicated by in-line spectrometry [22].

4.14.3
Oxidation of Arsenous Acid – Combined Dushman/Roebuck Reaction

Proceedings: [68, 145].

4.14.3.1 **Drivers for Performing Arsenous Acid Oxidation in Micro Reactors**
For arsenous acid oxidation, fundamental studies on the interplay of flow and reaction were made. By means of capillary-flow investigations, spatio-temporal concentration patterns were monitored which stem from the interaction of a specific complex reaction and transport of reaction species by molecular diffusion [68]. One prominent class of these patterns is propagating reaction fronts. By external electrical stimulus, electromigration of ionic species can be investigated.

4.14.3.2 **Beneficial Micro Reactor Properties for Arsenous Acid Oxidation**
To study effects of molecular diffusion and formation/destabilization of reaction fronts, it is advised to rely on small flow-through chambers such as capillaries or cells of sheet-type cross-section [68]. These micro reactors simply provide the small-scale environment needed for such laboratory inverstigations.

4.14.3.3 **Arsenous Acid Oxidation Investigated in Micro Reactors**
Organic synthesis 92 [OS 92]: Oxidation of arsenous acid by iodate

$$IO_3^- + 5\,I^- + 6\,H^+ \longrightarrow 3\,I_2 + 3\,H_2O$$

$$I_2 + H_3AsO_3 + H_2O \longrightarrow 2\,I^- + 3\,H_3AsO_4 + 2\,H^+$$

The arsenous acid–iodate reaction is a combination of the Dushman and Roebuck reactions [145]. These reactions compete for iodine and iodide as intermediate products. A complete mathematical description has to include 14 species in the electrolyte, seven partial differential equations, six algebraic equations for acid–base equilibriums and one linear equation for the local electroneutrality.

This reaction undergoes conversion in one sequence of consecutive elementary reaction steps and so only one propagating front is formed in a spatially distributed system [68]. Depending on the initial ratio of reactants, iodine as colored and iodide as uncolored product, or both, are formed [145].

4.14.3.4 **Experimental Protocols**
[P 72] No details on the protocol are given in [68] or [145].

4.14.3.5 **Typical Results**
Weak electrical field stimulus
[OS 92] [R 32] [P 72] At weak electrical field, the propagation velocity of a reaction front in a capillary-flow reactor could be increased or decreased depending on the mutual orientation of the electrical field and the reaction zone propagation [68]. The movement of two reaction fronts was given by optical images in [68].

Figure 4.99 Images of radial movement of reaction zones [68].

[OS 92] [R 32] [P 72] At weak electrical field, the propagation velocity of a reaction front in a capillary-flow reactor could be increased or decreased depending on the mutual orientation of the electrical field and the reaction zone propagation [68]. The movement of two reaction fronts (Figure 4.99) was given by optical images in [68].

Strong electrical field stimulus

[OS 92] [R 32] [P 72] The iodate–arsenous acid reaction proceeds to one of two stationary states in different parts of the capillary when an electrical field of specific strength is applied [68]. Accordingly, a spatially inhomogeneous distribution of reaction products is generated along the capillary.

Wave splitting by electrical stimulus

[OS 92] [R 32] [P 72] For radial movement from a center position, wave splitting was found [68]. Two new reaction zones were formed from a part of the circular reaction zone (Figure 4.100).

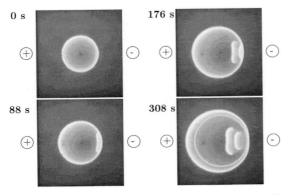

Figure 4.100 Wave splitting of the reaction zone (white circle) in an electric field. Top view of the monitored area (3 × 3 cm). Numbers show the time intervals after the electric field was switched on ($E = 4.06$ V cm^{-1}) [68].

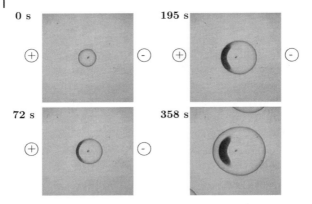

Figure 4.101 Formation of zones due to the change of reaction mechanism by applying an electrical field during the oxidation of arsenous acid by iodate ($E = 2.0$ V cm^{-1}). Numbers show the time intervals after the electric field was switched on. Intermediate product iodine (dark) and iodide (white) [68].

Electric field-induced switching between iodide and iodine

[OS 92] [R 32] [P 72] Iodide and iodine are autocatalysts. When an electrical field is applied, iodide moves to the positive electrode and, by this means, the propagation of the reaction is accelerated [145]. As a net result, more iodide is generated for fronts which approach the positive electrode; in turn, iodine is formed favorably for fronts propagating to the negative electrode. On switching the field off, the system comes back to the prior state.

Electric field-induced switching between iodide and iodine

[OS 92] [R 32] [P 72] A mathematical model based on transport and reaction was able to reproduce the experimental findings described above [145].

Changes of reaction mechanism by electrical stimulus

[OS 92] [R 32] [P 72] When switching on an electrical field, the reaction mechanism changes [68]. At zero field, iodine is formed as intermediate and converted to iodide. At non-vanishing field, iodine is the product, as evidenced by large colored zones. On switching off the field, uncolored iodide is formed again (Figure 4.101).

4.14.4
Landolt Reaction

Proceedings: [145].

4.14.4.1 Drivers for Performing Landolt Reactions in Micro Reactors
See Section 4.14.3.1.

4.14.4.2 Beneficial Micro Reactor Properties for Landolt Reactions
See Section 4.14.3.2.

4.14.4.3 Landolt Reactions Investigated in Micro Reactors
Organic synthesis 93 [OS 93]: Modified Landolt reaction

$$BrO_3^- + 3\,HSO_3^- \longrightarrow Br^- + 3\,SO_4^{2-} + 3\,H^+$$

This reaction is autocatalytic concerning the H+ ions [145]. The concentration of H^+ ions increases within the reaction front. For this reason, acid–base indicators can monitor the front propagation.

4.14.4.4 Experimental Protocols
[P 73] No details on the protocol are given in [145].

4.14.4.5 Typical Results
Change of homogeneous radial front propagation by unidirectional electrical field
[OS 93] [R 31] [P 73] When applying a uni-directional electrical field, migration of H^+ ions induces propagation of the reaction front towards the negative electrode and supresses the propagation towards the negative electrode [145]. Accordingly, the speed of the reaction front depends on the orientation to the electrode axis. As a result, the formerly homogeneously radial moving front is distorted towards the positive electrode.

Modeling the effect of axial dispersion
[OS 93] [R 31] [P 73] Using a simplified modeling approach, it was shown that axial dispersion changes the direction and shape of moving reaction fronts and also affects the interplay between dispersion and migration in an electrical field [145].

4.14.5
Transition Metal–Ligand Complex Formation – Co(II) Complexes

Peer-reviewed journals: [28].

4.14.5.1 Drivers for Performing Co(II) Complex Formations in Micro Reactors
Co(II) complex formation is the essential part of copper wet analysis. The latter involves several chemical unit operations. In a concrete example, eight such operations were combined – two-phase formation, mixing, chelating reaction, solvent extraction, phase separation, three-phase formation, decomposition of co-existing metal chelates and removal of these chelates and reagents [28]. Accordingly, Co(II) complex formation serves as a test reaction to perform multiple unit operations on one chip, i.e. as a chemical investigation to validate the Lab-on-a-Chip concept.

4.14.5.2 **Beneficial Micro Reactor Properties for Co(II) Complex Formations**

Micro reactors, in particular chip-based systems, offer a high degree of integration of diverse processing units in a confined space, typically not exceeding a few centimeters length scale (for width and length of chip). By flow-through operation, diverse unit operations can be performed on a time scale just long enough for completion of performance and be directly combined with the next operational step. As a result, operations that may require many manual steps at the macro-scale may here be carried out at once just by using pumping or electrical action for feeding.

4.14.5.3 **Co(II) Complex Formations Investigated in Micro Reactors**
Organic synthesis 94 [OS 94]: Complex formation between Co(II) and 2-nitroso-1-naphthol

4.14.5.4 **Experimental Protocols**

[P 74] Aqueous solutions of Co(II) with concentrations of 0–1.5×10^{-7} M in the presence of 1.0×10^{-6} M Cu(II) were introduced into a three-feed contactor [28]. One other feed was charged with an aqueous mixture solution of 2-nitroso-1-naph-thol and NaOH. The two aqueous phases merged, induced the chelating reaction to give a colored complex, and were contacted with *m*-xylene fed via the third line. The aqueous and xylene phases formed a bi-layer in one micro channel with spe-cial structuring of the channel bottom supporting flow guidance of continuous streams and avoiding intermixing of phases. After such extraction of the metal ions, the xylene stream was encompassed by two aqueous streams. One stream contained NaOH and cleaved the cupper complex, while not affecting the cobalt complex. Thereby, copper ions were released and moved to the second organic phase containing HCl. Analysis of concentration was done optically with a thermal lens microscope (TLM).

4.14.5.5 **Typical Results**
Conversion/selectivity/yield

[OS 94] [R 13] [P 74] For admixture of samples with varying concentrations of Co(II) and Cu(II), the respective changes in the Co(II) chelate complex concentration as a function of contact time were optically derived [28]. Analysis was performed in the reaction/extraction area and also in the decomposition/removal area (Figure 4.102). As expected, more complex is formed in the reaction/extraction area with increas-ing contact time. Also, more complex results when increasing the Co(II) concen-tration at constant Cu(II) concentration. This proves that mass transfer is efficient (as high concentrations can also be handled) and that no interference from other analytes falsifies the measurement. As a result, calibration curves were derived.

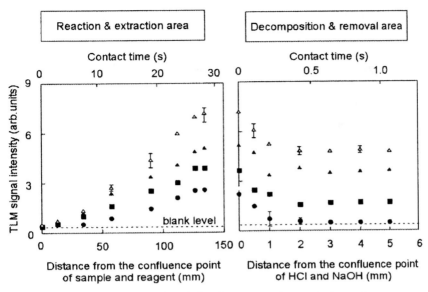

Figure 4.102 Dynamic evolution of Co chelate formation in the reaction/extraction and decomposition/removal areas at various concentrations: (△) 1.5×10^{-7}; (▲) 1.0×10^{-7}; (■) 5.0×10^{-8}; (●) 0 M Co(II) [28].

Analysis time
[OS 94] [R 13] [P 74] The analysis time for one sample, less than 1 min, is considerably faster than that of a conventional system, which needs about 2 h [28].

4.14.6
Transition Metal–Ligand Complex Formation – Nickel–Pyridine Complexes

4.14.6.1 Drivers for Performing Ni–Pyridine Complex Formations in Micro Reactors
The reversible formation of a complex by Ni^{2+} ions and the bidentate ligand pyridine-2-azo-*p*-dimethylaniline is a simple and thus reliable reaction, not accompanied by side reactions [17]. Kinetic rate law and rate constants for the reaction are known. The time demand of the reaction fits the short time scales typical for micro reactors. The strong absorption and the strong changes by reaction facilitate analysis of dynamic and spatial concentration profiles.

Accordingly, the above-mentioned complex formation was used as a test reaction to gather physico-chemical parameters and to validate quantitatively predictions of the spatial and temporal evolutions of concentrations [17].

4.14.6.2 Beneficial Micro Reactor Properties for Ni-Pyridine Complex Formations
Owing to its nature as a test reaction, not the reaction itself but rather the microchannel flow was the focus of the investigations. Hence nothing is to be said here on the beneficial micro reactor properties for the reaction applied.

4.14.6.3 **Ni–Pyridine Complex Formations Investigated in Micro Reactors**
Organic synthesis 95 [OS 95]: Complex formation from Ni^{2+} ions and
pyridine-2-azo-*p*-dimethylaniline

Yellow (λ_{max} = 450 nm)   Purple (λ_{max} = 550 nm)

The changes in absorption spectra due to the complex formation are given in [17].

4.14.6.4 **Experimental Protocols**
[P 75] The protocol relies on sequential filling of selected channels or parts of them in a chip micro reactor [17]. Thus, a short description of the micro reactor flow configuration is needed to understand details of the protocol.

The chip micro reactor comprises a long micro channel connected to two vertically positioned shorter channels at each end, which lead to two reservoirs (this sequence of three channels is named in the following the 'main channel') [17]. These shorter channels are oriented in opposite directions so that a Z-type flow configuration results. In the front section of the long channel, two other Z-types oriented shorter channels ('ligand insertion channels') are also attached, thereby defining a channel segment of the long channel ('ligand slug segment'). These channels are each connected to a liquid reservoir.

Before the start of the reaction, all channels were filled with 30 vol.-% ethanol in water containing 0.05 M Tris buffer [17]. The same solvent mixture with 2.1 mM Ni^{2+} was loaded into the respective reservoir using a micro-syringe. By setting the voltage properly, the main channel of the micro device was filled with the Ni^{2+} solution by electrophoresis without entering the two ligand insertion channels, adjacent to the main channel. This was termed 'flow mode'. Thereafter, a solvent mixture with 2.1 mM pyridine-2-azo-*p*-dimethylaniline was loaded into the respective reservoir. By EOF, the two ligand insertion channels encompassing the slug segment of the main channel were filled. This was termed 'insertion mode'. Mixing and reaction were initiated by restarting the flow mode.

4.14.6.5 **Typical Results**
Flow mode

[OS 95] [R 5] [P 75] Plugs of pyridine-2-azo-*p*-dimethylaniline were inserted in a continuous Ni^{2+} ion stream. Thereafter, spatial concentration profiles were monitored in the micro channels as a result of interpenetrating flow fronts due to varying mobility of the species in an electrical field [17].

Depending on what species port was 'activated', i.e. which reactant stream was moved, different flow modes were available, deliberately changing the concentration profiles in a predetermined manner [17]. These flow modes were termed 'flow mode', 'inject mode' and 'restarted flow mode', corresponding to Ni^{2+} ion channel filling, ligand slug injection and Ni^{2+} ion movement after slug insertion, respectively.

At the Ni^{2+} ion solution/ligand slug interface, the complex is formed, as evidenced by the color change [17]. Since the formation is reversible, decomposition of the complex can also be observed, on reducing the content of pyridine-2-azo-*p*-dimethylaniline.

Comparison with modeling

[OS 95] [R 5] [P 75] Numerical calculations were performed which were based on electroosmotic flow, electrophoresis, diffusion and chemical reaction [17]. The concentrations of Ni^{2+} ions, pyridine-2-azo-*p*-dimethylaniline and the respective complex as functions of both time and channel position were determined, describing mixing and reaction in a slug-insertion mode governed by electrophoretic mobility (see also [14]). The calculations were mainly in accordance with the experimental results. In particular, this refers to the formation of a relatively narrow peak of the complex. In turn, the concentration profiles at the edges of the ligand slug were not adequately described.

4.14.7
Diverse Inorganic Reactions

4.14.7.1 **Ionic Chemical Systems for Electrolyte Diode and Transistors**
Simple chemical systems with several components (HCl, KOH, KCl in hydrogel) were used for modeling mass and charge balances coupled with equations for electric field, transport processes and equilibrium reactions [146]. This served for demonstrating the chemical systems' function as electrolyte diodes and transistors, so-called 'electrolyte-microelectronics'.

References

1 NIELSEN, C. A., CHRISMAN, R. W., LAPOINTE, E., MILLER, T. E., *Novel tubing microreactor for monitoring chemical reactions,* Anal. Chem. **74**, 13 (**2002**) 3112–3117.

2 HASWELL, S. J., O'SULLIVAN, B., STYRING, P., *Kumada–Corriu reactions in a pressure-driven microflow reactor,* Lab. Chip **1** (**2001**) 164–166.

3 KUNZ, U., KIRSCHNING, A., *A new microreactor for the solution-phase synthesis of potential drugs,* in MATLOSZ, M., EHRFELD, W., BASELT, J. P. (Eds.), *Microreaction Technology – IMRET 5: Proc. 5th International Conference on Microreaction Technology,* pp. 424–445, Springer-Verlag, Berlin (**2001**).

4 WOOTTON, R. C. R., FORTT, R., DE MELLO, A. J., *On-chip generation and reaction of unstable intermediates – monolithic nanoreactors for diazonium chemistry: azo dyes,* Lab. Chip **2** (**2002**) 14N–21N.

5 WATTS, P., WILES, C., HASWELL, S. J., POMBO-VILLAR, E., *Solution phase synthesis of β-peptides using micro reactors,* Tetrahedron **58**, 27 (**2002**) 5427–5439.

6 SKELTON, V., GREENWAY, G. M., HASWELL, S. J., STYRING, P., MORGAN, D. O., *Micro-reactor synthesis: synthesis of cyanobiphenyls using a modified Suzuki coupling of an aryl halide and aryl boronic acid,* in EHRFELD, W. (Ed.), *Microreaction Technology: 3rd International Conference on Microreaction Technology, Proc. of IMRET 3,* pp. 235–242, Springer-Verlag, Berlin (**2000**).

7 GREENWAY, G. M., HASWELL, S. J., MORGAN, D. O., SKELTON, V., STYRING, P., *The use of a novel microreactor for high troughput continuous flow organic synthesis,* Sens. Actuators B: Chemical **63**, 3 (**2000**) 153–158.

8 WILES, C., WATTS, P., HASWELL, S. J., POMBO-VILLAR, E., *1,4-Addition of enolates to α,β-unsaturated ketones within a micro reactor,* Lab. Chip **2** (**2002**) 62–64.

9 GARCIA-EGIDO, E., WONG, S. Y. F., *A Hantzsch synthesis of 2-aminothiazoles performed in a microreactor system,* in RAMSEY, J. M., VAN DEN BERG, A. (Eds.), *Micro Total Analysis Systems,* pp. 517–518, Kluwer Academic Publishers, Dordrecht (**2001**).

10 GARCIA-EGIDO, E., WONG, S. Y. F., WARRINGTON, B. H., *A Hantzsch synthesis of 2-aminothiazoles performed in a heated microreactor system,* Lab. Chip **2** (**2002**) 31–33.

11 SANDS, M., HASWELL, S. J., KELLY, S. M., SKELTON, V., MORGAN, D., STYRING, P., WARRINGTON, B., *The investigation of an equilibrium dependent reaction for the formation of enamines in a microchemical system,* Lab. Chip **1** (**2001**) 64–65.

12 HASWELL, S. J., *Miniaturization – what's in it for chemistry,* in VAN DEN BERG, A., RAMSAY, J. M. (Eds.), *Micro Total Analysis System,* pp. 637–639, Kluwer Academic Publishers, Dordrecht (**2001**).

13 SKELTON, V., GREENWAY, G. M., HASWELL, S. J., STYRING, P., MORGAN, D. O., H., WARRINGTON, B. H., WONG, S., *Micro reaction technology: synthetic chemical optimisation methodology of Wittig synthesis enabling a semi-automated micro reactor for combinatorial screening,* in *Proceedings of the 4th International Conference on Microreaction Technology, IMRET 4,* 5–9 March 2000, pp. 78–88, AIChE Topical Conf. Proc., Atlanta, GA (**2000**).

14 FLETCHER, P. D. I., HASWELL, S. J., POMBO-VILLAR, E., WARRINGTON, B. H., WATTS, P., WONG, S. Y. F., ZHANG, X., *Micro reactors: principle and applications in organic synthesis,* Tetrahedron **58**, 24 (**2002**) 4735–4757.

15 WILES, C., WATTS, P., HASWELL, S. J., POMBO-VILLAR, E., *The aldol reaction of silyl enol ethers within a micro reactor,* Lab. Chip **1** (**2001**) 100–101.

16 SKELTON, V., HASWELL, S. J., STYRING, P., WARRINGTON, B., WONG, S., *A micro-reactor device for the Ugi four component condensation (4CC) reaction,* in RAMSEY, J. M., VAN DEN BERG, A. (Eds.), *Micro Total Analysis Systems,* pp. 589–590, Kluwer Academic Publishers, Dordrecht (**2001**).

17 FLETCHER, P. D. I., HASWELL, S. J., ZHANG, X., *Electrokinetic control of a chemical reaction in a lab-on-a-chip micro-*

reactor: measurement and qualitative modeling, Lab. Chip **2** (**2002**) 102–112.

18   FERNANDEZ-SUAREZ, M., WONG, S. Y. F., WARRINGTON, B. H., *Synthesis of a three-member array of cycloadducts in a glass microchip under pressure driven flow*, Lab. Chip **2** (**2002**) 170–174.

19   WILSON, N. G., MCCREEDY, T., *On-chip catalysis using a lithographically fabricated glass microreactor – the dehydration of alcohols using sulfated zirconia*, Chem. Commun. **9** (**2000**) 733–734.

20   GARCIA-EGIDO, E., SPIKMANS, V., WONG, S. Y. F., WARRINGTON, B. H., *Synthesis and analysis of combinatorial libraries performed in an automated micro-reactor system*, Lab. Chip **3** (**2003**) 67–72.

21   FLOYD, T. M., JENSEN, K. F., SCHMIDT, M. A., *Towards integration of chemical detection for liquid phase microchannel reactors*, in *Proceedings of the 4th International Conference on Microreaction Technology, IMRET 4*, 5–9 March 2000, pp. 461–466, AIChE Topical Conf. Proc., Atlanta, GA (**2000**).

22   FLOYD, T. M., LOSEY, M. W., FIREBAUGH, S. L., JENSEN, K. F., SCHMIDT, M. A., *Novel liquid phase microreactors for safe production of hazardous specialty chemicals*, in EHRFELD, W. (Ed.), *Micro-reaction Technology: 3rd International Conference on Microreaction Technology, Proc. of IMRET 3*, pp. 171–180, Springer-Verlag, Berlin (**2000**).

23   KIKUTANI, Y., HORIUCHI, T., UCHIYAMA, K., HISAMOTO, H., TOKESHI, M., KITAMORI, T., *Glass microchip with three-dimensional microchannel network for 2 × 2 parallel synthesis*, Lab. Chip **2** (**2002**) 188–192.

24   KIKUTANI, Y., HISAMOTO, H., TOKESHI, M., KITAMORI, T., *Fabrication of a glass micro-chip with three-dimensional channel network and its application to a single-chip combina-torial synthetic reactor*, in RAMSEY, J. M., VAN DEN BERG, A. (Eds.), *Micro Total Ana-lysis Systems*, pp. 161–162, Kluwer Aca-demic Publishers, Dordrecht (**2001**).

25   MITCHELL, M. C., SPIKMANS, V., BESSOTH, F., MANZ, A., DE MELLO, A., *Towards organic synthesis in microfluidic devices: multicomponent reactions for the construction of compound libraries*, in VAN DEN BERG, A., OLTHUIS, W., BERGVELD,

P. (Eds.), *Micro Total Analysis Systems*, pp. 463–465, Kluwer Academic Publishers, Dordrecht (**2000**).

26   KANNO, K., MAEDA, H., IZUMO, S., IKUMO, M., TAKESHITA, K., TASHIRO, A., FUJII, M., *Rapid enzymatic transclyco-sylation and oligosaccharide synthesis in a microchip reactor*, Lab. Chip **2** (**2002**) 15–18.

27   KANNO, K., FUJII, M., *Microreactor: new device for organic and enzymatic synthesis*, Yuki-Gosei-Kagagu-Kyokai-shi **60**, 7 (**2002**) 701–707 (in Japanese).

28   TOKESHI, M., MINAGAWA, T., UCHIYAMA, K., HIBARA, A., SATO, K., HISAMOTO, H., KITAMORI, T., *Integration of chemical processin on a microchip*, in RAMSEY, J. M., VAN DEN BERG, A. (Eds.), *Micro Total Analysis Systems*, pp. 533–534, Kluwer Academic Publishers, Dordrecht (**2001**).

29   UENO, K., KITAGAWA, F., KITAMURA, N., *Photocyanation of pyrene across an oil/water interface in a polymer microchannel chip*, Lab. Chip **2** (**2002**) 231–234.

30   WAN, Y. S. S., CHAU, J. L. H., GAVRIILIDIS, A., YEUNG, K. L., *Design and fabrication of zeolite-based micro-reactors and membrane microseparators*, Micropor. Mesopor. Mater. **42** (**2001**) 157–175.

31   BURNS, J. R., RAMSHAW, C., *Development of a microreactor for chemical production*, Trans. Inst. Chem. Eng. **77**, 5/A (**1998**) 206–211.

32   SCHWESINGER, K., FRANK, T., *A modular microfluid system with an integrated micromixer*, in *Proceedings of the MME '95*, pp. 144–147, Copenhagen (**1995**).

33   SCHWESINGER, W., FRANK, T., *A static micromixer built up in silicon*, in *Proceedings of Micromachining and Microfabrication*, pp. 150–155, SPIE, Austin TX (**1995**).

34   SCHWESINGER, N., FRANK, T., *Device for mixing small quantities of liquids*, WO 96/30113, Merck Patent GmbH, Darmstadt (**1995**).

35   SCHWESINGER, N., FRANK, T., WURMUS, H., *A modular microfluid system with an integrated micromixer*, J. Micromech. Microeng. **6** (**1996**) 99–102.

36   SCHWESINGER, N., *Mikrostrukturierte modulare Reaktionssysteme für die*

chemische Industrie, F & M, Feinwerk-technik, Mikrotechnik, Meßtechnik **110**, 4 (**2002**) 17–21.

37 ANTES, J., TUERCKE, T., MARIOTH, E., SCHMID, K., KRAUSE, H., LOEBBECKE, S., *Use of microreactors for nitration processes*, in *Proceedings of the 4th International Conference on Microreaction Technology, IMRET 4*, 5–9 March 2000, pp. 194–200, AIChE Topical Conf. Proc., Atlanta, GA (**2000**).

38 ANTES, J., TÜRCKE, T., MARIOTH, E., LECHNER, F., SCHOLZ, M., SCHNÜRER, F., KRAUSE, H. H., LÖBBECKE, S., *Investigation, analysis and optimization of exothermic nitrations in microreactor processes*, in MATLOSZ, M., EHRFELD, W., BASELT, J. P. (Eds.), *Microreaction Technology – IMRET 5: Proc. 5th International Conference on Microreaction Technology*, pp. 446–454, Springer-Verlag, Berlin (**2001**).

39 EHRFELD, W., GOLBIG, K., HESSEL, V., LÖWE, H., RICHTER, T., *Characterization of mixing in micromixers by a test reaction: single mixing units and mixer arrays*, Ind. Eng. Chem. Res. **38**, 3 (**1999**) 1075–1082.

40 HESSEL, V., HARDT, S., LÖWE, H., SCHÖNFELD, F., *Laminar mixing in different interdigital micromixers – Part I: experimental characterization*, AIChE J. **49**, 3 (**2003**) 566–577.

41 LÖWE, H., EHRFELD, W., HESSEL, V., RICHTER, T., SCHIEWE, J., *Micromixing technology*, in *Proceedings of the 4th International Conference on Microreaction Technology, IMRET 4*, 5–9 March 2000, pp. 31–47, AIChE Topical Conf. Proc., Atlanta, GA (**2000**).

42 EHRFELD, W., HESSEL, V., LÖWE, H., *Microreactors*, Wiley-VCH, Weinheim (**2000**).

43 LÖWE, H., SCHIEWE, J., HESSEL, V., DIETRICH, T., FREITAG, A., *Verfahren und statischer Mikrovermischer zum Mischen mindestens zweier Fluide*, DE 10041823, IMM Institut für Mikrotechnik Mainz GmbH, mgt mikroglas technik AG (**2000**).

44 FREITAG, A., DIETRICH, T. R., *Glass as a material for microreaction technology*, in *Proceedings of the 4th International Conference on Microreaction Technology,*

IMRET 4, 5–9 March 2000, pp. 48–54, AIChE Topical Conf. Proc., Atlanta, GA (**2000**).

45 FREITAG, A., DIETRICH, T. R., SCHOLZ, R., *Glass as a material for microreaction technology*, in **Proceedings of the VDE World Microtechnologies Congress, MICRO.tec 2000**, 25–27 September 2000, pp. 355–359, VDE Verlag, Berlin, EXPO Hannover (**2000**).

46 HERWECK, T., HARDT, S., HESSEL, V., LÖWE, H., HOFMANN, C., WEISE, F., DIETRICH, T., FREITAG, A., *Visualization of flow patterns and chemical synthesis in transparent micromixers*, in MATLOSZ, M., EHRFELD, W., BASELT, J. P. (Eds.), *Microreaction Technology – IMRET 5: Proc. 5th International Conference on Microreaction Technology*, pp. 215–229, Springer-Verlag, Berlin (**2001**).

47 DIETRICH, T. R., EHRFELD, W., LACHER, M., KRÄMER, M., SPEIT, B., *Fabrication technologies for microsystems utilizing photoetchable glass*, Micoelectron. Eng. **30** (**1996**) 497–504.

48 HESSEL, V., LÖWE, H., HOFMANN, C., SCHÖNFELD, F., WEHLE, D., WERNER, B., *Process development of fast reaction of industrial importance using a caterpillar micromixer/tubular reactor set-up*, in *Proceedings of the 6th International Conference on Microreaction Technology, IMRET 6*, 11–14 March 2002, pp. 39–54, AIChE Pub. No. 164, New Orleans (**2002**).

49 EHRFELD, W., HESSEL, V., KIESEWALTER, S., LÖWE, H., RICHTER, T., SCHIEWE, J., *Implementation of microreaction technology in process engineering*, in EHRFELD, W. (Ed.), *Microreaction Technology: 3rd International Conference on Microreaction Technology, Proc. of IMRET 3*, pp. 14–34, Springer-Verlag, Berlin (**2000**).

50 IMM, unpublished results.

51 KRAUT, M., NAGEL, A., SCHUBERT, K., *Oxidation of ethanol by hydrogen peroxide*, in *Proceedings of the 6th International Conference on Microreaction Technology, IMRET 6*, 11–14 March 2002, pp. 352–356, AIChE Pub. No. 164, New Orleans (**2002**).

52 TAGHAVI-MOGHADAM, S., KLEEMANN, A., GOLBIG, K. G., *Microreaction technology as a novel approach to drug design, process de-*

velopment and reliability, Organic Process Res. Dev. **5 (2001)** 652–658.

53 GOLBIG, K., TAGHAVI-MOGHADAM, S., BORN, P., *CYTOS^{TM}-technology – microreaction technology in practical sense*, in *Proceedings of the 6th International Conference on Microreaction Technology, IMRET 6, 11–14 March 2002*, pp. 131–134, AIChE Pub. No. 164, New Orleans **(2002)**.

54 TAGHAVI-MOGHADAM, S., GOLBIG, K., *Microreactors: application of CYTOS^{TM}-technology from laboratory to production scale*, MST News **3 (2002)** 36–38.

55 WILLE, C., AUTZE, V., KIM, H., NICKEL, U., OBERBECK, S., SCHWALBE, T., UNVERDORBEN, L., *Progress in transferring microreactors from lab into production – an example in the field of pigments technology*, in *Proceedings of the 6th International Conference on Microreaction Technology, IMRET 6, 11–14 March 2002*, pp. 7–17, AIChE Pub. No. 164, New Orleans **(2002)**.

56 RICHTER, T., EHRFELD, W., HESSEL, V., LÖWE, H., STORZ, M., WOLF, A., *A flexible multi component microreaction system for liquid phase reactions*, in EHRFELD, W. (Ed.), *Microreaction Technology: 3rd International Conference on Microreaction Technology, Proc. of IMRET 3*, pp. 636–644, Springer-Verlag, Berlin **(2000)**.

57 RICHTER, T., EHRFELD, W., WOLF, A., GRUBER, H. P., WÖRZ, O., *Microreactors in highly corrosion- and heat-resistant materials fabricated by various micro-EDM techniques*, in *Proceedings of Microreaction Technology, 1st International Conference on Microreaction Technology, IMRET 1, 23–25 February 1997, Frankfurt/M.* **(1997)**.

58 RICHTER, T., EHRFELD, W., GEBAUER, K., GOLBIG, K., HESSEL, V., LÖWE, H., WOLF, A., *Metallic microreactors: components and integrated systems*, in EHRFELD, W., RINARD, I. H., WEGENG, R. S. (Eds.), *Process Miniaturization: 2nd International Conference on Microreaction Technology, IMRET 2, Topical Conf. Preprints*, pp. 146–151, AIChE, New Orleans **(1998)**.

59 RICHTER, T., EHRFELD, W., WOLF, A., GRUBER, H. P., WÖRZ, O., *Fabrication of microreactor components by electro discharge machining*, in EHRFELD, W. (Ed.), *Microreaction Technology – Proc. of the 1st International Conference on Microreaction Technology, IMRET 1*, pp. 158–168, Springer-Verlag, Berlin **(1997)**.

60 WOLF, A., EHRFELD, W., LEHR, H., MICHEL, F., RICHTER, T., GRUBER, H., WÖRZ, O., *Mikroreaktorfertigung mittels Funkenerosion*, F & M, Feinwerktechnik, Mikrotechnik, Meßtechnik **105**, 6 **(1997)** 436–439.

61 WÖRZ, O., JÄCKEL, K.-P., RICHTER, T., WOLF, A., *Microreactors – a new efficient tool for reactor development*, Chem. Eng. Technol. **24**, 2 **(2001)** 138–143.

62 WÖRZ, O., JÄCKEL, K.-P., RICHTER, T., WOLF, A., *Mikroreaktoren – Ein neues wirksames Werkzeug für die Reaktorentwicklung*, Chem. Ing. Tech. **72**, 5 **(2000)** 460–463.

63 RICHTER, T., WOLF, A., JÄCKEL, J.-P., WÖRZ, O., *Mikroreaktoren, ein neues wirksames Werkzeug für die Reaktorentwicklung*, Chem. Ing. Tech. **71 (1999)** 973–974.

64 WEBER, M., TANGER, U., KLEINLOH, W., *Method and device for production of phenol and acetone by means of acid-catalyzed, homogeneous decoposition of cumene-hydroperoxid*, WO 01/30732, Phenolchemie GmbH **(1999)**.

65 MATLOSZ, M., VALLIERES, C., *Micro-sectioned electrochemical reactors for selective partial oxidation*, in EHRFELD, W., RINARD, I. H., WEGENG, R. S. (Eds.), *Process Miniaturization: 2nd International Conference on Microreaction Technology, IMRET 2, Topical Conf. Preprints*, pp. 54–59, AIChE, New Orleans **(1998)**.

66 SUGA, S., OKAJIMA, M., FUJIWARA, K., YOSHIDA, J.-I., *Cation Flow method: a new approach to conventional and combinatorial organic syntheses using electrochemical microflow systems*, J. Am. Chem. Soc. **123**, 32 **(2001)** 7941–7942.

67 SUGA, S., OKAJIMA, M., FUJIWARA, K., YOSHIDA, J., *New electrochemical micro flow sytem for CationFlow method*, in *Proceedings of the 6th International Conference on Microreaction Technology, IMRET 6, 11–14 March 2002*, pp. 29–31,

AIChE Pub. No. 164, New Orleans (2002).

68 SEVCIKOVA, H., SNITA, D., MAREK, M., *Reactions in microreactors in electric fields*, in EHRFELD, W. (Ed.), *Microreaction Technology – Proceedings of the 1st International Conference on Microreaction Technology, IMRET 1*, pp. 47–54, Springer-Verlag, Berlin (1997).

69 ZIOGAS, A., LÖWE, H., KÜPPER, M., EHRFELD, W., *Electrochemical micro-reactor: a new approach in microreaction technology*, in EHRFELD, W. (Ed.), *Microreaction Technology: 3rd International Conference on Microreaction Technology, Proc. of IMRET 3*, pp. 136–156, Springer-Verlag, Berlin (2000).

70 MENEGAUD, V., BAGEL, O., FERRIGNO, R., GIRAULT, H. H., HAIDER, A., *A ceramic electrochemical microreactor for the methoxylation of methyl-2-furoate with direct mass spectrometry coupling*, Lab. Chip 2 (2002) 39–44.

71 FERRIGNO, R., REID, V., GIRAULT, H. H., *Single flow electrochemical microreactor application to furan methoxylation*, in EHRFELD, W. (Ed.), *Microreaction Technology: 3rd International Conference on Microreaction Technology, Proc. of IMRET 3*, pp. 294–301, Springer-Verlag, Berlin (2000).

72 LU, H., SCHMIDT, M. A., JENSEN, K. F., *Photochemical reactions and on-line UV detection in microfabricated reactors*, Lab. Chip 1 (2001) 22–28.

73 JACKMAN, R. J., FLOYD, T. M., GHODSSI, R., SCHMIDT, M. A., JENSEN, K. F., *Microfluidic systems with on-line UV detection fabricated in photodefinable epoxy*, J. Micromech. Microeng. 11 (2001) 263–269.

74 LU, H., SCHMIDT, M. A., JENSEN, K. F., *Photochemical reactions and online product detection in microfabricated reactors*, in MATLOSZ, M., EHRFELD, W., BASELT, J. P. (Eds.), *Microreaction Technology – IMRET 5: Proc. 5th International Conference on Microreaction Technology*, pp. 175–184, Springer-Verlag, Berlin (2001).

75 BREMUS-KÖBBERLING, E., GILLNER, A., KÖBBERLING, J., ENDERS, D., BRANDTNER, S., *Development of a microreactor for solid phase synthesis*, in

MATLOSZ, M., EHRFELD, W., BASELT, J. P. (Eds.), *Microreaction Technology – IMRET 5: Proc. 5th International Conference on Microreaction Technology*, pp. 455–463, Springer-Verlag, Berlin (2001).

76 BREMUS-KÖBBERLING, E., GILLNER, A., KÖBBERLING, J., *Microreactor design for solid phase applications*, in *Proceedings of the VDE World Microtechnologies Congress, MICRO.tec 2000*, 25–27 September 2000, pp. 769–773, VDE Verlag, Berlin, EXPO Hannover (2000).

77 BREMUS, E., GILLNER, A., HELLRUNG, D., HÖCKER, H., LEGEWIE, F., POPRAWE, R., WEHNER, M., WILD, M., *Laser processing for manufacturing microfluidic devices*, in EHRFELD, W. (Ed.), *Microreaction Technology: 3rd International Conference on Microreaction Technology, Proc. of IMRET 3*, pp. 80–89, Springer-Verlag, Berlin (2000).

78 WERNER, B., DONNET, M., HESSEL, V., HOFMANN, C., JONGEN, N., LÖWE, H., SCHENK, R., ZIOGAS, A., *Specially suited micromixers for processes involving strong fouling*, in *Proceedings of the 6th International Conference on Microreaction Technology, IMRET 6*, 11–14 March 2002, pp. 168–183, AIChE Pub. No. 164, New Orleans (2002).

79 HESSEL, V., LÖWE, H., *Mikroverfahrenstechnik: Komponenten – Anlagenkonzeption – Anwenderakzeptanz – Teil 3*, Chem. Ing. Tech. 74, 4 (2002) 381–400.

80 HESSEL, V., LÖWE, H., *Micro chemical engineering: components – plant concepts – user acceptance: Part III*, Chem. Eng. Technol. 26, 5 (2003) 531–544.

81 SCHENK, R., HESSEL, V., WERNER, B., ZIOGAS, A., HOFMANN, C., DONNET, M., JONGEN, N., *Micromixers as a tool for powder production*, Chem. Eng. Trans. 1 (2002) 909–914.

82 HESSEL, V., personal communication (2002).

83 Schwalbe, T., Autze, V., Wille, G., *Chemical synthesis in microreactors*, Chimia 56, 11 (2002) 636–646.

84 GARCIA, E., FERRARI, F., GARCIA, T., MARTINEZ, M., ARACIL, J., *Use of micro-reactors in biotransformation processes: study of the synthesis of diglycerol mono-laurate ester*, in *Proceedings of the 4th*

International Conference on Microreaction Technology, IMRET 4, 5–9 March 2000, pp. 201–214, AIChE Topical Conf. Proc., Atlanta, GA (2000).

85 SCHWESINGER, N., MARUFKE, O., QIAO, F., DEVANT, R., WURZIGER, H., *A full wafer silicon microreactor for combinatorial chemistry,* in EHRFELD, W., RINARD, I. H., WEGENG, R. S. (Eds.), *Process Miniaturization: 2nd International Conference on Microreaction Technology, IMRET 2, Topical Conf. Preprints,* pp. 124–126, AIChE, New Orleans (1998).

86 WATTS, P., WILES, C., HASWELL, S. J., POMBO-VILLAR, E., *Investigation of racemization in peptide synthesis within a micro reactor,* Lab. Chip 2 (2002) 141–144.

87 WATTS, P., WILES, C., HASWELL, S. J., POMBO-VILLAR, E., STYRING, P., *The synthesis of peptides using micro reactors,* Chem. Commun. (2001) 990–991.

88 WATTS, P., WILES, C., HASWELL, S. J., POMBO-VILLAR, E., STYRING, P., *The synthesis of peptides using micro reactors,* in MATLOSZ, M., EHRFELD, W., BASELT, J. P. (Eds.), *Microreaction Technology – IMRET 5: Proc. 5th International Conference on Microreaction Technology,* pp. 508–519, Springer-Verlag, Berlin (2001).

89 HASWELL, S. T., WATTS, P., *Green chemistry: synthesis in micro reactors,* Green Chem. 5 (2003) 240–249.

90 JÄHNISCH, K., HESSEL, V., LÖWE, H., BAERNS, M., *Chemie in Mikrostrukturreaktoren,* Angew. Chem. 44, 4 (2003) in press.

91 BRIVIO, M., OOSTERBROEK, R. E., VERBOOM, W., GOEDBLOED, M. H., VAN DEN BERG, A., REINHOUDT, D. N., *Surface effects in the esterification of 9-pyrenebutyric acid within a glass micro reactor,* Chem. Commun. (2003) 1924–1925.

92 YOSHIDA, J.-I., NAGAKI, A., SUGA, S., *Highly selective reactions using microstructured reactors,* in *Proceedings of the Micro Chemical Plant – International Workshop,* 4 February 2003, pp. L2 (3–9), Kyoto (2003).

93 BÖKENKAMP, D., DESAI, A., YANG, X., TAI, Y.-C., MARZLUFF, E. M., MAYO, S. L., *Microfabricated silicon mixers for sub-millisecond quench flow analysis,* Anal. Chem. 70 (1998) 232–236.

94 DUMMANN, G., QUITMANN, U., GRÖSCHEL, L., AGAR, D. W., WÖRZ, O., MORGENSCHWEIS, K., *The capillary-microreactor: A new reactor concept for the intensification of heat and mass transfer in liquid–liquid reactions,* Catalysis Today, Special edition – 4th International Symposium on Catalysis in Multiphase Reactors, CAMURE IV 78–79, 3 (2002) 433–439.

95 BURNS, J. R., RAMSHAW, C., BULL, A. J., HARSTON, P., *Development of a microreactor for chemical production,* in EHRFELD, W. (Ed.), *Microreaction Technology – Proc. of the 1st International Conference on Microreaction Technology, IMRET 1,* pp. 127–133, Springer-Verlag, Berlin (1997).

96 BURNS, J. R., RAMSHAW, C., HARSTON, P., *Development of a microreactor for chemical production,* in EHRFELD, W., RINARD, I. H., WEGENG, R. S. (Eds.), *Process Miniaturization: 2nd International Conference on Microreaction Technology, IMRET 2, Topical Conf. Preprints,* pp. 39–44, AIChE, New Orleans (1998).

97 BURNS, J. R., RAMSHAW, C., *A micro-reactor for the nitration of benzene and toluene,* in *Proceedings of the 4th International Conference on Microreaction Technology, IMRET 4, 5–9 March 2000,* pp. 133–140, AIChE Topical Conf. Proc., Atlanta, GA (2000).

98 LÖBBECKE, S., ANTES, J., TÜRCKE, T., MARIOTH, E., SCHMID, K., KRAUSE, H., *The potential of microreactors for the synthesis of energetic materials,* in *Proceedings of the 31st Int. Annu. Conf. ICT, Energetic Materials – Analysis, Diagnostics and Testing,* 27–30 June 2000, Karlsruhe (2000).

99 GAVRIILIDIS, A., ANGELI, P., CAO, E., YEONG, K. K., WAN, Y. S. S., *Technology and application of microengineered reactors,* Trans. IChemE. 80/A, 1 (2002) 3–30.

100 HESSEL, V., LÖWE, H., *Micro chemical engineering: components – plant concepts – user acceptance: Part I,* Chem. Eng. Technol. 26, 1 (2003) 13–24.

101 Burns, J. R., Ramshaw, C., *The intensification of rapid reactions in multiphase systems using slug flow in capillaries*, Lab. Chip **1** (2001) 10–15.

102 Kasarekar, R. B., Ramakrishna, M., Juvekar, V. A., *Effect of intensity of agitation on selectivity of aromatic nitration*, Chem. Eng. Technol. **20** (1997) 282–284.

103 Doku, G. N., Haswell, S. J., McCreedy, T., Greenway, G. M., *Electric field-induced mobilisation of multiphase solution systems based on the nitration of benzene in a micro reactor*, Analyst **126** (2001) 14–20.

104 Hisamoto, H., Saito, T., Tokeshi, M., Hibara, A., Kitamori, T., *Fast and high conversion phase-transfer synthesis exploiting the liquid–liquid interface formed in a microchannel chip*, Chem. Commun. (2001) 2662–2663.

105 deWitt, S., *Microreactor for chemical synthesis*, Curr. Opin. Chem. Biol. **3** (1999) 350–356.

106 Geipel-Kern, A., *Vor dem Sprung in die Produktion – Trenbeitrag Mikroreaktionstechnik*, Chem. Produktion (2002) 28–30.

107 Salimi-Moosavi, H., Tang, T., Harrison, D. J., *Electroosmotic pumping of organic solvents and reagents in microfabricated reactor chips*, J. Am. Chem. Soc. **119** (1997) 8716.

108 Koch, M., Wehle, D., Scherer, S., Forstinger, K., Meudt, A., Hessel, V., Werner, B., Löwe, H., *Verfahren zur Herstellung von Aryl- und Alkyl-Bor-Verbindungen in Mikroreaktoren*, DE 10140857, Clariant GmbH, Frankfurt (2001).

109 Hessel, V., Hardt, S., Löwe, H., *Chemical processing with microdevices: device/plant concepts, selected applications and state of scientific/commercial implementation*, in Proceedings of the 6th Italian Conference on Chemical and Process Engineering, ICheaP-6, 8–11 June 2003, Pisa, Chem. Engin. Trans. **3**, pp. 479–484 (2003).

110 de Bellefon, Tanchoux, N., Caravieilhes, S., Grenouillet, P., Hessel, V., *Microreactors for dynamic high throughput screening of fluid–liquid molecular catalysis*, Angew. Chem. **112**, 19 (2000) 3584–3587.

111 de Bellefon, C., Abdallah, R., Lamouille, T., Pestre, N., Caravieilhes, S., Grenouillet, P., *High-throughput screening of molecular catalysts using automated liquid handling, injection and microdevices*, Chimia **56**, 11 (2002) 621–626.

112 de Bellefon, C., Caravieilhes, S., Grenouillet, P., *Application of a micromixer for the high throughput screening of fluid–liquid molecular catalysts*, in Matlosz, M., Ehrfeld, W., Baselt, J. P. (Eds.), *Microreaction Technology – IMRET 5: Proc. 5th International Conference on Microreaction Technology*, pp. 408–413, Springer-Verlag, Berlin (2001).

113 de Bellefon, C., *Application of microdevices for the fast investigation of catalysis*, in Proceedings of the Micro Chemical Plant International Workshop, 4 February 2003, pp. L3 (9–17), Kyoto (2003).

114 Hessel, V., Löwe, H., *Micro chemical engineering: components – plant concepts – user acceptance: Part II*, Chem. Eng. Technol. **26**, 4 (2003) 391–408.

115 Hessel, V., Löwe, H., *Mikroverfahrenstechnik: Komponenten – Anlagenkonzeption – Anwenderakzeptanz – Teil 2*, Chem. Ing. Tech. **74**, 3 (2002) 185–207.

116 Haverkamp, V., Ehrfeld, W., Gebauer, K., Hessel, V., Löwe, H., Richter, T., Wille, C., *The potential of micromixers for contacting of disperse liquid phases*, Fresenius' J. Anal. Chem. **364** (1999) 617–624.

117 Caravieilhes, S., de Bellefon, C., Tanchoux, N., *Dynamic methods and new reactors for liquid phase molecular catalysis*, Catal. Today **66** (2001) 145–155.

118 Taghavi-Moghadam, S., Kleemann, A., Overbeck, S., *Implications of microreactors on chemical synthesis*, in Proceedings of the VDE World Micro-technologies Congress, MICRO.tec 2000, 25–27 September 2000, pp. 489–491, VDE Verlag, Berlin, EXPO Hannover (2000).

119 Ehrfeld, W., Hessel, V., Haverkamp, V., *Microreactors*, in *Ullmann's Encyclopedia of Industrial Chemistry*, Wiley-VCH, Weinheim (1999).

120 Fukuyama, T., Shinmen, M., Nishitani, S., Sato, M., Ryu, I., *A cop-*

per-free Sonogashira coupling reaction in ionic liquids and its application to a microflow system for efficient catalyst recycling, Org. Lett. **4**, 10 (**2002**) 1691–1694.

121 DE BELLEFON, C., PESTRE, N., LAMOUILLE, T., GRENOUILLET, P., *High-throughput kinetic investigations of asymmetric hydrogenations with microdevices*, Adv. Synth. Catal. **345**, 1+2 (**2003**) 190–193.

122 WAN, Y. S. S., CHAU, J. L. H., GAVRIILIDIS, A., YEUNG, K. L., *TS-1 zeolite microengineered reactors for 1-pentene epoxidation*, Chem. Commun. (**2002**) 878–879.

123 WAN, Y. S. S., CHAU, J. L. H., GAVRIILIDIS, A., YEUNG, K. L., *Design and fabrication of zeolite-containing microstructures*, in MATLOSZ, M., EHRFELD, W., BASELT, J. P. (Eds.), *Microreaction Technology – IMRET 5: Proc. 5th International Conference on Microreaction Technology*, pp. 94–102, Springer-Verlag, Berlin (**2001**).

124 MENGEAUD, V., FERRIGNO, R., JOSSERAND, J., GIRAULT, H. H., *Towards a 'plant-on-a-chip' for the methoxylation of methyl-2-furoate: Electrochemical experiments and mixing simulations*, in MATLOSZ, M., EHRFELD, W., BASELT, J. P. (Eds.), *Microreaction Technology – IMRET 5: Proc. 5th International Conference on Microreaction Technology*, pp. 350–358, Springer-Verlag, Berlin (**2001**).

125 BAYER, T., PYSALL, D., WACHSEN, O., *Micro mixing effects in continuous radical polymerization*, in EHRFELD, W. (Ed.), *Microreaction Technology: 3rd International Conference on Microreaction Technology, Proc. of IMRET 3*, pp. 165–170, Springer-Verlag, Berlin (**2000**).

126 PYSALL, D., WACHSEN, O., BAYER, T., WULF, S., *Verfahren und Vorrichtung zur kontinuierlichen Hestellung von Polymerisaten*, DE 19816886, Aventis Research & Technologies GmbH (**1998**).

127 WÖRZ, O., *Microreactors as tools in chemical research*, in MATLOSZ, M., EHRFELD, W., BASELT, J. P. (Eds.), *Microreaction Technology – IMRET 5: Proc. 5th International Conference on Microreaction Technology*, pp. 377–386, Springer-Verlag, Berlin (**2001**).

128 HESSEL, V., LÖWE, H., *Mikroverfahrenstechnik: Komponenten – Anlagenkonzeption – Anwenderakzeptanz – Teil 1*, Chem. Ing. Tech. **74**, 2 (**2002**) 17–30.

129 UGI, I., ALMSTETTER, M., GRUBER, B., HEILINGBRUNNER, M., *MCR XII. Efficient development of new drugs by online-optimization of molecular libraries*, in EHRFELD, W. (Ed.), *Microreaction Technology – Proc. of the 1st International Conference on Microreaction Technology, IMRET 1*, pp. 190–194, Springer-Verlag, Berlin (**1997**).

130 UGI, I., ALMSTETTER, M., GRUBER, B., DÖMLING, A., *MCR X. Important aspects for automating preparative chemistry*, in EHRFELD, W. (Ed.), *Microreaction Technology – Proc. of the 1st International Conference on Microreaction Technology, IMRET 1*, pp. 184–189, Springer-Verlag, Berlin (**1997**).

131 GRUBER, B., ALMSTETTER, M., HEILINGBRUNNER, M., *MCR XI. Microreactors as processors for chemical computers*, in EHRFELD, W. (Ed.), *Microreaction Technology – Proc. of the 1st International Conference on Microreaction Technology, IMRET 1*, pp. 195–202, Springer-Verlag, Berlin (**1997**).

132 UGI, I. K., *New aspects of natural and preparative of one-pot multicomponent reactions and their libraries*, in EHRFELD, W., RINARD, I. H., WEGENG, R. S. (Eds.), *Process Miniaturization: 2nd International Conference on Microreaction Technology, IMRET 2, Topical Conf. Preprints*, pp. 111–117, AIChE, New Orleans (**1998**).

133 DÖMLING, A., UGI, I., *The seven component reaction*, Angew. Chem. Int. Ed. **32** (**1993**) 563–564.

134 KRUMMRADT, H., KOPP, U., STOLDT, J., *Experiences with the use of microreactors in organic synthesis*, in EHRFELD, W. (Ed.), *Microreaction Technology: 3rd International Conference on Microreaction Technology, Proc. of IMRET 3*, pp. 181–186, Springer-Verlag, Berlin (**2000**).

135 MARCH, J., *Advanced Organic Chemistry*, 4th ed., Wiley-Interscience, New York (**1992**).

136 SKELTON, V., GREENWAY, G., HASWELL, S., STYRING, P., MORGAN, D.,

WARRINGTON, B., WONG, S., *The design of a continuous flow combinatorial screening micro-reactor system with on-chip detection,* in VAN DEN BERG, A., OLTHUIS, W., BERGVELD, P. (Eds.), *Micro Total Analysis Systems,* pp. 59–62, Kluwer Academic Publishers, Dordrecht (**2000**).

137 MCCREEDY, T., WILSON, N. G., *Micro fabricated reactors for on-chip heterogeneous catalysis,* in *Proceedings of the Microreaction Technology – IMRET5: Proceedings of the 5th International Conference on Microreaction Technology,* 27–30 May 2001, Strasbourg (**2001**).

138 WILSON, N. G., MCCREEDY, T., *Microporous silica structures for the immobilization of catalysts and enhancement of electroosmotic flow (EOF) in micro reactors,* in EHRFELD, W. (Ed.), *Microreaction Technology: 3rd International Conference on Microreaction Technology, Proc. of IMRET 3,* pp. 346–352, Springer-Verlag, Berlin (**2000**).

139 LÖWE, H., KÜPPER, M., ZIOGAS, A., *Reaktor sowie Verfahren zur Durchführung elektrochemischer Umsetzungen,* DE 19841302, IMM Institut für Mikrotechnik Mainz GmbH (**1998**).

140 ANDREEV, V. V., *Microreactor for optical dissociation of non-linear molecules,* in *Proceedings of the VDE World Microtechnologies Congress, MICRO.tec 2000,* 25–27 September 2000, pp. 779–781, VDE Verlag, Berlin, EXPO Hannover (**2000**).

141 WÖRZ, O., JÄCKEL, K. P., *Winzlinge mit großer Zukunft – Mikroreaktoren für die Chemie,* Chem. Techn. **26,** 131 (**1997**) 130–134.

142 WÖRZ, O., JÄCKEL, K. P., RICHTER, T., WOLF, A., *Microreactors, a new efficient tool for optimum reactor design,* in EHRFELD, W., RINARD, I. H., WEGENG, R. S. (Eds.), *Process Miniaturization: 2nd International Conference on Microreaction Technology, IMRET 2, Topical Conf. Preprints,* pp. 183–185, AIChE, New Orleans (**1998**).

143 WÖRZ, O., JÄCKEL, K. P., RICHTER, T., WOLF, A., *Microreactors, new efficient tools for optimum reactor design,* in *Microtechnologies and Miniaturization, Tools, Techniques and Novel Applications for the BioPharmaceutical Industry,* IBC Global Conferences, London (**1998**).

144 WÖRZ, O., *Wozu Mikroreaktoren?* Chem. Unserer Zeit **34,** 1 (**2000**) 24–29.

145 SNITA, D., LINDNER, J., SEVCIKOVA, H., KOSEK, J., MAREK, M., *Microreactors for ionic reactions in liquids and gels in electric field,* in EHRFELD, W., RINARD, I. H., WEGENG, R. S. (Eds.), *Process Miniaturization: 2nd International Conference on Microreaction Technology, IMRET 2, Topical Conf. Preprints,* pp. 140–145, AIChE, New Orleans (**1998**).

146 PACES, M., HAVLICA, J., LINDNER, J., KOSEK, J., SNITA, D., MAREK, M., *Electrolyte microelectronics,* in *Proceedings of the VDE World Microtechnologies Congress, MICRO.tec 2000,* 25–27 September 2000, pp. 783–785, VDE Verlag, Berlin, EXPO Hannover (**2000**).

5
Gas/Liquid Reactions

5.1
Micro Reactors for Gas/Liquid Reactions

5.1.1
Gas/Liquid Micro Flow Contactors

These devices have gas and liquid streams which do not merge, i.e. are not fed into each other. Rather, by use of separate ports both phases are passed in their own encasing, e.g. their own micro channel. Thereby, dispersion of the phases is prevented and only one large specific interface is created for mass transfer (different from dispersive micro devices, see, e.g., Section 5.1.2). The separate guidance of the phases allows to heat the phases separately, usually from the side most distant from the gas/liquid interface.

The advantage of the two-phase micro flow contacting concept is easy phase separation, as the phases are never inter-mixed. However, in view of the normally facile separation of gases and liquids, this is not of major impact. A real large benefit stems from operating with gas and liquid layers of defined geometry with a known, defined interface, unlike most disperse systems having a size distribution of their bubbles in the continuous liquid.

As a further advantage, the two-phase contacting approach facilitates having the same flow patterns when operating in many parallel micro channels. The key for proper numbering-up is flow equipartition. Since for the contacting approach only single-phase distribution is needed for both phases, it stands to reason that this is generally much more easily achieved as for multi-phase distribution given in dispersive mixers (see Section 5.1.2). Accordingly, numbering-up seems to be practicable for two-phase contactors and, indeed, existing falling film micro reactors already comprise an array of parallel-operated micro channels (see [R 1]).

The disadvantage of the two-phase contacting principle is related to the technical expenditure of realizing phase separation throughout the complete reactor passage. Special measures have to be taken to prevent phase inter-mixing. Also, this has to be controlled during the process. Hence inspection windows are essential (for the first prototype; they may be eliminated later).

Figure 5.1 Photograph of a falling film micro reactor [1].

5.1.1.1 Reactor 1 [R 1]: Falling Film Micro Reactor

The falling film principle utilizes the wetting of a surface by a liquid stream, governed by gravity force, which thus spreads to form an expanded thin film.

The falling film micro reactor (Figure 5.1) transfers this well-known macro-scale concept to yield films of a few tens of micrometers thickness [1–3]. For this reason, the streams are guided through micro channels. To obtain a reasonable throughput, many micro channels are operated in parallel.

The liquid enters the micro channel device via a large bore that is connected to a micro channel plate via a slit (Figure 5.2). The slit acts as a flow restrictor and serves for equipartition of the many parallel streams [1, 3, 4]. The liquid streams are re-collected via another slit at the end of the micro structured plate and leave the device by a bore. The gas enters a large gas chamber, positioned above the micro channel section, via a bore and a diffuser and leaves via the same type of conduit.

By changing the flow direction of gas and liquid streams, a co-flow and counter-flow guidance is in principle possible; however, owing to the low gas velocities usually employed, this makes no practical difference [5].

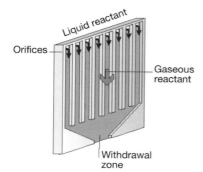

Figure 5.2 Principle of contacting liquid and gaseous reactants in a falling film micro reactor [5].

Internal heat exchange is realized by heat conduction from the microstructured reaction zone to a mini channel heat exchanger, positioned in the rear of the reaction zone [1, 3, 4]. The falling film micro reactor can be equipped, additionally, with an inspection window. This allows a visually check of the quality of film formation and identification of flow misdistribution. Furthermore, photochemical gas/liquid contacting can be carried out, given transparency of the window material for the band range of interest [6]. In some cases an inspection window made of silicon was used to allow observation of temperature changes caused by chemical reactions or physical interactions by an IR camera [4, 5].

The slit is made by μEDM; the micro channels are etched in the plate [1–3]. The gas chamber and the two diffusers are made by μEDM. The heat exchange channels are manufactured by micro milling.

Reactor type	Falling film micro reactor	Mini heat exchange channel width; depth	1500 μm; 500 μm
Housing and reaction plate material	Steel	Pressure stability	10 bar (20 bar without inspection window)
Heat exchange plate material	Copper	Temperature stability	Up to 180 °C
Outer dimensions (without connectors)	$120 \times 76 \times 40$ mm^3	Residence time	17 s (at 25 ml h^{-1})
Outer dimensions (with connectors)	$120 \times 128 \times 40$ mm^3	Interfacial area of liquid films	20 000 m^2 m^{-3} (at 25 ml h^{-1})
Number of micro channels	64	Active inner volume	100 mm^3 (per plate)
Micro reaction channel width; depth	300 μm; 100 μm	Total inner surface	1690 mm^2 (per plate)

5.1.1.2 Reactor 2 [R 2]: Continuous Two-phase Contactor with Partly Overlapping Channels

Solute transfer is performed here between immiscible phases each flowing in separate adjacent micro channels, only having as conduit a small, stable fluid interface [7]. This flow configuration is realized by having one micro channel each in two plates which are connected to a reactor sandwich. The position of these channels is so that their open channel sides do not completely overlap (as usual), but are displayed to result in partial overlap, covering more of the open area than releasing it as conduit. For this alignment, a special jig based on a mask aligner stage was applied, having a precision of positioning of ±2 μm.

The two plates were not manufactured via the same route and were not made of the same material [7]. Typically, rectangular channels in silicon are realized by sawing, whereas semi-circular channels are made in glass by wet-chemical etching. Such glass/silicon plates are joined by anodic bonding.

Several publications have referred to calculating the correct fluid layer thickness for efficient mass transfer and to determine the area of the interface sufficient to build up enough pressure to stabilize the continuous flow of the two phases and to prevent intermixing [8]. The first quantity should be below 100 µm; the channel opening should be about 20 µm.

Meanwhile, a numbered-up module was developed using the partly overlapping channels [7]; 120 single micro channels are operated here in parallel.

Reactor type	Continuous two-phase contactor with partly overlapping channels	Silicon micro channel: width; depth	Both typically 50–80 µm
Micro channels: materials	Silicon; glass	Channel opening	20 µm
Glass micro channel: diameter	35 µm	Number of micro channels per device	1–120

5.1.2
Gas/Liquid Micro Flow Dispersive Mixers Generating Slug and Annular Patterns

These devices have gas and liquid streams which merge, i.e. are fed into each other, e.g. by special dual-feed, triple-feed or multiple feed arrangements. From then, both phases are passed in the same encasing, e.g. the same micro channel. As a result of phase instability, fragmentation of the gas jet occurs under given conditions, forming a dispersion. When very small bubbles, below the characteristic length of the micro device, are generated in this way, bubbly flow is observed. More commonly, slug flow composed of alternating gas and liquid segments is achieved. At very large gas flow rates, annular flow is obtained, i.e. a thin liquid shell stream surrounding a large gas core. This pattern is similar to contacting gas and liquids in a non-dispersive manner as described in Section 5.1.1. However, the two-phase flow achieved here should be not as stable, probably creating more wavy structures, and spray phenomena (partial dispersion of the flow) are also known [9, 10]. Slug and annular flow provide very large specific interfaces for mass transfer; the latter gives even larger ones than the first.

The advantage of the dispersing principle is related to the relatively low technical expenditure to achieve dispersion, i.e. the simplicity of the concept. However, as flow patterns may change and are not known for new systems, they have to be identified, documented as flow-pattern maps and controlled. Thus, some analytical characterization has to be done in advance of the experiment. Hence inspection windows again are essential (for the first prototype; they may be eliminated later).

A slight disadvantage of the concept is phase separation, as the phases are thoroughly inter-mixed. In contrast to liquid/liquid dispersion, the gas/liquid separation should be, however, not nearly as troublesome. Another more serious drawback stems from the disperse nature of the systems involving a size distribution of the initial bubbles in the continuous liquid, which can be rather broad. By this means, slugs of varying length may be produced. The interface is not as defined as for two-phase continuous reactors, as described in Section 5.1.1.

As a further disadvantage, it is known concerning operation in many parallel micro channels that mixed flow patterns and even drying of the channels can occur [9, 10]. This comes from phase maldistribution to the channels. To overcome this problem, first solutions for phase equipartition have been proposed recently, but so far have not been applied for the mixers described here, but instead for mini-packed reactors, having feed sections similar to the mixers [11, 12]. Nevertheless, numbering-up of dispersive-acting micro devices generally seems to be more complicated than for two-phase contactors (see Section 5.1.1).

5.1.2.1 Reactor 3 [R 3]: Micro Bubble Column

The micro bubble column uses dispersion of gas in a liquid stream (Figure 5.3) (and to a minor extent dispersion of liquid in a gas, as e.g. given for spray flow) [2, 3, 9]. The naming of the micro channel device stems from the prevailing flow pattern related to the guidance bubbles through a continuous liquid medium [3, 9, 10]. On the micro scale, the *slug flow* pattern, comprising bubbles of a diameter approaching that of the micro channel (Taylor bubbles) and segmented by liquid slugs, has a large range of stability; the flow pattern *bubbly flow*, which is on the macro-scale a dominating flow pattern, is also found on the micro scale, but with limited stability [9, 10]. In addition, other flow patterns known from the macro scale such as *annular-* or *spray-flow* patterns were also identified in the micro bubble column [9, 10]. The annular-flow pattern has the largest internal surface, as one uniform thin liquid film is formed. This is notably different from slug flow with co-existence of thin liquid wall-wetting films and thick liquid bubble segmenting slugs.

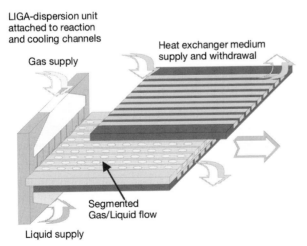

LIGA-dispersion unit
attached to reaction
and cooling channels

Gas supply

Heat exchanger medium
supply and withdrawal

Segmented
Gas/Liquid flow

Liquid supply

Figure 5.3 Schematic of contacting liquid and gaseous reactants in a micro bubble column [3].

Figure 5.4 Micro bubble column (redesigned version) [IMM, unpublished].

The micro bubble column (Figure 5.4) is composed of a four-piece housing [3, 9, 10]. Two main pieces carry the micro mixing unit and the micro channel plate and are closed by two end-caps. The mixing unit comprises an interdigital feed structure with very different hydraulic diameters for gas and liquid feed. This comes from the different demands for microstructure adaptation to achieve a good pressure barrier for equipartition. Separate micro-scale gas and liquid films enter in one reaction micro channel each, which is on a separate reaction plate. By this means, a specific flow pattern is generated, determined by the process parameters, in particular the gas and liquid velocities. The gas and liquid streams merge to be removed from the micro channel section.

Reactor type	Micro bubble column	Mini heat exchange channel width; depth; length	3000 μm; 500 μm; 40 mm
Housing, heat exchange and reaction plate material	Steel	Pressure stability	30 bar
Micro mixer plate material	Nickel	Temperature stability	Up to 180 °C
Outer dimensions (without connectors)	95 × 50 × 36 mm³	Residence time	0.14–0.56 s (at 10 ml h⁻¹ liquid flow; 600–3300 ml h⁻¹ gas flow)
Outer dimensions (with connectors)	143 × 102 × 36 mm³	Interfacial area of liquid films	20 000 m² m⁻³ (at 25 ml h⁻¹)
Number of reaction micro channels	32	Active inner volume	15.3 mm³ (per plate)
Micro reaction channel width; depth; length	200 μm; 70 μm; 60.5 mm	Total inner surface	860 mm² (per plate)
Micro mixer channel width; depth; length; number (gas- and liquid site)	6, 5, 10 μm; 20 μm; 600, 2000 μm; 64; 10, 20 μm; 600 μm; 64		

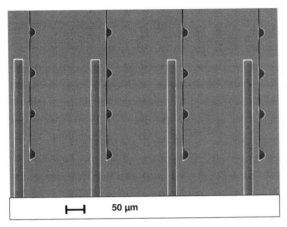

Figure 5.5 SEM of the interdigital gas and liquid feed channels [3].

The micro bubble column comprises internal cooling via heat conduction from the reaction zone to a mini channel heat exchanger [3, 9, 10]. Either two such heat exchange plates can encompass the reaction plate, or only one. In the latter case, the free position is occupied by an inspection window which allows direct observation of the quality of the flow patterns.

The micro mixing unit (Figure 5.5) is fabricated by UV lithography followed by replication via electroforming [3, 9, 10]. The reaction channels are made by wet-chemical etching. The housing is made by µEDM fabrication. The heat exchange channels are manufactured by micro milling.

Meanwhile, a redesign of the micro bubble column has been made with improved flow distribution (Figure 5.4).

5.1.2.2 Reactor 4 [R 4]: Dual-Micro Channel Chip Reactor

This reactor is based on two parallel micro channels that are separated by a wall. In front of the micro-channel section, one outlet hole is placed for liquid feed, followed by two holes for gas feed [13, 14]. The liquid feed enters in line with the wall long axis, while the gas feeds have the position of the two channels. Consequently, the liquid flow has to split.

The silicon chip reactor was compressed between a top plate, for direct observation of the flows, gaskets with punched holes and a base plate with all fluid connections [13, 14]. Thermocouples inserted between the two plates were located next to the micro reactor. A third inlet served for reaction quenching by introducing an inert gas such as nitrogen. Generally, heat removal is facilitated by the special reactor arrangement acting as a heat sink.

The reaction channels were made in silicon by several photolithographic steps, followed by potassium hydroxide etching [13, 14]. Silicon oxide was thermally grown over the silicon. Nickel thin films were vapor-deposited. Pyrex was anodically bonded to such a modified microstructured silicon wafer.

Reactor type	Dual-channel micro reactor	Base plate material of interfacing chip housing	Silicon stainless steel
Reactor material	Silicon; Pyrex	Gasket material	Kalrez®; hybrid Kalrez®/graphite
Micro channel: width; depth; length	435 µm × 305 µm × 20 mm (triangular cross-section)	Top plate material of interfacing chip housing	Plexiglas
Channel hydraulic diameter	224 µm	Surface-to-volume ratio of micro channels	18 000 m^2 m^{-3}
Volume of the reactor	2.7 µl	Silicon wafer: diameter; thickness	100 mm; 525 µm
Distance liquid and gas ports	3 mm	Silicon oxide layer: thickness	200–500 µm
Distance upstream end of wall to gas port	0.5 mm	Nickel layer: thickness	> 200 nm

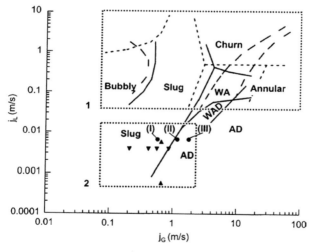

Figure 5.6 Flow pattern map for a gas/liquid flow regime in micro channels. Annular flow; wavy annular flow (WA); wavy annular-dry flow (WAD); slug flow; bubbly flow; annular-dry flow (AD). Transition lines for nitrogen/acetonitrile flows in a triangular channel (224 µm) (solid line). Transition lines for air/water flows in triangular channels (1.097 mm) (dashed lines). Region 2 presents flow conditions in the dual-channel reactor (●), with the acetonitrile/nitrogen system between the limits of channeling (I) and partially dried walls (III). Flow conditions in rectangular channels for a 32-channel reactor (150 µm) (▼) and single-channel reactor (500 µm) (▲) [13].

A flow-pattern map for gas/liquid flow (nitrogen/acetonitrile) was derived [13]. Bubbly, slug, churn and annular flows were found (Figure 5.6). In addition, wavy annular and wavy annular-dry flows were detected with a less extended stability region. Generally, the annular-flow processing was favored owing to the high specific interface and the simple concept relying on an inner gas core and surrounding liquid only.

5.1.2.3 Reactor 5 [R 5]: Single-/Tri-channel Thin-film Micro Reactor

This micro reactor (Figure 5.7) contains a three-plate structure forming a single micro channel and the conduits [15, 16]. The first plate is a thin frame for screw mounting and provides an opening for visual inspection of the single micro-channel section. The second plate serves as top plate shielding the micro-channel section and comprising the fluid connections. This plate also has a seal function and is transparent to allow viewing of the flow patterns in the single micro channel. The bottom plate comprises the micro channel which is made by cutting a groove in a metal block. The metal plate is highly polished to ensure gas tightness [16].

The solution containing the reactant is fed at one end of the micro channel and runs through the single micro channel for a certain passage to adapt to temperature. Then the gas stream is introduced in the flowing liquid via a second port in rectangular flow guidance [15, 16]. Thereby, the gas/liquid flow pattern is derived and the reaction initiated. After a reaction flow passage, the product mixture leaves the micro reactor via a third port.

The micro reactor was specially made for fluorination reactions. Before carrying out the fluorination reactions, passivity of the micro reactor has to be ensured by exposure of the micro channel to increasing concentrations of fluorine in nitrogen [16].

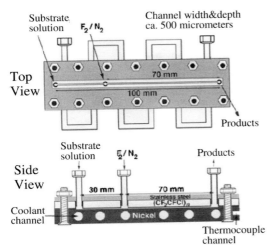

Figure 5.7 Schematic of the single-channel thin-film micro reactor [15].

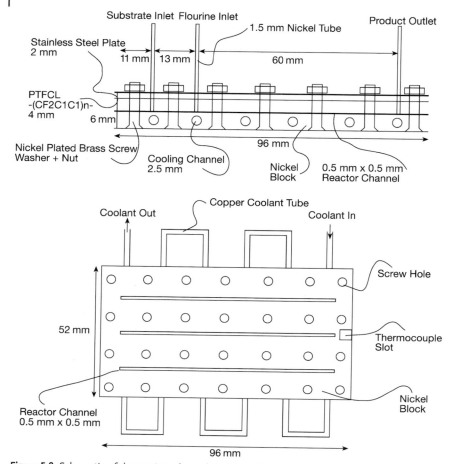

Figure 5.8 Schematic of three-micro-channel version of the single-channel micro reactor [16].

A coolant channel is guided through the metal block in a serpentine fashion [15]. Hence reactant and coolant flows are orthogonal. A thermocouple measures the temperature at the product outlet of the single-channel thin-film micro reactor.

The micro reactor was initially made as single-micro-channel version [15] and later as numbered-up (scale-out) three-micro-channel version (Figure 5.8) [16]. The data for both micro devices are given in the following.

Reactor type	Single-channel thin-film micro reactor	Bottom plate (metal block) material	Nickel (or copper)
Frame plate material	Stainless steel	Micro channel: width; depth; length	500 μm; 500 μm; 70 mm
Top plate material	Polytrifluorochloroethylene		

Reactor type	Three-channel thin-film micro reactor	Micro device outer dimensions	$96 \times 52 \times 12$ mm^3
Frame plate material	Stainless steel	Micro channel: width; depth; length	500 µm; 500 µm; 60 mm
Top plate material	Polytrifluorochloro-ethylene	Inlet and outlet ports	Nickel tubing
Bottom plate (metal block) material	Nickel (or copper)	Nickel tubing: internal diameter; length	500 µm; 100 mm
Frame plate: thickness	2 mm	Distance liquid/gas entries	13 mm
Top plate: thickness	4 mm	Cooling channel: internal diameter	2.5 mm
Bottom plate: thickness	6 mm		

5.1.2.4 Reactor 6 [R 6]: Modular Multi-plate-stack Reactor

A modular reactor concept was developed to meet the typical demands of laboratory reactors, which are flexibility, easiness of handling and fast change of parameters (Figure 5.9). It is based on five different assembly groups, namely micro structured platelets, a cylindrical inner housing, two diffusers and a cylindrical outer shell with a flange [17] (see also [18, 19]). The micro structured platelets are inserted in a recess of the bottom part of the inner housing, which is a rectangular mill cut. Cylindrical tube connectors guide the flow from the reactor inlet via the diffuser to the platelet stack in the mill cut. The flange and cylindrical outer housing are bolted by six 5 mm screws and tightened via insertion of a copper gasket. The platelets are fabricated by means of thin-wire µEDM.

Even for operation of the micro reactor at 480 °C, platelet exchange can be performed rapidly, needing only 15–30 min for cooling from operational to ambient temperature [18, 19]. Heat production rates of about 30 W can be achieved without the need for external cooling [19].

A nanoporous structure on the surface of the micro channels can be realized via anodic oxidation, thereby considerably enlarging the catalyst surface [17]. Catalysts

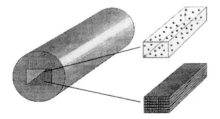

Space for fixed bed catalyst (catalysts WCh and FB)

Stack of micro-structured wafers (catalysts MCh and MEl)

Figure 5.9 The reactor module can be equipped either with a stack of micro structured catalyst wafers or with a 'mini' fixed-bed [17].

can either be deposited by electrochemical means (e.g. palladium from $PdSO_4$ solution; 298 K; 7.5 V AC; 50 Hz; 3 min) or a chemical method can be used. The micro structured platelets were placed in 10% formalin aqueous solution, then the micro channels were exposed to a solution 200 mg of $PdCl_2$ in 40 ml of distilled water for 5 h. The $PdCl_2$ species was reduced to elemental Pd by the formalin was still present in the nanopores of the oxide layer generated one step before. This was followed by calcination. The whole procedure of impregnation was repeated several times to increase the catalyst load.

(a)

Reactor type	Multi-plate-stack in cylindrical housing	Platelet material	Aluminum
Inner housing: outer dimensions	$10 \times 10 \times 50$ mm^3	Reaction channel: width; depth; length	300 µm; 700 µm; 40 mm
Tube connectors: inner diameter	1/8 in	Total number of microstructured platelets	6
Operating pressure	3 bar		

(b)

Reactor type	Multi-plate-stack in cylindrical housing	Platelet material	Aluminum
Inner housing: outer dimensions	$10 \times 10 \times 70$ mm^3	Reaction channel: width; depth; length	300 µm; 300 µm; 70 mm
Tube connectors: inner diameter	1/8 in	Total number of microstructured platelets	140
Operating pressure	50 bar		

5.1.2.5 Reactor 7 [R 7]: Micro-channel Reactor in Disk Housing

Only a rough description of this micro reactor was given (Figure 5.10), not disclosing all details [20]. A pair of iron plates coated with Pd catalyst is inserted in disk-type holders. Such a supported micro channel device is encased in a housing.

Reactor type	Micro-channel reactor in disk housing	Catalyst material	Pd
Micro channel: material	Iron	Catalyst layer thickness	5 µm
Micro channel: depth	100–200 µm	Total catalyst material	20 mg
Disk holder: material	Polypropylene	Catalyst specific surface area	3.6 ± 0.4 m^2 g^{-1}
Outer casing: material	Stainless steel		

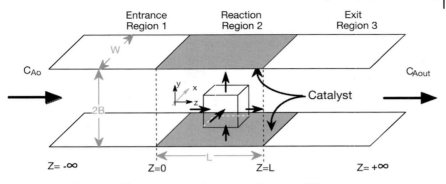

Figure 5.10 Schematic of the micro-channel reactor configuration [20].

5.1.2.6 Reactor 8 [R 8]: Photochemical Single-channel Chip Micro Reactor

This micro-chip reactor comprises a liquid inlet port which splits into two channels of equal passage [21]. These split channels merge with a third channel, which is connected to a second port for gas feed, in such way that the two liquid streams encompass the gas stream. This triple-stream feed section is followed by a long serpentine channel passage which ends in a third outlet port (Figure 5.11).

The microstructure is part of a bottom plate; a top plate serves as a cover [21]. Direct-write laser lithography and wet-chemical etching were employed for microfabrication of the bottom plate. Holes were drilled in the top plate to give conduits for the inlet and outlet ports. The top and bottom plates were bonded thermally.

Reactor type	Photochemical single-channel chip micro reactor	Reaction flow-through chamber: width; depth; length	150 μm; 50 μm; 50 mm
Channel material	Glass	Device outer dimensions	50×20 mm^2
Cover plate material	Glass		

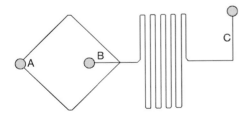

Figure 5.11 Schematic of the single-channel chip micro reactor. Divergent inlet channels (A); secondary inlet channel (B); serpentine irradiation channel (center); outlet channel (C) [21].

5.1.3
Gas/Liquid Micro Flow Dispersive Mixers Generating Bubbly Flows and Foams

These devices have gas and liquid streams which merge, i.e. are fed into each other, e.g. by special dual-feed, triple-feed or multiple-feed arrangements. Both phases are then passed in the same encasing, e.g. the same micro channel. From there, fragmentation of the gas stream occurs, forming a dispersion.

However, in contrast to the dispersive mixers forming slug and annular flow patterns, here the flow-through channel is much larger than the typical dimensions of the dispersed phase (compare with Section 5.1.2). As a result, bubbly flows and foams are the common flow patterns.

The most striking advantage of the concept is its simplicity. The micro mixers described below are commercially available, inexpensive and comparatively simple to operate.

As a further advantage, large flows are provided by these devices, even when operating with a single device. Pilot-scale operation, with one or a few micro devices, is easily feasible. Especially in the case of foam formation, parallel operation of many devices having the same flow pattern seems to be possible.

The need for analytical (flow-pattern) characterization in advance of the experiment is less than for the dispersive mixers forming slug and annular flow patterns, because the dispersion typically is formed in an attached tube. This tube is commonly made of glass and mostly of larger inner diameter. Hence visual inspection by the operator is routinely possible.

However, coalescence of the foam may occur. In aqueous systems, this may be prevented by adding surfactants to lower the surface tension. With organic solvents, this is not as facile. Hence there may be limits to applicability. For unstable gas/liquid dispersions, the micro devices described here may only be used for short-term contacting.

For systems of reduced coalescence, there is a need for phase separation, as the phases are thoroughly inter-mixed. In contrast to liquid/liquid dispersion the gas/liquid separation should be, however, much less troublesome.

Another major drawback stems from the disperse nature of the system itself involving a size distribution of the bubbles in the continuous liquid, which can be broad. The interface is not as defined as for two-phase continuous reactors, as described in Section 5.1.1. However, in the case of making foams, regular micro flow structures, such as hexagon flow, were described [22].

5.1.3.1   Reactor 9 [R 9]: Interdigital Micro Mixer
Interdigital micro mixers comprise feed channel arrays which lead to an alternating arrangement of feed streams generating multi-lamellae flows [23–26]. By proper shaping of the attached flow-through chamber, secondary effects can assist the diffusion mixing of the multi-lamellae. So-called triangular mixers geometrically focus the lamellae; slit-shaped mixers use jet mixing in addition to multi-lamination. In a rectangular mixer, solely multi-lamination takes place.

The interdigital feed can be fed in a counter-flow or co-flow orientation; the first principle is realized in metal/stainless-steel devices [23, 25] and the latter in glass devices [24]. Glass mixers allow observation of hydrodynamics, e.g. for process control during reaction. To prolong residence time and/or to increase temperature, tubes are usually attached to interdigital micro mixers. These comprise millimeter dimensions or below, if necessary.

For a more detailed description of interdigital micro mixers and their images see the corresponding section in Section 4.1.

Reactor type	Interdigital micro mixer	Triangular chamber: initial width; focused width; depth; focusing length; mixing length; focusing angle	3.25 mm; 500 μm; 150 μm; 8 mm; 19.4 mm; 20°
Mixer material	Metal/stainless steel; silicon/stainless steel; glass	Slit-type chamber: initial width; focused width; depth; focusing length; expansion width; expansion length; expansion angle	4.30 mm; 500 μm; 150 μm; 300 μm; 2.8 mm, 24 mm; 126.7°
Metal/silicon mixer feed channel width; depth	40 μm; 300 μm	Slit depth in steel housing	60 μm
Glass mixer feed channel width; depth	60 μm; 150 μm	Tubing attached to slit: diameter	500 μm
Type of flow-through chamber	Rectangular; triangular; slit-type	Device outer dimensions: diameter; height	20 mm; 16.5 mm
Rectangular chamber: width; depth; length	3250 μm; 150 μm; 27.4 mm		

5.1.3.2 Reactor 10 [R 10]: Caterpillar Mini Mixer

The caterpillar mixer acts by distributive mixing using the split–recombine approach which performs multiple splitting and recombination of liquid compartments [25–28]. A ramp-like microstructure splits the incoming flow into two parts which are lifted up and down. Thereafter, the two new streams are reshaped separately in such a way that the two new cross-sections combined restore the original one. Then, the streams are recombined, again by the lifting up and down procedure using ramps. This procedure is repeated multiple times, yielding a multi-lamellae flow configuration – in the ideal case.

A caterpillar steel mini mixer can be connected to conventional tubing, either stainless steel or polymeric, to prolong the residence time. The caterpillar mixer as all types of split–recombine mixers, profits from high volume flows (e.g. $100 \, \mathrm{l \, h^{-1}}$ and more at moderate pressure drops) at favorable pressure drop (not exceeding 5 bar) as its internal microstructures can be held large [25–28].

For a more detailed description of the hydrodynamics of the caterpillar mixer, respective images and the performance of a second-generation caterpillar device see the corresponding section in Section 4.1.

Reactor type	Caterpillar mini mixer–tube reactor, 1st generation	Microstructure in one plate: initial depth; maximum depth	600 µm; 850 µm
Mini mixer material	Stainless steel	Mini mixer stage: length	2400 µm
Number of plates needed to form mini mixer channel	2	Number of mixing stages	8
Mini mixer channel: initial width; maximum width	1200 µm; 2400 µm	Total length of caterpillar mini mixer	19.2 mm
Mini mixer channel (both plates): initial depth; maximum depth	1200 µm; 1700 µm	Device outer dimensions	$50 \times 50 \times 10$ mm^3

5.1.3.3 Reactor 11 [R 11]: Fork-like Chip Micro Mixer – Tube Reactor

Central part of this reaction unit is a split–recombine chip micro mixer made of silicon based on a series of fork-like channel segments [29–33]. Standard silicon micro machining was applied to machine these segments into a silicon plate which was irreversibly joined to a silicon top plate by anodic bonding.

The fork-type chip mixer was used in connection to conventional tubes. PTFE tubing has been applied [34, 35].

Reactor type	Fork-like chip micro mixer–tube reactor	Characteristic structure of the mixing stage	'G structure'
Brand name	accoMix (µRea-4 formerly)	Inner device volume	70 µl
Micro mixer material	Silicon	Outer device dimensions	$40 \times 25 \times 1.3$ mm
Inlet channel width	1000 µm	Tube material	PTFE
Number of parallel mixing channels to which the inlet flow is distributed	9	Tube diameter	Not reported
Number of mixing stages within one mixing channel	6; 3 in top plate, 3 in bottom plate	Tube length	Not reported
Channel width; depth of micro mixer	360 µm; 250 µm (triangular-shaped)		

5.1.4
Gas/Liquid Micro Flow Packed-bed or Trickle-bed Reactors

These devices have gas and liquid streams which merge, i.e. are fed into each other, e.g. by special dual-feed, triple-feed or multiple-feed arrangements. From there, both phases are passed in the same encasing, e.g. the same micro channel.

However, in contrast to the two classes of dispersive mixers mentioned before, the attached flow-through channel contains a packed bed of particles which may carry a catalyst. This chamber is much larger than the typical dimensions of the inlet channels (e.g. compare with Section 5.1.2). The packed bed and its interstices influence the gas/liquid flow patterns, e.g. a trickle-bed operation may be established.

The most striking advantage of the concept is the resemblance of industrially applied principles for gas/liquid/solid operation. Since the packed bed may contain commercial particles, the concept is also flexible and close to the needs of applications. As a further advantage, large flows are provided, even when operating with a single device. Parallel operation of many devices has also been demonstrated [11].

Analytical (flow-pattern) characterization is more difficult as the particle bed is not transparent and covers most of the flow-through chamber. Another drawback stems from the size distribution of the particles of the catalyst bed, giving interstices which vary in typical dimensions. Here, however, today's considerable efforts in nano- and micro-material research may provide regular, mono-sized particles in the near future which will allow one to create much improved micro flow-packed beds.

5.1.4.1 Reactor 12 [R 12]: Multiphase Packed-Bed Reactor

The design of this micro reactor was motivated by achieving high dispersion between the liquid and gas phases and minimizing pressure drop (see especially [12, 36] for a detailed description, but also [11]). The reactor was made as two similar devices towards this end. The first version comprises a single-channel reactor with a flow-through chamber to be filled with conventional porous catalyst particles or powders. The second version has ten parallel channels, each having a staggered array of 50 μm diameter columns within to provide both catalyst support and static mixing. For the first single-channel version, a standard porous catalyst particles are inserted in a mini flow-through chamber (Figure 5.12). An inlet manifold feeds this reaction chamber. Its purpose is to introduce alternately arranged gas and liquid streams to achieve a high degree of dispersion. A filter at the outlet serves for retaining the catalyst particles (Figure 5.13). This filter is composed of regularly arranged microstructured columns. Perpendicular to the reaction channel are inlet channels for feeding the catalyst slurry.

In the multi-channel version comprising 10 packed-bed reactors (Figure 5.14), the gas flow is distributed by star-type manifolds to the 10 reaction units [11, 12].

Such a micro reactor is compressed between a cover plate, a gasket and a base plate [11]. In the cover plate a cartridge heater is inserted. The base plate provides

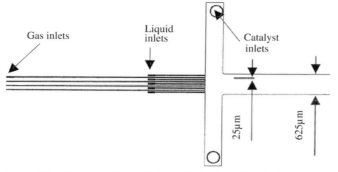

Figure 5.12 Schematic of a multiphase single-channel packed-bed reactor [36].

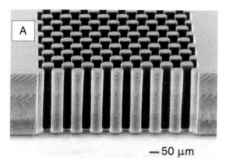

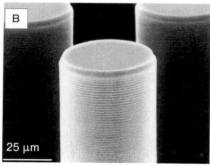

Figure 5.13 SEM images of the microstructured catalyst support in the multi-channel reactor version. Overall view (channel view) (A) and detailed view of one column (B) [12].

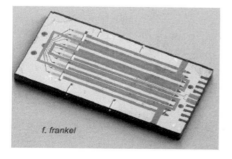

Figure 5.14 Image of a 10-channel chemical micro reactor [12].

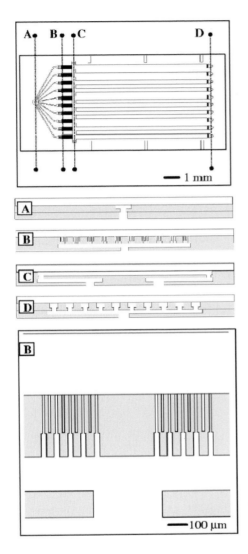

Figure 5.15 Schematic of the multiphase packed bed reactor. Gas inlet (A); liquid distributor (B); catalyst inlets (C); exit port manifold (D) [11].

conduits to the micro reactor (Figure 5.15). The outlets are standard high-pressure fittings. Thermocouples are inserted in the slurry feed channels.

Microfabrication involves multiple photolithographic and etch steps, a silicon fusion bond and an anodic bond (see especially [12] for a detailed description, but also [11]). A time-multiplexed inductively coupled plasma etch process was used for making the micro channels. The microstructured plate is covered with a Pyrex wafer by anodic bonding.

A hydrodynamic characterization of the micro reactor is given in [12]. A flow-pattern map reveals the existence of dispersed flow, annular flow, slug-dispersed flow and slug-annular flow. The highest specific interface measured amounts to $16\,000\;\mathrm{m^2\,m^{-3}}$.

A porous surface structure ($100\;\mathrm{cm^2}$) in the reaction channel can be generated by an SF_6 plasma etch process with silicon nitride masking [12].

Strategies for enhanced heat control resulting in a new micro reactor design are briefly mentioned in [37].

Reactor type	Multiphase packed-bed reactor	Gasket material	Viton
Reactor material	Silicon; Pyrex	Catalyst material	Platinum on alumina
Reaction flow-through chamber: width; depth; length	625 µm; 300 µm; 20 mm	Catalyst particle diameter	50–75 µm
Gas inlet flow channels: width	25 µm	Catalyst surface area	$0.57\;\mathrm{m^2\,g^{-1}}$
Slurry inlet flow channels: width	400 µm	Catalyst weight	40 mg
Silicon substrate: diameter, thickness	100 mm; 500 µm	Catalyst loading density	$0.8\text{–}1.0\;\mathrm{g\,cm^{-3}}$
Cover plate material	Aluminum	Column diameter	50 µm
Base plate material	Stainless steel	Gap between columns	25 µm

5.2
Aromatic Electrophilic Substitution

5.2.1
Halodehydrogenation – Fluorination of Aromatic Compounds

Peer-reviewed journals: [13, 38]; proceedings: [3, 13, 14, 37, 39, 40]; sections in reviews: [26, 41–48].

5.2.1.1 Drivers for Performing Aromatic Fluorination in Micro Reactors

Aromatic fluorination certainly is one of the best candidate reactions for micro reactors. Fluorinated compounds are of high industrial interest. For instance, they find wide application as pharmaceuticals, dyes, liquid crystals and crop-protection agents [3, 13, 16, 38]; about every third drug contains a fluorine moiety. The introduction of fluorine moieties in molecules has unique effects on biological activity (see original citations in [16]).

So far, complex synthesis routes have been followed to realize these compounds, leading to a reduction in selectivity and being associated with high waste generation (see the discussion on this topic in [GL 1, below]). A direct route using the elemental material would thus be highly favorable.

This direct route demands absolutely precise temperature control because over-heating of the reaction media increases radical formation [3, 13, 14, 16, 37, 38]. The number of radicals formed determines which reaction path dominates, radical or electrophilic. Another issue refers to increasing mass transfer. To utilize the potential of the extremely fast fluorination reaction, fluorine has to cross over the interface from the gas to the liquid phase. Since fluorine is hardly soluble in any organic solvent, the size of the interface becomes of special importance. In turn, control over this quantity with regard to time and space, besides heat setting, allows one to handle the delicate direct fluorination reaction. This is accompanied by a demand for precise residence time setting as the fluorination reaction is extremely fast and too long exposure to the aggressive fluorine results in secondary reactions such as multi-fluorination, C–C bond cleavage by additions, and polymerizations.

5.2.1.2 Beneficial Micro Reactor Properties for Aromatic Fluorination

The drivers for performing direct fluorinations in a micro-channel environment directly relate to the elemental advantages of micro reactors. Their large internal surface areas and interfaces facilitate mass and heat transport by building up large concentration and temperature gradients [3, 13, 14, 16, 37, 38]. Residence times can be controlled much more precisely than, e.g., in a stirred-tank reactor. Finally, the small hold-up volumes make the process safe, even if very high fluorine concentrations and hence large conversion rates are employed [16].

5.2.1.3 Aromatic Fluorinations Investigated in Micro Reactors

Gas/liquid reaction [GL 1]: Direct fluorination of toluene using elemental fluorine

Direct fluorinations with elemental fluorine still are not feasible on an industrial scale today; they are even problematic when carried out on a laboratory-scale [49–53]. This is caused by the difficulty of sustaining the electrophilic substitution path as the latter demands process conditions, in particular isothermal operation, which can hardly be realized using conventional equipment. As a consequence, uncontrolled additions and polymerizations usually dominate over substitution, in many cases causing large heat release which may even lead to explosions.

For this reason, industrial fluorinations of aromatics are performed by other routes, mostly via the Schiemann or Halex reaction [54, 55]. As these processes are multi-step syntheses, they suffer from low total selectivity and waste production and demand high technical expenditure, i.e. a need for several pieces of apparatus.

Accordingly, for decades scientific investigations have been carried out to achieve the direct fluorination which would be attractive as a one-step synthesis alternative. Although the early reports concerned gas-phase direct fluorinations, the most relevant work in the last three decades was based on contacting fluorine gas with the aromatic compound dissolved in a liquid phase. These attempts at gas/liquid processing focused on coping with the problem of removing the reaction heat. Mostly this was accomplished by extreme dilution or working at very low temperature or a combination of both [49–52]. By this means, it could be shown that the reaction proceeds predominantly via the electrophilic path, i.e. gives the ortho- and para-substitution patterns on the aromatic ring (see also [56]). Apart from demonstrating feasibility, the investigations could be extended to kinetic and mechanistic analysis. Using aromatic substrates of different electron density distribution, the varying reactivities could be described by the known equations in organic chemistry [51, 52].

Gas/liquid reaction 2 [GL 2]: Direct fluorination of 4-nitrotoluene using elemental fluorine

This fluorination was carried out in formic acid/acetonitrile mixtures and in acetonitrile at 5 °C [16].

5.2.1.4 Experimental Protocols

[P 1] The liquid volume flow to the micro reactor is controlled by an HPLC pump [38]. The gas flow was set by mass flow controllers. Temperature was monitored by resistance thermometers.

(a) Falling film micro reactor experiments (10% fluorine): the temperature was set to −15 to −42 °C [38] (see also [3]). The molar ratio of fluorine to toluene spans the range from 0.20 to 0.925; hence under-stochiometric fluorine contents were employed to favor mono-fluorination. The concentration of toluene in the solvent was 1.1 mol l^{-1}. As liquid volume flows 11.1 ml h^{-1}, 11.6 ml h^{-1} or 19.6 ml h^{-1} were applied. Acetonitrile or methanol was taken as solvent for the aromatic compound. In the gas phase, 10% fluorine in nitrogen was used.

(b) Falling film micro reactor experiments (> 10% fluorine): the temperature was set to −16 °C [38]. The molar ratio of fluorine to toluene spans the range from 0.40 to 2.0; hence under- and over-stochiometric fluorine contents were employed. The concentration of toluene in the solvent was 1.1 mol l^{-1}. As liquid volume flow always 19.6 ml h^{-1} was applied. Acetonitrile was taken as solvent for the aromatic compound.

(c) Micro bubble column experiments (50 × 50 μm^2 reaction channels): the temperature was set to −15 °C [38] (see also [3]). The molar ratio of fluorine to toluene spans the range from 0.20 to 0.83; hence under-stochiometric fluorine contents were employed. The concentration of toluene in the solvent was 1.1 mol l^{-1}. As liquid volume flow always 13 ml h^{-1} was applied. Acetonitrile was taken as solvent for the aromatic compound. In the gas phase, 10% fluorine in nitrogen was used. The gas volume flow was varied from 12.1 to 50.0 ml min^{-1}.

(d) Micro bubble column experiments (300 μm × 100 μm reaction channels): The temperature was set to −15 °C [38] (see also [3]). The molar ratio of fluorine to toluene spans the range from 0.20 to 0.83; hence under-stochiometric fluorine contents were employed. The concentration of toluene in the solvent was 1.1 mol l^{-1}. As liquid volume flow always 13 ml h^{-1} was applied. Acetonitrile was used as solvent for the aromatic compound. In the gas phase, 10% fluorine in nitrogen was used. The gas volume flow was varied from 12.1 ml min^{-1} to 50.0 ml min^{-1}.

(e) Laboratory bubble column experiments: the temperature was set to −17 °C. The molar ratio of fluorine to toluene spans the range from 0.40 to 1.00 [38]; hence under-stochiometric fluorine contents were employed. The concentration of toluene in the solvent was 1.1 mol l^{-1}. As liquid batch volume always 20 ml was applied. Acetonitrile or methanol was taken as solvent for the aromatic compound. In the gas phase, 10% fluorine in nitrogen was used. The gas volume flow was either 20 or 50.0 ml min^{-1}.

[P 2] A 0.1 M (and 1.0 M in one case) toluene solution was fluorinated in various solvents such as acetonitrile, methanol and octafluorotoluene using a gas mixture of 25 vol.-% fluorine in nitrogen [13]. Liquid solutions were dried with molecular sieves before use. The reactor was primed prior to operation with dry nitrogen gas and anhydrous solvent. First the reaction solution was entered, then the fluorine/nitrogen mixture. Toluene solution flow was fed by a syringe pump; the fluorine mixture was delivered by a mass-flow controller. The temperature was monitored by thermocouple. Heat removal was mainly achieved by heat conduction through the micro reactor to the top and base plate materials, acting as a heat sink.

Reactions were carried out at room temperature in the annular-dry flow regime (gas superficial velocity, 1.4 m s^{-1}; liquid superficial velocity, 5.6 · 10^{-3} m s^{-1}) [13]. The number of fluorine equivalents (to toluene) was varied; the gas and liquid flow velocities were kept constant to maintain the same flow pattern for all experiments. Liquid products were collected in an ice-cooled round-bottomed glass flask containing sodium fluoride to trap the hydrogen fluoride. The flask is connected to a cooling condenser to recover the solvent. Samples were typically collected for 1 h. Waste gases were scrubbed in aqueous 15% potassium hydroxide solution. Samples were degassed with nitrogen and filtered before analysis.

[P 3] The reactant solutions were injected into a three-micro channel thin-film reactor via a syringe or a syringe pump [16]. The gaseous reactant was fed directly from a small cylinder by a mass-flow controller. Gas and liquid flows were started simultaneously [16]. The flow rates were so set that an annular-flow regime results, comprising a gas inner core stream surrounded by a cylindrical liquid film which wets the micro-channel's surface. This flow regime was chosen for its large specific gas/liquid interfaces and the good temperature control achievable by the thin films.

After leaving the reactant zone, the product stream enters a 0.5 in diameter FEP tube cooled by either a salt–ice bath or acetone–carbon dioxide slush bath [16]. The gas mixture was scrubbed in a soda-lime tower. Hydrogen fluoride was trapped by adding sodium fluoride to the reaction mixture or simply adding water. Then, the product solution was extracted with dichloromethane, washed with aqueous

NaHCO$_3$ solution and dried over MgSO$_4$. Thereafter, the solvent was evaporated, leaving the crude product.

A gaseous mixture of 10% fluorine in nitrogen at 10 ml min^{-1} was used [16]. The reaction was carried out at 0 °C, 5 °C and room temperature. Formic acid/acetonitrile mixtures ranging from 1 : 1 and 3 : 2 were used in addition to pure acetonitrile at single-channel flow rates of 1.0–2.0 ml h^{-1} [15]. The fluorine-to-substrate ratios were 1.7 : 1 and 3.0 : 1.

5.2.1.5 Typical Results
Conversion/selectivity/yield
[GL 1] [R 1] [P 1a] Using acetonitrile as solvent, the conversions ranged from 14 to 50% at selectivities of 33–57% [38] (see also [3]). This corresponds to yields of 5–20%. The highest yield was found for a liquid volume flow of 11.6 ml h^{-1} using a 1.1 mol l^{-1} toluene concentration at –20 °C. The fluorine/toluene molar ratio was 0.925.

Using methanol as solvent, the conversions ranged from 12 to 42% at selectivities of 9–58% [38]. This corresponds to yields of 3–14%. Hence the performance of the direct fluorination in methanol is generally worse than that in acetonitrile. The highest yield was found for a liquid volume flow of 11.1 ml h^{-1} using a 1.1 mol l^{-1} toluene concentration at –17 °C. The fluorine/toluene molar ratio was 0.925.

[GL 1] [R 1] [P 1b] Using acetonitrile as solvent, the conversions ranged from 7 to 76% at selectivities of 31–43% [38]. This corresponds to yields of 3–28%. The highest yield was found for a liquid volume flow of 19.6 ml h^{-1} using a 10% toluene/acetonitrile molar ratio at –20 °C. The fluorine/toluene molar ratio was 2.0.

[GL 1] [R 3] [P 1c] Using acetonitrile as solvent and 50 × 50 μm reaction channels, the conversions ranged from 4 to 28% at selectivities of 21–75% [38]. This corresponds to yields of 3–11%. The highest yield was found for a liquid volume flow of 13.0 ml h^{-1} and a gas volume flow of 50.0 ml h^{-1} using a 1.1 mol l^{-1} toluene concentration at –15 °C. The fluorine/toluene molar ratio was 0.54.

[GL 1] [R 3] [P 1d] Using acetonitrile as solvent and 300 × 100 μm reaction channels, the conversions ranged from 9 to 41% at selectivities of 22–28% [38]. This corresponds to yields of 2–11%. The highest yield was found for a liquid volume flow of 13.0 ml h^{-1} and a gas volume flow of 50.0 ml h^{-1} using a 1.1 mol l^{-1} toluene concentration at –15 °C. The fluorine/toluene molar ratio was 0.83.

[GL 1] [R 4] [P 2] Selectivities of up to 36% at 33% conversion were achieved using acetonitrile as solvent (1.0 fluorine-to-toluene equivalent) [13]. When including multi-fluorinated toluenes and chain-fluorinated toluenes, in addition to the mono-fluorinated toluenes, in the selectivity balance, the value increases to 49%. The remainder is lost in other side reactions such as additions or polymerizations.

Complete toluene conversion is achieved when using 5.0 fluorine-to-toluene equivalents [13], but at a much decreased selectivity of 11%.

The highest yield of 14% was found at 2.5 fluorine-to-toluene equivalents (58% conversion; 24% selectivity) [13]. This yield was obtained using acetonitrile as solvent; slightly lower yields were obtained for methanol (Figure 5.16). The selectivities were as high as for acetonitrile, the conversion being lower. Still lower yields (7%)

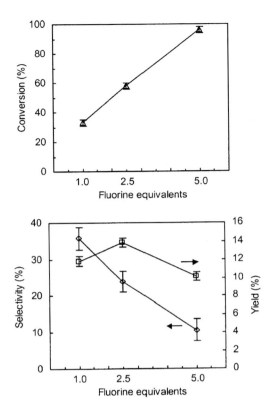

Figure 5.16 Influence of the fluorine-to-toluene equivalents (0.1 M toluene in acetonitrile) on conversion, selectivity and yield [13].

were achieved in octafluorotoluene for the same reason as for methanol, the selectivity decreasing substantially.

[GL 1] [R 4] [P 2] Conversions from 17 to 95% were achieved using methanol as solvent (1.0–10.0 fluorine-to-toluene equivalents; 0.1 M toluene; room temperature; 10 ml min^{-1} gas flow; 100 µl min^{-1} methanol) [14]. The respective selectivities ranged from 37 to 10%. Taking into account also the difluorotoluenes and trifluorotoluenes gives a selectivity of about 45%. The yields passed through a maximum at 18%.

Conversions of 35 and 52% were achieved using methanol as solvent at higher toluene concentration and with lower fluorine-to-toluene equivalents (0.5–1.0 fluorine-to-toluene equivalents; 1.0 M toluene; room temperature; 10 ml min^{-1} gas flow; 100 µl min^{-1} methanol) [14]. The respective selectivities were 20 and 17%. The yields amounted to 7 and 9%. The lower performance of the high-concentration processing compared with the more dilute 0.1 M processing is explained by a larger temperature rise leading to more pronounced radical formation causing side reactions. This is in line with calculations on heat transport for the micro reactor.

[GL 2] [R 5] [P 3] Using formic acid/acetonitrile mixtures, conversions of 44–77% and yields of 60–78% were obtained for different contents of the solvents and different flow rates [16].

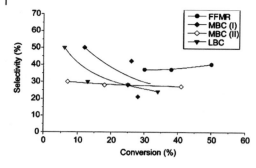

Figure 5.17 Comparison of performance of a typical laboratory column (LBC) with those of micro-reactor devices: falling film micro reactor (FFMR); micro bubble column (MBC I and MBC II) [38].

a)

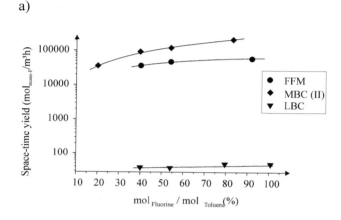

b)

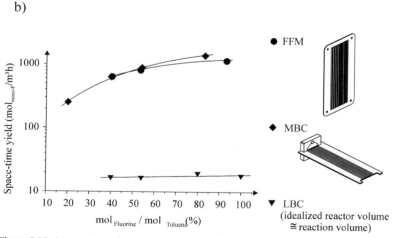

Figure 5.18 Comparison of space–time yields of direct fluorination of toluene for the falling film micro reactor (FFMR), micro bubble column (MBC) and laboratory bubble column (LBC) referred to the reaction volume (a) and referred to an idealized reactor geometry (b) [38].

Benchmarking to laboratory bubble column

[GL 1] [R 1] [R 3] [P 1e] The performance of a typical laboratory bubble column was tested and benchmarked against the micro reactors (Figure 5.17). Using acetonitrile as solvent, the conversion of the laboratory bubble column ranged from 6 to 34% at selectivities of 17–50% [3, 38]. This corresponds to yields of 2–8%. Hence the yields of the laboratory tool are lower than those of the micro reactors, mainly as a consequence of lower selectivities.

The laboratory and the micro bubble column show decreasing selectivity with increasing conversion. The falling film micro reactor shows a near-constant selectivity–conversion relationship [3, 38].

Benchmarking of the micro reactors themselves – slug flow vs. falling film

[GL 1] [R 1] [R 3] [P 1e] The falling film micro reactor has a better selectivity–conversion performance than the two micro bubble columns tested (Figure 5.18) [3, 38]. The micro bubble column with narrow channels has a better behavior at large conversion than the version with wide channels. The behavior of the falling film micro reactor and the micro bubble column with narrow channels is characterized by a nearly constant selectivity with increasing conversion, while the bubble column with wide channels shows notably decreasing selectivity with conversion (similar to the laboratory bubble column).

Ratio of *o*-, *m*- and *p*-isomers – substitution pattern

[GL 1] [R 1] [R 3] [P 1a–d] A ratio of ortho-, meta- and para-isomers for monofluorinated toluene amounting to 5 : 1 : 3 was found for the falling film micro reactor and the micro bubble column at a temperature of –16 °C [3, 38]. This is in accordance with an electrophilic substitution pathway. The relatively high amount of ortho-isomers is due to the small size of the fluorine moiety as the ortho position is amenable to steric effects.

[GL 1] [R 4] [P 2] A ratio of ortho-, meta- and para-isomers for monofluorinated toluene amounting on average to 3.5 : 1 : 2 was found in the dual-channel micro reactor at room temperature, using acetonitrile as solvent [13] (see also [14]). Ortho- and para products were the main products of the reaction mixture, unless the fluorine equivalents < 5 were used. Compared with the result cited above, the meta-isomer content was slightly higher, maybe as a consequence of using a higher reaction temperature.

Using methanol as solvent, the ratio is on average 5.5 : 1 : 2.4. Hence more products referring to an electrophilic substitution were formed [13].

An increase in toluene concentration, from 0.1 to 1.0 M, did not affect the substitution pattern when using acetonitrile as solvent [13].

Side-chain fluorination

[GL 1] [R 4] [P 2] A small amount of side-chain fluorination was reported [13]. Benzyl fluoride was formed to about the same extent as the meta isomer.

Multiple fluorination

[GL 1] [R 4] [P 2] Small amounts of difluoro- and trifluorotoluenes and also unidentified high-boiling compounds were detected [13]. Benzyl fluoride was formed to about the same extent as the meta isomer.

Space–time yield

[GL 1] [R 1] [R 3] [P 1a–d] Space–time yields higher by order of magnitude were found for the falling film micro reactor and the micro bubble column as compared with the laboratory bubble column [38]. The space–time yields for the micro reactors ranged from about 20 000 to 110 000 mol monofluorinated product $m^{-3} h^{-1}$. The ratio with regard to this quantity between the falling film micro reactor and the micro bubble column was about 2. The performance of the laboratory bubble column was of the order of 40–60 mol monofluorinated product $m^{-3} h^{-1}$.

The above-mentioned space–time yields were referred solely to the reaction volume, i.e. the micro channel volume. When defining this quantity via an idealized reactor geometry, taking into account the construction material as well, naturally the difference in space–time yield of the micro reactors from the laboratory bubble column becomes smaller. Still, the performance of the micro reactors is more than one order of magnitude better [38]. The space–time yields for the micro reactors defined in this way ranged from about 200 to 1100 mol monofluorinated product $m^{-3} h^{-1}$.

Fluorine-to-toluene ratio – fluorine equivalents

[GL 1] [R 1] [R 3] [P 1a–d] When the fluorine-to-toluene ratio is increased, conversion increases in a linear fashion [38]. This basically means that transport resistance would most likely not prohibit using still higher fluorine contents, thereby further possibly increasing the productivity (space–time yield) of the reactor (Figure 5.19).

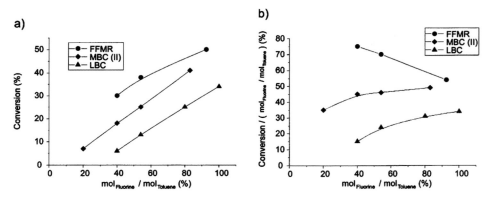

Figure 5.19 Conversion of the direct fluorination of toluene in different reactor types as a function of the molar ratio of fluorine to toluene (a) and efficiency of these reactors, defined as conversion normalized by the molar ratio of fluorine to toluene, as a function of the molar ratio of fluorine to toluene (b). Falling film micro reactor (FFMR); micro bubble column (MBC); laboratory bubble column (LBC) [38].

Content of fluorine consumed

[GL 1] [R 1] [R 3] [P 1a–d] For micro-channel processing, an analysis of the content of fluorine actually consumed as a function of the fluorine-to-toluene ratio was made [38]. The curves for two micro reactors and one laboratory bubble column do not show the same trend; a decrease of converted fluorine with increasing ratio results for the falling-film micro reactor, whereas the micro and laboratory bubble columns show increasing performance. The two micro reactors use about 50–75% of all fluorine offered, whereas the laboratory tool has an efficiency of only 15%.

[GL 1] [R 4] [P 2] A linear increase in conversion was found on increasing the number of fluorine-to-toluene equivalents (i.e. the above-mentioned ratio) from 1.0 to 5.0 [13]. By this means, the conversion increases from 33% to 96%, whereas the selectivity drops from 36% to 11% (0.1 M toluene, 25% fluorine, ambient temperature). The yield passes through a maximum.

Temperature

[GL 1] [R 1] [R 3] [P 1a–d] On increasing the temperature for micro-channel processing, conversion for the direct fluorination rises, as expected [38]. For the falling film micro reactor, conversion is increased from 15 to 30% on going from –40 to –15 °C. The selectivity varies widely between 30 and 50% without a clear tendency for this temperature range. The origin of this fluctuation is not understood.

Variation of solvent

[GL 1] [R 1] [R 3] [P 1a–d] A comparison of yield and selectivity for the direct fluorination in micro reactors when using the polar, protic solvent methanol and the polar, aprotic solvent acetonitrile was made [38]. Generally, the performance when using acetonitrile was much better; for instance, yields ranging from 20 to 28% were obtained in the falling film micro reactor. For methanol, the best yield was only 14%. This is somehow in contrast to literature findings, showing an increase in product yield with increasing solvent polarity (see citations in [38]). So far, no explanation for the differing micro channel performance has been given.

[GL 1] [R 4] [P 2] For the dual-channel micro reactor, the highest yield of 14% was found using acetonitrile (58% conversion; 24% selectivity) [13]. Slightly lower yields were obtained for methanol. The selectivities were as high as for acetonitrile, the conversion being lower. Still lower yields (7%) were achieved in octafluorotoluene for the same reason as for methanol, selectivity decreasing considerably.

[GL 1] [R 4] [P 2] A more detailed study on the methanol performance is given in [14]. Conversions from 17 to 95% were achieved (1.0–10.0 fluorine-to-toluene equivalents; 0.1 M toluene; room temperature; 10 ml min^{-1} gas flow; 100 µl min^{-1} methanol). The respective selectivities ranged from 37 to 10%. Taking into account also the difluorotoluenes and trifluorotoluenes gives a selectivity of about 45%. The yields passed through a maximum at 18%.

At higher toluene concentration and using lower fluorine-to-toluene equivalents (0.5–1.0 fluorine-to-toluene equivalents; 1.0 M toluene) [14], a lower performance is observed, which is explained by a larger temperature rise leading to more pronounced radical formation causing side reactions.

[GL 1] [R 4] [P 2] Variation of solvent affects also the substitution pattern to a certain extent [13]. A ratio of ortho-, meta- and para-isomers for mono-fluorinated toluene amounting on average to 3.5 : 1 : 2 was found in the dual-channel micro reactor at room temperature, using acetonitrile as solvent [13]. Using methanol as solvent, the ratio was on average 5.5 : 1 : 2.4. Hence more products referring to an electrophilic substitution were formed [13].

Increasing solvent polarity

[GL 2] [R 5] [P 3] By addition of formic acid, the polarity of the solvent can be enhanced, which is known to favor the electrophilic pathway. Using formic acid/acetonitrile mixtures, conversions of 44–77% and yields of 60–78% were obtained for different contents of the solvents and different flow rates [16]. The performance in pure acetonitrile was much lower (conversion, 15%; yield, 71%) and was accompanied by fouling of the micro device due to insufficient liquid reactant solubility.

Variation of residence time

[GL 1] [R 1] [R 3] [P 1a–d] In [38], only two residence times were applied, hence no large data sets were available. The yield generated for both experiments were the same, indicating that the reaction still might be much faster compared with the already short residence times in the two micro reactors.

[GL 2] [R 5] [P 3] By increasing the flow rate from 1.0 ml h^{-1} to 2.0 ml h^{-1}, the conversion increases from 44 to 53% [16]. It should be noted that the change in liquid flow rate may also have affected the flow pattern and hence the mass transfer, in addition to changing the residence time.

Residence time distribution

[GL 1] [R 1] [P 1a] The residence time distribution between the individual flows in the various micro channels on one reaction plate of a falling film micro reactor was estimated by analysing the starting wetting behavior of an acetonitrile falling film [3]. For a flow of 20 ml h^{-1}, it was found that 90% of all streams were within a 0.5 s interval for an average residence time of 17.5 s.

Experimental film thickness determination

[GL 1] [R 1] [P 1a] By autofocus laser imaging, the average position of the liquid surface in all micro channels of a reaction plate of a falling film micro reactor was determined [3]. It was found that very thin films of the order of 20–25 μm were formed for total volume flows of 20–80 ml h^{-1}. The thickness of the films in the various channels differed, but by no more than 30% on average. At high flows, e.g. > 180 ml h^{-1}, flooding of the channels occurs.

Fluorine and substrate concentration

[GL 1] [R 1] [R 3] [P 1a–d] The fluorine content in the gas phase of a falling film micro reactor was varied at 10, 25 and 50% [38]. A nearly linear increase in conversion results at constant selectivity. The substitution pattern, i.e. the ratio of ortho- to para-isomers, is strongly affected by this.

When using pure toluene instead of a dilute solution (1.1 mol l^{-1}), only a very low yield at high selectivity was found [38].

Estimation of temperature increase

[GL 1] [R 4] [P 2] A temperature rise of 0.4 K was estimated for a typical experiment in a dual-channel micro reactor based on assuming reaction rates and heat conductivity of the medium [13]. However, there are experimental indications that the real value is higher.

Corrosion

[GL 1] [R 4] [P 2] The stability of vapor-deposited protection coatings made from nickel depends on the process conditions, particularly on the concentrations of toluene and fluorine [14]. Nickel-coated silicon micro reactors were operated for several hours for the reaction conditions given. The nickel films lose to a certain extent their adhesion to the reaction channel with ongoing processing.

Safety

[GL 1] [R 1] [R 3] [P 1a–d] Fluorine contents in the gas phase of a falling film micro reactor as high as 50% could be handled safely while performing direct fluorination experiments [38].

5.3
Free Radical Substitution

5.3.1
Halodehydrogenation – Fluorination of Aliphatics and Other Species

Peer-reviewed journals: [15, 16]; sections in reviews: [26, 42–48].

5.3.1.1 Drivers for Performing Aliphatics Fluorination in Micro Reactors

Aliphatics fluorination certainly is one of the best candidate reactions for micro reactors. Fluorinated compounds are of high industrial interest. For instance, they find wide application as pharmaceuticals, dyes, liquid crystals and crop-protection agents [3, 13, 16, 38]; about every third drug contains a fluorine moiety. The introduction of fluorine moieties in molecules has unique effects on biological activity (see original citations in [16]).

So far, complex synthesis routes have been followed to realize these compounds, leading to a reduction in selectivity and being associated with high waste generation [15]. Instead, a direct route using the elemental material would be highly favorable. However, scale-up of such a direct route in conventional reactors most likely would suffer from problems with maintaining temperature control and safe handling, as fluorinations are among the most exothermic organic reactions.

Accordingly, this direct route demands absolutely precise temperature control because overheating of the reaction media increases radical formation in an uncontrolled way [3, 13, 14, 16, 37, 38]. Another issue refers to increasing mass transfer. To utilize the potential of the extremely fast fluorination reaction, fluorine has to cross over the interface from the gas to the liquid phase. Since fluorine is hardly soluble in any organic solvent, the size of the interface becomes of special importance. In turn, control over this quantity with regard to time and space, besides heat setting, allows one to handle the delicate direct fluorination reaction. This is accompanied by a demand for precise residence time setting as the fluorination reaction is extremely fast and too long exposure to the aggressive fluorine results in secondary reactions such as multi-fluorination, C–C bond cleavage, and polymerizations.

5.3.1.2 Beneficial Micro Reactor Properties for Aliphatics Fluorination

The drivers for performing direct fluorinations in a micro-channel environment directly relate to the elemental advantages of micro reactors. The large internal surface areas and interfaces facilitate mass and heat transport by building up large concentration and temperature gradients [3, 13–16, 37, 38]. Residence times can be controlled much more precisely than in a stirred-tank reactor. Finally, the small hold-up volumes make the process safe, even if very high fluorine concentrations and hence large conversion rates are employed [15, 16].

5.3.1.3 Aliphatics Fluorination Investigated in Micro Reactors

Gas/liquid reaction 3 [GL 3]: Fluorination of ethyl 3-oxobutanoate (ethyl acetoacetate) [15, 16]

This reaction is an example of fluorination of the methylene-group of β-dicarbonyl compounds [15].

Gas/liquid reaction 4 [GL 4]: Fluorination of ethyl 2-chloro-3-oxobutanoate (ethyl 2-chloroacetoacetate) [15, 16]

Gas/liquid reaction 5 [GL 5]: Fluorination of ethyl 2-methyl-3-oxobutanoate (ethyl-2-methylacetoacetate) [16]

Gas/liquid reaction 6 [GL 6]: Fluorination of 3-acetyl-3,4,5-trihydrofuran-2-one [16]

Gas/liquid reaction 7 [GL 7]: Fluorination of 2-acetylcyclohexan-one [16]

Gas/liquid reaction 8 [GL 8]: Fluorination of ethyl 2-oxocyclohexane carboxylate [16]

Gas/liquid reaction 9 [GL 9]: Fluorination of di(*m*-nitrophenyl) disulfide to the sulfur pentafluoride derivative [15, 16]

Gas/liquid reaction 10 [GL 10]: Fluorination of trifluorothio-*m*-nitrobenzene to the sulfur pentafluoride derivative [15, 16]

Sulfur pentafluoride derivatives are of considerable industrial interest [15]. The solubility of the di(*p*-nitrophenyl) disulfide in acetonitrile is too low to undergo a similar pathway as given in [GL 9]. Therefore, the better soluble trifluorothio-*m*-nitrobenzene was used, leading to the further fluorination of this substrate to the pentafluoro derivative as given here.

Gas/liquid reaction 11 [GL 11]: Perfluorination of tetrahydrofuran derivatives [15]

Gas/liquid reaction 12 [GL 12]: Perfluorination of cyclohexane derivatives [15]

5.3.1.4 Experimental Protocols
General procedure for all experiments
The reactant solutions were injected via a syringe or a syringe pump [15, 16]. The gaseous reactant was fed directly from a small cylinder by a mass-flow controller. Gas and liquid flows were started simultaneously [16]. The flow rates were set so that an annular-flow regime results, comprising a gas inner core stream surrounded by a cylindrical liquid film which wets the micro-channel surface. This flow regime was chosen for its large specific gas/liquid interfaces and the good temperature control achievable by the thin films. Flows in the range $0.5–5.0$ ml h^{-1} were used.

After leaving the reactant zone, the product stream entered a 0.5 in diameter PTFE tube cooled either by salt–ice bath or acetone–carbon dioxide slush bath [15, 16]. The gas mixture was scrubbed in a soda-lime tower. Hydrogen fluoride was trapped by adding sodium fluoride to the reaction mixture or simply adding water. Then, the product solution was extracted with dichloromethane, washed with aqueous NaHCO$_3$ solution and dried over MgSO$_4$. Thereafter, the solvent was evaporated, leaving the crude product.

[P 4] 10% fluorine in nitrogen at 10 ml min^{-1}; 5 °C; formic acid as solvent at 0.5 ml h^{-1} [15].

[P 5] 10% fluorine in nitrogen at 10 ml min^{-1}; 5 °C; formic acid as solvent at 0.25 ml h^{-1} [15].

[P 6] 10% fluorine in nitrogen at 10 ml min^{-1}; room temperature; acetonitrile as solvent at 0.5 ml h^{-1} [15].

[P 7] 50% fluorine in nitrogen at 15 ml min^{-1}; room temperature; no solvent; 0.5 ml h^{-1} liquid flow [15].

[P 8] 50% fluorine in nitrogen at 15 ml min^{-1}; room temperature, then 280 °C; no solvent; 0.5 ml h^{-1} liquid flow [15].

[P 9] 20% fluorine in nitrogen at 20 ml min^{-1}; room temperature; no solvent; 0.5 ml h^{-1} liquid flow [15].

[P 10] 50% fluorine in nitrogen at 10 ml min^{-1}; room temperature, then 50 °C; no solvent; 0.5 ml h^{-1} liquid flow [15].

[P 11] 50% fluorine in nitrogen at 15 ml min^{-1}; room temperature, then 280 °C; no solvent; 0.5 ml h^{-1} liquid flow [15].

5.3.1.5 Typical Results
Conversion/selectivity/yield
[GL 3] [R 5] [P 4] Yields of 73/71% were achieved at 99/98% conversion [15, 16], respectively. The amount and nature of difluorinated products were also specified in [16].

[GL 4] [R 5] [P 5] A yield of 62% was achieved at 90% conversion [15]. The amount and nature of difluorinated products was also specified in [16]. Besides this performance in single micro-channels, conversion and yield for a three-micro-channel reactor given [16].

[GL 5] [R 5] [P 5] A yield of 49% was achieved at 52% conversion [15]. The amount and nature of difluorinated products was also specified in [16].

[GL 6] [R 5] [P 5] A yield of 95% was achieved at 66% conversion [15]. The amount and nature of difluorinated products was also specified in [16].

[GL 7] [R 5] [P 5] A yield of 78% was achieved at 93% conversion [15]. The amount and nature of difluorinated products was also specified in [16].

[GL 8] [R 5] [P 5] A yield of 76% was achieved at 86% conversion [15].

[GL 9] [R 5] [P 6] A yield of 75% was achieved [15].

[GL 10] [R 5] [P 6] A yield of 44% was achieved [15]. An improved yield of 56% is reported in [16].

[GL 11] [R 5] [P 7] [P 8] A yield of 91% of the perfluorinated product was achieved (52% recovery) when using an additional heating stage to complete the reaction [15].

[GL 12] [R 5] [P 9] [P 10] [P 11] A yield of 70% of the perfluorinated product was achieved (82% recovery) when using an additional heating stage to complete the reaction [15].

Catalytic effect of metal surface

[GL 3] [R 5] [P 4] A catalytic effect of the fluorinated metal surface of the micro channel was clearly determined [15].

[GL 12] [R 5] [P 9] [P 10] [P 11] A high yield of 70% of the perfluorinated product was achieved without using the conventionally used cobalt trifluoride [15].

See also the next section for the importance of the surface on enol formation.

Importance of enol formation for β-keto ester fluorination

[GL 4] [R 5] [P 5] The rate of the fluorination of β-keto esters is usually correlated with the enol concentration or the rate of enol formation as this species is actually fluorinated [15, 16]. For the fluorination of ethyl 2-chloroacetoacetate in a micro reactor, much higher yields were found as expected from such relationships and as compared with conventional batch processing which has only low conversion. Obviously, the fluorinated metal surface of the micro channel promotes the enol formation.

Temperature

[GL 12] [R 5] [P 9] [P 10] [P 11] Significantly lower temperature can be used to achieve a yield of the perfluorinated product as high as when employing the conventionally used cobalt trifluoride process with traditional reactors [15].

Single- vs. three-micro-channel processing

[GL 3] [GL 4] [GL 5] [GL 6] [GL 7] [R 5] [P 5] No clear difference in the performance of a single-channel micro reactor and a numbered-up three-channel micro reactor was detected [16]. However, many differences in the results were found which were

probably due to fluctuations in the micro-channel hydrodynamics in general, rather than being a consequence of the numbering-up (e.g. due to flow maldistribution). Overall, this means that, despite partly unstable processes, the numbering-up concept was successful, i.e. no performance was lost during scale-out.

Safety
[GL 12] [R 5] [P 9] [P 10] [P 11] Hazardous perfluorination processes with high yield can be carried out safely in micro reactors [15].

5.3.2
Halodehydrogenation – Chlorination of Alkanes

Peer-reviewed journals: [6]; proceedings: [40]; sections in reviews: [47].

5.3.2.1   Drivers for Performing Alkane Chlorination in Micro Reactors
Side-chain photochlorination of toluene isocyanates yields important industrial intermediates for polyurethane synthesis, one of the most important classes of polymers [6]. The motivation for micro-channel processing stems mainly from enhancing the performance of the photo process. Illuminated thin liquid layers should have much higher photon efficiency (quantum yield) than given for conventional processing. In turn, this may lead to the use of low-intensity light sources and considerably decrease the energy consumption for a photolytic process [6] (see also [21]).

Owing to the planar layer structure of most micro reactors, uniform illumination is yielded in addition, which can be maintained on increasing throughput by numbering-up [6]. Here, the individual reaction units are assembled in parallel again on a plane, but a larger one.

5.3.2.2   Beneficial Micro Reactor Properties for Alkane Chlorination
Considering the drivers outlined above, the relevant micro reactor property is the capability to provide thin liquid layers. This is advantageous both for increasing the photon efficiency (quantum yield) and for the mass transfer across the gas/liquid interface [6]. As another result, the residence time may be considerably reduced, thereby decreasing side and follow-up reactions. Overheating as a consequence of using high-intensity light sources at low efficiency can be avoided, thus reducing thermally induced side reactions. The use of low-intensity light sources may change the technical expenditure needed for photoreactions and result in a completely different cost-investment scenario when building a new plant.

5.3.2.3   Alkane Chlorination Investigated in Micro Reactors
Gas/liquid reaction 13 [GL 13]: Photochlorination of toluene-2,4-diisocyanate
Chlorine molecules are cleaved at high temperatures by photoinduced radical formation. By this means, a gas/liquid reaction can be performed in the side chain of alkyl aromatics quite selectively. The electrophilic ring substitution, instead, is favored using Lewis catalysts in polar solvents at low temperature.

Via such a gas/liquid reaction, toluene-2,4-diisocyanate reacts with chlorine to give 1-chloromethyl-2,4-diisocyanatobenzene [6]. As a ring-substituted side product, toluene-5-chloro-2,4-diisocyanate is formed in minor quantities.

Typically, the reaction mechanism proceeds as follows [6]. By photoreaction, two chlorine radicals are formed. These radicals react with the alkyl aromatic to yield a corresponding benzyl radical. This radical, in turn, breaks off the chlorine moiety to yield a new chlorine radical and is substituted by the other chlorine, giving the final product. Too many chlorine radicals lead to recombination or undesired secondary reactions. Furthermore, metallic impurities in micro reactors can act as Lewis catalysts, promoting ring substitution. Friedel–Crafts catalyst such as $FeCl_3$ may induce the formation of resin-like products.

5.3.2.4 Experimental Protocols

[P 12] A falling film micro reactor was applied for generating thin liquid films [6]. A reaction plate with 32 micro channels of channel width, depth and length of 600 µm, 300 µm and 66 mm, respectively, was used. Reaction plates made of pure nickel and iron were employed. The micro device was equipped with a quartz window transparent for the wavelength desired. A 1000 W xenon lamp was located in front of the window. The spectrum provided ranges from 190 to 2500 nm; the maximum intensity of the lamp is given at about 800 nm.

For the range of flows applied (0.12, 0.23, 0.38 and 0.57 ml min^{-1}), the average film thickness was calculated to be 21–36 µm [6]. The corresponding residence times and specific interfacial areas amount to 4.8–13.7 s and 28 000–48 000 m^2 m^{-3}, respectively. A pump was used for liquid feed. The gas flow was controlled by a mass-flow controller, ranging from 14–56 ml min^{-1}. This provides an equimolar ratio of toluene-2,4-diisocyanate and chlorine. A solution of 0.1 mol of toluene-2,4-diisocyanate (14.4 ml) in 30 ml of tetrachloroethane was used (concentration: 3.3 mol l^{-1}). The reaction temperature of 130 °C was achieved by guiding a heating medium through the mini-channel heat exchanger of the device, heating up the back side of the reaction plate. Temperatures were controlled at the product inlet and outlet. The product stream was actively removed by another liquid pump to avoid overfilling of the outlet bore, which may affect the falling film.

5.3.2.5 **Typical Results**

Conversion/selectivity/yield – by-product analysis – benchmarking

[GL 13] [R 1] [P 12] Conversions from 30 to 81% at selectivities from 79 to 67% and yields from 24 to 54%, were found when using a falling film micro reactor (4.8–13.7 s; 130 °C) [6].

A selectivity–conversion plot for the product, the ring-substituted by-product and other, so far unidentified, by-products is given in [6] (Figure 5.22). While the content of the ring-substituted product decreased with increasing conversion (12–5%), the other by-products were formed in much larger amount (8–29%). This also indicates that the other by-products are formed by consecutive reactions at longer residence times.

Control experiments in a batch reactor (30 ml reaction volume) at a 30 min reaction time resulted in a conversion of 65% at 45% selectivity, hence having a selectivity which is higher by about a factor of 2 [6]. Interestingly, the selectivity of the ring-substituted by-product is 54%, different from the more dominant resin-formation in the micro reactor (Figure 5.20). The superior performance of the micro reactor is explained by the better photon yield stemming from the use of very thin liquid films. The low penetration of light in the conventional batch reactor generates largely reaction conditions which are not photoinduced, hence favoring ring chlorination.

Flow rate

[GL 13] [R 1] [P 12] When using lower flow rates, conversion is increased as a consequence of both increasing residence time and decreasing film thickness [6]. Reducing the flow rate from 0.57 ml min^{-1} to 0.12 ml min^{-1} leads to an increase in conversion from 30 to 81%. Selectivity is nearly unaffected, with the exception of the smallest flow rate.

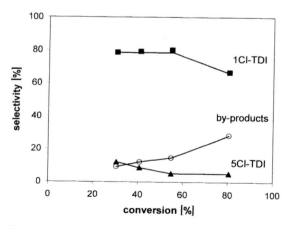

Figure 5.20 Selectivities of main and side product as a function of toluene-2,4-diisocyanate conversion when using a nickel-plate equipped micro reactor (Reactant ratio 1 : 1; 130 °C) [6].

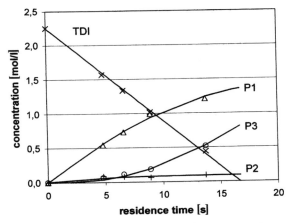

Figure 5.21 Comparison of reaction product concentrations (symbols) and model (solid line) as a function of residence time given in [6]. Toluene-2,4-diisocyanate (TDI); 1-chloromethyl-2,4-diisocyanatobenzene (P_1); toluene-5-chloro-2,4-diisocyanate (P_2); high-molecular-weight consecutive products (P_3) [6].

Residence time
[GL 13] [R 1] [P 12] As a function of residence time, conversion increases linearly from 30 to 81% at selectivities from 79 to 67% [6]. The associated yield increase is non-linear and seems to approach a plateau (Figure 5.21). Hence residence times much larger than 14 s are not suited to increase reactor performance.

Use of iron plate – formation of Lewis acids
[GL 13] [R 1] [P 12] By using an iron plate (instead of a non-active nickel plate), the impact of Lewis acid formation on the reaction course could be tested [6] (Figure 5.21). As a result, the selectivity to the target product decreases drastically, e.g. at a conversion of 80% it decreases to 50% (from 67% using a nickel plate). Interestingly, the content of ring-substituted isomer is not enhanced (actually it is reduced), but probably resin-type condensation products are formed instead yielding the product (Figure 5.22).

Space–time yield – benchmarking
[GL 13] [R 1] [P 12] By using a nickel plate, space–time yields up to 401 mol l^{-1} h^{-1} were achieved in the falling film micro reactor [6]. Control experiments in a batch reactor at a 30 min reaction time resulted in a space–time yield of only 1.3 mol l^{-1} h^{-1}, hence orders of magnitude smaller. By using an iron plate, space–time yields up to 346 mol l^{-1} h^{-1} were achieved in the falling film micro reactor.

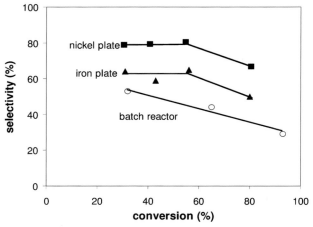

Figure 5.22 Influence of different reactor materials on selectivity for 1-chloromethyl-2,4-diisocyanatobenzene and toluene-2,4-diisocyanate conversion [6].

Kinetics – reaction modeling

[GL 13] [R 1] [P 12] Using a reaction model assuming plug-flow behavior and taking into account consecutive elemental reactions, the dynamic development of the species' concentrations can be predicted [6]. This fits well with experimental data when a reaction order of 0.1 is used (Figure 5.23). This is in contrast to comparable literature approaches relying on a first-order reaction with respect to chlorine and toluene. Also, a selectivity–conversion plot can be drawn for the target product, by-product and consecutive product.

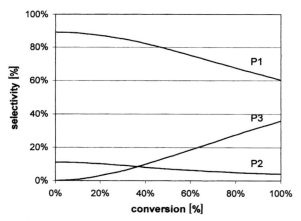

Figure 5.23 Simulated selectivity versus conversion of toluene-2,4-diisocyanate (TDI). P_1 is the target product 1-chloromethyl-2,4 diisocyanatobenzene; P_2 is the side-product 5-toluene-5-chloro-2,4-diisicyanate; P_3 represents high-molecular-weight consecutive products [6].

5.3.3
Halodehydrogenation – Chlorination of α-Keto Compounds

Patents: [57]; sections in reviews: [47].

5.3.3.1 Drivers for Performing Chlorination of α-Keto Compounds in Micro Reactors
The motivation of an industrial development was to increase selectivity for mono-chlorination of acetic acid to give chloroacetic acid [57]. This product is amenable under suitable reaction conditions by further chlorination to give dichloroacetic acid by consecutive reaction. The removal of this impurity is not simple, but rather demands laborious and costly separation. Either crystallization has to be performed with high technical expenditure or an expensive hydrogen reduction at a Pd catalyst is needed.

5.3.3.2 Beneficial Micro Reactor Properties for Chlorination of of α-Keto Compounds
The consecutive reaction will be triggered by too long exposure of already chlorinated product in an environment with a high density of chlorine radicals. Accordingly, controls over residence time, concentration profiles and efficient heat transfer have the potential to cope with such a problem.

5.3.3.3 Chlorination of of α-Keto Compounds Investigated in Micro Reactors
Gas/liquid reaction 14 [GL 14]: Chlorination of acetic acid

This process is carried out on an industrial scale in bubble columns [57]. Acetic acid and acetic anhydride are fed together with a recycle solution composed of acetic acid, acetyl chloride, monochloroacetic acid, dichloroacetic acid and hydrogen chloride. Under these conditions, acetic anhydride and hydrogen chloride give acetyl chloride spontaneously.

This mixture is fed into bubble columns and contacted with chlorine gas at 3.5 bar and 115–145 °C [57]. A typical reaction mixture has a composition of 38.5% acetic acid, 11.5% acetic anhydride and 50% chlorine gas. The crude product is first purified by distillation. Thereafter, either crystallization or hydrogen reduction at a Pd catalyst is conducted to separate the monochlorinated from the dichlorinated product.

5.3.3.4 Experimental Protocols
[P 13] Micro channels of 1500 µm width and 300 µm depth, separated by fins of 150 µm width, were employed [57].

Acetic acid and 10, 15, or 20% acetyl chloride were fed as a mixture into a modified falling film micro reactor (also termed micro capillary reactor in [57]) at a massflow rate of 45 g min^{-1} and a temperature of 180 or 190 °C [57]. Chlorine gas was fed at 5 or 6 bar in co-flow mode so that a residual content of only 0.1% resulted after reaction. The liquid product was separated from gaseous contents in a settler and collected. By exposure to water, acetyl chloride and acetic anhydride were converted to the acid. The hydrogen chloride released was removed.

5.3.3.5 Typical Results

Yield/selectivity/conversion – benchmarking to industrial bubble-column processing
[GL 14] [R 1] [P 13]* A yield of 85% was obtained by micro flow processing similarly to large-scale bubble column processing [57]. Selectivity was much better since less than 0.05% dichloroacetic acid was formed, whereas conventional processing typically gives 3.5%.

[R 1*] is a modified [R 1] micro device, the exact design of which has not been disclosed.

Numbering-up – benchmarking to industrial bubble-column processing
[GL 14] [R 1] [P 13]* Three falling-film units were operated in parallel in one device [57]. A yield of 85% and < 0.1% dichloroacetic acid resulted, exceeding the performance of conventional processing.

Temperature/pressure
[GL 14] [R 1] [P 13]* Increasing both temperature and pressure slightly increases the yield from 85% to 90% [57]. The content of dichloroacetic acid was below 0.05%.

5.3.4
Hydrodehalogenation – Dechlorination of Aromatics

Proceedings: [20].

5.3.4.1 Drivers for Performing Dechlorination of Aromatics in Micro Reactors
Micro reactors may be used for the removal of chlorinated organic compounds such as found in stockpiles of mixed waste [20]. On-site use of micro reactors may benefit from eliminating the need for waste transport, reduces the risk of exposure, could have lower investment and processing costs and may reduce the generation of secondary waste. These advantages seem to be clear, but so far there is no documentation in the literature giving experimental evidence.

5.3.4.2 Beneficial Micro Reactor Properties for Dechlorination of Aromatics
The above-mentioned drivers stem from the fact that micro reactors may be constructed light-weight and so are potentially mobile [20]. This holds particularly when integration of many components into one system can be performed. In addition, micro reactors can generally have high mass transfer efficiency, especially when

considering multi-phase processes as they provide large specific interfaces. This is needed when small-volume reactors are required to process large volumes of reactant solutions.

Reductions of investment and processing costs are general benefits that have been predicted for a long time; so far, this has not yet been demonstrated. This is strongly coupled with an onset of mass fabrication, proposed to have similar benefits as for microelectronics, which has not happened so far.

5.3.4.3 Dechlorination of Aromatics Investigated in Micro Reactors
Gas/liquid reaction 15 [GL 15]: Dechlorination of *p*-chlorophenol to phenol

This process involves a series of reactions, including dissolution, hydrogen reactions and chlorine withdrawal [20]. The second type of reactions include reduction of protons at the catalyst by electron transfer yielding hydrogen radicals that are consumed by reaction or give elemental hydrogen otherwise.

As catalyst, a Pd/Fe system is used, having finely dispersed Pd clusters (< 1 µm) on the Fe surface [20] (see also original citations in [20]). A considerable portion of the surface remains uncovered, exposing Fe for reaction.

5.3.4.4 Experimental Protocols
[P 14] A solution containing *p*-chlorophenol was fed from a 250 ml flask to the micro reactor by a syringe pump [20]. Flow rates from 10 to 65 ml min^{-1} were applied. The reaction temperature was set to either 20 or 40 °C.

The micro reactor consisted of a pair of iron plates coated with Pd catalyst placed on disk-type polypropylene holders [20]. This micro channel device was encased in a stainless-steel housing. The thickness of the catalyst layer was approximately 5.0 µm; 10 mg catalyst was deposited per plate, so giving 20 mg in total. The specific surface area of the catalyst was 3.6 ± 0.4 m^2 g^{-1}.

5.3.4.5 Typical Results
Dynamic behavior of catalyst
[GL 15] [R 7] [P 14] During a 24 h period, only minimal deactivation could be detected [20]. Only about 10–15% of the *p*-chlorophenol was converted.

Reaction rates
[GL 15] [R 7] [P 14] The reaction rate constant K''_w ranged from $3.15 \cdot 10^{-7}$ to $7.86 \cdot 10^{-7}$ m$_{reactor}$ s^{-1} depending on the flow rate (0.10–0.63 ml min^{-1}), channel depth (100, 200 µm) and temperature (20, 40 °C) [20]. Only about 10–15% of the *p*-chlorophenol was converted.

5.4
Addition to Carbon–Carbon Multiple Bonds

5.4.1
Dihydro Addition – Cycloalkene Hydrogenation

Peer-reviewed journals: [11, 12]; proceedings: [58, 59]; reactor description: [11, 12, 36, 58]; sections in reviews: [43, 46].

5.4.1.1 Drivers for Performing Cycloalkene Hydrogenation in Micro Reactors

The cyclohexene hydrogenation is a well-studied process especially in conventional trickle-bed reactors (see original citations in [11, 12]) and thus serves well as a model reaction. In particular, flow-pattern maps were derived and kinetics were determined. In addition, mass transfer can be analysed quantitatively for new reactor concepts and processing conditions, as overall mass transfer coefficients were determined and energy dissipations are known. In lieu of benchmarking micro-reactor performance to that of conventional equipment such as trickle-bed reactors, such a knowledge base facilitates proper, reliable and detailed comparison.

Cyclohexene hydrogenation is a fast process so mass transfer limitations are likely [12]. Processing at room temperature and atmospheric pressure reduces the technical expenditure for experiments.

Owing to the exothermic nature of hydrogenations, the avoidance of hot spots causing reductions in selectivity and catalytic activity is a second driver [11].

5.4.1.2 Beneficial Micro Reactor Properties for Cycloalkene Hydrogenation

Especially the favorable mass transfer of micro reactors is seen to be advantageous for cyclohexene hydrogenation [11]. As one key to this property, the setting and knowledge of flow patterns are mentioned. Owing to the special type of micro-reactor used, mixing in a mini trickle bed (gas/liquid flows over a packed particle bed) and creation of large specific interfaces were investigated in detail.

In addition, the good temperature control of micro reactors is mentioned as a further advantage [11].

5.4.1.3 Cycloalkene Hydrogenation Investigated in Micro Reactors
Gas/liquid reaction 16 [GL 16]: Catalytic hydrogenation of cyclohexene

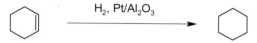

The reaction is carried out using a Pt/Al_2O_3 catalyst [11, 12]. Information on this reaction when conducted in trickle-bed reactors is available, comprising flow-pattern maps, kinetic data, mass transfer data and energy dissipation data (see original citations in [11]).

5.4.1.4 Experimental Protocols

[P 15] Cyclohexene was purified prior to use in a micro reactor experiment [11]. As catalyst, standard platinum supported on alumina powder with platinum contents of 1 or 5 wt.-% was used. For the Pt/Al_2O_3 catalyst, the surface area was 0.57 m^2 g^{-1}. The catalyst powder was sieved and separated into fractions with a range of particle sizes. A typical weight of 40 mg of catalyst was employed, corresponding to a catalyst loading density of 0.8–1.0 g cm^{-3}. All experiments were carried out at room temperature. The pressure was 0.01–0.25 MPa, depending on the flow rates. Steady co-current flow was utilized, avoiding pulsating flow, which leads to drying out of the catalyst. In this flow pattern, the liquid wets the wall of the micro channel prior to entering the packed bed. A stable gas/liquid interface near the feed entrance region was so achieved. A typical combination of gas and liquid flow rates was 5 standard cm^3 (sccm) min^{-1} and 10 μl min^{-1}, respectively. By visual check, it was ensured that an even flow distribution of the liquid to the catalyst particles was achieved.

5.4.1.5 Typical Results

Conversion/selectivity/yield

[GL 16] [R 12] [P 15] Conversions ranging from 2.8 to 16.0% were determined for a mini trickle-bed reactor [11] (see also [58]). The best result was obtained at a liquid flow rate of 15 mg min^{-1} and a gas flow rate of 5.0 sccm min^{-1}.

Average reaction rates

[GL 16] [R 12] [P 15] Average reaction rates ranging from 8.6 · 10^{-4} to 1.4 · 10^{-3} mol min^{-1} g$_{cat}^{-1}$ were determined for a mini trickle-bed reactor [11] (see also [58]). The best result was obtained both at a liquid flow rate of 153 mg min^{-1} and a gas flow rate of 5.0 sccm min^{-1} and at a liquid flow rate of 55 mg min^{-1} and a gas flow rate of 10.0 sccm min^{-1}. The intrinsic reaction rate was estimated to be 3.4 · 10^{-3} mol min^{-1} g$_{cat}^{-1}$, hence the best measured one is not in reach.

The initial reaction rate of a non-porous single-channel micro reactor, filled with porous catalyst particles, was 2.0 · 10^{-5} mol min^{-1} [12]. The same value of the porous 10-channel micro reactor was about three times larger. When normalized for the metal content of the device, the reaction rates of the porous reactor and the particle-containing reactor become similar, 6.5 · 10^{-5} and 4.5 · 10^{-5} mol min^{-1} m^{-2}, respectively.

Kinetics – volumetric rate constant

[GL 16] [R 12] [P 15] As excess of cyclohexene was used, the kinetics were zero order for this species' concentration and first order with respect to hydrogen [11]. For this pseudo-first-order reaction, a volumetric rate constant of 16 s^{-1} was determined, considering the catalyst surface area of 0.57 m^2 g^{-1} and the catalyst loading density of 1 g cm^{-3}.

Gas flow rate

[GL 16] [R 12] [P 15] On increasing the gas flow rate from 3.2 to 6.6 sccm min^{-1} at a constant liquid flow rate of 75 mg min^{-1}, the conversion and average reaction rate increase [11] (see also [58]). This is a hint for mass transfer limitations.

Liquid flow rate

[GL 16] [R 12] [P 15] On increasing the liquid flow rate from 15 to 153 mg min^{-1} at a constant liquid flow rate of 5.0 sccm min^{-1}, the average reaction rate increases,

while the conversion decreases [11] (see also [58]). This is a hint for mass transfer limitations.

Reactor performance with porous catalyst particles and porous walls that are catalyst coated

[GL 16] [R 12] [P 15] By a plasma etch process (see description in [R 12]), a highly porous surface structure can be realized which can be catalyst coated [12]. The resulting surface area of 100 m^2 is not far from the porosity provided by the catalyst particles employed otherwise as a fixed bed. In one study, a reactor with such a wall-porous catalyst was compared with another reactor having the catalyst particles as a fixed bed. The number of channels for both reactors was not equal, which has to be considered in the following comparison.

The initial reaction rate of the non-porous single-channel micro reactor, filled with porous catalyst particles, was $2.0 \cdot 10^{-5}$ mol min^{-1} [12]. The same quantity of the porous 10-channel micro reactor was about three times larger. When normalized for the metal content of the device, the reaction rates of the porous reactor and the particle-containing reactor become similar, $6.5 \cdot 10^{-5}$ and $4.5 \cdot 10^{-5}$ mol min^{-1} m^{-2}, respectively.

Mass transfer

[GL 16] [R 12] [P 15] Using a simple thin-film model for mass transfer, values for the overall mass transfer coefficient $K_L\,a$ were determined for both micro-channel processing and laboratory trickle-bed reactors [11]. The value for micro-reactor processing ($K_L\,a = 5\text{–}15$ s^{-1}) exceeds the performance of the laboratory tool ($K_L\,a = 0.01\text{–}0.08$ s^{-1}) [11, 12]. However, more energy has to be spent for that purpose (see the next section).

A lower-bound approximation for the overall mass transfer coefficient, $K_L\,a = 2\text{–}6$ s^{-1}, is also given in [11].

Energy dissipation

[GL 16] [R 12] [P 15] Increasing interfacial area is paid for by more flow resistance, which means energy dissipation. The energy dissipation factor, the power unit per reactor volume, of the micro-reactor process was higher ($\varepsilon_V = 2\text{–}5$ kW m^{-3}) as compared with the laboratory trickle-bed reactors ($\varepsilon_V = 0.01\text{–}0.2$ kW m^{-3}) [11]. Considering the still larger gain in mass transfer (see the previous section), an overall gain in performance for the micro reactor, nevertheless, can be stated.

5.4.2
Dihydro Addition – Alkene Aromatic Hydrogenation

Proceedings: [36]; reactor description: [11, 12, 36, 58]; sections in reviews: [26].

5.4.2.1   Drivers for Performing Alkene Aromatic Hydrogenation in Micro Reactors

As a second process, the hydrogenation of α-methylstyrene is a standard process for elucidating mass transfer effects in catalyst pellets and in fixed-bed reactors (see [36] and original literature cited therein). The physical properties of α-methyl-styrene are well described. The intrinsic kinetics are known. As the reaction is moderately fast, it fulfils one basic criterion for micro-channel processing. Owing to the exothermic nature, heat management is also known to play a role.

5.4.2.2 Beneficial Micro Reactor Properties for Alkene Aromatic Hydrogenation

Especially the favorable mass transfer of micro reactors is seen to be advantageous for cyclohexene hydrogenation [11]. As one key to this property, the setting and knowledge of flow patterns is mentioned. Owing to the special type of micro-reactor used, mixing in a mini trickle bed (gas/liquid flows over packed particle bed) and creation of large specific interfaces were investigated in detail.

5.4.2.3 Alkene Aromatic Hydrogenation Investigated in Micro Reactors
Gas/liquid reaction 17 [GL 17]: Catalytic hydrogenation of α-methylstyrene

The reaction was carried out using a palladium catalyst supported by activated carbon [36]. It is moderately fast at room temperature with 1 atm hydrogen. In the micro-reactor processing, however, operation at 50 °C was used. The reactor is first order with respect to hydrogen and zero order with respect to α-methylstyrene.

5.4.2.4 Experimental Protocols
[P 16] An ethanol slurry of activated carbon-supported palladium catalyst particles was introduced into a packed-bed micro reactor [36]. The fraction of 50–75 μm sized particles was used. The reaction was carried out with 1 atm hydrogen at 50 °C.

5.4.2.5 Typical Results
Conversion/selectivity/yield
[GL 17] [R 12] [P 16] Conversions ranging from 20 to 100% were determined for a mini trickle-bed reactor [36].

Initial reaction rates
[GL 17] [R 12] [P 16] The initial reaction rates were close to 0.01 mol min^{-1} per reaction channel without prior activation of the catalyst [36]. This is in agreement with literature data on intrinsic kinetics.

5.4.3
Dihydro Addition – Nitro Group Hydrogenation

Peer-reviewed journals: [60]; proceedings: [17, 61, 62]; sections in reviews: [47].

5.4.3.1 Drivers for Performing Nitro Group Hydrogenation in Micro Reactors

Nitro aromatics owe their great importance in organic synthesis for being intermediates for the generation of the respective anilines by hydrogenation [17, 61]. For instance, pharmaceuticals are produced via that route [61].

The reaction rates cannot be set as high as intrinsically possible by the kinetics, because otherwise heat removal due to the large reaction enthalpies (500–550 kJ mol^{-1}) will become a major problem [17, 60, 61]. For this reason, the hydrogen supply is restricted, thereby controlling the reaction rate. Otherwise, decomposition of nitrobenzene or of partially hydrogenated intermediates can occur [60].

The reaction involves various elemental reactions with different intermediates which can react with each other [60]. At short reaction times, the intermediates can be identified, while complete conversion is achieved at long reaction times. The product aniline itself can react further to give side products such as cyclohexanol, cyclohexylamine and other species.

5.4.3.2 Beneficial Micro Reactor Properties for Nitro Group Hydrogenation

Their favorable heat-transfer characteristics render micro reactors tools of interest for conducting nitro group hydrogenation via a gas/liquid process [17, 60].

However, in view of the large specific surface area of catalysts used in conventional fixed-bed reactors for this reaction, attempts have to be made to realize catalyst coatings of similar porosity in micro channels [17].

5.4.3.3 Nitro Group Hydrogenation Investigated in Micro Reactors
Gas/liquid reaction 18 [GL 18]: Catalytic hydrogenation of *p*-nitrotoluene [60]

The reaction is typically performed via a gas/liquid process on palladium-based catalysts, e.g. deposited on carbon [17]. Ethanol may be chosen as solvent.

5.4.3.4 Experimental Protocols

[P 17] In order to have a catalyst with a sufficiently high specific surface area, pretreatment of the micro channels made of aluminum was necessary [17]. Following a cleaning procedure, an oxide layer with a regular system of nanopores was generated by anodic oxidation (1.5% oxalic acid; 25 °C; 50 V DC; 2 h exposure using an aluminum plate cathode followed by calcination).

The active catalyst component was introduced by different ways:

- First, following a cleaning procedure, electrochemical deposition was carried out using an aqueous solution with $PdSO_4$ electrolyte, citric acid, and boric acid (25 °C; 7.5 V AC; 50 Hz; 3 min).
- Second, following a cleaning procedure with formalin, wet-chemical impregnation with palladium was performed [17]. The micro channels were exposed to a solution of 200 mg of $PdCl_2$ in 40 ml of distilled water for 5 h. The $PdCl_2$ was reduced to elemental Pd by formalin still present in the nanopores of the oxide layer which were generated one step before. This was followed by calcination. The whole procedure of impregnation was repeated several times to increase the catalyst load.

Typical catalyst loads were 21–14 mg [17].

Hydrogenation experiments were conducted in a flow apparatus (Figure 5.24) at 97 °C using a pressure of 2 MPa [17]. A 10% solution of p-nitrotoluene in 2-propanol was the liquid phase; as gas hydrogen (5.0 purity) was applied. The nitrotoluene flows normalized per unit area were 0.013 and 0.045 $g\,h^{-1}\,cm^{-2}$. The residence times were either 85 or 280 s. The recycle ratio was 21 or 43.

[P 18] A 0.5% $Pd/\gamma\text{-}Al_2O_3$ catalyst was used in the case of elemental hydrogen [61]. When using hydrogen donors, Pd/coal and Pd black were taken. As solvents, 2-propanol and ethanol were employed for elemental hydrogen and hydrogen donors, respectively. The reaction temperatures were 90 and 78 °C, respectively. In comparative experiments, a mini fixed bed of 10 mm inner diameter and 70 mm height was employed.

The surface of the micro channels was anodically oxidized to create a pore structure and thereafter wet-chemically impregnated [61]. The liquid reaction solution was fed by an HPLC pump; hydrogen was metered by a mass-flow controller. Pressure was kept constant.

Hydrogenation with hydrogen donors was carried out in a batch under refluxing conditions at ambient pressure [61].

[P 19] The nitrobenzene solution (0.04 or 1.0 mol l^{-1}) was fed into the falling film micro reactor at flow rates of 0.2–0.5 ml min^{-1} at 60 °C and 1–4 bar [60].

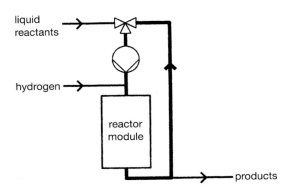

Figure 5.24 Schematic of the experimental set-up for the hydrogenation of p-nitrotoluene [17].

Catalyst coatings on the reaction plate of a falling film micro reactor were prepared by four routes and tested [60].

(a) Sputtering of palladium: a 100 nm thin layer of palladium was coated on the entire side of the reaction plate [60].

(b) UV decomposition of palladium acetate: a solution of this salt was introduced into the micro channels and then irradiated, inducing decomposition to yield elemental palladium [60].

(c) Wet impregnation: an γ-alumina layer was deposited via a slurry/washcoating route and subsequent calcination [60]. A 10 mm thick film with a surface area of 58 m^2 g^{-1} was obtained. By wet impregnation with a solution of palladium(II) nitrate (Pd 12–16%, w/w) and subsequent calcination, the final catalyst was obtained.

(d) Incipient wetness: as reported under (c), an γ-alumina layer was deposited via a slurry/washcoating route and subsequent calcination [60]. The same solution of palladium(II) nitrate was used to cover the micro channels. After drying, calcination completed catalyst preparation.

5.4.3.5 Typical Results
Conversion/selectivity/yield – benchmarking to fixed-bed reactor
[GL 18] [R 6a] [P 17] About 100% selectivity was achieved for the hydrogenation of p-nitrotoluene [17], with conversions of 58–98%. The conversion for the electro-deposited catalyst was 58%, whereas the impregnated catalyst gave a 58–98% conversion, depending on the process conditions (see Table 5.1).

Table 5.1 Results for hydrogenation of p-nitrotoluene using different micro-channel reactors: comparison of the results with those obtained by applying a conventional fixed-bed catalyst (rr-recycle ratio) [17].

Catalyst	M_{Pd} (mg)	T_{calc} (K)	$M_{p\text{-nitrotoluene}}/A_{geom}$ (g h^{-1} cm^{-2})	τ (s)	rr	Conversion (%)
MEI[a]	23	773	0.013	280	43	58
MCh[b]	24	773	0.013	280	43	96
MCh[b]	21	903	0.013	280	43	98
MCh[b]	21	903	0.045	85	21	58
WCh[c]	20	903	0.045	260	21	89
FB[d]	10	773	$1.7 \cdot 10^{-6}$	90	21	85

a Electrochemically deposited Pd.
b Chemically deposited Pd using formalin.
c Chemically deposited Pd on anodized aluminum wires using formalin.
d Conventional fixed-bed catalyst.

[GL 18] [R 6b] [P 18] Using a stoichiometric amount of hydrogen and operating in the slug-flow mode, it was shown that the yield of a micro reactor exceeds that of a mini fixed-bed reactor (LHSV = 60 h^{-1}) [61]. A maximum yield of 30% was obtained for the micro reactor for the range of pressure investigated (10–35 bar).

In advance, comparative fixed-bed measurements were undertaken. It was ensured that the performance of a plug-flow operation with both flows having the same direction is superior to trickle-bed operation, using counter-flow instead. The plug flow was assumed to model the slug-flow behavior in the micro reactor.

[GL 18] [R 1] [P 19a–d] Errors in judging selectivity came from problems of closing the carbon balance, rather than from intrinsic analytical fluctuations and sampling errors [60, 62]. The formation of species in solution not identified by the analytics used (GC) could be ruled out; instead, it was assumed that the loss of carbon is due to carbon deposition on the catalyst. The maximum loss of carbon amounted to about 20%, i.e. it was large.

Benchmarking to pure liquid-phase hydrogenation – use of hydrogen donors

[GL 18] [R 6b] [P 18] Using hydrogen donors, a pure liquid-phase operation, without any gas phase present, can be realized [61]. As a result, no mass transfer limitations across the gas/liquid boundary are present any longer. Despite this advantage in processing, it was found that the hydrogen-donor reaction takes several hours (up to 17 h at 11% yield) and gave only moderate yields (best yield: 27%); by contrast, gas/liquid processing took only a few minutes. Hence this option was disregarded by the authors for further experiments.

Method of catalyst preparation

[GL 18] [R 6a] [P 17] CFD calculations were performed to give the Pd concentration profile in a nanopore of the oxide catalyst carrier layer [17]. For wet-chemical deposition most of the catalyst was deposited in the pore mouth, in the first 4 μm of the pore. Hence most of the hydrogenation reaction is expected to occur in this location. For electrochemical deposition, large fractions of the catalyst are located in both the pore mouth and base. Since the pore base is not expected to contribute to large extent to hydrogenation, a worse performance was predicted for this case. This was corroborated by experimental evidence. Higher conversions were found for the wet-chemically prepared catalyst.

Catalysts prepared by different routes

[GL 18] [R 1] [P 19a] For a sputtered palladium catalyst, low conversion and substantial deactivation of the catalyst were found initially (0.04 mol l^{-1}; 60 °C; 4 bar; 0.2 ml min^{-1}) [60, 62]. Selectivity was also low, side products being formed after several hours of operation (Figure 5.25). After an oxidation/reduction cycle, a slightly better performance was obtained. After steep initial deactivation, the catalyst activity stabilized at 2–4% conversion and about 60% selectivity. After reactivation, the selectivity approached initially 100%. As side products, all intermediates except phenylhydroxylamine were identified.

In addition, it was assumed that too high reactivation temperatures were used, as a comparison with literature protocols reveals [60, 62]. Cracks in the plate and void areas with catalyst loss seem to corroborate this assumption.

[GL 18] [R 1] [P 19b] For a UV-decomposed palladium catalyst (Figure 5.26), a slightly higher conversion was found as for the sputtered catalyst (see above)

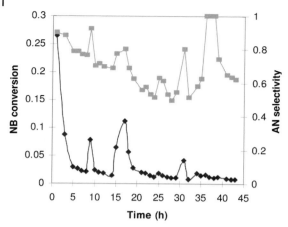

Figure 5.25 Catalytic activity of a sputtered palladium catalyst.
Nitrobenzene conversion (♦); aniline selectivity (■) [60].

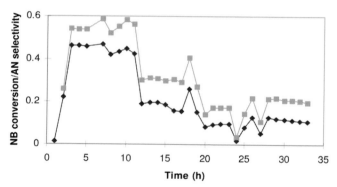

Figure 5.26 Catalytic activity of a UV-decomposed palladium acetate catalyst.
Nitrobenzene conversion (♦); aniline selectivity (■) [60].

(0.04 mol l^{-1}; 60 °C; 4 bar; 0.2 ml min^{-1}) [60, 62]. Steps in activity as a function of
time were found when the micro reactor was shut down. A similar spectrum of
side products as for the sputtered catalyst was found. The dynamic change in selectivity followed the trend of the conversion. Reactivation of the catalyst was hardly
successful.

Carbon species covered partly the catalyst [60, 62]. Also, cracks in the plate and
void areas with catalyst loss were observed which are responsible for catalyst deactivation.

[GL 18] [R 1] [P 19c] For an impregnated palladium catalyst, complete conversion
was found and maintained for 6 h (0.04 mol l^{-1}; 60 °C; 4 bar; 0.35 ml min^{-1}) [60, 62].
Selectivity decreased with time, but still remained high. Several reactivation routes
were tested, mainly based on removing organic residues by dichloromethane
(0.04 mol l^{-1}; 60 °C; 1 bar; 0.35 ml min^{-1}). The initial activity was recovered, but

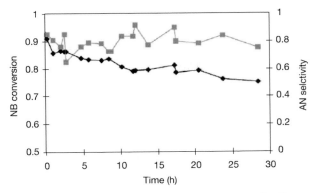

Figure 5.27 Comparison of nitrobenzene conversion and aniline selectivity as a function of reaction time for the incipient-wetness catalyst. Nitrobenzene conversion (♦); aniline selectivity (■) [60].

deactivation was more rapid than before. Reactivation by heating in air, however, led to a much improved performance. Complete conversion was regained and deactivation was slower. The time to the next reactivation step was shorter each time until finally a stable performance was achieved (0.04 mol l^{-1}; 60 °C; 1 bar; 0.5 ml min^{-1}). This is an indication of structural changes in the palladium and/or leaching.

[GL 18] [R 1] [P 19d] For an incipient-wetness palladium catalyst (Figure 5.27), the highest conversion was found of all catalysts investigated (the three other catalysts are described above) (0.04 mol l^{-1}; 60 °C; 1 bar; 0.5 ml min^{-1}) [60, 62]. Starting from a value of more than 90%, the performance settled at 82%, finally reaching 75%. Selectivity was on average 80%.

Leaching off of catalyst was observed; this was so substantial that even by visual inspection the channels appeared to be deeper than after initial coating. Black deposits were found at the channel end [60, 62].

[GL 18] [R 1] [P 19a–d] A comparison of all four catalysts evidences the importance of catalyst loading [60, 62]. The larger the loading, the longer the catalysts could be used before reactivation, as expected. When classified by their preparation, the four catalysts had the following sequence of lifespan and activity: wet impregnation >> incipient wetness > UV decomposition of precursors > sputtering. Selectivity followed a similar trend.

Mechanistic analysis/intermediates

[GL 18] [R 1] [P 19] For a sputtered palladium catalyst, all intermediates except phenylhydroxylamine were identified [60]. Their relative amounts allowed one to judge the route by which the hydrogenation proceeds. As a result, it was concluded that species containing nitroso, azo and azoxy groups have a strong interaction with the catalyst and so are preferably involved in the reaction course. In contrast, reduction of the hydrazo species was hindered. These assumptions are in line with literature reports.

Catalyst deactivation

[GL 18] [R 6a] [P 17] By applying a higher calcination temperature, the catalyst should be stabilized against hydrothermal dissolution [17]. Indeed, a slight increase in conversion from 96 to 98% was found on doing so.

[GL 18] [R 1] [P 19a–d] Catalyst longevity depends on the catalyst loading, as determined for four catalysts prepared by different preparation routes and hence having different loading [60, 62]. The larger the loading, the longer the catalysts could be used before reactivation, as expected. When classified by their preparation, the four catalysts had the following sequence of lifespan and activity: wet-impregnation >> incipient wetness > UV decomposition of precursors > sputtering. Selectivity followed a similar trend.

[GL 18] [R 1] [P 19a–d] High-molecular-weight deposits are assumed to form on the catalyst and can be removed by heating to 130 °C [60, 62]. The activity loss is fast, but recoverable. In contrast, palladium loss, another cause of deactivation, is not as fast, but is irrecoverable. This loss decreases gradually in the course of processing, in some cases reaching near-constant behavior.

Nitrobenzene concentration

[GL 18] [R 1] [P 19d] Using a high concentration of nitrobenzene, deleterious effects on the catalysts could be determined (1.0 mol l^{-1}; 60 °C; 1 bar; 0.5 ml min^{-1}) [60, 62]. Both conversion and selectivity were lower compared with processing at 0.04 mol l^{-1} concentration, being 41 and 68%, respectively. The catalyst surface turned dark after prolonged operation and particles settled (8.5 h). As a positive result, the (low) activity remained fairly stable over this period.

Separation of phases – benchmarking to fixed-bed reactor

[GL 18] [R 6a] [P 17] Using the same experimental conditions and catalysts with the same geometric surface area, the performance of micro-channel processing was compared with that of a fixed-bed reactor composed of short wires [17]. The conversion was 89% in the case of the fixed bed; the micro channels gave a 58% yield. One possible explanation for this is phase separation, i.e. that some micro channels were filled with liquids only, and some with gas. This is unlikely to occur in a fixed bed. Another explanation is the difference in residence time between the two types of reactors, as the fixed bed had voids three times larger than the micro channel volume. It could not definitively be decided which of these explanations is correct.

Reaching proper hydrodynamics/slug flow instead of channeling

[GL 18] [R 6b] [P 18] Scouting experiments were made in a micro reactor to look for proper conditions leading to slug-flow patterns [61]. Initially, the setting of flow rates was oriented on the process conditions of a fixed-bed reactor which gave maximum selectivity. The hydrogen supply was set three times higher than stoichiometrically needed. This corresponds to a volumetric ratio of liquid flow to gas flow of 1 : 2 for the conditions applied. However, this was expected to result in 'channeling', conducting liquid and gas in separate micro channels, since the pressure of a mixed gas/liquid flow will be higher than for the pure gas-filled channels.

Accordingly, the hydrogen supply was reduced to the stoichiometrically needed amount [61]. After ensuring that high yields can be achieved in a mini fixed-bed reactor (albeit needing high pressure of at least 30 bar) even under these conditions, the following gas and liquid flow rates were proposed: 1 : 1.3 at 20 bar and 1.0 : 0.65 at 40 bar. It was believed that these conditions should result in plug-flow patterns, although this was not experimentally confirmed.

Catalyst activity – benchmarking to fixed-bed reactor

[GL 18] [R 6a] [P 17] A sol–gel deposited catalyst used in a fixed-bed reactor gave higher conversion than a micro-channel catalyst impregnated on a porous alumina layer [17]. This was due to the higher geometric surface area of the sol–gel deposited catalyst.

Use of recycle loop

[GL 18] [R 6b] [P 18] A recycle loop was established by feeding six of seven parts of the product flow back to the reactor inlet [61]. The corresponding liquid-to-gas flow ratio was 1.0 : 0.37 which was assumed still to result in slug-flow behavior when using a micro reactor. For both a micro and a fixed-bed reactor, it was found that using such a recycle loop the same yield can be obtained at about half the pressure used in similar processing without recycling. By this means, a yield of 46% at a pressure of 10 bar could be achieved.

5.4.4
Dihydro Addition – Conjugated Alkene Hydrogenation

Peer-reviewed journals: [26, 63, 64]; proceedings: [65–67]; sections in reviews: [26, 43, 46, 48].

5.4.4.1 Drivers for Performing Conjugated Alkene Hydrogenation in Micro Reactors

The hydrogenation of a methyl cinnamate was investigated as a model reaction for demonstrating the benefits of a new screening technique, based on transient serial screening of multi-phase reactions [63, 64, 66]. Besides ensuring proper pulse formation and reducing pulse broadening, one main issue concerns having large mass transfer by use of micro reactors to carry out reactions in a kinetically controlled manner [63, 66].

Despite affecting conversion, mass transfer is known to impact enantio- and regioselectivity for many reactions [63]. For this reason, conventional micro-titration apparatus, typically employed in combinatorial chemistry of single-phase reactions, also often suffer from insufficient mixing when dealing with multi-phases [63, 66].

The hydrogenation of a cinnamate was also investigated as a first step to determine kinetics and finally to come to a quantitative determination of kinetic models and parameters in asymmetric catalysis [64]. The enantiomeric excess of enantioselective catalytic hydrogenations is known to be dependent on pressure, chiral additives and mixing. Such dependences are often due to kinetics, demanding appropriate studies.

In another investigation, process development and optimization studies were undertaken [67]. In the framework of a study conducted as contract research for industry, the general applicability of micro reactors to organic synthesis for pharmaceutical applications was tested, ultimately aiming at performing combinatorial chemistry by micro flow processing [67] (see a more detailed description in [26]). The scouting studies focused on determining suitable reaction parameters and monitoring yields as a function of time.

5.4.4.2 Beneficial Micro Reactor Properties for Conjugated Alkene Hydrogenation

Micro reactors are capable of generating large specific interfaces between two liquids, thereby enhancing mass transfer. For example, interdigital micro mixers are known to produce fine emulsions [68] and fine foams [22]. By using small samples, they also allow efficient screening. In addition to parallel approaches, serial screening is increasingly recognized as an efficient way particularly for multi-phase micro-channel processing [63, 64, 66]. Detailed investigations concerning test-throughput frequency, sample consumption and significance of screening experiments have been reported, e.g. [69, 70].

In addition, the interdigital micro mixers used have a very low inventory of catalytic material and have the potential for automation [64].

The hydrodynamics of multi-phase micro flow in a serial screening apparatus was investigated in detail [71, 72]. It was found that by optimizing the conditions, significantly smaller pulse broadening can be achieved, resulting in higher test-throughput frequencies.

Besides applications in serial screening, the process development and optimization studies mentioned above referred to the general advantages of micro flow processing in terms of enhanced heat and mass transfer [67] (see a more detailed description in [26]).

5.4.4.3 Conjugated Alkene Hydrogenation Investigated in Micro Reactors
Gas/liquid reaction 19 [GL 19]: Hydrogenation of Z-(α)-acetamidocinnamic methyl ester

By homogeneous reaction of the conjugated double bond system selectively the C=C double bond is hydrogenated [63–66]; the ester function is not affected. Moreover, by action of the chiral catalyst, a chiral hydrogenated product is created with good enantioselectivity.

Gas/liquid reaction 20 [GL 20]: Hydrogenation of methyl cinnamate

By heterogeneous reaction of methyl cinnamate, the C=C double bond is selectively hydrogenated [26] and methyl 3-phenylpropionate is formed; the ester function is not affected.

5.4.4.4 **Experimental Protocols**

[P 20] The reaction was carried out using ethylene glycol/water (60 : 40 wt.-%) and hydrogen [63, 66]. To stabilize the gas/liquid interface, sodium dodecyl sulfate was added as surfactant. By this means, a foam stable for at least 6 min at 60 °C was achieved. Bubbles of a typical size of 200 μm were formed. The liquid content in the foam amounted to 20%.

The catalyst was of Rh/diphos type, using as diphos ligands sulfonated (*S*,*S*)-1,2-bis(diphenylphosphanylmethyl)cyclobutane and sulfonated (2*S*,4*S*)-2,4-bis(diphenylphosphanyl)pentane [63, 66].

Pulses of 200 μl volume containing methyl (*Z*)-α-acetamidocinnamate at 0.05–0.10 M concentration and the catalyst at 1–4 mM concentration were injected via high-pressure liquid pumps and injection valves into an interdigital slit-type micro mixer [63]. The flow rates were 1–4 ml min^{-1} for the gas phase and 0.3–1.0 ml min^{-1} for the liquid phase. After passing through the micro mixer, the foam was transported in a reaction glass tube (2.85 mm inner diameter; 1.56 m length). The pulses moved as a reacting segment downstream of this reactor tube. The residence time was 3–6 min when leaving the tubular reactor (Figure 5.28) [64].

The reaction was carried out at 40–60 °C at atmospheric pressure [63, 66].

[P 21] Palladium on alumina was employed as catalyst [26]. Hydrogen and organic reactant were mixed in the micro mixer and fed to a Merck Superformance HPLC column of 100 mm length and 5 mm inner diameter, which was used as a hydrogenator. No further details are given in [67] or [26].

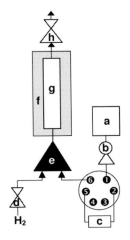

Figure 5.28 Schematic of the experimental set-up. Water/ethylene glycol/SDS reservoir (a); high-pressure liquid pumps (b); catalyst/substrate HPLC injection valve with 200 μl sample loop (c); hydrogen supply, equipped with mass flow controller (d); micro mixer (e); heating jacket (f); tubular glass or quartz reactor (g); back-pressure regulator (h) [64].

5.4.4.5 Typical Results
Conversion/selectivity/yield
[GL 20] [R 11] [P 21] By HPLC analysis it could be shown that the hydrogenation product was exclusively formed [67] (see a more detailed description in [26]).

Catalyst concentration
[GL 19] [R 9] [P 20] The rate of reaction is proportional to increasing catalyst concentration [63, 66]. This result is an indication of operation in a chemical regime. At higher temperature, larger rates are found [66].

Surfactant concentration
[GL 19] [R 9] [P 20] The rate of reaction is proportional to decreasing surfactant (sodium dodecyl sulfate) concentration [63]. No change in the enantiomeric excess was observed. These results are an indication of operation in a chemical regime.

Benchmarking to batch operation
[GL 19] [R 9] [P 20] Enantiomeric excess data for the micro reactor and batch operation are in agreement when being performed under similar conditions [63].

Catalyst consumption
[GL 19] [R 9] [P 20] Down to 0.2 μmol of the metal (rhodium) was consumed in a typical test [63].

Activation energy
[GL 19] [R 9] [P 20] An activation energy of 67 kJ mol^{-1} was determined (substrate, 0.05 kmol m^{-3}; 1 bar; gas flow rate, 10^{-4} m^3 h^{-1} (NPT); liquid flow rate, 1 cm^3 min^{-1}; 225 s) [66]. Hence the micro flow test is under chemical control.

Foam stability
[GL 19] [R 9] [P 20] Foam stability increases with hydrogen pressure [64, 70]. The foams remain stable for up to 12 min at 70 °C; no coalescence occurs within this period.

Reproducibility of conversion and enantiomeric excess
[GL 19] [R 9] [P 20] A moderate reproducibility of conversion was determined, lying in the range from 76 to 94% (0.001 M Rh; 0.05 M ester; 40 °C; 0.3 MPa; 3.2 min) [64]. This was explained by variations in the pressure, as control by the back-pressure regulator was insufficient. If one data point was excluded, conversions ranged only from 86 to 94%, giving better agreement.

The enantiomeric excess was very reproducible (19.6–20.6%) [64].

Surfactant influence on enantiomeric excess
[GL 19] [R 9] [P 20] The surfactant had no influence on the enantiomeric excess [64].

Enantiomeric excess distribution
[GL 19] [R 9] [P 20] The enantiomeric excess distribution was fairly narrow; 90% of all data were within 40–48% [64]. This is in line with first-order kinetics.

Share of rejected experiments
[GL 19] [R 9] [P 20] About 20% (44 of 214) experiments were rejected since their conversions were too low (< 3%) to provide any useful information [64]. After other experiments had also been rejected, finally 66% of all experiments could be used. In a classical mini-batch test, 71% of all tests were employed, using the same reaction and processing conditions.

Fit to empirical kinetic models
[GL 19] [R 9] [P 20] Experimental data were fitted to several empirical models from a mechanistic model [64]. By an iterative fitting process, a statistical model with first-order kinetics with respect to hydrogen was derived. With this model, a parity diagram was given, showing that 29 (17%) experiments of 170 had to be rejected; the others were adequately described by the model. All rejected data had higher conversion than theoretically predicted.

Residence-time distribution
[GL 19[R 9] [P 20] Axial dispersion, probably due to back-mixing, takes places in a vertically positioned tube [64, 70]. Horizontal placing gives better performance.

Kinetic constant – activation energy – benchmarking to mini batch
[GL 19] [R 9] [P 20] Both the kinetic constant and the activation energy were lower for a micro-reactor test unit than for a batch reactor (micro/batch: $k_{323} = 9.3$ vs $19.1 \ m^3 \ kmol^{-1} \ MPa^{-1} \ min^{-1}$); $E_a = 31$ vs $40 \ kJ \ mol^{-1}$) [64, 70]. This was assigned to the very low inventory of material. Also, poor control of residence time and temperature in the micro flow test was noted [64].

Enantioselectivity – benchmarking to mini batch
[GL 19] [R 9] [P 20] Data on enantioselectivity were in reasonable agreement for a micro-reactor test unit and a batch reactor (ee = 40–45%) [66, 70].

Reaction Volume – Benchmarking to Mini Batch
[GL 19] [R 9] [P 20] Reaction volumes required for testing are 1% of the volume needed for batch processing [70]. Whereas for a typical batch 10 cm^3 are demanded, only 0.1 cm^3 are needed for micro flow processing (see Table 5.2).

Noble metal and chiral ligand consumption – benchmarking to mini batch
[GL 19] [R 9] [P 20] The consumption of noble metal and chiral ligands per test was orders of magnitude lower for a micro-reactor test unit than a mini batch (noble metal: micro reactor, 5–20 µg; mini batch, 500–1000 µg; chiral ligand: micro reactor, 10 µmol; mini batch, 0.1 µmol) [70].

Table 5.2 Comparison of reaction conditions and results for hydrogenation of methyl (Z)-α-acetamidocinnamate in mini batch, micro liquid/liquid and micro gas/liquid reactors [70].

Feature	Mini batch	Micro	Micro
Reaction volume (ml)	10	0.1	0.1
Average amount of Rh/experiment (µg)	500–1000	5–20	5–20
Typical amount of ligand per experiment (µmol)	10	0.1	0.1
Temperature range (°C)	20–100	20–80	20–80
Pressure range (bar)	1–100	1–11	1–11
Residence time (min)	> 10	1–30	1–30
Average TTF achieved during study (d^{-1})	2	15	15
Maximum actual achievable TTF (d^{-1})	3	40	40
Range of solvents	Large	Restricted	Restricted
Automation of reagents/catalyst injection	No	Yes	Yes
Automation of sample collection	No	Yes	Yes

Temperature and pressure range – benchmarking to mini batch
[GL 19] [R 9] [P 20] The temperature range of a micro-reactor test unit is similar to that of a mini batch reactor, whereas the range for pressure operation is different (temperature: micro reactor, 20–80 °C; mini batch, 20–100 °C; pressure: micro reactor, 1–11 bar; mini batch, 1–100 bar) [70]. This is caused by the choice of glass as tube material for the micro reactor and may be overcome in the future by choosing stainless steel.

Residence time – benchmarking to mini batch
[GL 19] [R 9] [P 20] A micro-reactor test unit allows operation at short and medium residence times (1–11 min), whereas a mini batch has to be used for reactions longer than this period [70].

Range of solvents – benchmarking to mini batch
[GL 19] [R 9] [P 20] The range of solvents to be used for a micro-reactor test unit is limited as a stable foam has to be established. In turn, this range is large for a mini batch reactor [70].

Automation of injection and sample collection – benchmarking to mini batch
[GL 19] [R 9] [P 20] Such automation is (in principle) achievable for micro flow operation, but cannot be done for mini batch operation [70].

Time demand for a test – benchmarking to mini batch
[GL 19] [R 9] [P 20] A typical test on gas/liquid transient serial processing lasts 3–5 min [63]. By extrapolation, test-throughput frequencies of 500 per day are seen

as possible. To reach this performance, fast analysis is demanded as this is the rate-limiting step in combinatorial and related techniques. More than 20 tests per day have been reported [66].

[GL 19] [R 9] [P 20] A micro-reactor test unit runs 15 and 40 tests per day on average and at maximum, respectively, whereas only two and three such tests can be processed in a mini batch unit [70]; 214 tests were conducted in one investigation [64].

5.4.5
Dihydro Addition – First-order Model Reaction

Proceedings: [73]; sections in reviews: [26].

5.4.5.1 Drivers for Modeling First-order Model Reactions in Micro Reactors
The study referred to here was using a hypothetical hydrogenation reaction; however, of common nature so that, at least, qualitative conclusions can be drawn to optimize some of the reactions mentioned above [73].

5.4.5.2 First-order Model Reactions Modeled in Micro Reactors
Gas/liquid reaction 21 [GL 21]: Model reaction with hydrogen

$$H_2 + A + \text{solvent} \rightarrow A + H_2 + B + D + \text{solvent}$$

This model reaction was assumed to be first order with regard to hydrogen [73].

5.4.5.3 Modeling Protocols
[P 22] A flat-plate micro reactor was modeled where hydrogen was diffusing from the top into the micro channel [73]. At this border, hydrogen is at its solubility limit. The liquid solution is flowing in a counter-flow mode to the hydrogen feed. A catalyst coating is placed on the wall opposite the hydrogen feed.

The model reaction was based on hydrogen and a 5 wt.-% reactant solution; the liquid reactant did not need to be specified [73]. Assuming efficient heat transfer, the reaction was regarded as isothermal. A catalyst of zero thickness was considered. As a base case, a channel of a depth of 100 μm and of a length of 20 mm, was used. A velocity of 10 mm s^{-1} was assumed (Re = 0.04–0.78). Thereafter, these properties were varied during the modeling.

5.4.5.4 Typical Results
Concentration profiles
[GL 21] [no reactor] [P 22] Depending on the reaction velocity, different concentration profiles were simulated [73]. At 10 mm s^{-1}, reaction occurs over the full length and is not completed. At 1 mm s^{-1}, the reaction, reaction starts much earlier.

Channel depth
[GL 21] [no reactor] [P 22] The channel depth has a strong influence on the calculated conversion [73]. Smaller channels have enhanced conversion.

Reaction mixture velocity – residence time
[GL 21] [no reactor] [P 22] Higher reaction velocities have notably reduced conversion [73]. Corresponding residence times above 10 s were predicted to give above 20% conversion.

Diffusion coefficient
[GL 21] [no reactor] [P 22] A linear increase in conversion with increasing diffusion coefficient was observed [73]. This shows that liquid transport of hydrogen to the catalyst has a dominant role.

Reaction rate constant
[GL 21] [no reactor] [P 22] A constant conversion is approached on increasing the reaction rate constant [73]. This shows that liquid transport of hydrogen to the catalyst has a dominant role. In turn, this means that a higher catalyst loading should have not too much effect.

5.5
Addition to Carbon–Heteroatom Multiple Bonds

5.5.1
S-Metallo, *C*-Hydroxy Addition – Carbon Dioxide Absorption

Proceedings: [5]; PhD thesis [1]; sections in reviews: [26, 48].

5.5.1.1   Drivers for Performing Carbon Dioxide Absorption in Micro Reactors
This reaction serves as a known model reaction to characterize mass transfer efficiency in micro reactors [5]. As it is a very fast reaction, solely mass transfer can be analyzed. The analysis can be done simply by titration and the reactants are inexpensive and not toxic (although caustic).

5.5.1.2   Beneficial Micro Reactor Properties for Carbon Dioxide Absorption
Carbon dioxide absorption is used for the above-mentioned purpose, hence an analytical tool for judging micro-reactor performance [5].

5.5.1.3   Carbon Dioxide Absorption Investigated in Micro Reactors
Gas/liquid reaction 22 [GL 22]: Acid–base reaction between carbon dioxide and sodium hydroxide

$$2\,Na^+ + 2\,OH^- + CO_2\,(g) \rightarrow 2\,Na^+ + CO_3^{2-} + H_2O$$

5.5.1.4   Experimental Protocols
[P 23] Aqueous NaOH solutions of 0.1, 1.0 and 2.0 M were used, fed by pumps to the micro devices [5]. Carbon dioxide was supplied as a mixture with nitrogen, the flow rate being set by a mass-flow controller. Liquid samples were taken and subjected to carbonate analysis (see original citation in [5]).

Since many different gas/liquid contactors were used, the experimental conditions differed for each device and the reader is referred to the listing in the original reference [5]. The liquid flows range from 10 to 1042 ml h^{-1} and the gas volume flows from 180 to 25 020 ml h^{-1}. The corresponding residence times were 0.01–19.58 s. The ratio of carbon dioxide to sodium hydroxide was fixed at 0.4.

[P 24] Aqueous NaOH solutions of 0.1, 1.0 and 2.0 M were used, fed by pumps to the micro devices [5]. Carbon dioxide was supplied as a mixture with nitrogen, the flow rate being set by a mass-flow controller. Liquid samples were taken and subjected to carbonate analysis (see original citation in [5]).

The carbon dioxide volume content was varied from 0.8 to 100 vol.-%; the gas velocity changes from 0.1 to 42.9 mm s^{-1} [5]. The residence time varied from 0.1 to 9.7 min; 64 single streams of a liquid film thickness of 65 μm were used at a total volume flow of 50 ml h^{-1}. The ratio of carbon dioxide to sodium hydroxide was fixed at 0.4.

5.5.1.5 **Typical Results**
Mass transfer efficiency by conversion analysis – benchmarking to mixing tee
[GL 22] [R 3] [R 9] [R 10] [P 23] The mass transfer efficiency of different gas/liquid contactors as a function of residence time was compared qualitatively (Figure 5.29), including an interdigital micro mixer, a caterpillar mini mixer, a mixing tee and three micro bubble columns using micro channels of varying diameter [5].

The two micro bubble columns comprising the smaller micro channels reached nearly 100% conversion [5]. The micro bubble column with the largest hydraulic diameter reached at best 75% conversion. The curve obtained displays the typical shape, passing through a maximum due to the antagonistic interplay between residence time and specific interfacial area.

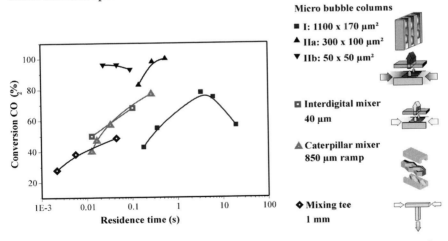

Figure 5.29 Special-type multi-purpose micro devices and mixing tee used for investigation of CO_2 absorption. Comparison of their reactor performance as a function of the residence time. Micro bubble columns (■) 1100 μm × 170 μm, (▲) 300 μm × 100 μm and (▼) 50 μm × 50 μm; interdigital mixer (□) (40 μm); caterpillar mixer (△) (850 μm ramp); mixing tee (◊) (1 mm) [5].

All other devices showed only the increasing part of such a dependence, i.e. the highest performance measured was obtained at the longest residence time [5]. The best conversions with the interdigital micro and caterpillar mini mixers (~78 and ~70%, respectively) still exceed considerably the performance of a conventional mixing tee (1 mm inner diameter).

Mass transfer efficiency by conversion analysis for the falling film micro reactor

[GL 22] [R 1] [P 23] The mass transfer efficiency of the falling film micro reactor as a function of the carbon dioxide volume content was compared quantitatively (Figure 5.30) [5]. The molar ratio of carbon dioxide to sodium hydroxide was constant at 0.4 for all experiments, i.e. the liquid reactant was in slight excess.

In a first set of experiments, the impact of the sodium hydroxide concentration (0.1, 1.0, 2.0 M) and gas-flow direction (co-current, counter-flow) was analysed (50 ml h^{-1} liquid flow, 65 μm film thickness) [5]. The higher the base concentration, the higher is the conversion of carbon dioxide. For all concentrations, complete absorption is achieved, but at different carbon dioxide contents in the gas mixture. The higher the carbon dioxide content, the higher is the gas flow velocity and the larger must be the sodium hydroxide concentration for complete absorption. The gas flow direction had no significant effect on carbon dioxide absorption as the gas velocities were still low, so that no pronounced co- or counter-flow operation was realized.

It is remarkable that the falling film micro reactor achieved complete conversion for all process variations applied [5]. This is unlike conventional reactor operation reported for this reaction, displaying pronounced mass transfer resistance.

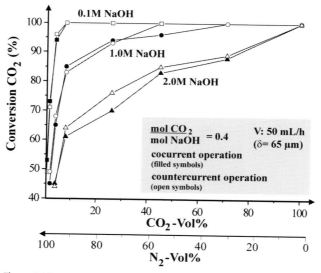

Figure 5.30 Conversion of CO$_2$ as a function of reactant concentration and flow direction of the gaseous phase [5].

In a second set of experiments, the height of the gas chamber was varied (2.5, 5.5 mm) in order to analyse the origin for some mass-transfer limitations for the micro-channel operation found in the first set of experiments (sodium hydroxide concentration, 1.0 M; 200 ml h^{-1} liquid flow; 100 μm film thickness) [5]. A large impact of the carbon dioxide content in the gas phase on the absorption in the alkaline solution was found; the importance of the gas velocity for this conversion was less.

Mass transfer efficiency – benchmarking of micro devices to literature data
[GL 22] [R 1] [R 3] [P 23] The mass transfer efficiencies of the falling film micro reactor and the micro bubble column were compared quantitatively with literature reports on conventional packed columns (see Table 5.3) [5]. The process conditions were chosen as similar as possible for the different devices. The conversion with the packed columns was 87–93%; the micro devices gave conversions of 45–100%. Furthermore, the space–time yield was compared. Here, the micro devices gave an order of magnitude larger values (the best results for the falling film micro reactor and the micro bubble column were 84 and 816 mol m^{-3} s^{-1}, respectively, compared to the conventional packed bed reactor having 0.8 mol m^{-3} s^{-1}.

Table 5.3 Comparison of space–time yields for CO_2 absorption when using micro devices and conventional packed columns [5].

Reactor type	NaOH (mol l^{-1})	CO$_2$ (vol.-%)	Molar ratio CO$_2$/NaOH	Conversion CO$_2$ (%)	Space–time yield (mol m^{-3} s^{-1})
Packed column	1.2	12.5	0.41	87	0.61
Packed column	2.0	15.5	0.43	93	0.81
Falling film micro reactor (65 μm)	1.0 (50 ml h^{-1})	8.0	0.40	85	56.1
Falling film micro reactor (65 μm)	2.0 (50 ml h^{-1})	8.0	0.40	61	83.3
Falling film micro reactor (100 μm)	1.0 (200 ml h^{-1})	8.0	0.40	45	83.7
Micro bubble column (300 × 300 μm)	2.0 (10 ml h^{-1})	8.0 (2400 ml h^{-1})	0.33	100	227
Micro bubble column (300 × 300 μm)	2.0 (50 ml h^{-1})	8.0 (12 000 ml h^{-1})	0.33	72	816

Temperature distribution in micro reactor
[GL 22] [R 1] [R 3] [P 23] In a first scouting experiment to monitor temperature profiles, a very uniform temperature profile was measured in the falling film micro reactor under non-reacting conditions, using 2-propanol as falling film [5]. The temperature deviations along the whole reaction plate were less than 0.3 °C at an average temperature of 30 °C. Hence isothermal conditions were given, provided that no large reaction terms were present.

5.6
Oxidations and Reductions

5.6.1
C,O-Dihydro Elimination – Oxidation of Alcohols to Aldehydes

Proceedings: [58, 59]; reactor description: [11, 12, 36, 58]; sections in reviews: [43, 46, 48].

5.6.1.1 Drivers for Performing Oxidation of Alcohols in Micro Reactors
The oxidation of benzyl alcohol to benzaldehyde is carried out for elucidating mass transfer effects in a mini trickle-bed reactor [58].

5.6.1.2 Beneficial Micro Reactor Properties for Oxidation of Alcohols
Especially the favorable mass transfer of micro reactors is seen to be advantageous for the oxidation of benzyl alcohol [58]. As one key to this property, the setting and knowledge on flow patterns are mentioned. Owing to the special type of micro-reactor used, mixing in a mini trickle bed (gas/liquid flows over a packed particle bed) and creation of large specific interfaces are special aspects of the reactor con-cept. In addition, temperature can be controlled easily and heat transfer is large, as the whole micro-reactor construction acts as a heat sink.

5.6.1.3 Oxidation of Alcohols Investigated in Micro Reactors
Gas/liquid reaction 23 [GL 23]: Oxidation of benzyl alcohol to benzaldehyde

The reaction is carried out using a palladium catalyst supported by activated carbon [58].

5.6.1.4 Experimental Protocols
[P 25] A slurry of activated carbon-supported 30 wt.-% palladium catalyst particles was introduced into a single-channel packed-bed micro reactor [36]. A fraction of 53–75 μm sized particles is used. The reaction is carried out at up to 8 atm pressure hydrogen and of up to 140 °C.

5.6.1.5 Typical Results
Conversion/selectivity/yield – benchmarking
[GL 23] [R 12] [P 16] Conversions near 70% were determined for a mini trickle-bed reactor (flow rate 20 mg min^{-1}) [36]. The corresponding reaction rate was 10 times larger than in typical batch operation on a laboratory-scale, which is restricted to milder conditions.

5.6.2
Cycloadditions – Photo-Diels–Alder Reactions Using Oxygen

Peer-reviewed journals: [21]; proceedings: [40]; patents: [74]; sections in reviews: [45, 47].

5.6.2.1 Drivers for Performing Photooxidation of Dienes in Micro Reactors

The photooxidation of cyclopentadiene by singlet oxygen is one step of an industrial process to make 2-cyclopentene-1,4-diol [40]. Hence the driver is a commercial one, namely to develop a continuous synthesis of this molecule.

In addition, particularly micro-channel processing is demanded here as the reaction is dangerous owing to the explosive nature of the endoperoxide formed as intermediate. This handling of explosive or toxic compounds for Diels–Alder reactions is also described in [21]. Owing to the use of only small volumes, the hazardous potential can be substantially minimized. This was exemplarily shown for the addition of singlet oxygen.

In another publication, attention is drawn on the difficulties in achieving a high oxygen saturation of during photooxidations with singlet oxygen and the aid of sensitizers [21]. Oxygen-rich organic solutions are always subject to explosions and hence pose a considerable risk during scale-up. Owing to the absorption of the latter species, problems also exist in achieving a sufficiently high illumination of the reaction sample. As a consequence, high-power irradiation is employed, which reduces selectivity by the formation of dimers and higher aggregates. In addition, the large light power leads to overheating of the sample, demanding collimators or active cooling.

5.6.2.2 Beneficial Micro Reactor Properties for Photooxidation of Dienes

Apart from showing the feasibility of an industrial synthesis by the micro reactor and looking for benefits with regard to mass and heat transfer, a further property relates to the numbering-up concept when carrying out the photooxidation of cyclopentadiene. By continuous processing, even in the falling film micro reactor [40], medium quantities of the product can already be prepared, which may be sufficient in the case of precious fine chemicals. A further increase in productivity can be achieved by parallel operation of many such micro devices which are all run under the same conditions. Favorably, this is done by internal numbering-up, e.g. by parallel feed of many plates. Also, a certain scale-up, i.e. increase in characteristic dimensions, can be tolerated. In the case of the falling film micro reactor, this means using a larger micro channel depth, which allows an increase in flow of about an order of magnitude.

Concerning safety issues, micro reactors are beneficial as they efficiently remove the reaction heat and also may intrinsically prevent explosions by terminating the radical chains. This has been impressively shown for the reaction between hydrogen and oxygen, widely known as being very dangerous [75, 76].

Owing to the small length scales in micro reactors, e.g. 50 μm, high concentrations of a sensitizer may be used [21]. As these materials typically have high costs, recycle loops with low inventory can be employed to consume only a small overall amount of sensitizer. The sensitizer absorption, despite the large molar extinction coefficient, is not over the tolerable limit since only small optical paths are employed. It is assumed that molecules in thin liquid layers face a broadly similar photon flux, unlike macro-scale photo processing.

Low-intensity light sources should give efficient irradiation of thin liquid layers [21]. Sample heating is reduced and so is radical recombination. In addition, oxygen enrichment of solutions before and after micro reactor passage can be handled differently and is no longer a major safety problem [21].

5.6.2.3 Photooxidation of Dienes Investigated in Micro Reactors
Gas/liquid reaction 24 [GL 24]: Oxidation of cyclopentadiene by singlet oxygen to 2-cyclopentene-1,4-diol

This reaction, of industrial interest, utilizes singlet oxygen generated by irradiation in the presence of Rose Bengal [40]. An endoperoxide is formed as intermediate which is converted to 2-cyclopentene-1,4-diol by reduction with thiourea.

Gas/liquid reaction 25 [GL 25]: [4 + 2] Cycloaddition of singlet oxygen to α-terpinene

Using a single-channel chip micro reactor, singlet oxygen is generated by photochemical means in presence of catalytic amounts of Rose Bengal [21]. By [4 + 2] cycloaddition of this oxygen species to α-terpinene, the product ascaridole is obtained.

5.6.2.4 Experimental Protocols
[P 26] A 4 ml volume of cyclopentadiene and 100 mg of Rose Bengal as photosensitizer were dissolved in 250 ml of methanol [40]. This solution was passed as a falling film through the micro channels in the reaction plate of the falling film micro reactor at a volume flow of 1 ml min^{-1}. Oxygen was guided in co-current flow to this reaction stream at a velocity of 15 l h^{-1}. The temperature was set to 10–15 °C using a cryostat. Via a quartz glass window in the housing, light from a xenon lamp was passed through the reaction zone. The reacted stream was passed directly into a solution

of 2.5 g of thiourea in 60 ml of methanol, cooled to 10 °C. To isolate the product quantitatively, the micro reactor was, in addition to the collection of the continuous stream, purged with 20 ml of methanol. The solvent was evaporated from the solution, leaving the crude product, which was extracted with 20 ml of acetone, filtered and purified by column chromatography [eluent: chloroform/methanol (9 : 1)]. A 0.95 g amount of 2-cyclopentene-1,4-diol was obtained in this way.

[P 27] Processing in the micro chip was safe only as small volumes were applied in the reaction zone and any larger volumes before and after micro-reactor processing could be prevented from containing oxygen-rich solutions, which are of potential danger [21]. The small volumes do not need to be pre-saturated and can be efficiently degassed with nitrogen after passage through the micro reactor.

The channel section of the micro chip was irradiated at a 10 cm distance at full power by an unfiltered 20 W (6 V) overhead tungsten lamp on an inverted microscope stage [21]. A solution of sensitizer dye (0.1 g) and terpinene (85%, 0.6 ml) with methanol (20 ml) as solvent was introduced at a flow rate of 1 μl min^{-1} through one port. Pure oxygen was fed through the other port at a flow rate of 15 μl min^{-1}. Hamilton PHD 2000 syringe pumps were used for liquid and gas feed. Both streams merged at the beginning of the micro channel section. The reactant solution was diluted with a 10-fold excess of diethyl ether and passed through a silica plug. The solvent was evaporated by a stream of nitrogen.

Since the sensitizer Rose Bengal is recyclable, relatively high concentrations (5 · 10^{-3} M) can be used without raising cost issues and optical detection problems [21]. Under these conditions, it was shown that molecules at any position in the micro channel have a similar photon flux.

5.6.2.5 Typical Results
Conversion/selectivity/yield
[GL 25] [R 8] [P 27] After irradiation for only 5 s, high conversions (> 80%) were obtained [21]. This is explained as being due to both the high absorption coefficient of the photosensitizer Bengal Rose and the high local number density of photons within the micro channel as a result of the large specific surface area.

[GL 24] [R 1] [P 26] For the oxidation of cyclopentadiene by singlet oxygen to 2-cyclopentene-1,4-diol, a yield of 19.5% was found [40].

Safe micro flow processing of explosive intermediates and on-site conversion to valuable, stable products
[GL 24] [R 1] [P 26] The feasibility of safely carrying out the oxidation of cyclopentadiene by singlet oxygen to 2-cyclopentene-1,4-diol was demonstrated [40]. The explosive intermediate endoperoxide was generated and without isolation used on-site for a subsequent hydration reaction. By reduction with thiourea the pharmaceutically important product 2-cyclopentene-1,4-diol was so obtained.

Safe micro flow processing of explosive mixtures
[GL 25] [R 8] [P 27] Continuous micro flow processing avoiding enrichment of the explosive endoperoxide ascaridole was established [21].

5.6.3
Oxidation of Aldehydes to Carboxylic Acids – Addition of Oxygen

Proceedings: [10]; PhD thesis: [9]; sections in reviews: [77, 78].

5.6.3.1 Drivers for Performing Oxidation of Aldehydes to Carboxylic Acids in Micro Reactors

The homogeneously catalyzed oxidation of butyraldehyde to butyric acid is a well-characterized gas/liquid reaction for which kinetic data are available. It thus serves as a model reaction to evaluate mass transfer and reactor performance in general for new gas/liquid micro reactors to be tested. This reaction was particularly used to validate a reactor model for a micro reactor [9, 10].

5.6.3.2 Beneficial Micro Reactor Properties for Oxidation of Aldehydes to Carboxylic Acids

The homogeneously catalyzed oxidation of butyraldehyde to butyric acid was used to analyse reactor performance for different flow patterns (or for different Weber numbers) [9, 10]. Hence it relates to the possibility of setting various flow patterns in gas/liquid micro devices and hence controlling mass transfer.

5.6.3.3 Oxidation of Aldehydes to Carboxylic Acids Investigated in Micro Reactors
Gas/liquid reaction 26 [GL 26]: Homogeneously catalyzed oxidation of butyraldehyde to butyric acid

Butyraldehyde is oxidized to butyric acid in the presence of air using manganese acetate as catalyst [9, 10].

5.6.3.4 Experimental Protocols
[P 28] The liquid feed was introduced by a pump and the gas feed using a mass-flow controller [10]. The reaction was carried out using liquid flows of 20.7–51.8 ml h^{-1} and gas flows of 1.7–173 ml min^{-1}. The gas and liquid velocities amounted to 0.02–1.2 and 0.03–3.0 m s^{-1}, respectively. The reaction was performed in mixed flow regimes, including bubbly, slug and annular patterns. The specific interfacial areas amounted to about 5000–15 000 m^2 m^{-3}. The reaction was conducted at room temperature.

5.6.3.5 Typical Results
Conversion/selectivity/yield – benchmarking
[GL 26] [R 3] [P 28] Conversions from 2 to 42% were found for the oxidation of butyraldehyde [10]. The highest conversions were obtained for large gas and liquid flows. On increasing the ratio of gas and liquid superficial velocities from 5 to 53, an increase in conversion from 10 to 41% resulted.

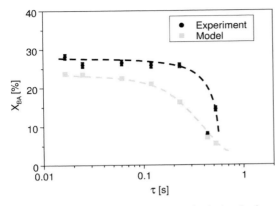

Figure 5.31 Experimental results (●) and calculated values
(■) for conversion of butyraldehyde as a function of
residence time [10].

Conversion normalized by residence time (Figure 5.31) increased nearly linearly
with the Weber number owing to an increase in specific interfacial area [10]. The
normalization is needed because by increasing gas and liquid velocities both interfacial area and time are affected in an antagonistic manner.

Reactor model of micro bubble column performance
[GL 26] [R 3] [P 28] A simple reactor model was developed assuming isothermal
behavior, confining mass transport to only from the gas to the liquid phase, and a
sufficiently fast reaction (producing negligible reactant concentrations in the liquid phase) [10]. For this purpose, the Hatta number has to be within given limits.

Using this reactor model, conversions as a function o residence time were modeled
and compared with experimental data [10]. The model describes qualitatively the
behavior of the experiment, showing at first near-constant behavior and then a
more notable decrease in conversion with increasing residence time (due to decreasing specific interface).

The reactor model is also able to describe the dependence of conversion on the
specific interfacial area (Figure 5.32) which passes through a maximum owing to
the antagonistic role of increasing interfacial area at the expense of reducing residence time [10]. For a liquid volume flow of 50 ml h^{-1}, optimum conversion was
achieved at a specific interfacial area of 12 000 m^2 m^{-3} and at a residence time of
0.093 s.

Residence time
[GL 26] [R 3] [P 28] See the discussion of results in the previous section [10].

Specific interfacial area
[GL 26] [R 3] [P 28] See the discussion of results in the section Reactor model of
micro bubble column performance, above [10].

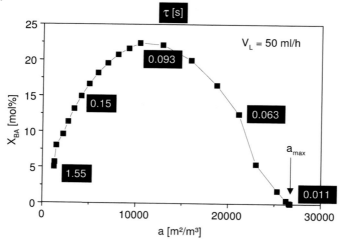

Figure 5.32 Calculated values for conversion of butyraldehyde as a function of specific interfacial area for a micro bubble column [10].

5.7
Inorganic Reactions

5.7.1
Sulfite Oxidation

Proceedings: [5, 10]; PhD thesis: [9].

5.7.1.1 Drivers for Performing Sulfite Oxidation in Micro Reactors
The sulfite oxidation is a recommended test reaction for determining the size of the specific interfacial area in gas/liquid systems, in particular as expressed by the $K_L a$ value [9, 10].

5.7.1.2 Beneficial Micro Reactor Properties for Sulfite Oxidation
The sulfite reaction is used for the above-mentioned purpose and hence is an analytical tool for judging micro-reactor performance [5, 9, 10]. The sulfite oxidation as a chemical method provides complementary information to optical analysis of the specific interfacial area.

5.7.1.3 Sulfite Oxidation Investigated in Micro Reactors
Gas/liquid reaction 27 [GL 27]: Oxidation of sulfite to sulfate

$$SO_3^{2-} \xrightarrow{\quad O_2/(Cu^{2+} \text{ or } Co^{2+}) \quad} SO_4^{2-}$$

The reaction is carried out by Cu^{2+} or Co^{2+} catalysis in an aqueous medium [9, 10]. The reaction mechanism has not been identified clearly so far. Most likely, a radical-chain reaction is involved, having Co^{2+}, a sulfite anion radical and an SO_5 anion radical as intermediate species.

5.7.1.4 Experimental Protocols

[P 29] The reaction was performed at 25 °C using a sulfite concentration of 0.68–0.8 kmol m^{-3}, a Co^{2+} concentration of $1 \cdot 10^{-6}$–$1 \cdot 10^{-3}$ kmol m^{-3} and air as oxygen source at a pH of 7–9 (an alternative protocol uses 15–60 °C with a sulfite concentration of 0.4–0.8 kmol m^{-3} and a Co^{2+} concentration of $3 \cdot 10^{-6}$–$5 \cdot 10^{-3}$ kmol m^{-3} at a pH of 7.5–8.5) [9]. Reaction velocities of $2 \cdot 10^5$–$1 \cdot 10^8$ m^3 $kmol^{-1}$ s^{-1}) resulted. Using the above-mentioned parameter set, the reaction had a second-order dependence on oxygen and a zero-order dependence on cobalt. With other sets, however, this may change considerably.

The catalyst concentration can be varied in a wide range for the above-mentioned parameter set, without changing the reaction kinetics [9]. Since gas/liquid micro reactors span a broad range of residence times, typically much shorter than for conventional apparatus, this allows a flexible adaptation of the test procedure to the needs of micro flow characterization.

5.7.1.5 Typical Results
Specific interfacial area determination

[GL 27] [R 3] [P 29] By means of sulfite oxidation, the specific interfacial area of the fluid system nitrogen/water was determined at Weber numbers ranging from 10^{-4} to 10^{-2} [10]. In this range, the interface increases from 4000 m^2 m^{-3} to 10 000 m^2 m^{-3}. The data are – with exceptions – in accordance with optically derived analysis of the interface and predictions from calculations. At still larger Weber number up to 10, the specific interfacial area increases up to 17 000 m^2 m^{-3}, which was determined optically.

[GL 27] [R 3] [P 29] By means of sulfite oxidation, the specific interfacial areas of the fluid system nitrogen/2-propanol were determined for different flow regimes [5]. For two types of micro bubble columns differing in micro-channel diameter, interfaces of 9800 and 14 800 m^2 m^{-3}, respectively, were determined (gas and liquid flow rates: 270 and 22 ml h^{-1} in both cases). Here, the smaller channels yield the multi-phase system with the largest interface.

5.7.2
Brønsted Acid–Base Reactions – Ammonia Absorption

Proceedings: [7]; sections in reviews: [26].

5.7.2.1 Drivers for Performing Ammonia Absorption in Micro Reactors

This reaction serves as a known model reaction to characterize mass transfer efficiency in micro reactors [7]. As it is a very fast reaction, mass transfer can be analysed solely. The analysis can be done simply by monitoring color changes from pH indicators. Ammonia absorption in aqueous acidic solutions generally is even faster than carbon dioxide absorption in alkaline solutions.

5.7.2.2 Beneficial Micro Reactor Properties for Ammonia Absorption

Ammonia absorption is used for the above-mentioned purpose and hence is an analytical tool for judging micro-reactor performance [7].

5.7.2.3 Ammonia Absorption Investigated in Micro Reactors
Gas/liquid reaction 28 [GL 28]: Acid–base reaction between ammonia and Brønsted inorganic acids

$$NH_3 + H^+/H_2O \rightarrow NH_4^+/H_2O$$

5.7.2.4 Experimental Protocols

[P 30] A 120 parallel micro-channel device was employed (see [R 2] for a description of the corresponding single-channel device) [7]. The total flow rate was $1–10 \text{ ml h}^{-1}$. The contact length was 14 mm; the channel cross-section was 3000 μm^2. A residence time of 2–20 s resulted.

Liquid feed was fed by hydrostatic means; gas feed was accomplished from a reservoir with the aid of a syringe pump [7]. The gas pressure was held nearly constant by passing a gas stream into a non-absorbing liquid. Analysis was performed both by visual means using a microscope and camera and by chemical analysis of the liquid output solution (Figure 5.33).

5.7.2.5 Typical Results
Mass transfer efficiency

[GL 28] [R 2] [P 30] Ammonia absorption in dilute acidic solution containing Cresol Purple indicator was rapid, as expected [7]. By appropriate choice of processing parameters, neutralization was achieved close to the gas/liquid contacting zone or distributed over the full contact length. This is evidence for having controls by both solution and gas-phase transport.

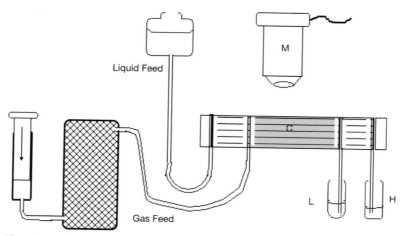

Figure 5.33 Schematic of the experimental set-up. 120 parallel micro channels (C); microscope (M); analysis of output solution (L); measurement of non-absorbed gas (H) [7].

References

1 WILLE, C., *Entwicklung und Charakterisierung eines Mikrofallfilm-Reaktors für stofftransportlimitierte hochexotherme Gas/Flüssig-Reaktionen*, PhD Thesis, Fakultät für Berbau, Hüttenwesen und Maschinenbau, University Clausthal-Zellerfeld (2002).

2 LÖWE, H., EHRFELD, W., GEBAUER, K., GOLBIG, K., HAUSNER, O., HAVERKAMP, V., HESSEL, V., RICHTER, T., *Microreactor concepts for heterogeneous gas phase reactions*, in EHRFELD, W., RINARD, I. H., WEGENG, R. S. (Eds.), *Process Miniaturization: 2nd International Conference on Microreaction Technology, IMRET 2, Topical Conf. Preprints*, pp. 63–74, AIChE, New Orleans (1998).

3 HESSEL, V., EHRFELD, W., GOLBIG, K., HAVERKAMP, V., LÖWE, H., STORZ, M., WILLE, C., GUBER, A., JÄHNISCH, K., BAERNS, M., *Gas/liquid microreactors for direct fluorination of aromatic compounds using elemental fluorine*, in EHRFELD, W. (Ed.), *Microreaction Technology: 3rd International Conference on Microreaction Technology, Proc. of IMRET 3*, pp. 526–540, Springer-Verlag, Berlin (2000).

4 WILLE, C., EHRFELD, W., HERWECK, T., HAVERKAMP, V., HESSEL, V., LÖWE, H., LUTZ, N., MÖLLMANN, K.-P., PINNO, F., *Dynamic monitoring of fluid Equipartition and heat release in a falling film microreactor using real-time thermography*, in *Proceedings of the VDE World Microtechnologies Congress, MICRO.tec 2000, 25–27 September 2000*, pp. 349–354, VDE Verlag, Berlin, EXPO Hannover (2000).

5 HESSEL, V., EHRFELD, W., HERWECK, T., HAVERKAMP, V., LÖWE, H., SCHIEWE, J., WILLE, C., KERN, T., LUTZ, N., *Gas/liquid microreactors: hydrodynamics and mass transfer*, in *Proceedings of the 4th International Conference on Microreaction Technology, IMRET 4, 5–9 March 2000*, pp. 174–186, AIChE Topical Conf. Proc., Atlanta, GA (2000).

6 EHRICH, H., LINKE, D., MORGENSCHWEIS, K., BAERNS, M., JÄHNISCH, K., *Application of microstructured reactor technology for the photochemical chlorination of alkylaromatics*, Chimia **56** (2002) 647–653.

7 SHAW, J., TURNER, C., MILLER, B., HARPER, M., *Reaction and transport coupling for liquid and liquid/gas microreactor systems*, in EHRFELD, W., RINARD, I. H., WEGENG, R. S. (Eds.), *Process Miniaturization: 2nd International Conference on Microreaction Technology, IMRET 2, Topical Conf. Preprints*, pp. 176–180, AIChE, New Orleans (1998).

8 ROBINS, I., SHAW, J., MILLER, B., TURNER, C., HARPER, M., *Solute transfer by liquid/liquid exchange without mixing in micro-contactor devices*, in EHRFELD, W. (Ed.), *Microreaction Technology – Proc. of the 1st International Conference on Microreaction Technology, IMRET 1*, pp. 35–46, Springer-Verlag, Berlin (1997).

9 HAVERKAMP, V., *Charakterisierung einer Mikroblasensäule zur Durchführung stofftransportlimitierter und/oder hochexothermer Gas/Flüssig-Reaktionen (in Fortschritt-Bericht VDI, Reihe 3, Nr. 771)*, PhD Thesis, University Erlangen (2002).

10 HAVERKAMP, V., EMIG, G., HESSEL, V., LIAUW, M. A., LÖWE, H., *Characterization of a gas/liquid microreactor, the micro bubble column: Determination of specific interfacial area*, in MATLOSZ, M., EHRFELD, W., BASELT, J. P. (Eds.), *Microreaction Technology – IMRET 5: Proc. 5th International Conference on Microreaction Technology*, pp. 202–214, Springer-Verlag, Berlin (2001).

11 LOSEY, M. W., SCHMIDT, M. A., JENSEN, K. F., *Microfabricated multiphase packed-bed reactors: characterization of mass transfer and reactions*, Ind. Eng. Chem. Res. **40** (2001) 2555–2562.

12 LOSEY, M. W., JACKMAN, R. J., FIREBAUGH, S. L., SCHMIDT, M. A., JENSEN, K. F., *Design and fabrication of microfluidic devices for multiphase mixing reaction*, J. Microelectromech. Syst. **11**, 6 (2002) 709–717.

13 DE MAS, N., GÜNTHER, A., SCHMIDT, M. A., JENSEN, K. F., *Microfabricated multiphase reactors for the selective direct*

fluorination of aromatics, Ind. Eng. Chem. Res. **42**, 4 **(2003)** 698–710.

14 DE MAS, N., JACKMAN, R. J., SCHMIDT, M. A., JENSEN, K. F., *Microchemical systems for direct fluorination of aromatics,* in MATLOSZ, M., EHRFELD, W., BASELT, J. P. (Eds.), *Microreaction Technology – IMRET 5: Proc. 5th International Conference on Microreaction Technology,* pp. 60–67, Springer-Verlag, Berlin **(2001)**.

15 CHAMBERS, R. D., SPINK, R. C. H., *Microreactors for elemental fluorine,* Chem. Commun. **(1999)** 883–884.

16 CHAMBERS, R. D., HOLLING, D., SPINK, R. C. H., SANDFORD, G., *Elemental fluorine. Part 13. Gas–liquid thin film reactors for selective direct fluorination,* Lab. Chip **1 (2001)** 132–137.

17 FÖDISCH, R., HÖNICKE, D., XU, Y., PLATZER, B., *Liquid phase hydrogenation of p-nitrotoluene in microchannel reactors,* in MATLOSZ, M., EHRFELD, W., BASELT, J. P. (Eds.), *Microreaction Technology – IMRET 5: Proc. 5th International Conference on Microreaction Technology,* pp. 470–478, Springer-Verlag, Berlin **(2001)**.

18 KURSAWE, A., HÖNICKE, D., *Epoxidation of ethene with pure oxygen as a model reaction for evaluating the performance of microchannel reactors,* in *Proceedings of the 4th International Conference on Microreaction Technology, IMRET 4, 5–9 March 2000,* pp. 153–166, AIChE Topical Conf. Proc., Atlanta, GA **(2000)**.

19 KURSAWE, A., HÖNICKE, D., *Comparison of Ag/Al- and Ag/α-Al$_2$O$_3$ catalytic surfaces for the partial oxidation of ethene in microchannel reactors,* in MATLOSZ, M., EHRFELD, W., BASELT, J. P. (Eds.), *Microreaction Technology – IMRET 5: Proc. 5th International Conference on Microreaction Technology,* pp. 240–251, Springer-Verlag, Berlin **(2001)**.

20 JOVANOVIC, G., SACRITTICHAI, P., TOPPINEN, S., *Microreactors systems for dechlorination of p-chlorophenol on palladium based metal support catalyst: theory and experiment,* in *Proceedings of the 6th International Conference on Microreaction Technology, IMRET 6, 11–14 March 2002,* pp. 314–325, AIChE Pub. No. 164, New Orleans **(2002)**.

21 WOOTTON, R. C. R., FORTT, R., DE MELLO, A. J., *A microfabricated nano-reactor for safe, continuous generation and use of singlet oxygen,* Org. Proc. Res. Dev. **60 (2002)** 187–189.

22 HESSEL, V., EHRFELD, W., GOLBIG, K., HAVERKAMP, V., LÖWE, H., RICHTER, T., *Gas/liquid dispersion processes in micro-mixers: the hexagon flow,* in EHRFELD, W., RINARD, I. H., WEGENG, R. S. (Eds.), *Process Miniaturization: 2nd International Conference on Microreaction Technology, IMRET 2, Topical Conf. Preprints,* pp. 259–266, AIChE, New Orleans **(1998)**.

23 EHRFELD, W., GOLBIG, K., HESSEL, V., LÖWE, H., RICHTER, T., *Characterization of mixing in micromixers by a test reaction: single mixing units and mixer arrays,* Ind. Eng. Chem. Res. **38**, 3 **(1999)** 1075–1082.

24 HESSEL, V., HARDT, S., LÖWE, H., SCHÖNFELD, F., *Laminar mixing in different interdigital micromixers – Part I: Experimental characterization,* AIChE J. **49**, 3 **(2003)** 566–577.

25 LÖWE, H., EHRFELD, W., HESSEL, V., RICHTER, T., SCHIEWE, J., *Micromixing technology,* in *Proceedings of the 4th International Conference on Microreaction Technology, IMRET 4, 5–9 March 2000,* pp. 31–47, AIChE Topical Conf. Proc., Atlanta, GA **(2000)**.

26 EHRFELD, W., HESSEL, V., LÖWE, H., *Microreactors,* Wiley-VCH, Weinheim **(2000)**.

27 HESSEL, V., LÖWE, H., HOFMANN, C., SCHÖNFELD, F., WEHLE, D., WERNER, B., *Process development of fast reaction of industrial importance using a caterpillar micromixer/tubular reactor set-up,* in *Proceedings of the 6th International Conference on Microreaction Technology, IMRET 6, 11–14 March 2002,* pp. 39–54, AIChE Pub. No. 164, New Orleans **(2002)**.

28 EHRFELD, W., HESSEL, V., KIESEWALTER, S., LÖWE, H., RICHTER, T., SCHIEWE, J., *Implementation of microreaction technology in process engineering,* in EHRFELD, W. (Ed.), *Microreaction Technology: 3rd International Conference on Microreaction Technology, Proc. of IMRET 3,* pp. 14–34, Springer-Verlag, Berlin **(2000)**.

29 SCHWESINGER, K., FRANK, T., *A modular microfluidic system with an integrated micromixer*, in *Proceedings of MME '95*, pp. 144–147, Copenhagen (**1995**).

30 SCHWESINGER, W., FRANK, T., *A static micromixer built up in silicon*, in *Proceedings of Micromachining and Microfabrication*, pp. 150–155, SPIE, Austin TX (**1995**).

31 SCHWESINGER, N., FRANK, T., *Device for mixing small quantities of liquids*, WO 96/30113, Merck Patent GmbH, Darmstadt (**1995**).

32 SCHWESINGER, N., FRANK, T., WURMUS, H., *A modular microfluidic system with an integrated micromixer*, J. Micromech. Microeng. **6** (**1996**) 99–102.

33 SCHWESINGER, N., *Mikrostrukturierte modulare Reaktionssysteme für die chemische Industrie*, F & M, Feinwerktechnik, Mikrotechnik, Meßtechnik **110**, 4 (**2002**) 17–21.

34 ANTES, J., TUERCKE, T., MARIOTH, E., SCHMID, K., KRAUSE, H., LOEBBECKE, S., *Use of microreactors for nitration processes*, in *Proceedings of the 4th International Conference on Microreaction Technology*, *IMRET 4*, 5–9 March 2000, pp. 194–200, AIChE Topical Conf. Proc., Atlanta, GA (**2000**).

35 ANTES, J., TÜRCKE, T., MARIOTH, E., LECHNER, F., SCHOLZ, M., SCHNÜRER, F., KRAUSE, H. H., LÖBBECKE, S., *Investigation, analysis and optimization of exothermic nitrations in microreactor processes*, in MATLOSZ, M., EHRFELD, W., BASELT, J. P. (Eds.), *Microreaction Technology – IMRET 5: Proc. 5th International Conference on Microreaction Technology*, pp. 446–454, Springer-Verlag, Berlin (**2001**).

36 LOSEY, M. W., SCHMIDT, M. A., JENSEN, K. F., *A micro packed-bed reactor for chemical synthesis*, in EHRFELD, W. (Ed.), *Microreaction Technology: 3rd International Conference on Microreaction Technology, Proc. of IMRET 3*, pp. 277–286, Springer-Verlag, Berlin (**2000**).

37 DE MAS, N., *Heat effects in a microreactor for direct fluorination of aromatics*, in *Proceedings of the 6th International Conference on Microreaction Technology*, *IMRET 6*, 11–14 March 2002, pp. 184–185, AIChE Pub. No. 164, New Orleans (**2002**).

38 JÄHNISCH, K., BAERNS, M., HESSEL, V., EHRFELD, W., HAVERKAMP, V., LÖWE, H., WILLE, C., GUBER, A., *Direct fluorination of toluene using elemental fluorine in gas/liquid microreactors*, J. Fluorine Chem. **105**, 1 (**2000**) 117–128.

39 JÄHNISCH, K., BAERNS, M., HESSEL, V., HAVERKAMP, V., LÖWE, H., WILLE, C., *Selective reactions in microreactors – fluorination of toluene using elemental fluorine in a falling film microreactor*, in *Proceedings of the 37th ESF/EUCHEM Conference on Stereochemistry*, 13–19 April 2002, Bürgenstock (**2002**).

40 JÄHNISCH, K., EHRICH, H., LINKE, D., BAERNS, M., HESSEL, V., MORGENSCHWEIS, K., *Selective gas/liquid-reactions in microreactors*, in *Proceedings of the International Conference on Process Intensification for the Chemical Industry*, 13–15 October 2002, Maastricht (**2002**).

41 BASELT, J. P., FÖRSTER, A., HERMANN, J., *Microreaction technology: focusing the German activities in this novel and promising field of chemical process engineering*, in EHRFELD, W., RINARD, I. H., WEGENG, R. S. (Eds.), *Process Miniaturization: 2nd International Conference on Microreaction Technology, IMRET 2, Topical Conf. Preprints*, pp. 13–17, AIChE, New Orleans (**1998**).

42 DEWITT, S., *Microreactor for chemical synthesis*, Curr. Opin. Chem. Biol. **3** (**1999**) 350–356.

43 HESSEL, V., LÖWE, H., *Mikroverfahrenstechnik: Komponenten – Anlagenkonzeption – Anwenderakzeptanz – Teil 2*, Chem. Ing. Tech. **74**, 3 (**2002**) 185–207.

44 SCHWALBE, T., AUTZE, V., WILLE, G., *Chemical synthesis in microreactors*, Chimia **56**, 11 (**2002**) 636–646.

45 HASWELL, S. T., WATTS, P., *Green chemistry: synthesis in micro reactors*, Green Chem. **5** (**2003**) 240–249.

46 HESSEL, V., LÖWE, H., *Micro chemical engineering: components – plant concepts – user acceptance: Part II*, Chem. Eng. Technol. **26**, 4 (**2003**) 391–408.

47 JÄHNISCH, K., HESSEL, V., LÖWE, H., BAERNS, M., *Chemie in Mikrostrukturreaktoren*, Angew. Chem. **44**, 4 (**2004**) in press.

48 GAVRIILIDIS, A., ANGELI, P., CAO, E., YEONG, K. K., WAN, Y. S. S., *Technology and application of microengineered reactors*, Trans. IChemE. **80**/A, 1 **(2002)** 3–30.

49 GRAKAUSKAS, V., *Direct liquid-phase fluorination of halogenated aromatic compounds*, J. Org. Chem. **34**, 10 **(1969)** 2835–2839.

50 GRAKAUSKAS, V., *Direct liquid-phase fluorination of aromatic compounds*, J. Org. Chem. **35**, 3 **(1970)** 723–728.

51 CACACE, F., WOLF, A. P., *Substrate selectivity and orientation in aromatic substitution by molecular fluorine*, J. Am. Chem. Soc. **100**, 11 **(1978)** 3639–3641.

52 CACACE, F., GIACOMELLO, P., WOLF, A. P., *Substrate selectivity and orientation in aromatic substitution by molecular fluorine*, J. Am. Chem. Soc. **102**, 10 **(1980)** 3511–3515.

53 PURRINGTON, S. T., KAGEN, B. S., *The application of elemental fluorine in organic chemistry*, Chem. Rev. **86** (1986) 997–1018.

54 BALZ, G., SCHIEMANN, G., Ber. Dtsch. Chem. Ges. **60** (1927) 1186.

55 Schiemann, G., Cornils, B., *Chemie und Technologie cyclischer Fluorverbindungen*, pp. 188–193, Ferdinand Enke Verlag, Stuttgart **(1969)**.

56 CONTE, L., GAMBARETTO, G. P., NAPOLI, M., FRACCARO, C., LEGNARO, E., *Liquid-phase fluorination of aromatic compounds by elemental fluorine*, J. Fluorine Chem. **70** (1995) 175–179.

57 WEHLE, D., DEJMEK, M., ROSENTHAL, J., ERNST, H., KAMPMANN, D., TRAUTSCHOLD, S., PECHATSCHEK, R., *Verfahren zur Herstellung von Mono-chloressigsäure in Mikroreaktoren*, DE 10036603 A1 **(2000)**.

58 LOSEY, M. W., ISOGAI, S., SCHMIDT, M. A., JENSEN, K. F., *Microfabricated devices for multiphase catalytic process*, in *Proceedings of the 4th International Conference on Microreaction Technology, IMRET 4*, 5–9 March 2000, pp. 416–424, AIChE Topical Conf. Proc., Atlanta, GA **(2000)**.

59 JENSEN, K. F., AJMERA, S. K., FIRE-BAUGH, S. L., FLOYD, T. M., FRANZ, A. J., LOSEY, M. W., QUIRAM, D., SCHMIDT, M. A., *Microfabricated chemical systems for product screening and synthesis*, in

HOYLE, W. (Ed.), *Automated Synthetic Methods for Specialty Chemicals*, pp. 14–24, Royal Society of Chemistry, Cambridge **(2000)**.

60 YEONG, K. K., GAVRIILIDIS, A., ZAPF, R., HESSEL, V., *Catalyst preparation and deactivation issues for nitrobenzene hydrogenation in a microstructured falling film reactor*, Catal. Today **81**, 4 **(2003)** 641–651.

61 FÖDISCH, R., RESCHETILOWSKI, W., HÖNICKE, D., *Heterogeneously catalyzed liquid-phase hydrogenation of nitro-aromatics using microchannel reactors*, in *Proceedings of the DGMK Conference on the Future Role of Aromatics in Refining and Petrochemistry*, pp. 231–238, Erlangen **(1999)**.

62 YEONG, K. K., GAVRIILIDIS, A., ZAPF, R., HESSEL, V., *Effect of catalyst preparation methods on the performance of a microstructured falling film reactor in nitrobenzene hydrogenation*, in *Proceedings of the 7th International Conference on Microreaction Technology, IMRET 7*, 7–10 September 2003, Lausanne, submitted for publication.

63 DE BELLEFON, TANCHOUX, N., CARAVIEILHES, S., GRENOUILLET, P., HESSEL, V., *Microreactors for dynamic high throughput screening of fluid–liquid molecular catalysis*, Angew. Chem. **112**, 19 **(2000)** 3584–3587.

64 DE BELLEFON, C., PESTRE, N., LAMOUILLE, T., GRENOUILLET, P., *High-throughput kinetic investigations of asymmetric hydrogenations with microdevices*, Adv. Synth. Catal. **345**, 1+2 **(2003)** 190–193.

65 DE BELLEFON, C., *Application of micro-devices for the fast investigation of catalysis*, in Proceedings of Micro Chemical Plant – International Workshop, 4 February 2003, pp. L3 (9–17), Kyoto **(2003)**.

66 DE BELLEFON, C., CARAVIEILHES, S., GRENOUILLET, P., *Application of a micromixer for the high throughput screening of fluid–liquid molecular catalysts*, in MATLOSZ, M., EHRFELD, W., BASELT, J. P. (Eds.), *Microreaction Technology – IMRET 5: Proc. 5th International Conference on Microreaction Technology*, pp. 408–413, Springer-Verlag, Berlin **(2001)**.

67 SCHWESINGER, N., MARUFKE, O., QIAO, F., DEVANT, R., WURZIGER, H., *A full wafer silicon microreactor for combinatorial chemistry*, in EHRFELD, W., RINARD, I. H., WEGENG, R. S. (Eds.), *Process Miniaturization: 2nd International Conference on Microreaction Technology, IMRET 2, Topical Conf. Preprints*, pp. 124–126, AIChE, New Orleans (**1998**).

68 HAVERKAMP, V., EHRFELD, W., GEBAUER, K., HESSEL, V., LÖWE, H., RICHTER, T., WILLE, C., *The potential of micromixers for contacting of disperse liquid phases*, Fresenius' J. Anal. Chem. **364** (**1999**) 617–624.

69 CARAVIEILHES, S., DE BELLEFON, C., TANCHOUX, N., *Dynamic methods and new reactors for liquid phase molecular catalysis*, Catal. Today **66** (**2001**) 145–155.

70 DE BELLEFON, C., ABDALLAH, R., LAMOUILLE, T., PESTRE, N., CARAVIEILHES, S., GRENOUILLET, P., *High-throughput screening of molecular catalysts using automated liquid handling, injection and microdevices*, Chimia **56**, 11 (**2002**) 621–626.

71 PENNEMANN, H., HESSEL, V., KOST, H.-J., LÖWE, H., DE BELLEFON, C., *Investigations on pulse broadening for transient catalyst screening in gas/liquid systems*, AIChE J. (**2003**) 34.

72 PENNEMANN, H., HESSEL, V., KOST, H.-J., LÖWE, H., DE BELLEFON, C., PESTRE, N., Lamouille, T., Grenouillet, P., *High-throughput experimentation with a micromixer based, automated serial screeening apparatus for polyphasic fluid reactions*, in *Proceedings of the International Conference on Process Intensification for the Chemical Industry*, 13–15 October 2002, Maastricht, in press.

73 ANGELI, P., GOBBY, D., GAVRIILIDIS, A., *Modeling of gas-liquid catalytic reactions in microchannels*, in EHRFELD, W. (Ed.), *Microreaction Technology: 3rd International Conference on Microreaction Technology, Proc. of IMRET 3*, pp. 253–259, Springer-Verlag, Berlin (**2000**).

74 JÄHNISCH, K., BAERNS, M., *Verfahren zur Photooxygenierung von Olefinen*, Aktenzeichen DD 10257239.9 (**2002**).

75 VESER, G., FRIEDRICH, G., FREYGANG, M., ZENGERLE, R., *A modular microreactor design for high-temperature catalytic oxidation reactions*, in EHRFELD, W. (Ed.), *Microreaction Technology: 3rd International Conference on Microreaction Technology, Proc. of IMRET 3*, pp. 674–686, Springer-Verlag, Berlin (**2000**).

76 VESER, G., *Experimental and theoretical investigation of H_2 oxidation in a high-temperature catalytic microreactor*, Chem. Eng. Sci. **56** (**2001**) 1265–1273.

77 HESSEL, V., LÖWE, H., *Mikroverfahrenstechnik: Komponenten – Anlagenkonzeption – Anwenderakzeptanz – Teil 1*, Chem. Ing. Tech. **74**, 2 (**2002**) 17–30.

78 HESSEL, V., LÖWE, H., *Micro chemical engineering: components – plant concepts – user acceptance: Part I*, Chem. Eng. Technol. **26**, 1 (**2003**) 13–24.

Index